HANDBOOK OF BIOMEDICAL TELEMETRY

IEEE Press
445 Hoes Lane
Piscataway, NJ 08854

HANDBOOK OF BIOMEDICAL TELEMETRY

Edited by

KONSTANTINA S. NIKITA

IEEE Engineering in Medicine
and Biology Society, *Sponsor*

IEEE Press Series in Biomedical Engineering
Metin Akay, *Series Editor*

IEEE PRESS

WILEY

For general information on our other products and services or for technical support, please contact our Customer Care Department within the United States at (800) 762-2974, outside the United States at (317) 572-3993 or fax (317) 572-4002.

Wiley also publishes its books in a variety of electronic formats. Some content that appears in print may not be available in electronic formats. For more information about Wiley products, visit our web site at www.wiley.com.

Library of Congress Cataloging-in-Publication Data:

Handbook of biomedical telemetry / edited by Konstantina S. Nikita.
p. ; cm.
Includes bibliographical references.
Summary: "Handbook of Biomedical Telemetry describes the main components of a typical biomedical telemetry system as well as its technical challenges. Author K. S. Nikita addresses technologies for biomedical sensing and design of biomedical telemetry devices with special emphasis on powering/integration issues and materials for biomedical telemetry applications. Covering biomedical telemetry devices, biosensor network topologies and standards, clinical applications, wearable and implantable devices, and the effects on the mobile healthcare system, this compendium is a must have for professional engineers as well as researchers and graduate students"–Provided by publisher.
ISBN 978-1-118-38861-7 (cloth)
I. Nikita, Konstantina S., editor of compilation. II. Institute of Electrical and Electronics Engineers.
[DNLM: 1. Telemetry. 2. Biosensing Techniques. QT 34]
R857.M3
610.28′4–dc23

2013051203

10 9 8 7 6 5 4 3 2 1

This book is dedicated to
Maria and Elias

CONTENTS

PREFACE

The word telemetry is derived from the Greek words "tele" = "remote", and "metron" = "measure", and means data measurements allowed to be made at a distance. In other words, data are measured in situ, and further transmitted remotely to a receiving station. Biomedical telemetry permits the measurement of physiological signals at a distance. Physiological signals are obtained by means of appropriate transducers, post-processed, and eventually transmitted to an exterior monitoring and/or control device. The exterior device can be placed on or at a close distance from the subject's body, but can also further communicate with a distant hospital or physicians' station with the help of communication technologies and infrastructure.

The "Handbook of Biomedical Telemetry" aims to form a complete book in this emerging scientific field, addressing all arising scientific issues and technologies. It includes a comprehensive set of topics, sources and resources and contains generic information, detailed scientific analyses, as well as example applications. To accurately cover the broad range of topics arising within biomedical telemetry, the book comprises 23 chapters, divided into three parts.

The **First Part** of the book addresses technologies for the design of biomedical telemetry devices, biomedical sensing, and powering techniques for biomedical telemetry applications.

The **Second Part** of the book deals with Propagation and Communication Issues in Biomedical Telemetry, and covers numerical and experimental techniques for body area electromagnetics, inductive coupling, antennas and RF communication, intra-body communication, optical communication, biosensor communication technologies and standards, context-aware sensing and multi-sensor fusion, security and privacy issues, connection between biomedical telemetry and telemedicine, as well as safety issues and exposure assessment to high frequency biotelemetry fields.

The **Third Part** of the book presents selected examples of wearable, implantable and ingestible devices, stimulators, integrated mobile healthcare systems, and advanced material-based sensing paradigms for monitoring and therapeutic intervention.

The material in the book is written for persons at a number of levels. Much of it is introductory for an engineer, but serves to link engineering principles with living systems, such as the material related to sensing and propagation of physiological signals over time and space. Even though the book is expected to enable a diverse group of readers to become acquainted with biomedical telemetry, it also intends to provide highly specialized comprehensive knowledge in the field. As a result, it is mainly intended for persons who are familiar with conventional engineering and biomedical engineering principles. It can be addressed to a wide audience—from advanced undergraduate and postgraduate students pursuing education/degrees in bioengineering / biomedical / communications engineering to practicing engineers, engineering professors, researchers and industry specialists in the fields of communications and biomedical engineering sciences.

Several chapters have been written by experts in the respective fields, and a great effort has been devoted to secure chapter-to-chapter cohesion and writing style uniformity. As a result, the book contains more perspectives, analyzes more data, incorporates knowledge and research from more disciplines, and uses a wider variety of methodologies.

An important feature of the work is the breadth and depth of treatment given to the component parts of a biomedical telemetry system and their illustration through numerous application case studies. An overview of the commercially (off-shelf) available equipment related to particular described scenarios is also presented. Human phantoms, spectrum regulations, safety, standards and interoperability issues are presented and supported by relevant resource listings, while major technical challenges related to advanced materials, miniaturization and biocompatibility issues, are also included.

Konstantina S. Nikita

National Technical University of Athens
Athens, Greece

ACKNOWLEDGMENTS

In getting to the end of this venture, I wish to express my sincere thanks to colleagues, students, and friends for their support and help.

I would like to begin by thanking all the contributors for the quality of their work, their patience in the review and editing of the book, and their valuable efforts to make the completion of this book possible.

I owe a special note of thanks to Asimina Kiourti who was involved in this project from conception to completion while doing her doctoral thesis with me. Dr. Kiourti was enthusiastically engaged in every aspect of this project, from preliminary discussions on the content of the book to suggesting ways to improve manuscript material. The time and energy invested by Dr. Kiourti were critical, for this book to come to fruition. Moreover, particular gratitude is due to Dr. Ioannis Andreadis for his valuable help in editing and proofreading the manuscript.

I am indebted to several anonymous reviewers who have read over portions of this book and have provided thorough and constructive comments. They did a superb job of shaping the book and suggesting ways to improve the draft manuscripts.

My thanks to the editorial staff of Wiley-IEEE Press and my appreciation to Mary Hatcher for her invaluable efforts in the publishing process of this book. I also wish to thank and formally acknowledge the dedicated work by Jayashree Saishankar, Project Manager, and Sheeba Ravi, Production Team Leader, at Laserwords. The time invested by Mrs. Saishankar and Mrs. Ravi in editing the manuscript has ensured accuracy of the material.

CONTRIBUTORS

María Teresa Arredondo Waldmeyer, Life Supporting Technologies, Universidad Politécnica de Madrid, Madrid, Spain

Nikolaos Bourbakis, Wright State University, Assistive Technologies Research Center, Dayton, OH

Carlos J. Cela, North Carolina State University, Raleigh, NC

Amparo Callejón-Leblic, University of Seville, Seville Spain

Chi-Chih Chen, ElectroScience Laboratory, Ohio State University, Columbus, OH

Dominik Cirmirakis, Department of Electronic and Electrical Engineering, University College London, London, United Kingdom

Andreas Demosthenous, Department of Electronic and Electrical Engineering, University College London, London, United Kingdom

Michael D. Dickey, North Carolina State University, Raleigh, NC

Javier Espina, Philips Research High Tech Campus, Eindhoven, the Netherlands

Miguel Á. Estudillo-Valderrama, University of Seville, Seville Spain

Thomas Falck, Philips Research High Tech Campus, Eindhoven, the Netherlands

Giuseppe Fico, Life Supporting Technologies, Universidad Politécnica de Madrid, Madrid, Spain

Alessio Fioravanti, Life Supporting Technologies, Universidad Politécnica de Madrid, Madrid, Spain

Robert Foster, Queen Mary University of London, London, United Kingdom

Stavros Georgakopoulos, Florida International University, Miami, FL

Maysam Ghovanloo, Georgia Institute of Technology, Atlanta, GA

Keyoor C. Gosalia, North Carolina State University, Raleigh NC

Alejandra Guillén Arredondo, Medtronic Ibérica, Spain

Yang Hao, Queen Mary University of London, London United Kingdom

Gerard J. Hayes, North Carolina State University, Raleigh, NC

Stefan Hey, KIT Karlsruhe Institute of Technology, Karlsruhe, Germany

Koichi Ito, Chiba University, Chiba, Japan

Raghudeep Kannavara, Wright State University, Assistive Technologies Research Center, Dayton, OH

Alexandros Karargyris, Wright State University, Assistive Technologies Research Center, Dayton, OH

Yoshihiro Kawahara, Georgia Institute of Technology, Atlanta, GA

Mehdi Kiani, Georgia Institute of Technology, Atlanta, GA

Asimina Kiourti, School of Electrical and Computer Engineering, National Technical University of Athens, Athens, Greece

Sangkil Kim, Georgia Institute of Technology, Atlanta, GA

Richard M. Kwasnicki, Imperial College London, United Kingdom

Vasileios Lakafosis, Georgia Institute of Technology, Atlanta, GA

Gianluca Lazzi, North Carolina State University, Raleigh, NC

Jan-Paul Leuteritz, Institute for Human Factors and Technology, Management at the University of Stuttgart, Stuttgart, Germany

Taoran Le, Georgia Institute of Technology, Atlanta, GA

Gil-Young Lee, ElectroScience Laboratory, Ohio State University, Columbus, OH

Hoseon Lee, Georgia Institute of Technology, Atlanta, GA

Athanasios Lioumpas, School of Electrical and Computer Engineering, National Technical University of Athens, Athens, Greece

Kostas Marias, Computational Medicine Laboratory, Institute of Computer Science, Foundation for Research and Technology–Hellas (FORTH). Heraklion, Crete, Greece

Sagar Mukala, Georgia Institute of Technology, Atlanta, GA

David Naranjo, University of Seville, Seville, Spain

Konstantina S. Nikita, School of Electrical and Computer Engineering, National Technical University of Athens, Athens, Greece

Georgia Ntouni, School of Electrical and Computer Engineering, National Technical University of Athens, Athens, Greece

Alexandros Pantelopoulos, Wright State University, Assistive Technologies Research Center, Dayton, OH

Alejandro González Patón, Life Supporting Technologies, Universidad Politécnica de Madrid, Madrid, Spain

Konstantinos A. Psathas, School of Electrical and Computer Engineering, National Technical University of Athens, Athens, Greece

Dimitris Psychoudakis, ElectroScience Laboratory, Ohio State University, Columbus, OH

Anil Kumar RamRakhyani, University of Utah, Salt Lake City, UT

Javier Reina-Tosina, University of Seville, Seville, Spain

Laura M. Roa, University of Seville, Seville, Spain

Vangelis Sakkalis, Computational Medicine Laboratory, Institute of Computer Science, Foundation for Research and Technology–Hellas (FORTH). Heraklion, Crete, Greece

Lars Schmitt, Philips Research High Tech Campus, Eindhoven, the Netherlands

Koichi Shimizu, Hokkaido University, Sapporo, Japan

Shruthi Soora, North Carolina State University, Raleigh, NC

Emmanouil G. Spanakis, Computational Medicine Laboratory, Institute of Computer Science, Foundation for Research and Technology–Hellas (FORTH). Heraklion, Crete, Greece

Masaharu Takahashi, Chiba University, Chiba, Japan

Toshiyo Tamura, Osaka Electro-Communication University, Neyagawa, Japan

Manos M. Tentzeris, Georgia Institute of Technology, Atlanta, GA

Anya Traille, Georgia Institute of Technology, Atlanta, GA

Manolis Tsiknakis, Computational Medicine Laboratory, Institute of Computer Science, Foundation for Research and Technology–Hellas (FORTH). Heraklion, Crete, Greece; Biomedical Informatics Laboratory, Department of Informatics Engineering, TEI Crete, Heraklion, Crete, Greece

John L. Volakis, ElectroScience Laboratory, Ohio State University, Columbus, OH

Rushi Vyas, Georgia Institute of Technology, Atlanta, GA

Dong Wang, Philips Research High Tech Campus, Eindhoven, the Netherlands

Wei Wei, Georgia Institute of Technology, Atlanta, GA

Guang-Zhong Yang, Imperial College London, United Kingdom

Li Yang, Georgia Institute of Technology, Atlanta, GA

1

INTRODUCTION TO BIOMEDICAL TELEMETRY

Konstantina S. Nikita
School of Electrical and Computer Engineering, National Technical University of Athens, Athens, Greece

1.1 WHAT IS BIOMEDICAL TELEMETRY?

The word *telemetry* is derived from the Greek words *tele* = "remote" and *metron* = "measure" and allows data measurements to be made at a distance. In other words, data are measured in situ and further transmitted remotely to a receiving station. Typically, telemetry systems have been used in the testing of moving vehicles such as cars, aircraft, and missiles.

Biomedical telemetry permits the measurement of physiological signals at a distance. Physiological signals are obtained by means of appropriate transducers, postprocessed, and eventually transmitted to an exterior monitoring and/or control device. The exterior device can be placed onto the patient's body or at a close distance next to the patient but can also further communicate with a distant hospital or physicians' station with the help of telemedicine technologies and infrastructure.

The principal purpose of biomedical telemetry is to take advantage of the recent advances in wired and wireless communication technologies in order to address the growing demands of the health care community. The goal is to take advantage of the recent improvements in electronics and communications in order to develop a new generation of medical devices with incorporated biomedical telemetry functionalities. Medical devices can be defined as any physical device which is useful for preventive, diagnostic, monitoring, or therapeutic functions. Such devices are expected to support

Handbook of Biomedical Telemetry, First Edition. Edited by Konstantina S. Nikita.

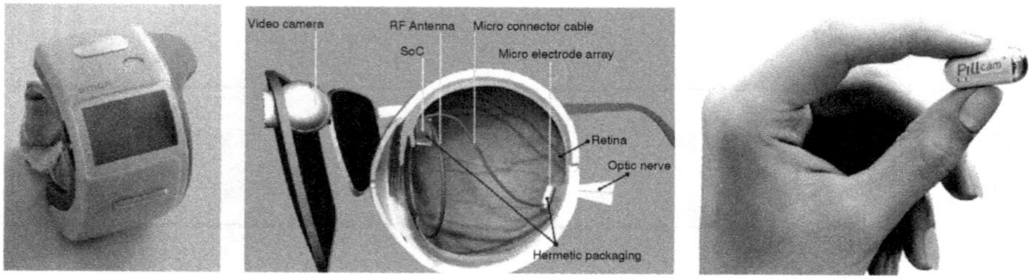

(*a*) (*b*) (*c*)

Figure 1.1 (*a*) Wearable [advanced care and alert portable telemedical monitor, AMON (Anliker et al., 2004)], (*b*) implantable [epiretinal prosthesis (Sivaprakasam et al., 2005)], and (*c*) ingestible [PillCam (Mc Caffrey et al., 2008)] medical devices.

an expanding variety of medical applications and have the potential to revolutionize medicine. Even though prevention is perhaps the most desirable goal for medical devices, early diagnosis, effective treatment, and accurate monitoring of diseases can also be considered as the cornerstones of an effective biomedical telemetry system.

There exist three categories of medical devices, according to their location on or inside the patient's body:

1. *Wearable devices* can be worn by the patient as an accessory or embedded into clothing with the help of textile and flexible technologies (e.g., Figure 1.1*a*). They can be used to monitor several physiological parameters (e.g., glucose or cardiac events), assist the movement of artificial limbs, and work as receivers for the collection and retransmission of various vital signals.
2. *Implantable devices* can be implanted inside the patient's human body by means of a surgical operation (e.g., Figure 1.1*b*) (Kiourti and Nikita, 2012*a,b*, 2013). Example applications are heart rate control, artificial retina, cardiac pacemakers, cochlear implants, hypertension monitoring, functional electrical stimulation, and intracranial pressure monitoring.
3. *Ingestible devices* are integrated into capsules or pills and can be swallowed by the patient (e.g., Figure 1.1*c*). Main focus is on their use for gastrointestinal track and drug use monitoring.

Since medical devices are used on human beings, with at least a theoretical potential for misuse or harmful side effects, they must first meet the criteria established by government-operated regulations before they can be designated as medical devices and enter the market. For example, in the United States, medical devices are primarily regulated via the Department of Health and Human Services (HHS) of the Food and Drug Administration (FDA).

Historically, wired links have been the most prevalent method of biomedical telemetry. To overcome the inherent drawbacks of restricted communication range as well as patient discomfort and limited activity level, research is nowadays mostly oriented toward wireless technologies. Wireless biomedical telemetry offers the advantage of obtaining accurate physiological signal measurements from freely moving patients and has significantly risen in the last decades thanks to the explosive growth in Internet traffic, the commercial success of digital cellular communication systems, and the scaling of integrated circuits (ICs) at a manageable cost, power, and size (Rappaport et al., 2002).

Recent global focus on health care issues has stimulated research and development of innovative technologies which address many unsustainabilities of the current health care provision models. Several health care organizations are seeking new techniques to deliver quality health care in a timely, cost-effective, and efficient manner. Biomedical telemetry can be considered as an important technological innovation toward freeing hospital resources, improving patient care, and rendering health care affordable for all. The utmost aim is to enhance the patients' quality of life by encouraging and maintaining their independence. With rising health care costs, an increasing average age of populations in the occidental world, a significant presence of wireless communications in our daily lives, and recent advances in electronics and information and communication technologies (ICTs), biomedical telemetry devices are attracting significant scientific interest in both academia and industry.

1.2 SIGNIFICANCE OF AREA

Exploitation of ICT assists in a fundamental redesign of the health care processes based on the use and integration of communication technologies at all levels. Recent advances in ICT enable cost-effective and efficient health care delivery in home, hospital, assisted-living, and nursing home settings to promote disease management and wellness (Nikita et al., 2012).

Disease management programs aim to support patient-specific care plans and the provider–patient relationship via evidence-based guidelines while focusing on prevention of deteriorations and/or complications. Aiming at citizen empowerment, the paradigm of disease management can be extended to wellness management, where the focus is on disease prevention, maintenance, and improvement of the health status of any individual.

Continuous and remote monitoring of patients in the comfort of their own home rather than inside a hospital or clinic environment offers a number of benefits, including continuous medical monitoring of the disease progression or fluctuation, patient convenience, sophisticated monitoring capabilities, and lower health care costs (Lin and Nikita, 2010; Nikita et al., 2011). Example applications are:

- Monitoring of patients with chronic diseases (e.g., diabetes or hypertension) by means of a single medical device
- Development of "smart" body sensor networks where physiological data are collected from multiple on/in body sensors, preferably with context-aware sensing capabilities
- Drug delivery feedback loops which continuously monitor a drug's effect and adjust its delivery from drug pumps

TABLE 1.1 Ten Targets for Wireless Medicine

Disease	Number Affected (millions)	Monitoring
Alzheimer's	5	Vital signs, location, activity, balance
Asthma	23	Respiratory rate (RR), peak-flow breathing volume (FEV1), air quality, oximetry, pollen count
Breast cancer	3	Ultrasound, self-examination
Chronic obstructive pulmonary disease (COPD)	10	RR, FEV1, air quality, oximetry
Depression	21	Medicine compliance, activity, communication
Diabetes	24	Glucose, hemoglobine A1C, activity
Heart failure	5	Cardiac pressures, weight, BP, fluid status
Hypertension	74	Continous BP, medical compliance
Obesity	80	Smart scales, glucose, caloric in/out, activity
Sleep disorder	40	Sleep phases, quality, apnea, vital signs

- Rehabilitation for the elderly
- Measuring medical parameters at the scene of an accident and providing surveillance during transport to the hospital

A top 10 list for conditions and diseases that are already benefiting from wireless health services or soon will is shown (in alphabetical order) in Table 1.1 (Topol, 2012).

Biomedical telemetry has the potential to empower patients and support a transition from a role in which the patient is the passive recipient of care services to an active role in which the patient is informed, has choices, and is involved in the decision-making process. However, these modern healthcare systems set some additional critical requirements and challenges compared to traditional networks.

1.3 TYPICAL BIOMEDICAL TELEMETRY SYSTEM

A schematic of a typical biomedical telemetry system is shown in Figure 1.2. A number of medical devices are worn on, implanted into, or ingested by the patient to perform measurements of the intended physiological signals to be monitored.

Measured data are further transmitted to a remotely located receiving device, which is most commonly placed onto the patient's body or at a close distance next to the patient. This central receiving device can take several actions, such as:

- Postprocess the signals and make decisions based on several factors (e.g., a change in heart rate, lack of motion, and sudden increases in blood oxygen saturation could be used together to identify the onset of a stroke).
- Warn the patient of an important event.

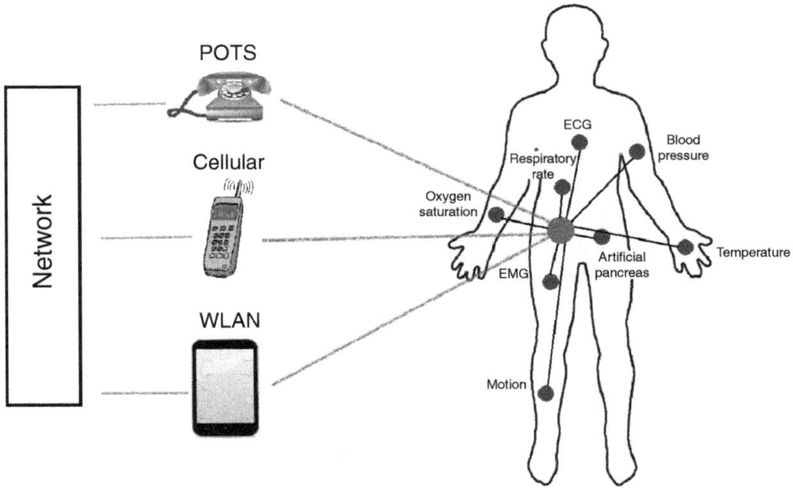

Figure 1.2 Schematic of typical biomedical telemetry system with example sensor/actuator applications.

- Notify other medical devices to administer drugs (e.g., insulin for diabetics).
- Make emergency calls.
- Store the signals locally for medical postprocessing.

In addition to the above, the central receiving device also has the ability to acquire the measured signals and retransmit them into the data network and communication infrastructure for long-distance transmission (telemedicine). Telemedicine systems are mostly constructed using a combination of off-the-shelf equipment and services as well as specialized terminals and software applications. Most of these systems rely on the Plain Old Telephone Service (POTS) for data communication, but, recently, emphasis is mostly given on the use of wireless systems [cellular phones, devices with wireless local area network (WLAN) capabilities, etc. (Berezdivin et al., 2002)].

Considering a more complicated scenario, medical devices can communicate with several on-body or external units as well as with other medical devices carried by the patient. What is important to note is that application requirements can vary widely. For example, average data rates for typical biomedical telemetry applications span over five orders of magnitude, as shown in Table 1.2.

1.4 CHALLENGES IN BIOMEDICAL TELEMETRY

1.4.1 Spectrum Regulations

Demand on radio spectrum for use in wireless biomedical telemetry systems is currently on the increase. This demand is driven by a rapid increase in the use of

TABLE 1.2 Average Data Rates for Typical Biomedical Telemetry Applications

Application	Average Data Rate (kbps)
Glucose monitoring	0.01–0.1
Blood pressure monitoring	0.01–0.1
Electroencephalogram (EEG)	10–100
Electrocardiogram (ECG)	10–100
Electromyogram (EMG)	10–100

medical devices, advancements in wireless communication technologies, and the need to improve quality, reliability, and delivery of health care. Some of the most commonly used frequency bands for biomedical telemetry systems include the Medical Implant Communications Systems (MICS) band, the Wireless Medical Telemetry Service (WMTS) bands, the industrial, scientific, and medical (ISM) bands, and the ultrawide band (UWB). Regardless of the band chosen, constant management is highlighted as crucial for reducing the probability of interference from other transmitting devices. The most important regulations governing the aforementioned frequency bands are summarized below.

Medical Implant Communications Systems In 1998, the International Telecommunication Union—Radiocommunication (ITU-R) outlined the use of the 402–405-MHz frequency band for MICS (ITU-R, 1998). The MICS band is currently regulated by the U.S. Federal Communications Commission (FCC) (MICS Federal Register, 1999) and the European Radiocommunications Committee (ERC) (ERC Recommendation, 1997) and is expected to become a true global standard within several years. Two fields of application are indicated for this standard: communication between an implantable medical device and an exterior receiving station and communication between medical devices implanted within the same human body. MICS devices can use up to 300 kHz of bandwidth at a time for the complete session. Equivalently, separate transmitter and receiver bands, each with a bandwidth of 300 kHz, may be adopted as long as they are not used simultaneously. Assuming a full-duplex solution in which the system uses two separate frequencies for up- and downlink transmission, the two link bandwidths should not exceed 300 kHz. The range is typically 2 m, and the maximum power limit is set to 25 μW of equivalent radiated power (ERP), that is, the maximum field strength in any direction should be equal to or lower than what a resonant dipole would give in its maximum direction at the same distance, with the dipole being fed with a power of 25 μW. No licensing is required for MICS, but equipment must be certified and operated by or under the direction of a physician or a medical professional.

Wireless Medical Telemetry Service The FCC has allocated the frequency bands of 608–614, 1395–1400, and 1427–1432 MHz for WMTS in the United States

(FCC, 2003). These bands are very advantageous for biomedical telemetry because they allow a relatively large bandwidth for communication (e.g., four 1.5-MHz-wide channels are allowed in the 608–614-MHz WMTS band). Furthermore, WMTS bands are solely reserved for biomedical telemetry, meaning that medical devices which operate at these frequencies are protected from interference caused by other sources. The American Society for Healthcare Engineering (ASHE) maintains a database of WMTS transmitters and is responsible for notifying users of potential frequency conflicts. However, there is currently no indication that the WMTS bands would be allotted in other parts of the world, meaning that devices cannot be marketed or used freely in countries other than the United States. What is more, the WMTS bands are considered to be narrow for high-data-rate applications, such as video or voice transmission.

Industrial, Scientific, and Medical The ISM bands were originally reserved internationally for noncommercial use of radio frequency (RF) electromagnetic fields. They are defined by the ITU-R, but individual countries' use of the bands differs due to variations in national radio regulations. The 902–928- and 2400.0–2483.5-MHz frequency bands are used in the United States and are defined by the FCC, whereas the European countries use the 433.1–434.8- and 868.0–868.6-MHz frequency bands, which are defined by the Electronic Communications Committee (ECC). The ISM bands offer users the advantage of increased bandwidth, thus enabling video and voice transmissions. Furthermore, since government approval is not required, the ISM bands are nowadays being used by a wide variety of commercial standards. However, the ISM bands are not exclusive to biomedical telemetry equipment, meaning that transmission of sensitive medical data in these bands is susceptible to interference from other devices.

Ultrawide Band UWB systems are spread-spectrum communication systems or, equivalently, systems in which the bandwidth of the transmitted signal is considerably wider than the frequency content of the original information. More specifically, UWB is defined by the FCC as any communication system which has a spectral occupation of greater than 20% or occupies an instantaneous bandwidth of more than 500 MHz. The band of 3.1–10.6 GHz, which has been authorized by the FCC for unlicensed use, is nowadays receiving the most attention by standardization bodies. Extremely short pulses are transmitted, and high data rates are thus achieved. Despite the fact that UWB medical devices are currently only allowed in the United States and Singapore, regulatory efforts are already underway in Europe and Japan.

1.4.2 Sensing Technologies

Biomedical telemetry requires the communication with sensors providing physical, chemical, and biological data for continuous monitoring of the physiological state. Advances in microelectromechanical system (MEMS) and biological, chemical, electrical, and mechanical sensor technologies have led to a wide range of medical devices, such as pressure sensors, silicon microphones, accelerometers, gyroscopes, optical MEMS and image sensors, microfluidic chips, microdispensers for drug delivery, flowmeters, infrared (IR) temperature sensors, radio frequency identification (RFID) tags, and strain sensors.

Wearable medical devices monitor electrical, physical, and physiological parameters in a noninvasive way. For example, ECG data can be collected by using smart shirts (Lee and Chung, 2009) or flexible polymeric dry-potential electrodes (Jung et al., 2012). Pulse oximetry allows the monitoring of hemoglobin saturation through the transmission of light of two different wavelengths to a photodetector, while motion analysis can be performed with the help of several types of motion sensors and systems, such as accelerometer gyro sensor, magnetoresisitive sensors, eletromagnetic tracking systems (ETSs), textile sensors, force sensors, and electromyography sensors. Most of the invasive (implantable and ingestible) medical devices monitor human-generated chemical products. For example, MEMS-based capacitive pressure sensors fabricated using silicon micromachining techniques may be used for implantable blood pressure monitoring (Cong et al., 2010), while electrochemical reactions may be identified for implantable pH and blood glucose monitoring (Zeng and Grimes, 2007). In the field of capsule endoscopy, silicon diodes, ion-selective field effect transistors (ISFETs), direct-contact gold electrodes, and electrochemical cells are most commonly used for temperature, pH, conductivity, and dissolved oxygen monitoring, respectively.

1.4.3 Advanced Materials

The explosive growth of biomedical telemetry and biomedical devices is accentuating the need for miniaturized, high-efficiency conformal materials that can operate over a wide range of frequencies and can be integrated in lightweight configurations. Advanced materials are, thus, targeted for the design and fabrication of the different components of medical devices in an attempt to tackle all the subtle but essential problems that hinder their performance or even their realization. Examples of current research focus on advanced material technologies are:

(a) *Inkjet Printing on Paper Substrates with Conductive Inks Which Form Conductive Traces When Cured* Inkjet printing has a significant smaller environmental impact as compared to other methods of electronic device fabrication, such as etching, while the use of paper allows for cheaper production cost of electronics (Lakafosis et al., 2010).

(b) *Electromagnetic Band Gap (EBG) Structures for Wearable Antennas to Be Integrated on Wearable Medical Devices Which Help Eliminate Effects of Lossy Human Body* Advanced techniques consider an ultrathin completely vialess, split-ring-resonator-based EBG design for both on-metal and on-body mounted rugged applications featuring low-cost, low-fabrication-complexity and flexibility characteristics.

(c) *Liquid Ionic Antennas* Liquid antennas were first reported by Ida et al., (2002) and are based on aqueous-salt solutions operating in microwave frequencies (around 1.7 GHz). The liquid must be encapsulated in a flexible, noncorrosive, biocompatible, nontoxic container. As compared to traditional antennas, liquid antennas exhibit significantly improved range and efficiency characteristics as well as a much simpler and flexible fabrication mechanism.

(d) *Inkjet-Printed Substrate Integrated Waveguides (SIWs)* SIW structures can form a transition between microstrip and dielectric-filled waveguides and can be considered as a promising approach for the planar microwave and millimeter wave components. Implementation of SIW structures on paper using inkjet printing technology reduces the overall cost, allows for arbitrary geometries, enhances the flexibility of the SIW components, and offers the possibility for multilayer miniature structures (Bozzi et al., 2011).

1.4.4 Data and Power Circuits

Picked-up physiological signals need to be amplified before being fed to any other component. Designing application-specific, low-noise, low-power amplifiers has been the focus of several research works. For example, a bioamplifier for EEG signals with a power dissipation of 0.9 μW has been proposed and tested by Harrison and Charles (2003). Amplified signals are then digitized and fed to a transceiver which will code and modulate the data and finally lead them to the antenna. The AMIS-52100 is an example of such a low-cost, ultralow-power commercially available transceiver for MICS applications.

Several systems use inductive techniques for data transmission (Figure 1.3*a*), but improvements in transceiver design are enabling devices with integrated antennas (Figure 1.3*b*) which transmit higher data rates at longer distances. Intrabody (Figure 1.3*c*) and optical communications (Figure 1.3*d*) have also been suggested. Improvements in transceiver architectures, data mining, and data compression are used to provide high chip rates at increased distances using low-power biomedical telemetry.

All aforementioned components require power. Batteries are suitable for wearable medical devices as well as for implantable or ingestible devices with a short lifespan (Warty et al., 2008). The reason is that batteries supply power for only a limited period of time, even when conservationist techniques are applied. External power transmission for medical implants, in terms of electromagnetic induction between an exterior-transmitting and an implantable-receiving coil, has been proposed as an alternative (Kendir et al., 2005). Ultimately, the user should not have to replace or recharge the medical devices frequently. It is, thus, desirable to get rid of the battery entirely and rely on energy scavenged. A number of power-scavenging sources have been suggested, including motion, vibration, air flow, temperature difference, light, and infrared radiation. For example, a vibration-based generator for wearable and implantable medical devices capable of delivering 2 μJ/cycle has been designed (Mitcheson et al., 2004), while ambient electromagnetic energy harnessing has recently been investigated (Lakafosis et al., 2010).

The sensor, transmitting coils or antenna, signal processing electronics, and energy source will need to be integrated in a single heterogeneous platform. Ideally, the

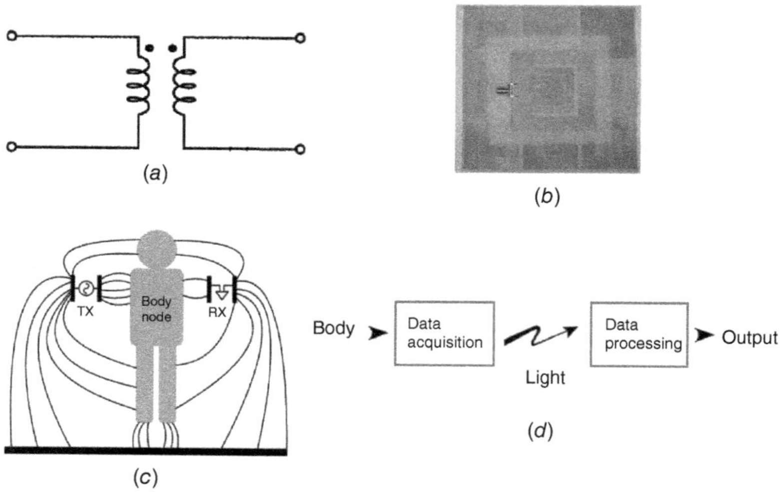

Figure 1.3 Data transmission techniques in biomedical telemetry systems: (*a*) inductive coupling, (*b*) antennas and RF communication (dual-band wearable textile antenna (Zhu and Langley, 2009), (*c*) intrabody communication (Cho et al., 2007), and (*d*) optical communication.

overall medical device should occupy a small size so that the user does not feel discomfort or inconvenience. This implies the use of miniature components as well as advanced integration and packaging technologies. Driven by the effect of Moore's law, the chip size can be divided by 2 every 18 months while preserving the same performance. Fortunately, this miniaturization often also reduces power consumption.

The last decade has witnessed a rapid surge of interest in new sensing and monitoring medical devices. To sum up, the following challenges can be considered as crucial toward the future development of advanced biomedical telemetry systems:

- Improved medical sensor design
- Miniaturization and low power consumption of components
- Development of new materials and methods to allow miniaturized communication systems to be seamlessly integrated within the medical device itself
- Context awareness and multisensor fusion
- Use of power conservation and power-scavenging techniques

1.4.5 Biocompatibility Issues

Biocompatibility can be defined as the capability of a material to exist in harmony with a biological tissue environment. In other words, biocompatibility refers to the ability of a medical device to perform its intended function without eliciting any undesirable biological effects to the surrounding human tissues.

Biocompatibility plays a key role in the development of implantable and ingestible devices and is an important input requirement for their design. Such devices must be biocompatible in order to preserve patient safety and prevent rejection of the implant. Furthermore, human tissues are conductive and would short circuit the implantable or ingestible components if they were allowed to be in direct contact with their metallization. Biocompatibility and prevention of undesirable short circuits are especially significant in the case of devices that are intended for long-term implantation.

It is important to highlight that medical devices consist of a variety of materials. Therefore, in order to assess the biocompatibility of the device, one must consider the applied materials one by one as well as the complete medical device as a whole. Furthermore, biocompatibility of a medical device depends on the time that it is exposed to the human body as well as its specific location inside the body. In any case, the designer of the medical device is responsible for its biocompatibility and safety, rather than the physician.

1.4.6 Standardization and Interoperability

The development of standards for biomedical telemetry communications is necessary in order to provide connectivity for a variety of services in a vast range of communication scenarios. In other words, standards are required in order to enable communication between medical devices and other types of equipment and networks or, equivalently, interoperability. In a strict sense, interoperability means the capability of two devices or systems (of different type, model, and/or manufacturer) to cooperate and communicate.

Wireless communication standards are in general being defined by standardization bodies such as the Institute of Electrical and Electronics Engineers (IEEE), the International Telecommunication Union (ITU), and the European Telecommunication Standards Institute (ETSI). Technologies such as Bluetooth, Zigbee, Wireless Fidelity (WiFi), Worldwide Interoperability for Microwave Access (WiMAX), Global System for Mobile Communications (GSM), General Packet Radio Service (GPRS), and Universal Mobile Telecommunications System (UMTS) are available for short-, medium-, and long-range communications, allowing a wide coverage area and offering the possibility of ubiquitous worldwide wireless mobility of medical devices with telemetry functionalities (Foster and Hao, 2008).

A wireless personal area network (WPAN) is a network of devices centered around an individual's workspace (typical range of approximately 10 m). The original standard for WPANs has been Bluetooth, also known as the IEEE 802.15.1 standard, which offers short-range communications at low data rates under 1 Mbps. The Bluetooth technology has been the basis of a new family of standards, the IEEE 802.15, which includes seven task groups (IEEE 802.15.1 to IEEE 802.15.7). For example, in 2003, Zigbee, known as the IEEE 802.15.4 standard, was ratified, targeting a similar market as Bluetooth but at lower data rates (20–250 kbps), lower cost, and lower power consumption (IEEE, 2003). For higher speed communications, the high-rate WPAN (HR WPAN), known as the IEEE 802.15.3 standard, has been developed, which targets at rates up to 55 Mbps over short ranges. The IEEE 802.15.6 task group focuses on wireless body area networks (WBANs) and aims at low-power and low-frequency short-range communications. WBANs are a more confined version of WPANs and consist of a number of medical devices placed in proximity of or inside the human body, with communication being based on infrared light, microwave radio, or even near-field coupling through skin conduction. The IEEE 802.15.6 task group is responsible for the development of a standard for WBANs, and the purpose of the group is to tackle the main challenges and constraints, such as power consumption and quality of service (QoS).

As technology continues to evolve, existing standards will keep being updated, while new standards will very likely emerge. Standards-based connectivity of medical devices to information and communications technology networks is crucial toward forming a standard interoperable framework, which will allow advanced clinical solutions to be safely and efficiently incorporated. The optimal solution, therefore, would be the implementation of a device integration system that could achieve interoperability today, and that, when combined with any future standards, would form a robust and scalable integration solution, possible to meet any evolving connectivity needs.

1.4.7 Privacy and Security

Issues related to privacy and security of medical devices with telemetry functionalities are considered to be highly critical.

The challenge in data privacy is to share data while protecting patient-identifiable information. Medical data are considered to be highly sensitive in nature and should, thus, be encrypted in order to preserve patient privacy. Although it is important that the physician or the patient can have access to this information, medical data should not be broadcast in public. Biometrics has been used for automated authentication of people to systems for over a decade, making the authentication more convenient as it does not require memorizing passwords or PIN codes (Ivanov and Yu, 2010). Furthermore, laws need to be established in order to control and regulate access to such patient-identifiable information.

Security typically involves protecting the data and the network from theft or destruction. A compromise, such as eavesdropping or tampering by a malicious third party, may result in identity theft, incorrect diagnosis and treatment, and even death. The problem is most commonly decomposed into two subproblems: (a) securing the biomedical telemetry link between the patient and the receiving station and (b) securing the telemedicine link between the receiving station and the network. Radio systems can be made secure with cryptographic strategies to prevent any sort of interception. However, even though it may seem advantageous to encrypt all of the data, it is important to additionally take into account the required computational resources.

1.4.8 Biomedical Telemetry Toward Telemedicine

Telemedicine is defined as the use of ICTs for the exchange of medical information (images, data, audio, etc.) between remote locations which are located far away from

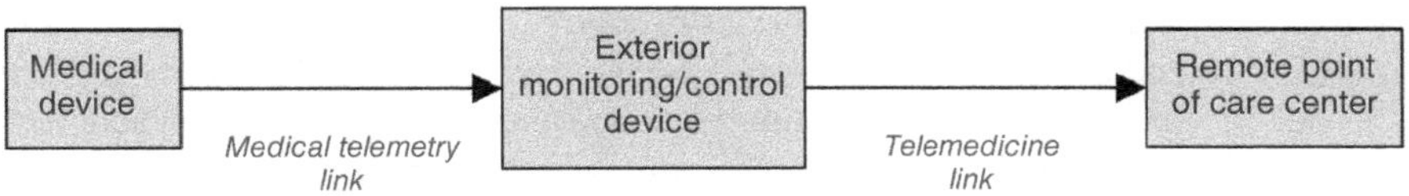

Figure 1.4 Block diagram of biomedical telemetry–telemedicine-integrated system.

each other. Telemedicine is a rapidly growing application of wireless technologies, giving a significant boost in the provision and improvement of health care services. Advances in communications and medical technology have led to increasing deployment of telemedicine systems and services around the world. The aim of such systems is to increase the accessibility of physicians and caregivers, improve the quality of patient care, and reduce the overall cost of health care.

The development of advanced ICTs allows the integration of biomedical telemetry with telemedicine, thus enabling the perfusion of biomedical telemetry into the entire health care supply network. This will influence the architecture of medical devices to accommodate remote-programming facilities as well as the potential of telemedicine itself.

A typical scheme of a system with integrated biomedical telemetry and telemedicine functionalities is shown in Figure 1.4. Biomedical telemetry data sensed by the medical device are obtained at the exterior monitoring or control unit and further forwarded to a remote patient care center for decoding, storage, and analysis. Data forwarding in the telemedicine part of the system is performed by either the telephone (mobile-Health, mHealth) or the Internet (electronic Health, eHealth). Some wireless sensor platforms for pervasive health care monitoring have been designed in order to improve the quality of human life and minimize restrictions on daily activities. Examples include platforms with wearable/on-body sensors (Yang, 2006; Hall et al., 2006) as well as polysomnographic supervision and surveillance systems (Miles, 1999; Penzel et al., 2002). Telemedicine issues related to protocol selection, bandwidth limitations, interference mitigation and interoperability placed within the limitations and requirements imposed by biomedical telemetry are highly challenging.

1.4.9 Patient Safety

Before medical devices can be widely accepted, the public needs to be convinced of their safety. People's perception of electromagnetic radiation is generally fearful and safety standards need to be established in order to quantify biological damage and preserve patient safety. It was not until recently that research on the biological effects of medical devices with telemetry functionalities started being carried out.

The approach that is currently in use for establishing such safety standards is through animal experimentation (Johnson and Guy, 1972). The actual fields, current density, and absorbed energy density which cause biological damage inside the tissues of the animal are recorded and further extrapolated to human beings. The specific absorption rate (SAR), which is defined as the rate of energy deposited per unit mass of tissue, is generally accepted as the most appropriate dosimetric measure. Limits for the United States and Europe are based on recommendations from the IEEE (1999, 2005) and the International Committee on Non-Ionizing Radiation Protection (ICNIRP, 1998), respectively. Actual regulations may vary according to the scenario under study and legislation of each country.

Device manufacturers must show that their products do not introduce higher SAR values than the specified limits. This is most commonly accomplished through in vivo experimental investigations or numerical computations. It is worth noting that the actual SAR values not only are determined by the medical device itself but also depend on any device in the close vicinity of the human body that could influence the fields inside the human tissues. Increasing the power incident in the medical device to improve its communication range may result in the device exceeding the regulations for maximum power absorption inside the body and must be taken into account by manufacturers.

1.5 COMMERCIAL MEDICAL TELEMETRY DEVICES

A number of commercial medical telemetry devices have already been reported for wearable, implantable, and ingestible applications. Indicative examples can be summarized as follows.

1.5.1 Wearable Devices

Disposable wireless diagnostic skin patches that can be read directly by cell phones or computers can be customized by GENTAG to revolutionize mobile or home remote diagnostics worldwide (Figure 1.5*a*) (Gentag, 2011). There exist four types of patches, that is, the glucose monitoring patch, the fever monitoring patch, the post orthopedics surgery skin patch, and the drug delivery skin patch.

Corventis has designed the AVIVO Mobile Patient Management (MPM) System to provide continuous insight into the health status of ambulatory patients, such as those living with heart failure or fluid management problems (Avivo, 2009). The goal is to

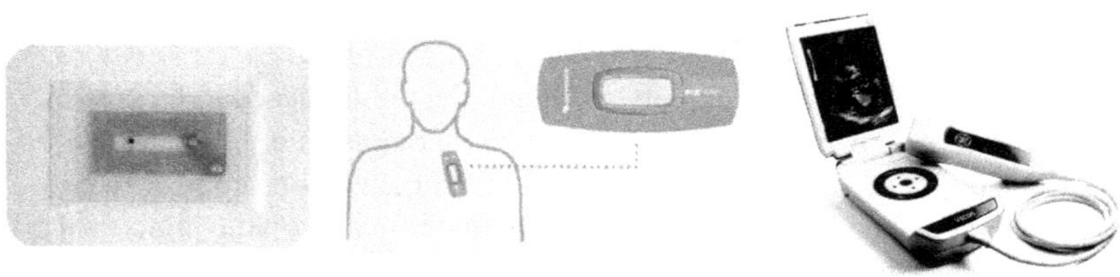

(*a*) (*b*) (*c*)

Figure 1.5 Commercial wearable devices: (*a*) Gentag diagnostic skin patch [Copyright © Gentag, Inc. (Gentag, 2011)]; (*b*) Corventis AVIVO Mobile Patient Management System [Copyright © 2010 Corventis, Inc. (AVIVO, 2009)]; (*c*) GE Healthcare VScan (GE Healthcare, 2009).

help health care providers proactively identify concerning trends (fluid status, heart rate, respiration rate, activity, posture, etc.) and intervene before problems progress. Patients are continuously monitored via a wearable device (PiiX), an unobtrusive, water-resistant device designed to support patient compliance (Figure 1.5*b*). Health information is automatically collected while patients go about their daily activities. This information is then wirelessly transmitted via a small hand-held device (zLink) to Corventis. Clinical reports containing physiological trends are then delivered and made available at the Corventis website to prescribing physicians. Physicians may also be contacted by the Corventis Monitoring Center directly when health care conditions that meet predefined criteria are detected.

Recently, GE Healthcare announced the commercial release of a new, smartphone-size imaging tool, known as the VScan, which lets physicians carry ultrasound technology in their pockets (Figure 1.5*c*) (GE Healthcare, 2009). The VScan imaging device is cleared as a prescription device for ultrasound imaging, measurement, and analysis in the clinical applications of abdominal, cardiac (adult and pediatric), urological, fetal, pediatric, and thoracic/pleural motion and fluid detection. A wand attached to the device sends and receives sound wave data upon contact with the body. This is then translated into an image for analysis by the physician.

1.5.2 Implantable Devices

Biotronik has recently proposed a small battery-powered electrical impulse generator to be implanted in patients who are at risk of sudden cardiac death due to ventricular fibrillation and ventricular tachycardia (Lumax 540 DR-T) (Figure 1.6*a*) (Biotronik, 2012). The process of implantation of an implantable cardioverter–defibrillator (ICD) is similar to implantation of a pacemaker. Similar to pacemakers, these devices typically include electrode wire(s) that pass through a vein to the right chambers of the heart, usually lodging in the apex of the right ventricle. The difference is that pacemakers are more often temporary and are generally designed to correct bradycardia, while ICDs are often permanent safeguards against sudden arrhythmias.

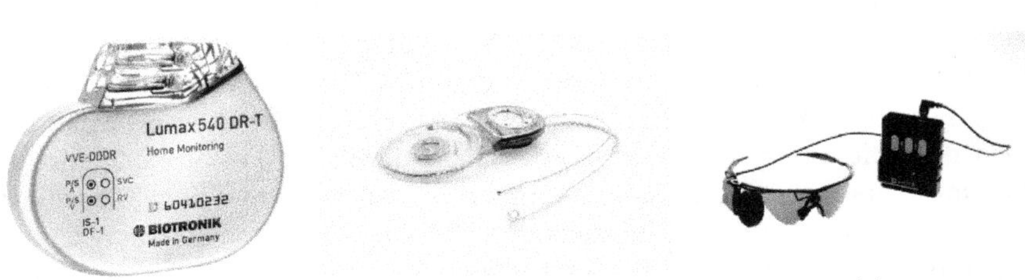

(*a*) (*b*) (*c*)

Figure 1.6 Commercial wearable devices: (*a*) Biotronik defibrillator [Copyright © Biotronik, Inc. (Biotronik, 2012)]; (*b*) Nucleus Freedom Cochlear Implant [Image courtesy of Cochlear Americas (Nucleus Freedom, 2010)]; (*c*) Second Sight Argus II Retinal Prosthesis [Copyright © 2013 Second Sight Medical Products, Inc. (Second Sight, 2012)].

The Medtronic Adapta with MVP pacing system offers managed ventricular pacing (MVP), atrial therapy, ventricular capture, and remote cardiac telemetry (Medtronic, 2010a). The new standard of care in pacing is to reduce unnecessary right ventricular pacing to as close to zero as possible. Mounting evidence suggests that right ventricular pacing is associated with a variety of detrimental effects, including risk of heart failure hospitalization and atrial fibrillation. On the other hand, the Medtronic Revo MRI SureScan pacing system is magnetic resonance (MR) conditional designed to allow patients to undergo magnetic resonance imaging (MRI) under the specified conditions of use (Medtronic, 2011a).

The Nucleus Freedom cochlear implant includes a sound processor which is worn behind the ear and a cochlear implant which is placed under the skin, behind the ear (Figure 1.6*b*) (Nucleus Freedom, 2010). The sound processor captures sounds, digitizes them, and sends the digital code to the implant. The implant converts the digitally coded sound to electrical impulses and sends them along an electrode array to further stimulate the cochlea's hearing nerve. Hearing may be managed via a remote assistant or directly from the sound processor.

The Medtronic SynchroMed Pump is a drug infusion system which provides precise drug delivery for chronic therapy of severe spasticity (Medtronic, 2012). The pump is part of the SynchroMed II programmable drug infusion system which provides precise drug delivery for chronic therapy for severe spasticity. In addition to the implanted pump, the SynchroMed II infusion system uses a catheter to deliver programmed amounts of intrathecal baclofen (a muscle relaxant and antispasticity agent) directly to the intrathecal space and cerebrospinal fluid.

The Argus II retinal implant was approved by the FDA's Ophthalmic Devices Advisor Panel in 2012 (Figure 1.6*c*) (Second Sight, 2012). It includes a video camera, a transmitter mounted on a pair of eyeglasses, a video processing unit, and a 60-electrode implanted retinal prosthesis that replaces the function of degenerated cells in the retina. Although it does not fully restore vision, this setup can improve a patient's ability to perceive images and movement using the video processing unit to transform images from the video camera into electronic data that are wirelessly transmitted to the retinal prosthesis.

Finally, implantable glucose monitoring systems appear as a promising treatment for diabetes on a continuous basis [e.g., Medtronic Guardian REAL-Time (Medtronic, 2010b), Medtronic MiniMed Paradigm Veo (Medtronic, 2011b), Dexcom SEVEN Plus (Dexcom, 2008), and Abbott FreeStyle Navigator (Abbott, 2011)]. For some patients, who either have very brittle diabetes and experience sudden dramatic changes in their blood glucose levels or do not experience the symptoms of highs and lows, testing several times a day using traditional finger prick tests may not provide sufficient information to be able to prevent potentially harmful levels of blood glucose. Therefore, implantable glucose monitoring systems appear as highly promising for continuous monitoring. A tiny sensor is inserted under the skin to measure glucose levels and further transmit this information to an exterior monitor via radio waves, as shown in Figure 1.7.

Challenges to be addressed by commercial implantable devices which will be employed in the market in the years to come are shown in Figure 1.8. The first

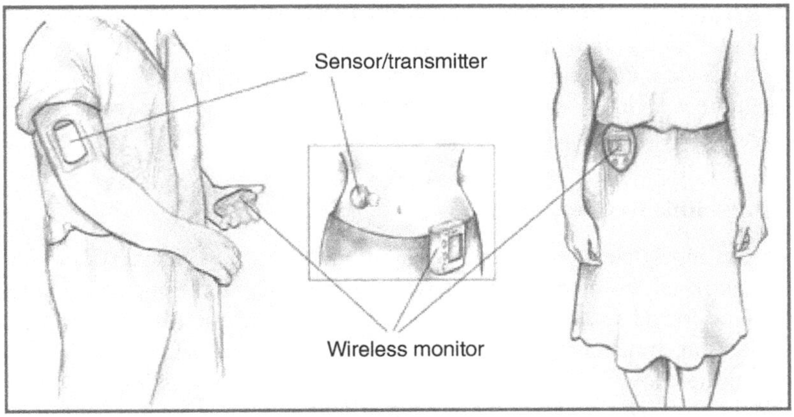

Figure 1.7 Schematic diagram of implantable systems used for continuous glucose monitoring.

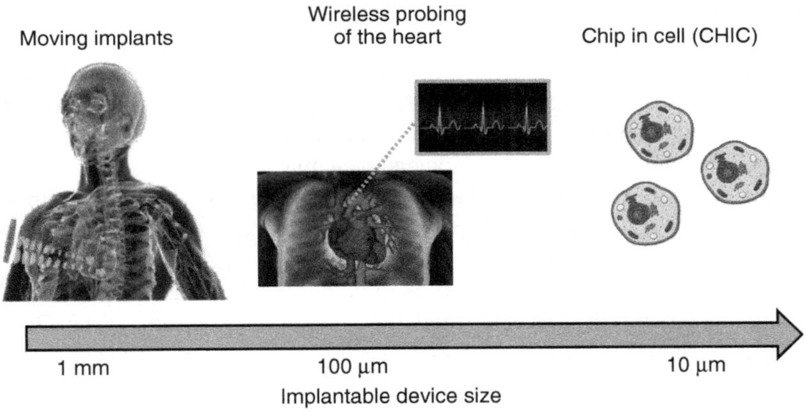

Figure 1.8 Future challenges for commercial implantable medical devices.

challenge involves the design of moving implants with a size of the order of 1 mm. The idea is similar to the paddle in kayaking: Asymmetrical shape produces asymmetrical drag forces, with alternate direction of electromagnetic torque resulting in a net forward force. Wireless endocardial pacing and sensing are emphasized as two of the most favorable applications of such moving implants. The second challenge involves the design of miniature (order of 100 μm) implants for wireless intracardiac electrogram transmission, wireless probing of the heart, and wireless epicardial mapping. Finally, the chip-in-cell technology, also known as CHIC technology, is set as the third challenge of future commercial implantable devices. The goal is to reduce the size of the implant down to a few micrometers in order to design autonomous sensors which will perform active and continuous monitoring of the cellular activity. It is important to highlight that the choice of cell is what will actually determine the size of the device. For example, the Xenopus Oocytes cell has a size of 1 mm and would require a device size of 50 μm, whereas the Plant Protoplasts cell has a size of 60 μm and would require a device size of 10–15 μm.

1.5.3 Ingestible Devices

Commercial ingestible medical devices are used for gastrointestinal (GI) endoscopy and sensing of physiological parameters within the GI (pH, temperature, pressure), which allow for direct and noninvasive examination of the GI tract. Images and sensed data are transmitted from a disposable, ingestible wireless video capsule and are further downloaded for review. In 2000, the introduction of low-power, complementary metal–oxide–semiconductor-based (CMOS-based) image sensors and application-specific integrated circuits (ASICs) made the video capsule possible. Capsules are mainly composed of a CMOS image sensor, light emission diodes (LEDs) for illumination, a miniature video transceiver of sufficient output power, a microcontroller unit (MCU), a cell battery, and optional sensors. They travel through the small intestine via normal muscle contractions, with pictures and data being wirelessly transmitted to a patient-worn receiving device.

Up until 2007, wireless endoscopic capsules were only developed by Given Imaging (2012a). The vitamin-sized capsule (PillCam) provides a way to visualize, monitor, and diagnose small-bowel abnormalities including abnormalities associated with obscure GI bleeding (OGIB), iron deficiency anemia (IDA), and Crohn's disease. It captures a broad mucosal area per image with a 156° field of view at 4.5 mm working distance. In addition, advanced optics and automatic light control provide optimal image quality and illumination.

After 2007, other companies, such as Olympus (2012) and IntroMedic (2012) made significant improvements in their own endoscopic capsules. The Olympus Endocapsule (Figure 1.9*a*) is a small-bowel endoscopy system. The capsule travels

(*a*) (*b*)

Figure 1.9 Commercial ingestible devices: (*a*) Olympus Endocapsule (Olympus, 2012) and (*b*) IntroMedic MiroCam (IntroMedic, 2012).

through the small intestine via normal muscle contractions, taking thousands of pictures that are transmitted to a recorder worn around the waist. The IntroMedic MiroCam (Figure 1.9*b*) is a capsule endoscope based upon HBC (human body communication), a state-of-the-art patented technology utilizing the human body as a communication medium.

Based on the above, most of the capsules are intended for visualization, monitoring, and diagnosis of small-bowel abnormalities. However, there also exist capsules with integrated sensor technologies which aim at physiological parameter sensing within the GI tract. For example, the Bravo pH Monitoring System (Medtronic, 2010c) is a catheter-free way to measure pH. The Bravo system involves a pH capsule that is temporarily attached to the wall of the esophagus. Throughout the 24- or 48-hr study period, the capsule measures pH levels in the esophagus and transmits readings via radio telemetry to a receiver worn on the patient's belt or waistband. Another example is the Smartpill Wireless Mobility Capsule (Given Imaging, 2012b), which uses sensor technology to measure pH, pressure, and temperature from within the entire GI tract. Once the patient ingests the capsule, it transits the GI tract collecting data and sending them to a receiver worn by the patient. The single-use capsule is capable of transmitting data continuously for more than five days and is excreted naturally from the body. Recently, Philips marketed the Intellicap ingestible device, which performs targeted delivery of pharmaceutical drugs and biologicals to the GI tract (Philips, 2008).

1.6 OVERVIEW OF BOOK

Biomedical telemetry is a highly modern scientific field, and no coordinated effort has been performed so far toward writing a book that will address all arising scientific issues and technologies. The *Handbook of Biomedical Telemetry* aims to form a complete book in the field and includes generic information, detailed scientific analyses, as well as example applications. To accurately cover the broad range of topics arising within biomedical telemetry, the book comprises 23 chapters, divided into three parts, where each delves, in both breadth and depth, onto the corresponding subject addressed.

The structure of the book is shown in Figure 1.10. Themes being addressed are summarized as follows:

The first part of the book (Chapters 2–5) is entitled Biomedical Telemetry Devices and addresses technologies for the design of biomedical telemetry devices, biomedical sensing, and power techniques for biomedical telemetry applications.

- Chapter 2 (Design Considerations of Biomedical Telemetry Devices) focuses on the system and circuit design of biomedical telemetry devices from the transmitter, receiver, and communication link point of view.
- Chapter 3 (Sensing Principles for Biomedical Telemetry) reviews and analyzes the various different recognition and detection principles of biosensors,

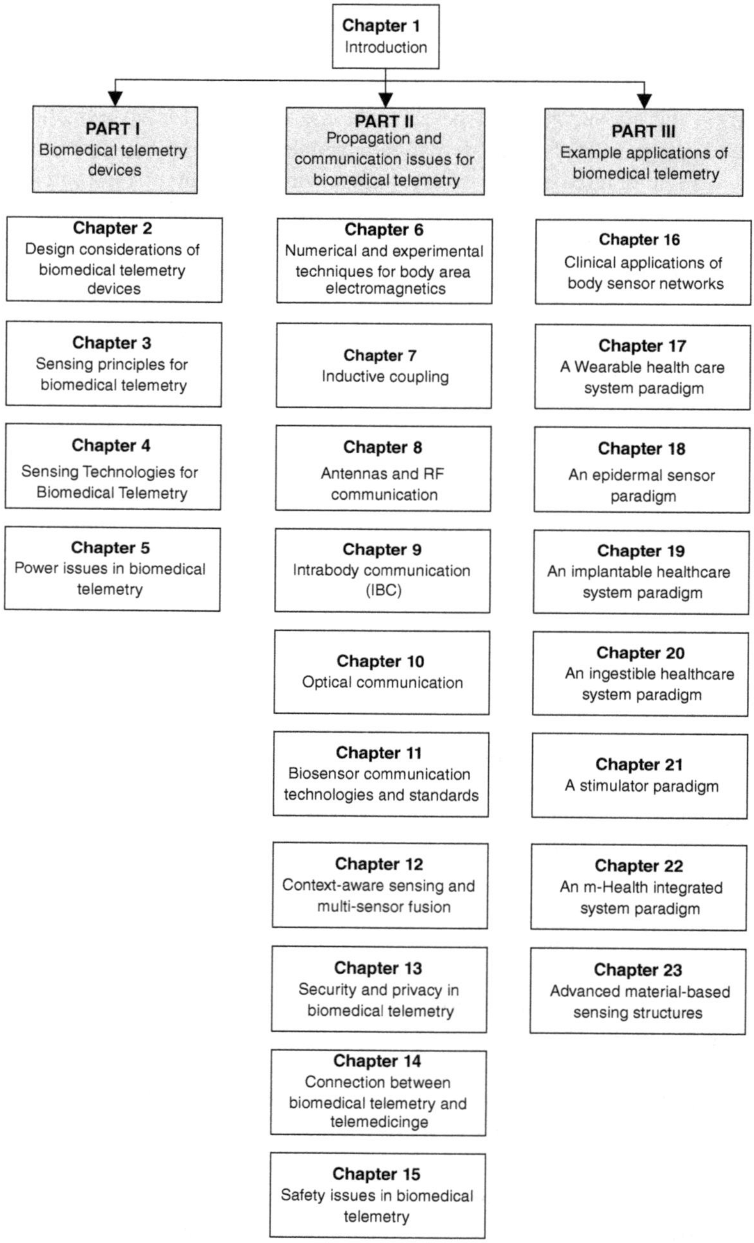

Figure 1.10 Structure of the book.

providing insight into the design challenges and the applications to which they are applied.

- Chapter 4 (Sensing Technologies for Biomedical Telemetry) concentrates on the sensor operation, as required to provide physical, chemical, and biological data for continuous monitoring of physiological parameters.
- Chapter 5 (Power Issues in Biomedical Telemetry) reviews the status and the major challenges concerning the powering and integration of biomedical telemetry systems.

The second part of the book (Chapters 6–15) deals with propagation and communication issues for biomedical telemetry and covers numerical and experimental techniques for body area electromagnetics, inductive coupling, antennas and RF communication, intrabody communication, optical communication, biosensor communication technologies and standards, context-aware sensing and multisensor fusion, security and privacy issues, connection between biomedical telemetry and telemedicine, and safety issues and exposure assessment to high-frequency biotelemetry fields.

- Chapter 6 (Numerical and Experimental Techniques for Body Area Electromagnetics) makes an overview of the (numerical and experimental) bioelectromagnetics modeling tools and methods to be used in biomedical telemetry.
- Chapter 7 (Inductive Coupling) introduces the methods and challenges in transmitting wide-band data in addition to power across transcutaneous inductive links.
- Chapter 8 (Antennas and RF Communication) addresses wireless biomedical telemetry performed by means of antennas for wearable, implantable, and ingestible applications.
- Chapter 9 (IntraBody Communication) offers varied material that ranges from IBC basics for the generally knowledgeable individual working in the biomedical telemetry field to more specialized topics on modeling and simulation of IBC.
- Chapter 10 (Optical Biotelemetry) discusses the technical considerations related to the use of light as a transmission medium for biomedical telemetry.
- Chapter 11 (Biosensor Communication Technology and Standards) addresses the communication and networking aspects of biomedical telemetry systems.
- Chapter 12 (Context-Aware Sensing and Multisensor Fusion) describes systems and methods for context-aware sensing and multisensor fusion in the field of body sensor network applications.
- Chapter 13 (Security and Privacy in Biomedical Telemetry) provides an overview on information security, host computer security, and network methodologies and biometrics used for authentication/authorization and for protection of health information in networking.

- Chapter 14 (Connection between Biomedical Telemetry and Telemedicine) defines ways in which biomedical telemetry and telemedicine integrated environments can deliver seamless, personalized, and nonobtrusive health care services to people.
- Chapter 15 (Safety Issues in Biomedical Telemetry) discusses concerns and safety guidelines for unintended effects of electromagnetic fields on the human body.

The third part of the book (Chapters 16–23) addresses example applications of biomedical telemetry and presents an introduction to clinical applications of body sensor networks (BSNs) as well as selected examples of wearable, implantable, and ingestible devices, stimulators, integrated mobile health care systems, and advanced material-based sensing paradigms for monitoring and therapeutic intervention.

- Chapter 16 (Clinical Applications of Body Sensor Networks) introduces the concept of biotelemetry in health care, defining its key aspects, and reviews technological advances central to facilitating scalable and cost-effective biotelemetry.
- Chapter 17 (Wearable Health Care System Paradigm) presents examples of wearable wireless physiological measurement systems in specific application areas.
- Chapter 18 (Epidermal Sensor Paradigm) introduces a new method for in vivo determination of the human tissue dielectric properties.
- Chapter 19 (Implantable Health Care System Paradigm) discusses the paradigm of an implantable antenna for a cardiac pacemaker designed to operate in the MICS band.
- Chapter 20 (Ingestible Health Care System Paradigm for Wireless Capsule Endoscopy) provides an overview of the methodologies used on wireless endoscopic imaging and presents the design of an ingestible medical device with telemetry functionalities.
- Chapter 21 (Stimulator Paradigm) provides an introduction to the artificial retina system and provides design strategies and constraints for a suitable telemetry system.
- Chapter 22 (An mHealth Integrated System Paradigm) describes the METABO system, an integrated mobile diabetes management tool, by summing up the state of the art concerning mobile diabetes mellitus management systems and integrated mHealth systems in general.
- Chapter 23 (Advanced Material-Based Sensing Structures) presents advanced material technology used for biomedical sensing and discusses the major integration challenges.

REFERENCES

Abbott. 2011. FreeStyle Navigator. Available: https://www.abbottdiabetescare.com/; last accessed: April 29, 2013.

Anliker U, Ward JA, Lukowicz P, Tröster G, Dolveck F, Baer M, Keita F, Schenker EB, Catarsi F, Coluccini L, Belardinelli A, Shklarski D, Alon M, Hirt E, Schmid R, Vuskovic M. 2004. AMON: A wearable multiparameter medical monitoring and alert system. *IEEE Trans. Inf. Technol. Biomed.*, 8(4):415–427.

Avivo. 2009. Mobile Patient Management System. Available: http://www.corventis.com/us/avivo.asp; last accessed: April 29, 2013.

Berezdivin R, Breinig R, Topp R. 2002. Next-generation wireless communications concepts and technologies. *IEEE Comm. Mag.*, 40(3):108–116.

Biotronik. 2012. Lumax. Available: http://www.biotronik.com/wps/wcm/connect/en_de_web/biotronik/sub_top/healthcareprofessionals/products/tachyarrhythmia_therapy/; last accessed: April 29, 2013.

Bozzi M, Georgiadis A, Wu K. 2011. Review of substrate-integrated waveguide circuits and antennas. *IET Microw. Antennas Propag.*, 5(8):909–920.

Cho N, Yoo J, Song SJ, Lee J, Jeon S, Yoo HJ. 2007. The human body characteristics as a signal transmission medium for intrabody communication. *IEEE Trans. Microwave Theory Tech.* 55(5):1080–1086.

Cong P, Ko WH, Young DJ. 2010. Wireless batteryless implantable blood pressure monitoring microsystem for small laboratory animals. *IEEE Sensors J.*, 10(2):243–254.

Dexcom. 2008. Seven Plus. Available: http://www.dexcom.com/seven-plus; last accessed: April 29, 2013.

European Radiocommunications Commission (ERC). 1997. Recommendation 70–03 relating to the use of short range devices (SRD). Conf. Eur. Postal Telecomm. Admin. (EPT), CEPT/ERC 70–03, Annex 12.

Federal Communications Commision (FCC). 2003. Code of Federal Regulations, Title 47 Part 95, WMTS Band Plan. Available: http://www.fcc.gov.

Foster R, Hao Y. 2008. Wireless body sensor networks for health-monitoring applications. *Physiol Meas.*, 29(11):27–56.

GE Healthcare. 2009. VScan. Available: https://vscan.gehealthcare.com/gallery/a-quick-look-at-vscan; last accessed: April 29, 2013.

Gentag. 2011. Diagnostic Skin Patch. Available: http://www.gentag.com/applications.html; last accessed: April 29, 2013.

Given Imaging. 2012a. PillCam. Available: http://www.givenimaging.com/en-int/Innovative-Solutions/Capsule-Endoscopy/Pages/default.aspx; last accessed: April 29, 2013.

Given Imaging. 2012b. Smart Pill. Available: http://www.smartpillcorp.com/; last accessed: April 29, 2013.

Hall PS, et al. 2006. *Antennas Propagation for Body-Centric Wireless Communications*. London: Artech House.

Harrison RR, Charles C. 2003. A low-power low-noise CMOS amplifier for neural recording applications. *IEEE J. Solid-State Circuits*, 38:958–965.

Ida I, Sato J, Sekizawa T, Yoshimura H, Ito K. 2002. Dependence of the efficiency-bandwidth product on electrical volume of small dielectric loaded antennas. *IEEE Trans. Antennas Propag.*, 50(6):821–826.

Institute of Electrical and Electronics Engineers (IEEE). 1999. Standard for safety levels with respect to human exposure to radiofrequency electromagnetic fields, 3 kHz to 300 GHz, *IEEE* C95.1-1999.

Institute of Electrical and Electronics Engineers (IEEE). 2005. Standard for safety levels with respect to human exposure to radiofrequency electromagnetic fields, 3 kHz to 300 GHz, *IEEE* C95.1-2005.

Institute of Electrical and Electronics Engineers (IEEE). 2003. Wireless medium access control (MAC) and physical layer (PHY) specifications for low-rate wireless personal area networks (LR-WPANs), *IEEE* 802.15.4.

International Commission on Non-ionizing radiation Protection (ICNIRP). 1998. Guidelines for limiting exposure to time-varying electric, magnetic, and electromagnetic fields (up to 300 GHz).

International Telecommunications Union-Radiocommunications (ITU-R). 1998. Recommendation ITU-R SA.1346.

International Telecommunications Union-Radiocommunications (ITU-R). Radio regulations, Section 5.138 and 5.150. Available: http://www.itu.int/home.

IntroMedic. 2012. MiroCam. Available: http://www.intomedic.com/eng/sub_products_2.html; last accessed: April 29, 2013.

Ivanov VI, Yu PL. 2010. Securing the communication of medical information using local biometric authentication and commercial wireless links. *Health Inf. J.*, 16(3):211–223.

Johnson CC, Guy AW. 1972. Nonionizing electromagnetic wave effects in biological materials and systems. *Proc. IEEE*, 60(6):692–718.

Jung HC, Moon JH, Baek DH, Lee JH, Choi YY, Hong JS, Lee SH. 2012. CNT/PDMS composite flexible dry electrodes for long-term ECG monitoring. *IEEE Trans. Biomed. Eng.*, 59(5):1472–1479.

Kendir GA, Liu W, Wang G, Sivaprakasam M, Bashirullah R, Humayun MS, Weiland JD. 2005. An optimal design methodology for inductive power link with class-E amplifier. *IEEE Trans. Circuits Syst.*, 52:857–866.

Kiourti A, Nikita KS. 2012a. A review of implantable patch antennas for biomedical telemetry: Challenges and solutions. *IEEE Antennas Propag. Mag.*, 54(3):210–228.

Kiourti A, Nikita KS. 2012b. Recent advances in implantable antennas for medical telemetry. *IEEE Antennas Propag. Mag.*, 54(6):190–199.

Kiourti A, Nikita KS. 2013. Design of implantable antennas for medical telemetry: Dependence upon operation frequency, tissue anatomy, and implantation site. *Int. J. Monit. Surveillance Technol. Res.*, 1(1):16–33.

Lakafosis V, Rida A, Vyas R, Li Y, Nikolaou S, Tentzeris MM. 2010. Progress towards the first wireless sensor networks consisting of inkjet-printed, paper-based RFID-enabled sensor tags. *IEEE Proc.*, 98:1601–1609.

Lee YD, Chung WY. 2009. Wireless sensor network based wearable smart shirt for ubiquitous health and activity monitoring. *Sensors Actuators B: Chem.*, 140(2):390–395.

Lin J, Nikita KS. 2010. *Wireless Mobile Communication and Healthcare*. Springer LNICST, p. 55.

McCaffrey C, Chevalerias O, O'Mathuna C, Twomey K. 2008. Swallowable-capsule technology. *Pervasive Comput.* 7(1):23–29.

Medical Implant Communications Service (MICS) 1999. Rules Reg, *Fed. Reg.* 64: 69926–69934.

Medtronic. 2010a. Adapta with MVP Pacing System. Available: http://www.medtronic.com/for-healthcare-professionals/products-therapies/cardiac-rhythm/pacemakers/adapta-with-mvp-pacing-system/; last accessed: April 29, 2013.

Medtronic. 2010b. Guardian REAL-Time. Available: http://www.medtronic-diabetes-me.com/Guardian-REAL-Time.html; last accessed: April 29, 2013.

Medtronic. 2010c. Bravo pH Monitoring System. Available: http://www.medwow.com/used-gastro-esophageal-monitoring/medtronic/bravo-ph-system/395666699.item; last accessed: April 29, 2013.

Medtronic. 2011a. Revo MRI SureScan. Available: http://www.medtronic.com/for-healthcare-professionals/products-therapies/cardiac-rhythm/pacemakers/revo-mri-pacing-system/; last accessed: April 29, 2013.

Medtronic. 2011b. MiniMed Paradigm Veo. Available: http://www.medtronic-diabetes.co.uk/product-information/paradigm-veo/index.html; last accessed: April 29, 2013.

Medtronic. 2012. SynchroMed II Pump. Available: http://www.medtronic.com/patients/severe-spasticity/therapy/itb-therapy/synchromed-ii-pump/; last accessed: April 29, 2013.

Miles LE. 1999. World-wide real-time remote polysomnographic supervision, surveillance and control using an Internet virtual sleep laboratory. 16th International Symposium on Biotelemetry.

Mitcheson PD, Green TC, Yeatman EM, Holmes AS. 2004. Architectures for vibration-driven micropower generators. *IEEE J. Microeletromech. Sys.*, 13:429–440.

Nikita KS, Lin JC, Fotiadis DI, Arredondo Waldmeyer MT. 2011. *Wireless Mobile Communication and Healthcare*. Springer LNICST, p. 83.

Nikita KS, Lin JC, Fotiadis DI, Arredondo Waldmeyer MT. 2012. Special issue on mobile and wireless technologies for healthcare delivery. *IEEE Trans. Biomed. Eng.*, 59(11):3083–3270.

Nucleus Freedom. 2010. Cochlear Nucleus System. Available: http://www.cochlear.com/wps/wcm/connect/us/recipients/recipient-support/nucleus; last accessed: April 29, 2013.

Olympus. 2012. EndoCapsule. Available: http://www.olympusamerica.com/msg_section/endocapsule/; last accessed: April 29, 2013.

Penzel T, Brandenburg U, Peter JH, Otto R, Hundemer HP, Lledo A, Wetter TC, Trenkwalder C. 2002. A new design of a polysomnography-based multi-center treatment study for the restless legs syndrome. *Clin. Neurophysiol.*, 113:571–578.

Philips. 2008. Intellicap. Available: http://www.research.philips.com/initiatives/intellicap/index.html; last accessed: April 29, 2013.

Rappaport TS, Annamalai A, Beuehrer RM, Tranter WH. 2002. Wireless communications: Past events and a future perspective. *IEEE Comm. Mag.*, 40(5):148–161.

Second Sight. 2012. Argus II Retinal Prosthesis System. Available: http://2-sight.eu/en/product-en; last accessed: April 29, 2013.

Sivaprakasam M, Liu W, Humayun MS, Weiland JD. 2005. A variable range bi-phasic current stimulus driver circuitry for an implantable retinal prosthetic device. 40(3):763–771.

Topol E. 2012. *The Creative Destruction of Medicine: How the Digital Revolution Will Create Better Health Care*. Basic Books, New York.

Warty R, Tofighi MR, Kawoos U, Rosen A. 2008. Characterization of implantable antennas for intracranial pressure monitoring: Reflection by and transmission through a scalp phantom. *IEEE Trans. Microw. Theory Tech.*, 56:2366–2376.

Yang GZ. 2006. *Body Sensor Networks*. New York: Springer.

Zeng K, Grimes CA. 2007. Wireless magnetoelastic physical, chemical, and biological sensors. *IEEE Trans. Magn.*, 43(6):2358–2363.

Zhu S, Langley R. 2009. Dual-band wearable textile antenna on an EBG substrate. *IEEE Trans. Antennas Propag.*, 57(4):926–935.

PART I

BIOMEDICAL TELEMETRY DEVICES

2

DESIGN CONSIDERATIONS OF BIOMEDICAL TELEMETRY DEVICES

DOMINIK CIRMIRAKIS AND ANDREAS DEMOSTHENOUS

Department of Electronic and Electrical Engineering, University College London, London, United Kingdom

2.1 INTRODUCTION

The specifications for an implantable medical device (IMD) are provided by the application but share a common set of constraints in size, power, and functionality. The interplay between these factors determines the wireless communication link range and bandwidth, battery lifetime, and electronics technology. The size restrictions limit the power consumption of an IMD and as a result functionality is limited. Moreover, the size dictates the dimensions of the antennas and their radiation efficiency.

Powering of IMDs is usually performed by an inductive link (Harb et al., 2004), but other types of power delivery have been reported (Olivo et al., 2011). Besides power, in many applications there is a need for bidirectional communication through the same set of coils or another type of communication link [e.g., high-frequency radio frequency (RF) transmission] as shown in Figure 2.1. Different modulation techniques have been used to transmit data to and from IMDs. The most important factor in the design process is to keep the power consumption in the implant as low as possible. Another consideration is the frequency and power transmitted by the carrier, which is limited due to the absorption of the higher frequency RF signals by the tissue (Olivo et al., 2011). This results in a limitation to high data rates, which require higher carrier frequencies. It is also desirable to have a simple demodulator for the data link from the transmitter to the implant (downlink) which is placed in

Handbook of Biomedical Telemetry, First Edition. Edited by Konstantina S. Nikita.

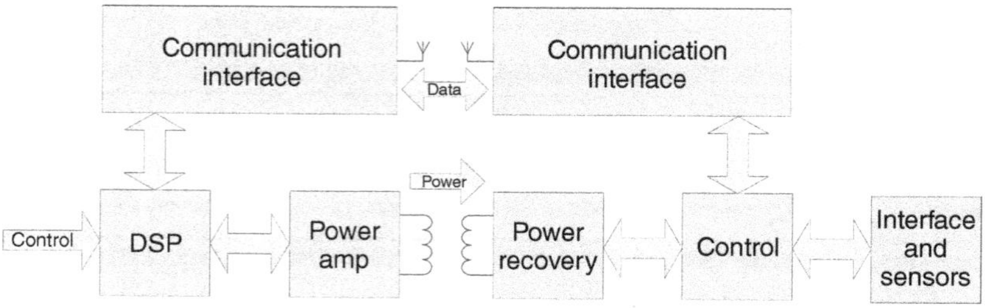

Figure 2.1 Basic architecture of biomedical telemetry devices.

the implant and should provide low power consumption and high reliability [low bit error rate (BER)]. The same concept is used for communication from the implant to the transmitter (uplink), but the complexity of the circuitry is often shifted to the transmitter so that the modulator can take advantage of lower power consumption and simple implementation.

This chapter provides a brief overview of the design of biomedical telemetry devices and in particular IMDs using inductive links. The basic system architecture and inductive link fundamentals are discussed, as well as methods and circuits for power transfer and data communication. Safety issues relating to overheating of body tissue as a result of increase of temperature are also examined.

2.2 ENERGY TRANSFER TYPES

There are several methods of harvesting energy for IMDs such as kinetic, thermoelectric, fuel cells, infrared radiation, low-frequency magnetic fields, and inductive links.

Kinetic energy can be converted to electric via electromagnetic, electrostatic, or piezoelectric energy harvesters. Electromagnetic energy harvesters generate energy due to the change of an external magnetic flux through a closed-loop conductor. Electrostatic energy harvesters are built from capacitors where an external force changes the position of its plates. The change in the position of the plates causes the voltage across the capacitor to change. This technique is suitable for low-power systems like microelectromechanical systems (MEMSs) as it is hard to achieve high power. In many cases precharging is also required. Piezoelectric energy harvesters use piezoelectric materials where mechanical deformation generates an electric field. This technique generates very low power in the range of 40–60 μW with footprints of 1 cm^3 (Elfrink et al., 2009), which makes it unattractive for IMDs.

Thermoelectric energy harvesters are based on the Seedback effect. A voltage drop is generated between two materials with different temperatures (thermocouple). The power generated is in the range of a few hundred microwatts with a size of 0.19 cm^3 (Stark and Stordeur, 1999).

In fuel cells, current is generated by the electromechanical reaction between two types of chemical. It was reported by in vivo experiments in Malachesky et al. (1972)

that fuel cells abiotycally catalyzed can generate up to 50 μW/cm^2. The drawback of this method is a limited lifetime of one month.

The infrared power-harvesting method is based on optical transmission of infrared light, which can generate current in photodiodes (Murakawa et al., 1999; Goto et al., 2001). The size of photodiodes needed for efficient power transfer and the limitation in distances between transmitter and IMD make them unsuitable.

A low-frequency magnetic field may be used to rotate an implanted rotor and thereby generate power (Suzuki et al., 1999, 2002). The high level (few watts) of transferred power makes it attractive but the dimensions of the rotor as well as the maintenance of mechanical components might be challenging for implantation.

Finally, the most widely used technique for power (and data) transfer in IMDs is the inductive link approach. In this method, one coil works as a transmitter while another coil works as a receiver. The coils are coupled by a magnetic field generated by the external primary coil. High power transfer of up to 250 mW and the possibility of bidirectional communication via the same inductive link make this method the more suitable for IMDs. The communication to the implant is achieved by modulation of the magnetic field. Communication from the implant is achieved by modulation of the secondary coil load that is reflected in the primary and can be detected by current changes in the primary coil.

2.3 ARCHITECTURE OF INDUCTIVELY COUPLED BIOMEDICAL TELEMETRY DEVICES

Wireless inductive telemetry systems can be grouped into two categories, active and passive transmission methods, as shown in Figure 2.2. In active transmission, the implant uses a local oscillator that generates the RF waveform carrying data. In passive transmission, the implant does not generate RF but interacts with a waveform generated by the external circuit. In passive transmission the implant is placed under the skin because power attenuation is high and relatively large antennas (coils)

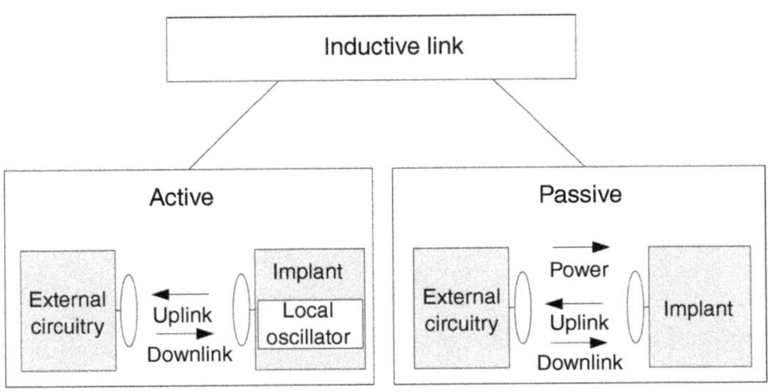

Figure 2.2 Inductive link transmission methods.

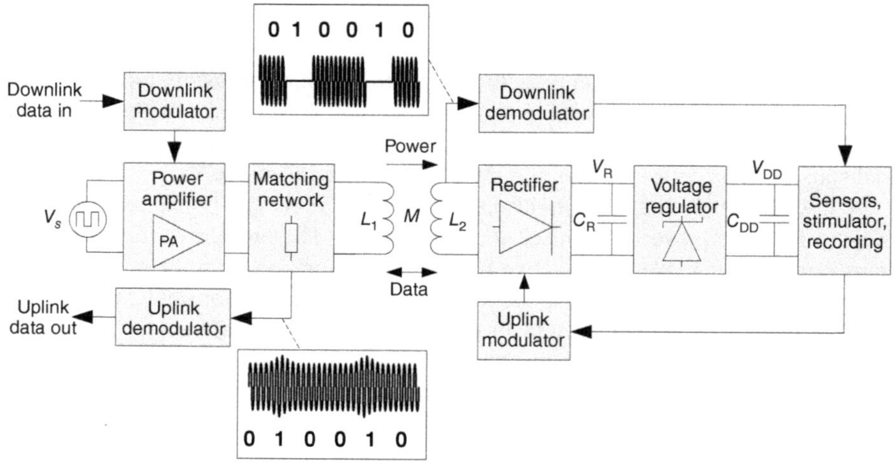

Figure 2.3 General block diagram of a wireless inductive telemetry system.

are needed. Active transmission is usually used in small implants, for example, in ingestible devices where small antennas are used.

A simplified block diagram of a passive, inductively coupled biomedical telemetry system is shown in Figure 2.3. The external transmitter usually consists of a class D or class E power amplifier capable of providing large currents in the tuned primary coil (L_1) from a relatively low voltage (Donaldson and Perkins, 1983; Terman, 1943; Zierhofer and Hochmair, 1996; Jow and Ghovanloo, 2007; Kumar et al., 2011; Ghovanloo and Najafi, 2004a). In the implant the induced voltage that appears across the secondary coil (L_2, tuned by a capacitor) is then rectified and regulated to provide a power supply for the electronics. The purpose of tuning is to boost power transfer efficiency. The data link from the transmitter to the implant, called the *downlink*, is usually achieved by modulating the envelope of the power carrier to create detectable changes across the secondary coil. The data link from the implant to the external circuit, called the *uplink*, is implemented by load modulation techniques, which utilize the property of the mutual coupled coils in which a change in the load of the secondary circuit is reflected back as changing (reflected) impedance in the primary through their mutual inductance. Due to the risk of tissue heating, the upper operational frequency is limited to about 10 MHz (Donaldson and Perkins, 1983; Zierhofer and Hochmair, 1996; Jow and Ghovanloo, 2007; Ko et al., 1977) because the magnetic field can easily penetrate the biological media (Galbraith et al., 1987). Power typically ranges from 10 to 250 mW while practical data rates of about 1 Mbps are achievable.

2.3.1 Inductive Link Fundamentals

The link properties related to power transfer between air-coupled coils have been studied in several applications (Donaldson and Perkins, 1983; Terman, 1943;

Zierhofer and Hochmair, 1996; Jow and Ghovanloo, 2007; Kumar et al., 2011; Ghovanloo and Najafi, 2004; Ko et al., 1977; Suzuki et al., 1999, 2002). The coils' quality factors and mutual inductance are the most important parameters in coil design for IMDs. Donaldson and Perkins (1983), Foster (1981), Hochmair (1984), and Jow and Ghovanloo (2007) developed a design procedure to adjust the coils' parameters for a particular "working" position of the critical coupling in order to minimize variation of the receiver voltage and maximize power transfer. Ko et al. (1977) have found a relationship between the known coil diameters and their relative positions, Zierhofer and Hochmair (1996) recommended distributed winding to improve coupling between coils, and Dormer et al. (1980) introduced the use of a magnet to minimize lateral displacement and hold the transmitter coil in place. Donaldson and Perkins (1983) showed a relationship between the numbers of turns in the coils for optimal operation. Jow and Ghovanloo (2007) introduced a design procedure for printed spiral coils. Another recent approach by Kumar et al. (2011) introduced the concept of four coils to improve power transfer. Additionally, Simard et al. (2010) used three coils in an advanced arrangement to reduce coupling between them and separate the communication and the power link.

The term "link gain" can refer to the voltage gain, current gain, transimpedance gain, and transconductance gain depending on the arrangement of the primary and secondary circuits. The circuit arrangement of Figure 2.4 refers to the voltage gain. As the coils move relative to each other, their gain changes. This implies that there is a strong dependence between the gain and their relative position.

The circuit model in Figure 2.4 consists of two resonant circuits: the primary and the secondary. The primary circuit of the link is tuned in series in order to provide a low-impedance load to the driving transmitter. At resonance, the voltages related with C_1 and L_1 cancel each other and thus the primary circuit requires small voltage swings at its inputs. Therefore, the primary circuit loads the secondary circuit with a small load. On the other hand, the secondary is a parallel LC circuit. The topology of the secondary circuit is tuned in parallel in order to amplify the induced voltage enough and to drive a nonlinear rectifier load (Sawan and Gosselin, 2008). In contrast, this resonating configuration results in an increasing current flowing through the secondary coil and hence to an increased power dissipation.

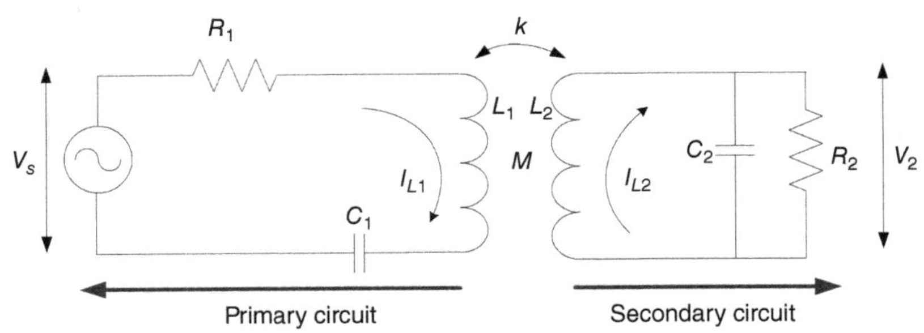

Figure 2.4 Schematic of inductive link.

In the Laplace domain the voltage gain transfer function of the inductive link is given by (Sawan and Gosselin, 2008)

$$\frac{V_2(s)}{V_s(s)} = \frac{sMR_2}{\left(sL_2 + s^2R_2L_2C_2 + R_2\right)\left[R_1 + 1/\left(sC_1\right) + sL_1\right] - s^2M^2\left(1 + sR_2C_2\right)} \tag{2.1}$$

where R_1 and C_1 are respectively the series resistance and tuning capacitance of the primary coil L_1 and R_2 and C_2 are respectively the equivalent resistance (parallel loss resistance combined with load resistance caused by the implanted circuits) and tuning capacitance of the secondary coil L_2. The rest of the symbols have their usual meaning.

Additionally, assuming that the operating frequency is low enough, the geometry of the coils in the primary and secondary circuits can be simplified into three parameters: their mutual inductance M and their self-inductances L_1 and L_2. The coupling coefficient k is given by (Sawan and Gosselin, 2008)

$$k = \frac{M}{\sqrt{L_1L_2}} \qquad 0 \le k \le 1 \tag{2.2}$$

The coupling coefficient shows the amount of magnetic flux from the primary coil coupled to the secondary coil. It varies with the coils' size, geometry, and mutual separation and the properties of the medium separating the coils. Hence, their mutual separation and the lateral and angular displacements affect the link gain as a result of the mutual coupled coils (Soma et al., 1987). Because of these constraints, the coupling coefficient can only be calculated for simple configurations. Considering the case of air-cored parallel coils, center aligned, the coupling coefficient between the mutually coupled coils can be empirically defined as (Sauer et al., 2005)

$$k = \frac{r_2^2 r_1^2}{\sqrt{r_2 r_1}\left(\sqrt{x^2 + r_1^2}\right)^3} \tag{2.3}$$

Thus, the coupling coefficient for the case of parallel and center-aligned coils depends on r_1, r_2, that is, the radii of coils L_1 and L_2, and on x, the distance between them. For air-cored coils, typical values of k are between 0.01 and 0.1 (Sauer et al., 2005). When k is zero, the coils are said to be fully decoupled. When k is 1, all the magnetic flux generated by coil L_1 is fully coupled to coil L_2. For the case of equation (2.3), k can be equal to 1 when x is zero and when the radii of the coils are the same.

Solving equation (2.2) for M and substituting it into equation (2.1), a new expression for the voltage gain can be obtained. Also, further rearrangement is needed since equation (2.1) is given in the Laplace domain. Thus,

$$\frac{V_2(j\omega)}{V_s(j\omega)} = -\frac{\omega^2}{\omega_0^2} \times \frac{k\sqrt{L_2/L_1}}{k^2 + 1/\left(Q_1Q_2\right) - \left(1 - \omega^2/\omega_0^2\right) + j\left(1 - \omega^2/\omega_0^2\right)\left(1/Q_1 + 1/Q_2\right)} \tag{2.4}$$

where $Q_1 = \omega L_1/R_1$ is the quality factor of the primary coil L_1, $Q_2 = \omega L_2/(R_2+R_L)$ is the quality factor of the secondary coil L_2, and ω is the source's operating frequency. The primary and secondary circuits are tuned at the same resonant frequency, given by

$$\omega_0 = \frac{1}{\sqrt{L_1 C_1}} = \frac{1}{\sqrt{L_2 C_2}} \tag{2.5}$$

From equation (2.4) the maximum gain, that is, the highest transferred function of the inductive link, is achieved when k becomes equal to the critical coupling coefficient k_c given by (Sawan and Gosselin, 2008)

$$k_c = \frac{1}{\sqrt{Q_1 Q_2}} \tag{2.6}$$

In the concept of biomedical implants this statement is of significant importance since it implies that the link transfers maximum power when the resistance in the primary is equivalent with the reflected secondary resistance assuming that reactive components are being canceled. Therefore, at resonant frequency and critical coupling, half of the energy in the primary coil is transferred to the secondary circuit. This implies that efficiency η at critical coupling is equal to 50%. The efficiency is calculated by (Sawan and Gosselin, 2008; Donaldson and Perkins, 1983)

$$\eta = \frac{k^2}{k_c^2 + k^2} \tag{2.7}$$

When relatively high Q factors are maintained, the link behaves as a narrow-band bandpass filter. Assuming critical coupling and that the quality factors of the secondary and primary coils are equal (i.e., $Q = Q_1 = Q_2$), the bandwidth of the link can be estimated using equation (2.4) and is found to be

$$B = \frac{\omega_0}{2\pi Q}\sqrt{2} \tag{2.8}$$

Typically, low coupling coefficients are achieved ($k < 0.1$) since the primary and secondary coils are weakly coupled together. This happens because, despite the fact that the coils have to be small, they are also separated by a skin layer of a few centimeters. Further to this, the inductance of the primary L_1 and secondary L_2 coils is small and it can be maintained using coils of few metal loops. This also shows the main difference with transformers where high coupling coefficients can be achieved. Although the inductive links differ from transformers, the basic principles for the operation of transformers can still be applied.

Power transfer (gain factor) optimization and data transfer (bit rate) optimization have contradicting requirements. To illustrate this, the circuit in Figure 2.4 was simulated in ADS software with $R_1 = 1\ \Omega$, $L_1 = 1\ \mu H$ tuned by C_1 at 1 MHz and $L_2 = 1\ \mu H$, $R_2 = 1\ k\Omega$ tuned by C_2. As shown in Figure 2.5, the optimum power transfer is at the resonant frequency (1 MHz) and coupling coefficient of around 0.05.

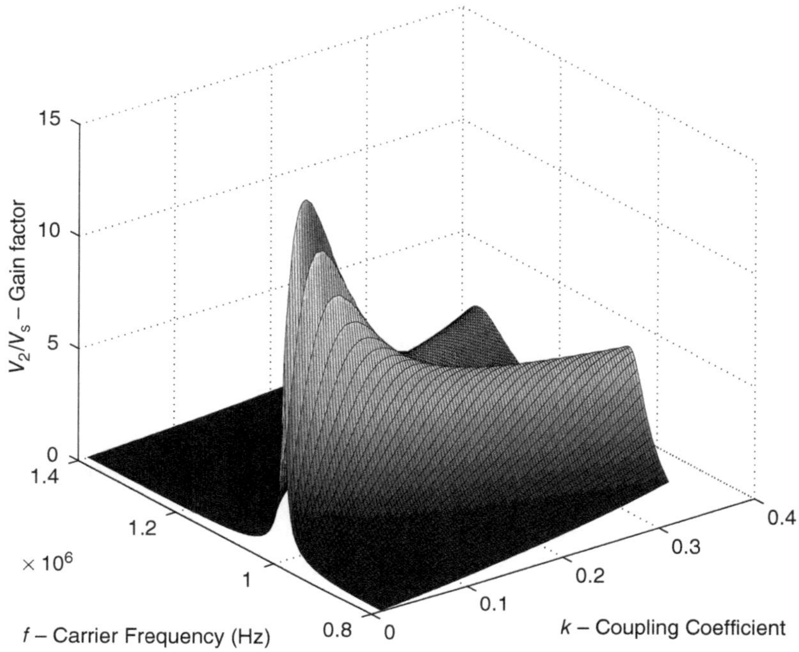

Figure 2.5 Inductive link gain factor versus coupling coefficient and carrier frequency simulated in ADS for circuit model in Figure 2.4.

The power transfer can be optimized (increased) by increasing the quality factor Q. The bandwidth of the inductive link can be increased by lowering the Q. Moreover, the bandwidth of the link increases as the coupling coefficient increases, but it is important that the link be designed for the worst-case scenario. Often power and data links are separated in frequency and space because it allows independent optimization to maximize performance (Simard et al., 2010).

2.3.2 Coupling Compensation

A number of closed-loop techniques have been developed to keep the RF amplitude at a minimum (Wang et al., 2005; Zierhofer and Hochmair, 1990; Si et al., 2008; Galbraith et al. 1987; Ghovanloo and Atluri, 2007) and to keep the voltage amplitude across the secondary coil constant. These techniques require some kind of a feedback loop in order for the transmitter to sense the amplitude received in the implant. The simplest method requires a change of the supply voltage of the power amplifier according to the measured coupling at a known load. A second method is to change the carrier frequency while the inductive link is tuned to one resonant frequency. This causes a change of the amplitude across the resonant circuit, and the amplitude on the implant side can be controlled. The third method is based on stagger tuning (Galbraith et al., 1987). Both the transmitter and receiver resonant frequencies are different from the carrier frequency, but they are on separate sides of the carrier frequency. The change in the position does not change the receiver amplitude because the transfer factor is compensated by the transmitter or receiver tuning network. It requires an accurate design of resonating frequencies for the transmitter and receiver. Recently, an investigation of closed-loop power telemetry has been presented (Kiani and Ghovanloo, 2010). In this method, an increase or decrease of the supply voltage of the amplifier in the transmitter is controlled by the measured direct-current (DC) level of the rectified voltage in the implant. This requires a feedback mechanism in the uplink protocol to be implemented in order for the transmitter to know whether the receiver's rectifier voltage is too high or too low. The key for stability using this method is the frequency of updates of the rectified voltage in the transmitter. Nevertheless, when design constraints are followed, it can be very efficient in terms of power transfer.

2.3.3 Rectification and Voltage Regulation

Power conversion from RF to implant internal DC power supply is achieved by rectification. Of prime importance is the power dissipated in the rectifier. There are several different topologies that can be used. The choice of the circuit is a compromise between limitations of the foundry process technology, power dissipation, area, and input loading. Figure 2.6 shows different topologies of rectifiers where the rectification elements are presented as diodes but could as well be any element with directional conductivity.

The rectifiers pass current only during a small period of the carrier cycle, and it can be assumed that the forward voltage drop is maximal at that point. If the load current is $I_{DC,}$ then the power dissipated in circuits with diodes can be expressed by

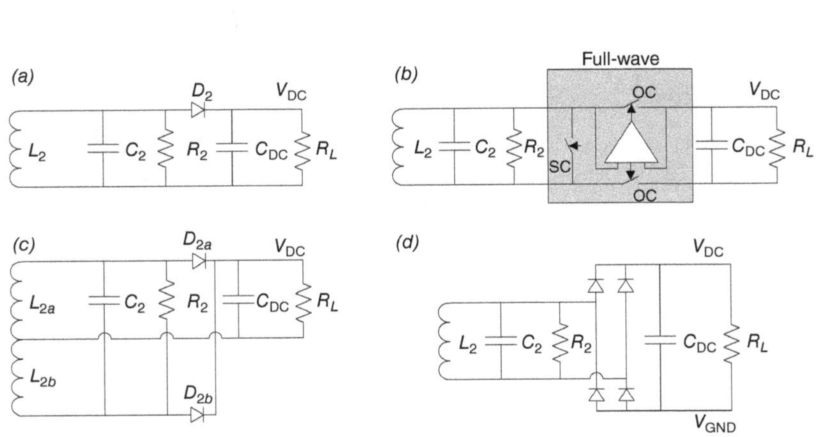

Figure 2.6 Different arrangements of rectifier configurations: (*a*) half-wave; (*b*) full-wave with built-in back load shift keying (LSK) telemeter (Bawa and Ghovanloo, 2008)]; (*c*) full-wave rectifier, and (*d*) full-wave bridge rectifier.

$$P = nI_{\text{DC}}V_F \tag{2.9}$$

where n is the number of rectification elements and V_F is the forward drop.

The efficiency can be expressed as

$$\eta = \frac{I_{\text{DC}}V_{\text{DC}}}{P_D + I_{\text{DC}}V_{\text{DC}}} = \frac{V_{\text{DC}}}{nV_F + V_{\text{DD}}} \tag{2.10}$$

In IMDs, data and power links are often not separated, so the nature of the load seen by the coil can affect the data transmission. The problem is created by nonlinear loads and the most harmful effect is intermodulation. Harmonic distortion can be suppressed when a simple envelope detector is used. The half-wave rectifier is seen by the inductive link as an asymmetric load and as a result even-order distortion products appear in the signal and direct conversion performance is degraded (Abidi, 1995; Binkley et al., 1997). The other rectifier examples in Figure 2.6 are symmetrical and odd-order distortion dominates.

In an inductive link, the voltage needs to be stabilized after rectification. Low-dropout-voltage (LDO) regulators are an obvious choice (Hu et al., 2003; Cirmirakis et al., 2013) because of their high efficiency (better than shunt regulators), simple implementation, small area, and good transient response if designed correctly.

The input current in the voltage regulator is always higher than the output current. The current efficiency of the regulator presented in Figure 2.7 can be represented by

$$\eta_I = \frac{I_{\text{DDL}}}{I_{\text{DDH}}} = \frac{1}{\alpha + 1} - \frac{I_{\text{int}}}{(\alpha + 1)\,I_{\text{int}} + (\alpha + 1)^2 I_{\text{DDL}}} \qquad I_{\text{DDH}} = I_{\text{int}} + (\alpha + 1)\,I_{\text{DDL}} \tag{2.11}$$

The current I_{int} should be as small as possible to increase efficiency as it is the internal current of the regulator. To minimize the transient load noise and other transients (mainly from digital blocks in the load), a large off-chip capacitor is usually placed at the output node of the regulator.

2.3.4 Transmitter Power Amplifier

To transfer power to the receiver, a power amplifier is required in the output stage of the transmitter. The transmitter coil needs relatively high alternating current (AC) to generate an electromagnetic field that will generate current flow in the receiver coil.

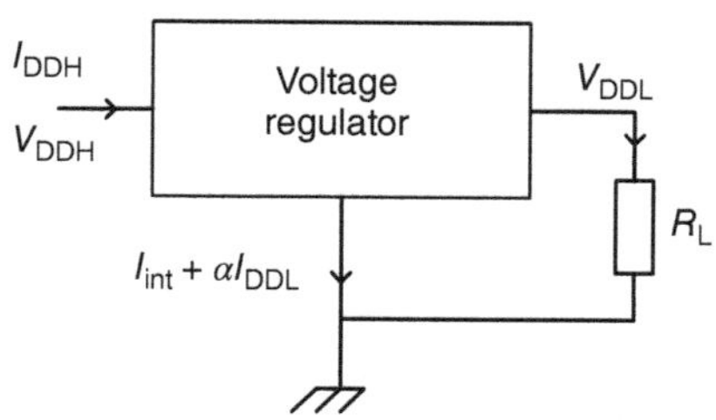

Figure 2.7 Simple model of regulator for current efficiency calculation.

The most common method of driving the transmitter coil is a class E power amplifier. This topology was invented in 1975 by Sokal and Sokal (1977). Detailed analysis on this class of amplifier can be found in Raab (1977). The basic circuit contains a single power transistor which switches current through the *LC* network. The efficiency of this amplifier is close to 100%, which makes it very attractive for designers. However, there are some difficulties in implementing back-telemetry schemes because this topology is very sensitive to rapid load variations.

Another method of driving the transmitter coil is a class D power amplifier. This topology consists of two power transistors that are alternately switched on and off, generating current flow in the loaded *LC* network. The efficiency of this amplifier is above 90%, and this topology is more stable as it is less sensitive to rapid load variations. Power transistors used in the output stage have special packages to minimize parasitic effects and increase the power dissipative abilities of the device (large pads).

2.4 DATA TRANSMISSION METHODS

As mentioned earlier, data communication is called the downlink from the transmitter to the implant, and the uplink from the implant to the transmitter. Due to dual data and power transfer, the carrier amplitude is greater than or equal to the supply voltages. In regular RF communication usually the first block is an amplifier, but in IMDs it is often an attenuator. High-efficiency expectations in the transmitter set some limitations on modulation techniques. As a result, frequency modulation (FM) and pulse amplitude modulation (PAM) are never used. In contrast, amplitude modulation (AM), phase modulation (PM), and quadrature modulation (QAM) are commonly used because of better power efficiency. Another important factor is BER, but a comparison based on signal-to-noise ratio (SNR) cannot be used as, in inductively coupled systems, power levels are too large. Coils used as antennas for these systems are in near field; therefore the operational distance is not great.

For uplink communication, when one set of coils is used, load modulation is always employed. There is a danger of disruption in power delivery if low impedance (short) is connected for too long to the implant coil.

Because the carrier is supplying power to the implant, long periods of low amplitude should not occur. When the communication mode is in idle, the link should be optimized for maximum power transfer. Another approach is to fix the power associated with each bit to the level where the implant can safely operate. In inductively coupled systems, communication can work only in the half-duplex mode as downlink can affect uplink and vice versa.

2.4.1 Downlink

A number of different modulation schemes have been reported for inductively coupled IMDs. Binary amplitude shift keying (BASK), binary frequency shift keying (BFSK), and binary phase shift keying (BPSK) are the most common. The simplest in implementation is BASK. However, it is harmful to amplitude noise and as a result the data transmission rate is typically limited to about 10% of the carrier frequency (Hu and Sawan, 2005). In BFSK modulation the binary data are represented by constant amplitude using two different frequencies, where logic 1 is assigned to one frequency and logic 0 is assigned to the other. Compared to BASK, BFSK can provide data-to-carrier frequency ratio as high as 67% (Ghovanloo and Najafi, 2003, 2004b, 2005). However, the BFSK scheme requires a wide pass band in the inductive link to allow different frequencies, which limits the power transfer due to high filtration by *RLC* components. The advantage of BPSK over BASK and BFSK is the use of a carrier with fixed amplitude and fixed frequency (Luo and Sonkusale, 2008), which allows for efficient and stable transfer of power. Unfortunately, the disadvantage of the BPSK is that it requires a much more complex demodulator in the IMD, implemented by some kind of phase-locked loop (PLL) [usually a Costas loop (Hu and Sawan, 2005)]. Another important factor is the amount of power sent via the RF link as excessive power can damage the tissue around the IMD. The power level should be kept slightly above operational level and never exceed safety standards [International Organization for Standardization (ISO) 14708-1 part 17].

If sidebands are in the bandwidth of the link, the modulation of the carrier can be achieved by amplitude, frequency, and phase modulation and any derivatives of these methods, namely, ASK, ASK on–off keying (ASK-OOK), BPSK, offset phase shift keying (OPSK), quadrature phase shift keying (QPSK), and offset quadrature phase shift keying (OQPSK). The most commonly used is ASK-OOK because it is simple and transmission power efficient. Moreover, demodulation can be implemented by a simple envelope detector.

ASK Demodulator The ASK demodulator is the simplest to implement and consists of only three stages, envelope detector, low-pass filter and level detector, as shown in Figure 2.8*a*. The time constant of the peak-detection *RC* elements should be set to a value larger than the worst-case time rise of the carrier across the implant coil (the coil element is not shown). The drawback of this method is its sensitivity to noise and relatively slow speed. To increase the BER, wide Schmitt hysteresis is used and as a result high modulation indexes and long “low” carrier time must be used to transmit data. The approach in Figure 2.8*b* is more resistant to errors and noise but dissipates more power (mainly in triggers). In this method two Schmidt triggers (one with wide hysteresis and another with narrow hysteresis) are used for the detection of amplitude modulation. When a low carrier occurs, a constant logic high appears at the output of the triggers. However, an increase/decrease of the carrier is not instant, and therefore a narrow trigger is used for sampling the second trigger output.

BPSK Demodulator BPSK is a modulation method where the input signal is shifted between 0° and 180° to represent logic levels 0 and 1. The nature of the inductive link of filtering harmonics and introducing transient response does not attenuate the baseband tone which can still be demodulated. An example of a BPSK demodulator for IMDs is shown in Figure 2.9 (Hu and Sawan, 2004).

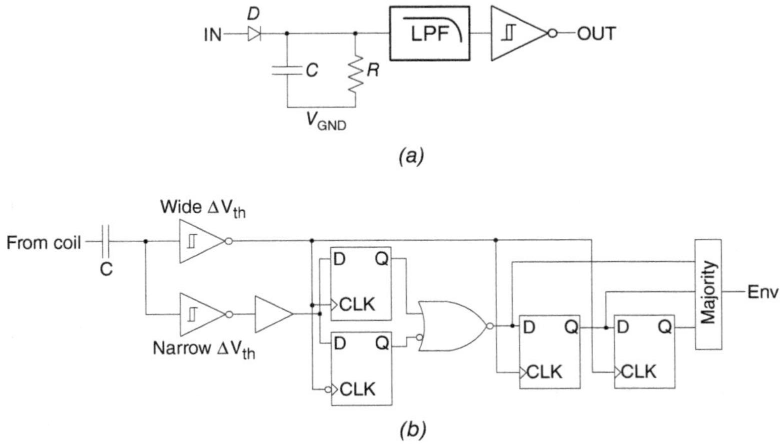

Figure 2.8 ASK demodulator circuits: (*a*) simple ASK based on envelope detection (Coulombe and Sawan, 2004); (*b*) fast ASK demodulator working in differential mode (Coulombe et al., 2003).

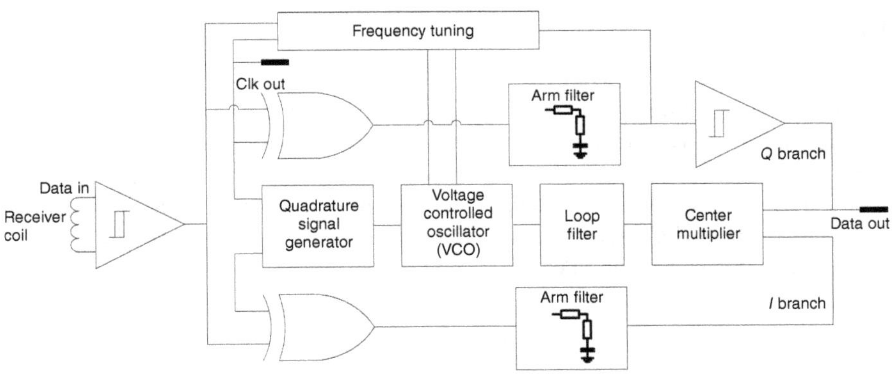

Figure 2.9 Block diagram of BPSK demodulator (Hu and Suwan, 2004).

The Costas loop is the most commonly used technique for demodulation of BPSK. Two parallel PLLs, one for the in-phase (*I*) and one for the quadrature (*Q*) branches, are used. The multiplication product of both is used for controlling the voltage-controlled oscillator (VCO) to track carrier frequency. The digitized *Q* branch follows the demodulated data. The advantage of this method is that the carrier can continuously supply power to the implant even during the time that the modulation is performed.

FSK Demodulator Common techniques of FSK demodulation such as FM discriminator, PLL, or quadrature detector circuits need low-pass filtering in the signal path

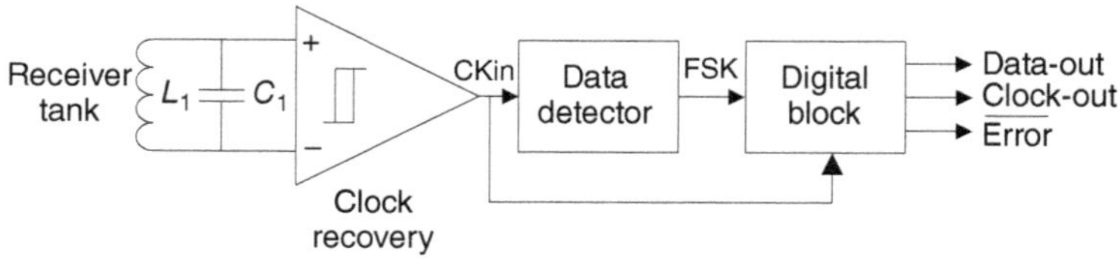

Figure 2.10 Block diagram of FSK demodulator (Ghovanloo and Najafi, 2004).

where a large capacitor is required and would consume a large chip area. Figure 2.10 shows a low-power implantable FSK demodulator (Ghovanloo and Najafi, 2004a) where the FSK carrier is treated as a baseband signal to avoid mixing or filtering. The demodulation technique is based on measuring the period of each carrier cycle for two different frequencies. If the period is higher than a specified threshold, logic 1 is detected, otherwise logic 0 is detected. The clock recovery block digitizes the sinusoidal carrier from the receiver tank. The detector block detects short and long carrier periods and feeds it into the digital block where the data stream and data clock are recovered.

2.4.2 Uplink

The most commonly used technique for communication between the implant and the transmitter is LSK (Tang et al., 1995). This modulation is based on the reflection of the implant's load to the transmitter via the inductive link. A typical implementation is shown in Figure 2.11. The data stream shorts the coil, and the change in impedance is reflected to the transmitter because the implant load is much bigger than the on resistance of the switch transistor. The impedance of the secondary at resonance is expressed as (Donaldson and Perkins, 1983)

$$Z_2 = \frac{L_2}{R_2 C_2} \tag{2.12}$$

and the impedance of the primary at resonance with reflected $Z_2' = (\omega M^2)/Z_2$ is given by

$$Z_1 = R_1 + \omega^2 k^2 L_1 R_2 C_2 \tag{2.13}$$

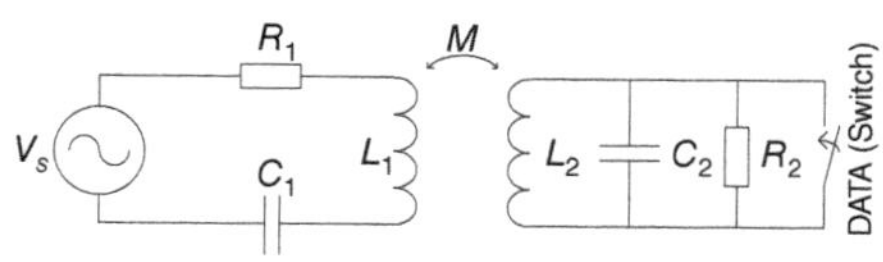

Figure 2.11 Implementation of LSK.

The voltage across L_1 is

$$V_{L1} = \left(\frac{\omega L_1}{Z_1}\right) V_s = \left(\frac{\omega L_1}{R_1 + \omega^2 k^2 L_1 R_2 C_2}\right) V_s \tag{2.14}$$

When the load impedance changes, R_2 changes to R_2', the voltage across L_1 changes from V_{L1} to V_{L1}'. The modulation index is expressed as

$$m = \left| \frac{V_{L1} - V_{L1}'}{V_{L1} + V_{L1}'} \right| \tag{2.15}$$

The bandwidth of this modulation is limited by the coupling, the parameters of the coils, and the transient response of the inductive link. The inductive link bandwidth due to the coil's quality factors sets the maximum data rate at around $1.414(f_{\text{carrier}}/Q)$ at critical coupling (about 200 kbps) (Djemouai and Sawan, 1999).

Another approach is to use BPSK modulation (Simard et al., 2010; Chaimanonart et al., 2010). In this technique, the modulation is performed by changing the tuning of the implant *RLC* network in the imaginary domain. This requires adding or extracting extra capacitance from the tuning capacitor in the receiver *RLC* network. The drawback of using this method is a decrease in power transfer, but the advantage is a faster data rate related to the carrier and speed of the demodulator. Unfortunately, to reach a speed of around 30% of the carrier, a separate inductive link for the power supply is required.

A more advanced technique called passive phase shift keying (PPSK) modulation (Zhou and Donaldson, 2003; Cirmirakis et al., 2012) allows speeds of 10% of the carrier without significant distortion to the power link. Data from the implant are sent by closing or not closing the switch DATA in Figure 2.12 at regular intervals. The switching must be aligned with carrier periods and have a width of $n + 0.5$ carrier periods. Figure 2.12 shows an example of the shape of the voltage and current waveforms during steady state. Switch DATA closes when the voltage V_{L2} crosses zero and the current I_{L2} is at its peak, shorting L_2 and holding this current at its peak. The voltage across C_2 is zero. After half a cycle switch DATA opens in antiphase and the current I_{L2} is forced to go back in phase with the carrier. Part of the current stored in the inductor L_2 charges C_2, while the rest of it (through the inductive link) increases I_{L1} after a short transient and V_{C1} recovers to its steady state. Binary 1 is represented by modulation and binary 0 by steady state (no modulation). In this arrangement, serial communication can be maintained. Power delivery to the implant is only slightly affected as V_{L2} quickly resumes its steady state (Zhou and Donaldson, 2003; Donaldson and Perkins, 1983). The small power loss is proportional to the number of transferred 1's. The drawback of this method is the heavy dependence of the modulation index on the variance of the load.

Recently, a new approach has been reported based on pulse harmonic modulation (PHM) (Inanlou et al., 2011, 2010). The data rate achieved using this method could reach up to 10.2 Mbps, and it is the fastest communication method currently known via inductively coupled coils. Unfortunately, to implement this method, a separate set of coils is required for powering as this method is carrierless.

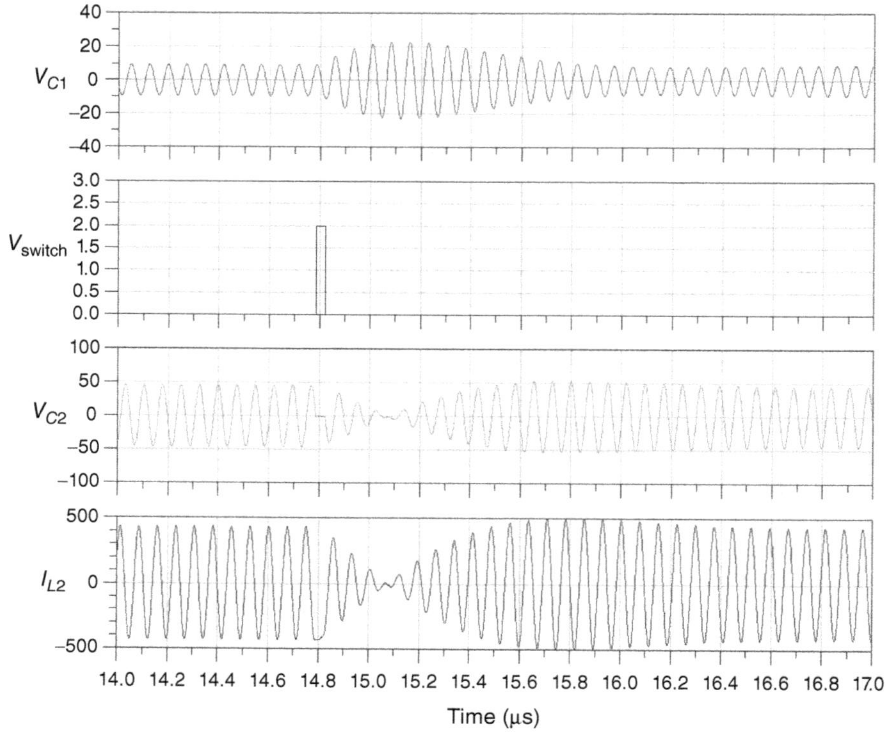

Figure 2.12 PPSK; simulated transient responses in ADS of voltages and currents after modulation pulse with $k = 0.07$. With reference to Figure 2.10, I_{L2} is the current in L_2, V_{C2} is the voltage across C_2, and V_{switch} is the digital pulse applied to the switch.

2.5 SAFETY ISSUES

The major consideration during the operation of any IMD is the possibility of the device overheating the body part where it is located, causing a rise in the temperature which could damage the surrounding tissue (Gabriel et al. 1996). The ISO imposed limitations of the temperature changes in IMDs in standard ISO 14708-1 part 17. This states that an IMD should not create an increase of more than 2°C above the normal body temperature of 37°C during normal operation or during any single fault condition (ISO 14708-1). The time and amplitude of electromagnetic waves exposed to the human tissues at various frequencies are regulated by the Federal Communications Commission (FCC). The American National Standards Institute (ANSI) standard C95.1-1982 describes electromagnetic field strength limits for the general public for frequencies between 300 kHz and 100 GHz (ANSI C95.1-1982; Sawicki, 2011). The ANSI standard C95.1-1982 is overruled by the Institute of Electrical and Electronics Engineers (IEEE) standard C95.1-1991, which defines the electric and the magnetic field strength limits for the general public for frequencies between

TABLE 2.1 IEEE Standard C95.1-1991 (IEEE C95.1)

Frequency Range (MHz)	Electric Field Strength, E (V/m)	Magnetic Field Strength, H (A/m)	Power Density, S (mW/cm^3) E Field	H Field	Averaging Time $\|E\|^2$, $\|H\|^2$ (min)
0.003–0.1	614	163	100	10^6	6
0.1–3	614	$16.3/f$	100	$10{,}000/f$	6
3–30	$1842/f$	$16.3/f$	$900/f$	$10{,}000/f$	6
30–100	61.4	16.3/f	1	$10{,}000/f$	6
100–300	61.4	0.163	1		6
300–3000	—	—	$f/300$		6
3000–15,000	—	—	10		6
15,000–300,000	—	—	10		$616{,}000/f$

Note: f = frequency (MHz).

3 kHz and 300 GHz (IEEE C95.1-1991). Table 2.1 summarises the IEEE standard C95.1-1991.

Heating in an IMD occurs due to three separate mechanisms, namely electronics heating, electromagnetic heating, and eddy and circular current heating.

2.5.1 Implant Heating

Electronics heating has the biggest impact on temperature rise. It has been investigated in depth by researchers at Case Western Reserve University (Saidel, 2001; Davies et al., 1994; Liu et al., 2003; Seese et al., 1998; Okazaki et al., 1997). The experimental data related power consumption to surface temperature increase. For contact with muscle tissue, the power density of 60–80 mW/cm^2 increases the contact temperature by 4–7°C. For 40 mW/cm^2 the rise was in the range of 2–3°C, and for 10 mW/cm^2 a maximum of 1°C rise is allowed in the government safety standard (Sawan and Gosselin, 2008). This indicates that the size of the device is related to the power dissipation. Another heating mechanism is eddy and circular current heating. When magnetic fields pass through conductive material (packaging) such as titanium alloy, eddy and circular currents are generated, which results in increased heat because of I^2R losses in the conductor (Geselowitz et al., 1992). The last mechanism for heat increase is a result of the direct absorption of electromagnetic energy by tissue called electromagnetic heating. The absorption is described by the specific absorption rate (SAR) (IEEE C95.1) and can be expressed as

$$\mathrm{SAR} = \frac{d}{dt}\left(\frac{dW}{dm}\right) = \frac{d}{dt}\left(\frac{dW}{\rho dV}\right) \tag{2.16}$$

where W is the energy delivered to the sample, m is the mass of the sample, ρ is the sample material density, and V is a sample volume.

SAR is used to determine the amount of radiation that human tissue absorbs and is expressed in units of energy per unit time per unit mass. The relationship to temperature rise (Ruggera et al., 2003) or the induced field and material parameters (Tang et al., 2005; Litvak et al., 2002) can be evaluated as

$$\mathrm{SAR} = \frac{d}{dt}\left(\frac{dT}{dt}\right)c = \frac{\sigma |E|^2}{\rho} \quad \mathrm{W/kg} \tag{2.17}$$

where σ is the sample electrical conductivity, E is the root-mean-square (RMS) of the electric field, T is the temperature, and t is the time.

According to IEEE (2005) standard C95.1, for a frequency range of 1–30 MHz, exposure should be limited to 16.3/*f*(MHz) (A/m). It is an RMS magnetic field strength averaged over a vertical cross section of the human body. The European Union Standardisation Committee (CENELEC) specifies SAR limits to the exposure of the public within the EU, giving an average whole-body SAR limit of 0.08 W/kg and localized SAR limit of 2 W/kg for the frequency range 100 kHz–10 GHz. Another limitation from CENELEC for a human head is "in cases where the orbit is in the field of a small local RF transmit coil, care should be taken to ensure that the temperature rise is limited to 1°C."

2.5.2 Transmission to Human Body

Investigation of how a human head absorbs a radiated wave from a transmitter coil (diameter 25 mm; two turns) and the temperature increase that the absorbed radiation causes was conducted by finite-element modeling software, COMSOL version 4.0, and is presented in Figures 2.13 and 2.14. The model of human head geometry was provided by the software and is the SAM phantom provided by IEEE, IEC, and CENELEC from their standard specification of SAR value measurements. Adjustment had to be made on the conductivity and relative permeability of the brain due to the close relation to the frequency of these parameters. Figure 2.13 shows the parameters used for simulations at 13.56 MHz (IEEE, 2005; Ruggera et al., 2003; Tang et al., 2005; Litvak et al., 2002; Gabriel et al., 1996). The modelers in COMSOL used two-dimensional magnetic resonance imaging (2D MRI) pictures to create the 3D human head model used in these simulations. The carrier frequency of 13.56 MHz was chosen as one of the highest ISM available frequencies. Generally, the higher the frequency, the larger the SAR value. A carrier current amplitude through the transmitter inductor of 200 mA was used.

The results show that the maximum tissue temperature due to electromagnetic heating at 13.56 MHz is well below the limits (Figure 2.14*a*; the colors represent different temperatures). SAR results show maximum absorption in the range of 0.01 mW/kg, which is eight times less than the limit specified by CENELEC.

2.5.3 Transmission from Human Body

In small IMDs like ingestible capsules (Wheeler, 1947; Yu et al., 2007) size is determined by the functional constraints. The power loss due to tissue absorption

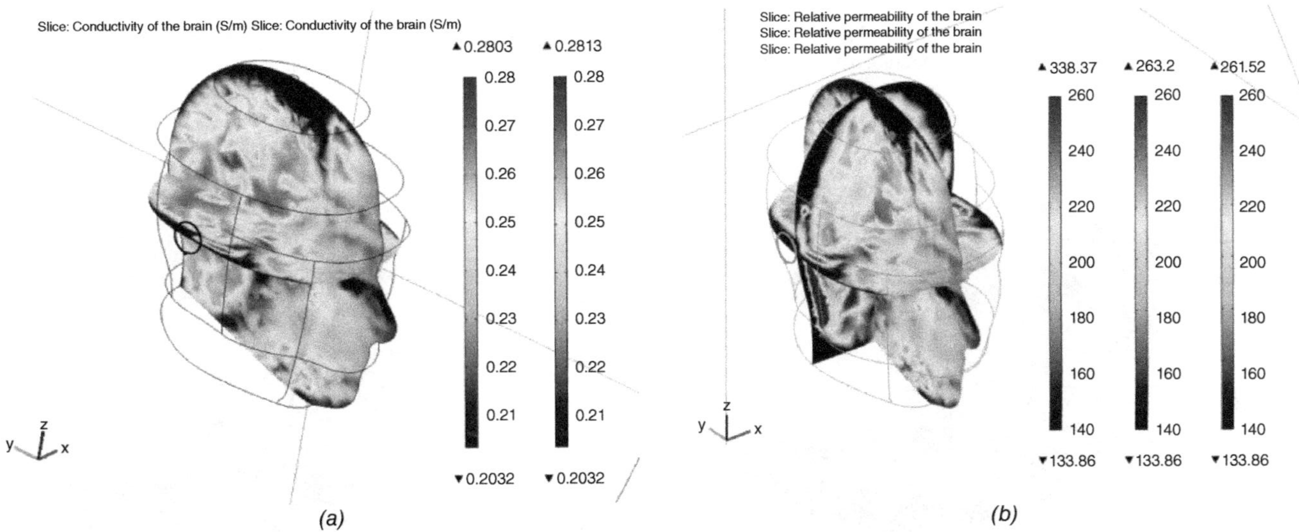

Figure 2.13 Model of human head at 13.56 MHz in COMSOL simulations: (*a*) conductivity of brain tissue; (*b*) relative permeability. (*See insert for color representation of the figure.*)

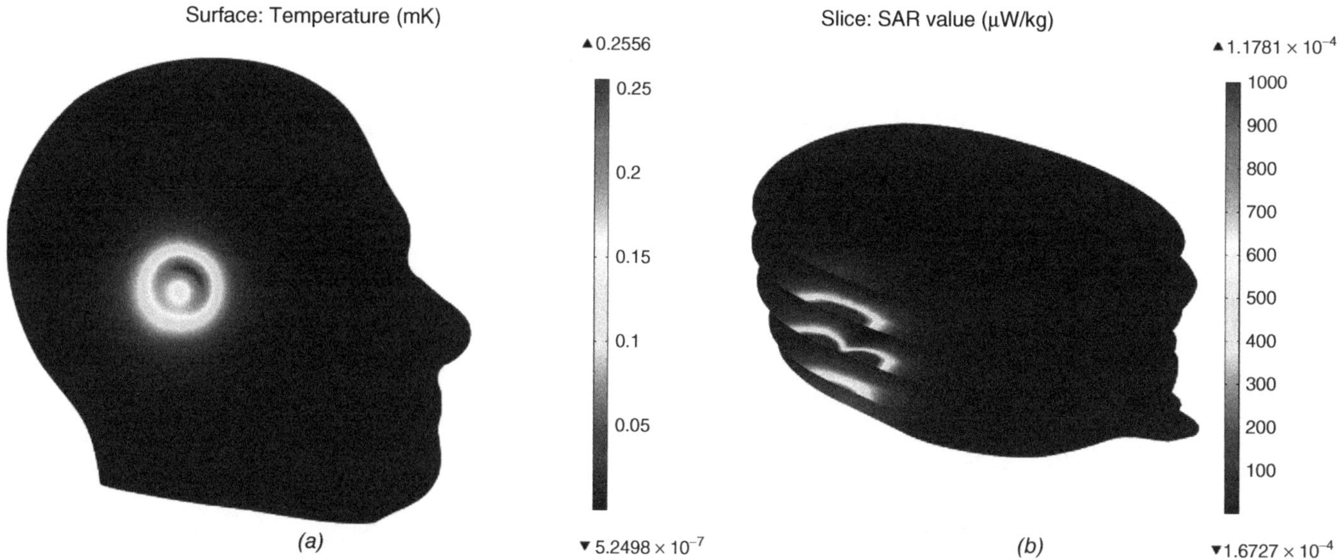

Figure 2.14 Result of simulation at 13.56 MHz with current of 200 mA through coil: (*a*) surface temperature increase due to electromagnetic heating of head; (*b*) SAR in human head. (*See insert for color representation of the figure.*)

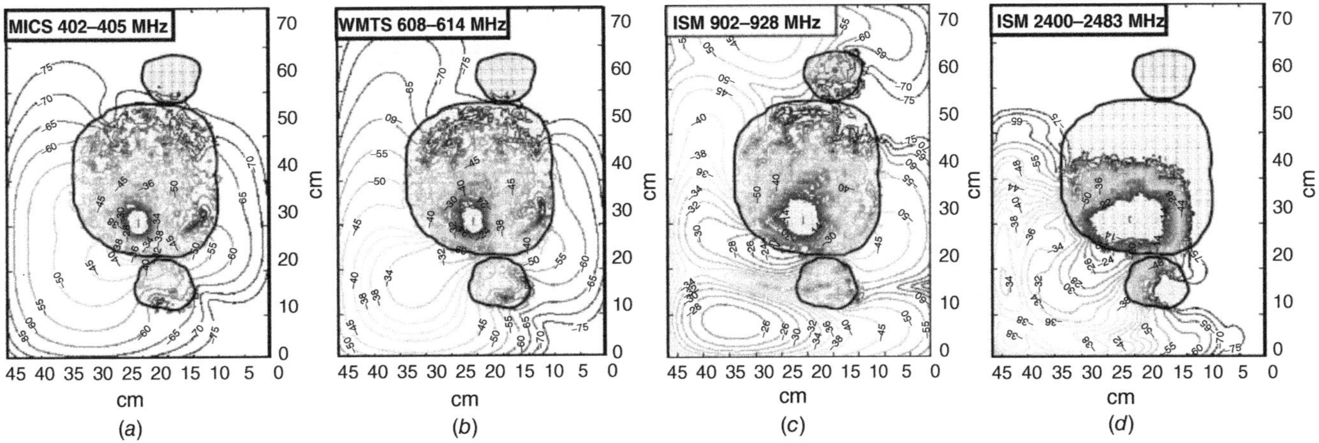

Figure 2.15 Normalized contours of radiated fields from capsule antenna inside human body torso (cross sections shown) operating at (*a*) MICS 402–405 MHz, (*b*) WMTS 608–614 MHz, (*c*) ISM 902–928 MHz, and (*d*) ISM 2.4–2.483 GHz FCC frequency bands (Yu et al., 2009). (*See insert for color representation of the figure.*)

TABLE 2.2 Comparison of Various Telemetry Systems

System	Carrier Frequency (MHz)	Downlink Modulation	Downlink Speed (kbps)	Uplink Modulation	Uplink Speed (kbps)	Separate Power	Number of Coils
Xu et al., 2009; Sonkusale and Luo, 2008	13.56	BPSK	1690	LSK	100	Yes	2
Ghovanloo and Najafi, 2004a	5/10	DFSK	2500	OOK	2000	Yes	3
Sodagar et al., 2009	4/8		2000		at 70–200 MHz		
Hu and Sawan, 2005	10	BPSK	1120	LSK	Not available	No	1
Coulombe et al., 2003	13.56	ASK	1130	LSK	200	No	1
Deng et al., 2006; Lu and Sawan, 2008; Simard et al., 2010	13.56	OQPSK	4160	OQPSK	4160-	Yes	3
Inanlou et al., 2010, 2011	N/A	PHM	10200	PHM	10200	Yes	2/3

(Gabriel et al., 1996) determines the operational frequency and has a significant impact on the power budget of the implant (Yu et al., 2009). A finite-difference time-domain tool from REMCOM Inc. was used to determine radiation characteristics (Yu et al., 2009). The study with a zig-zag dipole antenna placed inside an average male stomach shows that the optimal carrier frequency is near 900 MHz, as shown Figure 2.15. The antenna's radiation capabilities are limited by antenna size at lower frequencies and tissue absorption at higher frequencies. The optimal frequency is determined by the antenna size and the position of the implant in the body.

2.6 CONCLUSION

IMDs have become one of the biggest drivers for scientific and commercial research. With constant development of microelectronics, nanotechnology, sensors, and micropackaging, these biomedical devices will be used in an ever wider range of applications aiming to improve the quality of life. This chapter presented a brief overview of the architecture of biomedical telemetry using inductive links. The challenges are a trade-off between implant size, power dissipation, and system functionality. Wireless powering of IMDs is generally performed by low-frequency inductively coupled links to minimize power absorption by tissue. Data communication link parameters conflict with power transfer optimization and are often separated. The data link optimization ranges from antenna and frequency optimization to choice of modulation techniques. A comparison of telemetry systems is presented in Table 2.2. Safety issues relating to overheating of body tissue as a result of increases in temperature have also been discussed together with a study followed by simulations of how a human head absorbs a radiated wave from a transmitter coil and the temperature increase that the absorbed radiation causes.

REFERENCES

Abidi AA. 1995. Direct-conversion radio transceivers for digital communications. *IEEE J. Solid-State Circuits*, SC-30:1399–1410.

American National Standards Institute (ANSI). 1982. American National Standard safety levels with respect to human exposure to radio frequency electromagnetic fields, 300 kHz to 100 GHz. ANSI Std C95.1-1982.

Bawa G, Ghovanloo M. 2008. Active high power conversion efficiency rectifier with built-in dual-mode back telemetry in standard CMOS technology. *IEEE Trans. Biomed. Circuits Syst.*, 2(3):184–192.

Binkley DM, Rochelle JM, Swann BK, Clonts LG, Goble RN. 1997. A micropower CMOS direct-conversion VLF receiver chip for magnetic-field wireless applications. Proc. Custom Integrated Circ. Conf., pp. 415–418.

Chaimanonart N, Suster MA, Young DJ. 2010. Two-channel passive data telemetryWith remote RF powering for high-performance wireless and batteryless strain sensing microsystem applications. *IEEE Sensors J.*, 10(8):1375–1382.

Cirmirakis D, Jiang D, Demosthenous A, Donaldson N, Perkins T. 2012. A fast passive phase shift modulator for inductively coupled implanted medical devices. Proc 2012 European Solid-State Circuits Conference (ESSCIRC 2012), pp. 301–304.

Cirmirakis D, Demosthenous A, Saeidi N, Donaldson N. 2013. Humidity-to-frequency sensor in CMOS technology with wireless readout. *IEEE Sensors J.*, 13(3):900–908.

Coulombe J, Gervais JF, Sawan M. 2003. A cortical stimulator with monitoring capabilities using a novel 1 Mbps ASK data link. Proceedings of the 2003 International Symposium on Circuits and Systems ISCAS, 5:v-53–v-56.

Coulombe J, Sawan M. 2004. An implant for a visual cortical stimulator. CDEN Biomimetics Symposium.

Davies CR, Fukumura F, Fukamachi K, Muramoto K, Himley SC, Massiello A, Chen J, Harasaki H. 1994. Adaptation of tissue to a chronic heat load. *ASIAO J.*, 40(3):514–517.

Deng S, Hu Y, Sawan M. 2006. A high data rate QPSK demodulator for inductively powered electronics implants. IEEE International Symposium on Circuits and Systems ISCAS.

Djemouai DA, Sawan M. 1999. Prosthetic power supplies. In *Wiley Encyclopedia of Electrical and Electronics Engineering*. New York: Wiley, Vol. 17, pp. 413–421.

Donaldson N, Perkins T. 1983. Analysis of resonant coupled coils in the design of radio frequency transcutaneous links. *Med. Biol. Eng. Comput.*, 21(5):612–627.

Dormer KJ, Richard GPE, Hough JVD, Heweft T. 1980. The cochlear implant (auditory prosthesis) utilizing rare earth magnets. *Am. J. Otol.*, 2(1):22–27

Elfrink R, Kamel TM, Goedbloed M, Matova S, Hohlfeld D, van Andel R, van Schaijk R. 2009. Vibration energy harvesting with aluminum nitride-based piezoelectric devices. *J. Micromech. Microeng.*, 19(9):4005.

Federal Communications Commission (FCC). 1996. Guidelines for evaluating the environmental effects of radio frequency radiation.

Foster IC. 1981. Theoretical design and implementation of a transcutaneous, multichannel stimulator for neural prosthesis applications. *J. Biomed. Eng.*, 3:107–120.

Gabriel C, Gabriel S. 1996. Compilation of the dielectric properties of body tissues at RF and microwave frequencies. Available: http://niremf.ifac.cnr.it/docs/DIELECTRIC/Title.html.

Gabriel S, Lau RW, Gabriel C, 1996. The dielectric properties of biological tissues. Measurements in the frequency range 10 Hz to 20 GHz. *Phys. Med. Biol.*, 41:2251–2269.

Galbraith DC, Soma M, White RL. 1987. A wide-band band efficent inductive transdermal power and data link with coupling insensitive gain. *IEEE Trans. Biomed. Eng. BME*, 34:265–275.

Geselowitz DB, Hoang QTN, Gaumond RP. 1992. The effects of metals on a transcutaneous energy transmission system. *IEEE Trans. Biomed. Eng.*, 39(9):928–934.

Ghovanloo M, Atluri S. 2007. A wide-band power-efficient inductive wireless link for implantable microelectronic devices using multiple carriers. *IEEE Trans. Circuits Syst. I: Reg. Papers*, 54(10):2211–2221.

Ghovanloo M, Najafi K. 2003. A high data-rate frequency shift keying demodulator chip for the wireless biomedical implants. *IEEE Int. Symp. Circuits Syst. Proc.*, 5:45–48.

Ghovanloo M, Najafi K. 2004a. A wideband frequency-shift keying wireless link for inductively powered biomedical implants. *IEEE Trans. Circuits Syst. I*, 51(12):2374–2383

Ghovanloo M, Najafi K. 2004b. A modular 32-site wireless neural stimulation microsystem. *IEEE J. Solid-State Circuits*, 39(12):2457–2466.

Ghovanloo M, Najafi K. 2005. A tri-state FSK demodulator for asynchronous timing of high-rate stimulation pulses in wireless implantable microstimulators. Proceedings, 2nd Intl. IEEE IEMBS Conf. on Neural Engineering, pp. 116–119.

Goto K, Nakagawa T, Nakamura O, Kawata S. 2001. An implantable power supply with an optical rechargeable lithium battery. *IEEE Trans. Biomed. Eng.*, 48(7):830–833.

Harb A, Hu Y, Sawan M. 2004. Low-power CMOS interface for recording and processing very low amplitude signals. *J. Analog Integr. Circuits Signal Process*, 39(1):39–54.

Hochmair E. 1984. System optimization for improved accuracy in transcutaneous signal and power transmission. *IEEE Trans. Biomed. Eng.*, 31(2):177–186.

Hu Y, Sawan M. 2004. A fully-integrated low-power BPSK based wireless inductive link for implantable medical devices. *The 47th IEEE International Midwest Symposium on Circuits and System*, Vol. 3, pp. 25–28.

Hu Y, Sawan M. 2005. A fully integrated low-power BPSK demodulator for implantable medical devices. *IEEE Trans. Circuits Syst. I. Reg. Papers*, 52(12):2552–2562.

Hu Y, Sawan M, Ei-Gamal MN. 2003. A power recovery strategy dedicated to implantable applications. *Proc. 10th IEEE Int. Conf. Electronics Circuits Syst. (ICECS)*, Vol. 3, pp. 1212–1215.

Inanlou F, Kiani M, Ghovanloo M. 2010. A novel pulse-based modulation technique for wideband low power communication with neuroprosthetic devices. 32nd Annual International Conference of the IEEE EMBS, pp. 5326–5329.

Inanlou F, Kiani M, Ghovanloo M. 2011. A 10.2 Mbps pulse harmonic modulation based transceiver for implantable medical devices. *IEEE J. Solid-State Circuits*, 46(6):1296–1306.

Institute of Electrical and Electronics Engineers (IEEE). 1992. IEEE standard for safety levels with respect to human exposure to radio frequency electromagnetic fields, 3 kHz to 300 GHz. IEEE Std C95.1-1991.

Institute of Electrical and Electronics Engineers (IEEE). 2005. Standard for safety levels with respect to human exposure to radio frequency electromagnetic fields, 3 kHz to 300 GHz. IEEE std C95.1.

International Organization for Standardization (ISO). 2000. Implants for surgery—active implantable medical devices—part 1: General requirements for safety, marking, and for information to be provided by the manufacturer, ISO 14708–1.

Jow U, Ghovanloo M. 2007. Design and optimization of printed spiral coils for efficient transcutaneous inductive power transmission. *EEE Trans. Biomed. Cir. Syst.*, 1(3):193–202

Kiani M, Ghovanloo M. 2010. An RFID-based closed-loop wireless power transmission system for biomedical applications. *IEEE Trans. Circuits Syst.: Express Briefs*, 57(4):260–264.

Ko WH, Liang SP, Fung CDF. 1977. Design of radio-frequency powered coils for implant instruments. *Med. Biol. Eng. Comput.*, 15(6):634–640.

Kumar A, Rakhyani R, Mirabbasi S, Chiao M. 2011. Design and optimization of resonance-based efficient wireless power delivery systems for biomedical implant. *IEEE Trans. Biomed. Cir. Syst.*, 5(1):48–63.

Litvak E, Foster KR, Repacholi MH. 2002. Health and safety implications of exposure to electromagnetic fields in the frequency range 300 hz to 10 mhz. *Bioelectromagnetics*, 23(1):68–86.

Liu EH, Saidel GM, Harasaki H. 2003. Model analysis of tissue response to transient and chronic heating. *Ann. Biomed. Eng.*, 31(8):814–820.

Lu Z, Sawan M. 2008. An 8 Mbps data rate transmission by inductive link dedicated to implantable devices. IEEE International Symposium on Circuits and Systems ISCAS, pp. 3057–3060.

Luo Z, Sonkusale S. 2008 A novel BPSK demodulator for biological implants. *IEEE Trans. Circuits Syst. I: Reg. Papers*, 55(6):1478–1484.

Malachesky P, Holleck G, McGovern F, Devarakonda R. 1972. Parametric studies of implantable fuel cell. Proceedings of the 7th Intersociety Energy Conversion Engineering Conference, pp. 727–732.

Murakawa K, Kobayashi M, Nakamura O, Kawata S. 1999. A wireless near-infrared energy system for medical implants. *IEEE Eng. Med. Biol. Mag.*, 18(6):70–72.

Okazaki Y, Davies CR,. Matsuyoshi T, Fukamachi K, Wika KE, Harasaki H. 1997. Heat from an implanted power source is mainly dissipated by blood perfusion. *ASIAO J.*, 43(5):585–588.

Olivo J, Carrara S, De Micheli G. 2011. Energy harvesting and remote powering for implantable biosensors. *IEEE Sensors J.*, 11(7):1573–1586.

Raab FH. 1977. Idealized operation of the class E tuned power amplifier. *IEEE Trans. Circuits Syst.*, 24(12):725–735.

Ruggera PS, Witters DM, Maltzahn G, Bassen HI. 2003. In vitro assessment of tissue heating near metallic medical implants by exposure to pulsed radio frequency diathermy. *Phys. Med. Biol.*, 48(17):2919–2928.

Saidel GM, Davies CR, Liu EH, Harasaki H. 2001 Temperature and perfusion responses of muscle and lung tissue during chronic heating in vivo. *Med. Biol. Eng. Comput.*, 39(1):126–133.

Sauer C, Stanacevic M, Cauwenberghs G, Thakor N. 2005. Power harvesting and telemetry in CMOS for implanted devices. *IEEE Trans. Circuits Syst.*, 52(12):2605–2613.

Sawan M, Gosselin D. 2008. CMOS circuits for biomedical implantable devices. *In VLSI Circuits for Biomedical Applications*, 1st edn. K. Iniewski, London, Chapter 3, pp. 45–57.

Sawicki DS. 2011. Police Traffic SPEED RADAR Handbook: A Comprehensive Guide to Speed Measuring Systems. *Includes Microwave and Laser Radar*. CreateSpace Independent Publishing Platform (January 31, 2011).

Seese TM, Harasaki H, Saidel GM, Davies CR. 1998 Characterization of tissue morphology, angiogenesis, and temperature in the adaptive response of muscle tissue to chronic heating. *Lab. Invest.*, 78(12):1553–1562.

Si P, Hu AP, Malpas S, Budgett D. 2008. A frequency control method for regulating wireless power to implantable devices. *IEEE Trans. Biomed. Circuits Syst.*, 2(1):22–29.

Simard G, Sawan M, Massicotte D. 2010. High-speed OQPSK and efficient power transfer through inductive link for biomedical implants. *IEEE Trans. Biomed. Cir. Syst.*, 4(3):192–200.

Sodagar AM, Wise KD, Najafi K. 2009. A wireless implantable microsystem for multichannel neural recording. *IEEE Trans. Microw. Theory Tech.*, 57(10): 2565–2573.

Sokal NO, Sokal AD. 1977. Class E—a new class of high-efficiency tuned single-ended switching power amplifiers. *IEEE J. Solid-State Circuits, SC*, 10(3):168–176.

Soma M, Galbraith DC, White RL. 1987. Radio-frequency coils in implantable devices: Misalignment analysis and design procedure. *IEEE Transaction on Biomedical Engineering, BME*, 34(4):276–282.

Sonkusale S, Luo Z. 2008. A complete data and power telemetry system utilizing BPSK and LSK signaling for biomedical implants. 30th Annual International IEEE EMBS Conference, pp. 3216–3219.

Stark I, Stordeur M. 1999. New micro thermoelectric devices based on bismuth telluride-type thin solid films. 18th International Conference on Thermoelectric, pp. 465–472.

Suzuki S, Katane T, Saotome H, Saito O. 1999. A proposal of electric power generating system for implanted medical devices. *IEEE Trans. Magn.*, 35(5):3586–3588.

Suzuki S, Katane T, Saotome H, Saito O. 2002. Electric powergenerating system using magnetic coupling for deeply implanted medical electronic devices. *IEEE Trans. Magn.*, 38:3006–3008.

Tang Z, Smith B, Schild JH, Peckham PH. 1995. Data transmission from an implantable biotelemeter by load-shift keying using circuit configuration modulator. *IEEE Trans. Biomed. Eng.*, 42(5):524–528.

Tang Q, Tummala N, Kumar S, Gupta S. 2005. Communication scheduling to minimize thermal effects of implanted biosensor networks in homogeneous tissue. *IEEE Trans. Biomed. Eng.*, 52(7):1285–1294.

Terman FE. 1943. Radio Engineer's Handbook. McGraw-Hill Book Company, inc; 1st edition, 1943 New York and London http://www.itermoionici.it/letteratura_files/Radio-Engineers-Handbook.pdf

Wang G, Liu W, Sivaprakasam M, Kendir GA. 2005 Design and analysis of an adaptive transcutaneous power telemetry for biomedical implants. *IEEE Trans. Circuits Syst. I: Reg.* Papers, 52(10):2109–2117.

Wheeler HA. 1947. Fundamental limitations of small antennas. *Proc. IRE*, 35(12):1479–1488.

Xu W, Luo Z, Sonkusale S. 2009. Fully digital BPSK demodulator and multilevel LSK back telemetry for biomedical implant transceivers. *IEEE Trans. Circuits Syst.: II Express Briefs*, 56(9):714–718.

Yu H, Irby GS, Peterson DM, Nguyen MT, Flores G, Euliano N, Bashirullah R. 2007. Printed capsule antenna for medication compliance monitoring. *Electron. Lett.*, 43(22):1179–1181.

Yu H, Tang CM, Bashirullah R. 2009. An asymmetric RF tagging IC for ingestible medication compliance capsules. Proc. IEEE Radio Frequency Integrated Circuits (RFIC) Symp., pp. 101–104.

Zhou L, Donaldson N. 2003. A fast passive data transmission method for ENG telemetry. *Neuromod.: Tech. Neural Interf.*, 6(2):116–121.

Zierhofer CM, Hochmair ES. 1990. High-efficiency coupling-insensitive transcutaneous power and data transmission via an inductive link. *IEEE Trans. Circuits Syst. I: Regu. Papers*, 37(7):716–722.

Zierhofer CM, Hochmair ES. 1996. Coil design for improved power transfer efficiency in inductive links. Proceedings of the 18th Annual International Conference of the IEEE, Vol. 4, pp. 1538–1539.

3

SENSING PRINCIPLES FOR BIOMEDICAL TELEMETRY

Athanasios Lioumpas, Georgia Ntouni, and Konstantina S. Nikita

School of Electrical and Computer Engineering, National Technical University of Athens, Athens, Greece

3.1 INTRODUCTION

During the last decades, the advances in technology have had significant impact on monitoring applications in fields such as the food industry, environmental monitoring, e-health, clinical diagnosis, and biomedical telemetry. Particularly for biomedical applications, besides other challenges, such as the demanding wired or wireless communication requirements, the successful development of these applications is highly related with the sensing devices, that is, the biosensors. These are charged with the challenging task of providing the physical, chemical, and biological data for the monitoring, which includes among others heart rate, blood oxygen saturation, physical activity, intracorporeal (e.g., intracranial) pressure, gastrointestinal parameters, and lumen visualization.

According to the International Union of Pure and Applied Chemistry (IUPAC), *a biosensor is an independently integrated receptor transducer device, which is capable of providing selective quantitative or semiquantitative analytical information using a biological recognition element (biochemical receptor)* (Thevenot et al., 1999). The biosensors aim to provide accurate and reliable information about the sensed parameter rapidly and in real time. The first biosensor was presented by Clark and Lyons in 1962, who utilized external electrode systems to record, quantitatively, blood pH,

Handbook of Biomedical Telemetry, First Edition. Edited by Konstantina S. Nikita.

oxygen, and carbon dioxide tensions and contents and intravascular electrodes for hydrogen detection and oxygen tension changes (Clark and Lyons, 1962).

Over the last years, novel sophisticated tools and new materials have made it possible to construct sensing devices with increased reliability and accuracy with respect to the sensed information. Based on the transduction process, biosensors can be divided into several categories, such as electrical, electrochemical, optical, piezoelectric, and thermal/calorimetric biosensors, while based on the receptor type, they can be characterized as enzymatic biosensors, genosensors, immunosensors or DNA-based sensors.

This chapter focuses on the operation principles of the biosensors in terms of both the receptor type and the transduction process. The different recognition and detection principles are reviewed and analyzed, providing insight into the biosensor design challenges.

3.2 BIOSENSOR STRUCTURE

Although there are numerous different types of biosensors and various operation principles, the fundamental elements of the biosensors are the following (Monošík et al., 2012):

1. The *bioreceptor*, which is an immobilized biological element able to recognize the target analyte, that is, the compound whose concentration is to be determined (e.g., enzyme substrate, complementary DNA, antigen). Among the biosensing elements that are utilized, enzymes are most commonly used.
2. The *transducer*, which is used to convert (translate) the (bio)chemical signal resulting from the interaction of the analyte with the bioreceptor, that is, the recognition event, into a measurable electric signal into one single sensor. The concentration of the analyte is directly related with the intensity of the generated signal. Based on the transducer's detection principle, they can be classified into electrochemical, gravimetric, calorimetric, or optical transducers with the electrochemical ones being widely used because of their low cost and small size.

The uniqueness of a biosensor is that the two components are integrated into one single device (Figure 3.1), a characteristic that distinguishes them from a bioanalytical system, which requires additional processes, such as a reagent (Sassolas et al., 2012). Some examples of usually employed bioreceptors and transducers are given in Table 3.1 (Grieshaber et al., 2008).

3.2.1 Design Constraints

Toward the construction of a biosensor able to measure the concentration of an analyte or interact with a specific conformation, several design principles should be taken into account, such as the choice for the biological element that will be used for the

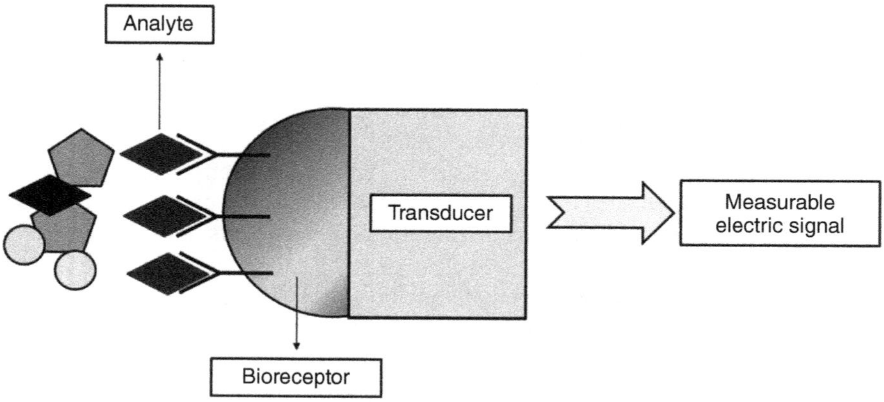

Figure 3.1 Components of a biosensor.

TABLE 3.1 Elements and Selected Components of Typical Biosensor

Samples	Bioreceptors	External System	Transducer	Interface
Analytes (e.g., glucose, lactate)	Enzyme Antibody Microorganism	Signal amplifier + signal processor + display	pH change	Semiconductor pH electrode
			Heat	Thermistor
			Light	Photon counter
			Mass change	Piezoelectric device
			Electroactive substance	Electrode

detection of the analyte, the chemical nature of the transducer, and the position into the protein to introduce the signal (Serra, 2011; Perumal and Hashim, 2013; Wang et al., 2009). The biological component is most usually a protein or nucleic acid, able to provide the stereospecificity that is necessary for the desired interaction with the analyte. The periplasmic binding proteins (PBPs) are able to bind a large number of analytes, such as carbohydrates, amino acids, ions, hormones, and heme groups (Medintz and Deschamps, 2006). The chemical nature of the transducer is a very critical parameter for the construction of a biosensor, and it must be carefully selected since the physiochemical properties of the transducer determine the quality of the generated signal (e.g., some signals are sensitive to the polarity of the solvent or the pH). Another important factor to be considered for the construction of a biosensor is its ability and efficiency to provide fast real-time reproducible measurements (Plaxco and Soh, 2011).

The design constraints are highly related to the biomedical application and the particular use of the biosensor. The most stringent constraints are imposed by those applications involving an "invasive" sensing, where a penetration into the body is required. In such applications, the size, the power source, and the energy efficiency of the implanted device play a decisive role. Apart from these issues, the most critical issue for the implanted biosensors is their biocompatibility with the individual patient, since the biosensor could be attacked by the immune system and lead to serious complications (Iniewski, 2012). Besides, the tissues conductivity may prevent the normal operation of the implanted device in case of their direct contact. In this sense, the used materials as well as the packaging of the biosensor and the implanted device play a key role in the design. Usually, the implanted or ingestible devices are encapsulated in a biocompatible material, such as porous silicon, thermal silicon oxide, polysilicon, silicon nitrate, titanium, and SU-8 (Guiseppi-Elie et al., 2005; Lopez-Martinez et al., 2009). Moreover, the electrical current sent to the device should be taken into account (e.g., through biocompatible wire sheathing) in order to avoid unintentional discharge or unnecessary current heating effects. Furthermore, proper biocompatible packaging and filtering could prevent incorrect signal detections because of the noise created by the compounds within the body (Guiseppi-Elie et al., 2005).

"Noninvasive" sensing applications, where the biosensor is attached to a wearable device, do not pose the same challenges as the invasive ones, since larger device sizes can be tolerated and conventional power sources can be utilized. The most challenging task is to effectively detect the biosignals on the exterior of the body and encounter the increased probability to sense undesired signals because of the noise and interference (Guiseppi-Elie et al., 2005). Wearable sensing devices usually employ advanced filtering and interference-mitigating techniques in order to achieve sufficient reliability.

3.3 ELECTROCHEMICAL BIOSENSORS

The principle of the first enzyme electrode with immobilized glucose oxidase was presented in 1962 by L. C. Clark, while the first commercial biosensor was produced by Yellow Spring Instruments in 1972 and was applied to the fast glucose assay in blood samples from diabetics (Grieshaber et al., 2008). The operation of a typical electrochemical biosensor is based on the presence of an appropriate enzyme in the biorecognition layer (baroreceptor), which is able to provide those electroactive substances to the physicochemical transducer in order to detect a measurable signal. Native enzyme can be used as the biorecognition element or enzymes can be additionally used as labels bound to antibodies, antigens, and oligonucleotides with a specific sequence, thus providing affinity-based sensors (Bakker, 2004). Depending on the analyte, different enzymes are employed for its detection, for example, glucose oxidase and glucose dehydrogenase for glucose assays, urease for urea, cholesterol oxidase coimmobilized with cholesterol esterase for the cholesterol assay, NADH-dependent lactate dehydrogenase, and lactate–cytochrome *c* oxidoreductase for lactate (Grieshaber et al., 2008). Electrochemical biosensors are classified to different categories, according to the transducer's operation principle, with the most common ones being the potentiometric, amperometric, and impedimetric transducers (Pohanka and Skladal, 2008).

3.3.1 Amperometric Electrochemical Biosensors

The operation of the amperometric electrochemical biosensors is based on the reaction between an enzyme immobilized at the surface of an amperometric electrode and a substrate. Clark oxygen electrodes represent the basis for the amperometric biosensors, which consist of a central Pt (platinum) cathode and an Ag/AgCl (silver) anode. When a reference potential is applied between the cathode and the anode, a current produced in proportion to the oxygen concentration is measured by the reduction of oxygen at the Pt electrode with respect to a Ag/AgCl reference electrode. If the current is measured at a constant potential, it is referred as *amperometry*, while if the current is measured during controlled variations of the potential, this is referred to as *voltammetry* (Chaubey and Malhotra, 2002). Figure 3.2 depicts the basic structure of an amperometric biosensor. The platinum cathode and the silver anode are placed in a solution (e.g., saturated potassium chloride) which is separated from the biocatalyst (e.g., glucose oxidase) by a membrane permeable to O_2, that is, a product after the chemical reaction between the enzyme and the analyte (e.g., glucose). The biocatalyst is separated from the analyte by another membrane permeable to specific products of the chemical reaction. For this specific example the chemical reactions that take place

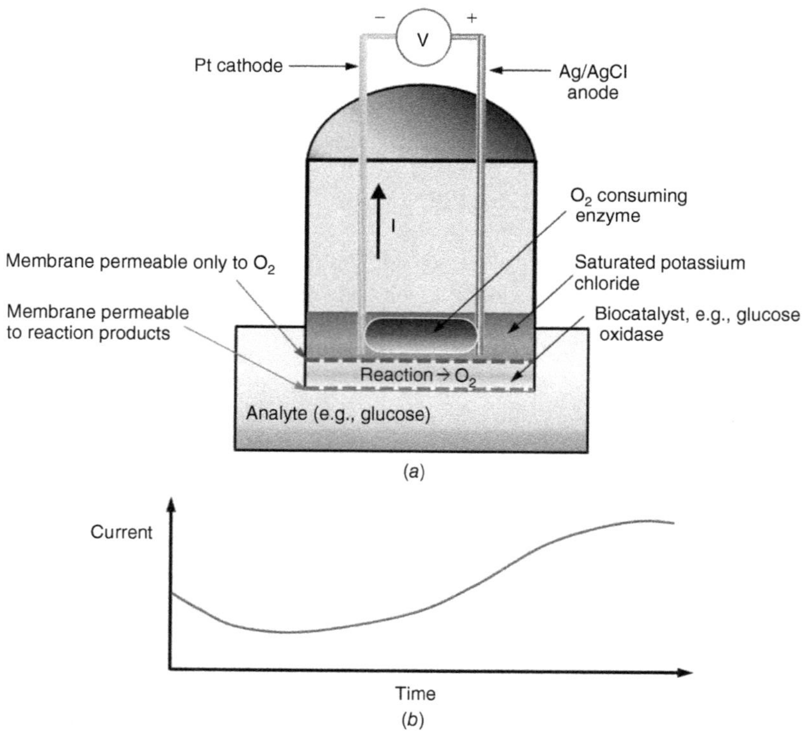

Figure 3.2 (*a*) Structure of simple amperometric biosensor and (*b*) current at electrodes after reaction between analyte and oxygen-consuming enzyme.

can be expressed as follows:

Silver anode: $4Ag^0 + 4Cl^- \rightarrow 4AgCl + 4e^-$

Platinum cathode: $O_2 + 4H^+ + 4e^- \rightarrow 2H_2O$

The amperometric biosensors that utilize two electrodes have some disadvantages with the most important being the limited control of the potential when higher currents occur, resulting in a shortened linear range. A solution to this problem involves the introduction of a third auxiliary electrode, so that the potential is applied between the reference and the working electrodes, while the current flows between the working and the auxiliary one.

The simple structure and operation principles of the amperometric biosensors have allowed their mass production and have made them the most commonly used sensors for biomedical applications. Several commercial amperometric biosensors exist, with the glucose biosensors being the most known.

Furthermore, several attempts have been made toward the further simplification of amperometric biosensors (Belluzo et al., 2008). For example, an outstanding approach has been proposed to determine glucose using glucose oxidase directly "plugged" to the electrode taking advantage of the controlled electron transference through a carbon nanotube (Patolsky et al., 2004). Moreover, biosensor research is highly motivated by the need for miniaturized clinically useful devices where microscale biosensor devices react with the analyte and generate a measurable response within the same compact body. A good example is the enzymatic biosensor to detect lactate in saliva in a one-step analysis (Schabmueller et al., 2006).

3.3.2 Potentiometric Electrochemical Biosensors

The operation of the potentiometric electrochemical biosensors is based on potentiometry, which detects the ion activity during an electrochemical reaction. The biosensor measures the cumulative potential between two electrodes in an electrochemical cell when the current flowing through the electrodes is equal to or near zero. Usually, ion-selective electrodes (ISEs) and ion-sensitive field effect transistors (ISFETs) are employed, with the output signal resulting from the ions that are accumulated at the ion-selective membrane. The relationship between the ion concentration and the potential is governed by the *Nernst equation*, also referred to as the electromotive force (EMF), that is,

$$\text{EMF or } E_{\text{cell}} = E^0_{\text{cell}} - \frac{RT}{nF} \ln Q \tag{3.1}$$

where E_{cell} represents the observed cell potential at zero current, E^0_{cell} a constant potential contribution to the cell, R the universal gas constant, T the absolute temperature in degrees Kelvin, n the charge number of the electrode reaction, F the Faraday constant, and Q the ratio of ion concentration at the anode to ion concentration at the cathode (Buerk, 1993).

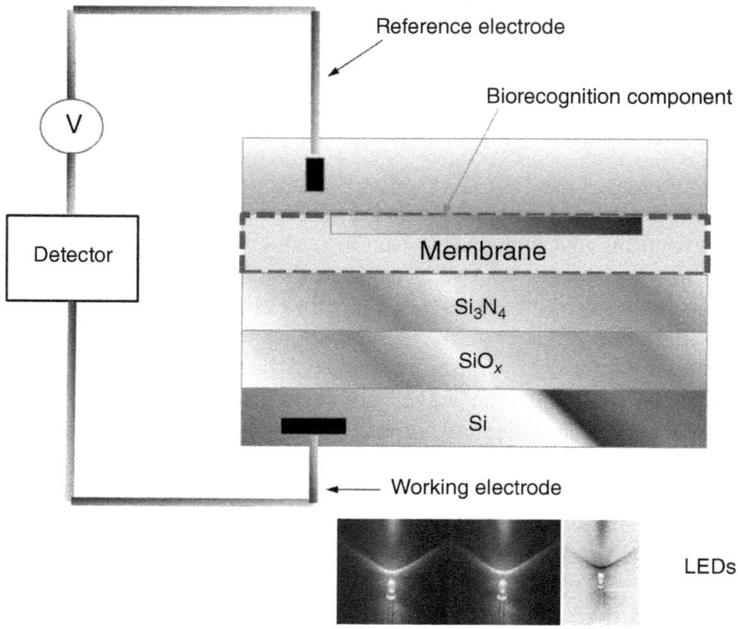

Figure 3.3 Structure of simple potentiometric biosensor.

Semiconductor-based physicochemical transducers are among the most commonly used potentiometric biosensors and more particularly the ISFETs and light addressable potentiometric sensor (LAPS).

In potentiometric biosensors employing the LAPS, a light-emitting diode (LED) activates the semiconductor (Yoshinobu et al., 2005). Figure 3.3 illustrates a simple biosensor based on the LAPS principle. In such structures an n-type Si is coated with 30 nm of SiO_x, 100 nm of Si_3N_4, and indium–tin oxide (ITO). The LAPS measures a voltage change as a function of medium pH in the LED-activated zone. A commercial biodetector device (Smiths Detection, Warrington, UK) based on the LAPS-type biosensor is found in mobile laboratories for automated eight-channel analysis of biological agents (Pohanka and Skladal, 2008).

3.3.3 Impedimetric Electrochemical Biosensors

The impedimetric biosensors measure the ability of an analyte to conduct an electrical current between electrodes or reference nodes and follow either impedance (Z) or its component resistance (R) and capacitance (C). The impedance biosensor is commonly a functional part of the Wheatstone bridge. They are also called conductometric devices, because of the inverse value of resistance, that is, the conductance. These devices are usually related to enzymes, where the ionic strength of a solution between two electrodes changes as a result of an enzymatic reaction.

A common application of the impedimetric biosensors is the assay of urea using urease as a biorecognition component. In general, impedimetric biosensors are not as popular as the potentiometric or amperometric biosensors. Nevertheless, they have contributed to important research studies, such as the hybridization of DNA fragments previously amplified by a polymerase chain reaction, or to the monitoring of microorganism growth due to the production of conductive metabolites.

3.4 OPTICAL BIOSENSORS

The demanding requirements for fast and accurate detection of any type of substances have accelerated the development of a large variety of biosensors. For most applications it is desirable to have a compact biosensor with high sensitivity, fast response, and the ability to perform measurements in real time. Optical methods for the detection of biomolecular interactions and their biomedical applications have attracted great research interest, showing that they meet these requirements. Moreover, the cost reduction of high-quality fibers and optoelectronic components has played a key role for the development of optical biosensors (Velasco-Garcia, 2009). These biosensors have many advantages that stem from the optical technology, such as the immunity to external electromagnetic interference, while they are characterized by the increased speed of biodetection. Furthermore, thanks to the increased bandwidth at the optical frequencies, they are suitable for biomedical applications with a large volume of information. Besides, the optical may directly detect a target of interest or indirectly through optically labeled probes. Most usually the operation of the optical biosensors is based on the detection of changes in (Gauglitz, 2005; Velasco-Garcia, 2009):

- Absorbance
- Fluorescence/phosphorescence
- Chemiluminiscence
- Reflectance
- Light-scattering or refractive index

The optical methods for detection are based on fluorescence spectroscopy, surface plasmon resonance, interferometry, and spectroscopy of guided modes in optical waveguide structures (Passaro et al., 2007; Fan et al., 2008). The advantages of optical sensors become more visible with the utilization of photonic integration, where multiple photonic functions are integrated on a single optical circuit, thus improving the functionality, sensitivity, and resistance of the biosensor and enabling the ability for mass production and reduced costs.

Various technologies are available for the construction of photonic biosensors, with the silicon technology being one of the most practical and promising tools. In silicon photonics the construction of the devices is based on silicon materials using microelectronics technologies in order to integrate all elements on a single chip. The propagation of light for biosensor applications is based on the total internal reflection

(TIR) in flat or lateral waveguides, concave waveguides, antiresonant reflecting optical waveguides (ARROWs), or waveguides apertures.

3.4.1 Integrated Optical Biosensors

A fiber is a waveguide that transmits light along its axis by the process of total internal reflection, and it consists of a dielectric material (the core) surrounded by another dielectric material with a lower refractive index (the cladding). Although most of the light power is confined within the waveguide, a small amount of this power is radiated and can interact with the environment. This type of optical sensor has important advantages, such as small size, high sensitivity, and low cost. In a large number of integrated optic sensors that have been proposed, the presence of the analyte either causes a change in the refractive index of the medium that covers the waveguide (homogeneous detection) or determines the thickness of the molecular layer that is placed over the interface between the waveguide and the cover medium (surface detection). Both phenomena affect the propagating optical mode effective index, measured in various ways, depending on the architecture of each biosensor.

3.4.2 Interferometric Architectures

The interaction between the sample (analyte) and the optical signal that propagates through the sensor causes a change in the propagating optical mode effective index and therefore in the phase. The transduction of this phase change to amplitude change width is based on the interferometric architectures. Among these architectures, the Mach–Zehnder approach is the most commonly used to determine the relative phase shift variations between two collimated beams and ensures high sensitivity. In this kind of integrated optical sensor (Figure 3.3), the optical input signal is separated into two signals by a Y splitter. These two signals propagate in the reference and sensing arms, and the interaction between the sample and the optical signal is carried in the sensing arm. Thereafter, the two optical signals acquire a phase shift $\Delta\phi$. If the input power P_{in} is unequally divided between the two arms [the sensing and reference arms receive an optical power equal to γ_1 and $(1 - \gamma_1)P_{\text{in}}$], the output power will be the sum of the optical power at the sensing arm multiplied by γ_2 and the optical power at the reference arm multiplied by $1 - \gamma_2$, resulting in an output power–input power ratio that is expressed as (Passaro et al., 2007)

$$\frac{P_{\text{out}}}{P_{\text{in}}} = \gamma_1\gamma_2 + (1 - \gamma_1)(1 - \gamma_2) + 2\sqrt{\gamma_1\gamma_2(1 - \gamma_1)(1 - \gamma_2)}\ \cos(\Delta\phi) \qquad (3.2)$$

where $\Delta\phi = 2\pi(L/\lambda)(n^S_{\text{eff}} - n^R_{\text{eff}})$ with L being the length of the reference and sensing arm and λ the optical signal wavelength and n^S_{eff} and n^R_{eff} are the propagating optical mode effective index at the sensing and reference arms respectively.

During the last two decades, a large number of integrated optical biosensors have been proposed and implemented based on the Mach–Zehnder interferometer (MZI)

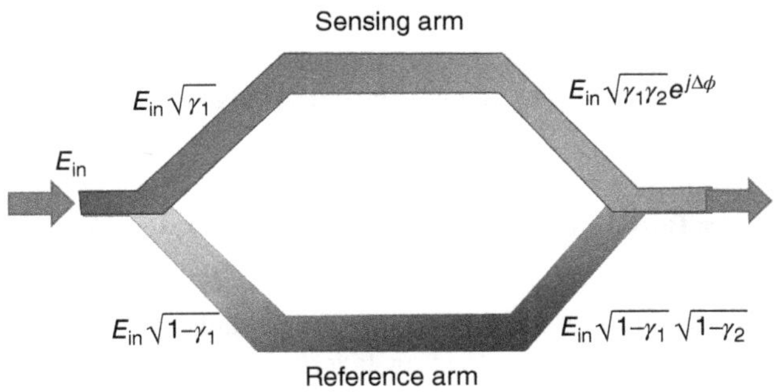

Figure 3.4 Mach–Zehnder interferometer.

architecture (Figure 3.4). The technologies that are employed for the construction of such biosensors is complementary metal–oxide–semiconductor (CMOS) compatible (guiding field on silicon, silicon nitrides, or silicon oxynidrite), but also glass and III–V semiconductor material have been proposed.

The sensitivity S of the MZI-based biosensors can be expressed as

$$S = -\frac{2\pi}{\lambda} L S_W \tag{3.3}$$

where S_W is the waveguide sensitivity, which for the case of homogeneous sensing is defined as

$$S_W = -\frac{\partial n_{\text{eff}}}{\partial n_C} \tag{3.4}$$

while for the case of surface sensing as

$$S_W = -\frac{\partial n_{\text{eff}}}{\partial \rho} \tag{3.5}$$

where ρ is the thickness of the molecular layer, which is placed at the interface between the guiding film and the cover medium of the wavelength.

Furthermore, the sensitivity of the MZI-based integrated biosensors depends on the length, L, of the sensing arm and therefore can be reduced by the unavoidable changes of the propagating optical mode effective index (e.g., because of temperature variations). Prieto et al. (2003) implemented an optical MZI biosensor capable of detecting the immune reactions between antibodies–antigens with detection limit equal to 7×10^{-6} with respect to the minimum detectable change of the refractive index of the wavelength cover medium. A detection limit of 1.5×10^{-6} was achieved by Drapp et al. (1997) using a wavelength with glass substrate BGG36.

Other types of interferometric optical sensors include the Young interferometers with four channels, which achieve the simultaneous and independent detection of up to three antigen–antibody reactions (Ymeti et al., 2003). In this device, the interference pattern is monitored by a charge-coupled device (CCD) camera and is analyzed with a fast Fourier transform algorithm, resulting in a detection limit equal to 8.5×10^{-8}.

3.4.3 Biosensors Based on Antiresonant Reflecting Optical Waveguides

The ARROW is a five-layer waveguide where the optical confinement is based on the total internal reflection at the air–core interface and a very high reflectivity, of 99.96%, at the two interference cladding layers underneath the core (Figure 3.5) (Prieto et al., 2000).

The ARROW waveguides are usually designed with CMOS-compatible technology and the material commonly used is silicon for the substrate, oxide silicon for the core, and the second cladding layer and nitride silicon for the first cladding layer. ARROW structures exhibit low loss, allowing greater dimensions than conventional waveguides based on TIR and are characterized by their high sensitivity. These features are quite attractive for sensoring applications.

3.4.4 Biosensors Based on Surface Plasmon Resonance

The principle behind surface plasmon resonance (SPR) is the interaction between light and a thin metallic film (e.g., Au or Ag) coated on a transparent medium. When the opposite surface of the substrate is coated with a thin metallic layer, there is an angle θ_{SPR} greater than the critical angle θ_C (i.e., the angle of incidence at which the light is totally reflected back) where the light, instead of being totally internally reflected back, is "coupled" into the metallic film, resulting in a minimum in the reflected light intensity (Figure 3.6). This angle is called the SPR angle and is the result of the oscillations of the surface electrons that propagate along the boundary of the metal film and the dielectric layer, causing the surface plasmon wave.

It has been shown that θ_{SPR} depends not only on the physical properties of the metal boundary layer but also on the dielectric properties of the medium, which is

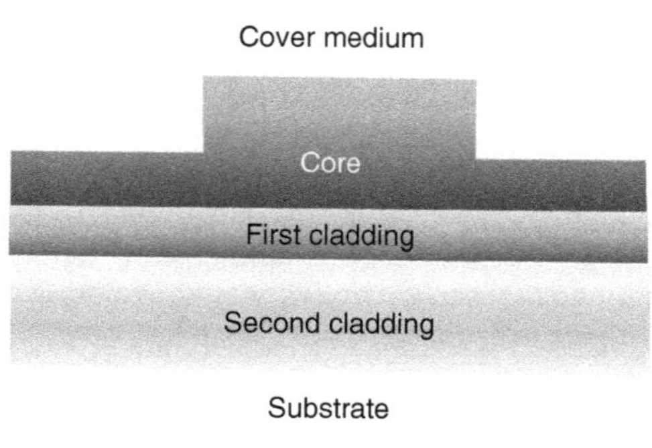

Figure 3.5 ARROW waveguide.

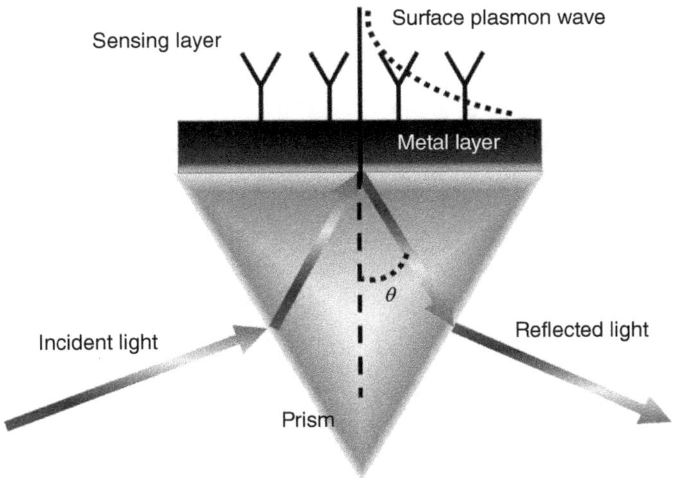

Figure 3.6 Structure of biosensor based on SPR.

directly in contact with the metal film. SPR can be used to study the biomolecular interactions and measure the concentration levels of analytes in complex samples (Hoa et al., 2007). The development of SPR biosensors is motivated by the need for compact, low-cost, sensitive devices. Prism coupling or Kretchmann's configuration has been found to be very suitable for sensing and is the most widely used geometry in SPR sensors. Biacore AB is one of the companies that have successfully commercialized prism-based SPR sensor systems (Homola, 2003). Nevertheless, because of the size constraints of the prism, it is not always easy to integrate in compact devices, and therefore optical fibers and waveguides have been proposed as good alternatives to transmit the surface plasmon excitation and reflection light.

Waveguide coupling is a robust and simple way to control the optical path in the sensing layer and is easier to integrate with other optical and electrical components, offering extra benefits like miniaturized devices and remote sensing (Velasco-Garcia, 2009). The excitation of a surface plasmon wave in an optical waveguide-based SPR sensor is similar to that in the Kretschmann approach, since the light propagates in a waveguide through total internal reflection and generates an evanescent field at the waveguide–metal interface. The surface plasmon wave is excited if the phase velocities of the waveguide mode and that of the surface plasmon match. Recently, microfabrication has been employed in order to improve the coupling of light to the surface plasmon mode.

3.5 THERMAL/CALORIMETRIC BIOSENSORS

A property that all biological reactions taking place inside living organisms share is the absorption or evolution of heat, reflected to the surrounding environment as a change in the temperature. Thermal biosensors exploit this property and constitute a promising tool with various applications (e.g., detecting pathogenic bacteria), especially since the enzyme thermistor (ET) was invented (Danielsson and Mosbach, 1988). Among their advantages are their small size, long-term and high stability, and non–chemical contact measuring (Syam et al., 2012).

Calorimetry involves measuring the heat following a biochemical reaction. Defining Q as the total heat absorbed or produced during a chemical reaction, n_p as the amount of moles produced, ΔH as the molar enthalpy change, C_p as the heat capacity of the system, and ΔT as the temperature shift caused by the heat evolution or absorption, the first law of thermodynamics, which deals with energy conservation, is expressed as

$$Q = -\, n_p \sum \Delta H \tag{3.6}$$

$$Q = C_p \, \Delta T \tag{3.7}$$

The basic part of a thermal biosensor is a temperature sensor that measures the temperature and immobilized enzyme molecules. When these enzymes come into contact with the analyte, a biochemical reaction takes place and the temperature shift is measured. The enthalpy change that a biochemical reaction will cause depends on the specific enzyme-catalyzed reaction, namely on the enzyme and the substrate used, and the heat capacity of the system on the organic solvent of which the surrounding environment consists. These can be specified by tables available in the literature. Thus, the amount of product molecules created can be calibrated against the temperature change (Lammers and Scheper, 1999).

The temperature sensor most commonly used for measuring the temperature in the reaction medium is an ET. (See Figure 3.7.) It consists of an external thermostated box surrounded by a foam which insulates it from the environment and the internal cylinder as well. The cylinder inside the box has two cavities for two columns which are inserted inside it. Each column has a probe attached to it, where the immobilized enzymes can be placed. If different enzymes are placed on the probes, two different analytes can be measured by the device, but the thermistor can also be used with an immobilized enzyme on the first column and a reference column with an inactivated enzyme or just support material. Two separate small thermistors connected to a Wheatstone brigde are fixed on each of the two columns respectively measuring the temperature of the sample inserted in the probe connected to its column.

The buffer certifies the current flow of a stream inside the cylinder to which samples are injected and pumped to the thermistor. When the thermistor probe where the immobilized enzymes are placed reacts with the analyte, the temperature change caused is detected by the measuring thermistor. The differential temperature signal of the measuring and reference thermistor is the temperature shift, which according to equations (3.6)–(3.7), corresponds to a substrate concentration.

The characteristic that distinguishes thermistors from other temperature sensors is that they work over a relatively small temperature range. As resistance decreases exponentially with the increase of temperature, thermistors are considered highly

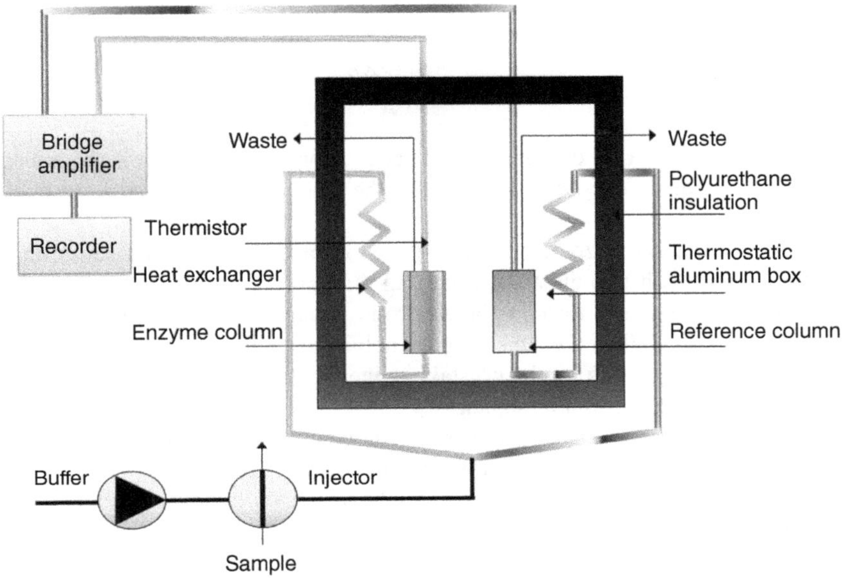

Figure 3.7 Schematic setup of enzyme thermistor.

sensitive devices within their working range, which constitutes another advantage accompanying their low cost and small size.

Thermal biosensors were recently applied as devices for detecting pesticides and pathogenic bacteria (Syam et al., 2012), and of course they are still used as detectors for the presence of particular substances or for measuring biological parameters.

3.6 PIEZOELECTRIC BIOSENSORS

The fundamental principle on which piezoelectric biosensors base their operation is the piezoelectric effect connected with their basic component, namely the piezoelectric crystal. Piezoelectricity is the electric charge in crystals and biological matter (e.g, DNA, proteins) resulting from changes in pressure. In our case, when the substance on the surface of the piezoelectric crystal reacts with a substrate, its mass changes, resulting in an oscillation of the crystal's characteristic reasonant frequency. Thus, a piezoelectric biosensor can be defined as a device that measures pressure by converting it to an electrical charge utilizing the piezoelectric effect.

Assuming a crystal with a thin, uniform, and purely elactic surface, Sauerbrey (1959) derived the basic equation describing the relationship between the reasonant frequency of an oscillating piezoelectric crystal (f in Hertz) and the amount of mass deposited or removed from the crystal's surface (Δm in grams) as follows:

$$\Delta f = \frac{Kf^2 \ \Delta m}{A} \tag{3.8}$$

where K is a constant of the crystal which depends on its structure, A is the crystal's surface area in square centimeters, and Δf is the change of the reasonant frequency in Hertz.

The most common piezoelectric sensor used in biomedical applications consists of a thin crystal with two electrodes fixed to the two opposite sides. An electrical field applied to the crystal through the electrodes makes the crystal oscillate near its resonant frequency. When the substance immobilized on the crystal's surface reacts with the analyte, the mass on the crystal changes, causing a frequency shift that is easily detected by the oscillator signal.

Choosing a crystal with proper reasonant frequency is critical for the detection of the desired biomolecule. Especially for small molecules, such as antigens, amplification and other types of signal processing may be necessary in order to obtain a direct and observable signal if the reasonant frequency is not the appropriate one.

Apart from being small and inexpensive, piezoelectric biosensors have many advantages. Crystal materials are not very sensitive and offer long-term stability as well as almost excellent temperature behavior. Above all they are capable of giving a rapid response signal, which makes them especially attractive when compared to other classical devices. The major drawback of these sensors is that they cannot be used for static measurements as they only detect changes of a variable.

Piezoelectric biosensors can be used as part of a flow injection analysis (FIA) system (Figure 3.8), but they have other applications as well. They are devices that can detect substances in aqueous solutions or directly and continuously monitor their

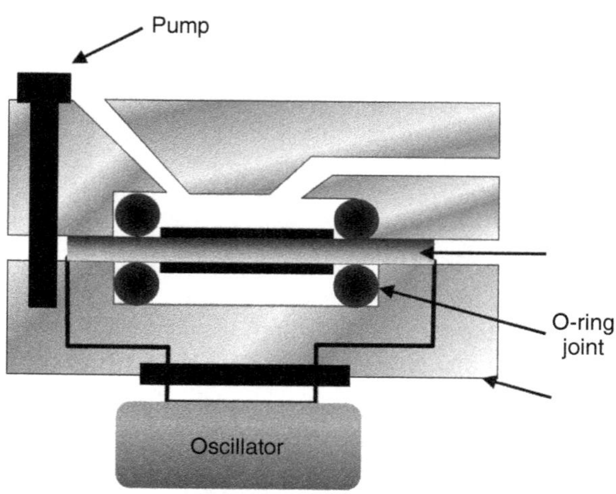

Figure 3.8 Schematic setup of piezoelectric biosensor.

concentration. In one of their recent and most attractive applications they were used for point mutation detection in human DNA (Del'Atti et al., 2006).

3.7 OTHER TYPES OF BIOSENSORS

3.7.1 Magnetic Biosensors

Magnetic biosensors were used in the past but have been disregarded mainly because of their size and power consumption. The magnetic biosensors generated recently overcome these problems by the use of giant magnetoresistance material (GMR) and promise detection of biomolecules in small samples with low concentration with great sensitivity, selectivity, speed, and economy (Wang and Guanxiong, 2008).

The sensing technique is based on magneto-nanodetection of magnetic nanoparticles that biomolecules have. The sensors consist of the GMR detector, conductors carrying the current by which the permanent magnetic field is created, and immobilized microbeads. (See Figure 3.9.) These microbeads develop a dipole field when magnetized by the external field and can be used in order to detect small fields. It should be noted that the sensitivity is maximized when the size of the sensor is matched to the microbead's size.

The fact that this technology is scalable makes magneto-nano-GMR biosensors very promising. They can be applied for molecular diagnostics of various diseases, such as cancer, cardiac problems, or infectious diseases.

3.7.2 Pyroelectric Biosensors

Pyroelectic biosensors were named after the basic substance on which their operation is based. Pyroelectric materials have the ability to generate voltage or create a current

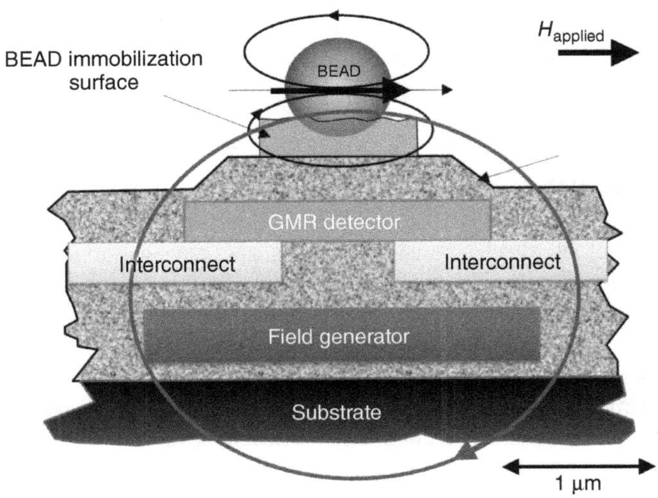

Figure 3.9 Schematic setup of magnetic biosensor.

whenever their temperature changes as the temperature shift modifies the position of their atoms, resulting in a polarization change. The sensing technique involves measuring the change in the pyroelectic material polarization, expressed as a generation of a voltage, when the temperature changes.

Assuming a pyroelectric material on which constant pressure (σ) and electric field (E) are applied, its pyroelectric coefficient (p) is expressed as

$$p = \left(\frac{\partial P}{\partial T}\right)_{E,\sigma} \tag{3.9}$$

where T stands for the temperature and P for the polarization.

3.7.3 Ion Channel Biosensors

Biological ion channels are membrane proteins whose main role is to regulate the flow of ions across the cell membrane and regulate the cell volume and its electrical and biochemical activities. The fact that they are present in the membranes of all cells makes these biosensors suitable for detecting molecules of interest, such as drugs with low molecular weight, large proteins, or microorganisms (Krishnamurthy et al., 2010).

A sensor of this type has a proper molecule fixed to a gold surface. When two nonconductive monomers align, forming a conductive dimer, the ions start flowing. The electrical charge generated when the analyte meets the immobilized molecule is the detection signal, which can also be used to feed an external device. Then it may be processed and provide analytical data, for example, about the concentration of the molecule of interest. The most important part of this type of biosensor is the switching mechanism used to disrupt the ion flow when the analyte will be detected.

This type of biosensor is very useful as it can be used with multiple immobilized molecules on it. Thus, it makes it possible for the sensor to detect a number of different analytes that contribute to ion channel signal delivery. One of its most common uses is for the detection of influenza and viruses (Syam et al., 2012).

3.8 CONCLUSIONS

The development of novel sophisticated tools and new materials as well as the advances in microelectronics over the last years have resulted in a major transformation of sensing devices, which are nowadays characterized by increased reliability and accuracy with respect to the sensed information. Several types of biosensors exist—electrical, electrochemical, optical, piezoelectric, and thermal/calorimetric—each with different advantages and disadvantages. The construction of a biosensor is a complex task where several design principles must be taken into account, including the selection of the biological element that will be used for the detection of the analyte, the chemical nature of the transducer, and the position in the protein to introduce the signal. Another important factor to be considered is the requirement for fast real-time and reliable reproducible measurements.

Modern biomedical applications, which require miniaturized medical devices to be implanted into human bodies, dictate even more stringent requirements, including the size, energy efficiency, and biocompatibility of the device with the individual patient. Biosensor research is particularly motivated by the need for miniaturized clinically useful devices where microscale biosensor devices react with the analyte and generate a measurable response within the same compact body. In the future the further miniaturization of the implantable devices is expected to enable on-organ monitoring and highly specific treatment delivery without prohibiting the normal functioning of surrounding organs and tissues. Furthermore, in addition to more sophisticated neuroprosthetics and artificial organs, the expected developments in brain–computer interfacing will enhance our ability to investigate and alter cognitive or sensori-motor functions in humans.

REFERENCES

Bakker E. 2004. Electrochemical sensors. *Anal. Chem.* 76:3285–3298.

Belluzo MS, Ribone MÉ, Lagier CM. 2008. Assembling amperometric biosensors for clinical diagnostics. *Sensors* 8(3):1366–1399.

Buerk D. 1993. *Biosensors. Theory and Applications*. Lancaster: Technomic Publishing.

Chaubey A, Malhotra BD. 2002. Mediated biosensors. *Biosensors Bioelectron.*, 17(6):441–456.

Clark LC, Lyons C. 1962. Electrode systems for continuous monitoring in cardiovascular surgery. *Ann. NY Acad. Sci.*, 102(1):29–45.

Danielsson B, Mosbach B. 1988. Enzyme Thermistors. *Methods Enzymol.*, 137:181–197.

Dell'Atti D, Tombelli S, Minunni M, Mascini M. 2006. Detection of clinically relevant point mutations by a novel piezoelectric biosensor. *Biosens Bioelectron.*, 21(10):1876–1879.

Drapp B, Piehler J, Brecht A, Gauglitz G, Luff BJ, Wilkinson JS, Ingenhoff J. 1997. Integrated optical Mach-Zehnder interferometers as simazine immunoprobes. *Sensors Actuators B: Chem.*, 39(1):277–282.

Fan X, White IM, Shopova SI, Zhu H, Suter JD, Sun Y. 2008. Sensitive optical biosensors for unlabeled targets: A review. *Anal. Chim. Acta*, 620(1):8–26.

Gauglitz G. 2005. Direct optical sensors: Principles and selected applications. *Anal. Bioanal. Chem.*, 381(1):141–155.

Grieshaber D, MacKenzie R, Vörös J, Reimhult E. 2008. Electrochemical biosensors—Sensor principles and architectures. *Sensors*, 8(3):1400–1458.

Guiseppi-Elie A, Brahim S, Slaughter G, Ward KR. 2005. Design of a subcutaneous implantable biochip for monitoring of glucose and lactate. *Sensors J., IEEE*, 5(3):345–355.

Hoa XD, Kirk AG, Tabrizian M. 2007. Towards integrated and sensitive surface plasmon resonance biosensors: A review of recent progress. *Biosensors Bioelectron.*, 23(2):151–160.

Homola J. 2003. Present and future of surface plasmon resonance biosensors. *Anal. Bioanal. chem.*, 377(3):528–539.

Iniewski K (ed.). 2012. *Biological and Medical Sensor Technologies*. Boca Raton, FL: CRC Press.

Krishnamurthy V, Monfared SM, Cornell B. 2010. Ion-channel biosensors—Part I: Construction, operation, and clinical studies. *IEEE Trans. Nanotechn.*, 9(3):303–312.

Lammers F, Scheper T. 1999. Thermal biosensors in biotechnology. In Scheper T. (ed.). *Thermal Biosensors, Bioactivity, Bioaffinitty*. Springer: Berlin/Heidelberg, pp. 35–67.

Lopez-Martinez MJ, Campo EM, Caballero D, Fernandez E, Errachid A, Esteve J, Plaza JA. 2009. Versatile micropipette technology based on deep reactive ion etching and anodic bonding for biological applications. *J. Micromech. Microeng.*, 19(10):105013.

Medintz IL, Deschamps JR. 2006. Maltose-binding protein: A versatile platform for prototyping biosensing. *Curr. Opin. Biotechnol.*, 17(1):17–27.

Monošík R, Stre'anský M, Šturdík E. 2012. Biosensors—classification, characterization and new trends. *Acta Chim. Slov.*, 5(1):109–120.

Passaro V, Dell'Olio F, Casamassima B, De Leonardis F. 2007. Guided-wave optical biosensors. *Sensors*, 7(4):508–536.

Patolsky F, Weizmann Y, Willner I. 2004. Long-range electrical contacting of redox enzymes by SWCNT connectors. *Angew. Chem. Int. Ed.*, 43(16):2113–2117.

Perumal V. Hashim U. 2013. Advances in biosensors: Principle, architecture and applications. *J. Appl. Biomed.*, 11.

Plaxco KW, Soh HT. 2011. Switch-based biosensors: A new approach towards real-time, in vivo molecular detection. *Trends Biotechnol.*, 29(1):1–5.

Pohanka M, Skladal P. 2008. Electrochemical biosensors—principles and applications. *J. Appl. Biomed.*, 6(2):57–64.

Prieto F, Llobera A, Calle A, Lechuga LM. 2000. Design and analysis of silicon antiresonant reflecting optical waveguides for evanscent field sensor. *J. Lightwave techn.*, 18(7):966.

Prieto F, Sepulveda B, Calle A, Llobera A, Domínguez C, Abad A, Lechuga LM. 2003. An integrated optical interferometric nanodevice based on silicon technology for biosensor applications. *Nanotechnology*, 14(8):907.

Sassolas A, Blum LJ, Leca-Bouvier BD. 2012. Immobilization strategies to develop enzymatic biosensors. *Biotechn. Adv.*, 30(3):489–511.

Sauerbrey G. 1959. Verwendung von Schwingquarzen zur Wägung dünner Schichten und zur Mikrowägung. *Zeitschrift für Physik* 155(2):206–222.

Schabmueller CGJ, Loppow D, Piechotta G, Schütze B, Albers J, Hintsche R. 2006. Micromachined sensor for lactate monitoring in saliva. *Biosensors Bioelectron.*, 21(9):1770–1776.

Serra PA (ed.). 2011, Biosensors for health, environment and biosecurity. *InTech Open Access*.

Syam R, Davis KJ, Pratheesh MD, Anoopraj R, Surej Joseph B. 2012. Biosensors: A Novel Approach for Pathogen Detection. *Vetscan*, 7:14–18.

Thevenot DR, Toth K, Durst RA, Wilson GS. 1999. Electrochemical biosensors: Recommended definitions and classification. *Pure Appl. Chem.*, 71(12):2333–2348.

Velasco-Garcia MN. 2009. Optical biosensors for probing at the cellular level: A review of recent progress and future prospects. In *Seminars in Cell & Developmental Biology*, Vol. 20, No. 1, San Diego, CA: Academic Press, pp. 27–33.

Wang, SX, Guanxiong L. 2008. Advances in giant magnetoresistance biosensors with magnetic nanoparticle tags: Review and outlook, *IEEE Trans. Magnetics*, 44(7):1687–1702.

Wang H, et al. 2009. Recent progress in strategies for the creation of protein-based fluorescent biosensors. *Chembiochem*, 10(16):2560–2577.

Ymeti A, Kanger JS, Greve J, Lambeck PV, Wijn R, Heideman RG. 2003. Realization of a multichannel integrated Young interferometer chemical sensor. *App. optics*, 42(28):5649–5660.

Yoshinobu T, Iwasaki H, Ui Y, Furuichi K, Ermolenko Y, Mourzina Y, Schöning MJ. 2005. The light-addressable potentiometric sensor for multi-ion sensing and imaging. *Methods* (San Diego, Calif.), 37(1):94.

4

SENSING TECHNOLOGIES FOR BIOMEDICAL TELEMETRY

TOSHIYO TAMURA
Osaka Electro-Communication University, Neyagawa, Japan

4.1 INTRODUCTION

Biotelemetry systems use sensors which provide physical, chemical, and biological data for the continuous remote monitoring of human physiological parameters. Among other applications, biotelemetry is applied for the monitoring of high-risk patients through the monitoring of crucial physiological parameters such as heart rate, pulse waveform, respiratory rate, blood oxygen saturation, intracorporal pressure, gastrointestinal chemical parameters, or lumen visualization. In the late 1950s wireless communication technologies began to be applied for measuring the pH and the temperature from internal cavities and also for applications involving subcutaneous and implantable sites, though with several limitations (Farrar et al., 1957). Later, during the 1970s, several systems were developed, such as devices for fetal monitoring, electrocardiogram telemetry, and gastrointestinal pressure monitoring (Ko and Neuman, 1967; Neuman et al., 1970; Ko et al., 1971). Hewlett Packard developed a portable electrocardiograph in 1972, the HP71800A, for ECG monitoring on a 24-h basis. Since then, several biotelemetry devices have been developed. A critical parameter for all the biomedical telemetry applications is the sensing operation and especially the power feed of the wearable or implantable device. Other challenges include the size, the sampling rate, and the reliability of measurements. The hardware of these sensors usually comprises a microcontroller, an ultralow-power RF transceiver, antennas, analog signal conditioning circuitry, data converters, and

Handbook of Biomedical Telemetry, First Edition. Edited by Konstantina S. Nikita.

a powering module. Wireless technology is employed to safely isolate the human body from the AC power supply and ground. Microelectromechanical systems and commercial available wireless technologies such as Bluetooth and Zigbee encourage the development of biotelemetry devices (Patel et al., 2012). MEMSs include various types of devices such as pressure sensors, silicon microphones, accelerometers, gyroscopes, optical MEMS and image sensors, microfluidic chips, microdispensers for drug delivery, flowmeters, and IR temperature sensors, strain sensors, and energy-harvesting devices. Usually, the output of the sensor is a digital signal which allows the direct connection with a microcontroller. Accelerometers are classified as analog or digital according to their output. Digital accelerometers usually use the pulse width modulation (PWM), with a square wave of a certain frequency, and amplitude proportional to the measured acceleration. The advantages of digital accelerometers include the low complexity and low power consumption. On the other hand, they usually require strict synchronization as well as computationally intensive division operations.

This chapter focuses on the sensor interface of various biomedical telemetry applications for both humans and animals. The sensors applied to humans could be divided into two groups: noninvasive sensors and invasive and implantable sensors. Wireless sensors also include ingestible pills for the wireless monitoring of pH, blood pressure, and so on.

4.2 NONINVASIVE SENSORS AND INTERFACES

4.2.1 Sensors Using Electrophysiological Signals

Electrocardiography Electrocardiography is widely used in the diagnosis and management of many cardiac diseases. A typical ECG monitoring system consists of the electrodes, amplifier, and transmitter. More recently wireless, mobile, and remote technologies have been applied to enhance the functionalities and usability of ECG. Since the first ECG monitoring systems (Lopez and Richardson, 1969), significant advances have taken place, including dry noncontact electrodes with capacitive coupling, leadless ECG monitoring (Scherr et al., 2008; Melhorn et al., 2009; Casson et al., 2010; Chi et al., 2010), the transition from traditional 12-lead to 3-lead ECG (Hsieh and Lo, 2010; Atoui et al., 2010; Bsoul et al., 2011), flexible polymeric dry-potential electrodes (Baek et al., 2008; Jung et al., 2012), and a microfabricated electrode with hollow microneedles (Yu et al., 2009). The carbon nanotube (CNT)/polydimethylsiloxane (PDMS) composite-based dry ECG electrode showed its long-term wearable monitoring capability and robustness to motion and sweat (Jung et al., 2012). Head movement or sweating can cause electrodes to shift or fall off, making them rather unsuitable for long-term monitoring or for demanding applications under strict conditions. The belt-type electrode with a transmitter has been developed by Polar USA (Polar 2012), while current ECG approaches include wearable textile-based monitoring systems, where ECG data can be collected using smart shirts (Finlay el al., 2008; Lee and Chung, 2009).

Electroencephalography Electroencephalography (EEG) has long been used to record and study the electrical activity of the outermost layer of the brain–the cerebral cortex. The International 10–20 system is a method to describe and apply the location of scalp electrodes in the context of an EEG test or experiment. The system is based on the relationship between the location of an electrode and the underlying area of cerebral cortex. The electrode is an active electrode which consists of a unity-gain amplifier placed next to the electrode on the scalp. The development of wireless ambulatory EEG is crucial for achieving long-term monitoring of a patient in their everyday environment.

Recently an EEG-based brain–machine interface (BMI) was proposed where the oscillatory EEG components were used as input signals for the BMI (Lin et al., 2010; Verma et al. 2010; Xu et al., 2011), a direct communication pathway between the brain and an external device, which is used for assisting, augmenting, or repairing human cognitive or sensorimotor functions. To accomplish this, researchers have been studying ways to measure brain activity and convert the data into input for computer systems. Furthermore, wireless EEG devices, headsets (Figure 4.1), and headbands have been developed for BMI. Over the years, different applications have had success with restoring vision to blind patients, control of prosthetic limbs, control of robotics (Zhang et al., 2010; Oh et al., 2012), speech generation (Guenther et al., 2012), gaming, and cognitive imaging.

An important goal is to make BMIs more flexible and less invasive, progressing from neuronal implants in the brain to headsets using EEG technology towards an integrated hardware and software solution for acquisition and real-time EEG analysis to monitor indexes of alertness, cognition, and memory (Berka et al., 2004). Another example is a system integrated into a headband with small size, low power consumption of only 0.8 mW, and autonomous operation using the body heat dissipated naturally from the forehead as the power source (Strickland and Chartier, 1997).

Electromyography Electromyography (EMG) signals generated during muscle contraction are often used in rehabilitation devices because of their distinct output

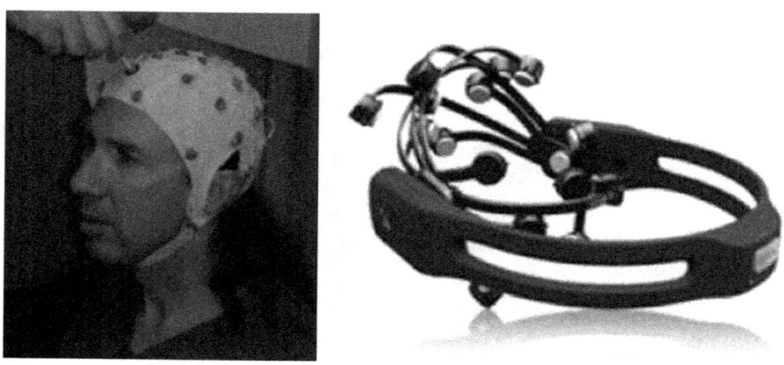

Figure 4.1 Headset-type of EEG monitoring.

characteristics compared to other biosignals (Moon et al., 2005). EMG can be used to detect abnormal electrical activity of muscle that may occur in many diseases and conditions, including muscular dystrophy, inflammation of muscles, pinched nerves, peripheral nerve damage (damage to nerves in the arms and legs), and amyotrophic lateral sclerosis (ALS). It consists of surface electrodes, reference electrodes, an amplifier, a wireless transmitter, and batteries. In commercial devices, most electrodes are active and equipped with a small unit for signal processing and transmission. Additionally, the dual electrode is very popular for detecting the most useful EMG signals. Fabric electrodes are usually applied for ambulatory monitoring (Akita et al., 2008).

Besides the detection of abnormality, EMG signals have been used for recognizing isometric muscular activity: activity related to very subtle or no movement at all. The EMG devices worn on an armband around the bicep are able to reliably recognize a motionless gesture without calibration or training across the user with different muscle volumes. Thus EM-based input devices can provide an effective solution for designing mobile interfaces.

To solve the complicated wires and battery maintenance problems in the application of human–machine interface (HMI) or human–computer interface (HCI), EMG using conductive fabric to power supply and electrical shield for noise reduction has been developed. It enables precise EMG measurement with a wearable system.

In another application, HCI is used to control an electric-powered wheelchair via combinations of left-, right and both-shoulder elevation gestures. The proposed wearable HCI hardware consists of two active surface electrodes, a high-speed micro controller, a Bluetooth module, and a battery. Experimental results using the wearable EMG-based HCI and the electric-powered wheelchair developed showed that wearable EMG-based HCI is feasible for users with severe motor disabilities. Wearable EMG also aims to control smartphones, notebooks, and other gadgets. The wearable electromyography-based controller uses sensors to interpret the electrical signals generated by a user's muscles and then communicates with the wearer's computer via a wireless connection. This can be available as an armband equipped with sensors, a shirt, eyeglasses, and even nodes attached directly to the user's body. No matter the setup, a calibration process allows the system to locate specific sensors and collect information based on specific gestures or movements.

Wearable electroretinographs (ERGs), Electrooculograms (EOG) (Bulling et al., 2008, 2009), and galvanic skin reflex (GSR) (Ouchi et al., 2004; Sun et al., 2012) have been also developed. Wearable goggles introduced the analysis of eye motion as a new input modality for activity recognition, context awareness, and mobile HCI applications.

4.2.2 Photoplethysmogram Sensor

A photoplethysmogram (PPG) is an optically obtained plethysmogram, a volumetric measurement of an organ. A PPG (Figure. 4.2) consists of the light-emitting diode (LED) usually of infrared light and the photodetector (Allen, 2007). A conventional

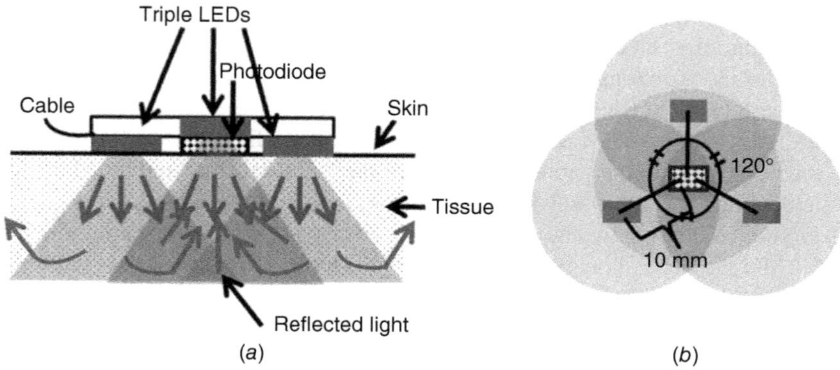

Figure 4.2 Conceptual diagrams of PPG sensor with three light Sources: (*a*) frontal view; (*b*) transverse view (Baek et al., 2009).

pulse oximeter monitors the perfusion of blood to the dermis and subcutaneous tissue of the skin. Because the skin is richly perfused, it is relatively easy to detect the pulsatile component of the cardiac cycle. The DC component of the signal is attributable to the bulk absorption of the skin tissue, while the AC component is directly attributable to change in blood volume in the skin caused by the pressure pulse of the cardiac cycle. The height of AC component of the PPG is proportional to the pulse pressure, the difference between the systolic and diastolic pressure in the arteries. The change in volume caused by the pressure pulse is detected with each cardiac cycle appearing as a peak. Because blood flow to the skin can be modulated by multiple other physiological systems, the PPG can also be used to monitor breathing, hypovolemia, and other circulatory conditions (Reisner et al., 2008). The sensor is simple to handle and just attached to the measuring site. However, the motion artifact is a challenging issue to solve, especially during exercise. Additionally, the shape of the PPG waveform differs from subject to subject and varies with the location and manner in which the pulse monitor is attached. Mostly these solutions are mathematical compensations such as fast Fourier transform (FFT) and wavelet analyses. The other solution is the changes in light wavelength. The green light has been used to eliminate the artifact during movement (Maeda et al., 2011a,b). Furthermore, a second derivative of the pulse wave estimated the stiffness of blood vessel (Takazawa et al., 1998).

The heart rate monitor by plethysmography in conjunction with respiratory rate has also been developed for wheelchair users (Postolache et al., 2009). The wireless respiratory monitor is not so common and the signal of acceleration can be used (Bates et al., 2010). MEMS accelerometers worn on the torso can measure inclination changes due to breathing, from which a respiratory rate can be obtained. The triaxial accelerometer data from a wireless sensor device tracks the axis of rotation and obtains angular rates of breathing motion. The resulting rates are validated against gyroscope measurements and show high correlation to flow rate measurements using a nasal cannula. This is a potential method for estimating absolute air flow rate and tidal volume.

4.2.3 Pulse Oximeter

Pulse oximetry is a noninvasive method allowing the monitoring of the saturation of a patient's hemoglobin. A sensor is placed on a fingertip or earlobe or, in the case of an infant, across a foot. Light of two different wavelengths is passed through the patient to a photodetector. The changing absorbance at each of the wavelengths is measured, allowing determination of the absorbances due to the pulsing arterial blood alone, excluding venous blood, skin, bone, muscle and fat, as shown in Figure 4.3. It is possible to measure both oxygenated and deoxygenated hemoglobin on a peripheral

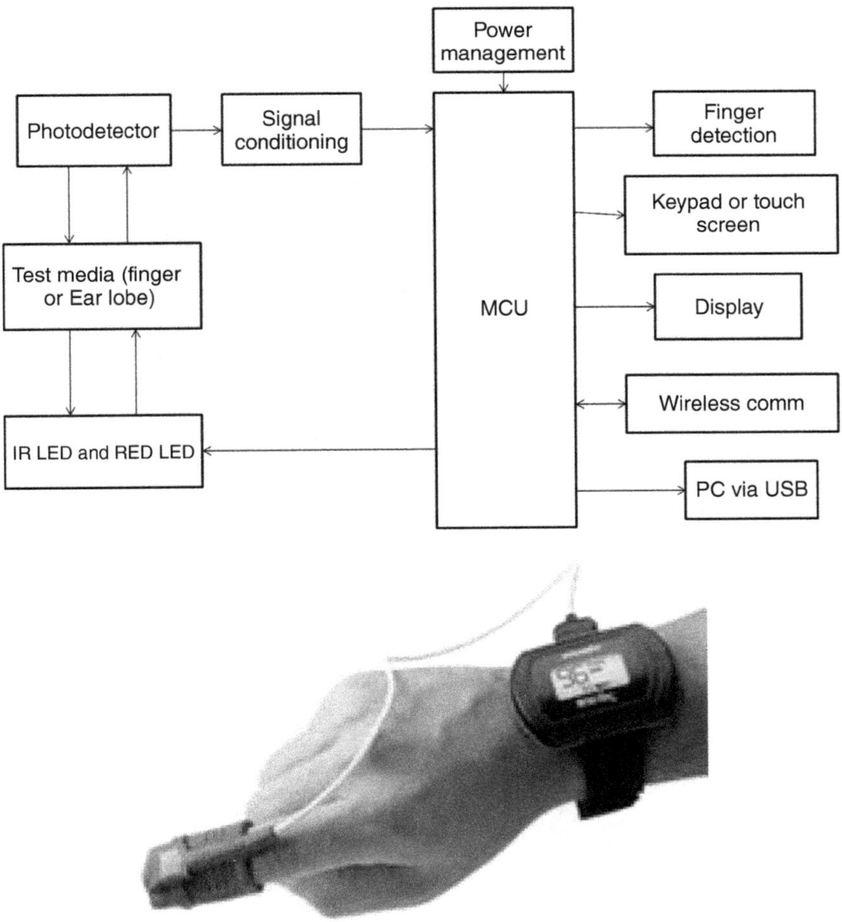

Figure 4.3 Pulse oximetry.

scale. Reflectance pulse oximetry may be used as an alternative to transmissive pulse oximetry. This method does not require a thin section of the patient's body and is therefore well suited to more universal application such as the feet, forehead, and chest, but it also has some limitations.

Wireless pulse oximeters are now in the market, with only a few companies producing these devices. Most current wireless pulse oximeters use the Bluetooth protocol to communicate sensor readings. However, there are significant disadvantage when using the Bluetooth specification for this application, most notably the battery life. The Zigbee specification is much more suited to wireless patient monitoring, and although it lacks the cross-integration of Bluetooth, it has the advantage of significantly increased battery life, a simpler secure transmission scheme, and lower system cost than Bluetooth.

The main operation of a pulse oximeter is the determination of oxygen saturation. Arterial oxygen saturation is the percentage of arterial hemoglobin that is oxygenated. Oxygen saturation measured with a pulse oximeter is called SpO_2. It is based on estimation, peripheral measurements, and the assumption that only HbO_2 and Hb are present in the blood, that is,

$$SpO_2 = \frac{HbO_2}{Hb + HbO_2} \tag{4.1}$$

Oxygenated and reduced hemoglobin differ in their absorption of light, a fact that pulse oximetry exploits to find the relative levels of the two hemoglobins. The most common pulse oximetry uses a red light-emitting diode (LED) and infrared LED to shine light through a person's finger (Figure 4.4). A photodetector is used on the other side of the finger to measure the transmitted red and infrared light. The red and infrared LEDs are generally pulsed in alternating fashion, so that one photodetector can be used to measure the light intensity of both LEDs. With a known measurement of red and infrared lights transmitted through the finger, an estimation of the ratio between oxygenated hemoglobin and reduced hemoglobin can be determined based on absorption curves at the various wavelengths of transmitted light. A typical oximeter works with 660-nm red light and 940-nm infrared light.

At 660 nm reduced hemoglobin absorbs about 10 times as much light as does oxygenated. A normal finger has light absorbed from bloodless tissue, venous blood,

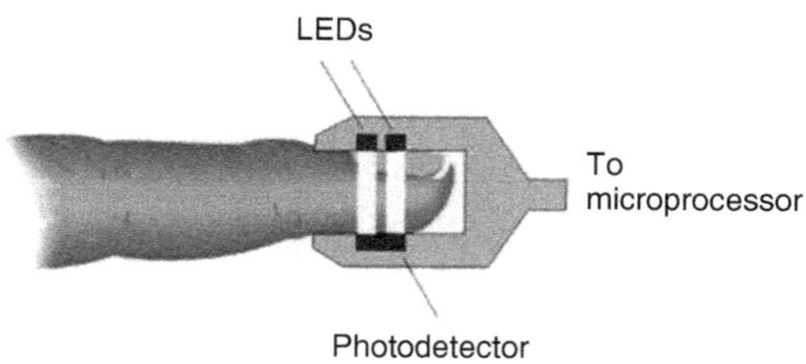

Figure 4.4 Measuring site of pulse oximeter.

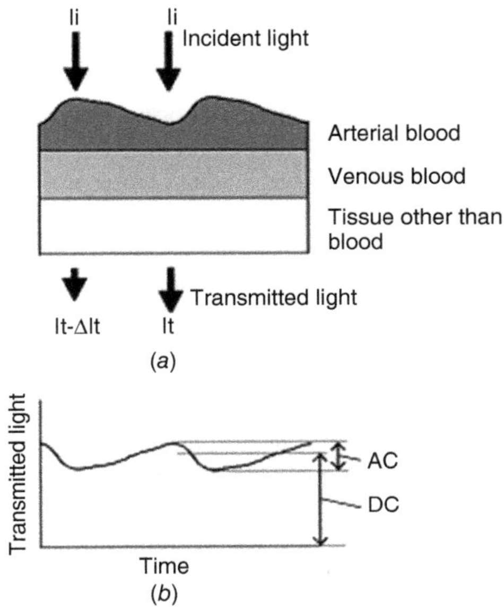

Figure 4.5 Principle of pulse oximeter: (*a*) incident light (Ii) and transmitted light (It) with arterial blood variation; (*b*) change in amount of transmitted light with time.

and arterial blood. The volume of arterial blood changes in pulse, so the absorption of light also changes. The photodetector will therefore see a large DC signal representing the residual arterial blood, venous blood, and bloodless tissue. A small amount of the detected signal will be an AC signal representing the arterial pulse. Because this is the only AC signal, the arterial part of the signal can be differentiated as shown in Fig. 4.5. This AC signal is separated with simple filtering and a root-mean-square (RMS) value can be calculated.

An intermediate value, known as the norm ϕ, represents the ratio of reduced to oxygenated arterial hemoglobin:

$$\phi = \frac{\text{ACrms 660 nm/DC 660 nm}}{\text{ACrms 940 nm/DC 940 nm}} \tag{4.2}$$

Using this value, a value of oxygen saturation is calculated based on empirical data.

4.2.4 Wireless Pressure Monitor

Wearable pressure monitors are available for measuring blood pressure, intraocular pressure, and so on. For blood pressure monitoring, the American Heart Association recommends home monitoring for all people with high blood pressure to help the health care provider determine whether treatments are working. Home monitoring is a substitute for regular visits to the physician. For this, an ambient blood pressure monitor which consists of a cuff, inflator, and pressure sensor has been developed. In wireless blood pressure monitoring, the system operates wirelessly to inflate the cuff and display the values. The other option is the smartphone-based blood pressure monitor which provides mobile means for monitoring blood pressure. The device is connected to smartphone as a display.

A cuffless, approach is now popular to estimate blood pressure. Pulse transit time (PTT) is a proven, simple-to-measure marker of blood pressure that could potentially permit continuous, noninvasive, and cuffless BP monitoring after an initial calibration (Smith et al., 1999; Chen et al., 2000; Nashitz et al., 2004; Foo et al., 2005; Yoon et al 2009; Nakamura et al., 2011). The PTT or pulse transit velocity (PTV) is monitored by the time between the PPG signal and the R-wave of ECG. The device transmits its data via radio or WiFi (Asada et al., 2003, 2005).

The speed of the pulse wave depends on the tension of the arterial walls. When the blood pressure is high, the arterial walls are tense and hard and the pulse wave travels faster. When the blood pressure is low, the arterial walls have less tension and the pulse wave travels slower. Although the actual blood pressure itself cannot be determined from the speed of the pulse wave, a *change* in blood pressure is indicated by a change in the speed of the pulse wave. Therefore, PTT is used to detect changes in pressure. PTT is defined as a time of propagation of a pulse to travel in artery and based on preejection time (PEP) and aortic pulse wave (aPTT). A-PTT as shown in Fig. 4.6, is the time it takes the pulse wave to travel from the aorta to a peripheral artery and is directly related to blood pressure, but unfortunately, it cannot be measured directly. PEP is the period just before the blood is pumped into the aorta. In general, PEP change over short periods of time is negligible in most cases so we can assume that PTT corresponds to a-PTT and therefore to blood pressure.

However, vasoactive and other drugs can cause significant changes in PEP and affect the correlation between PTT and blood pressure. Generally in most cases, we can say that PTT corresponds to a-PTT and blood pressure (Shaltis et al., 2006; Wang et al., 2004; Zhang et al., 2011).

Also the blood pressure in an extremity changes depending on the height above/below the heart, and the device contains accelerometers that can keep track of hand position and adjust accordingly (Shaltis et al., 2006).

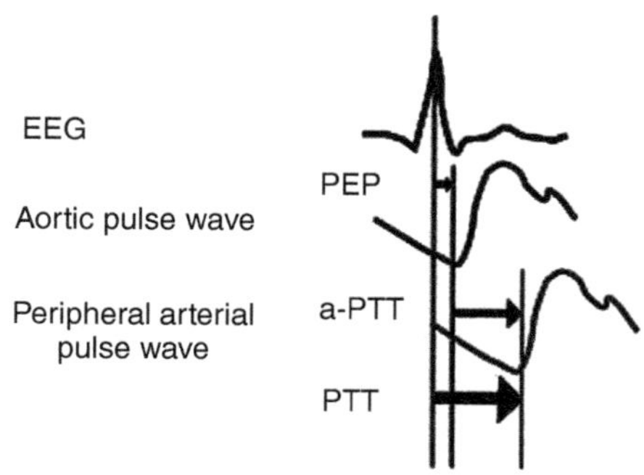

Figure 4.6 Principle of pulse transit time.

Difference in pulse transit time (PTT-D) is calculated as the timing variations between the toe and finger peripheral sites so that a more accurate PTT can be obtained.

Another interesting MEMS sensor development is a digital MEMS barometer. Worn by patients, the device essentially determines the altitude—wherein higher altitudes demand higher levels of oxygen and vice versa—to conserve oxygen and energy in ventilator systems. It can be used as a smart bandage for negative-pressure wound therapy, employing differential pressure measurements (Figure 4.7).

Sensimed's system, called dubbed Triggerfish, includes a contact lens that uses an embedded MEMS strain gauge that measures the curvature of the eye over a time period. It is made on a flexible substrate. It can measure, monitor, and control intraocular pressure levels for patients as well as catch early cases of glaucoma. It monitors this pressure for 24 hr and then provides a record to the attending physician (Figure 4.8). The contact lens has an antenna, a dedicated processing unit, and a transmitter. The lens is powered by radio waves so it does not need to be connected to a battery. The lens transmits information such as eye measurements to a receiver that is worn around the patient's neck. Data transmitted from the lens provide

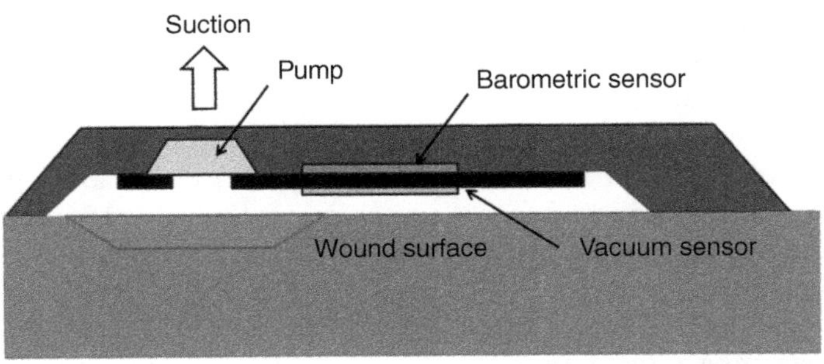

Figure 4.7 A MEMS digital barometer, such as Freescale Semiconductor's MPL115A, can be used as a smart bandage for wound therapy by taking advantage of negative differential pressure measurements.

Figure 4.8 To catch early cases of glaucoma, the Triggerfish ocular implant measures, monitors, and controls intraocular pressure levels for patients over a 24-hr period. Sensimed and STMicroelectronics codeveloped the implant.

information that cannot be discovered using traditional ophthalmic equipment. The standard test for glaucoma is via an instrument called a tonometer that measures eye pressure. The problem is that the tonometer usually finds issues after glaucoma has progressed too far.

Circumferential fluctuations in the area of the corneoscleral junction, directly correlated in intraocular pressure readings, are measured. This information is subsequently transmitted from a recorder via wireless communications. Cumbersome glaucoma tests, which require a visit to an ophthalmologist, could soon be history thanks to a test that provides earlier and more accurate detection of the malady. An easy-to-use self-test probe allows patients to gently rub the eyelid in the comfort of their own home. Although the probe concept appears simple, the technology behind it is rather complex. It involves a system of microforce sensors, specially designed microchips, and math-based procedures programmed into the probe.

4.2.5 Motion Sensors

Motion analysis uses several types of motion sensors and systems, such as the accelerometer Gyrosensor, magnetoresisitive sensor, flexible goiometer, electromagnetic tracking system (ETS), textile sensor, force sensor, and EMG. Multiple sensors may be used for various gait analyses, physical activities, and daily activites with different diseases. The wireless and wearable motion sensors beside EMG are desribed in this section. Common types of MEMS accelerometers are available, namely piezoelectric, piezoresistive, and capacitive accelerometers.

The wristwatch type of inertial sensor is also commercially available, such as MotoinPods. It clips onto a strap for easy attachment to the body or can patch directly onto the body and be networked to gather information simultaneously from different parts of the body.

Accelerometer and Gyrosensor The use of acceleration signals was discussed in the late 1980s. With the development of MEMS, accelerometer sizes and cost decreased and several rehabilitation studies were performed using them. The commercial off-the-shelf, 2.4-GHz, wireless-transmitting MotionPod comes in a fully integrated printed-circuit board (PCB) module measuring 33 × 22 × 15 mm and weighing 14 g.

Acceleration signals at the low back, which is close to the center of gravity, were monitored, and different signals were found in patients with different diseases (Bouten et al., 1997; Bonato, 2005; Culhane et al., 2005; Tao et al., 2012). It is designed to support high-resolution motion studies of patients being treated for neuromotor conditions such as Parkinson's disease (Patel et al., 2009), stroke (Hester et al., 2006), and epilepsy (Sung et al., 2005).

A prototype of the acceleration platform for long-term motion analysis in a home setting consists of a number of wearable sensors and a base station installed in the patient's home. Each sensor samples multiple channels of accelerometer, gyrosensor, and/or physiological data and stores raw signals to a local flash drive. Sensors also perform *feature extraction* on the raw signals, which may involve expensive on-board

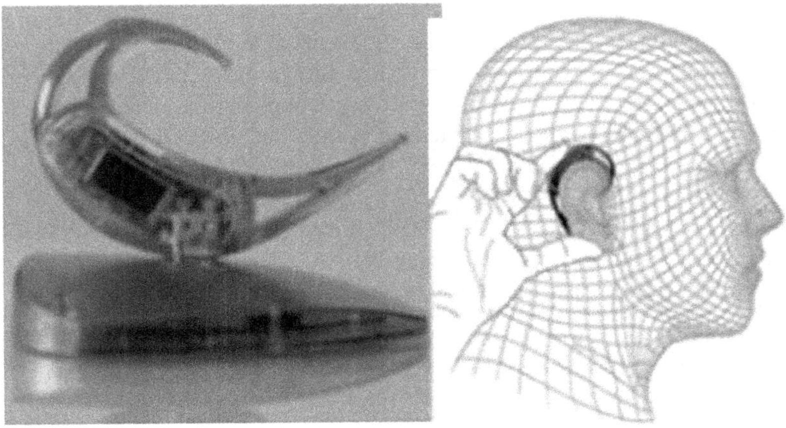

Figure 4.9 e-AR accelerometer.

computation. The body sensor network performs data transfer to the base station, based on the quality of the radio link to each sensor and the remaining battery capacity. Each node dynamically tunes the number of data transfers and degree of computation applied to the sampled signal to meet a target lifetime like 12 or 24 hr. Nodes also save energy by dropping down to a low-power state when the sensor is not moving.

A small wireless sensor platform that can record and transmit physiological and kinematic data in realtime is now availabe (SHIMMER 2012). Another type of accelerometer, e-AR, is shown in Figure 4.9. It is a low-power, miniaturized ear-worn activity recognition sensor. The e-AR sensor is equipped with a MEMS three-axis accelerometer capable of detecting acceleration in three dimensions (up and down, left and right, front and back). This device also allows detecting the gait cycle, seat locomotion, and accelaration (Atallah et al., 2012)

The wireless gyrosensor has been applied for the measurement of the motion and posture of human segments during walking by measuring the angular velocity and angle (Jovanov et al., 2005, Bamberg et al., 2008, Gouwanda and Senanayake, 2010).

Magnetroresisitive Sensors Magnetroresistive sensors are based on the magnetoresistive effect. If magnetic flux is applied, a Lorentz force proportional to the magnetic flux density will deflect the current path. This sensor can estimate changes in the orientation of the body segment (Dai et al., 1996; O'Donovan et al., 2007).

Wireless Goniometer The goniometer measures the changes in the angle. An electronic optical fiber and flexible goniometer can be used to measure relative rotation between two body segments (Tracher Freedom Goniometer 2012).

Electromagnetic Tracking System An electromagnetic tracking system is a magnetic sensor system which uses an electromagnetic field to detemine the position and orientation of a remote object. Wireless electromagnetic tracking devices have been developed for computed tomography environments.

Textile Sensor Textile sensors are a combination of fabric and sensing technologies, where electric components are attached to the fiber and yarns. The textile sensor is more flexible and comfortable to measure human posture and movement. The obtained signals are transmitted by wireless communication. The sensing fabrics are piezoresistive, piezoelectric, or piezocapacitive materials (Tognetti et al., 2007; Helmer et al., 2008; Munro et al., 2008; Lorussi et al., 2009).

Force Sensor The force sensor is embedded into shoes to monitor ambulation of grand force and center of pressure during the gait. The piezoelectric, strain gauge, and capacitive types are feasible to use (Chin, 2005; Bamberg et al., 2008).

4.2.6 Temperature Sensor

Wireless temperature monitoring is common in environmental applications, but for human use, not many applications have been developed. One example is a paste sensor connected to the abdomen with the data transmitted to a smartphone. For thermal therapy, a wireless thermometer used a silicon bipolar transistor voltage-controlled oscillator (Si VCO) based on three-dimensional monolithic-microwave integrated-circuit (MMIC) technology. The VCO's oscillation frequency (around 4.35 GHz) is extremely linear against temperature in the range from 30°C to 50°C. Tests using a phantom material show that a prototype thermometer can be detected at depths of more than 8 cm of soft tissue (Kotsuka et al., 1999).

4.2.7 Wireless and Wearable Chemical Sensor

Applying Soft-MEMS techniques onto functional polymers, flexible chemical sensors have been developed. The structure of the wearable oxygen sensor involves gas-permeable membrane-coated electrodes, membranes with KCl solution, and non-permeable membranes same as the original amperometric principle (Mitsubayashi et al 2003, Iguchi et al. 2005) (Figure 4.10)

Wearable glucose sensors, constructed by immobilizing glucose oxidase onto the sensing region were reported. Considering the progress toward contact-lens-type sensors, the optical-transparent sensor was developed. On the other hand, the biocompatible polymer was introduced. Both sensors were sensitive enough to measure glucose concentration in tear fluids (0.14 mmol/L). Recent researches on wearable oxygen sensors, wearable glucose sensors, and their applications are reported (Yao et al., 2011). Thick-film technology was applied to monitor transcutaneous blood oxygen. The principle is amperometry with incorporate an integral heating element to enhance transcutaneous diffusion of oxygen at 44°C. (Lam and Atkinson 2007).

The principle of measurement involves an electrochemical reaction that is activated with the aid of an enzyme. Glucose oxidase converts glucose into hydrogen peroxide (H_2O_2) and other chemicals whose concentration can be measured with a potentiostat. This measurement is used for calculating the glucose level. The chip, measuring just 0.5 × 2.0 mm, can fit more than just the nanopotentiostat itself.

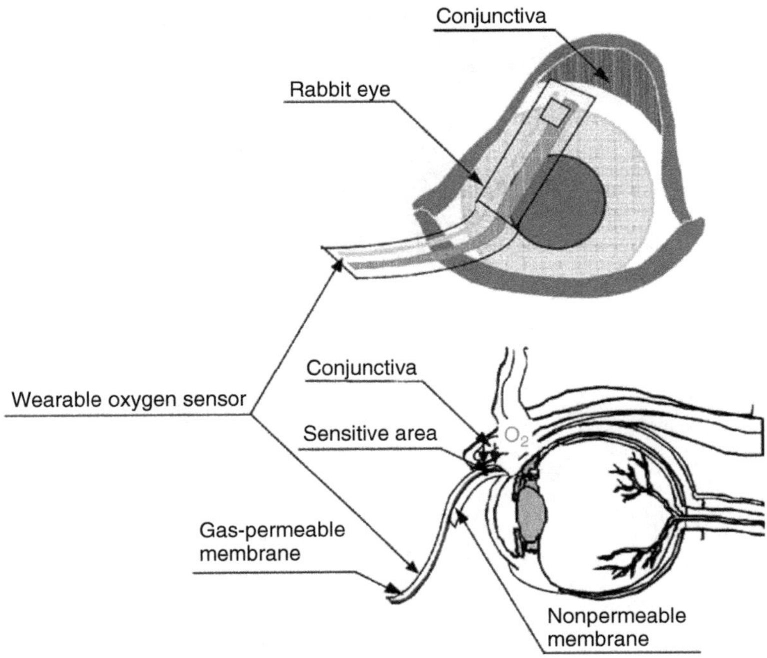

Figure 4.10 Schematic diagram of attachment of sensor for rabbit conjunctival oxygen monitoring (Iguch et al., 2005 with permission).

4.2.8 Capsule Sensor and Endoscopic Camera

Telemetry capsules have existed since the 1960s (Jacobson, 1963) and were used to measure temperature, pH, or pressure inside the gastrointestinal (GI) tract. It was hoped that these capsules would replace invasive techniques in the diagnosis of function disorders in the GI tract. However, problems such as signal loss and uncertainty of the pill position limited their use in a clinical setting. MEMS technology, modern techniques for energy transmission, and thick-film technology (TFT) for the fabrication of a wireless physiological sensing microsystem offer solutions to these problems.

Capsules are noninvasive devices, and MEMS technology has contributed in constructing miniaturized small drinkable capsules for easy swallowing. New sophisticated radio-telemetry capsules based on a 30-year-old technology have been specifically developed for clinical use, but their applications have the potential to reach only basic research laboratories beyond the hospital.

The most common sensors are silicon diode, an ion-selective field effect transistor (ISFET), a pair of direct-contact gold electrodes, and a three-electrode electrochemical cell for monitoring temperature, pH, conductivity, and dissolved oxygen in a solution, respectively.

Figure 4.11 Capsule endoscope. 1. Optical dome, 2. Lens holder, 3. Lens, 4. LEDs, 5. CMOS imager, 6. Battery, 7. ASIC transmitter 8. Antenna.

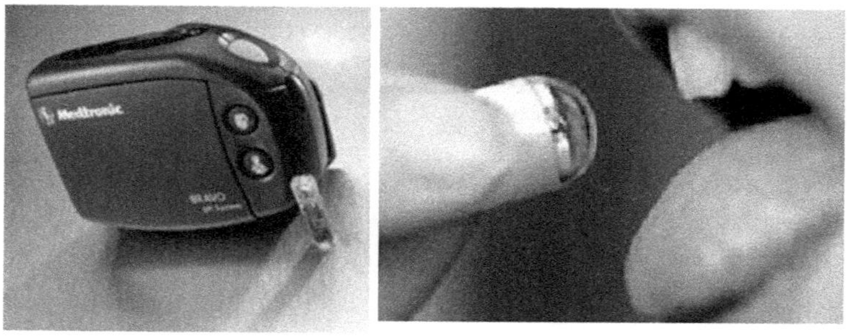

Figure 4.12 Medtronic Bravo pH system.

Small-bowel endoscopy with capsules offers painless endoscopic imaging of the whole small bowel (Figure 4.11). The capsule has its own lens and light source in order to make it possible for us to view the images captured on a video monitor (Iddan et al., 2000; Swain, 2003).

Catheter-based esophageal pH monitoring has limitations, which are discomfort, inconvenience, and interference with normal activity. An alternative to conventional pH monitoring is the wireless Medtronic Bravo pH system (Figure 4.12). The system has already been evaluated in terms of safety, performance, and tolerability (des Varannes et al., 2005; Tseng et al., 2005; Wong et al., 2005).

SmartPill is designed to travel through the entire gastrointestinal tract, where it measures temperature, pressure, and pH. Besides that, it provides temporal–spatial analysis and differentiates between normal and abnormal transit times (Figure 4.13). As illustrated in Figure 4.13, the SmartPill capsule contains a power switch, a battery, an antenna, a transmitter, and a microprocessor in addition to pH, temperature, and pressure sensors. The data from the pill are transmitted to a receiver for later analysis by the physician. To determine the location of the pill, the use of pills combined with a catheter is proposed (Pandolfino et al., 2003) (Figure 4.14).

Designed for human use, the ingestible core body temperature sensor consists of a silicon diode and wirelessly transmits core body temperature as it travels through the digestive tract. The silicon diode is used to measure the body core temperature

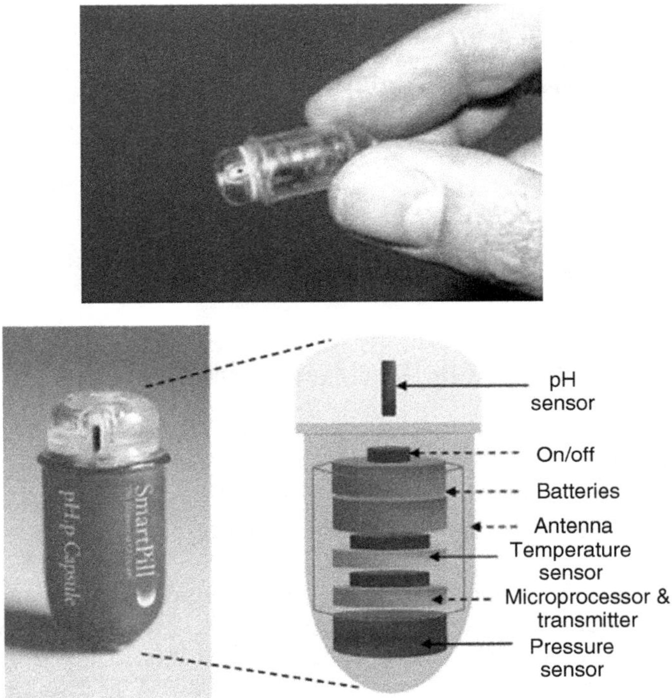

Figure 4.13 Image and schematic diagram of components of the SmartPill®. (Image reproduced and schematic adapted with permission from SmartPill® Corporation, Buffalo, NY.)

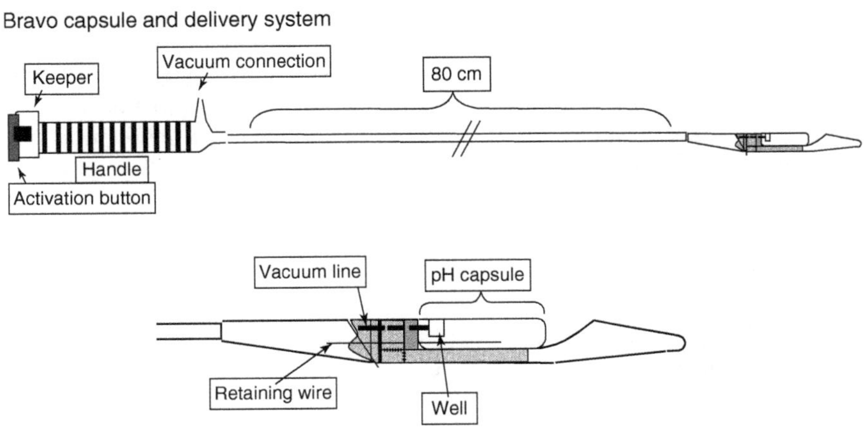

Figure 4.14 Prepacked assembly incorporating both a delivery catheter and the capsule. (Reproduced with permission from AJG, Figure 4.1.)

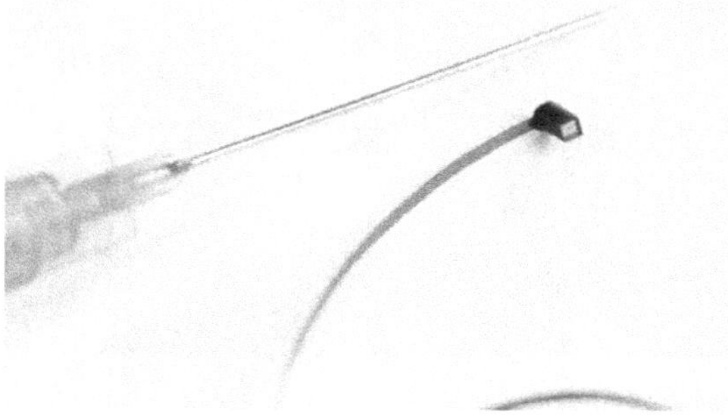

Figure 4.15 NanEye camera from Awaiba.

and also identify local changes associated with tissue inflammation and ulcers. The other temperature sensor is Mutemp HTC, which changes its magnetic permeability rapidly when changes in temperature occur (Connell et al., 1963).

Endoscopic Camera Minimally invasive robotic surgeries are becoming simpler and easier to perform thanks to new endoscopic device development. A wafer-level digital complementary metal–oxide–semiconductor (CMOS) image sensor customizable to low-power medical needs has been developed. The Nan Eye camera measures a mere 0.5 × 0.5 mm—roughly the size of a matchstick's tip—and features a resolution of 140 × 140 pixels at a rate of 40 frames/secr, which makes it suitable for endoscopic examinations and robotic surgeries (Figure 4.15).

The camera lens, based on B33 (Borofloat) glass, is designed in a way that the surface toward the object to be viewed is flat, minimizing the influence caused by the medium between the lens and the object. Therefore, only the opening angle of the lens is reduced when the system operates in contact with body fluids. Replete with a 3-μm-pitch 250 × 250-pixel rolling shutter, the camera provides clear and sharp color images by utilizing Bayer pattern filters. Finally, a low-power 1.8-V battery-operated version is available that dissipates only 600 μA.

4.3 INVASIVE AND IMPLANTABLE SENSORS

Invasive and implantable wireless technology is common for biotelemetry. There are several implantable devices which were constructed using MEMS technology. Implantable sensors are developed to monitor pressure, motility, pH, and temperature (Cumming et al., 2006). In addition to this, the integration of an ECG device into a chip is already well known in the field of implantable devices, such as pacemakers and defibrillators, and an implantable EEG device has already been developed for animal study.

4.3.1 Pressure Sensors

MEMS-based capacitive pressure sensors are typically fabricated using silicon micromachining techniques (Allen, 2005; Palasagaram and Ramadoss, 2006; Cong et al., 2010).

A novel, wireless, batteryless, implantable real-time blood pressure monitoring microsystem for small laboratory animals employs an instrumented elastic circular cuff which is wrapped around a blood vessel. The elastic circular cuff is made of soft biocompatible silicone material filled with biocompatible insulating fluid. It also contains an immersed MEMS pressure sensor and an integrated electronic system in order to detect a down-scaled vessel blood pressure waveform. This technique avoids vessel penetration and substantially minimizes vessel restriction due to the soft cuff elasticity. For this reason it becomes attractive for long-term monitoring. A wireless and batteryless monitoring microsystem is then implantable and characterized in a laboratory rat. The measured in vivo blood pressure waveform by the microsystem and a reference waveform recorded by a commercial catheter tip transducer are closely matched in shape with a constant scaling factor. In this way they demonstrate a blood pressure signal with high fidelity that can be wirelessly obtained by the implantable monitoring microsystem. The overall implant dissipates 300 μW and is powered by an external adaptive RF powering source (Cong et al., 2010).

Passive wireless pressure sensors, originally developed for use in harsh environments, have been adapted for use in the human body. They are mostly applied as monitors of endovascularly repaired abdominal aortic aneurysms. For this application the device is permanently implanted deep within the body and remains functional for the remainder of the patient's life.

MEMS technologies have been utilized to fabricate sensors with sizes and form factors suitable for endovascular delivery and permanent implantation. The sensors are interrogated with an external measurement antenna and in this way a real-time waveform of the pressure environment is extracted.

The telemetrically powered Implantable Telemetric Endo-System (ITES) consists of a surface micromachined capacitive-type absolute pressure sensor fabricated with an eight-mask MOS-like process and containing two low-power ASICs for capacity change read-out and telemetric data and energy transmission (Eggers et al., 2000).

Special integrated circuits (ICs) were designed and fabricated using the CMOS process to enable wireless telemetry, remote powering, voltage regulation, and processing of pressure measurements from MEMS capacitive sensors with pressure resolution of 0.5 mm Hg over a range of 0–50 mm Hg. The following biocompatible materials are most commonly used in these cases: low-temperature cofired ceramic (LTCC) parylene, liquid crystal polymer (LCP), silicon, and alumina (Chow et al., 2010).

Diaphragm and strain gauge pressure sensors are also available. A tiny passive MEMS *LC* resonator lies at the core of CardioMEMS's Champion implantable device (Figure 4.16) for monitoring and treating aneurisms, a leading cause of heart failure.

Figure 4.16 CardioMEMS an implantable MEMS passive *LC* resonator.

It is smaller than and as thin as a U.S. dime and also wirelessly addressed and powered by an external source.

The two-component device will consist of the biocompatible sensor—a pressure-sensitive MEMS capacitor with a chip inductor/radiator—that will be implanted internally in the wall of a cardiac patient's major artery and an external hand-held reader that patients will hold in front of their body. The internal sensor will send a RF signal that indicates real-time blood pressure data to the external reader. The external device will then wirelessly transmit the data via the telephone or the Internet to the doctor or hospital. Thus, the device will enable the doctor to monitor the blood pressure after the patient goes home.

The pressure sensor and its wireless antenna are inserted near the heart with a catheter during a procedure that takes only a few minutes. The RF-addressed wireless diaphragm-type pressure sensor needs no batteries, as it is powered by external inductive coupling. Pressure changes deflect the transducer's diaphragm and change the *LC* circuit's resonant frequency, which can be monitored externally. Blood pressure readings are sent to a wireless scanner. When blood pressure readings taken over several days remain outside a desired range, doctors can be notified by phone for further action.

To investigate the motility of intestine in human, improved silicon strain gauge pressure senor has been developed. The probe consists of six ultraminiature strain gauge pressure sensors placed 10 cm apart from one another and encased in a flexible polyurethane sheath. The probe is small [2.67 mm (8F) diameter], is easily passed, and can be implanted to patients for prolonged recording periods (24–36 hr) without discomfort. The initial semiconductor and catheter material designed for use in the cardiovascular system proved to be easily damaged by gastric acid and enzymes. After improvement of this probe, they can now collect recordings from more than 100 patients in approximately 6000 hr of recording time. The improved probe is a durable recording device that facilitates the investigation of motility of the small intestine in humans either in health or in disease (Mathias et al., 1985).

Figure 4.16 show an implantable intracranial pressure (ICP) sensor. ICP provides a strong indication of brain health and is a good indicator in the monitoring

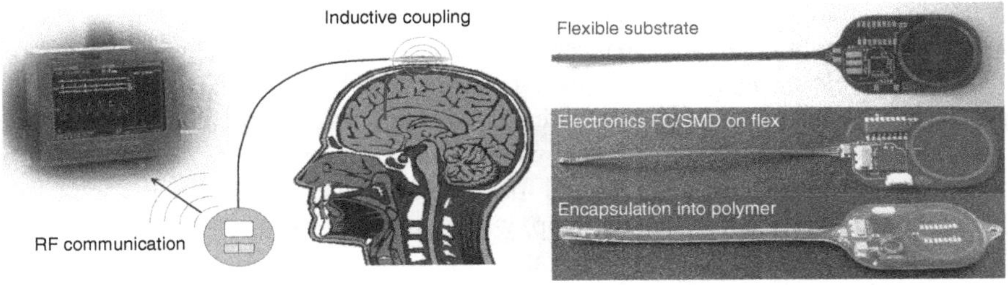

Figure 4.17 Implantable intracranial pressure sensor.

of conditions affecting the central nervous system. A fully implantable wireless ICP sensor would allow to monitor ICP without invasive surgeries. A passive implantable sensor consisting of an integrated series (variable) capacitor, inductor, and resistor has been developed. When implanted, the resonant frequency of the circuit varies according to intraocular pressure. Internally powered LC and MEMS sensors have also been produced. The intracranial pressure can be monitored using a passive implantable sensor with a capacitive pressure sensor (Fig. 4.17).

4.3.2 Chemical Sensor

Chemical implantable sensors are used to monitor mainly pH and glucose in either humans or animals. The principle on which they are based is an electrochemical reaction. In the following a remotely powered implantable microsystem for continuous blood glucose monitoring is presented.

The microsystem consists of a MEMS glucose biosensor flip-chip bonded to a transponder chip and has a volume of 32 mm^3. The transponder chip is inductively powered by an external reader with a 13.56 MHz carrier and is used for measuring the output signal of the glucose biosensor and transmitting the measured data back to the external reader using load shift keying (LSK).

The procedures for the microfabrication of the glucose sensor and the assembly of the microsystem are described along with the description of the circuit blocks of the transponder chip. The transponder chip has been fabricated with the TSMC 0.18-mum CMOS process and has a total area of 1.3×1.3 mm^2. The chip can measure the sensor output current ranging from 1 nA to 1 μA with less than 0.3% nonlinearity error, provided that the amplitude of the received RF signal is higher than 2.6 V. Finally, the circuit consumes a total current of about 110 μA (Ahmadi and Jullien, 2009).

A prepared wireless enzyme sensor system used to determine blood glucose levels in fish is also presented. Periodic checks of fish health and the rapid detection of abnormalities are necessary at fish farms. Several studies indicate that blood glucose levels closely correlate to stress levels in fish and represent the state of respiratory or nutritional disturbance. These can be rapidly and conveniently monitored using the newly developed needle-type enzyme sensor, consisting of a Pt–Ir wire, Ag/AgCl paste, and glucose oxidase. To prevent the effects of interfering anionic species, such as uric acid and ascorbic acid, on the sensor response, the Pt–Ir electrode was coated with Nafion, and then glucose oxidase was immobilized on the coated electrode.

The principle of measurement involves an electrochemical reaction that is activated with the aid of an enzyme. Glucose oxidase converts glucose into hydrogen peroxide (H_2O_2) and other chemicals whose concentration can be measured with a potentiostat. This measurement is used for calculating the glucose level. The special feature of this biosensor is that the chip, with dimensions of just 0.5 × 2.0 mm, can fit to more than just the nanopotentiostat.

The fluid of the inner sclera of the fish eyeball (EISF) was a suitable site for sensor implantation to obtain the glucose sample. Glucose concentrations in fish blood could be monitored in free-swimming fish in an aquarium for three days. The calibration curve of the glucose concentration was linear, from 0.18 to 144 mg/dL, and the detection limit was 0.18 mg/dL. It should be noted that the sensor-calibrated glucose levels and actual blood glucose levels were in excellent agreement. Thus, there was a close correlation between glucose concentrations in the EISF and those in the blood.

A wireless label-free detection of disease-related C-reactive proteins (CRPs) was reported (Chen et al., 2009). The level of CRP rises when there is inflammation throughout the body. The sensor is based on a 200-mm-long MEMS microcantilever housed in a 7 × 7–mm^2 reaction chamber. The deflection of the microcantilever due to specific CRP-antiCRP binding is detected using a position-sensitive detector. The CRP concentrations that can be detected range from 1 to 500 mg/mL, while the assay duration ranges from about 30 min to 3 hr, depending on the desired accuracy. The converted biosignal is transmitted by a custom-designed wireless amplitude shift keying (ASK) transceiver IC fabricated in an 0.18-mm CMOS process. Instead of traditional bases/acids, a 0.2 Hz 1 V AC signal is applied to the bio-MEMS sensor in order to unbind the CRP from the microcantilever and certify that it has a safe reusable feature.

4.3.3 Electroencephalography Sensor

A continuous in vivo EEG and single-unit electrical activity from untethered rodents using an inductively powered and implantable wireless neural recording device have been developed. The device uses an IC to amplify, modulate, and transmit neural signals. The IC transmits neural signals (15 μV–15 mV) of a behaving test animal at 3.2 GHz to a receiver located outside the environment with an input–output correlation better than 90% (Irazoqui-Pastor et al., 2003; Mohseni et al., 2005; Harrison et al., 2007; Sodagar et al., 2007).

An implantable brain stimulator is under development to treat epilepsy. The responsive brain neurostimulation (RNS) is significantly reducing the frequency of seizures among people who have a common form of epilepsy which is otherwise difficult to treat with medication. The device consists of a stimulator, a microprocessor, and an EEG monitor.

The therapy uses an implantable pacemaker-like device to stimulate a targeted region of the brain with electrical impulses, blocking or correcting abnormal nerve signals that cause the tremor. The researchers are also using deep-brain stimulation surgery to treat essential tremor. Essential tremor, which affects the hands, head, and voice, is three times more prevalent than Parkinson's disease. This largely hereditary neurological condition can cause uncontrollable shaking that interferes with normal daily activities.

4.3.4 Magnetoelastic Sensor

Magnetoelastic sensors, now common for implantable sensors, are made of a piezoelectric sensor, amorphous metallic glass ribbons, or wires with a characteristic resonant frequency inversely proportional to length. The remotely detected resonance frequency of a magnetoelastic sensor shifts in response to different physical parameters, including stress, pressure, temperature, flow velocity, liquid velocity, magnetic field, and mass loading.

The sensors can be characterized in the time domain, where the resonance frequency is determined through analysis of the sensor transient response, or in the frequency domain, where the resonant frequency is determined from the frequency–amplitude spectrum of the sensor.

Reactions resulting in the precipitation of calcium oxalate and brushite formation are monitored using magnetoelastic sensors. These salts frequently occur in pathological precipitation in higher mammals. Calcium oxalate hydrates and brushite were formed by mixing anionic and cationic reactant solutions, with the resulting precipitation reaction starting immediately and mass loading the sensor. Magnetoelastic sensors operate on a remote query basis, using magnetic field telemetry, enabling the wireless, passive sensors to be monitored from within opaque reaction chambers. The mass sensitivity of the 17 mm × 6 mm × 28 μm sensor was found equal to 1.38 kHz/mg, suggesting applicability of the sensor for the real-time investigation of the nucleation and crystal growth kinetics of sparingly soluble salts.

The fabrication and application of a miniaturized array of four magnetoelastic sensors that enable the simultaneous remote query measurement of pH, temperature, and pressure from a passive, wireless platform are presented in the following (Jain et al., 2001; Grimes et al., 2002; Bouropoulos et al., 2005). The practical use of the sensors is facilitated with the smart microprocessor-based monitoring electronics that accurately and efficiently characterize the pulse sensor response. The low cost and remote query nature of the sensors make the sensor platform ideally suited for applications where the sensors have to be placed inside sealed, optically opaque containers and also when disposable use is desired. The applications of the sensors are focused on the quantification of endotoxin concentrations, blood coagulation monitoring, measurement of gastroesophageal pH, and glucose sensing (Zeng and Grimes, 2007).

4.3.5 Surface Acoustic Wave Sensors

Surface acoustic wave (SAW) sensors rely on the modulation of surface acoustic waves in order to sense a physical phenomenon. The sensor transduces an input electrical signal into a mechanical wave which can be easily influenced by physical phenomena. The device then transduces this wave back into an electrical signal. Thus, a typical acoustic wave device consists of two sets of interdigital transducers.

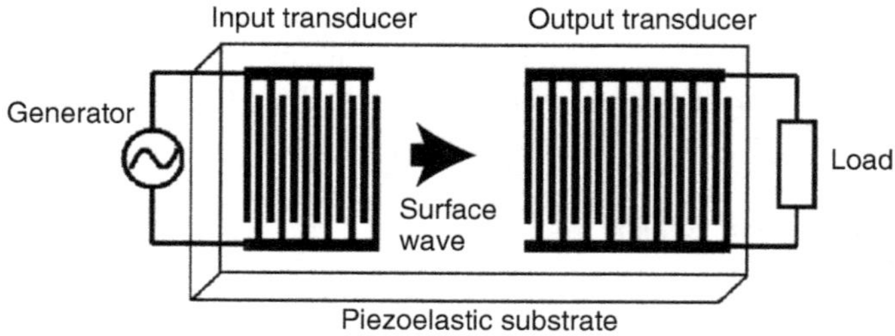

Figure 4.18 Typical acoustic wave device.

One transducer converts electric field energy into mechanical wave energy and the other converts the mechanical energy back into an electric field. Changes in amplitude, phase, frequency, or time delay between the input and output electrical signals can be used to measure the presence of the desired phenomenon.

The basic surface acoustic wave device consists of a piezoelectric substrate, an input interdigitated transducer (IDT) on one side of the surface of the substrate, and a second, output interdigitated transducer on the other side of the substrate (Figure 4.18). The space between the IDTs, across which the surface acoustic wave will propagate, is known as the delay line. This region is called the delay line because the signal, which is a mechanical wave at this point, moves much slower than its electromagnetic form, thus causing an appreciable delay.

The phenomena of pressure, strain, torque, temperature, and mass can be sensed by the basic device, consisting of two IDTs separated by some distance on the surface of a piezoelectric substrate. These phenomena can all cause a change in length along the surface of the device. A change in length will affect both the spacing between the interdigitated electrodes—altering the pitch—and the spacing between IDTs—altering the delay. This can be sensed as a phase shift, frequency shift, or time delay in the output electrical signal.

When a diaphragm is placed between the environment at a variable pressure and a reference cavity at a fixed pressure, the diaphragm will bend in response to a pressure differential. As the diaphragm bends, the distance along the surface in compression will increase. A surface acoustic wave pressure sensor simply replaces the diaphragm with a piezoelectric substrate patterned with interdigitated electrodes. Strain and torque work in a similar manner, as application to the sensor will cause a deformation of the piezoelectric substrate. A surface acoustic wave temperature sensor can be fashioned from a piezoelectric substrate with a relatively high coefficient of thermal expansion in the direction of the length of the device.

In the surface acoustic wave a MEMS gyrosensor and the SAW resonator are used to set up a stable reference vibration and the SAW sensor is used for the detection of the secondary SAW generated by the Coriolis force. Further to this resonator, strategically positioned metallic dots that form an array along the standing-wave antinode locations are subjected to the reference vibratory motion. These vibrating dot arrays through the Coriolis effect will generate a secondary SAW, which is picked up by the SAW sensor (Jose et al., 2002).

4.3.6 Energy- and Power-Harvesting Piezoelectric MEMS Device

A piezoelectric MEMS device generates 10 times more energy than conventional energy harvesters. It portends important implications for powering medical implants in the body as well as wireless sensor networks in automobiles.

The bulk MEMS device is packaged together with other tiny circuit elements, creating a complete vibration energy harvester in a tiny 27-mm^3 package. It can harvest vibrational energy between 14 and 155 Hz to produce about 200 μW from 1.5*g* of vibration. The device charges a supercapacitor to 1.5 V. The supercapacitor then powers up the wireless sensor that is in place of the battery.

The piezoelectric effect is also used in an ultrasonic pressure-sensing echo probe with an aluminum–nitride film that can noninvasively measure the tissues of living bodies (Figure 4.19). The sensor features mechanical strength and durability. This comes from the placement of a single internal electrode between a pair of thin-film external electrodes that have piezoelectric layers on the inner sides and by completely shielding the internal electrode from the outside between the external pair of electrodes

4.3.7 Microfluidic Sensors

Microfluidic technology is steadily becoming even more popular in implantable devices and for lab-on-a-chip technology. Many lab-on-a-chip developments focus on producing low-cost, high-accuracy, and rapid diagnostic tools that aim to detect cancer in the blood. In fact, this is the goal of the Miracle (Magnetic Isolation and Molecular Analysis of Single Circulating and Disseminated Tumor Cells on Chip) project for detecting cancer in the blood.

Microfluid technology also applies as implantable drug delivery systems for colonic disease. Figure 4.20 shows a typical implantable drug delivery device with

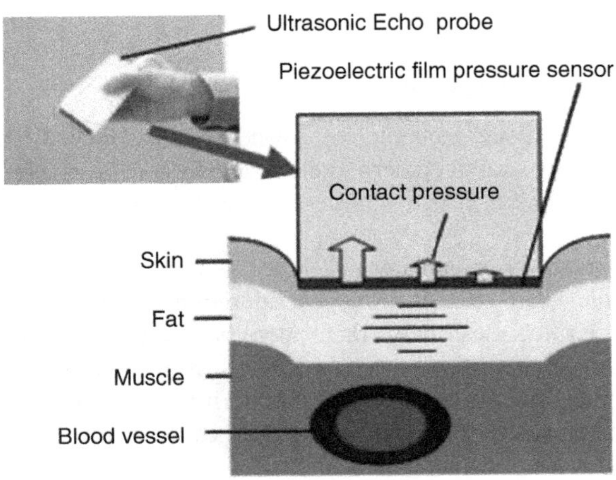

Figure 4.19 Ultrasonic measurement pressure-sensing probe.

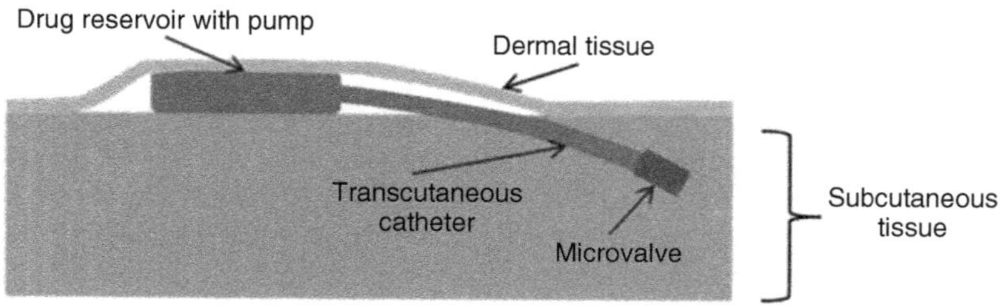

Figure 4.20 A example of drug delivery device (Tng et al., 2012), with permission.

a separated reservoir embedded under the skin. The drug was pumped by micro pump. It was flown through the catheter to the target organ where a microvavle was regulated the flow rate of drug.

One of the more notable microfluidic-based drug delivery mechanisms is the Jewel insulin pump from Debiotech (Jewlpump 2013), codeveloped with STMicroelectronics using Debiotech's microfluidic MEMS technology. The pump can be mounted on a disposable skin patch to provide continuous insulin infusion. It promises substantial improvement in the treatment efficiency and quality of life of diabetic patients.

Smart infusion pumps are complicated machines that require careful design considerations. A recent U.S. Food and Drug Administration (FDA) analysis of such pumps found that, of the 56,000 medical device reports it received relating to the use of infusion pumps (during a five-year period), more than half of the problems were caused by user errors, where software errors were found to be common. The FDA uncovered shortcomings in patient education on proper settings and other matters. However, the FDA also gave high marks to the technology behind such pumps and stated that problems were more likely to stem from user errors rather than device defects.

4.3.8 In-Stick Electrode Sensor

One of the most promising technologies nowadays is the in-stick electrode sensor (Figure 4.21). Putting on the epidermal electronic system is as easy as applying a temporary tattoo: The sensor is just placed on the skin and then the sheet is rubbed with water.

The miTag sensor is an already developed sensor of this type. It monitors a patient's location, heart rate, and oxygen saturation and relays this information over a wireless ad hoc mesh network to monitor stations. Thousands of patients can be monitored simultaneously over the miTags' resilient wireless network. Thus miTags can facilitate the tasks of patient monitoring and tracking, documentation, and coordination currently undertaken by medical care providers.

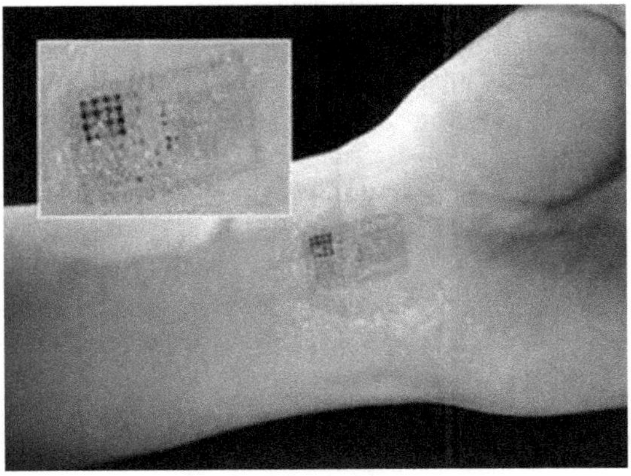

Figure 4.21 In-stick electrode.

4.4 CONCLUSION

Recent advances in wireless and embedded computing systems have led to the emergence of wireless sensor networks consisting of small, battery-powered devices with limited computation and radio communication capabilities. Sensor networks permit data gathering and computation to be deeply embedded in the physical environment. This technology has the potential to impact the delivery and study of resuscitative care by allowing vital signs to be automatically collected and fully integrated into the patient care record and used for real-time triage, correlation with hospital records, and long-term observation.

REFERENCES

Ahmadi MM, Jullien GA. 2009. A wireless implantable microsystem for continuous blood glucose monitoring. *IEEE Trans actions on Biomed. Circuits Syst.*, 3(3): 169–180.

Akita J, Shinmura T, Sakurazawa S, Yanagihara K, Kunita M, Toda M, Iwata K. 2008. Available: http://www.sciencedirect.com/science/article/pii/S0933365707001492-aff2. Wearable electromyography measurement system using cable-free network system on conductive fabric. *Artif. Intell. Med.*, 42 (2): 99–108.

Allen J. 2007. Photoplethysmography and it application in clinical physiological measurement. *Physiol. Meas.*, 28:R1–R39.

Allen, MG. 2005. Micromachined endovascularly-implantable wireless aneurysm pressure sensors; from concept to clinic. The 13th International Conference on Solid-State Sensors, Actuators and Microsystems. Digest of Technical Papers. *Transducers '05,* 1: 275–278.

Asada H, Reisner A, Shaltis P, McCombie D. 2005. Towards the development of wearable blood pressure sensors: A photo-plethysmograph approach using conducting polymer actuators. *Conf. Proc. IEEE Eng. Med. Biol. Soc.*, 4:4156–4159.

Asada H, Shaltis P, Reisner A, Sokwoo R, Hutchnison RC. 2003. Mobile monitoring with wearable photoplethymographic biosensor. *IEEE Eng. Med. Biol. Mag.*, 22:28–40.

Atallah L, Wiik A, Jones GG, Lo B, Cobb JP, Amis A, Yang GZ. 2012. Validation of an ear-worn sensor for gait monitoring using a force-plate instrumented treadmill. *Gait Posture*, 35(4):674–676.

Atoui H, Fayn J, Rubel P. 2010. A novel neural-network model for deriving standard 12-lead ECGs from serial three-lead ECGs: Application to self-care. *IEEE Trans. Inf. Technol. Biomed.*, 14(3):883–890.

Baek J-Y, An J-H, Cho J-M, Park K-W, Lee S-H. 2008. Flexible polymeric dry electrodes for the long-term monitoring of ECG. *Sensors Actuators A Phys.*, 143(2):423–429.

Baek J-H, Chung G-S, Kim K-K, Kim J-S, Park K-W. 2009. Photoplethysmogram measurement without direct skin-to-sensor contact using an adaptive light source intensity control. *IEEE Trans. Inf. Tech. Biomed.*, 13(6):1085–1088.

Bates A, Ling MJ, Mann J, Arvind DK. 2010. Respiratory rate and flow waveform estimation from tri-axial accelerometer data. International Conference on Body Sensor Networks (BSN), PP. 144–150.

Bamberg S, Benbasat AY, Scarborough DM, Krebs DE, Paradiso JA. 2008. Gait analysis using a shoe-integrated wireless sensor system. *IEEE Trans. Inform. Techn. Biomed.*, 12 (4):413–423.

Berka C, Levendowski DJ, Cvetinovic MM, Petrovic MM, Davis G, MN. Lumicao MN, Zivkovic VT, Popovic MV, Olmstead R. 2004. Real-time analysis of EEG indexes of alertness, cognition, and memory acquired with a wireless EEG headset. *Int. Jo. Human-Compute. Inter.*, 17(2):151–170.

Bonato P. 2005. Advances in wearable technology and applications in physical medicine and rehabilitation *J. NeuroEng. Rehab.* 2–2.

Bouropoulos N, Kouzoudis D, Grimes C. 2005. The real-time, in situ monitoring of calcium oxalate and brushite precipitation using magnetoelastic sensors. *Sensors Actuators B: Chem.* 109(2):227–232.

Bouten CVC, Koekkoek KTM, Verduin M, Koddel R, Jansen JD. 1997. A triaxial accelerometer and portable data processing unit for the assessment of daily physical activity. *IEEE Trans. Biomed. Eng.*, 44:136–147. Available: http://www.sciencedirect.com/science /article/pii/S0924424707008576 - aff1.

Bsoul M, Minn H, Tamil L. 2011. Apnea MedAssist: Real-time sleep apnea monitor using single-lead ECG. *IEEE Trans. Inf. Technol. Biomed.*, 15(3):409–415.

Bulling A, Roggen D, Tröster G. 2009. Wearable EOG goggles: Seamless sensing and context-awareness in everyday environments. *J. Ambient Intell. Smart Environ. Comput. Sci. Artif. Intell.*, 1(2):157–171.

Bulling A, Ward JA, Gellersen H, Tröster G. 2008. Robust recognition of reading activity in transit using wearable electrooculography pervasive computing. *Lecture Notes Comput. Sci.*, Vol. 5013, pp. 19–37.

Casson AJ, Yates DC, Smith SJM, Duncan JS, Rodriguez-Villegas E. 2010. Wearable electrocardiography. *IEEE Eng. Med. Biol. Mag.*, 29(3):44–56.

Chen C-H, Hwang R-Z, Huang L-S, Lin S-M, Chen H-C, Yang Y-C, Lin Y-T, Yu S-A, Lin Y-S, Wang Y-H, Chou N-K, Lu S-S. 2009. A wireless bio-MEMS sensor for C-reactive protein detection based on nanomechanics. *IEEE Trans. Biomed. Eng.*, 56(2):462–470.

Chen W, Kobayashi T, Ichikawa S, Takeuchi Y, Togawa T. 2000. Continuous estimation of systolic blood pressure using pulse arrival time and intermittent calibration. *Med. Biol. Eng. Comput.*, 38:569–574.

Chi YM, Jung TP, Cauwenberghts G. 2010. Dry-contact and noncontact biopotencial electrodes: Methological review. *IEEE Rev. Biomed. Eng.*, 3:106–118.

Chin EH. 2005. Introducing wearable force sensors in martial arts. *Pervasive comput.*, 47–53.

Chow EY, Chlebowski AL, Chakraborty S, Chappell WJ, Irazoqui PP. 2010. Fully wireless implantable cardiovascular pressure monitor integrated with a medical stent. *IEEE Trans. Biomed. Eng.*, 57(6):1487–1496.

Cong P, Ko WH, Young DJ. 2010. Wireless batteryless implantable blood pressure monitoring microsystem for small laboratory animals. *IEEE Sensors J.*, 10(2):243–254.

Connell AM, MacCall J, Misiewicz JJ, Rowlands EN. 1963. Observations on the clinical use of radio pills. *Br. Med. J.*, 771–774.

Culhane KM, O'Connor M, Lyons D, Lyons GM. 2005. Accelerometers in rehabilitation medicine for older adults. *Age Aging*, 34:556–560

Cumming DRS, Hammond PA, Wang L, Cooper JM, Johannessen EA. 2006. *Wireless sensor microsystem design: A practical perspective in body sensor networks.* In Yacoub M. Yang G-Z (eds), Body sensor networks, Springer-Verlag London UK, pp. 373–397

Dai R, Stein RB, Andrew BJ, James KB, Wieler M. 1996. Applicatoin of tilt sensors in functional electrical simulation. *IEEE Trans. Rehabil. Eng.*, 4:63–72.

des Varannes SB, Mion F, Ducrotté P, Zerbib F, Denis P, Ponchon T, Thibault R, Galmiche JP. 2005. Simultaneous recordings of oesophageal acid exposure with conventional pH monitoring and a wireless system (Bravo). *Gut*, 54:1682–1686.

Eggers T, Marschner C, Marschner U, Clasbrummel B, Laur R, Binder J. 2000. Advanced hybrid integrated low-power telemetric pressure monitoring system for biomedical applications. The Thirteenth Annual International Conference on Micro Electro Mechanical Systems MEMS, pp. 329–334.

Endo H, Yonemori Y, Hibi K, Ren H, Hayashi T, Tsugawa W, Sode K. 2009. Wireless enzyme sensor system for real-time monitoring of blood glucose levels in fish. *Biosensors Bioelectr.*, 24(5):1417–1423.

Farrar JT, Zworykin VK, Baum J. 1957. Pressure-sensitive telemetering capsule for study of gastrointestinal motility. *Science*, 126:975–976.

Finlay DD, Nugent CD, Donnelly MP, McCullagh PJ, Black ND. 2008. Optimal electrocardiographic lead systems: Practical scenarios in smart clothing and wearable health systems. *IEEE Trans. Inf. Technol. Biomed.*, 12(4):433–441.

Foo JYA, Wilson SJ, Williams G, Harris M-A, Cooper D. 2005. Pulse transit time as a derived noninvasive mean to monitor arterial distensibility changes in children *J. Human Hyperten.*, 19:723–729.

Grimes CA, Mungle CS, Zeng K, Jain MK, Dreschel WR, Paulose M, Ong KG. 2002. Wireless magnetoelastic resonance sensors: A critical review. *Sensors*, 2(7):294–313.

Gouwanda D, Senanayake SMNA. 2010. Application of gyroscopes in identifying gait symmetry in walking. *IFMBE Proc. 6th World Congr. Biomech. (WCB 2010)*. 31:1378–1381.

Guenther FH, Brumberg JS, Wright EJ, Nieto-Castanon A,Tourville JA, Panko M, Law R, Steven Siebert A, Bartels JL, Andreasen DS, Ehirim P, Mao H, Kennedy PR. 2012. A wireless brain-machine interface for real-time speech synthesis. *PLOS one* 4(12):e8218.

Harrison RR, Watkins PT, Kier RJ, Lovejoy RO, Black DJ, Greger B, Solzbacher F. 2007. A low-power integrated circuit for a wireless 100-electrode neural recording system. *IEEE J. Solid-State Circuits*, 42(1):123–133.

Helmer RJN, Mestrovic MA, Farrow D, Lucas S, Spratford W. 2008. Smart textiles: Position and motion sensing for sport, environment and rehabilitation. In Vincenzini P, Paradiso R (eds). *Smart Textiles. Advanes in Science and Technology*, Trans. Tech. Publications Limited, Dürnten Switzerland. Vol. 60, pp. 144–153.

Hester T, Hughes R, Sherrill DM, Knorr B, Akay M, Stein J, Bonato P. 2006. Using wearable sensors to measure motor abilities following stroke. International Workshop on Wearable and Implantable Body Sensor Networks, pp. 4–8.

Hsieh J-C, Lo H-C. 2010. The clinical application of a PACS-dependent 12-lead ECG and image information system in E-medicine and telemedicine. *J. Digital Imaging*, 23(4):501–513.

Iddan G, Meron G, Glukhovsky A, Swain P. 2000. Wireless capsule endoscopy *Nature* 405(6785), 417

Iguchi S, Mitsubayashi K, Uehara T, Ogawa M. 2005. A wearable oxygen sensor for transcutaneous blood gas monitoring at the conjunctiva. *Sensors Actuators B: Chem.*, 108:733–737.

Irazoqui-Pastor P, Mody I, Judy JW. 2003. In-vivo EEG recording using a wireless implantable neural transceiver. First International IEEE EMBS Conference on Neural Engineering, Conference Proceedings, pp. 622–625.

Jacobson B. 1963. Endoradiosonde techniques–a survey. *Med. Electron. Biol. Eng.*, 165–180.

Jain MK, Cai Q, Grimes CA. 2001. A wireless micro-sensor for simultaneous measurement of pH, temperature, and pressure. *Smart Mater. Struct.*, 10(2):347–353.

JewlPump http://www.jewelpump.com. access on 1 April.2014.

Jose KA, Suh WD, Xavier PB, Varadan VK, Varadan VV. 2002. Surface acoustic wave MEMS gyroscope. *Wave Motion* 36(4):367–381.

Jovanov E, Milenkovic A, Otto C, de Groen PC. 2005. A wireless body area network of intelligent motion sensors for computer assisted physical rehabilitation. *J NeuroEngin. Rehab.*, 2:6.

Jung HC, Moon JH, Baek DH, Lee JH, Choi YY, Hong JS, Lee SH. 2012. CNT/PDMS composite flexible dry electrodes for long-term ECG monitoring. *IEEE Trans. Biomed. Eng.* 59(5):1472–1479.

Kim D-H, Lu N. Ma R, CONTINOU TO Kim Y-S.

Kim Y-S, Kim R-H, Wang S, Wu J, Won S-M, Tao H, Islam A, Yu K-J, Kim T-I, Chowdhury R, Ming Ying M, Xu L, Li M, Chung H-J, Keum H, McCormick M, Liu P, Zhang Y-W, Omenetto FG, Huang Y, Coleman T, Rogers JA. 2011. *Epidermal Electro. Sci.*, 333 (6044):838–843.

Ko WH, Neuman MR. 1967. Implant biotemeletry and microelectronics. *Science*, 156:351–360,

Ko WH, Yon ET, Mabrouk S, Hynecek J. 1971. Taped on heart rate and electrocardiogram telemetry transmitters. *J. Assoc. Adv. Med. Instrum.*, 5:268–272.

Kotsuka Y, Orii K, Kojima H, Tanaka M. 1999. New wireless thermometer for RF and microwave thermal therapy using an MMIC in an Si BJT VCO type. *IEEE Trans. Microw. Theory Techn.*, 47(12):2630–2635.

Lam Y-Z, Atkinson JK. 2007. Biomedical sensor using thick film technology for transcutaneous oxygen measurement. *Med. Eng. Phys.*, 29:291–297.

Lee YD, Chung WY. 2009. Wireless sensor network based wearable smart shirt for ubiquitous health and activity monitoring. *Sensors Actuators B: Chemi.*, 140(2):390–395.

Lin C-T, Ko L-W, Chang M-H, Duann J-R, Chen J-Y, Su T-P, Jung T-P. (2010). Review of wireless and wearable electroencephalogram systems and brain-computer interfaces—A mini-review. *Gerontology*, 56:112–119.

Lopez A, Richardson PC. 1969. Capacitive electrocardiographis and bioelectric electrode. *IEEE Trans. Biomed. Eng.* 16:299–300.

Lorussi F, Galatolo S, De Rossi DE. 2009. Textile-based electrogoniometers for wearable posture and gesture capture systems. *IEEE Sensors J.*, 9 (9):1014–1024.

Maeda Y, Sekine M, Tamura T. 2011a. The advantages of wearable green reflected photoplethysmography. *J. Med. Syst.*, 35:829–834.

Maeda Y, Sekine M, Tamura T. 2011b. Relationship between measurement site and motion artifacts in wearable reflected photoplethysmography *J. Med. Syst.*, 35:969–976.

Mathias JR, Sninsky CA, Millar HD, Clench MH, Davis RH. 1985. Development of an improved multi-pressure-sensor probe for recording muscle contraction in human intestine. *Dig. Dis. Sci.*, 30(2):119–123

Melhorn K, Oehler M, Ling V, Schilling MA. 2009. Multichannel protable ECG system with capacitive sesnors. *Physiol. Meas.*, 29:783–793.

Mitsubayashi K, Wakabayashi Y, Murotomi D, Yamada T, Kawase T, Iwanaga S, Karube I. 2003. Wearable and flexible oxygen sensor for transcutaneous oxygen monitoring. *Sensors Actuators B: Chem.*, 95:373–377.

Mohseni P, Najafi K, Eliades SJ, Wang A. 2005. Wireless multichannel biopotential recording using an integrated FM telemetry circuit. *IEEE Trans. Neural Syst. Rehabil. Eng.*, 13(3):263–271.

Moon I, Lee M, Chu J, Mun M. 2005. Wearable EMG-based HCI for electric-powered wheelchair users with motor disabilities. Proceedings of the 2005 IEEE International Conference on Robotics and Automation, ICRA, pp. 2649–2654.

Munro BJ, Campbell TE, Wallace GG, Steele JR. 2008. The intelligent knee sleeve: A wearable biofeedback device. *Sensors Actuators B: Chem.*, 131(2):541–547.

Nakamura M, Nakamura J, Lopez G, Shuzo M, Yamada I. 2011. Collaborative processing of wearable and ambient sensor system for blood pressure monitoring. *Sensors*, 11:6760–6770.

Nashitz JE, Bezobchuk S, Renata Mussafia-Priselac R, Sundick S, Dreyfuss D, Igal Khorshidi I, Karidis A, Manor H, Nagar M, Peck ER, Peck S, Storch S, Rosner I, Gaitini L. 2004. Pulse transit time by r-wave-gated infrared photoplethysmography: Review of the literature and personal experience. *J. Clin. Monit.*, 333–342.

Neuman MR, Picconnatto J, Roux JF. 1970. A wireless radiotelemetry system for monitoring fetal heart rate and intrauterine pressure during labor and delivery. *Gynecol. Invest.*, 1 92–104.

O'Donovan KJ, Kamnik R, O'Keeffe DT, Lyons GM. 2007. An inertial and magnetic sesnor based technique for joint angle measurement. *J. Biomech.*, 40:2604–2611.

Oh S, Kumar PS, Kwon H, Varadan VK. 2012. Wireless brain-machine interface using EEG and EOG: Brain wave classification and robot control. Proc. SPIE 8344, Nanosensors, Biosensors, and Info-Tech Sensors and Systems 2012. 83440U doi:1117/12/918159.

Ouchi K, Suzuki T, Doi M. 2004. LifeMinder: A wearable healthcare support system with timely instruction based on the user's context. *IEICE Trans. Inf. Syst.*, E87-D (6):1361–1369.

Palasagaram JN, Ramadoss R. 2006. MEMS-capacitive pressure sensor fabricated using printed-circuit processing techniques. *IEEE Sensor J.*, 6(6):1374–1375.

Pandolfino JE, Richter JE, Ours T, Guardino JM, Chapman J, Kahrilas PJ. 2003. Ambulatory esophageal pH monitoring using a wireless system. *Am. J. Gastroenterol.*, 98:740–749.

Patel S, Lorincz K, Hughes R, Huggins N, Growden J, Standaert D, Akay M, Dy J, Welsh M, and Bonato P. 2009. Monitoring motor fluctuations in patients with Parkinson's disease using wearable sensors. *IEEE Trans. Inf. Tech. Biomed.*, 13(6):864–873.

Patel S, Park H, Bonato P, Chan L, Rodgers M. 2012. A review of wearable sensors and systems with application in rehabilitation. *J NeuroEng Rehab*, 9:21.

Polar Available: http://www.polar.com/en. access on 30 Oct, 2012

Postolache O, Girao PS, Mendes J, Postolache G. 2009. Unobstrusive heart rate and respiratory rate monitor embedded on a wheelchair. *IEEE International Workshop on Medical Measurements and Applications, MeMeA* 2009, pp. 83–88.

Reisner AT, Shaltis PA, McCombie D, Asada HH. 2008. Utility of the photoplethysmogram in circulatory monitoring. *Anesthesiology*, 108:950–958.

Robin P, Smith RP, Argod J, Pépin J-L, Levy PA. 1999. Pulse transit time: An appraisal of potential clinical applications. *Thorax*, 54:452–457.

Scherr D, Dalal D, Henrikson C, Spragg D, Berger R, Calkins H, Cheng A. 2008. Prospective comparison fo the diagnostic utility of a standard even monitor verus a "leadless" portanle ECG monitor in the evaluation of patients with palpitations. *J. Interv. Card. Electrophysiol.*, 22:39–44.

Shaltis PA, Reisner A, Asada HH. 2006. Wearable, cuff-less PPG-based blood pressure monitor with novel height sensor. *Conf Proc. IEEE Eng. Med. Biol. Soc.*, 1:908–911.

Shimmer. Available: http://www.shimmer-research.com/. access on 30 Oct. 2012.

Smith RP, Argod J, Pepin J-L, Levy P.A. 1999. Pulse transit time: an appraisal of potential clinical applications. Thorax 54:452–458.

Sodagar AM, Wise KD, Najafi K. 2007. A fully integrated mixed-signal neural processor for implantable multichannel cortical recording. *IEEE Trans. Biomed. Eng.*, 54(69):1075–1088.

Strickland D, Chartier D. 1997. EEG measurements in a virtual reality headset. *Presence: Teleoperat. Virt. Environ.*, 6(5):581–589.

Sun F-T, Kuo C, Cheng H-T, Buthpitiya S, Collins P, Griss M. 2012. Activity-aware mental stress detection using physiological sensors. Mobile Computing, Applications, and Services. Lecture Notes of the Institute for Computer Sciences. *Social Inform. Telecommun. Eng.*, 76:282–301.

Sung M, Marci C, Pentland A. 2005. Wearable feedback systems for rehabilitation. *J. Neuroeng. Rehabil.*, 2:17.

Swain P. 2003. Wireless capsule endscopy. *Gut*, 52(Supp IV):iv48–iv50.

Takazawa K, Tanaka N, Fujita M, Matsuoka O, Saiki T, Aikawa M, Tamura S, Ibukiyama C. 1998. Assesment of vasoactive agents and vascular aging by the second derivative of photoplethysmogram waveform. *Hypertension*, 32:365–370.

Tao W, Liu T, Zheng R, Feng H. 2012. Gait analysis using wearable sensors. *Sensors (Basel)*, 12(2):2255–2283.

Tng D.J.H, Hu R, Song P, Roy I. Yong K-T (2012). Approaches and Challenges of Engineering Implantable Microelectromechanical Systems (MEMS) Drug Delivery. Micromachines. 3:615–631. Systems for in Vitro and in Vivo Applications.

Tognetti A, Bartalesi R, Lorussi F, de Rossi D. 2007. Body segment position reconstruction and posture classification by smart textiles. *Trans. Inst. Measure. Control*, 29(3–4):215–253.

Tracher Freedom Goniometer. Available: http://www.jtechmedical.com/t5-goniometry. access on 29 Oct, 2012.

Tseng D, Rizvi AZ, Fennerty MB, Jobe BA, Diggs BS, Sheppard BC, Gross SC, Swanstrom LL, White NB, Aye RW, Hunter JG. 2005. Forty-eight-hour pH monitoring increases sensitivity in detecting abnormal esophageal acid exposure. *J. Gastrointest. Surg*., 9(8):1043–1052.

Verma N, Shoeb A, Bohorquez J, Dawson J, Guttag J, Chandrakasan AP. 2010. A micro-power EEG acquisition SoC with integrated feature. *IEEE J. Solid-State Circuits*, 45(4):804–816.

Wang YY, Jan MY, Wang GC, Bau JG, Wang WK. 2004. Pressure pulse velocity is related to the longitudinal elastic properties of the artery. *Physiol. Meas*., 25(6):1397–1403.

Wong W-M, Bautista J, Dekel R, Malagon IB, Tuchinsky I, Green C, Dickman R, Esquivel R. 2005. Feasibility and tolerability of transnasal/per-oral placement of the wireless pH capsule vs. traditional 24-h oesophageal pH monitoring—a randomized trial. *Aliment. Pharmacol. Ther*., 21(2):155–163.

Xu J, Yazicioglu RF, Grundlehner B, Harpe PJA, Makinwa KAA, Van Hoof C. 2011. A 160 uW 8-channel active electrode system for EEG monitoring. *IEEE Trans. Biomed. Circuits Syst*., 5(6):555–567.

Yao H, Shum AJ, Cowan M, Lahdesmaki I, Parviz BA. 2011. A congtact lens with embedded sensor for monitroing tear glucose level. *Biosens. Bioelectron*., 26(7):3290–3296.

Yoon Y, Cho J, Yoon G. 2009. Non-constrained blood pressure monitoring using ECG and PPG for personal healthcare. *J. Med. Syst*., 33:261–266.

Yu LM, Tay FEH, Guo DG, Xu L, Yap KL. 2009. A microfabricated electrode with hollow microneedles for ECG measurement. *Sensors and Actuators A: Phys*., 151:17–22.

Zeng K, Grimes CA. 2007. Wireless magnetoelastic physical, chemical, and biological sensors. *IEEE Trans. Magn*., 43(6):2358–2363.

Zhang B, Wang J, Fuhlbrigge, T. 2010. A review of the commercial brain-computer interface technology from perspective of industrial robotics. 2010 IEEE International Conference on Automation and Logistics (ICAL), pp. 379–384.

Zhang G, Gao M, Xu D, Oliver NB, Millamada R. 2011. Pulse arrival time is not an adequate surrogate for pulse transit time as a marker of blood pressure. *J. Appl. Physiol*., 111:(6)1681–1686.

5

POWER ISSUES IN BIOMEDICAL TELEMETRY

MANOS M. TENTZERIS, RUSHI VYAS, WEI WEI, YOSHIHIRO KAWAHARA, AND LI YANG

Georgia Institute of Technology, Atlanta, Georgia

STAVROS GEORGAKOPOULOS

Florida International University, Miami, Florida

VASILEIOS LAKAFOSIS, SANGKIL KIM, HOSEON LEE, TAORAN LE, SAGAR MUKALA, AND ANYA TRAILLE

Georgia Institute of Technology, Atlanta, Georgia

5.1 INTRODUCTION AND POWERING MECHANISMS

While sensors and wireless front ends are becoming pervasive for a number of biomedical applications from gas sensing to EKG monitoring as covered in the preceding, powering them on has proven to be quite a bottleneck in their wide-scale adoption. One of the reasons for limited adoption of wireless EKGs has been the increased cost and logistics involved in replacing batteries to power their wireless front ends and the microcontroller unit. Also the medical liability of missing a lethal EKG or gas reading has been proven to be another reason for the limited use of many of the wireless technologies. In addition to this, batteries add significant per-unit costs to the overall health of the wireless module, which can be a detriment in a number of health care IT applications. Energy harvesting or power scavenging, which is the process of capturing power from the ambience, can help in finding solutions for the powering issues of wireless and embedded motes for biomedical

Handbook of Biomedical Telemetry, First Edition. Edited by Konstantina S. Nikita.

TABLE 5.1 Power Levels of Various Energy-Harvesting Sources

Energy Source	Harvested Power
Vibration Motion	
Human	4 μW/cm^2
Industry	100 μW/cm^2
Temperature Difference	
Human	25 μW/cm^2
Industry	1–10 mW/cm^2
Light	
Indoor	10 μW/cm^2
Outdoor	10 mW/cm^2
Ambient RF	
Cellular	0.1 μW/cm^2
WiFi	0.001 mW/cm^2

applications. While the energy-harvesting devices do not have the energy density of batteries, their potentially endless supply coupled with smart embedded control of the application-specific operating duty cycle do make them useful for a number of biomedical applications. A study of the power density present in various energy sources in the ambience is presented in Table 5.1.

5.2 MOTION-POWERED RADIO FREQUENCY IDENTIFICATION (RFID) WIRELESS SENSORS

A potential energy source that all humans produce is kinetic energy. Comparing all human activities it can be found that the greaetest amount of energy is generated by the most basic activity of walking. The abundant energy produced by this basic daily activity can be exploited to power embedded sensors such as patient tracking, temperature, and EKG monitoring of patients, making it ideal in hospitals and assisted-living facilities. In this section, we present an advanced human-step-powered energy-harvesting mechanism to power a commercial wireless transmitter that can be used for RFID-based patient tracking applications. The basic system-level diagram of the energy-harvesting device is shown in Figure 5.1. Its individual building blocks are described in the subsequent Sections.

A piezoelectric pushbutton 35 mm long and 5 mm in diameter with a deflection of 4.5 mm at a maximum force of 15 N has been selected as a good compromise among compactness, simplicity, and cost of specific energy (joules per volume). Such devices are able to provide up to several thousand volts, while the output current is extremely small and may only cause a small electric shock. When the pushbutton is pressed, an inner spring is compressed, and when the pressure exceeds a fixed threshold, the

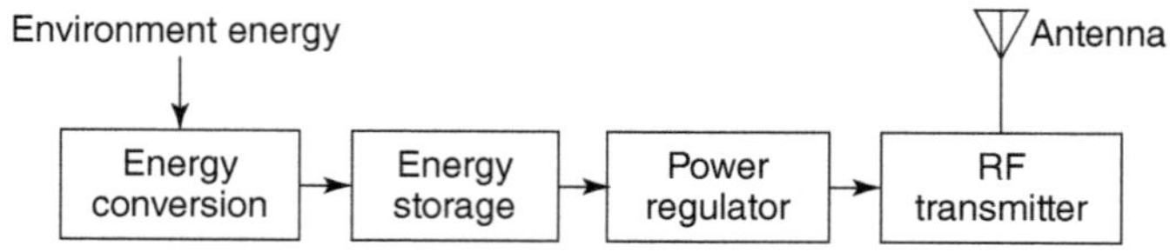

Figure 5.1 System-level diagram of human-step-powered energy-harvesting device (Orecchini et al., 2011).

spring-loaded hammer is released in order to deliver the dynamic mechanical force to the piezoelectric component. Once the hammer strikes the piezoelectric element, a pressure wave is generated and reflected a few times between the hammer and the element, creating a mechanical resonance (Tan et al., 2006). Consequently the output voltage that is generated will follow similarly to an AC signal course depending on the dynamic polarization of the element.

Since this radio frequency (RF) circuitry requires lower voltages at higher currents, a step-down transformer is used for better impedance matching to the following circuitry. An amorphous-core device with a 25:1 turn ratio can transform a peak of voltage close to a thousand volts down to tens of volts. Typically a RF transmitter takes tens of milliseconds to transmit one complete word while the piezoelectric push-button harvests energy in a transient time; therefore an energy storage device needs to collect the electrical energy and provide it to the transmitter even when the external power source is temporarily unavailable. The device that was chosen for energy storage is a 4.7-μF tank capacitor that can also act as a filter smoothing the voltage ripples in the raw harvested electrical energy.

An AC-to-DC full-wave diode bridge rectifier is used to convert the stepped-down AC voltage across the output windings of the transformer into an unregulated DC output which is then fed into a voltage regulator to provide a cleaner regulated 3 V of output. The rectified and regulated voltages at the output of the energy-harvesting source are shown in Figure 5.2. A low-leakage MAX666 low-dropout-voltage

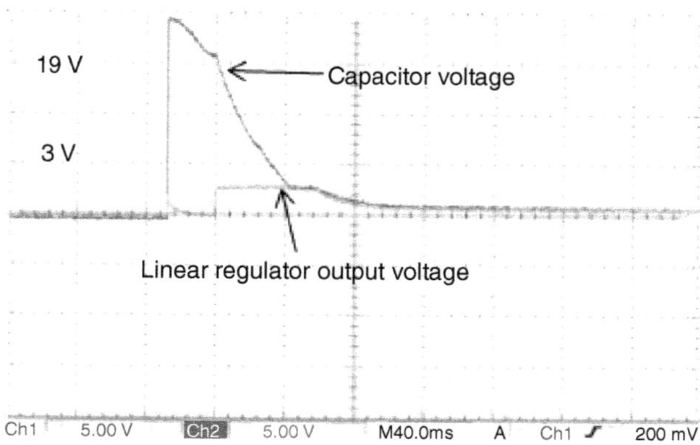

Figure 5.2 Voltage waveforms of rectifier linear regulator output (Orecchini et al., 2011).

regulator was used to provide a steady 3-V output supply to the end electronics as the energy stored across the tank capacitor from each human step discharged down. The end electronics used to carry out the RFID communication from the energy harvested from each step was a TXE-433-KH2wireless transmitter made by LINX that could be programmed to send out any patient data using an RFID-based scheme at 433 MHz.

As mentioned previously, body effects make it particularly challenging to make antennas that need to be worn. Significant effort was put into making antenna for the wireless transmitter that was unobtrusive while having a range of at least the length of a room. The antenna needed to be conformal to fit in the shoe's garment where no flat surfaces can be easily found. In addition, the 69-cm wavelength associated to the ultra high-frequency (UHF) RFID band at 433 MHz requires the antenna to be unobtrusive, low profile, and comfortable to wear while being far smaller than the wavelength. Finally the antenna needs to have robust performance in the presence of human body tissues because a ground place can hardly be incorporated into the show so the detuning effect of the human tissue must be accounted for in the antenna design (Orecchini et al., 2011). In order to meet all of the aforementioned conditions, a unique antenna in the form of a Nike logo was designed that could be wrapped around the heel cap of the shoe.

The antenna structure intrinsically works as a dipole with a center feed from the wireless transmitter, which along with energy-harvesting piezoelectric element and electronics was mounted in the heel of the shoe, as shown in Figures 5.3*a*–*c*. The radiating arms of the dipole form each of the Nike logo shown in Figure 5.3*c*. The conformity of the energy-harvesting prototype is made possible by depositing the silver conductor on a flexible paper-based substrate using an advanced inkjet printing process. The electronic components were mounted on using a multistep process with silver epoxy forming interconnects as outlined elsewhere (Vyas et al., 2009; Yang, 2007).

The wearable shoe mounted antenna performance as measured by a Rohde and Schwarz ZVA-8 vector network analyzer is shown in Figure 5.4. The simulated gain and simulated and measured return loss of the Nike logo antenna are shown in Figures 5.4*a*–*c* and show a resonance at 433 MHz, which is the frequency of transmission of the RFID tag. The effect of the human foot on the gain and return loss of the optimized antenna can be seen in Figures 5.5*a*–*c* respectively and exhibit a downward shift in the return loss resonance. However, the antenna optimization does result in a −10-dB return loss at the transmit frequency of 433 MHz. The body does parasitically assist in increasing the antenna directivity from around 2 to 2.66 dBi while maintaining close to an omnidirectional radiation pattern, as can be implied by the gain plots in Figures 5.4*b* and 5.5*b*.

The system-level performance of the energy-harvesting shoe was measured through the wireless link given out by the RFID tag as a result of each step during which the piezoelectric device embedded in the sole of the shoe is struck against the ground. Each step produces roughly 848.4 mJ of energy that gets stored in the 4.7-μF storage capacitor, which is enough for transmitting a packet and a half of the RFID data with the transmitter power level adjusted to 4 dBm. For less-data-intense vital sign monitoring such as body temperature, the low data rate and throughput of such an RFID-based communication is sufficient. The wireless packets generated by

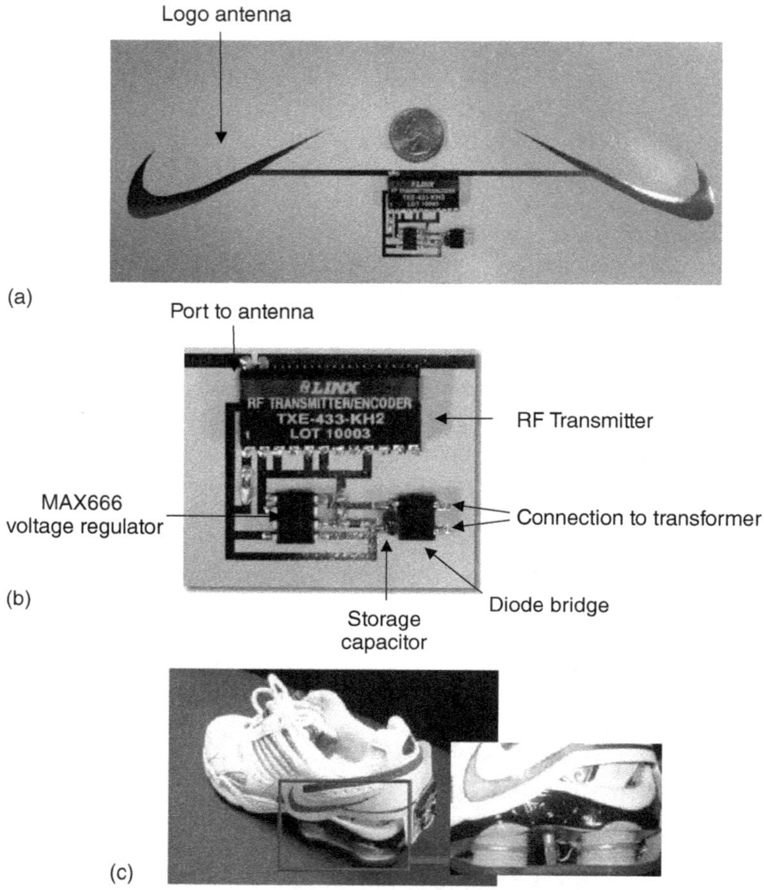

Figure 5.3 (*a*) Unfolded prototype of step-powered RFID system. (*b*) AC-to-DC converting electronics and wireless transmitter of step-powered RFID system. (*c*) Final prototype with piezoelectric device mounted on shoe.

each heel strike was measured remotely using the Tektronics RSA30408A spectrum analyzer and is shown in Figure 5.6. The energy link budget of the system is outlined in Table 5.2.

5.3 NONINVASIVE WIRELESS METHODS FOR POWERING ON SENSORS

Near-field magnetic coupling has been in use to noninvasively charge up implanted pacemakers (Ghovanloo and Najafi, 2004). It is also the primary mechanism used in smart cards and labels that medical patients and professionals use for access and

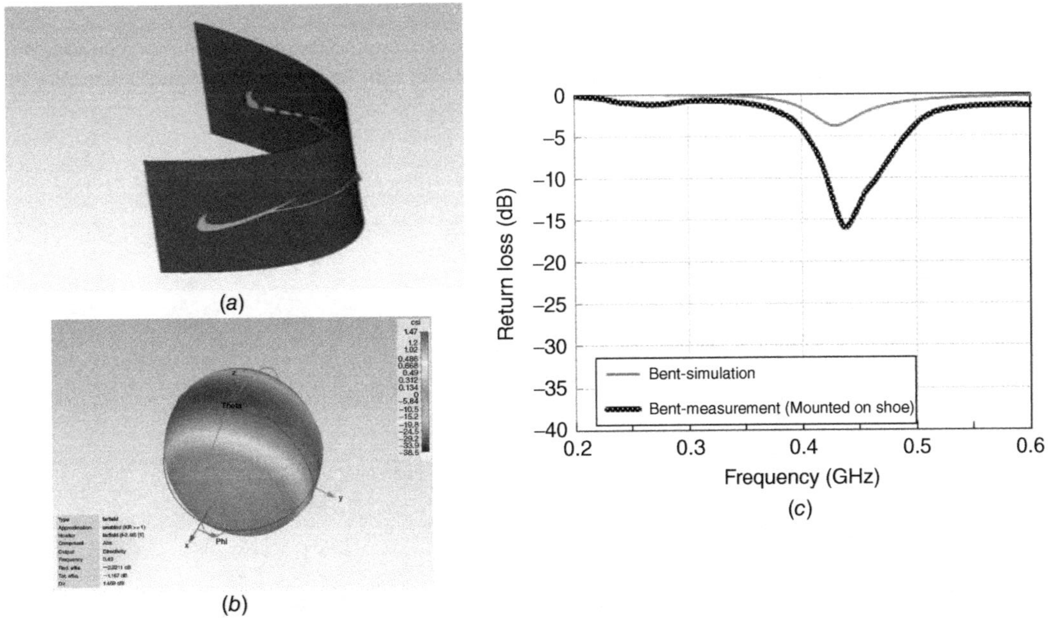

Figure 5.4 (*a*) Nike logo antenna without body effect, (*b*) Nike logo antenna simulated radiation pattern, and (*c*) Nike logo antenna simulated and measured return loss. (*See insert for color representation of the figure.*)

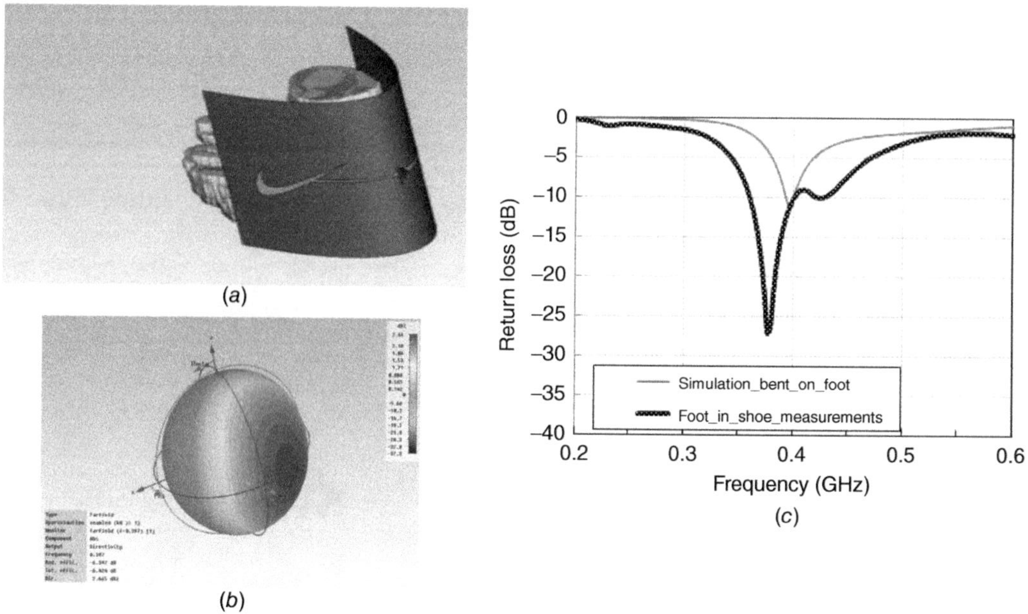

Figure 5.5 (*a*) Nike logo antenna with body effect, (*b*) Nike logo antenna simulated radiation pattern in presence of human tissue, and (*c*) Nike logo antenna simulated and measured return loss in presence of human tissue. (*See insert for color representation of the figure.*)

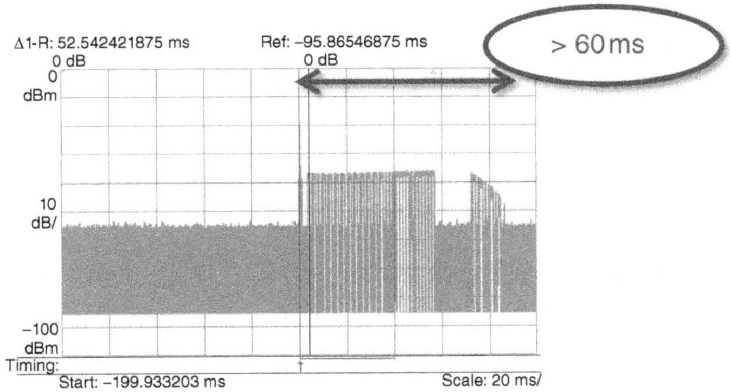

Figure 5.6 RFID data packets produced by each step. Total transmit time per step 60 msec.

TABLE 5.2 Energy Link Budget of Step-Powered Shoe

State	Energy/Power	Note
Energy Produced by Piezoelectric Device	848.4 μJ	Stored energy in the capacitor
Unused energy	17.1 μJ	Voltage across the capacitor is below 2.7V (The active RFID tag stops transmitting)
Available energy	831.3 μJ	–
Energy for single-word transmission	450 μJ	–
Required power for 50 mS operation	9 mW	–

tracking. In both these applications the user is still required to be within a distance of less than a tenth of the wavelength for maximum coupling efficiency (Ghovanloo and Najafi, 2004), which typically is a foot or less. Currently, wireless power transfer (WPT) via electromagnetic induction is mainly utilized in short transmission distance (tens of cm) while long-distance wireless power transfer via microwave frequencies is mainly restricted for use in space solar power transmission applications with long transmission distance (kilometers).

5.3.1 Inductive Coupling

Implantable medical devices (IMDs) play an important role in the diagnosis and treatment of disease due to their capability to monitor, stimulate, and regulate vital internal organs. Besides that, certain IMDs can communicate report data to an external host. IMDs have found applications in a wide range of areas, including pacemakers, physiological monitoring devices, pain relief devices, cochlear hearing implants, functional electrical simulators (FESs), left ventricular assist devices (LVADs) (Kim et al.,

2012), artificial hearts, bladder pressure monitoring devices (Lakafosis et al., 2010), and neurostimulators (Yang, 2007).

The power required by IMDs depends on the application and the typical range is from a few microwatts (Collins et al., 2000) to hundreds of microwatts (Kong et al., 2000). Conventional operation of most IMDs relies on continuous power supply from batteries (conventional or rechargeable), which due to their sizes and lifetime limits the capabilities of IMDs. Wireless powering of IMDs is a promising solution to this problem because it eliminates transcutaneous wiring and does not require medical procedures for the replacement of batteries.

In addition, power levels of IMDs are strictly restricted to avoid excessive heating of tissue that may result in significant tissue damage (Flujiwara et al., 2001). Specifically, the Federal Communication Commission (FCC) specifies that for biomedical devices the exposure to RF energy from wireless devices must be limited to a maximum specific absorption rate (SAR) value of 1.6 W/kg as averaged over 1 g of tissue, beyond which significant tissue heating and damage may occur (Shaozhen and Langley, 2009; Salonen and Rahmat-Samii, 2006). Therefore, such restrictions require that wireless powering systems for IMDs exhibit high power transmission efficiency in order to transfer substantial amount of power without using excessive transmitting power that will generate high-intensity electromagnetic (EM) fields on the human body (Marliere et al., 1999) and increase SAR.

Resonant inductive coupling and SCMR are currently the two most popular wireless powering techniques for IMDs. Resonant inductive coupling involves near-field wireless transfer of electrical energy between two devices that are tuned to resonate at the same frequency (Chopra et al., 2002). On the other hand, SCMR employs resonators to transmit power wirelessly and efficiently over midrange distances (Varghese, 2001). SCMR has also been applied to in vitro and in vivo experiments (Rais et al., 2009; Marliere et al., 1999).

Recent work on wireless power transfer using electromagnetic resonance has shown a range of between 1 and 2 m (Wei, 2012; Kurs et al., 2007), which can potentially be used for the purpose of charging up biomedical devices or environmental health monitoring sensors or patient tracking labels anywhere in the vicinity of the room as per the setup envisioned in Figure 5.7. To show proof of concept, a double-spiral resonator was fabricated as shown in Figure 5.8. Compared to other types of coils such as helical and spiral resonators, double-spiral resonators with optimized parameters (size, thickness, etc.) achieve higher transmission efficiency and longer coupling range (Figure 5.9). This is achieved by using two pieces of spiral resonators as a single coil. Resonators with configurations shown in Figure 5.9 were designed and fabricated to resonate optimally at 13.56 MHz to test out their individual performances. The coils were simulated using the ANSYS high-frequency structure simulator (HFSS), and design was prototyped by embedding the metallic resonators in a specially selected Styrofoam material with dielectric permittivity and magnetic permeability close to air. Dimensions of the final design are outlined in Figure 5.8*c*. The input impedance of the double spiral was found to vary between 1.35 and 2.78 Ω with inductances of between 12.03 and 31.35 μH depending on

Figure 5.7 Ubiquitous power transmission of electronics in vicinity of room using electromagnetically resonant inductive coupling.

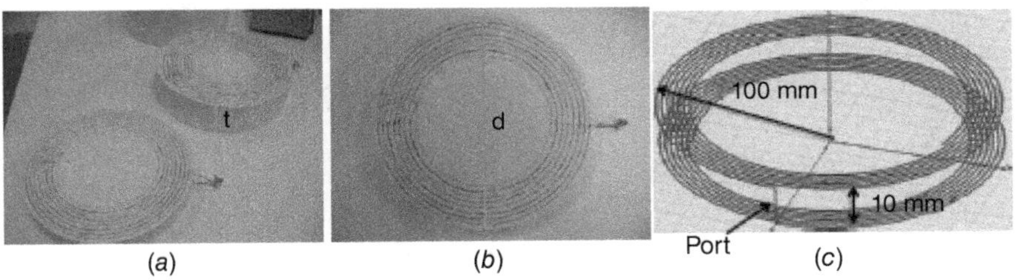

Figure 5.8 Double-spiral resonators to demonstrate electromagnetic resonance coupling.

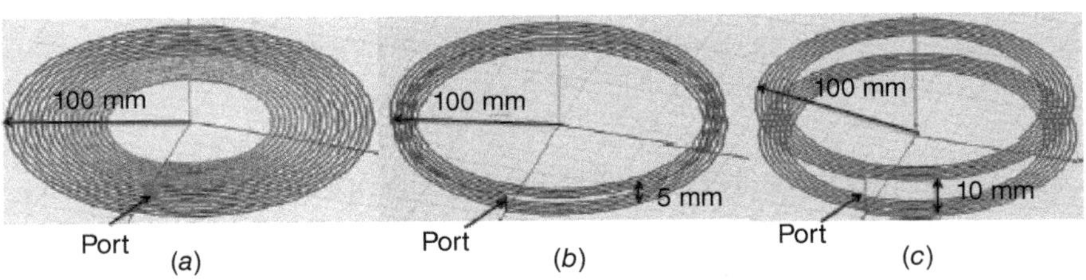

Figure 5.9 (*a*) Spiral resonator, (*b*) helical resonator, and (*c*) double-spiral resonator.

the conductor thickness used. Quality factors of between 758.99 and 960.68 were observed at the resonant frequency of 13.56 MHz.

Comparisons of the double spiral with the spiral and helical resonators were carried out using a vector network analyzer, as shown in Figure 5.10. Each of the coils was fed by a port on the vector network analyzer and the coupling and range were measured using the S parameter S_{21}. The double-spiral resonator showed 15 cm

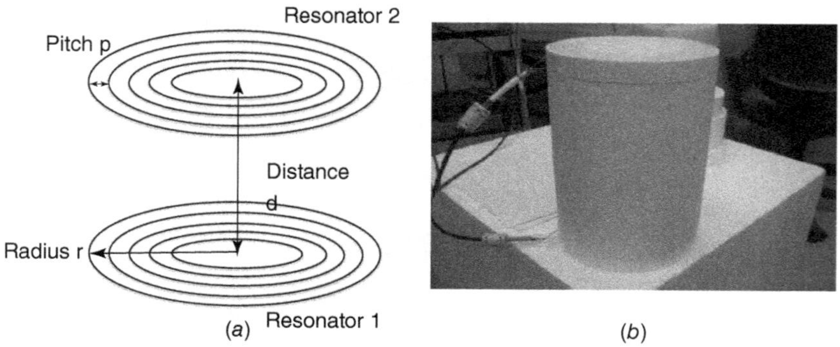

Figure 5.10 Measurement setup used to test range of resonators.

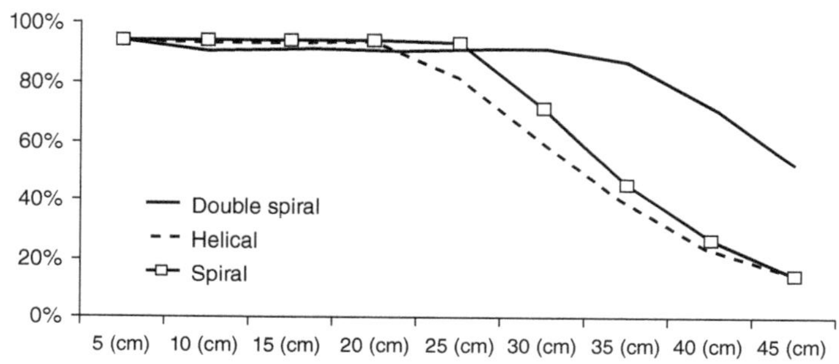

Figure 5.11 Efficiency and range comparison between resonators.

higher range and 40% higher efficiencies at longer distances compared to the spiral and helical resonators, as shown in Figure 5.11.

5.3.2 Conformal Strongly Coupled Wireless Powering of Biomedical Devices

Conventional SCMR requires that source and load resonators have certain distance from the TX and RX resonators, respectively. Therefore, conventional SCMR uses transmitter (TX) and receiver (RX) systems that are three dimensional and occupy significant volume. In this chapter, a conformal (i.e., planar) SCMR (CSCMR) scheme is presented. The efficiency of CSCMR is compared to the efficiency of standard SCMR through simulations and measurements. In addition, resonant inductive coupling SCMR and CSCMR systems with bifilar spiral resonators are presented for conformal wireless powering of IMDs. The efficiency and SAR of such systems are calculated and compared through simulations. It is shown that CSCMR achieves similar performance with conventional SCMR systems that are much bulkier.

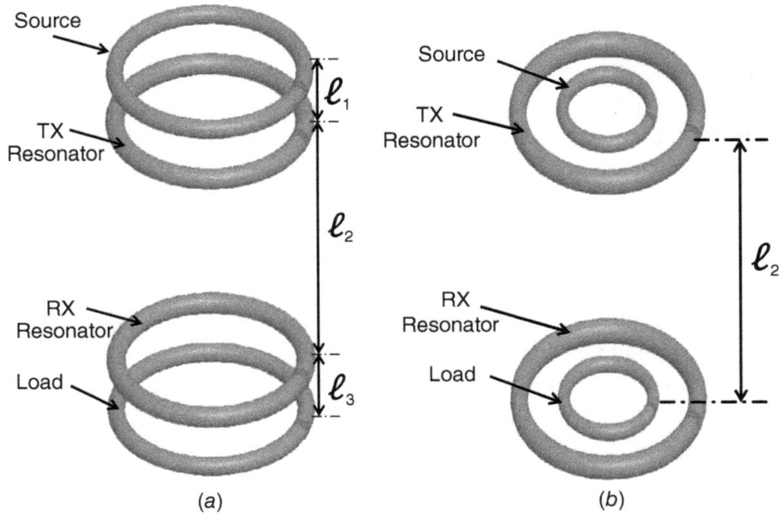

Figure 5.12 Schematic of SCMR system in air.

A typical SCMR system is formed by a TX and a TR resonator (self-resonant elements, e.g., helices and spirals, can be used or loops with capacitors can also be used) and two single loops acting as the load and source devices, as shown in Figure 5.12*a* (Varghese et al., 2001). In the setup of Figure 5.12*a*, ℓ_1 is the distance between source loop and the TX resonator, ℓ_2 is the distance between the TX and the RX resonator, and ℓ_3 is the distance between the load loop and the RX resonator. SCMR systems use resonant transmitters and receivers that are strongly coupled and also able to transfer energy efficiently (Varghese et al., 2001). In order for SCMR to achieve high efficiency, the TX and RX resonators (typically loops or coils) must be designed so that they resonate at the desired operational frequency that must coincide with the frequency at which the resonators exhibit a maximum Q factor. As seen in Figure 5.12*a*, a conventional SCMR system requires certain spacing (ℓ_1 and ℓ_3) between the source and load elements and the TX and RX resonators, respectively. It has been shown that for each SCMR design an optimum distance exists where the efficiency is maximum. In fact, previous work on SCMR has shown that if distances ℓ_1 and ℓ_3 are smaller than their optimum values, the efficiency decreases. These spacing requirements make TX (source and TX resonator) and RX (load and RX resonator) systems for SCMR systems three dimensional and bulky in volume. Here, a new SCMR method which flattens traditional SCMR systems is proposed. Figure 5.12*b* shows the proposed CSCMR that shrinks the source and load elements in size and places them on the same plane with the TX and RX resonators, respectively, thereby providing conformal TX and RX systems (Sievenpiper et al., 1999). CSCMR systems are more suitable for IMDs and wearable biosensors and devices since they can be implemented in planar circuits and flexible substrates, thereby minimizing their volume and yielding conformal devices. The TX and RX resonating elements for CSCMR must also be designed to operate at the resonant frequency where the elements naturally exhibit a maximum Q factor.

In what follows, the performance of CSCMR is compared through simulations and measurements with the performance of conventional SCMR. In order to achieve maximum efficiency for both SCMR and CSCMR systems, the TX and RX resonators are designed to resonate at the frequency where the loops exhibit a maximum Q factor. Previously established analytical quasi-static and simulation models (Andreuccetti et al., 2002) are used to calculate the frequency where the loops exhibit a maximum Q factor and fine tune their design. In this case, the resonant frequency is 46.5 MHz for the SCMR and CSCMR designs and the RX and TX loops are resonated at this frequency using a 68-pF capacitor. HFSS is used for all simulations. The measurement setups for SCMR and CSCMR are shown in Figure 5.13. All loops in Figures 5.12*a* and *b* have a cross-sectional radius of 2.2 mm, and the conductivity of copper is given as $\sigma = 5.8 \times 10^7$ S/m. The SCMR system of Figure 5.12*a* uses identical loops that all have a radius of 4.5 cm. The TX and RX resonators of the CSCMR system in Figure 5.12*b* have a radius of 4.5 cm, which is the same with the radius of the loops in the conventional SCMR system. The radius of CSCMR load and source loops is 2.5 cm. The distances between the loops are as follows: $\ell_2 = 10$ cm and $\ell_1 = \ell_3 = 1.5$ cm. Distances ℓ_1 and ℓ_3 were optimized using simulation to provide optimal efficiency in the SCMR system.

Figure 5.14 compares the S_{21} parameter and efficiency of the SCMR and CSCMR systems using simulations and measurements and shows that CSCMR achieves high efficiency similar in magnitude to the efficiency of SCMR for the same size resonators and the same distance between the resonators. In fact, both the measurements and the simulations show that CSCMR exhibits slightly larger efficiency than SCMR. This may be partially attributed to the fact that the source and load resonators for CSCMR are embedded into the TX and RX resonators, and therefore, they are physically closer

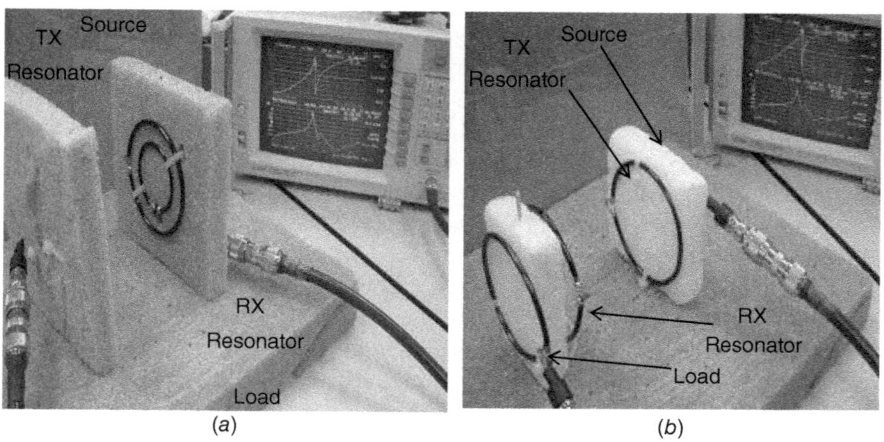

Figure 5.13 Loop-based wireless powering systems: (*a*) CSCMR, and (*b*) SCMR.

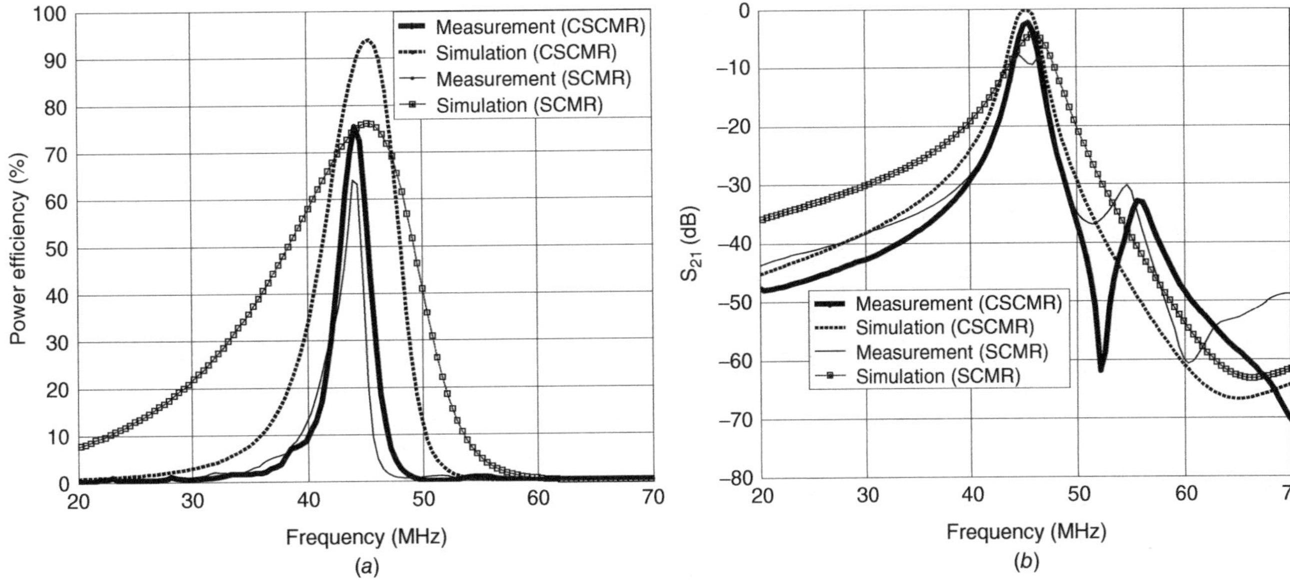

Figure 5.14 Performance comparison of SCMR and CSCMR: (*a*) efficiency and (*b*) S_{12}.

by an amount equal to distance $\ell_1 + \ell_3$ (i.e., the distance between the source and load for the SCMR system is $\ell_1+\ell_2+\ell_3$ whereas for the CSCMR system is ℓ_2).

CSCMR is a very promising wireless powering method because it achieves large efficiencies similar to SCMR and is conformal, thereby requiring minimal volume. Therefore, CSMR is more suitable than SCMR for wireless powering of IMDs and wearable devices as it can be implemented in planar circuits and flexible dielectrics. Finally, CSCMR also works in cases where the TX and RX resonators are placed within the source and load elements, respectively, which is the inverse of what is shown in Figure 5.1*b*. Such inverse CSCMR systems will be designed in the next section for wireless powering of devices that are implanted in the human body.

In this section, an advanced design of a CSCMR system suitable for wireless powering of biomedical devices is presented. Moreover, the performance of the proposed CSCMR system is compared to the performance of traditional resonant inductive coupling and SCMR. Here all three wireless powering systems are designed based on self-resonating bifilar spirals, as shown in Figure 5.15. Self-resonating TX and RX elements (i.e., helices and spirals) are preferred for SCMR systems because external capacitors cause the Q factor and thus the efficiency of SCMR systems to decrease. In fact, helices and spirals have been used to design and power SCMR systems for IMDs (Marliere et al., 1999; Rais et al., 2009). Bifilar resonators are used here because they can be excited at their center, thereby facilitating a direct comparison of the performance of the three wireless powering systems shown in Figure 5.15. The basic geometrical parameters of a bifilar spiral are N, W, S, T, and d_{out}, which are the number of turns, cross-sectional width, spacing between turns, thickness of the trace material, and outermost side length of the spiral, respectively (see Figure 5.16). In the following the three systems are briefly described and compared.

The resonant inductive coupling system of Figure 5.15*a* relies on the use self-resonant source and load elements (i.e., bifilar spirals) that are fed at their center. The SCMR system uses loops for the source and load elements and bifilar spirals for TX and RX resonators, as shown in Figure 5.15*b*. Finally, the CSCMR consists of source and load loop elements and bifilar TX and RX resonators. The bifilar TX and RX resonators are placed inside the source and load loops, respectively. It should

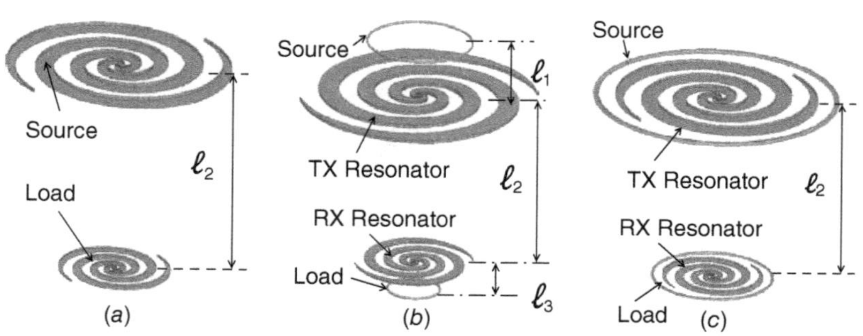

Figure 5.15 Models of three wireless powering systems: (*a*) resonant inductive coupling, (*b*) SCMR, and (*c*) CSCMR.

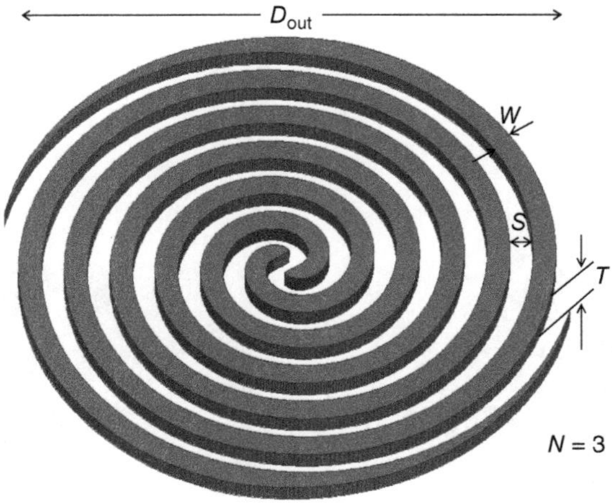

Figure 5.16 Bifilar spiral geometry.

be noted that the bifilar spirals used in the SCMR and CSCMR systems have their excitation gaps open since they are used as parasitic resonators. For both methods the source loop is connected to the power source and the load loop is connected to a 50-Ω load. All systems are designed to operate at 39 MHz.

The three systems of Figure 5.15 use the same two bifilar spirals, one in air (external) and one in tissue (implanted). The implanted spirals are embedded at a depth of 2.6 cm inside muscle tissue, as shown in Figure 5.17*a*. The geometrical parameters of the external TX bifilar spiral are as follows: $d_{out} = 98$ mm, $N = 8$, $T = 0.5$ mm, $W = 1.52$ mm, and $S = 1.52$ mm. The geometrical parameters of the implanted RX bifilar spiral are as follows: $d_{out} = 30$ mm, $N = 9$, $T = 0.3$ mm, $W = 0.48$ mm, and $S = 0.48$ mm. The distances in Figure 5.15 are as follows: $\ell_1 = 5$ mm, $\ell_2 = 50$ mm, and $\ell_3 = 4$ mm. The SCMR system shown in Figure 5.15*b* also uses (a) a source loop with radius of 30 mm and cross section of 1 mm and (b) a load loop with radius of 15 mm and cross section of 0.5 mm. The CSCMR system shown in Figure 5.15*c* also uses (a) a source loop with radius of 55 mm and cross section of 1 mm and (b) a load loop with radius of 17 mm and cross section of 0.1 mm. All elements are assumed to be made from copper (i.e., $\sigma = 5.8 \times 10^7$ S/m). In order to achieve miniaturization of the SCMR and CSCMR implanted RX resonator bifilar spirals, they were embedded in a dielectric with permittivity value of 10.2. This can be implemented in practice either by strip line circuits on the appropriate substrate or by encapsulating the bifilar spirals inside an appropriate dielectric material.

The performance of the three wireless powering systems of Figure 5.15 is compared through simulations. All simulations are performed using Ansoft HFSS and the top part of a 3D human body model in which the properties of the various tissue types are specified. For example, Figure 5.17*a* shows the part of the human phantom included in our simulations with the CSCMR system of Figure 5.15*c*.

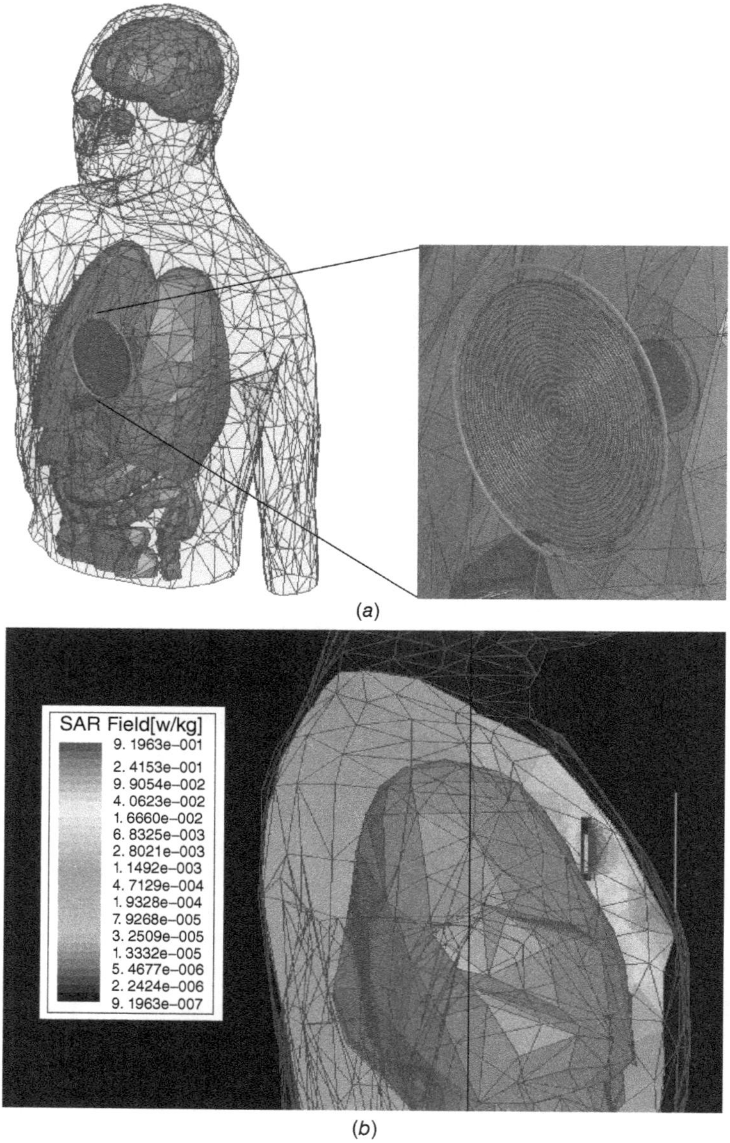

Figure 5.17 CSCMR in human body: (*a*) geometry of human body and (*b*) SAR distribution. (*See insert for color representation of b.*)

Power requirements of IMDs depend on the specific applications and are typically in the range of a few microwatts (Collins et al., 2000) to hundreds of milliwatts (Kong et al., 2000; Marliere et al., 1999). However, the more efficient a wireless powering method is, the less power needs to be used at the transmitter, thereby decreasing the

TABLE 5.3 Comparison of Different Wireless Powering Systems for Output Power of 385 mW

	Field Parameters and Efficiency			
Coupling Type	Input Power(W)	Maximum H Field (A/m)	Maximum SAR (W/kg)	Efficiency (%)
Resonant Coupling	97	50	35	0.46
SCMR	1.3	5.6	0.98	36.9
CSCMR	1	3.4	0.91	38.5

intensity of the generated fields as well as lowering the SAR. Therefore, it is expected that SCMR and CSCMR will outperform traditional resonant inductive coupling in terms of safety.

First, the conformal SCMR model is simulated and optimized in Ansoft HFSS and Nexxim with 1 W input power. An output power of 385 mW was achieved at $\ell_2 =$ 50 mm. In order to compare the three techniques for the same power delivered to the load, the input power of the resonant inductive coupling and the SCMR systems is adjusted so that an output power of 385 mW is delivered to the load. The performance of the three methods is compared in Table 5.3, where the input power, maximum H-field intensity, maximum SAR, and wireless powering efficiency are reported. As expected, it is clearly seen that SCMR and CSCMR significantly outperform resonant inductive coupling. Specifically, SCMR and CSCMR achieved efficiencies of 36.9 and 38.5%, respectively, versus resonant inductive coupling that achieved an efficiency of only 0.46%.

This in turn means that resonant inductive coupling must utilize a significantly greater amount of input power (i.e., 97 W) to achieve the same output power with SCMR and CSCMR. As a result, SCMR and CSCMR exhibit substantially lower levels of SAR (approximately 35 times lower maximum SAR) and magnetic field intensity. Therefore, SCMR and CSCMR are safer methods for wireless powering of implanted or wearable devices. Also, the SAR distribution for the CSCMR model is shown in Figure 5.17*b*.

In addition to achieving large efficiency and reducing SAR levels, CSCMR exhibits a compact size because it uses planar transmitter and receiver configurations, thereby minimizing the volume required. Minimization of volume is particularly important for IMDs and biomedical applications. Therefore, CSCMR is very well suited for wireless powering of wearable and implantable devices. A comparison of the dimensions, volume, and height of reported SCMR systems for biomedical applications is shown in Table 5.4 that clearly illustrates that the CSCMR system designed in this section requires very small volume and footprint.

5.3.3 Far-Field Wireless Power Harvesting

While far-field wireless power harvesting to date has been largely restricted to RFID EPC Gen-2 tags that utilize 860–960 MHz globally, low-power advancements in

TABLE 5.4 Comparison of Different SCMR Systems for IMD Applications

Papers	Dimension (r, H) (mm)	Volume (mm^3)	Volume (mm)	Frequency of Operation (MHz)
(Rais et al., 2009)	(85,50)	1, 134, 900	50	6.9
(Marliere et al., 1999)	(11, 2.5)	950.3	10	0.7
SCMR	(15, 4)	2827.4	4	39
CSCMR	(15, 0.35)	247.4	0.35	39

the semiconductor industry has resulted in additional applications starting to use long-range wireless power transfer in the far field. In addition to the near field in which a dedicated source is required, the pervasiveness of advanced communication mediums such as cellphones and digital TV is seeing higher amounts of power being broadcasted wireless through the air. Vyas et al. (2011) proposed an advanced method of utilizing RFID technology for the purpose of harvesting ambient power present in the air due to TV and cellular wireless signals. Ambient wireless spectrum measurements show significant amounts of wireless power present in the VHF and UHF bands due to radio, TV, and cellular signals, as shown in Figure 5.18.

By carefully designing wide-band antennas and RF–DC converting circuits to rectify and step up the low RF signals to a higher usable DC voltage, such a technology can be used to power sensors for biomonitoring or environmental health sensing.

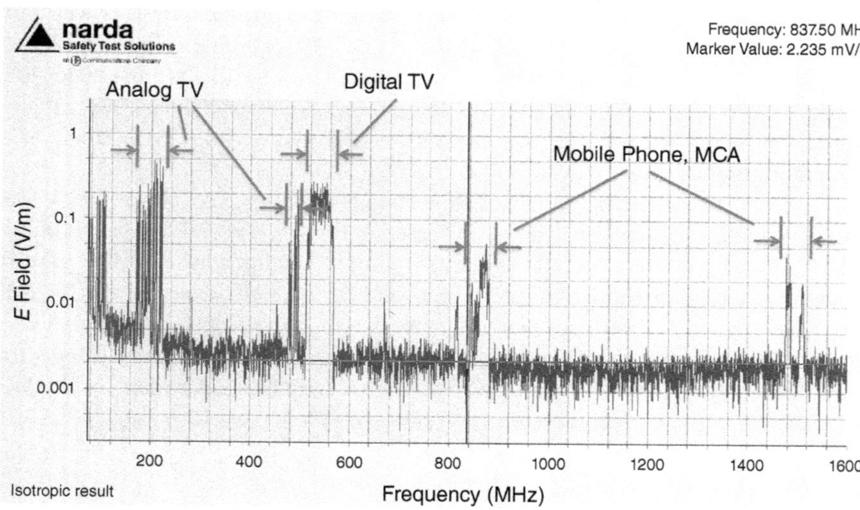

Figure 5.18 Measured ambient wireless spectrum showing radio, TV, and cellular signals in downtown Tokyo ~6.5 km from broadcast towers (Vyas et al., 2011).

The prototype used for harvesting far-field wireless signals is shown in Figure 5.19. Through careful design, the correct trade-off between sensitivity and bandwidth of far-field energy-harvesting systems can be obtained to power devices such as temperature sensors, embedded processors, and low-power radios. The antenna and the RF–DC converter were optimized to capture the greatest amount of power in the 500–700-MHz range since most of the wireless power was present in that spectrum. The antenna was optimized to fit on a conformal paper-based substrate with a peak gain of 2.0 dBi, as shown in Figure 5.20.

By using low-loss Schottky diodes and capacitors with low leakages, the RF–DC losses were kept to a minimum of between 1.0 and 1.2 Ω in the 500–700-MHz range, as shown by the measured RF–DC converter impedances in Figure 5.21. This allowed the far-field harvester to collect around 0.5 mJ of energy in under 3 min time with an output voltage of close to 3.0 V across a charge storage 100-mF capacitor. The energy was collected from sources of wireless broadcasts from a TV tower 6.5 km away. While the power may not be high enough to continuously power all types of sensors, it could be used for powering temperature sensors and pressure sensors for limited duty cycles occurring at intervals of 3–5 min.

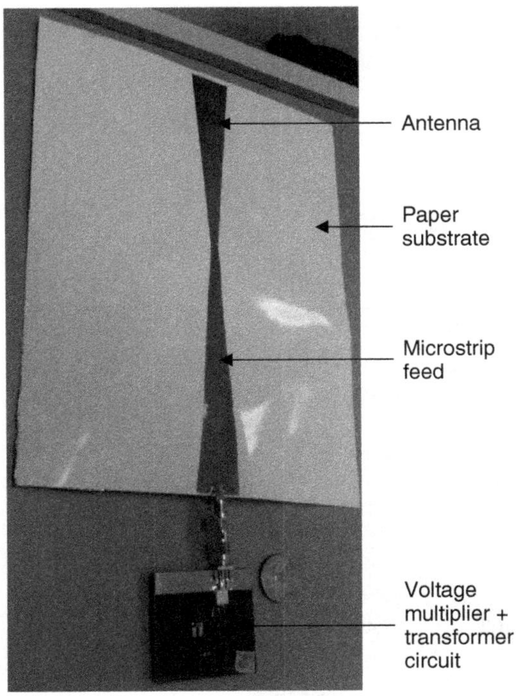

Figure 5.19 Far- wireless power-harvesting prototype (Vyas et al., 2012).

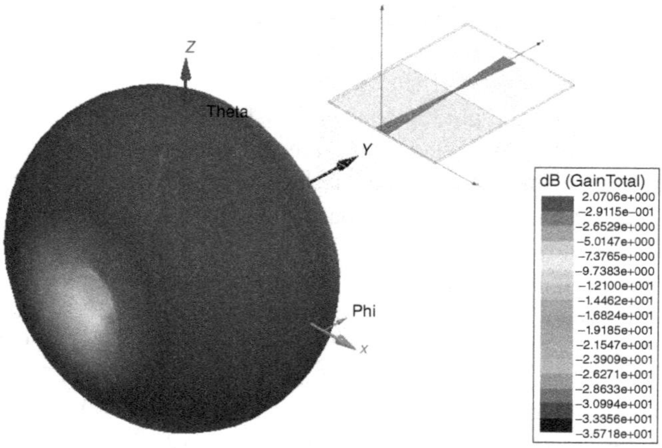

Figure 5.20 Simulated antenna radiation pattern and gain at frequency of 550 MHz (Vyas et al., 2012).

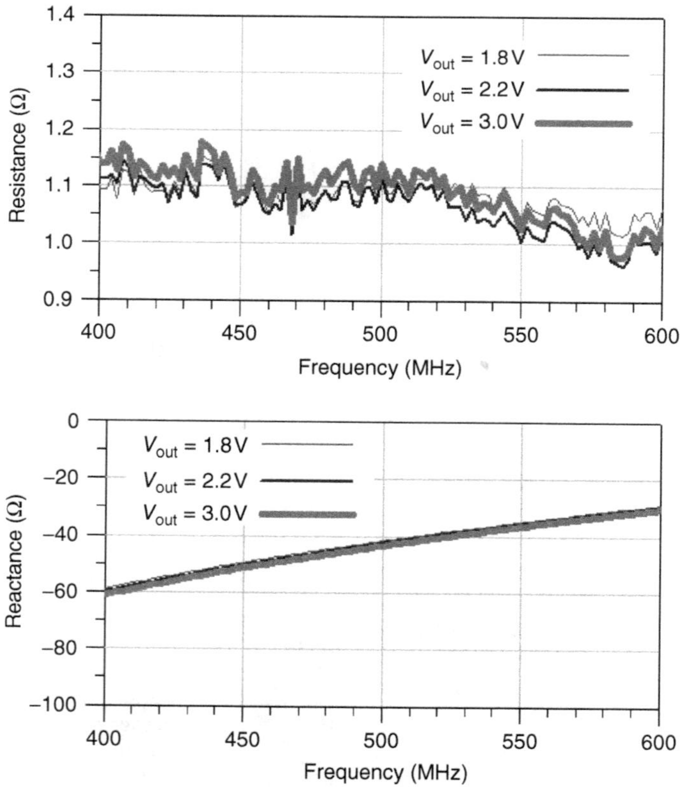

Figure 5.21 Measured impedance of RF–DC converter circuit. (*See insert for color representation of the figure.*)

5.4 CONCLUSION

This chapter introduced a new conformal SCMR method, that is, CSCMR, that achieves high efficiency and minimizes the volume of wireless powering systems. High efficiency enables CSCMR to deliver a significant amount of power to IMDs without excessive transmitter power. In fact, CSCMR outperformed resonant inductive coupling and performed similarly to SCMR in terms of the SAR levels for the same power delivered. CSCMR is also planar and therefore is more suitable for planar and flexible implantable or wearable circuits. Due to these important features, CSCMR is a very well suited and promising technique for wireless powering of implantable and wearable devices.

Moreover, a number of batteryless ways to power biomedical sensors were presented. While using piezoelectric devices energy of about 1 mJ can be produced from a single step, using specially designed electromagnetic resonating coils, the range of near-field powering mechanisms can be increased from tens of millimeters to well over a couple of feet with higher efficiencies. In addition to near field, far-field sources such as ambient TV and cellular signals can also be put in use to power less data-intense biomedical devices. Through proper wireless front-end designs between 0.1 and 1 mJ of energy can be harvested from the air. While not enough for continuously powering on data-intensive operations, the ambience can be used to generate enough power to run embedded processors and temperature sensors that nowadays consume only tens of microamperes.

REFERENCES

Andreuccetti D, Fossi R, Petrucci C. 2002. An Internet resource for the calculation of the dielectrtic properties of body tissues in the frequency range 10 Hz–100 GHz. http://niremf.ifac.cnr.it/tissprop/.

Chopra S, Pham A, Gaillard J, Parker A, Rao A. 2002. Carbon-nanotube-based resonant-circuit sensor for ammonia. *Appl. Phys. Lett.*, 80:4632.

Collins PG, Bradley K, Ishigami M, Zettl A. 2000. Extreme oxygen sensitivity of electronic properties of carbon nanotubes. *Science*, 287(5459):1801–1804.

Fujiwara A, Ishii K, Suematsu H, Kataura H, Maniwa Y, Suzuki S, Achiba Y. 2001. Gas adsorption in the inside and outside of single-walled carbon nanotubes. *Chemical Physics Letters*, 336(3):205–211.

Ghovanloo M, Najafi K. 2004. A wideband frequency-shift keying wireless link for inductively powered biomedical implants. *IEEE Trans. Circuits Syst. I: Reg. Papers*, 51(12):2374–2383.

Kim S, Tentzeris MM, Nikolaou S. 2012. Wearable biomonitoring monopole antennas using inkjet printed electromagnetic band gap structures. 6th European Conference on Antennas and Propagation (EUCAP), pp. 181–184.

Kong J, Franklin NR, Zhou C, Chapline MG, Peng S, Cho K, Dai H. 2000. Nanotube molecular wires as chemical sensors. *Science*, 287(5453):622–625.

Kurs A, Karalis A, Moffatt R, Joannopoulos JD, Fisher P, Soljacic M. 2007. Wireless power transfer via strong coupled magnetic resonances. *Science*, 317(5834):83–86.

Lakafosis V, Rida A, Vyas R, Li Y, Nikolaou S, Tentzeris MM. 2010. Progress towards the first wireless sensor networks consisting of inkjet-printed, paper-based RFID-enabled sensor tags. *Proc. IEEE*, 98:1601–1609.

Marliere C, Poncharal P, Vaccarini L, Zahab A. 1999. Effect of gas adsorption on the electrical properties of single-walled carbon nanotubes mats. In MRS Proceedings. 593(1). Cambridge University Press.

Orecchini G, Yang L, Tentzeris MM, Roselli L. 2011 Wearable battery-free active paper printed RFID tag with human-energy scavenger. *Microw. Symp. Dig. (MTT), 2011 IEEE MTT-S Int.*, pp.1–4, 5–10.

Rais NHM, Soh PJ, Malek F, Ahmad S, Hashim NBM, Hall PS. 2009. A review of wearable antenna. Antennas & Propagation Conference, LAPC 2009. Loughborough, pp. 225–228.

Raju M, Grazier M. 2008. Energy harvesting. ULP meets energy harvesting: a game-changing combination for design engineers. TI, http://focus.ti.com/lit/wp/slyy018/slyy018. pdf.

Salonen P, Rahmat-Samii Y. 2006. Textile antennas: Effects of antenna bending on input matching and impedance bandwidth. First European Conference on Antennas and Propagation, EuCAP. 2006, pp. 1–5.

Shaozhen Z, Langley R. 2009. Dual-band wearable textile antenna on an EBG substrate. IEEE Trans. Antennas Propag. , 57:926–935.

Sievenpiper D, Lijun Z, Broas RFJ, Alexopolous NG, Yablonovitch E. 1999. High-impedance electromagnetic surfaces with a forbidden frequency band. *IEEE Trans. Microw. Theory Techn.*, 47:2059–2074.

Tan Y, Hoe K, Panda S. 2006. Energy harvesting using piezoelectric igniter for self-powered radio frequency wireless sensor. IEEE International Conference on Industrial Technology, pp. 1711–1716.

Varghese PDKOK, Gong D, Ong KG, Dickey EC, Grimes CA. 2001. Gas sensing characteristics of multi-walled carbon nanotubes. *J. Sensors Actuators B. Chem.*, 81.

Vyas R, et al. 2009. "Paper-Based RFID-Enabled Wireless Platforms for Sensing Applications", IEEE Trans. Microwave Theory & Techniques, vol 57, pp 1370–1382.

Vyas R, Lakafosis V, Tentzeris MM, Hiroshi N, Kawahara Y, 2011 "A battery-less, wireless mote for scavenging wireless power at UHF (470–570 MHz) frequencies," Antennas and Propagation (APSURSI), IEEE International Symposium on, vol., no., pp.1069–1072, 3–8 July 2011.

Vyas R, Hiroshi N, Tentzeris MM, Kawahara Y, Tohru A 2012 "A battery-less, energy harvesting device for long range scavenging of wireless power from terrestrial TV broadcasts", Microwave Symposium Digest (MTT), IEEE MTT-S International, vol., no., pp.1–3, 17–22 June 2012.

Wei W. 2012. "Design and Analysis of Resonators for Electromagnetic Resonant Wireless Power Transmission Systems" University of Tokyo.

Yang L, Rida A, Vyas R, Tentzeris MM. 2007. RFID tag and RF structures on a paper substrate using inkjet-printing technology. IEEE Trans. *Microw. Theory Techn.*, 55:2894–2901.

PART II

PROPAGATION AND COMMUNICATION ISSUES FOR BIOMEDICAL TELEMETRY

6

NUMERICAL AND EXPERIMENTAL TECHNIQUES FOR BODY AREA ELECTROMAGNETICS

Asimina Kiourti and Konstantina S. Nikita
School of Electrical and Computer Engineering, National Technical University of Athens, Athens, Greece

6.1 INTRODUCTION

Recently, there is growing research activity on biomedical telemetry systems, while millions of people worldwide depend upon wearable, ingestible, and implantable biotelemetry devices to support and improve their quality of life. Such devices can be used to perform several monitoring, diagnostic, therapeutic, and interventional applications that range from cardiac pacemakers and defibrillators to emerging applications in visual prosthesis, brain computer interfaces, and monitoring of a variety of physiological parameters such as oxygen, glucose, pH level, pressure, and temperature.

Biotelemetry communication links are formed between biomedical devices placed on, inside, or in close proximity to the body and control/monitoring equipment placed at a short distance. Such links serve a variety of purposes including transmission of real-time or stored data and device parameter adjustment, with new opportunities constantly arising. Depending on the application scenario and performance requirements, biotelemetry links can be implemented in a number of ways:

- Inductive links formed between mutually coupled coils have widely been employed for data and power transcutaneous biomedical telemetry. The degree

Handbook of Biomedical Telemetry, First Edition. Edited by Konstantina S. Nikita.

of coupling is described in terms of the coils' mutual inductance, which indicates the proportion of the exterior (implanted) coil's field that is captured by the implanted (exterior) coil. However, low data rates as well as size, weight, and biocompatibility issues often prohibit the use of inductive biotelemetry. Inductive links for biomedical telemetry applications will be analyzed in Chapter 7.

- Wireless biotelemetry performed by means of antennas operating in predefined frequency bands can be used instead. Nowadays, design of wearable, implantable, and ingestible antennas is attracting high scientific interest to deal with challenges related to miniaturization, biocompatibility, and high-quality performance. Biomedical telemetry performed by means of antennas will be described in Chapter 8.
- The use of the human body itself as a transmission medium for electrical signals, known as intrabody communication, is recently serving as a data communication technique in biomedical telemetry. Transmission of low-power electrical signals through the human body, performed by means of either galvanic or capacitive coupling, is a promising approach for wireless intrabody data transmission between sensors. Intrabody communications will be discussed in Chapter 9.
- Finally, in order to answer many new demands for biotelemetry and to overcome various problems of radio telemetry, a technological method called optical biotelemetry has also been developed. Using light as a transmission medium, the bandwidth for signal transmission is greatly increased and many electromagnetic interference (EMI) problems can be solved. Realization of optical biotelemetry will be discussed in Chapter 10.

Non-ionizing EM fields in the RF and microwave (MW) regions of the EM spectrum have a strong potential in biomedical telemetry because of their ability to be transmitted, guided, and focused. As a result, significant scientific research has been reported and is currently being performed on diagnostic and therapeutic applications of EM fields in biotelemetry, including data and power transmission as well as induction or collection of biological responses.

The human body is an important part of the wireless biotelemetry channel. Propagation of EM fields inside and around the human body has been found to be highly complex and is strongly affected by a number of factors including the lossy, subject-dependant dielectric human tissues, and the geometry of the subject. Human subjects can be modeled either numerically (numerical models imported in numerical EM codes) or physically (physical models made from liquid, gel, or solid materials). In order to understand the basic ways in which EM waves interact with the body and, thus, identify optimal biotelemetry solutions, it is essential to study the EM properties of human tissues and (numerically or physically) model the body in a practical and realistic way.

Due to the risk of adverse health effects caused by the use of wireless biomedical telemetry devices adjacent to or implanted inside the human body, such studies are also significant toward minimizing interaction of these devices with biological tissues and preserving patient safety. Concerns and safety guidelines for unintended effects of fields induced by biomedical telemetry on the human body will be discussed in Chapter 11.

This chapter discusses the electrical properties of human tissues, as well as numerical and physical modeling of the human body, and corresponding numerical and experimental procedures. A brief outline of safety issues and conformance to (national and international) safety guidelines will also be presented.

6.2 ELECTRICAL PROPERTIES OF HUMAN BODY TISSUES

The human body is a heterogeneous medium consisting of several types of tissues (e.g., skin, muscle, fat, blood, and organs such as lung, liver, or heart). Human tissues contain insulating materials (lipids) and electrical charges (ions, electrically polarized molecules, etc.). Therefore, they can be viewed as a weakly conducting medium (dielectric).

Electrical properties of each type of tissue result from the interaction between the incident EM radiation and the tissue constituents at the cellular and molecular level and control the propagation, attenuation, reflection, and other behavior of EM fields inside and outside the human body. As a result, it is essential to know the dielectric properties of the lossy human tissues in order to understand the interaction of EM waves with the human body and enable a detailed analysis of RF transmission and absorption in biomedical telemetry systems.

Energy exchange in biological tissues occurs in terms of either free charges or dipolar molecules (e.g., water). When a temporal-variant EM field is applied, free charges are accelerated to produce current, resulting in resistive losses that depend on the conductivity of the tissue, while dipolar molecules are re-oriented accordingly (dipolar polarization). Adjustment is reached after a certain period of time, which is known as the relaxation time. The impact on the overall field strength is expressed in terms of the complex permittivity:

$$\varepsilon = \varepsilon_0 \varepsilon_r \tag{6.1}$$

where ε_0 is the permittivity of free space, and ε_r is the complex relative permittivity, defined as

$$\varepsilon_r = \varepsilon_r' - j\varepsilon_r'' \tag{6.2}$$

In equation (6.2), ε_r' is the relative permittivity of the material, and $\varepsilon_r'' = \sigma/\omega\varepsilon_0$ is the out-of-phase loss factor, where ω denotes the angular frequency of the EM field and σ the conductivity of the material.

In general, relative permittivity describes how the material is affected by electric fields, while conductivity describes attenuation of the EM wave as it transits the material. The loss tangent, defined as

$$\tan\delta = \frac{\varepsilon_r''}{\varepsilon_r'} \tag{6.3}$$

tells which component of the electrical properties dominates the effect on the field. At low frequencies, conductivity dominates the behavior of the field, while at high frequencies, the relative permittivity tends to dominate.

Based on equations (6.1) and (6.2), the complex permittivity of the lossy body tissues can, thus, be calculated as

$$\varepsilon = \varepsilon_0 \varepsilon_r' - j\frac{\sigma}{\omega} \tag{6.4}$$

Electrical properties of the human body are usually specified either in terms of ε_r' and ε_r'' or in terms of ε_r' and σ, as a function of frequency.

However, human tissues are heterogeneous dielectric materials and there exist several relaxation times. Relative permittivity and conductivity are a strong function of tissue type and frequency, with each tissue having a different frequency variation, as follows:

- Tissues with high water content (e.g., muscle, skin, or brain) feature higher permittivity and conductivity values than tissues with low water content (e.g., fat or bone) and are therefore more lossy. This behavior can be attributed to the resonant polar properties of water molecules.
- Relative permittivity decreases with frequency in three main steps known as the α, β, and γ dispersion regions, from values in the range of 10^5 at a few hundred hertz to less than 1 in the gigahertz range. Conductivity increases with frequency starting around 10^{-4} to above 1 in the same frequency range.

For example, Figure 6.1 shows relative permittivity data of two very extreme tissues, muscle and fat, in a range of frequencies. Conductivity values are presented in Figure 6.2. Results are obtained from a compilation presented by Gabriel et al. (1996), which covers several body tissues and provides values of the electrical properties at various frequencies.

High-order terms are necessary to mathematically describe the complex permittivity, and a model based on the summation of four Cole–Cole expressions is most

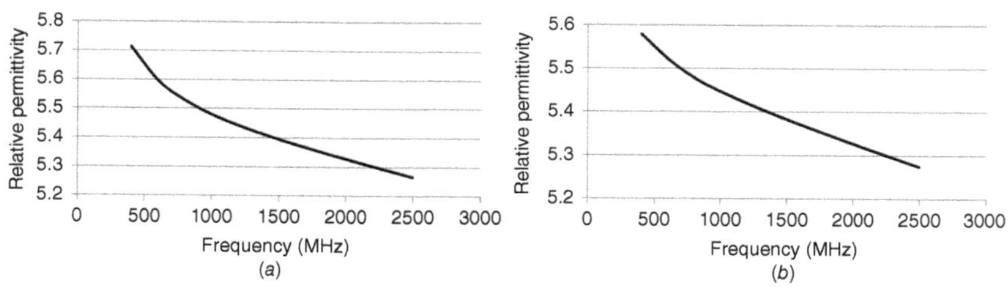

Figure 6.1 Dielectric permittivity values of (*a*) muscle and (*b*) fat tissues with respect to frequency.

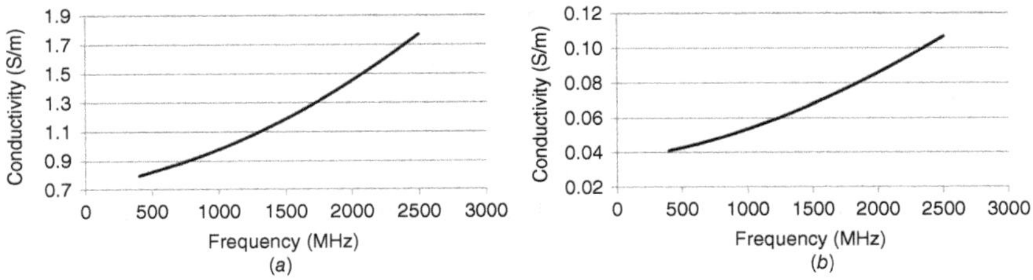

Figure 6.2 Conductivity values of (*a*) muscle and (*b*) fat tissues with respect to frequency.

commonly used (Gabriel et al., 1996):

$$\varepsilon = \varepsilon_{\infty} + \sum_{m=1}^{4} \frac{\Delta\varepsilon_m}{1 + (j\omega\tau_m)^{(1-a_m)}} + \frac{\sigma_j}{j\omega\varepsilon_0} \tag{6.5}$$

where ε_{∞} is the material permittivity at very high frequencies (of the order of terahertz), ε_0 is the permittivity of free space, σ_j is the conductivity, and ε_m, τ_m, α_m are material parameters for each dispersion region.

Finally, an important parameter toward understanding EM propagation in human tissues is the skin depth, δ. The skin depth is a measure of the distance over which the electric field has been attenuated by a factor of $^1/_e$ or 0.368 from its original value and is defined as

$$\delta = \frac{1}{\alpha} \tag{6.6}$$

where a is the attenuation constant obtained from the real part of the propagation constant, γ, as

$$\gamma = \alpha + j\beta = j\omega\sqrt{\mu\varepsilon}\left(1 + \frac{\sigma}{j\omega\varepsilon}\right)^{1/2} \tag{6.7}$$

where μ is the permeability of the material.

At low frequencies, where relative permittivity values are high and conductivity values low, skin depth is significant and the EM wave can penetrate well into the human body. For example, at 433 MHz, 69% of the field is transferred through 10 cm of fat, and 11% is transferred through 10 cm of muscle. In implantable biomedical telemetry devices, the use of lower operation frequencies (e.g., 402 or 433 MHz) is, therefore, recommended. On the other hand, penetration of incident EM fields into biological tissues decreases with frequency. At high frequencies, conductivity values become high, skin depth decreases, and propagation is limited around the surface of the body. For example, at 2.45 GHz, the penetration depth is limited to 113 and 21 mm for fat and muscle, respectively.

Currently, the most comprehensive, complete, and best known database of electrical properties of body tissues is based on the work of Gabriel et al. (1996). Before this work was published, most data on the dielectric properties of tissues were obtained from measurements of animals, such as pigs, sheep, or rabbits. Based on these measurements, a Cole–Cole analysis was performed to propose a parametric model for several body tissues in a wide frequency range (from 10 Hz to 20 GHz).

It is worth noting, however, that significant deviations are observed to these commonly used values from data produced by other works (Smith and Foster, 1985; Campbell and Land, 1992). Smith and Foster measured the complex permittivities of two low-water-content tissues, bone marrow and adipose tissue, between 1 kHz and 1 GHz. Comparison with previous measurements of dielectric properties for high-water-content tissues suggested that bone marrow and adipose tissues contain less motionally altered water per unit dry volume compared with the previously studied tissues with lower lipid fractions (Smith and Foster, 1985). Campbell and Land (1992) measured the dielectric properties of female breast tissues at 3.2 GHz and concluded that apart from the dielectric relaxation of tissue water, dielectric relaxation of bound water and the tail end of a beta dispersion also affect the complex pemittivities at this frequency.

Several other research efforts have been performed to assess the electrical properties of human tissues. Recently, Lazebnik et al. (2007) reported the results of a large-scale, multi-institutional study characterizing the dielectric properties of normal breast tissue samples obtained from reduction surgeries at the University of Wisconsin and University of Calgary hospitals. The analysis of 354 tissue samples revealed that there was a large variation in the dielectric properties of normal breast tissue due to substantial tissue heterogeneity. No statistically significant difference was observed between the within-patient and between-patient variability in the dielectric properties. Furthermore, O'Rourke et al. (2007) characterized electrical properties of in vivo and ex vivo human liver tissues between 0.5 and 20 GHz. It was observed that wideband dielectric properties of in vivo liver tissue are different from the wideband dielectric properties of ex vivo liver tissue, and that the in vivo data cannot be represented in terms of a Cole–Cole model.

Variability of electrical properties has also been studied. Bao et al. (1997) presented in vitro complex dielectric measurements of gray and white matter of rat brain tissue in the frequency range between 45 MHz and 26.5 GHz. Two empirical models were suggested to describe the experimental data: the first model contained two Cole–Cole functions, while the second utilized one Havriliak–Negami and one Cole–Cole function. Standard deviations of the order of ±4 to ±16% were reported (Bao et al., 1997). Electrical properties of animal tissues as a function of time after death have also been investigated (Schmid et al., 2003). Experiments were performed on pigs to investigate possible postmortem changes of the dielectric properties of brain gray matter in the frequency range of 800–1900 MHz. After being kept in stable anaesthesia for at least 45 min, animals were euthanatized, and measurements of dielectric properties were repeatedly performed from at least 45 min prior to death to 18 hr after euthanasia. A decrease by 4% in relative permittivity and by 10% in conductivity was observed in pig tissue 4 h after death. Results indicate that in vitro measurements of dielectric properties of brain tissue underestimate equivalent conductivity as well as permittivity of living tissue.

Furthermore, tissue electrical properties have been demonstrated to decrease with age for most tissues, as attributed to changes in water content and organic composition of the tissues. The dielectric properties of rat tissues at different ages have been measured (Peyman et al., 2001). Higher conductivity values were found for the brain and skull of newborn rats compared to adult rats at 900 MHz (16 versus 43%, respectively), while a lower increase of relative permittivity was also recorded (9.9 versus 33%, respectively). A significant dependence of the electrical properties of white matter and spinal cord on age has also been reported, while no age-related variation has been found for the gray matter (Peyman et al., 2007). Finally, a systematic evaluation of the age-dependent changes of the electrical properties of a large number of different tissues has recently been published (Peyman et al., 2009). In vitro dielectric properties of aging porcine tissues were measured in the frequency range of 50 MHz–20 GHz, and a statistically significant reduction with age in both permittivity and conductivity of 10 out of 15 measured tissues was observed. The establishment of a database for children's dielectric properties should be an essential and urgent task.

6.3 NUMERICAL MODELING

Numerical modeling provides an effective way of assessing and predicting the EM performance of biomedical telemetry systems in terms of radiation, propagation, and interaction with body tissues. Numerical modeling overcomes the inherent difficulties of experimental measurements and is considered of utmost importance in the calculation of field values inside body tissues as well as in the evaluation of implantable biomedical devices where it is impossible to make measurements in real operating scenarios.

Analytical methods can be applied to simplified canonical geometries modeling the human body (or parts of it), while numerical methods use body models ranging from very simple homogeneous models to millimeter resolution anatomic models, with the latter providing highly accurate results. Developing powerful, reliable, and subject-specific (in)homogeneous numerical phantoms is highly significant in reliably predicting the EM performance of biomedical telemetry systems.

6.3.1 Numerical Phantoms

Canonical Models Simple-shaped numerical phantoms, in the shape of a sphere or cube, are known as canonical models. These can be solved either numerically to provide an understanding of the exhibited EM performance through simplified estimations, or analytically for computer code checking.

Canonical models are computationally efficient with standard simulation resources and seem to be adequate for obtaining preliminary results. Inaccuracy comes from inaccuracy in modeling and not inaccuracy in the calculations. Furthermore, the choice of simple models provides the possibility to easily construct experimental phantoms complying with the numerical ones, which is important for validation purposes.

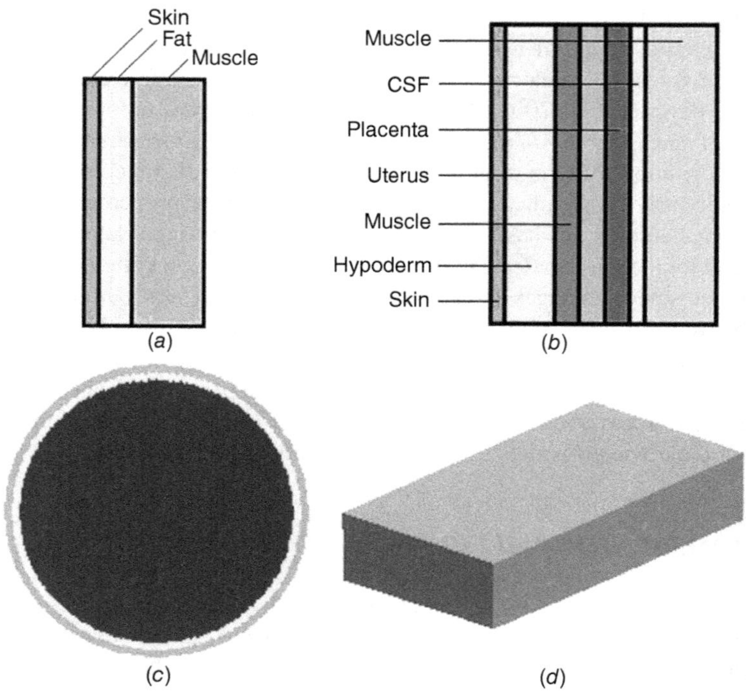

Figure 6.3 Whole-body canonical models used in (*a*) (Curto and Ammann, 2007) and (*b*) (Wiart et al., 2005), and canonical geometries modeling the human (*c*) head (Koulouridis and Nikita, 2004) and (*d*) torso (Kuhn et al., 2009).

Whole-body canonical models have largely been used in the literature to model the human body. A planar three-layer body model, consisting of a low-water-content tissue layer (fat) embedded between two high-water-content tissues (skin and muscle) has been proposed (Curto and Ammann, 2007) (Figure 6.3*a*). The study reported on the coupling mechanism of a half-wavelength dipole at 434 MHz, as a preliminary step toward analyzing antennas interacting with tissues in the near and farfields. Nine combinations of skin, fat, and muscle tissue with different thicknesses were analyzed to model the absorption in different parts of the body. A multi-layer structure composed of skin, hypoderm, muscle, uterus, placenta, amniotic fluid (considered as cerebrospinal fluid, CSF), and fetus (considered as muscle) has also been analyzed (Wiart et al., 2005) (Figure 6.3*b*). The study assessed the main parameters influencing the RF exposure of children's heads. The evolution of the head shape and the growth of specific parameters, such as the skull thickness, were taken into account and comparisons with corresponding results in adults were reported. Cylindrical phantoms can otherwise be applied for whole-body models.

Canonical geometries have also been used to model specific parts of the human body. For example, a three-layer spherical human head model consisting of skin, bone, and brain tissues has been used (Koulouridis and Nikita, 2004) (Figure 6.3*c*).

The aim was to study the interaction between normal-mode helical antennas and human head models, while emphasizing the comparative dosimetric assessment between adults and children head models. Rough approximations of the human head have also been obtained by using homogeneous or layered dielectric box models. Furthermore, a rectangular structure filled with tissue-emulating material has been used to model the human torso (Kuhn et al., 2009) (Figure 6.3*d*). The study evaluated the radiation emitted by mobile phones when used with wireless and wired hands-free kits in order to assess the significance of a dedicated compliance procedure and the extent to which the use of wired and wireless hands-free kits can reduce human exposure.

Regarding children models, it is most commonly assumed that a child model is perfectly proportional to an adult model. As such, children canonical models are obtained through uniform downscaling of the corresponding adult models. For example, homogeneous and multi-layer spherical children head models have been obtained through uniform deformation of spherical adult head models (Koulouridis and Nikita, 2004).

Example canonical models used in recent biomedical telemetry studies are shown in Figure 6.4:

- A spherical six-layer head model consisting of brain, CSF, dura, bone, fat, and skin has been used to characterize a dipole antenna implanted inside the human head and operating at the biomedical frequency band of 402–405 MHz (Figure 6.4*a*) (Kim and Rahmat-Samii, 2004). The human head was simplified as a lossy multi-layered sphere to allow analytical solving of the EM problem and facilitate parametric studies.

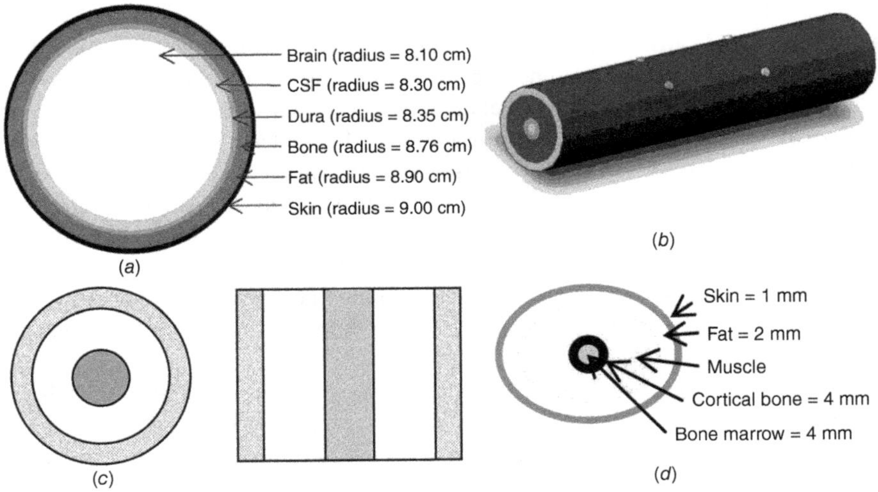

Figure 6.4 Canonical phantom models used to model the human (*a*) head (Kim and Rahmat; Samii, 2004), (*b*) upper arm (Wegmueller et al., 2007), (*c*) thigh (Weiss et al., 2009), and (*d*) arm (Lim et al., 2011) in recent biomedical telemetry studies.

- The geometry of the human upper arm has been modeled by concentric cylinders representing the skin, fat, muscle, cortical bone, and bone marrow layers in order to characterize it as a transmission medium for electrical current (galvanic coupling) (Figure 6.4*b*) (Wegmueller et al., 2007). Properties of specific types of tissue, as well as geometrical body variations were investigated, and different electrodes were compared.
- A thigh model consisting of concentric cylinders representing the bone, muscle, and fat has been presented to study the biotelemetry link for an artificial hip joint (Figure 6.4*c*) (Weiss et al., 2009). The accuracy of the cylindrically symmetrical model was verified through comparison of simulation results with measurements performed for a 433-MHz biomedical telemetry link in a 6-cm-thick porcine thigh.
- Finally, the human arm has been approximated as a conical frustum (tapered cone) of different radii and axis ratio as well as tissue compositions (single- or multiple-tissue layers and different thicknesses of tissue combinations) to report results on ultrawide-band (UWB) signal propagation along the human arm (Figure 6.4*d*) (Lim et al., 2011). The tissue types used included skin, fat, muscle, cortical bone, and bone marrow. The aim of this study was to assess the accuracy and efficiency and, thus, the suitability of such a simplified canonical model, as well as to investigate the effects of the model's structure complexity on the simulation results.

Anatomical Models To obtain more accurate simulation results, it is necessary to use realistically shaped numerical phantoms consisting of several types of tissues, known as anatomical models. The human body (or parts of it) is modeled by cubic cells (voxels) in which tissue electrical properties (relative permittivity and conductivity) are considered constant. By assigning the corresponding electrical properties to each voxel, one can easily model the anatomical tissues and organs.

Development of anatomical models is one of the most significant challenges in numerical bioelectromagnetics, and nowadays it is facilitated by the progress in medical imaging technologies and the increase in computing power. As computing power increases and computer resources get less expensive, there is a trend to refine the numerical description of the space modeled and move to more detailed anatomical structures. In contemporary models, the highest complexity used for modeling the whole human body is about 50 tissue types, and the finest resolution is about 1 mm. Advanced computer graphics techniques can be employed to pose such models in virtually any position.

In the majority of studies, the data for designing the anatomical body models are taken from magnetic resonance imaging (MRI) or computed tomography (CT) scans. MRI and CT scans provide gray-scale image data of the human body at several transverse slices at a designated spacing. The resolution in each slice is on the order of several millimeters. Several tissues can be distinguished; however, image segmentation is further required to convert the density mappings into tissue maps. This is generally a complex and time-consuming activity and is performed semi-manually, although automatic methods have been reported as well. Even if software for automatic identification is applied, manual verification or correction is required. MRI data are generally superior to CT data in identifying interior tissues because of the high contrast images of the soft tissues. It is worth noting that MRI or CT produced in different laboratories inevitably contain differing discretizations.

Furthermore, the resolution of the medical imaging techniques is, presently, too high for using their results directly in numerical modeling. Several anatomical models have been developed for use in a wide range of applications:

(a) Data acquired from CT scans of a cancer patient have been used to create a model consisting of approximately 35,000 10-mm-edge cubic cells (Sullivan, 1990). Simulations were conducted to assess the electric field distribution for various prototype applications used in deep regional hyperthermia, and comparison with measurements was performed to investigate the ability to simulate the patterns from the near field of the applicators.

(b) A whole-body model of 1.9 mm × 1.9 mm × 3 mm resolution based on MRI scans of an adult male (height 1.88 m, weight 64 kg), scaled to 71 kg, and segmented into 30 types of tissue has also been reported (Gandhi et al., 1996). The EM energy coupled to the head due to mobile phones operating at the frequencies of 835 and 1900 MHz was studied, and the effect of using simplified homogeneous models instead of the anatomical heterogeneous models was assessed.

(c) The Visible Human Project (VHP) (Ackerman, 1998) has made available a digital axial anatomical image library at 1-mm resolution of the body of a human male, named Hugo, based on data from a 38-year-old cadaver (height 1.86 m, weight 90 kg) (Figure 6.5*a*). This project involved the creation of complete, anatomically detailed, three-dimensional representations of the normal male and female human bodies, and its long-term goal was to produce a system of knowledge structures that would transparently link visual knowledge forms to symbolic knowledge formats such as the names of body parts. Hugo's body model is larger than that of an average male and

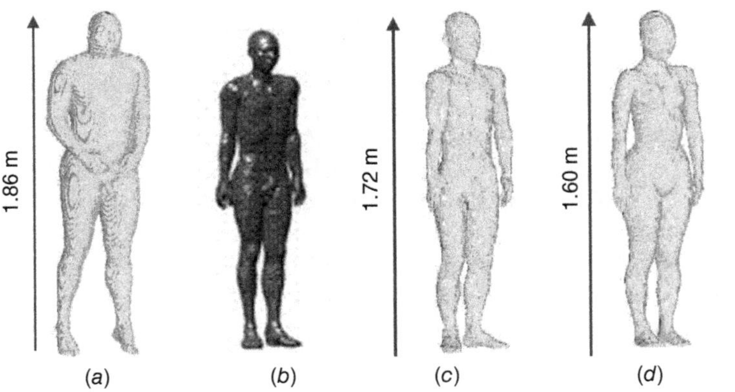

Figure 6.5 Anatomical models of (*a*) Hugo (Ackerman, 1998), (*b*) Norman (Dimbylow, 2002), (*c*) Japanese male (Nagaoka, 2004), and (*d*) Japanese female (Nagaoka, 2004).

consists of 38 tissues. Various studies have been performed with the VHP Man, and the model is nowadays being included in many commercially available EM simulation tools. Furthermore, the deviation of Hugo's dimensions (height and weight) from the average values has promoted development of other anatomically realistic models with average height and weight.

(d) The Norman model consists of 2-mm-edge cubic voxels based on MRI scans of a subject, scaled to the height and mass of a reference man, and segmented to 38 types of tissue (Dimbylow, 2002) (Figure 6.5b). The reduction in voxel size compared to previous works allowed dosimetric quantities to be calculated at higher frequencies, while calculations were extended down to 10 MHz to cover the whole-body resonance regions.

(e) Finally, two 2-mm-resolution whole-body Japanese models with hands placed at the side of the body have been developed using MRI (Nagaoka et al., 2004). These classified over 50 types of tissues and were based on images of a 22-year-old male (height 1.73 m, weight 65 kg) (Figure 6.5c) and a 22-year-old female (height 1.60 m, weight 53 kg) (Figure 6.5d). The reported female model was the first of its kind in the world, and both the male and female models were the first Asian voxel models (representing average Japanese), which enabled numerical evaluation of dosimetric quantities at frequencies of up to 3 GHz.

Due to the lack of children MRI or CT images, the first anatomical children models in the literature were based on uniform downscaling of adult models. However, the uniform downscaling approach does not take into account the different growth patterns of certain parts of the body. For example, head growth is age dependent. The volume of the brain, the skin, and skull thickness each grow at different rates. As a result, the scaled children models do not exactly reproduce the dimensions and anatomy of children. A more realistic approach suggests making a piecewise reduction of the adult model with respect to the main anatomical parameters. The adult model is divided in different parts and non-uniform downscaling is applied to each of these ("child-like" approach). However, inaccuracies in the size of the non-uniform downscaled child models are still present. As a result, recently, whole-body child models have been developed based on MRI or CT database of children. In Wiart et al., (2008), several MRI data sets of children at different ages from three different French hospitals were used to provide segmented versions of children head models. Six child head models at different ages (5, 6, 8, 9, 12, and 15 years old) were built using this approach, as illustrated in Figure 6.6.

Example anatomical models used in recent biomedical telemetry studies are shown in Figure 6.7. The 67-biological-tissue phantom file of a human body produced at Yale University (Zubal et al., 1994) has been translated into a 30-biological-tissue model to analyze the characteristics of a head-implantable dipole antenna (Figure 6.7*a*) (Kim and Rahmat-Samii, 2004). Implantable antennas have been analyzed inside a 31-tissue realistic model of the human shoulder derived from the University of Utah model (Figure 6.7*b*) (Soontornpipit et al., 2004), while an anatomical human head model comprising 24 biological tissues has been used to characterize the transmission of RF signals between implantable and free-space loop antennas (Figure 6.7*c*) (Chen et al., 2009). Finally, 2-mm-resolution human body models have been developed at

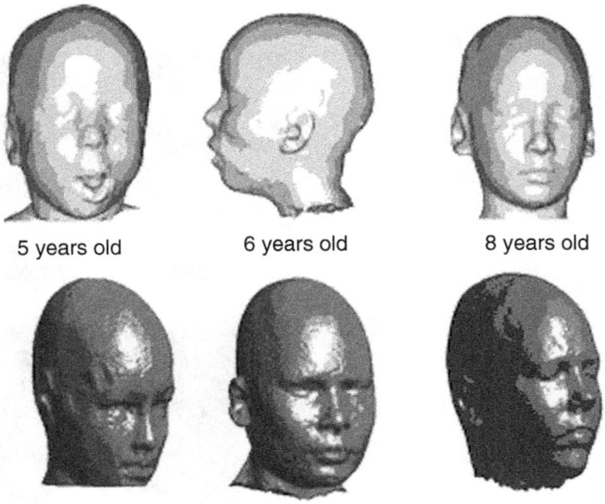

Figure 6.6 Child head models based on MRI data sets of children at different ages (Wiart et al., 2008).

different postures on the basis of average statistical values of the body parameters of Asian adults in order to investigate on-body UWB communication (Figure 6.7*d*) (Wang et al., 2009).

However, it is worth noting that even though heterogeneous models are more representative of actual coupling in human tissue, it has been observed that homogeneous models overestimate the absorbed energy and can be considered as worst-case approximations. This is why, although sophisticated body models have appeared in most recent studies, the homogeneous model case has almost always been present as well.

6.3.2 Computational Methods

Analytical Methods Analytical techniques are so called because, in contrast to numerical techniques, they consist of some solution to Maxwell's equation that is not based on a direct numerical solution and does not require the inversion of large matrices. Exact solutions to Maxwell's equations can be found in terms of mathematical formulas that describe the propagating EM fields. The main restriction consists in adopting a simplified canonical geometry for modeling the human body (or parts of it). The choice of such highly simplified geometries is essential due to the necessity of characterizing a structure resembling the human body (or parts of it) and having at the same time a closed form of the wave equation.

For example, body heating due to exposure to 150 MHz–10 GHz fields has been investigated by analyzing transmission and reflection at boundaries within

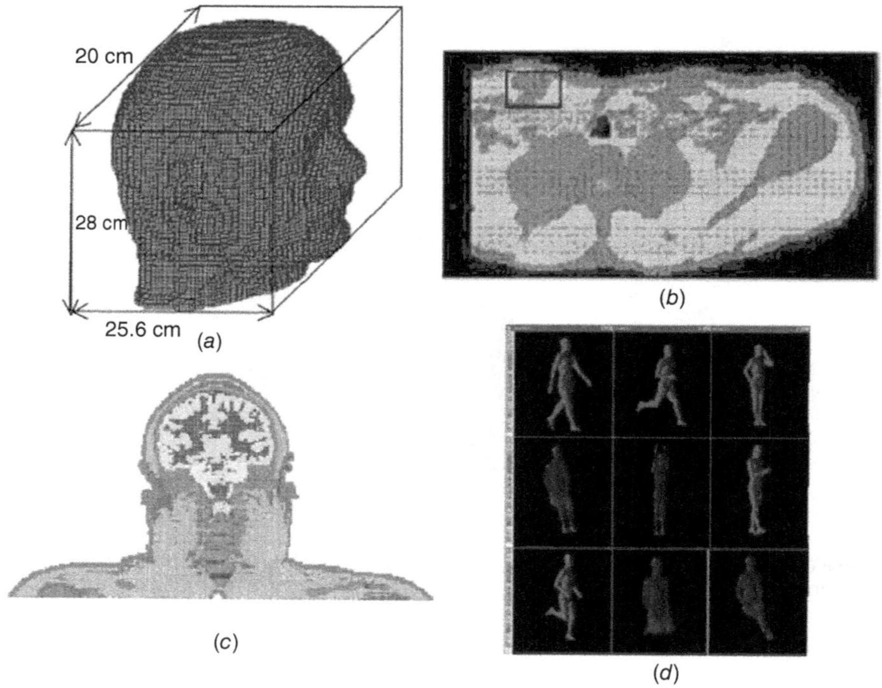

Figure 6.7 Anatomical phantom models used to model the human (*a*) head (Kim and Rahmat-Samii, 2004), (*b*) upper shoulder (Soontornpipit et al., 2004), (*c*) head with shoulders (Chen et al., 2009), and (*d*) body (Wang et al., 2009) in recent biomedical telemetry studies. (*See insert for color representation of c and d.*)

a three-layered (skin, fat, and muscle) planar model (Schwan and Li, 1956). Two quantities were considered to analyze EM radiation inside the human body and the resultant heat development: (a) the percentage of airborne EM energy absorbed by the body and (b) the distribution of heat sources in skin, subcutaneous fat, and deeper situated tissues. Analysis of the irradiation of spherical head models using spherical vector wavefunctions has also been reported (Weil, 1975). The head models consisted of a core of brain-like material surrounded by five outer layers of CSF, bone, fat, and skin-dura tissues, while their outer radii ranged in size from 2 to 12.5 cm. The distribution of internally deposited energy was also investigated for three basic spheres with radii of 3.3, 6, and 10 cm, to study the creation of hot spots (i.e., localized regions of strong heating).

The problem of interaction between a homogeneous sphere and a simple waveform (e.g., plane wave, short dipole, etc.) has been treated using analytical methods (Lin, 1976). The transmitted field strengths in homogeneous spherical models of human and animal heads were determined as a function of time and position using frequency analytic techniques. Constant conductivity was assumed for the electrical behavior of the brain matter. Zhou and Oosterom (1992) focused on the evaluation of the potential distribution inside spherical or spheroid volumes. A quasi-static approximation was used because of the low-frequency range of interest, but the method was applied to layered anisotropic media.

Nowadays, analytical methods are still used to find the exact solutions in simplified geometries and verify the accuracy of simulation results. For example, Kim and Sahmat-Ramii (2004) modeled the human head as a six-layer lossy dielectric sphere and implemented a spherical dyadic Green's function (DGF) code to characterize a dipole antenna implanted inside the human head. Gupta et al., (2008) used the DGF in a cylindrical human body model to obtain a simplified channel model. Four possible cases were considered, where the transmitter and receiver were either inside or outside the body, and an exact analytical expression was derived for the case where both the transmitter and receiver were placed out of the body.

Despite being restricted to very simple configurations, analytical methods significantly contribute to qualitative analyses. They are particularly useful to provide an insight into the physical mechanisms of EM propagation and interaction with biological tissues, identify the structure resonant frequencies that represent conditions of maximum power deposition inside the human body, evaluate the effect of dielectric and geometric parameters spread, and test the accuracy and performance of numerical codes.

Numerical Methods Numerical methods involve numerical solutions to Maxwell's equations subject to a set of initial or boundary values and are generally implemented on powerful computing platforms. They can handle complex geometries and provide a physical insight into the EM performance of the simulated systems. The constant evolution of computer systems (e.g., parallel systems) offers new possibilities for the execution of numerical codes with high computing requirements, thus facilitating more realistic and accurate modeling.

Several numerical techniques are available, with each using different forms of Maxwell's equations and employing different methods for their solution, thus each having its own advantages and drawbacks for particular applications. Among them, four have been most commonly used in bioelectromagnetics: the method of moments (MoM) (Harrington, 1968), the finite-element method (FEM) (Silvester and Ferrari, 1996), the finite-difference time-domain (FDTD) method (Yee, 1966), the transmission line matrix (TLM) method (Christopoulos, 1995), and the multiple-multipole (MMP) method (Hafner, 1990). Hybrid methods derived from the combination of these methods and other methods for EM propagation characterization are also used.

(a) Method of Moments The MoM was introduced by Harrington (1968) in a very general formulation. With this method, the problem is initially formulated in terms of integral equations obtained by using Green's functions. After appropriate choice of basis and weighting functions, these equations are reduced to a system of linear equations and solved.

The MoM is very efficient in modeling thin-wire structures and scattering from perfectly conducting objects. A significant advantage of this method is that only the structure in question (and not free-space) is discretized, thus limiting computational space, and boundary conditions do not have to be set.

On the other hand, the MoM is not effective for modeling arbitrarily shaped configurations, while memory requirements scale in proportion to the size of the problem, thus making EM characterization of physically and geometrically complex objects very difficult (or impossible) and time consuming. As a result, the MoM is not generally suitable for simulating the interactions between biomedical telemetry systems and the human body. The problem size grows so fast that the method does not allow a reasonably fine discretization of the human body. This also limits its application at frequencies higher than several hundred megahertz because smaller size blocks are required for higher frequencies.

Despite its drawbacks, MoM numerical solutions in bioelectromagnetics problems do appear in the literature. For example, Ito et al., (1992) and Chuang and Chen (1997) used the MoM to investigate the effect of the human body on small loop antennas in pager systems. Ito et al. (1992) approximated the human body as a conducting reflecting surface, for simplicity reasons. The influence of a reflector on the input impedance of a small loop antenna was examined both experimentally and theoretically, and results obtained by using the MoM were compared with experimental data to confirm the validity of the numerical method. Chuang and Chen (1997) adopted a crude squared model of the human body and assessed its effect on the performance of a circular-loop-wire antenna, which simulated the pager antenna. The MoM was employed to study the antenna characteristics and body absorption at 152, 280, and 400 MHz. More recently, Psychoudakis et al. (2008) considered a realistic homogeneous body model to analyze the performance of a body-worn diversity antenna for communications in the 225–380 MHz band. MoM techniques were employed to estimate optimum positioning and orientation for improved channel capacity, and results were validated through measurements for a human body phantom.

(b) Finite-Element Method The FEM was introduced in 1996 by Silvester and Ferrari (1996). It is based on the discretization, or, equivalently, meshing, of the EM problem into a number of elements of various shapes (usually tetrahedral) and sizes. Small elements are used to describe complex geometries, while larger elements are used in uniform regions.

The field equations are determined in terms of polynomials (interpolation functions) with unknown coefficients defined on the mesh nodes (corners of the elements) or along the element edges. In the edge-element technique the unknowns are associated with the fields that are tangential to the edges of the basic element. Within each element, vector expansion functions are used that are tangential to all element surfaces of which the edge is part and normal to all other surfaces. In contrast to the node-based FE technique, the edge-element technique possesses a direct way of controlling discretization errors in the solution by looking at the behavior of the normal components of the field between elements. Furthermore, the edge-element technique is more appropriate to model heterogeneous problems because there is no need to enforce any internal boundary conditions between elements with different electrical properties.

The major advantage of the FEM is that the electrical and geometric properties of each element can be defined independently. The FE mesh gives, in principle, very high flexibility in discretizing complicated geometries and is very efficient in modeling curved structures and arbitrarily shaped dielectric regions. As a result, the FEM has been widely used in bioelectromagnetics.

For example, Renhart et al. (1994) adopted the FEM to calculate eddy currents in the human body when magnetic resonance techniques were employed for diagnosis at the frequency of 64 MHz. A model was introduced that was feasible for FEM techniques without neglecting the most important organic influences on the eddy currents. The model was further analyzed, and the derived simulation results were compared with measurements obtained by MRI. Regarding functional electrical stimulation (FES), FEM models have frequently been used to find the parameters that have the biggest influence on nerve activation during stimulation. Kuhn and Keller (2006), used FE modeling combined with an active nerve model to assess the influence of the human anatomy and tissue properties on the electrical stimulation performance. Both the resistivity and permittivity of tissues were taken into account, and emphasis was given to the influence of muscle permittivity. Wegmueller et al. (2007) investigated the dependence of galvanic coupling on geometry and tissue parameters through FEM simulations with a mesh size of 150,000–200,000 elements. Simulation results showed that an increase in distance by 5 cm between transmitter and receiver increased the attenuation by 6–9 dB. Joints were found to increase the attenuation by up to 8 dB, with larger joints resulting in higher attenuation factors. Finally, Weiss et al. (2009) reported FE simulation results on the RF coupling of a 433 MHz near-field biotelemetry link for an artificial hip joint (titanium implant). Power losses were discussed as a function of patient size and antenna positioning, and accuracy of the simulation results was verified through experimental investigations in live porcine tissue.

The main drawback of the FEM is its relatively high computational complexity. Furthermore, the generation of meshes in three dimensions is a formidable task. While there are reasonably good techniques for the discretization of technical structures, the difficulty of generating FEMs for the typically very heterogeneous problems in dosimetry currently prevents wider use.

(c) Finite-Difference Time-Domain Method The FDTD is a powerful method for solving Maxwell's equations in three-dimensional space and time for the six-vector components of the electric and magnetic fields (Yee, 1966). Space is discretized in the form of cubic cells (dimensions of $\Delta x, \Delta y, \Delta z$), known as the Yee cells. Electric field (E) and magnetic field (H) components form the edges, and the normals to the faces of the cells, respectively, as shown in Figure 6.8. Simulation time is divided into time interval steps (Δt), and Maxwell's equations are solved iteratively in the cells as time is stepped forward. The electric and magnetic field components in the grid are updated in a leapfrog scheme using the finite-difference form of the curl operators on the fields that surround each component. The field updating process stops when the field quantities over the computational volume reach a steady state.

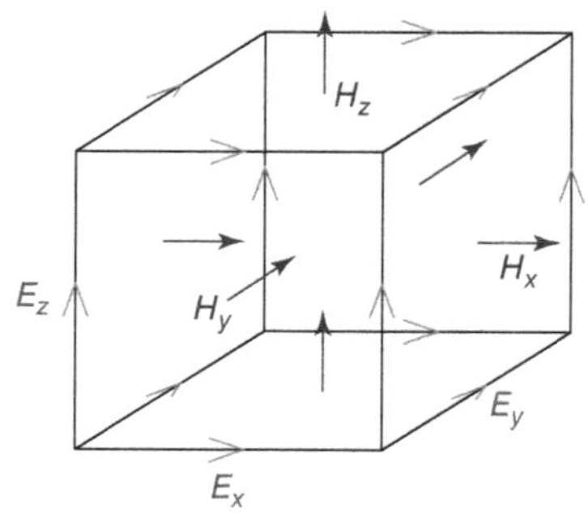

Figure 6.8 The Yee cell of the FDTD method.

The FDTD has the following main advantages:

- It is very versatile in analyzing complex geometries and can efficiently model arbitrary configurations of heterogeneous structures (e.g., the human body in various postures) with high spatial resolution.
- It demonstrates high computational efficiency because it requires only $O(N)$ computer complexity, where N is the number of the unknowns, and there is no need to invert large matrices. This means that the total memory storage required as well as the computational time are directly related to the number of field unknowns in the problem space (which in turn depends on the electrical size of the problem space and space resolution). For comparison, the MoM and the FEM need $O(N)^2$ computational storage.
- Because it is a time-domain technique, wideband results (i.e., results in a broad frequency range) can be obtained with a single simulation, simply by performing a fast Fourier transform on the time-domain response.
- There is no intrinsic upper bound to the number of unknowns it can solve.

On the other hand, the following points need to be carefully addressed in FDTD modeling:

- The entire computational domain has to be meshed. Cells must be smaller than the smallest wavelength divided by 10, and smaller than the smallest feature in the model. This greatly increases the number of cells for which calculations must be made, thus requiring a large amount of computer memory. In bioelectromagnetics, the smallest wavelength is generally found in the tissue with the highest water content at the highest frequency of interest.
- An upper bound on the time step (Δt) is required to ensure stability. Stability issues have thoroughly been discussed in Taflove and Hagness (2000). Once the cell size has been selected, the maximum time step is determined by the Courant

stability condition, which leads to a final upper bound given by

$$\Delta t \leq \left(c\sqrt{\frac{1}{(1/\Delta x)^2 + (1/\Delta y)^2 + (1/\Delta z)^2}} \right)^{-1} \tag{6.8}$$

- A very fine, dense, grid (and, therefore, a small time step) is required in modeling curved structures, sharp edges, or small geometric features, which in turn drastically increases computational time and memory required. Although non-uniform meshing and sub-gridding techniques can be applied, they may result in spurious solutions or suffer from instability (Monk and Suli, 1994).
- Difficulties may be encountered in modeling antenna structures not conforming to the used grid, such as helical antennas (Cavagnaro and Pisa, 1996; Nikita et al., 2000).

Nevertheless, the FDTD is being considered as the most suitable technique for modeling EM wave propagation inside and around the human body and is, currently, the most prevalent computational method in bioelectromagnetics. This is due to the fact that enhanced computer power has become available at lower cost, while intensive research work in the fields of source modeling, absorbing boundary conditions, and error checking has facilitated simulations with more confidence in the obtained results. The body model is discretized into Yee cells, and electrical properties are assigned to each cell to model human tissues before calculating the electric and magnetic fields as a function of time.

One of the first studies is that of Sullivan et al. (1987), who showed the capacity of the FDTD to compute EM energy absorption in human tissues. The FDTD method was evaluated by comparing its results to analytic solutions in two and three dimensions. The results obtained demonstrated that the FDTD method is capable of calculating internal dosimetric quantities with high accuracy. Since the early 1990s, the FDTD has been widely applied to the study of safety issues in mobile communications.

Regarding biomedical telemetry applications, the FDTD has more recently been used to study the radiation from implantable devices and evaluate the behavior of body-worn antennas. Chen et al. (2009) carried out FDTD investigations to characterize the transmission of RF signals between loop antennas placed in free space and implanted under the skull. The study assessed the effects of the human head on the transmission of RF signals between the implanted and exterior antennas, and simulation results were verified through measurements with a phantom. Computational models based on the FDTD were developed in Zhao et al. (2006) and Wang et al. (2009), with a single-cell Hertzian dipole being used as the excitation source, with the aim to provide a physical insight of the on-body radio channel. More specifically, Zhao et al. (2006) presented modeling for the ultrawideband on-body radio channel, and divided the frequency band between 3 and 9 GHz into 12 sub-bands to take into account the material frequency dispersion. Each subband was simulated separately, and a combination technique was subsequently employed (Zhao et al., 2006).

Wang et al. (2009) modeled on-body channels by simulating various body postures. Based on the FDTD simulation results, an on-body propagation model was derived and the model parameters for some representative transmission links on the human body were determined. Experimental measurements were performed to verify the simulation results for the chest-to-right-waist transmission link, and model parameters were found to highly agree between the two approaches (Wang et al., 2009). Finally, Kiourti et al. used the FDTD to study the resonance, radiation, and safety performance of a miniature antenna intended for integration in wireless head-implantable biomedical telemetry devices (Kiourti and Nikita, 2012, 2013; Kiourti et al., 2012).

(d) Transmission Line Matrix Method The TLM method provides discretization in both time and space (Christopoulos, 1995). Unlike other time-domain numerical methods, which are based on the direct discretization of Maxwell's equations, the TLM embodies Huygen's principle in discretized form. The method is based on the analogy between the electromagnetic field and a mesh of transmission lines, and, thus, considers the computational domain as a mesh of transmission lines, interconnected at nodes. Electric and magnetic fields are made equivalent to voltages and currents on the network, respectively. The mesh is initially excited by voltage impulses at specific points, which further propagate into the mesh while bouncing at the boundaries and being scattered by the nodes.

The TLM has proven to be one of the most powerful time-domain methods, especially for the computation of complex three-dimensional electromagnetic structures. Recently, this numerical method is being applied in the field of biomedical telemetry to assess the performance of advanced wearable liquid antennas, that is, antennas that are based on engineering the properties of liquids to communicate information about a person's health condition (body temperature, tissue damage, tumors, inflammation, and breathing). For example, in Traille et al. (2008) a bracelet-type liquid antenna consisting of an aqueous salt solution ($\varepsilon_r = 38$) inside a very thin plastic tubing container was characterized through numerous TLM calculations.

(e) Multiple-Multipole Method The MMP method is based on the generalized multipole technique (GMT) in which the fields are expanded as a linear superposition of basis functions (Hafner, 1990). As the name of the method suggests, the multipolar functions are the most used basis functions. These are obtained from the Helmholtz equation by separation of variables in spherical coordinates. The region where the fields are to be computed is divided into domains in which the material parameters are constant. The electric and magnetic fields are expanded as a linear superposition of known analytical solutions and are computed by minimizing the errors in the fulfillment of the boundary conditions.

The MMP method offers a high degree of freedom in the selection of the multipole basis functions, thus providing a high degree of accuracy. Furthermore, it offers the possibility of estimating the quality of the solution found, through evaluating the residual errors at the interfaces of the domains. Because of its efficiency, accuracy, and validation capability, the MMP method is especially suited to handle high-gradient fields in the vicinity of biological bodies. On the other hand, positioning of the multipolar function origins is difficult, giving rise to the development of several procedures for locating the multipoles (or auxiliary sources) in a systematic way. Moreover, MMP shows strong limitations for scatterers with complicated angular shapes or inhomogeneous bodies. In these cases, modeling of only the homogeneities that are of interest is generally preferred.

(f) Hybrid Methods Based on the above, none of the numerical methods and commercially available codes seems to be completely satisfactory for all cases of interest. A way to overcome the problems and drawbacks specific to each method is to use hybrid techniques. Combining existing methods into new hybrid approaches allows for highly efficient and robust numerical methods for application-specific scenarios, while keeping all the advantages and eliminating the disadvantages of component methods.

For example, a hybrid FEM/MoM method was employed by Meyer et al. (2003) to evaluate safety issues in the near field of a GSM base station antenna. The FEM was used to model the human phantom, while the MoM was used to model the metallic surfaces and wires of the antenna. The mathematical formulation of the hybrid technique was presented, and a discussion was performed on implementation details. The performance of the proposed hybrid FEM/MoM method was further validated by comparing results to MoM and FDTD solutions of human exposure problems.

Another typical hybrid technique is the combination of the MoM and the FDTD method. The method permits the computationally efficient FDTD method to model dielectrics, and the MoM, which represents conducting structures more accurately, to model antenna structures. It is, thus, particularly useful for analyzing complex problems that involve coupling between antennas and dielectric volumes (e.g., the human body). A hybrid FDTD/MoM technique was used by Chen et al. (1998) and Abd-Alhameed et al. (2005) to assess safety due to exposure to the radiation of an MRI transmit coil and base station antennas operating at 1800 MHz, respectively. Chen et al. (1998) computed the EM fields of shielded RF coils loaded with an anatomically human head model for high-frequency MRI applications and found that the proposed hybrid technique could accurately predict dosimetric quantitites excited by the RF coils. Finally, Abd-Alhameed et al. (2005) used a hybrid FDTD/MoM technique to verify the compliance of safety standards for mobile communications base stations and compared its performance with the traditional power density method.

(g) Boundary Conditions Most EM problems entail a structure whose behavior is to be studied in an infinite unbounded computational space. Because of the limited computer resources, infinite geometries cannot be simulated, and, thus, a method of terminating the computation domain is required. This is achieved by adding absorbing boundary conditions (ABCs) at the outside of the simulation space, which effectively simulate unbounded regions (infinite space) by minimizing reflections in the boundaries.

ABCs can be achieved in a number of ways. For example, Engquist and Majda (1977) described a one-way equation approach that numerically absorbs waves impinging on the boundaries. Mur (1981) described a finite-difference scheme, and Liao et al. (1984) described extrapolation of the wave fields in space and time using a backward difference polynomial. Another typical ABC is the perfectly matched layer (PML) in which layers of a material with both electric and magnetic loss (electric and magnetic conductivity) are introduced at the boundaries of the mesh to effectively suppress reflections and truncate the computational region (Taflove, 1995).

For example, Soontornpipit et al. (2004) discussed the design of a microstrip patch antenna for communication with medical implants. The FDTD method was used to evaluate the microstrip antenna design parameters. The edge of the cubic voxels was set to 1 mm, and Mur absorbing boundaries were placed 10 cells away from the antenna model. Kiourti et al. (2011, 2012, 2013; Kiourti and Nikita, 2012) conducted FDTD simulations to study the resonance and radiation performance of an implantable antenna operating in the medical implant communications service (MICS) band of 402–405 MHz for integration in wireless biotelemetry devices implanted in the human head. Free space surrounded the simulation set-ups by 200 mm, and Liao absorbing conditions were assumed at the boundaries to extend radiation infinitely far into space.

6.4 PHYSICAL MODELING

Numerical simulations do have some shortcomings. The computational time is usually very long, and it is generally not easy to simulate realistic conditions and include the effects of the surrounding environment. Measurements can be performed instead or as a means of validating the simulation results.

6.4.1 Physical Phantoms

In biomedical telemetry applications, experimental investigations can be performed with either real human subjects or physical phantoms. Physical phantoms are defined as surrogates of the human body that have electrical properties (relative permittivity and conductivity) equivalent to those of biological tissues.

Especially in safety testing where internal fields need to be measured, or in the case of wireless implants, measurements cannot be performed in human subjects by non-invasive methods, and physical phantoms become an essential tool. Furthermore, the use of phantoms can provide a stable and controllable EM environment, which cannot be easily realized with human subjects.

Tissue Material Formulation Physical phantoms consist of biological tissue-simulant materials, that is, materials in liquid, gel, or solid state that mimic the electrical properties of real biological tissue. Several recipes have been proposed to produce such materials accounting for different types of tissue. Furthermore, since it is not possible to produce a valid approximation for a wide frequency range by using a single formula, separate recipes are proposed for different operation frequency scenarios.

Liquid phantoms consist of an outer thin shell with very low RF absorption (usually made from fiberglass or other plastic material) that is further filled with a tissue-simulant liquid material. For frequencies in the range of 0.8–3 GHz, the relative permittivity of the outer shell should be less than 5, its loss tangent less than 0.05, and its thickness should be in the range of 2.0 ± 2 mm (Kanda et al., 2004). Liquid phantoms are very easy to prepare and their electrical properties are easily adjustable; most recipes are based on deionized water to which sugar and salt are added to adjust the relative permittivity and conductivity values, respectively.

Given their advantages, liquid phantoms have been widely used in the measurement of internal fields in safety studies, as well as in implantable antenna testing. For example, a liquid phantom made of water, sugar, salt and hydroxyethyl cellulose (HEC) (compound for adjusting the viscosity) was presented in Hartsgrove et al. (1987). A liquid UWB phantom material was developed in Hara and Kobayashi, (2005), and Liu et al. (2008) tested a skin tissue-implantable antenna immersed inside a skin-tissue-simulating fluid made of fruit sugar, salt, cellulose, and deionized water.

However, liquid phantoms suffer from the limited frequency range over which they have the desired electrical properties, the presence of the outer shell, their adequacy to simulate only high-water-content tissues, the instability of electrical properties (mainly due to water evaporation and growth of fungi), and the inaccuracy in human body modeling (since the internal structure is approximated as homogeneous).

Gel phantoms have the advantage of covering a wider frequency range than liquid phantoms, while they allow more accurate modeling of the human body by placing one on top of the other layers of gels mimicking different types of tissues. Similarly to liquid phantoms, they are very easy to prepare and their electrical properties are easily adjustable by appropriately adjusting the salt and sugar concentrations in the mixture.

Gel phantoms are formed by adding coagulants to the liquid solution. A common gelling agent is TX-150. Guy (1968) presented a recipe for a gel phantom composed of water, sodium chloride, TX-150, and polyethylene powder. Based on this recipe, Ito et al. (2001) developed a self-shaping phantom. Karacolak et al. (2008) added agarose (a linear polysaccharide) to a liquid mixture made of deionized water, sugar, and salt to fabricate skin-mimicking gels. The effects of sugar and salt on the relative permittivity and conductivity were further investigated, and it was found that the relative permittivity significantly decreases and conductivity slightly increases with increased sugar and salt concentration. Finally, Sani et al. (2010) added gelatin into liquid mixtures to convert them into gels and fabricate a three-layer phantom with layers representative of skin, fat, and muscle at 868 MHz.

However, similarly to liquid phantoms, gel phantoms are adequate only for high-water-content tissue modeling, and their electrical properties highly degrade over time (mainly due to water evaporation and growth of fungi).

Solid phantoms are made from materials that keep their shape over time, and they have the advantages of high accuracy in modeling the heterogeneous human body, fine mechanical stability, and minimized degradation over time (due to absence of water).

Several recipes have been proposed in the literature for solid phantoms. These include mixtures of Laminac 4110 (a polyester resin), acetylene black, and aluminum powder (Cheung and Koopman, 1976), flour, oil, and saline (Lagendijk and Nilsson, 1985), graphite powder and ceramic (Tamura et al., 1997), silicone rubber and carbon fiber (Nikawa et al., 1996), polyethyl methacrylate and carbon black (Chang et al., 2000), as well as agar, deionized water, polyethelene powder, sodium chloride, TX-151, and sodium dehydroacetate (Onishi et al., 2005). Solid phantoms are mainly used for analyzing propagation around the body; however, measurement of internal fields can also be performed by the method of thermography (Kobayashi et al., 1993).

On the other hand, fabrication of solid phantoms is generally more expensive and requires more complex and skilled production procedures as compared with liquid or gel phantoms.

Canonical Models Canonically shaped physical phantoms, that is phantoms with simplified geometries, have widely been used for biomedical telemetry applications in the literature, with special emphasis on the testing of tissue-implantable antennas.

Rectangular and cylindrical containers filled with skin- (Figure 6.9*a*) and muscle- (Figure 6.9*b*) tissue mimicking materials have been used to test prototypes of planar inverted-F and 3D spiral implantable antennas, respectively (Karacolak et al., 2009; Abadia et al., 2009). Karacolak et al. (2009) described the design and test of a miniature dual-band implantable antenna operating in the medical implant communications service (MICS) (402–405 MHz) and industrial, scientific, and medical

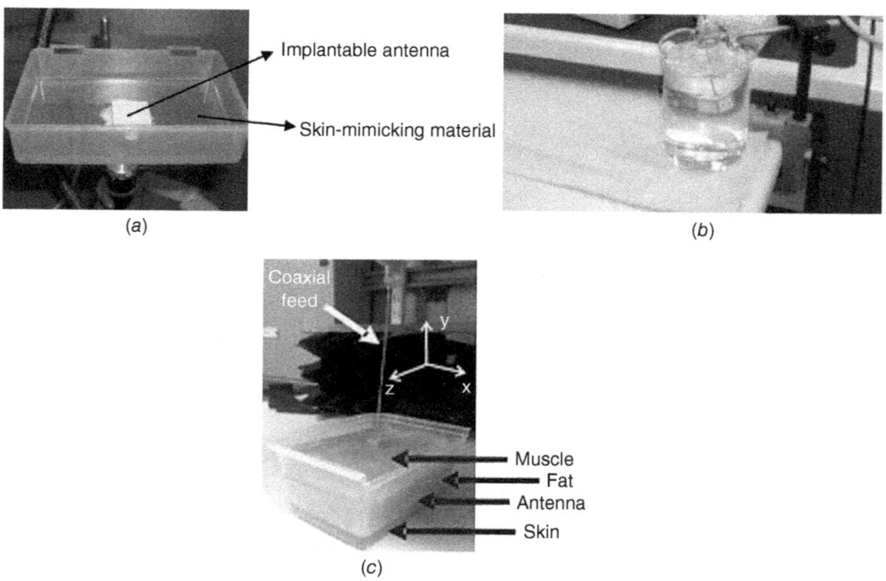

Figure 6.9 Canonical physical phantoms proposed in (*a*) Karacolak et al. (2009), (*b*) Abadia et al. (2009), and (*c*) Sani et al. (2010). (*See insert for color representation of the figure.*)

(ISM) (2.4–2.48 GHz) bands to be used in animal studies for medical research, while Abadia et al. (2009) presented the design and realization procedure of a MICS band 3D spiral radiator. Finally, a canonical three-layer phantom with layers representative of skin, fat, and muscle has been fabricated to investigate the radiation performance of an implantable antenna intended for RFID applications at 868 MHz (Figure 6.9*c*) (Sani et al., 2010).

Anatomical Models To provide realism in experimental modeling, several realistically shaped models have been developed that are filled with one or several different tissue-simulant liquids/gels to represent specific parts and tissues of the human body.

The most widely used homogeneous head phantom is the Specific Anthropomorphic Mannequin (SAM), which has been proposed by the Institute of Electrical and Electronics Engineers (IEEE, 2003) and the International Electrotechnical Commission (IEC, 2005) standards for compliance testing (Figure 6.10*a*). This phantom has also been adopted by the European Committee for Electrotechnical Standardization (CENELEC, 2001), the Association of Radio Industries and Businesses in Japan (ARIB, 2002), and the U.S. Federal Communications Commission (FCC, 1997). Prior to the introduction of the SAM phantom by the standardization bodies, the

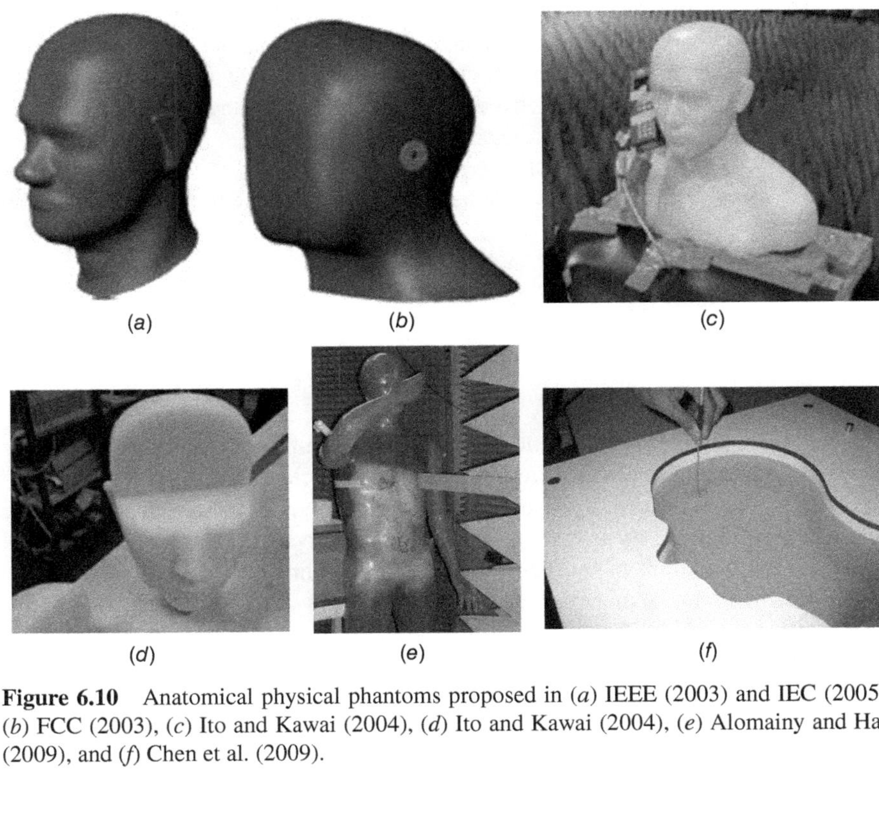

(*a*) (*b*) (*c*) (*d*) (*e*) (*f*)

Figure 6.10 Anatomical physical phantoms proposed in (*a*) IEEE (2003) and IEC (2005), (*b*) FCC (2003), (*c*) Ito and Kawai (2004), (*d*) Ito and Kawai (2004), (*e*) Alomainy and Hao (2009), and (*f*) Chen et al. (2009).

Generic Twin Phantom was used, and several cellular phone models were authorized based on measurements with this phantom (Figure 6.10*b*) (FCC, 2003). Figure 6.10*c* shows a realistic-shaped upper-half body phantom for evaluating a future mobile satellite phone at 2.6 GHz, while Figure 6.10*d* shows the phantom for evaluation of the internal specific absorption rate (SAR) distribution (Ito and Kawai, 2004). A human phantom of approximately 1.7 m in height and 0.35 m in average width, with animal organs used to represent human tissues, has also been used as a radio propagation measurement setup (Alomainy and Hao, 2009) (Figure 6.10*e*). Finally, an anatomical head phantom filled with gray-matter-simulant liquid has been used to measure the transmission coefficient between implantable and exterior loop antennas (Chen et al., 2009) (Figure 6.10*f*).

6.4.2 Experimental Equipment and Measurements

Experimental investigations for biomedical telemetry applications mainly include measurements of:

1. The physical phantom's electrical properties (relative permittivity and conductivity)
2. The antennas' resonance characteristics (reflection coefficient frequency response, or, equivalently, resonance frequency and bandwidth)
3. Parameters that characterize propagation, transmission, and quality of communication (e.g., transmission coefficient, channel frequency response, etc.) in the wireless channel (formed between the medical device and monitoring/control equipment placed at a short distance)

Some of the most commonly applications used in the literature to perform such measurements are outlined below.

Measurement of Electrical Properties Even when tissue-simulant recipes are followed exactly, complex dielectric properties (or, equivalently relative permittivity and conductivity) of the fabricated material must be measured in order to ensure agreement with the intended theoretical values for the tissue to be modeled at the desired operation frequency.

Measurement and characterization of the relative permittivity and conductivity has attracted significant scientific interest not only in bioelectromagnetics but also in several other research and development fields including material science and microwave circuit design. Several measurement techniques have been developed [e.g., coaxial probe, slotted line, transverse electromagnetic (TEM) line, etc.] with each applying to specific materials, frequency ranges, and applications (Burdette et al., 1980; IEEE, 2001).

The most commonly used technique to measure the complex dielectric properties of physical phantoms is the open-ended coaxial probe technique, which is a broadband, nondestructive and noninvasive measurement method. This technique is based on the fact that the reflection coefficient of an open-ended coaxial probe

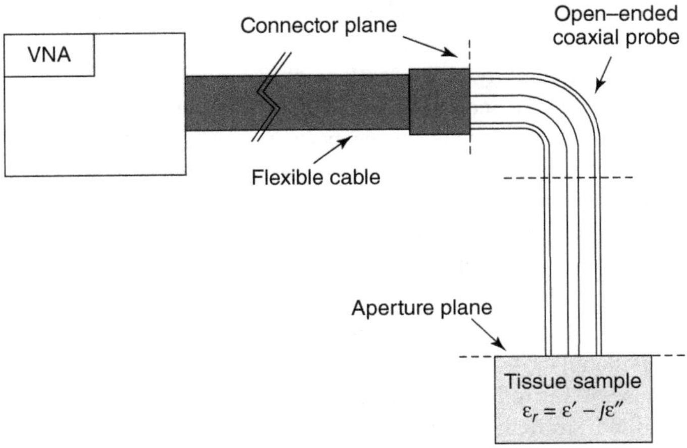

Figure 6.11 Open-ended coaxial probe technique for complex dielectric parameter measurement.

depends on the electrical parameters of the material that is attached to it. As shown in Figure 6.11, an open-ended coaxial probe attached to a vector network analyzer (VNA) is immersed into the liquid, or pressed against the solid specimen, and the reflection coefficient is measured. The VNA is calibrated at the probe aperture plane. Measured reflection coefficient values are subsequently postprocessed and converted into relative permittivity and conductivity values.

Calibration at the probe aperture plane can be performed in two ways:

- In the first way, reference liquids with known dielectric properties are used (e.g., water, saline, and methanol). The probe is directly calibrated at the aperture plane, by placing the short, open, and reference liquid standards at the end of the probe. This method is very direct and simple; however, it suffers from uncertainties in the selection and characterization of the reference liquids.
- In the second way, the VNA is initially calibrated at the connector plane by applying the short, open, and matched standards, and complex coefficient data referenced to the connector plane are recorded. A simulation (de-embedding) model of the open-ended probe is then used to compensate propagation in the probe and make the translation to the probe aperture. Measurement accuracy with this method highly depends on the physical characteristics of the probe's aperture. A rational function model (RFM) is finally applied at the probe aperture-referenced reflection coefficients to calculate the complex permittivity of the sample.

Recently, with the advance of new technologies, measurement of the complex dielectric parameters can easily be performed with commercially available dielectric probe kits. These consist of a VNA and an integrated software program to measure the complex reflection and transmission coefficients and directly convert them into electrical properties data. For example, Karacolak et al. (2008) performed dielectric constant and conductivity measurements of tissue-simulant gels using Agilent's 85070E dielectric probe kit and a network analyzer. The same technique was also used by the authors to measure the dielectric properties of 25 mm × 25 mm × 5 mm rat skin samples extracted from the middorsal area of rats (Karacolak et al., 2009). Samples were measured between 200 MHz and 20 GHz for 1800 frequency points, and 4 consecutive measurements were performed and averaged to ensure data reliability. The experimental set-ups are shown in Figure 6.12*a* and Figure 6.12*b*, respectively.

Measurement of Antenna Resonance Characteristics Impedance mismatch between an antenna and its feeding circuit causes reflection at the feed point of the antenna. Because of this reflection not all of the available power reaches the antenna, and thus the field strength of the radiated EM wave is reduced. The reflection coefficient (S_{11}) of an antenna describes this phenomenon and is defined as

$$S_{11} = \frac{Z_a - Z_0}{Z_a + Z_0} \tag{6.9}$$

where Z_a is the impedance of the antenna, and Z_0 is the impedance of the transmission feed line.

The reflection coefficient frequency response can easily be measured by means of a network analyzer. Before the measurements, the network analyzer should be calibrated for a suitable frequency range containing the operation frequency band of the antenna. To measure the reflection at the feed point of the antenna, a coaxial cable with an Sub-Miniature A (SMA) connector in one end is generally used. The cable is soldered to the feed point of the antenna, and the connector is connected to the

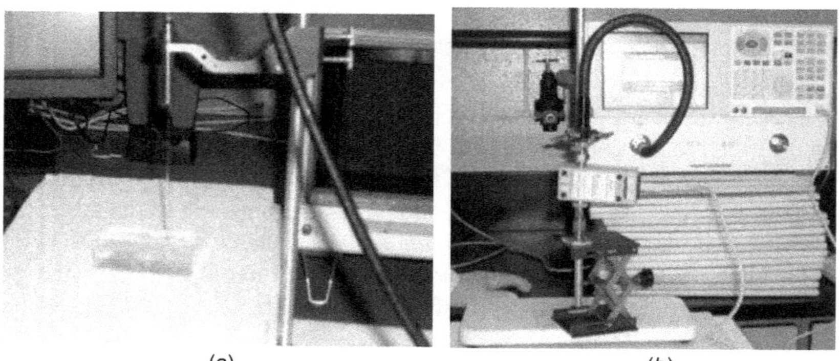

(*a*) (*b*)

Figure 6.12 Experimental setups for measuring the complex dielectric parameters of (*a*) skin-mimicking gels (Karacolak et al., 2008), and (*b*) rat skin samples (Karacolak et al., 2009) by means of a dielectric probe kit.

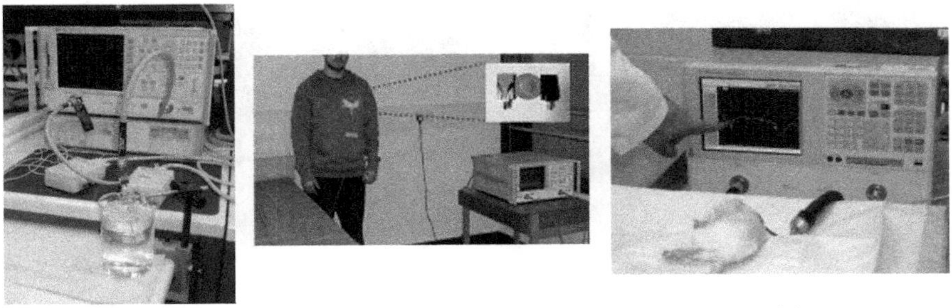

(a) (b) (c)

Figure 6.13 Experimental setups for measuring resonance characteristics of antennas (*a*) immersed in a tissue-simulant liquid (Abadia et al., 2009), (*b*) worn by a human subject (Alomainy et al., 2009), and (*c*) implanted inside a test animal (Karacolak et al., 2010).

network analyzer. In order to measure the reflection coefficient, the network analyzer transmits a small amount of power to the antenna and measures the reflected power.

For example, Abadia et al. (2009) built a prototype of their proposed muscle-implantable 3D spiral antenna and measured the magnitude of the reflection coefficient with a network analyzer (Figure 6.13*a*). A choke was used to reduce the current flow on the cable connecting to the network analyzer. Alomainy et al. (2009) measured UWB body-worn antennas placed directly on a tight cotton jumper worn by a human subject (Figure 6.13*b*). Finally, Karacolak et al. (2010) performed in vivo reflection coefficient measurements of implantable antennas using rats as model animals (Figure 6.13*c*). A stab wound allowed the coaxial cable attached to the antenna to exit the skin, and reflection coefficient measurements were performed immediately after the implantation surgery.

Specific Absorption Rate Measurements The dosimetric quantity most commonly used to determine the interaction of EM fields with human tissue is the specific absorption rate (SAR). The SAR represents power deposition in the body and is defined as

$$\mathrm{SAR} = \sigma \frac{|E|^2}{\rho} \tag{6.10}$$

where E is the root mean square (RMS) of the electric field (V/m), σ is the tissue conductivity (S/m), and ρ is the tissue density (kg/m^3). It is a measure of the power absorbed per unit mass of tissue and is expressed in watts per kilogram (W/kg).

The SAR can be experimentally measured by means of high-precision, multichannel exposition acquistion systems (EASY), as shown in Figure 6.14. Each measurement channel is characterized and processed individually, enabling high-precision isotropic measurements from lower MHz to above 10 GHz with a spatial resolution of a few millimeters. In the literature, there exists only a limited number of reported SAR measurements for biomedical telemetry systems. For example, Zhu and Langley (2009) performed SAR measurements for a dual-band textile wearable

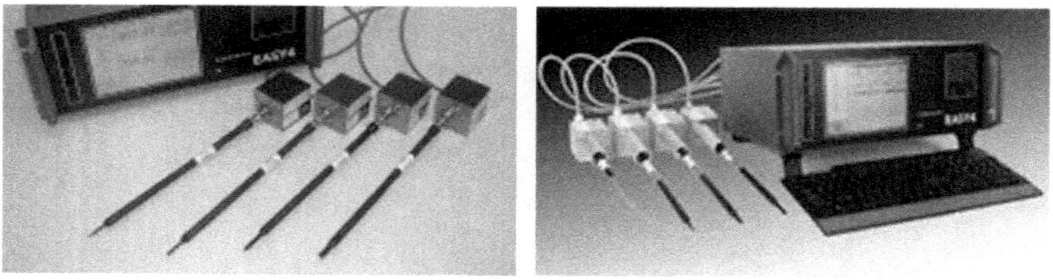

(a) (b)

Figure 6.14 (*a*) Portable four-channel exposure acquisition system and (*b*) portable four-channel exposure acquisition system hardened for use in hostile MRI environments up to 7 tesla.

antenna covering the 2.45-GHz and the 5-GHz wireless networking bands. Measured SAR values were found to be slightly higher than those of the simulations, which can be attributed to inaccuracies in phantom modeling and uncertainties in the realistic environment, as compared to that of the simulations (including the effects of connectors, cables, surrounding devices, and temperature).

Channel Measurements Several experimental investigations have been reported in the literature to assess propagation and transmission in wireless biomedical telemetry channels. Such channels are most commonly characterized by the transmission coefficient (S_{21}) between the antennas of the transmitter (biomedical telemetry device) and receiver (monitoring/control equipment placed at a short distance), which is defined so that

$$|S_{21}|^2 = \frac{P_r}{P_t} \tag{6.11}$$

where P_t is the power available at the transmitting port, and P_r is the power absorbed in a 50-Ω load terminating the receiving antenna.

Transmission coefficient measurements can be performed by connecting the transmitting and receiving antennas at the two ports of a VNA. The VNA measures the magnitude and phase of each frequency component of S_{21}, also known as the channel's frequency response. Such measurements often take place inside anechoic chambers, that is, chambers that are built from absorbers made of a high-quality, low-density lossy medium to provide a quiet zone needed to simulate the surrounding environment. Given the channel's frequency response, its time-domain response can subsequently be obtained by means of an inverse discrete fourier transform (IDFT), and time-domain channel characteristics (e.g., mean delay and RMS delay spread) can further be derived.

Several channel measurements have been reported in the literature for wearable, on-body antennas:

(a) UWB on-body propagation measurements have been performed in an anechoic chamber to characterize the channel between two pairs of different UWB antennas

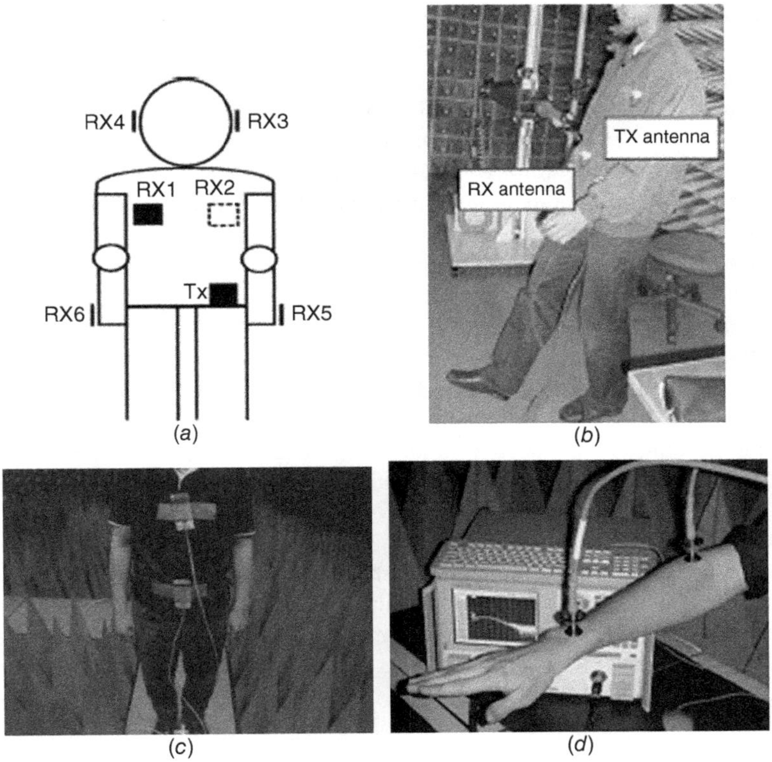

Figure 6.15 Channel measurement experimental setups found in (*a*) Alomainy et al. (2005), (*b*) Wang et al., (2009), (*c*) Takizawa et al. (2010), and (*d*) Lim et al. (2011).

[printed horn-shaped self-complementary antennas (HSCA) and planar inverted-cone antennas (PICA)] in terms of both frequency and time-domain response (Alomainy et al., 2005). Several positions between the transmitting (Tx) and receiving (Rx) antennas were investigated (Figure 6.15*a*).

(b) The chest-to-right-waist on-body link formed between two small-size, low-profile antennas mounted on the body has also been characterized (Wang et al., 2009). The transmitting and receiving antennas were fixed on the left side of the chest and the right side of the waist, respectively. The frequency-domain transfer function of the channel was measured in terms of a VNA and was further converted to the time domain by applying an IDFT. Measurements were conducted for 8 persons and 10 body postures per person, and the average delay profile of the channel was derived from the average of all 80 readings. Measurement antennas mounted on the body can be seen in Figure 6.15*b*.

(c) The performance of wearable wireless body area networks (WBANs) has been evaluated during walking motion (Takizawa et al., 2010). Received signal strength, packet error rate, and bit error rate were measured in an anechoic chamber and in an office room. Measurements were based on an Agilent Connected Solution using ADS Ptolemy. The signal waveform was calculated in the ADS Ptolemy installed in a laptop PC, and then the waveform was uploaded into a vector signal generator connected to a body-worn antenna. The receiving body-worn antenna received the transmitted signal, which was subsequently digitized at a spectrum analyzer and fed back to the ADS Ptolemy through ether cable (Figure 6.15*c*).

(d) Finally, propagation path loss measurements have been performed by considering two discone antennas attached to the arm of a volunteer (Lim et al., 2011). The transmitter was fixed on the wrist, while the receiver was positioned at different distances, as shown in Figure 6.15*d*. Antennas were connected to a calibrated VNA by two separate 50-Ω coaxial cables.

Channel measurements for ingestible and implantable biomedical telemetry systems have also been performed, as detailed below:

(a) The biotelemetric channel from ingestible implants has been measured by assuming a 4-cm-long monopole (without the ground plane) probe to be used as both the transmitting and receiving antennas (Figure 6.16*a*) (Alomainy and Hao, 2009). A near-field scanning system was used to obtain path loss data at different distances from the body.

(b) Transmission coefficient measurements have been conducted between an implantable antenna resting against a scalp phantom and a linearly polarized chip antenna used as a receiving (probing) antenna, by means of a full two-port calibrated VNA (Warty et al., 2008). A schematic of the experimental setup is shown in Figure 6.16*b*.

(c) Finally, measurements of $|S_{21}|$ in a mock biotelemetry link in porcine tissue have been performed to validate simulation results of a near-field biotelemetry link for an artificial hip joint. Both the implanted and external antennas were assumed to be stripped coax monopoles with a 3-cm-long exposed inner conductor (Weiss et al., 2009). The experimental setup is shown in Figure 6.16*c*.

6.5 SAFETY ISSUES

The biological effects of the emitted RF EM radiation can be divided into three categories: thermal, athermal, and nonthermal effects. Thermal effects cause tissue heating since the EM energy absorbed by the human tissues increases the molecular translational and rotational kinetic energy. In the case of athermal effects, even though the amount of absorbed EM energy is capable of heating the tissues, the temperature of the tissues does not increase because of the body thermoregulation mechanisms. Finally, non-thermal effects comprise complicated interactions between live cells and ions (calcium, potassium, etc.) and are related to the behavior of big molecules (proteins and DNA).

The effects of EM radiation in the human body depend not only on the field level but also on its operation frequency. Biomedical telemetry systems only produce nonionizing radiation. The photon energy of nonionizing radiation is low enough to break molecules and can only cause biological hazards in terms of thermal effects, that

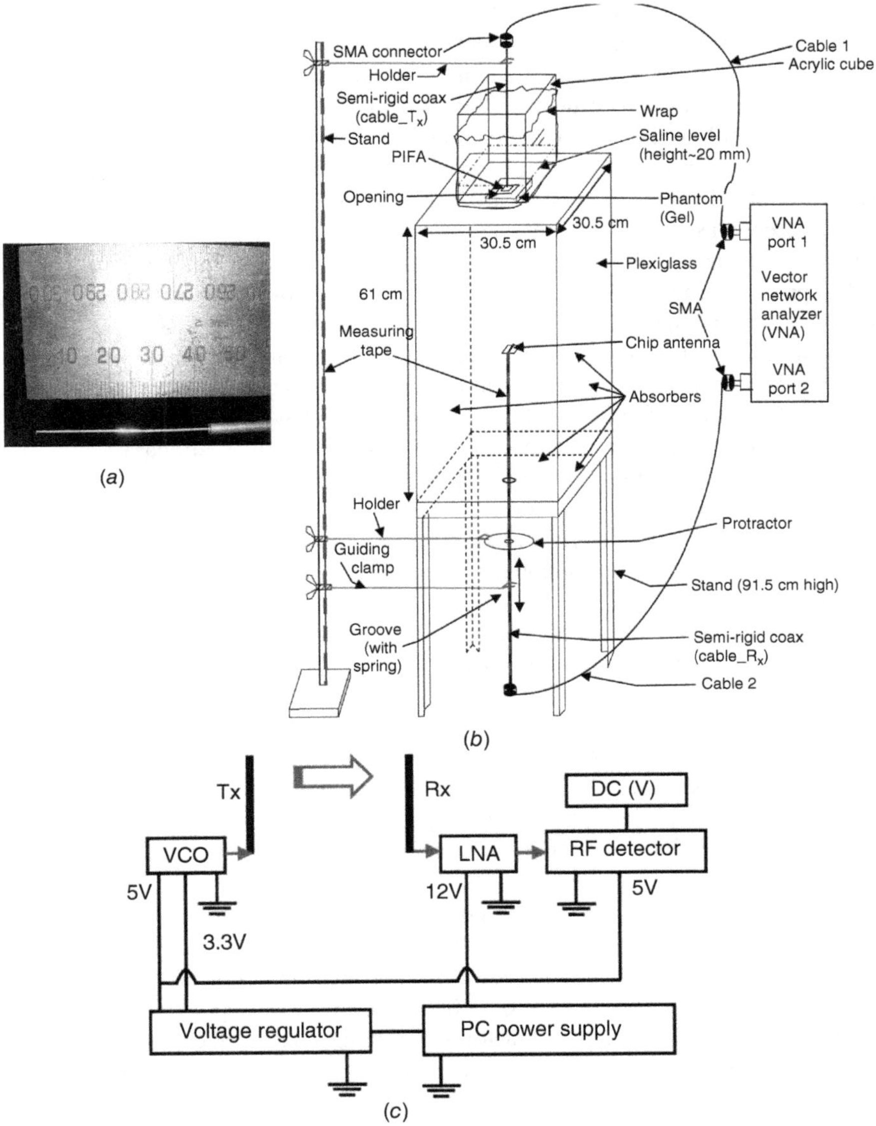

Figure 6.16 (*a*) Probe used as transmitter and receiver in the measurements of Alomainy and Hao (2009), and channel measurement experimental setups found in (b) Warty et al. (2008) and (*c*) Weiss et al. (2009).

is, tissue heating as energy of the photons is transformed into kinetic energy of the absorbing molecules. As EM waves propagate through the human body, their energy is absorbed by tissues and produces heating resulting from both ionic conduction and vibration of the dipole molecules of the water and the protein in tissue cells.

Dosimetry is necessary to "evaluate the dose" or, equivalently, to identify the dose metric that is closely related to the effect of concern. The proliferation of biomedical telemetry applications has given rise to national and international guidelines to determine exposure limits and ensure their safe use. Furthermore, considerable dosimetric research efforts have been devoted to assess the interactions between the EM radiation emitted by biomedical telemetry devices and the human body. These efforts have been motivated by three factors: (1) the need to evaluate potential health effects and compliance with standards (compliance testing), (2) the need to verify if existing protection standards are still adequate, and (3) the need to assess antenna performance and improve antenna design to minimize the energy absorbed in the human body while maximizing the radiated energy.

So far, all recommendations and regulations regarding the limits on allowable absorbed power in the body are based on quantitative short-term evaluation of the thermal effects caused by EM fields. The two major standards relating to RF radiation have been set up by the IEEE (1999, 2005) and the International Commission on Non-Ionizing Radiation Protection (ICNIRP, 1998), while several other standards exist as well [IEC, 2002; National Radiological Protection Board (NRPB), 2004; Health Protection Agency (HPA), 2008]. Countries can also set their own national standards, which are most commonly based on the guidelines set by international organizations.

Acceptable levels of radiation are typically expressed in terms of maximum permission exposures and SAR values averaged over specifically defined volumes of tissue. For example, by following the computational procedures recommended by IEEE, the average SAR is computed over 1-g (IEEE, 1999) or 10-g (IEEE, 2005) cubical volumes of voxels where no face of the averaging volume is external to the body. In cases where the cubical volume rule cannot be satisfied (e.g., at the surface of the body), special rules apply for setting the SAR value in a given voxel. On the contrary, the ICNIRP (1998) defines volumes of contiguous tissue, not necessarily in the shape of a cube.

Table 6.1 summarizes the most important SAR limits for uncontrolled exposure to mobile portable devices between 100 kHz and 6 GHz in the United States (IEEE, 1999, 2005) and Europe (ICNIRP, 1998). Uncontrolled environments are those where a person has no knowledge or control of his exposure (general public). Uncontrolled exposure is generally reduced by a factor of 5 from the controlled exposure. This happens because occupationally exposed population consists of adults who are trained to be aware of potential risks and to take appropriate precautions, while the general public comprises individuals of all ages and of varying health statuses, and may include particularly susceptible groups or individuals. A partial body average has been developed to deal with partial exposure to the sensitive head and eyes.

Finally, it is worth noting that new guidelines for limiting exposure to time-varying electric, magnetic, and electromagnetic fields have been developed to account for safety issues in the advanced biomedical telemetry technology of intrabody communications (currents propagating through the body). Table 6.2 summarizes ranges of threshold currents for the general public as set by the European Committee for Electrotechnical Standardization (ICNIRP, 2002). In general, it has been shown that

TABLE 6.1 IEEE and ICNIRP SAR Limits for Uncontrolled Exposure to Mobile Portable Devices between 100 kHz and 6 GHz

	IEEE C95.1-1999 (IEEE, 1999) (W/kg)	IEEE C95.1-2005 (IEEE, 2005) (W/kg)	ICNIRP (ICNIRP, 1998) (W/kg)
Whole-body average	0.08	0.08	0.08
Partial-body average	1.6	2	2
	Averaged over 1 g of tissue in the shape of a cube	Averaged over 10 g of tissue in the shape of a cube	Averaged over 10 g of contiguous tissue

TABLE 6.2 Ranges of Threshold Currents for General Public

	Threshold Current (mA)			
Effect	50/60 Hz	1 kHz	100 kHz	1 MHz
Touch perception	0.2–0.4	0.4–0.8	25–40	24–40
Pain on finger contact	0.9–1.8	1.6–3.3	33–55	28–50
Painful shock	8–16	12–24	112–224	
Severe shock	12–23	21–41	160–320	

Source: ICNIRP (2002).

threshold currents, which produce perception and pain, vary little over the frequency range of 100 kHz to 1 MHz, while they are unlikely to vary significantly over the frequency range up to about 110 MHz.

6.6 CONCLUSION

In this chapter, a summary was provided regarding the numerical and experimental techniques used to model and analyze EM wave propagation in biomedical telemetry systems. Canonical and anatomical models used for computational exposure assessment were presented, while phantoms used for experimental investigations were described. Safety issues were also addressed.

Analytical methods can be applied to simplified canonical geometries modeling the human body (or parts of it) and allow detailed calculation of the EM performance. However, numerical techniques implemented in computer codes running in powerful PCs prevail today to provide fast and accurate results, with the FDTD method dominating numerical modeling in bioelectromagnetics. Several anatomical models have been developed for subjects of varying age and ethnicity, with 1 mm resolution and higher being available in some cases.

In order to take into account the effects of the surrounding environment, or validate the simulation results, experimental investigations can also be performed. A range of liquid, gel, and solid canonical and anatomical physical phantoms have been developed for this purpose that account for specific types of tissue at given operation frequencies. Emphasis is most commonly given on measurements of electrical parameters, antenna reflection coefficient, and channel transmission coefficient, which can easily be performed by means of network analyzers and other commercially available equipment.

Finally, in order to ensure human safety, recommendations have been set by professional bodies and government agencies on the maximum allowable SAR levels. However, the health hazards occurring from long-term exposure to these new technologies are far from settled. All organizations agree that with any new technology it is a sensible precautionary approach to keep the situation under ongoing review.

REFERENCES

Abadia J, Merli F, Zurcher JF, Mosig JR, Skrivervik AK. 2009. 3D-spiral small antenna design and realization for biomedical telemetry in the MICS band. *Radioengineering*, 18(4):359–367.

Abd-Alhameed RA, Excell PS, Mangoud MA. 2005. Computation of specific absorption rate in the human body due to base-station antennas using a hybrid formulation. *IEEE Trans. Electromag. Compat.*, 47:374–381.

Ackerman MJ. 1998. The visible human project. *Proc IEEE*, 86(3):504–511.

Alomainy A, Hao Y. 2009. Modeling and characterization of biotelemetric radio channel from ingested implants considering organ contents. *IEEE Trans. Antennas Propag.*, 57(4):999–1005.

Alomainy A, Hao Y, Parini CG, Hall PS. 2005. Comparison between two different antennas for UWB on-body propagation measurements. *IEEE Antennas Wireless Propag. Lett.*, 4:31–34.

Alomainy A, Sani A, Rahman A, Santas JG, Hao Y. 2009. Transient characteristics of wear able antennas and radio propagation channels for ultrawideband body-centric wireless communications. *IEEE Trans. Antennas Propag.*, 57(4):875–884.

Association of Radio Industries and Businesses in Japan (ARIB). 2002. Specific Absorption Rate (SAR) Estimation for Cellular Phone, ARIB STD-T56.

Bao JZ, Lu ST, Hurt WD. 1997. Complex dielectric measurements and analysis of brain tissues in the radio and microwave frequencies. *IEEE Trans. Microw. Theory Tech.*, 45(10):1730–1741.

Burdette E, Cain F, Seals J. 1980. In vivo probe measurement technique for determining dielectric properties at VHF through microwave frequencies. *IEEE Trans. Microw. Theory Tech.*, 8 (4):414–427.

Campbell AM, Land DV. 1992. Dielectric properties of female human breast tissue measured in vitro at 3.2 GHz. *Phys. Med. Biol.*, 37:193–210.

Cavagnaro M, Pisa S. 1996. Simulation of cellular phone antennas by using inductive lumped elements in the 3D-FDTD Algorithm. *Microw. Opt. Technol. Lett.* 13(6):324–326.

Chang JT, et al. 2000. A conductive plastic for simulating biological tissue at microwave frequencies. *IEEE Trans. Electromagn. Compat.*, 42:76–81.

Chen J, Feng Z, Jin JM. 1998. Numerical simulation of SAR and B1-field inhomogeneity of shielded RF coils loaded with the human head. *IEEE Trans. Biomed. Eng.*, 45:650–659.

Chen ZN, Liu GC, See TSP. 2009. Transmission of RF signals between MICS loop antennas in free space and implanted in the human head. *IEEE Trans. Antennas Propag.*, 57(6):1850–1853.

Cheung AY, Koopman DW. 1976. Experimental development of simulated biomaterialsfor dosimetry studies of hazardous microwave radiation. *IEEE Trans. Microw. Theory Tech.*, 24:669–673.

Christopoulos C. 1995. *The Transmission Line Modeling Method: TLM*. Piscataway, NJ, IEEE Press.

Chuang H, Chen W. 1997. Computer simulation of the human-body effects on a circular loop-wire antenna for radio-pager communications at 152, 280, and 400 MHz. *IEEE Trans. Vehicular Tech.*, 46(3):544–559.

Curto S, Ammann MJ. 2007. Electromagnetic coupling mechanism in a layered human tissue model as reference for 434 MHz RF medical therapy applicators. Proc. IEEE Antennas and Propag. Soc. Int. Symp., Honolulu, pp. 3185–3188.

Dimbylow PJ. 2002. Fine resolution calculations of SAR in the human body for frequencies up to 3 GHz. *Phys. Med. Biol.*, 47:2835–2846.

Engquist B, Majda A. 1977. Absorbing boundary conditions for the numerical simulation of waves. *Proc. Natl. Acad. Sci.*, 74:1765–1766.

European Committee for Electrotechnical Standardization (CENELEC). 2001. Basic Standard for the Measurement of Specific Absorption Rate Related to Exposure to Electromagnetic Fields from Mobile Phones (300 MHz–3 GHz), EN 50361.

Federal Communications Commission (FCC). 1997. Federal Communications Commission (FCC) evaluating compliance with FCC guidelines for human exposure to radio frequency electromagnetic fields. Supplement C to *OET Bulletin 65* (Edition 9701), Washington, DC.

Federal Communications Commission (FCC). 2003. Equipment authorization system generic search. Available http://gullfoss2.fcc.gov/prod/oet/cf/eas/reports/GenericSearch.cfm.

Gabriel C, Gabriel S, Corthout E. 1996. The dielectric properties of biological tissues. *Phys. Med. Biol.*, 41:2231–2293.

Gandhi OP, Lazzi G, Furse CM. 1996. Electromagnetic absorption in the human head and neck for mobile telephones at 835 and 1900 MHz. *IEEE Trans. Microw. Theory Tech.* 44:1884–1897.

Gupta A, Abhayapala TD. 2008. Body area networks: Radio channel modelling and propagation characteristics. Australian Communications Theory Workshop, pp. 58–63.

Guy A. 1968. Analyses of electromagnetic fields induced in biological tissues by thermographic studies on equivalent phantom models. *IEEE Trans. Microw. Theory Tech.*, 19(2):205–214.

Hafner C. 1990. *The Generalized Multipole Technique for Computational Electromagnetic*. Boston: Artech House.

Hara D, Kobayashi T. 2005. Development of ultra wideband electromagnetic phantom materials for antennas and propagation studies. 13th Conf. Microw. Tech., pp. 64–67.

Harrington R. 1968. *Field Computation by Moment Method*. New York: Macmillan.

Hartsgrove G, Kraszewski A, Surowiec A. 1987. Simulated biological materials for electromagnetic radiation absorption studies. *Bioelectromagnetics*, 8: 29–36.

Health Protection Agency (HPA) 2008. *Protection of Patients and Volunteers Undergoing MRI Procedures*. Chilton: Health Protection Agency.

International Electrotechnical Commission (IEC). 2005. *IEC Standard 62209-1. Human Exposure to Radio Frequency Fields from Hand-Held and Body-Mounted Wireless Communication Devices–Human Models, Instrumentations, and Procedures*. IEC.

Institute of Electrical and Electronics Engineers (IEEE). 1999. C95.1-1999 IEEE Standard for Safety Levels with respect to Human Exposure to Radio Frequency Electromagnetic Fields, 3 kHz to 300 GHz. New York: IEEE.

Institute of Electrical and Electronics Engineers (IEEE). 2001. Draft: Recommended practice for determining the spatial-peak specific absorption rate (SAR) in the human body due to wireless communications devices: experimental techniques, p. 1528.

Institute of Electrical and Electronics Engineers (IEEE) Standard 1528 SCC34. 2003. IEEE recommended practice for determining the peak spatial-average specific absorption rate (SAR) in the human head from wireless communications devices: measurement techniques.

Institute of Electrical and Electronics Engineers (IEEE). 2005. C95.1-2005 IEEE Standard for Safety Levels with respect to Human Exposure to Radio Frequency Electromagnetic Fields, 3 kHz to 300 GHz. New York: IEEE.

International Commission on Non-Ionising Radiation Protection (ICNIRP). 1998. Guidelines for limiting exposure to time-varying electric, magnetic, and electromagnetic fields (up to 300 GHz). *Health Phys.* 74:494–522.

International Commission on Non-Ionizing Radiation Protection (ICNIRP). 2002. Guidelines for limiting exposure to time-varying electric, magnetic, and electromagnetic fields (up to 300 GHz), 513.

International Electrotechnical Commission (IEC). 2002. *Medical Electrical Equipment: Part 2. Particular Requirements for the Safety of Magnetic Resonance Equipment for Medical Diagnosis*. Geneva: IEC.

Ito K, Furuya K, Okano Y, Hamada L. 2001. Development and characteristics of a biological tissue-equivalent phantom for microwaves. *Electron Commun. in Japan* (Part I: Communications), 84(4).

Ito K, Ida I, Wu M. 1992. Body effect on characteristics of small loop antenna in pager systems. *IEEE Antennas Propag. Society Int. Symp.*, pp. 1081–1084.

Ito K, Kawai H. 2004. Phantoms for evaluation of interactions between antennas and human body. URSI EMTS International Symposium, pp. 1104–1106.

Kanda M, Ballen M, Salins S, Chou C, Balzano Q. 2004. Formulation and characterization of tissue equivalent liquids used for RF densitometry and dosimetry measurements. *IEEE Trans. Microw. Theory Tech.*, 52(8):2046–2056.

Karacolak T, Cooper R, Butler J, Fisher S, Topsakal E. 2010. In vivo verification of implantable antennas using rats as model animals. *IEEE Antennas Propag. Lett.*, 9(10):334–337.

Karacolak T, Cooper R, Topsakal E. 2009. Electrical properties of rat skin and design of implantable antennas for medical wireless telemetry. *IEEE Trans. Antennas Propag.*, 57(9):2806–2812.

Karacolak T, Hood AZ, Topsakal E. 2008. Design of a dual-band implantable antenna and development of skin mimicking gels for continuous glucose monitoring. *IEEE Trans. Microw. Theory Techn.*, 56(4):1001–1008.

Kim J, Rahmat-Samii Y. 2004. Implanted antennas inside a human body: Simulations, designs, and characterizations. *IEEE Trans. Microw. Theory Tech.*, 52(8):1934–1943.

Kiourti A, Christopoulou M, Nikita KS. 2011. Performance of a novel miniature antenna implanted in the human head for wireless biotelemetry. *IEEE APS Int. Symp. Antennas Propag.*, pp. 392–395.

Kiourti A, Costa JR, Fernandes CA, Santiago AG, Nikita KS. 2012. Miniature implantable antennas for biomedical telemetry: From simulation to realization. *IEEE Trans. Biomed. Eng.*, 59(11):3140–3147.

Kiourti A, Nikita KS. 2012. Miniature scalp-implantable antennas for telemetry in the MICS and ISM bands: design, safety considerations and link budget analysis. *IEEE Trans. Antennas Propag.*, 60(6):3568–3575.

Kiourti A, Nikita KS. 2013. Numerical assessment of the performance of a scalp-implantable antenna: Effects of head anatomy and dielectric parameters. *Wiley Bioelectrom.*, 34(3):167–179.

Kiourti A, Psathas KA, Costa JR, Fernandes CA, Nikita KS 2013. Dual-band implantable antennas for medical telemetry: A fast design methodology and validation for intracranial pressure monitoring. *Prog. Electrom. Res.*, 141:161–183.

Kobayashi T, Nojima T, Yamada K, Uebayashi S. 1993. Dry phantom composed of ceramics and its application to SAR estimation. *IEEE Trans. Microw. Theory Tech.*, 41(1):136–140.

Koulouridis S, Nikita KS. 2004. Study of the coupling between human heads and cellular phone helical antennas. *IEEE Trans. Electromagn. Compat.* 46:62–70.

Kuhn A, Keller T. 2006. The influence of capacitive properties on nerve activation in transcutaneous electrical stimulation. Proc. Int. Symp. Comput. Methods Biomech. Biomed. Eng., p. 7.

Kuhn S, Cabot E, Christ A, Capstick M, Kuster N. 2009. Assessment of the radio-frequency electromagnetic fields induced in the human body from mobile phones used with hands-free kits. *Phys. Med. Biol.*, 54:5493–5508.

Lagendijk JJW, Nilsson P. 1985. Hyperthermia dough: A fat and bone equivalent phantom to test microwave/radiofrequency hyperthermia heating systems. *Phys. Med. Biol.*, 30:709–712.

Lazebnik M, et al. 2007. A large-scale study of the ultrawideband microwave dielectric properties of normal breast tissue obtained from reduction surgeries. *Phys. Med. Biol.*, 52:2637–2656.

Liao Z P, Wong H L, Yang B, Yuan Y. 1984. A transmitting boundary for transient wave analysis. *Sci. Sin. A*, 27:1063–1076.

Lim HB, Baumann D, Li EP. 2011. A human body model for efficient numerical characterization of UWB signal propagation in wireless body area networks. *IEEE Trans. Biomed. Eng.*, 58(3):689–697.

Lin JC. 1976. Electromagnetic pulse interaction with mammalian cranial structures. *IEEE Trans. Biomed. Eng.* 23:61–65.

Liu WC, Chen SH, Wu, CM. 2008. Implantable broadband circular stacked PIFA antenna for biotelemetry communication. *J. Electromagn. Waves Appl.*, 22:1791–1800.

Meyer FJC, et al. 2003. Human exposure assessment in the near field of GSM base-station antennas using a hybrid finite element/method of moments technique. *IEEE Trans. Biomed. Eng.*, 50(2):224–233.

Monk P, Suli E. 1994. Error estimates for yee's method on non-uniform grids. *IEEE Trans. Magn.*, 30:3200–3203.

Mur G. 1981. Absorbing boundary conditions for the finite-difference approximation of the time-domain electromagnetic-field equations. *IEEE Trans. Electromag. Compat.*, 23:377–382.

Nagaoka T, Watanabe S, Sakurai K, Kunieda E, Watanabe W, Taki M, Yamanaka Y. 2004. Development of realistic high-resolution whole-body voxel models of Japanese adult males and females of average height and weight, and application of models to radio-frequency electromagnetic-field dosimetry. *Phys. Med. Biol.*, 49(4):1–15.

National Radiological Protection Board (NRPB). 2004. Advice on limiting exposure to electromagnetic fields (0–300 GHz). Doc. NRPB 15, 1–35.

Nikawa Y, Chino M, Kikuchi K. 1996. Soft and dry phantom modeling material using silicone rubber withcarbon fiber. *IEEE Trans. Microw. Theory Tech.*, 44(10):1949–1953.

Nikita KS, Cavagnaro M, Bernardi P, Uzunoglu NK, Pisa S, Piuzzi E, Sahalos JN, Krikelas GI, Vaul JA, Excell PS, Cerri G, Chiarandini S, De Leo R, Russo P. 2000. A study of uncertainties in modeling antenna performance and power absorption in the head of a cellular phone user. *IEEE Trans. Microw. Theory Tech.* 48:2676–2685.

Onishi T, Ishido R, Takimoto T. 2005. Biological tissue-equivalent agar-based solid phantoms and SAR estimation using the thermographic method in the range of 3–6 GHz. *IEICE Trans. Commun.*, E88-B(9):3733–3741.

O'Rourke APO, Lazebnik M, Bertram JM, Converse MC, Hagness SC, Webster JG, Mahvi DM. 2007. Dielectric properties of human normal, malignant and cirrhotic liver tissue: in vivo and ex vivo measurements from 0.5 to 20 GHz using a precision open-ended coaxial probe. *Phys. Med. Biol.*, 52:4707–4719.

Peyman A, Gabriel C, Grant EH, Vermeeren G, Martens L. 2009. Variation of the dielectric properties of tissues with age: The effect on the values of SAR in children when exposed to walkie-talkie devices. *Phys. Med. Biol.* 54:227–241.

Peyman A, Holden SJ, Watts S, Perrott R, Gabriel C. 2007. Dielectric properties of porcine brain tissue in the transition from life to death at frequencies from 800 to 1900 MHz. *Phys. Med. Biol.*, 52:2229–2245.

Peyman A, Rezazadeh A, Gabriel C. 2001. Changes in the dielectric properties of rat tissue as a function of age at microwave frequencies. *Phys. Med. Biol.*, 46:1617–1629.

Psychoudakis D, Lee G, Chen C, Volakis J. 2008. Body-worn diversity antennas for squad area networks (SAN). URSI General Assembly.

Renhart W, et al. 1994. Modeling and calculation of influences of RF-fields on the human body using the finite element method. *IEEE Trans. Mag.*, 30(5):3092–3095.

Sani A, Rajab M, Foster R, Hao Y. 2010. Antennas and propagation of implanted RFIDs for pervasive healthcare applications. *Proc IEEE*, 98(9):1648–1655.

Schmid G, Neubauner G, Udo MI, Francois A. 2003. Dielectric properties of porcine brain tissue in the transition from life to death at frequencies from 800 to 1900 MHz. *Bioelectromagnetics*, 24:413–422.

Schwan HP, Li K. 1956. Hazards due to total body irradiation by radar. *Proc. IRE*, 44:1572–1581

Silvester P, Ferrari R. 1996. *Finite Elements for Electrical Engineers*. Cambridge University Press.

Smith SR, Foster KR. 1985. Dielectric properties of low-water-content tissues. *Phys. Med. Biol.*, 30:965–973.

Soontornpipit P, Furse CM, Chung YC. 2004. Design of implantable microstrip antenna for communication with medical implants. *IEEE Trans. Microw. Theory Tech.*, 32(8):1944–1951.

Sullivan D. 1990. Three-dimensional computer simulation in deep regional hyperthermia using the finite-difference time-domain method. *IEEE Trans. Microw. Theory Tech.*, 38:204–211.

Sullivan DM, Borup DT, Gandhi OP. 1987. Use of the finite-difference time-domain method in calculating EM absorption in human tissues. *IEEE Trans. Biomed. Eng.*, 34:148–157.

Taflove A. 1995. *Computational Electrodynamics*. London: Artech House.

Taflove A, Hagness S. 2000. *Computational Electrodynamics: The Finite Difference Time Domain Method*. London: Artech House.

Takizawa K, Watanabe K, Kumazawa M, Hamada Y, Ikegami T, Hamaguchi K. 2010. Performance evaluation of wearable wireless body area networks during walking motions in 444.5 MHz and 2450 MHz. 32nd Annu Int Conf IEEE EMBS, pp. 370–373.

Tamura H, Ishikawa Y, Kobayashi T, Nojima T. 1997. A dry phantom material composed of ceramic and graphite powder. *IEEE Trans. Electromagn. Compat.*, 39(2):132–137.

Traille A, Yang L, Rida A, Tentzeris M. 2008. A novel liquid antenna for wearable bio-monitoring applications. *IEEE MTT-S*, 923–926.

Wang Q, Tayamachi T, Kimura I, Wang J. 2009. An on-body channel model for UWB body area communications for various postures. *IEEE J. Sel. Areas Commun.*, 57(4):991–998.

Warty R, Tofighi MR, Kawoos U, Rosen A. 2008. Characterization of implantable antennas for intracranial pressure monitoring: Reflection by and transmission through a scalp phantom. *IEEE Trans. Microw. Theory Tech.*, 56(10):2366–2376.

Wegmueller MS, Kuhn A, Froehlich J, Oberle M, Felber N, Kuster N, Fichtner W. 2007. An attempt to model the human body as a communication channel. *IEEE Trans. Biomed. Eng.*, 54(10):1851–1857.

Weil CM. 1975. Absorption characteristics of multilayered sphere models exposed to UHF/microwave radiation. *IEEE Trans. Biomed. Eng.*, 6:468–476.

Weiss MD, Smith JL, Bach J. 2009. RF coupling in a 433-MHz biotelemetry system for an artificial hip. *IEEE Antennas Wireless Propag. Lett.*, 8:916–919.

Wiart J, Hadjem A, Gadi N, Bloch I, Wong MF, Pradier A, Lautru D, Hanna VF, Dale C. 2005. Modeling of RF head exposure in children. *Bioelectromagnetics* 26:S19–S30.

Wiart J, Hadjem A, Wong M, Bloch I. 2008. Analysis of RF exposure in the head tissues of children and adults. *Phys. Med. Biol.* 53:3681–3695.

Yee KS. 1966. Numerical solution of initial boundary value problems involving Maxwell' s equations in isotropic media. *IEEE Trans. Antennas Propag.*, 14(3):302–307.

Zhao Y, Hao Y, Alomainy A, Parini C. 2006. UWB on-body radio channel modeling using ray theory and subband FDTD method. *IEEE Trans. Microw. Theory Tech.*, 54(4):1827–1835.

Zhou H, Oosterom van A. 1992. Computation of the potential distribution in a four-layer anisotropic concentric spherical volume conductor. *IEEE Trans. Biomed. Eng.*, 39:154–158.

Zhu S, Langley R. 2009. Dual-band wearable textile antenna on an EBG substrate. *IEEE Trans. Antennas Propag.* 57(4):926–935.

Zubal IG, Harrell CR, Smith EO, Rattner Z, Gindi G, Hoffer PB. 1994. Computerized three-dimensional segmented human anatomy. *Med. Phys.* 21(2):299–302.

7

INDUCTIVE COUPLING

MAYSAM GHOVANLOO AND MEHDI KIANI

GT-Bionics Lab, School of Electrical and Computer Engineering, Georgia Institute of Technology, Atlanta, Georgia

7.1 INTRODUCTION

Near-field power transmission is a viable technique to wirelessly power up devices, such as sensors and actuators, with a wide range of power requirements or recharge their batteries from a short range without any direct electrical contact between the energy source and the device. Moreover, it is possible to use the same short-range wireless link to establish wide-band bidirectional data communication with those devices. Wireless IMDs are good examples of where near-field power and data transmission links can be used effectively. IMDs have been significantly improved by going through many generations since the invention of the first implantable pacemaker in 1958, and their importance in several state-of-the-art medical treatments is on the rise (Zhou and Greenbaum, 2009). They have made it possible to treat a wide range of ailments and disabilities from bradycardia (Allan, 2003) and chronic back pain to epilepsy (Morrel, 2011; Fisher, 2011) and deafness (Zeng et al., 2008; Rauschecker et al., 2002). IMDs have the potential to alleviate more challenging types of disabilities such as blindness (Weiland and Humayun, 2008; Chen et al., 2010; Shire et al., 2009; Terasawa et al., 2008), paralysis (Nicolelis, 2003; Schwartz et al., 2006), and loss of limbs (Kuiken et al., 2007). These devices need to transmit and receive information wirelessly across the skin barrier since breaching the skin with interconnect wires would be a source of morbidity for the patient and significantly increases the risk of infection. They also increase the risk of damage to the IMD.

Handbook of Biomedical Telemetry, First Edition. Edited by Konstantina S. Nikita.

In sensory prosthetic devices, which interface with the central nervous system (CNS) to restore a sensory function such as hearing or vision, the quality of perception enhances with the number of stimulating sites and electrodes and the rate of stimulation (Fernandes et al., 2012; Theogarajan, 2012). These devices may stimulate the neural tissue by means of tens to thousands of stimulating channels and they generally require considerably more power and communication bandwidth than autonomous devices, such as pacemakers. State-of-the-art visual prostheses are currently targeting beyond one thousand sites to improve the quality of the visual functions, such as mobility without a cane, face recognition, and reading large fonts (Mathieson et al., 2012).

Inductive links can be utilized to power up wireless neural recording systems for freely behaving small animal subjects (Fernandes et al., 2012). A key limitation of the current wireless neural recording systems that are developed for neuroscience applications is the need for the animal subject to carry a large payload of batteries for continuous recording over several hours or even days (Lee et al., 2010; Yin and Ghovanloo, 2011; Chae et al., 2009). This may not be a major issue for larger animal subjects such as non-human primates (Gao et al., 2012; Rizk et al., 2009). However, such a payload can affect the behavior of smaller animals such as rats and mice, which are more common in behavioral neuroscience labs due to their lower cost and easier animal husbandry.

Radio frequency identification (RFID) takes advantage of inductive links to not only power up the ultralow-power RFID tags, which cannot have batteries due to their size, weight, and lifetime limitations, but also read the tags, stored information through back telemetry (Finkenzeller, 2003). Because of the significant growth of hand-held, wireless, and mobile electronic devices and gadgets, such as smartphones, tablets, and ultramobile laptops, in recent years, efficient and safe wireless energy transfer to wirelessly charge such devices and cut the last cord has gained considerable attention (Karalis et al., 2007; Fernandez and Borras, 2001).

7.2 INDUCTION PRINCIPLES

The main physical principle behind the operation of telemetry coils is Faraday's law, which states that when the total magnetic flux through a conductive loop—defined as the integral over the surface enclosed by the loop of the magnetic flux—varies with time, a current is induced in the loop itself. This, in turn, results in an electromotive force (EMF) induced in the loop (Sadiku, 2007). Thus, a primary loop will generate the varying magnetic field, which concatenates with the secondary loop, resulting in an induced current in the secondary loop.

7.2.1 Magnetic Fields

A magnetic field is associated with any moving charged particles that result in flow of current. The magnitude of the magnetic field regardless of the material properties of the medium can be described by the magnetic field strength, which integrated along a closed curve that encircles the moving charges results in the total current that passes through the curve. For designing a telemetry link, a number of parameters that are directly associated with the magnetic field and its strength must be considered, such as self- and mutual inductances.

7.2.2 Inductance and Inductive Coupling

Self-inductance is the ratio of the magnetic flux generated in an area enclosed by a conductor loop to the current passing through the loop. Under the condition of $r/R \ll 1$, where r and R are the radii of the wire and the circular loop that it is forming, self-inductance can be approximated by

$$L(R, r) \approx \mu_0 R \left[\ln \left(\frac{8R}{r} \right) - 2 \right] \tag{7.1}$$

where μ_0 is the permeability of free space (Zierhofer and Hochmair, 1996).

For the case of circular coils with N turns, if the coil length d is much smaller than R, the self-inductance is approximately equal to N^2L, where L is the self-inductance of a one-turn loop in (7.1). Whereas, for the case of planar spiral coils having N turns with different radii R_i ($i = 1, 2, \ldots, N$) the total self-inductance should be calculated from

$$L = \sum_{i=1}^{N} L(R_i, r) + \sum_{i=1}^{i=N} \sum_{j=1}^{j=N} M_{ij}(R_i, R_j, d_r = 0)(1 - \alpha_{i,j}) \tag{7.2}$$

where $\alpha_{i,j} = 1$ if $i = j$ and $\alpha_{i,j} = 0$ otherwise (Zierhofer and Hochmair, 1996).

7.2.3 Mutually Coupled Coils

The mutual inductance between two conductor loops, $M_{ij,}$ depends on the proportion of the magnetic flux generated by one loop that passes through the other loop (flux coupling). Therefore, it highly depends on their geometries, relative orientation, and magnetic properties of the medium. In a simplified case for two perfectly aligned parallel coaxial circular coils in the air separated by relative distance d_{12} and the radius and number of turns of (R_1, N_1) and (R_2, N_2) for the first and second coils, respectively,

$$M_{12}(R_1, R_2, d_{12}) = \frac{\pi \mu_0 N_1 R_1^2 N_2 R_2^2}{2\sqrt{(R_1^2 + d_{12}^2)^3}} \tag{7.3}$$

In order to more accurately calculate self- and mutual inductances of coils with various geometries, one should either use tabulated parameterized equations (Grover, 1946; Terman, 1943) or finite-element electromagnetic software, such as FastHenry, SONET, or HFSS (Ansoft, Pittsburgh, PA). A detailed overview of the numerical and experimental bioelectromagnetic modeling tools and methods used in biomedical telemetry is covered in Chapter 6.

In addition to the distance and geometry, alignment of the coils has a significant effect on their mutual inductance. The effects of coil misalignments have been analyzed elsewhere (Soma et al., 1987). For example, if one of the coils is tilted by an angle θ, their mutual inductance reduces by a factor of $\cos(\theta)$,

$$M_{12}(d_{12}, \theta) = M_{12}(d_{12}, 0) \cos(\theta) \tag{7.4}$$

We can normalize the mutual inductance between two coils to get a qualitative sense of how strongly they are coupled and compare the coupling between different pairs of coils. The coupling coefficient k_{12} between two coils with self-inductances L_1 and L_2 is defined as

$$k_{12} = \frac{M_{12}}{\sqrt{L_1 \times L_2}} \qquad 0 \leq k_{12} \leq 1 \tag{7.5}$$

Assuming $R_2 < R_1$, the coupling coefficient in this case can be approximated by (Finkenzeller, 2003)

$$k_{12}(d_{12}) = \frac{R_1^2 R_2^2}{\sqrt{R_1 R_2}\left(\sqrt{(R_1^2 + d_{12}^2)}\right)^3} \tag{7.6}$$

As a result of (7.4), k_{12} also depends on the coil orientation and alignment,

$$k_{12}(d_{12}, \theta) = k_{12}(d_{12}) \cos(\theta) \tag{7.7}$$

According to (7.7), in order to minimize k_{12} between two planar coaxial coils at a certain distance d_{12}, their planes should be orthogonal. On the other hand, parallel and perfectly aligned coaxial coils provide maximum k_{12}.

7.2.4 Equivalent Network Models

In an inductive link, shown in Figure 7.1*a*, a time-variant current $i_2(t)$ in the primary coil, L_2, generates a time-variant magnetic field, part of which passes through the secondary coil, L_3. This part of the time-varying magnetic field generates voltage $V_3(t)$ across L_3 and current $i_3(t)$ through the secondary loop due to its mutual inductance M_{23} with L_2. The time-domain relationship between these voltages and currents can be found from

$$\begin{aligned} V_2(t) &= R_2 \cdot i_2(t) + L_2 \cdot di_2(t)/dt + M_{23} \cdot di_3(t)/dt \\ V_3(t) &= M_{23} \cdot di_2(t)/dt + R_3 \cdot i_3(t) + L_3 \cdot di_3(t)/dt \end{aligned} \tag{7.8}$$

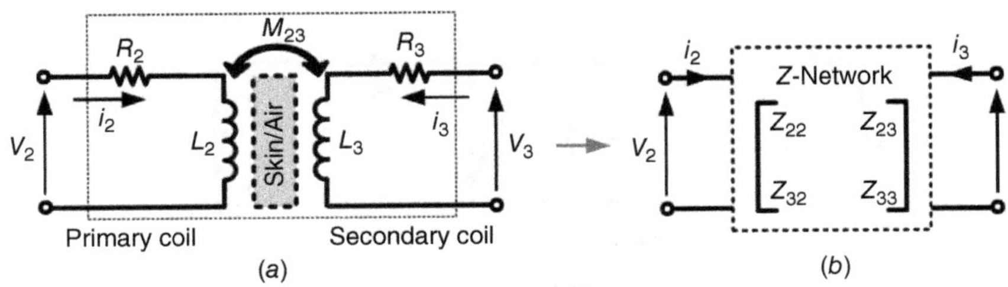

Figure 7.1 (*a*) Simplified circuit diagram of inductive link: R_2 and R_3 represent the ohmic losses of L_2 and L_3, respectively. (*b*) Equivalent *Z*-network model for inductive link.

where R_2 and R_3 are the ohmic losses of L_2 and L_3, respectively. In order to find the inductive link equivalent *Z*-network model, shown in Figure 7.1*b*, all voltages and currents in (7.8) should be represented in the Laplace domain,

$$V_2(s) = R_2 \cdot i_2(s) + L_2 \cdot si_2(s) + M_{23} \cdot si_3(s)$$
$$V_3(s) = M_{23} \cdot si_2(s) + R_3 \cdot i_3(s) + L_3 \cdot si_3(s) \tag{7.9}$$

Therefore, the equivalent *Z* matrix can be written as

$$\begin{bmatrix} V_2(s) \\ V_3(s) \end{bmatrix} = \begin{bmatrix} R_2 + L_2 s & M_{23}s \\ M_{23}s & R_3 + L_3 s \end{bmatrix} \begin{bmatrix} i_2(s) \\ i_3(s) \end{bmatrix} = Z \begin{bmatrix} i_2(s) \\ i_3(s) \end{bmatrix} \tag{7.10}$$

7.3 WIRELESS POWER TRANSMISSION

7.3.1 Resonant versus Nonresonant Inductive Links

In the most simplified case of power transmission across a nonresonant inductive link, shown in Figure 7.2, a time-variant current $i_2(t)$ in L_2 on the transmitter (Tx) side generates current $i_3(t)$ in L_3 on the receiver (Rx) side due to their mutual inductance M_{23}. Here, L_3 drives the resistive load R_L, which represents the target power consumption, by passing current $i_3(t)$.

For sinusoidal waveforms where $s = j\omega$, the voltage that appears across the load, V_3, can be found from the equivalent *Z* matrix in (7.10),

$$V_3(j\omega) = j\omega M_{23} \cdot i_2 + j\omega L_3 \cdot i_3 + R_3 \cdot i_3 \tag{7.11}$$

Substituting i_3 with $-V_3 / R_L$ gives V_3 as a function of i_2 and other circuit parameters,

$$V_3(j\omega) = \frac{j\omega M_{23} \cdot i_2}{1 + (j\omega L_3 + R_3)/R_L} \tag{7.12}$$

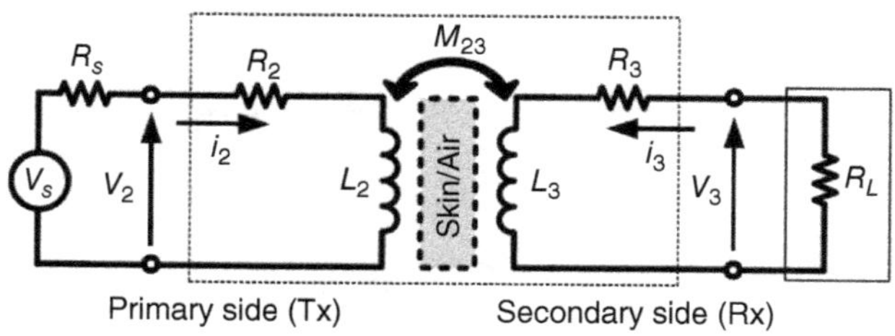

Figure 7.2 Simplified wireless link circuit diagram between two nonresonant inductively coupled coils: R_L represents the device loading on the receiver coil and R_s the loss of the power amplifier.

The received power at the load can be derived from

$$P_L = \frac{|V_3|^2}{2R_L} \tag{7.13}$$

Therefore, to maximize the delivered power to the load (PDL) with a constant R_L, V_3 should be increased.

In order to significantly increase V_3 at a certain frequency of interest, f_{res}, we should maximize the impedance across R_L at f_{res}. The best way to do this is to add a capacitor, C_3, in parallel with L_3 to form a parallel resonant *LC* tank circuit. The resonance frequency of the L_3C_3 tank circuit should be tuned to match that of the external transmitter carrier frequency, f_0, that is,

$$f_{res} = \frac{1}{2\pi\sqrt{L_3 \cdot C_3}} = f_0 \tag{7.14}$$

In choosing C_3 one should also consider the parasitic capacitance of L_3, interconnects, and the input capacitance of the stage following the L_3C_3 tank circuit, which is usually an AC-to-DC converter (e.g., a rectifier). If we represent the sum of all these parasitic capacitors by C_p, then $C_3 = C_p + C_L$, where C_L is the capacitance that is physically added across the load. Figure 7.3 shows the equivalent circuit of a resonant inductive link. By substituting R_L with $R_L \parallel C_3$ in (7.12),

$$V_3(j\omega) = \frac{j\omega M_{23} \cdot i_2}{1 + (j\omega L_3 + R_3)\left({}^1/_{R_L} + j\omega C_3\right)} \tag{7.15}$$

Figure 7.4 compares the magnitude of the received voltage for a nonresonant coil (7.12) and a resonance *LC* tank circuit (7.15) within a wide range of frequencies. It can be seen that at frequencies well below f_{res}, received voltages are in the same range. However, at resonance, the *LC* tank circuit produces a voltage across the load that is about one order of magnitude larger than a pure nonresonant inductive circuit.

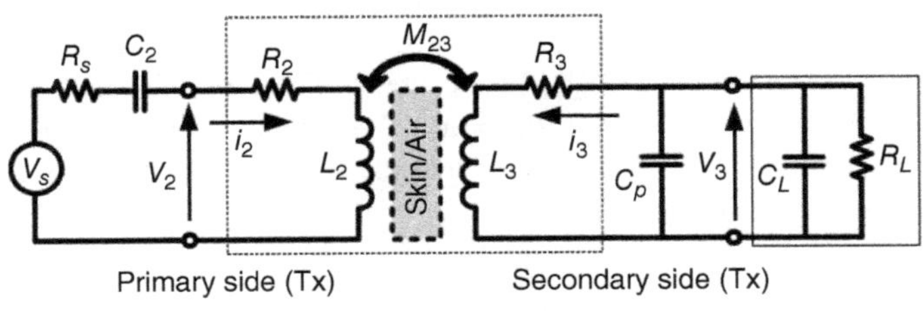

Figure 7.3 Wireless link circuit diagram with series and parallel resonance circuits on the transmitter and receiver sides, respectively: C_p represents parasitic capacitance and C_L is a physically added capacitor to form the Rx resonance circuit ($C_3 = C_p + C_L$); R_s is the output resistance of the power amplifier.

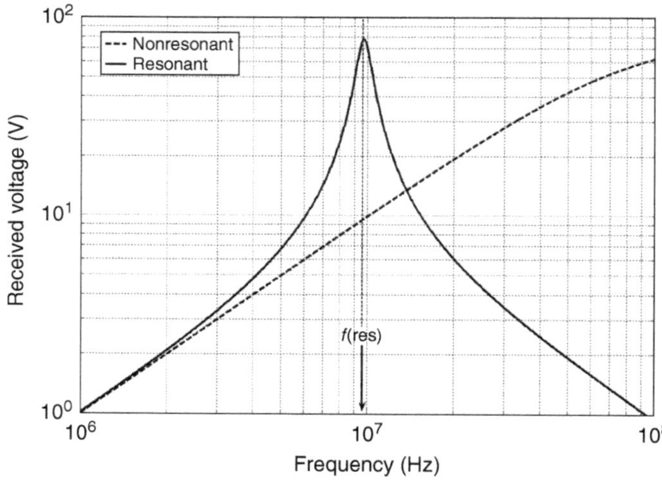

Figure 7.4 Comparison of received voltage magnitude across the load (V_3) between nonresonant Rx coil (Figure 7.2) and resonance *LC* tank circuit (Figure 7.3) when same current i_2 passes through primary coil (R_L = 5 kΩ, R_3 = 1 Ω, L_3 = 10 μH, C_3 = 27 pF) (Finkenzeller, 2003).

7.3.2 Power Transfer Efficiency

Power transfer efficiency is a key parameter in the design of wireless links which identifies the energy source specifications, heat dissipation, power transmission range, and possibility of interference with other electronic devices. The inductive link PTE is defined as the ratio between the power delivered to the load (PDL) and the power drained from the energy source and delivered to the primary loop in Figure 7.3 (Finkenzeller, 2003). To better understand the effects of inductive link circuit parameters in Figure 7.3 on the PTE, we define the Tx and Rx coil quality (Q) factors as $Q_2 = \omega_0 L_2/(R_2 + R_s)$ and $Q_3 = \omega_0 L_3/R_3$, respectively, where R_s represents the power amplifier (PA) loss.

To further simplify the circuit analysis, the secondary loop in Figure 7.3, including R_L, can be reflected on to the primary side with the reflected impedance, $L_{\text{ref}} \parallel C_{\text{ref}} \parallel R_{\text{ref}}$. To find R_{ref}, the secondary side in Figure 7.3 is modeled with only a parallel load resistance. The series loss resistance of L_3 that is, R_3, can be transformed to a parallel resistance equal to $R_{\text{P3}} = Q_2^3 R_3$ (Lee, 2004). Therefore, the equivalent parallel resistance in the secondary loop is $R_P = R_{P3} \parallel R_L$. Due to the mutual coupling of L_2 and L_3, the secondary load resistance and capacitance can be transferred to the primary side as shown in Figure 7.5*a*,

$$R_{\text{ref}} = k_{23}^2 \left(\frac{L_2}{L_3}\right) R_P = k_{23}^2 \omega_0 L_2 Q_{3L} \qquad C_{\text{ref}} = \left(\frac{L_3}{L_2}\right)\left(\frac{C_3}{k_{23}^2}\right) = \frac{1}{\omega_0^2 L_2 k_{23}^2} \quad (7.16)$$

where $Q_{3L} = R_P / \omega_0 L_3$ is referred to as the loaded quality factor of the secondary coil (Harrison, 2007). It should be noted that C_{ref} still resonates out with $k_{23}^2 L_2$ at f_0,

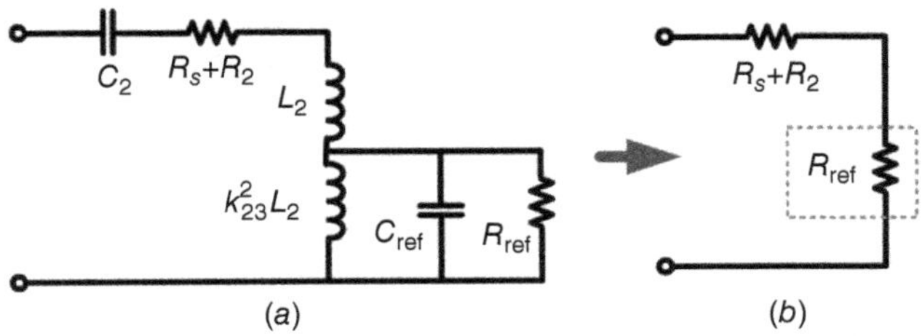

Figure 7.5 (*a*) Equivalent circuit of inductive link shown in Figure 7.3 reflected onto primary side; (*b*) C_{ref} and C_2 resonate out with $k_{23}^2L_2$ and L_2 at ω_0 and leave behind R_{ref} and $R_s + R_2$, respectively, and form simple resistive divider on primary side at carrier frequency, f_0.

that is, becomes an open circuit, leaving behind only a resistance, R_{ref}, in the primary loop, as shown in Figure 7.5*b*.

In the simplified circuit diagram shown in Figure 7.5*b*, L_2 and C_2, which form a series *LC* tank, are chosen such that they also resonate at f_0, ideally turning into a short circuit. Therefore, the input power provided by the source simply divides between R_s+R_2 and R_{ref}. The power absorbed by R_s+R_2 is dissipated as heat in the PA and the primary coil, while the power delivered to R_{ref} is that portion of the source power that is transferred across the link to the secondary loop in Figure 7.3. This power further divides between R_3 and R_L, which are the only power-consuming components on the Rx side. The power absorbed by R_3 is dissipated as heat in the secondary coil, and the rest is the useful power delivered to the load.

In order to calculate the PTE of the inductive link, we consider both of these power divisions (Harrison, 2007; Baker and Sarpeshkar, 2007; Kiani et al., 2011),

$$\eta_{2\text{-coil}} = \frac{R_{ref}}{R_s + R_2 + R_{ref}} \frac{R_{P3}}{R_{P3} + R_L} = \frac{k_{23}^2 Q_2 Q_{3L}}{1 + k_{23}^2 Q_2 Q_{3L}} \cdot \frac{Q_{3L}}{Q_L} \tag{7.17}$$

where $Q_{3L} = Q_3Q_L / (Q_3+Q_L)$ and $Q_L = R_L / \omega_0 L_3$ is often referred to as the load quality factor. It can be seen from (7.17) that k_{23}, Q_2, and Q_3 are the factors that should be increased to maximize the PTE. However, for a given set of Q_2, Q_3, and k_{23} values, there is an optimal load, $R_{L,\text{PTE}} = \omega_0 L_3 Q_{L,\text{PTE}}$, which can maximize the PTE for that particular arrangement, where $Q_{L,\text{PTE}}$ can be found by calculating the derivative of (7.17) with respect to Q_L,

$$Q_{L,\text{PTE}} = \frac{Q_3}{(1 + k_{23}^2 Q_2 Q_3)^{1/2}} \tag{7.18}$$

The amount of PDL, on the other hand, can be calculated by multiplying the power provided by V_s, that is, $V_s^2/2(R_s + R_2 + R_{ref})$, by the PTE from (7.17),

$$P_{L,2\text{-coil}} = \frac{V_s^{\,2}}{2(R_s + R_2)} \frac{k_{23}^2 Q_2 Q_{3L}}{(1 + k_{23}^2 Q_2 Q_{3L})^2} \cdot \frac{Q_{3L}}{Q_L} \tag{7.19}$$

Similar to the PTE, for a given set of Q_2, Q_3, and k_{23} values, there is an optimal load, $R_{L,\text{PDL}} = \omega_0 L_3 Q_{L,\text{PDL}}$, which can maximize the PDL at that particular arrangement, where $Q_{L,\text{PDL}}$ can be found by calculating the derivative of (7.19) with respect to Q_L,

$$Q_{L,\text{PDL}} = \frac{Q_3}{1 + k_{23}^2 Q_2 Q_3} \tag{7.20}$$

7.3.3 Multicoil Inductive Coupling

In order to achieve the optimal R_L condition for the PTE and PDL in (7.18) and (7.20), respectively, and also reduce the negative effect of R_s on Q_2, multicoil inductive links in the simple forms of three and four coils have recently been proposed (Kiani et al., 2011; Kurs et al., 2007; Hamam et al., 2009; RamRakhyani et al., 2011; Kiani and Ghovanloo, 2012). If the simple two-coil inductive link in Figure 7.3 is extended to an m-coil link, in which the first and mth coils are connected to the energy source and load, respectively, the reflected load from the (j+1)th coil to the jth coil can be found from

$$R_{\text{ref}\,j,j+1} = k_{j,j+1}^2 \omega_0 L_j Q_{(j+1)L} \qquad j = 1, 2, \ldots, m-1, \tag{7.21}$$

where $k_{j,j+1}$ is the coupling coefficient between the jth and (j+1)th coils and all coils are tuned at the same resonance frequency. The loaded quality factor of the (j+1)th coil, $Q_{(j+1)L,}$ can be found from

$$Q_{jL} = \frac{\omega_0 L_j}{R_j + R_{\text{ref}\,j,j+1}} = \frac{Q_j}{1 + k_{j,j+1}^2 Q_j Q_{(j+1)L}} \qquad j = 1,\ 2,\ \ldots,\ m-1 \tag{7.22}$$

where $Q_j = \omega_0 L_j / R_j$ and R_j are the unloaded quality factor and parasitic series resistance of the jth coil, L_j, respectively. It should be noted that for the last coil, which is connected to the load in series, $Q_{mL} = \omega_0 L_m/(R_m + R_L)$, and for the first coil, which is connected to the source in series, $Q_{1L} = \omega_0 L_1/(R_1 + R_s)$. Therefore, assuming that the coupling between nonneighboring coils is negligible, the PTE from the jth coil to (j+1)th coil can be written as

$$\eta_{j,j+1} = \frac{R_{\text{ref}\,j,j+1}}{R_j + R_{\text{ref}\,j,j+1}} = \frac{k_{j,j+1}^2 Q_j Q_{(j+1)L}}{1 + k_{j,j+1}^2 Q_j Q_{(j+1)L}} \tag{7.23}$$

Using (7.21), (7.22), and (7.23), the overall PTE in such a multicoil inductive link can be found from

$$\eta_{m\text{-coil}} = \prod_{j=1}^{m-1} \eta_{j,j+1} \cdot \frac{Q_{mL}}{Q_L} \tag{7.24}$$

and the PDL can be calculated from

$$P_{L,m\text{-coil}} = \frac{V_s^2}{2(R_s + R_1)} \frac{1}{1 + k_{12}^2 Q_1 Q_{2L}} \eta_{m\text{-coil}} \tag{7.25}$$

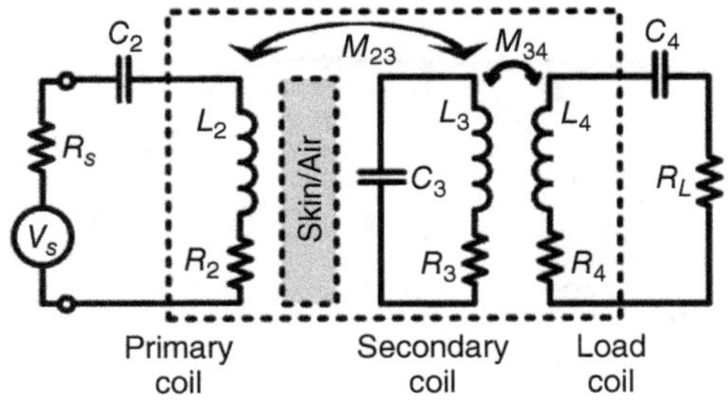

Figure 7.6 Lumped circuit model of three-coil inductive link.

For (7.24) and (7.25) to be valid, all coils should be tuned at the same resonance frequency to also achieve the highest PTE and PDL between each neighboring pair of coils and maximize the PTE and PDL of the multicoil link.

The three-coil inductive link circuit model, which comprises the primary coil, L_2, on the Tx side and the secondary and load coils (L_3 and L_4) on the Rx side, is shown in Figure 7.6. The PTE of this link can be found from

$$\eta_{3\text{-coil}} = \frac{(k_{23}^2 Q_2 Q_3)(k_{34}^2 Q_3 Q_{4L})}{(1 + k_{23}^2 Q_2 Q_3 + k_{34}^2 Q_3 Q_{4L})(1 + k_{34}^2 Q_3 Q_{4L})} \cdot \frac{Q_{4L}}{Q_L} = \eta_{23}\eta_{34} \tag{7.26}$$

where

$$\eta_{23} = \frac{k_{23}^2 Q_2 Q_{3L}}{1 + k_{23}^2 Q_2 Q_{3L}} = \frac{k_{23}^2 Q_2 Q_3}{1 + k_{23}^2 Q_2 Q_3 + k_{34}^2 Q_3 Q_{4L}}$$

$$\eta_{34} = \frac{k_{34}^2 Q_3 Q_{4L}}{1 + k_{34}^2 Q_3 Q_{4L}} \cdot \frac{Q_{4L}}{Q_L} \tag{7.27}$$

while k_{24} is ignored due to large separation between L_2 and L_4. Similarly, the three-coil link PDL can be found from (7.25)

$$P_{L,3\text{-coil}} = \frac{{V_s}^2}{2(R_s + R_2)} \frac{(k_{23}^2 Q_2 Q_3)(k_{34}^2 Q_3 Q_{4L})}{(1 + k_{23}^2 Q_2 Q_3 + k_{34}^2 Q_3 Q_{4L})^2} \cdot \frac{Q_{4L}}{Q_L} \tag{7.28}$$

The problem with the two-coil power transfer link, shown in Figure 7.3, was that for a given set of Q_2, Q_3, and k_{23}, the optimal PTE could only be achieved for a certain load, $R_{L,\text{PTE}}$, which can be far from the desired R_L that is more dependent on the application. In the three-coil power transfer inductive link, however, L_3, L_4, and M_{34} in Figure 7.6 can play the role of an impedance-matching circuit, which can convert any arbitrary R_L to $R_{L,\text{PTE}}$. This is equivalent to having a load quality factor $Q_L = Q_{L,\text{PTE}}$ in the secondary loop of a two-coil link, which was defined in (7.18). The optimal $Q_{L,\text{PTE}}$ is also adjustable with k_{34}, as shown in Figure 7.7, where the optimal PTE has been maintained for the three-coil link in a wide range of R_L (10 Ω–1 kΩ). On the other

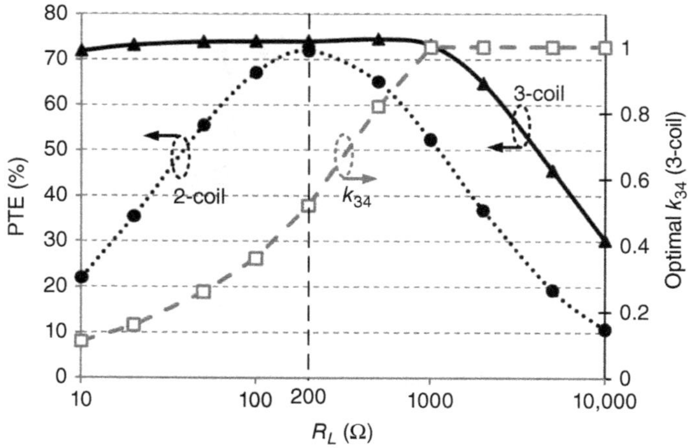

Figure 7.7 Adjustment of k_{34} to maintain optimal PTE in three-coil link vs. R_L. The two-coil link only reaches the optimal PTE for a specific $R_L = 200\ \Omega$ that satisfies (7.18) (parameter values: $L_2 = 0.9\ \mu$H, $L_3 = L_4 = 0.4\ \mu$H, $Q_2 = 255$, $Q_3 = Q_4 = 177$, $k_{23} = 0.03$).

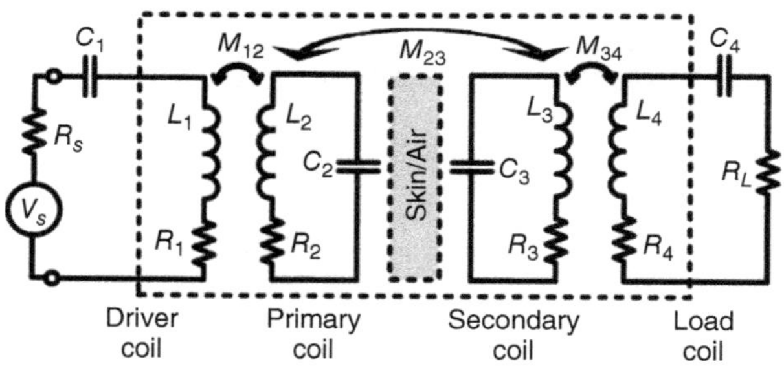

Figure 7.8 Lumped circuit model of four-coil inductive link.

hand, with a two-coil link the optimal PTE has been achieved in similar conditions only for a specific $R_{L,\mathrm{PTE}} = 200\ \Omega$ that satisfies (7.18). For short-distance inductive power transmission where k_{23} is large, which is the case in most transcutaneous IMD applications, the two-coil link requires smaller $Q_{L,\mathrm{PTE}}$ according to (7.18), and that is relatively easy to achieve. Therefore, a properly designed conventional two-coil inductive link can be very close to the optimal conditions.

The four-coil inductive link circuit model, which comprises the driver and the primary coils (L_1 and L_2) on the Tx side and the secondary and load coils (L_3 and L_4) on the Rx side, is shown in Fig. 7.8. Similar to the three-coil link, the PTE and PDL of this link can be found from

$$\eta_{4\text{-coil}} = \frac{(k_{12}^2 Q_1 Q_2)(k_{23}^2 Q_2 Q_3)(k_{34}^2 Q_3 Q_{4L})}{[(1 + k_{12}^2 Q_1 Q_2)(1 + k_{34}^2 Q_3 Q_{4L}) + k_{23}^2 Q_2 Q_3][1 + k_{23}^2 Q_2 Q_3 + k_{34}^2 Q_3 Q_{4L}]} \cdot \frac{Q_{4L}}{Q_L} \tag{7.29}$$

$$P_{L,4\text{-coil}} = \frac{V_s^2}{2(R_s + R_1)} \frac{(k_{12}^2 Q_1 Q_2)(k_{23}^2 Q_2 Q_3)(k_{34}^2 Q_3 Q_{4L})}{[(1 + k_{12}^2 Q_1 Q_2)(1 + k_{34}^2 Q_3 Q_{4L}) + k_{23}^2 Q_2 Q_3]^2} \cdot \frac{Q_{4L}}{Q_L} \tag{7.30}$$

The four-coil link also takes advantage of the load matching at the Rx side similar to the three-coil link using k_{34}. However, in the four-coil link a driver coil is also used to decouple the loosely coupled L_2-L_3 inductive link from R_s and, therefore, maximize Q_2 because the PTE between L_2 and L_3 is the dominant factor in determining the overall PTE of the four-coil link at a large coupling distance, d_{23}.

The total reflected impedance onto L_2, based on (7.21), reduces the L_2 quality factor from $Q_2 = \omega_0 L_2 / R_2$ to

$$Q_{2L} = \frac{Q_2}{1 + k_{23}^2 Q_2 Q_{3L}} \tag{7.31}$$

From (7.31) and (7.23) it can be inferred that a strong coupling between L_2 and L_3 (i.e., a high k_{23}) reduces Q_{2L} and consequently η_{12}, which is the PTE between L_1 and L_2. Therefore, η_{12} will significantly reduce at small coupling distances if k_{12} is not large enough. To avoid this problem, k_{12} should be kept large, which according to (7.21) results in a large loading reflected onto L_1. This can reduce the available power from the source, according to (7.25), unless V_s is increased. However, V_s can only be increased up to a certain extent before it raises safety issues in medical applications. This is a disadvantage of the four-coil arrangement for inductive power transfer to IMDs, particularly when a large PDL is required.

A comparison between PTE and PDL of two-, three-, and four-coil links versus R_s is demonstrated in Figure 7.9 at d_{23} = 10 mm. These curves are quite instructive to find which inductive link can provide the optimal PTE while proving sufficient PDL. For instance, a low-power driver with R_s = 5 Ω for a low-PDL application favors a four-coil inductive link to maintain a high PTE despite its low PDL. However, the same four-coil link at d_{23} = 20 mm will require V_s = 30 V with 100 V across a class-E PA transistor to deliver 250 mW to the load. On the other hand, a three-coil inductive link, driven by a class E PA with R_s = 0.5 Ω, will require V_s = 1.8 V and imposes only 6 V across the PA transistor in the same conditions (Kiani et al., 2011). It is quite clear from Figure 7.9 that the two-coil link performance is close to optimal when a large PDL is needed and, therefore, R_s is small.

7.3.4 Power Amplifiers

Power amplifiers are commonly utilized in wireless power transmission links to drive the primary coil by converting the DC power of an external source, often a battery, to an AC signal at the carrier frequency, f_0, represented by V_s in Figure 7.2. There are several well-known topologies for such PAs, designated by classes A to F. Each of these amplifier designs has certain advantages and limitations for specific applications. For transcutaneous power transmission, class-E PAs are the most popular ones, followed by class-D PAs, due to their high power efficiency (which can theoretically be close to 100%), self-oscillating capability, and small number of components.

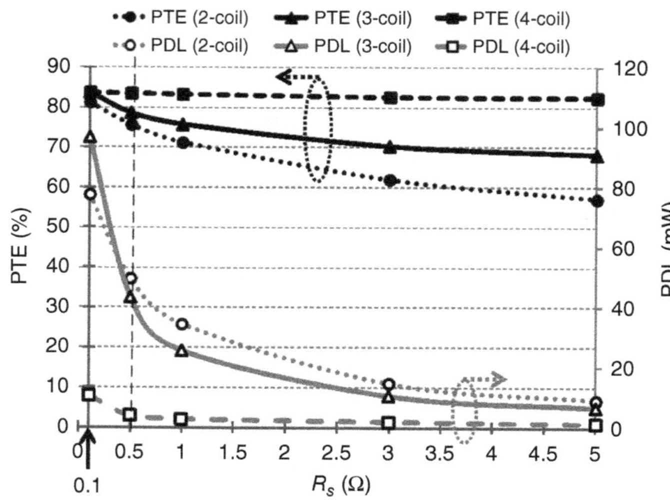

Figure 7.9 Comparison between calculated values of PTE and PDL versus R_s for two-, three-, and four-coil inductive links for IMD applications ($V_s = 1$ V, $R_L = 100\ \Omega$, $d_{23} = 1$ cm). The detailed coil geometries for the two-, three-, and four-coil links can be found in Kiani et al. (2011).

Fortunately all of the aforementioned PA classes, including class-E, have been widely covered in the literature (Sokal and Sokal, 1975; Raab and Sokal, 1978; Zierhofer and Hochmair, 1990; Kendir et al., 2005; Troyk and Schwan, 1992; Kazimierczuk and Puczko, 1987; Ziaie et al., 2001). It should be noted that the PTE equations for two-, three-, and four-coil links in (7.17), (7.26), and (7.29), respectively, include the PA losses as well, modeled by R_s in Figures 7.3, 7.6, and 7.8, to help designers calculate the overall PTE all the way from the battery to the load. One can exclude the PA loss to focus on the inductive link PTE by redefining Q_1 and Q_2 as $Q_1 = \omega_0 L_1 / R_1$ and $Q_2 = \omega_0 L_2 / R_2$ for the four-, and two-/three-coil links, respectively.

7.4 INDUCTIVE COUPLING FOR BIOMEDICAL TELEMETRY

A key design requirement in IMDs is to provide sufficient PDL while maintaining high PTE. High PTE is required to reduce (1) the heat dissipation within the coils; (2) tissue exposure to electromagnetic field, which can cause additional heat dissipation in the power transmission medium; (3) size and weight of the external energy source; and (4) interference with nearby electronics that is necessary to satisfy regulatory requirements.

7.4.1 Design Challenges and Possible Solutions

A simplified diagram of the transcutaneous power flow and various losses along the way is shown in Figure 7.10. The inductive link includes the Tx coil(s) attached to

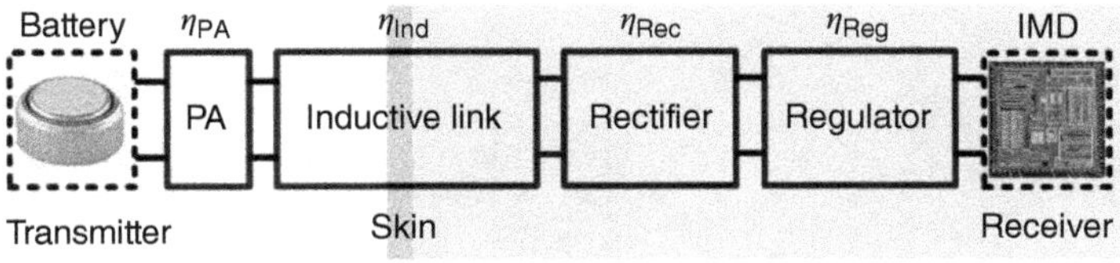

Figure 7.10 Power flow diagram in transcutaneous inductive power transmission link from external battery to IMD electronics.

the skin from outside of the body and the Rx coil(s) implanted under the skin flap often with the rest of the implant electronics. The inductive link follows the battery and the PA on the Tx side. On the Rx side, a highly efficient rectifier followed by a regulator provides a constant DC voltage for the IMD. Therefore, the overall power transfer efficiency is determined considering all the losses from,

$$\eta_{\text{link}} = \eta_{\text{PA}} \times \eta_{\text{Ind}} \times \eta_{\text{Rec}} \times \eta_{\text{Reg}} \tag{7.32}$$

It can be seen that maintaining a high efficiency in the PA, rectifier, and regulator is as important as in the inductive link. However, in terms of power efficiency, the inductive link is often the weakest link in this kind of application. Active rectifiers are good candidates for implementing highly efficient rectifiers on-chip. A 13.56-MHz active rectifier, which has achieved 80.2% power conversion efficiency when delivering 20 mW to the load, has been reported by Lee and Ghovanloo (2011).

One of the major problems with the power transfer link shown in Figure 7.10 is the received voltage fluctuations in the presence of coils coupling variations due to the changes in the coil relative distance or misalignment. Load variations can also result in received voltage fluctuations according to the PDL of two-coil links in (7.19). A closed-loop power transmission system, initially proposed by Kiani and Ghovanloo (2010), can be used to maintain the received voltage constant in such conditions. An RFID reader is a suitable candidate for establishing this closed-loop power control mechanism because it can both drive the primary coil and recover the backward telemetry data, containing the received voltage information as the feedback, both on the Tx side. Figure 7.11 shows a more detailed block diagram of such an RFID-based system, composed of two main blocks: power Tx and IMD (Kiani and Ghovanloo, 2010).

Key parameters that can affect the design of power transfer links aside from the coils are driver voltage (V_s), power required by the load (P_L), PA supply voltage (V_{DD}), PA loss (R_s), PA transistor breakdown voltage, and safety limits considering the application. In a class E PA, zero-voltage switching (ZVS) allows for high power efficiency with peak voltages across the primary coil and the PA transistor that are 1.07 and 3.56 times the V_{DD}, respectively (Kazimierczuk and Czarkowski, 1995). Therefore, when the application requires a large P_L, R_s should be reduced to make sufficient power available from the source, $P_{av} = V_s^2/8R_s$, at a reasonable V_s and V_{DD}. Achieving a high PTE while maintaining impedance matching at the PA output would not be possible because that reduces the PA efficiency to only 50%. Therefore, the

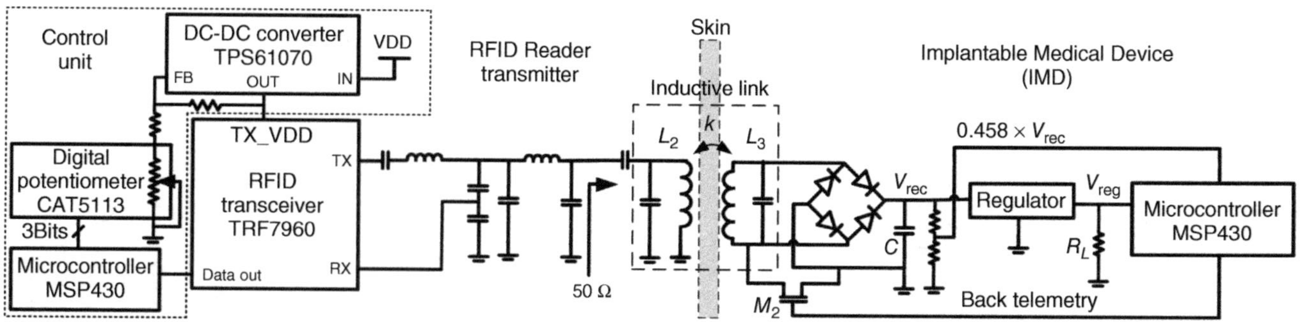

Figure 7.11 Block diagram of RFID-based closed-loop power transmission system operating at 13.56 MHz (Kiani and Ghovanloo, 2010).

power delivered to L_2 under high-PTE conditions would be much smaller than P_{av}, and increasing P_{av} is the only way to achieve the desired PDL. Utilizing large transistors to reduce R_s results in increased dynamic switching losses in the PA (Baker and Sarpeshkar, 2007). Therefore, the power transmission link should be designed in a way that it achieves high PTE and also provides sufficient PDL while considering practical limitations of the PA design.

7.4.2 Optimization of Coil Geometries

The inductive link PTE in (7.17) implies that the coil mutual inductance M_{23} and the quality factors Q_2 and Q_3 are the most important circuit parameters affecting the PTE. It is well known from electromagnetic theory that M_{23}, Q_2, and Q_3 are dependent on coil geometry, which in general can also include their relative distance, orientation, and number of turns. Therefore, one of the decisions that should be made early on is, for example, whether the coils should be wire wound or lithographically defined on a planar conductive surface, such as a printed circuit board (PCB) or micromachined surface.

Wire-wound coils are made of filament wires in the form of a single or multiple individually insulated strands twisted in a circular bundle. The latter type is also known as a Litz wire and helps reduce the resistive losses in the coil due to skin effects, especially at higher carrier frequencies, by increasing the circumferential area of the wire. The optimal number of turns for a coil made of Litz wire depends on the desired volume of the coil and Litz wire parameters such as the number and diameter of individual strands (Kendir et al., 2005; Sullivan, 1999; Tourkhani and Viarouge, 2001). Wire-wound coils, however, cannot be batch fabricated or reduced in size without the use of sophisticated machinery.

Printed spiral coils (PSCs), which are lithographically defined in one or multiple layers on rigid or flexible substrates, such as PCB, can be batch fabricated and offer more flexibility in optimizing their geometries and aspect ratios while occupying less space, deeming them attractive for IMD and RFID applications. Figure 7.12 shows a square-shaped PSC as an example. Geometric parameters of the PSCs that affect circuit parameters such as Q and k, and consequently the PTE, are the line

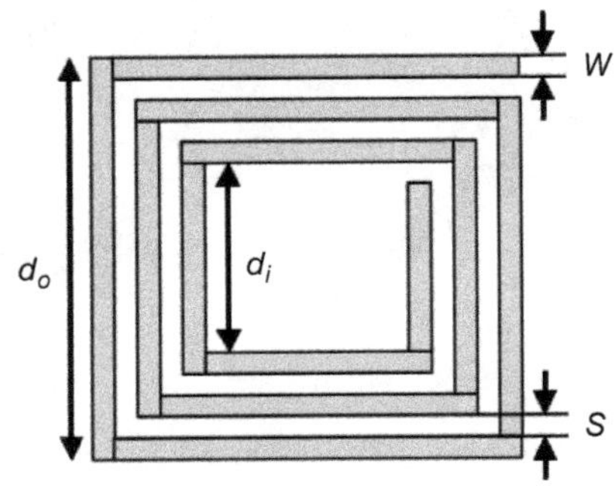

Figure 7.12 Geometric parameters of square-shaped PSC.

width (w), line spacing (s), outer diameter (d_o), inner diameter (d_i), and fill factor [$\varphi = (d_o - d_i)/(d_o + d_i)$], which have been described in more detail in (Jow and Ghovanloo, 2007).

The design procedure to find the optimal geometries of the primary and secondary coils, L_2 and L_3, in a two-coil inductive link should take into account the major design constraints imposed by regulations, application, and PSC fabrication technology. IMD applications often impose strict design constraints on the maximum implant size, in this case, the outer diameter d_{o3} and the coil relative distance d_{23}. Other important parameters affecting the coil geometries are the operating frequency f_0 and the load resistance R_L. The fabrication process indicates the minimum feature size that result in acceptable yield in mass manufacturing, such as the minimum line width w_{min}, minimum line spacing s_{min}, and minimum thickness, as well as substrate and conductor material properties.

An iterative design procedure, which starts with a set of design constraints and initial values, can result in the optimal PSC pair geometries (Jow and Ghovanloo, 2007). It can be seen from (7.6) that k increases by increasing the secondary coil radius in the simple case of two one-turn filament coils. Therefore, the largest possible d_{o3} that is allowed by the application is always chosen. Except for very high carrier frequencies, the minimum line spacing is often desired in order not to limit the maximum line width, which is an important parameter to increase the coil's Q factor (Jow and Ghovanloo, 2007). The nominal values for d_{23} and R_L are also identified from the application constraints. Then a set of initial values for L_2 and L_3 geometries, such as w_2, w_3, φ_2, φ_3, and d_{o2}, need to be selected before starting the iterative optimization process. A more detailed discussion about how to choose these initial values can be found elsewhere (Jow and Ghovanloo, 2007), a summary of which follows.

In the first step, d_{o2} and φ_2 are swept to maximize η_{23} in (7.17). Figure 7.13 shows an example of the 3D surface of η_{23} versus d_{o2} and φ_2 when $d_{o3} = 20$ mm, $\varphi_3 = 0.43$,

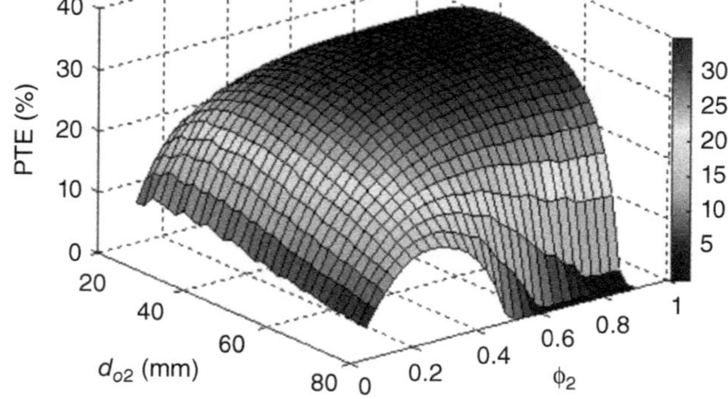

Figure 7.13 Power transfer efficiency (PTE) vs. outer diameter and fill factor of primary PSC when d_{o3} = 20 mm, $\varphi_3 = 0.43$, $s_{2,3} = w_{2,3} = 150$ µm, d_{23} = 10 mm, R_L = 500 Ω, and f_0 = 1 MHz.

$s_{2,3} = w_{2,3} = 150\ \mu\text{m}$, $d_{23} = 10$ mm, $R_L = 500\ \Omega$, and $f_0 = 1$ MHz. It can be seen that d_{o2} would be 48 mm and η_{23} almost flattens for $\varphi_2 > 0.8$. This indicates that the turns very close to the center of the PSC do not help with improving the efficiency (Jow and Ghovanloo, 2007).

In the second step, w_3 and φ_3 are swept to maximize η_{23} utilizing the temporary geometry of the primary PSC, which was resolved in the previous step. In step 3, we increase w_2 while providing room for this change by increasing d_{o2} and looking for the maximum η_{23}. If η_{23} increased for more than 0.1% in step 3, we return back to step 1 and continue the optimization until improvement in η_{23} per iteration is less than 0.1%, which indicates that the geometries are close to optimal. Since the closed-form equations governing the electromagnetic fields and parasitic components of the 3D PSC conductors become complicated rapidly, it becomes necessary to validate the design outcome by employing commercially available 3D field solvers, such as Fast-Henry and high frequency structural simulator (HFSS) (Ansoft, Pittsburgh, PA). The three- and four-coil link design procedures are similar and can be found elsewhere (Kiani et al., 2011; RamRakhyani et al., 2011).

IMDs need to be hermetically sealed in biocompatible materials and placed in a conductive environment with high permittivity (tissue), which can affect the implanted coil characteristics. Similarly, there are other inductive links that may need to operate in media other than air. A detailed model that considered the effects of the surrounding volume conductor on the PSC parasitic components and eventually on the power transfer efficiency has been proposed by Jow and Ghovanloo (2009). The measurement results have shown that, by including the volume conductor and PSC environment in the design and optimization process, the PTE can be improved by more than 40% compared to PSCs that are optimized for air but used in the tissue environment.

7.4.3 Power Absorption in Tissue

Coil designs for IMD applications should consider the electromagnetic power absorbed in the human body to ensure that it meets international safety standards and does not pose health hazards (Lin, 1986; Gosalia et al., 2004). Several guidelines for suggested absorption limits of electromagnetic energy are either expressed in terms of currents or *specific absorption rate* (SAR) of power induced in the human body, with the latter defined for a sinusoidal excitation in watts per kilogram as

$$\text{SAR}(x, y, z) = \frac{\sigma(x, y, z)E^2(x, y, z)}{2\rho(x, y, x)} \tag{7.33}$$

where ρ is the tissue density (in kg/m^3), σ is the conductivity (S/m), and E is the electric field amplitude (V/m) at point (x, y, z). Full-wave electromagnetic computational tools and experimental methods with phantoms filled with tissue simulants can be used to determine the field induced by telemetry or induction devices in the human body and compare them with standard safety limits.

TABLE 7.1 Maximum Permissible Exposure Limits to Human Body

Frequency Range (MHz)	Electric Field Strength, E (V/m)	Magnetic Field Strength, H (A/m)	Power Density E field, H field (mW/cm^2)
0.003–0.1	614	163	100, 1,000,000
0.1–3.0	614	$16.3 / f_0$	100, $10{,}000 / f_0^2$
3–30	$1842 / f_0$	$16.3 / f_0$	$900 / f_0^2$, $10{,}000 / f_0^2$
30–100	61.4	$16.3 / f_0$	1, $10{,}000 / f_0^2$

7.4.4 Safety Issues

Recommendations to prevent harmful effects in human beings exposed to electromagnetic energy have been issued by numerous organizations, such as the Institute of Electrical and Electronics Engineers (IEEE, 1999). For humans, the maximum permissible exposure (MPE) limits in terms of root-mean-square (RMS) electric (E) and magnetic (H) field strengths and power densities are given in Table 7.1 as functions of frequency. Table 7.1 clearly shows that lower frequencies are more appropriate for IMDs. The safety issues will be explained in more detail in Chapter 15.

7.5 INDUCTIVE DATA TRANSMISSION

Bidirectional wireless data transmission is essential for IMD and RFID systems to establish a short-range wireless communication between the Tx and Rx parts of the system. RFID readers use the same inductive link to not only power the passive RFID tags but also interrogate them (Finkenzeller, 2003). The majority of advanced IMDs have several adjustable parameters that can be fine tuned after implantation for every individual patient according to his or her specific needs. In addition to those parameters, research is underway to equip sensory devices with a flow of stimulation commands from the external artificial sensors and signal processing units to build closed-loop neuroprosthetic devices (Schwartz et al., 2006). Sending adjustment and control commands wirelessly from the external unit to the implanted unit is known as the *forward telemetry* or *downlink*. Moreover, the same devices often need to inform the external processing components about the IMD operating status, possible faults, and in some cases the neuronal response immediately after stimulation for proper adjustment of the stimulation parameters (Venkatraman et al., 2009; Lee et al., 2010) This direction of data flow that sends information from inside toward out of the body is often referred to as *backward telemetry* or *uplink*.

7.5.1 Forward Telemetry

A simple option for forward telemetry, which has been used in most of todays' IMD and RFID systems, is to modulate the same carrier that has been used for power transmission for transmitting data (Ghovanloo and Najafi, 2004; Arabi and Sawan,

1999; Boyer et al., 2000; Ziaie et al., 1997; Smith et al., 1998; Liu et al., 2000; Suaning and Lovell, 2001) The advantage of this method is less complexity on both external or reader and implant or transponder components of the system, which can result in lower power consumption and smaller size. In the following, three main modulation techniques for forward data telemetry through magnetic coupling are presented.

The majority of the IMD and RFID systems use *amplitude shift keying* (ASK) for forward telemetry due to its simplicity on both Tx and Rx sides (Arabi and Sawan, 1999; Boyer et al., 2000; Ziaie et al., 1997; Smith et al., 1998; Liu et al.,2000; Suaning and Lovell, 2001). In this method, shown in Figure 7.14*a*, the external data modulates the amplitude of the power carrier, known as the *carrier envelope*. This can be easily done by changing the supply voltage and consequently the output swing of the PA (V_s in Figure 7.2) based on the modulating signal. The ASK carrier frequency should be several times higher than the data rate to provide the Rx with enough cycles to detect the change in the envelope amplitude. The difference between data rate and carrier frequency also depends on the amplitude modulation index and primary

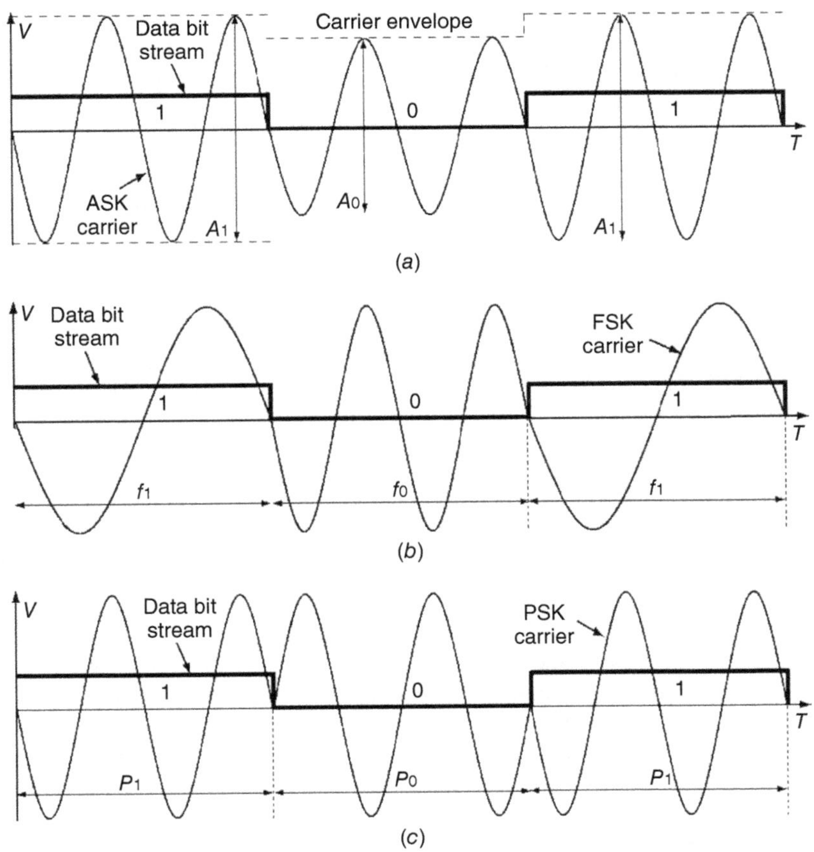

Figure 7.14 (*a*) Amplitude shift keyin, (*b*) frequency shift keying, and (*c*) phase shift keying.

and secondary coil quality factors. Increasing the modulation index helps with adding to the data rate at the cost of degrading the PTE. The higher the Q factor, the more the PTE improves; however, the longer it takes for the Tx or Rx LC tank circuits to follow the carrier amplitude and the smaller the data transmission bandwidth. Considering (7.17), it can be concluded that in single-carrier systems there is always a compromise between the inductive link power efficiency and data bandwidth.

One way to demodulate the ASK signal in the Rx is to use an envelope detector. The ASK signal passes through a low-pass filter (LPF), which has a cutoff frequency somewhere between the carrier frequency and data bandwidth. The filter rejects the higher frequency carrier from the carrier envelope, which can then be easily converted back to a serial data bit stream by passing through a high-pass filter (HPF) and a comparator. One factor that affects forward data rate in ASK is the percentage of change in carrier amplitude when transmitting logic 0 and logic 1. This parameter, which is known as the ASK modulation index, has a direct relationship with the data bandwidth. However, since carrier power is proportional to the second power of its amplitude, a higher modulation index means less average delivered power in the carrier signal.

Another limitation of the ASK method in IMD and RFID applications is its susceptibility to noise, interference, and motion artifacts, all of which mainly affect the carrier amplitude. Equation (7.3) shows that the mutual coupling between two coils highly depends on their relative distance. Therefore, the voltage across the Rx coil, V_3, is also strongly dependent on d_{23} ($V_3 \propto 1/d_{23}^3$). When a patient wears an inductively powered IMD and walks, the motion artifacts and vibrations affect d_{23} and consequently the amplitude of the received signal. Even when d_{23} is constant, any instantaneous changes in the IMD current consumption, due to stimulation for example, directly results in V_3 variations and deteriorates the quality of the ASK signal (Raker et al., 2001; Kaiser and Steinhaugen, 1995; Abrial et al., 2001). Hence, the ASK demodulator should be able to distinguish between the amplitude variations that result from noise, interference, and artifacts and those that represent the received data bits. Utilizing subcarriers is one of the methods to get around the aforementioned problems in ASK (Finkenzeller, 2003).

Frequency shift keying (FSK), a popular modulation technique in high-fidelity (HiFi) audio transmission as well as digital communications, has not been widely utilized in IMDs and RFID systems because of the complexities in implementation of the FSK modulation and demodulation circuits. In this method, shown in Figure 7.14*b*, the external data modulates the frequency of the power transmission carrier and the carrier amplitude remains constant. Therefore, logic 0 and logic 1 are transmitted by sinusoidal signals at frequencies f_0 and f_1, respectively. As a result, the carrier power stays constant regardless of the data contents, which is an advantage of the FSK compared to ASK. In addition to constant power, the superior robustness against various noise sources and interference of FM over AM has been known since the early days of radio engineering. In FSK, it is very unlikely that d_{23} or IMD current variations would affect the frequency of the induced signal. Phase noise is the main noise source in FSK, which depends on the stability of the FSK modulator on the Tx side.

For FSK modulation the PA input can be switched between two oscillators operating at f_0 and f_1 depending on the serial data bit stream. Another method is changing the oscillation frequency of a single oscillator by changing its time constant, which is usually done by a voltage-controlled oscillator (VCO), which is an essential component of many radio frequency (RF) circuits. The oscillation frequency of a VCO, which is often controlled by a voltage-dependent capacitor, called a varactor, drifts over time or changes with temperature variations. To stabilize the FSK modulator frequency, the VCO is often used in a closed loop and referenced to a local oscillator with low phase noise, which frequency is stabilized by a crystal. Such a circuit is called a phase-locked loop (PLL).

One limitation of the FSK technique is that its frequency spectrum occupies a wider bandwidth compared to ASK at the same data rate. Therefore, the Q factors of both Tx and Rx LC tank circuits should be lowered to provide enough bandwidth to pass major FSK carrier components at the expense of lowering the PTE. According to Carlson's rule, the bandwidth, BW, required to include 98% of the total power of an FM signal is

$$\mathrm{BW} \approx 2(\delta_{\mathrm{max}} + f_{i,\mathrm{max}}) \tag{7.34}$$

where δ_{max} is the maximum frequency shift caused by modulation, depending on the VCO gain and amplitude of the modulating signal, and $f_{i,\mathrm{max}}$ is the maximum frequency content of the modulating signal. To provide enough bandwidth and improve the inductive link robustness against coupling variations, a stagger tuned circuit is proposed by Galbraith et al. (1987). Also a variation of the class-E power amplifiers, which can switch the carrier frequency by switching the LC tank capacitive components, is proposed by Troyk and DeMichele (2003).

There are several traditional methods for FSK demodulation. One of the basic methods involves a limiter to eliminate noise and interference on the received signal amplitude, a discriminator to convert the FSK signal to an ASK signal, and an envelope detector to demodulate the ASK signal. PLLs can also be used for FSK demodulation. For high-bandwidth forward telemetry, however, these methods may require high-order analog filtering down the signal path, which would consume a large chip area or several off-chip components in the low-end RF application of interest (IMD).

To achieve a high data rate along with synchronization between the Tx and Rx without filtering, a phase-coherent FSK (pc-FSK) protocol, shown in Figure 7.14*b*, was proposed by Ghovanloo and Najafi (2004). In the pc-FSK protocol, binary symbols 1 and 0 are transmitted by one and two carrier cycles at f_1 and f_0, respectively. Choosing $f_0 = 2 \times f_1$ provides a constant bit length, which helps extract a synchronous sampling clock directly from the pc-FSK carrier. The FSK demodulator manages the received carrier as a baseband signal and directly measures the duration of each received carrier cycle. The measured duration is then compared with a predefined value to indicate its associated binary symbol. Therefore, every single carrier cycle can transfer a data bit, resulting in a data rate–carrier frequency ratio close to 1, which is higher than the same ratio in many wideband wireless communication techniques currently in use.

In *phase shift keying* (PSK), shown in Figure 7.14*c*, the serial data bit stream modulates the phase of the power transmission carrier and both carrier amplitude and frequency remain constant. Therefore, PSK has the highest spectral efficiency compared to the other two techniques, which means that, using PSK, it is possible to transmit higher data rates per unit available wireless link bandwidth. PSK is also the basis for vector modulation. In binary PSK (BPSK) each phase transition represents one bit, and logic 0 and logic 1 are 180° out of phase. This is equivalent to multiplying the original carrier with a bit stream of 1s and − 1s to represent logic 0 and 1, respectively (Sawan and Coulombe, 2005).

It is also possible to send more than one bit per phase transition by using smaller phase shifts. For example, in quadrature PSK (QPSK), by using four different phases that are 90° apart, it is possible to transmit four symbols, that is, two bits per phase. As a result of these capabilities, PSK is a popular modulation technique in wide-band digital communications and wireless local area networks (WLANs). This is the case especially when PSK is combined with ASK to further increase the number of bits per phase/amplitude transition. This method is yet another type of modulation, which is known as quadrature amplitude modulation (QAM). However, QAM has not been widely utilized in IMDs since it requires very stable and accurate local oscillators on both Tx and Rx sides, which can add to the volume and power consumption of the IMD.

PLLs can be used for both PSK modulation and demodulation. Traditional PLLs require accurate local oscillators, which in turn require crystals. The intense size constraints in many implantable devices and RFID transponders do not allow inclusion of crystals, which are relatively large and not scalable. Therefore, researchers have tried to either use specific types of PLLs that do not need crystals (Sawan and Coulombe, 2005) or extract the phase transitions directly from the incoming carrier signal (Marschner et al., 2002). The Rx complexity, which results in high power consumption, is one of the disadvantages of the PSK for use in IMD applications (Chen et al., 2010).

7.5.2 Backward Telemetry

Back telemetry can be implemented either passively by relying on the mutual coupling between the power coils or actively by adding a Tx and an antenna to the secondary side. Each method has its own advantages and limitations, which are explained in the following.

Load shift keying (LSK) is a common passive back-telemetry method in RFID applications (Finkenzeller, 2003; Parker et al., 2001; Kaiser and Steinhaugen, 1995; Abrial et al., 2001), which has also been used in many IMDs (Suaning and Lovell, 2001; Sawan et al., 2005; Tang et al., 1995; Zhou and Donaldson, 2003; Catrysse et al, 2004; Mandal and Sarpeshkar, 2008). In this method, also known as impedance modulation or load modulation, changing the loading of the secondary coil based on the modulating back-telemetry signal affects the current in the external primary coil due to their mutual coupling. Equation (7.16) shows that a change in R_L or C_3 in the secondary side can result in variations in the reflected resistance and capacitance and consequently changes in the primary current, especially if k_{23} is sufficiently large. By detecting these primary current variations, which also affect the primary coil voltage, the primary side of the system can demodulate and recover the back-telemetry signal. Therefore, the LSK signal on the primary coil should always be picked up and demodulated as an ASK signal, and our discussions in Section 7.6.1 applies to the LSK as well.

The main advantage of the LSK is its simplicity, especially on the IMD or transponder side, where size matters most. On the other hand, LSK affects the power transmission efficiency by disturbing the resonance circuit and reducing or completely cutting the received power to the main load for short periods of time. The back-telemetry data rate that is achievable through LSK highly depends on k_{23} and power carrier frequency f_0. It also depends on many other factors, including Q_2, Q_3, load variations during normal operation, sensitivity of the current or voltage sensing circuit on the external primary side, the amount of noise and interference in the primary coil, and the type of encoding technique that is usually combined with the LSK data (Mandal and Sarpeshkar, 2008).

There are three possible configurations for LSK as shown in Figure 7.15. Two of them are ohmic and the third one is capacitive. In the series ohmic load modulation, shown in Figure 7.15*a*, the secondary loading is changed between Z_L and the open circuit (infinity) by a series switch based on the back-telemetry data. In

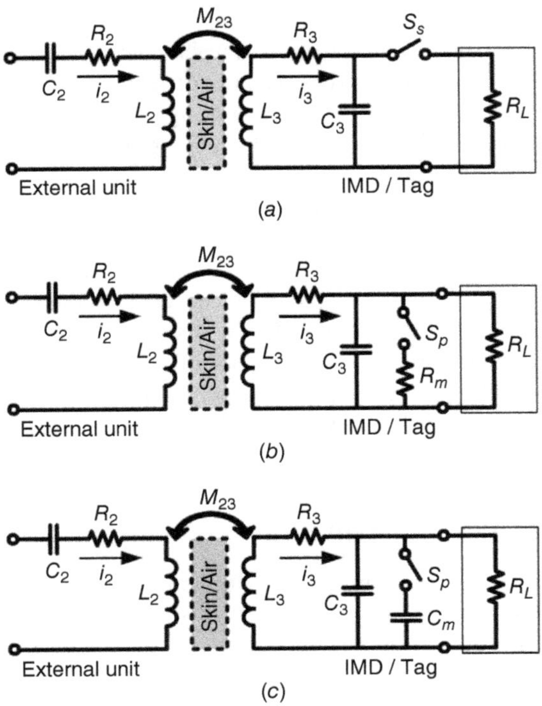

Figure 7.15 Passive back telemetry by LSK using (*a*) series ohmic load modulation, (*b*) parallel ohmic load modulation, and (*c*) parallel capacitive load modulation.

the parallel configuration, shown in Figure 7.15*b*, the secondary loading is changed between Z_L and the short circuit (close to zero) by a parallel switch. Since a larger change in the secondary loading results in easier detection of the back-telemetry signal in the primary coil, the series and parallel configurations are suitable for small and large loads (i.e., value of Z_L), respectively. Most RFID devices have very small power consumption and therefore use the parallel configuration (Finkenzeller, 2003; Raker et al., 2001; Kaiser and Steinhaugen, 1995; Abrial et al., 2001). However, implantable microstimulators may have much higher power consumption, especially when the stimulation is active. Therefore, the series ohmic configuration or a combination of both depending on the loading would be a better choice (Bawa and Ghovanloo, 2008). Finally, in capacitive load modulation, an additional capacitor, C_m in Figure 7.15*c*, is switched in and out in parallel to C_3 based on back-telemetry data bits. This action would result in detuning the Rx *LC* tank circuit from its original resonance frequency, f_0, which in turn affects the reflected capacitance onto the Tx and consequently i_2 according to (7.16).

Considering that neural signals have a bandwidth of about 10 kHz, a wide-band telemetry link on the order of several megahertz is needed to wirelessly record from a large number of sites simultaneously. Therefore, the small bandwidth provided by the passive back-telemetry method is not enough for IMDs that are dedicated to multichannel neural recording. These IMDs are usually equipped with low-power transmitters for active back telemetry and utilize a separate carrier that is significantly higher in frequency than the power carrier. The major challenges in active back telemetry are reducing the power consumption while achieving sufficient range, small size of the IMD, and efficient antenna design for an effective wireless link. Design of the external Rx would be less challenging due to more relaxed size and power constraints outside of the body.

Several research groups have implemented active back-telemetry links for neural recording systems using commercial components (Obeid et al., 2003; Hawley et al., 2002, Gosalia et al., 2004) or custom ASICs (Lee et al., 2010; Yin and Ghovanloo, 2011; Chae et al., 2009; Miranda et al., 2010; Rizk et al., 2009; Mohseni and Najafi, 2004). In most of these designs the IMD Tx is significantly simplified to reduce the size and power consumption at the expense of more complexity in the external Rx. Use of the high-frequency band known as the industrial, scientific, and medical (ISM) band along with inductive power has also been proposed (Neihart and Harrison, 2005).

7.5.3 Single Carrier versus Multicarrier

The main advantage of using a single carrier for both power and data transmission is the relatively robust coupling between power coils, which can lead to more reliable data transfer. Another advantage is the saving in space by reusing the power coils for multiple purposes. However, achieving high PTE and high data transmission bandwidth utilizing the same carrier would be challenging because of their conflicting requirements. It was shown in Sections 7.6.1 and 7.6.2 that modulating the power carrier in any form or direction complicates the power Tx circuitry and reduces the

PTE. Another important issue is the low frequency of the power carrier, which further limits the data transfer bandwidth in either direction to levels that are insufficient for advanced neuroprosthetic devices for sensory substitution and brain–computer interfacing (BCI).

As a result, the use of two or three carrier signals for power, downlink, and uplink has been proposed with each carrier having its own pair of coils or antennas in order to decouple the data transfer link bandwidth from the power transmission efficiency (Shire et al., 2009; Ghovanloo and Atluri, 2007; Zhou et al., 2008). Aside from the size overhead, the use of multiple-carrier signals within a space as small as an IMD introduces new challenges, the most important of which is the strong power carrier interference with much weaker data carriers. Several researchers have offered solutions such as using orthogonal symmetrical coils (Ghovanloo and Atluri, 2007; Jow and Ghovanloo, 2010), coaxial coils with differential phase shift keying (DPSK) (Shire et al., 2009), and shifted coplanar coils with offset quadrature phase shift keying (OQPSK) (Shire et al., 2009; Zhou et al., 2008). Nonetheless, the most effective way to reduce interference is to separate out the carrier frequencies and take advantage of the bandpass filtering (BPF) effect of the high-Q LC tank circuits at resonance.

In the case of the orthogonal coils, a pair of planar spiral coils (PSCs), shown in Figure 7.16*a*, which geometries have been optimized based on the power carrier frequency and tissue volume conductor, are used for transcutaneous power transfer (Jow and Ghovanloo, 2009). A second pair of vertical coils is wound symmetrically across the PSC pair to establish the data transfer link. Orthogonal orientation and symmetry lower the undesired cross-coupling between the two pairs without affecting the desired coupling within the pairs (Jow and Ghovanloo, 2010). This will minimize the power carrier interference on the data carrier, which can then benefit from any robust data modulation technique. There are also other symmetrical coil geometries, such as figure-8 shown in Figure 7.16*b*, which can attenuate the effects of external

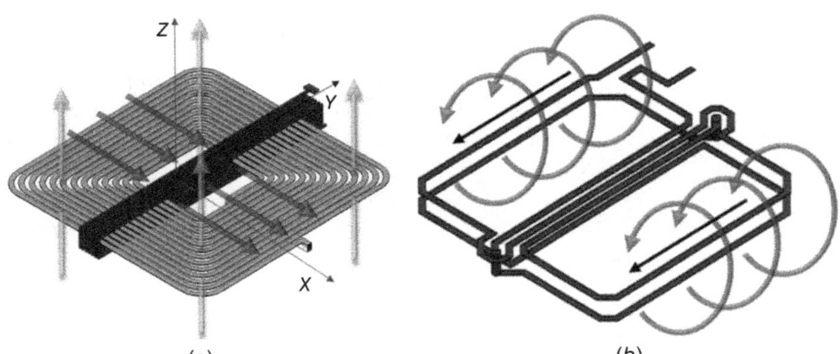

Figure 7.16 (*a*) Pair of PSCs used for power transfer. A second pair of coils can be wound symmetrically across the PSC pair for data transfer such that their fluxes are orthogonal and minimize the power carrier interference when the coils are perfectly aligned. (*b*) Symmetric figure-8 coil geometry to attenuate effects of strong common-mode magnetic fields due to power carrier interference (Jow and Ghovanloo, 2010).

common-mode magnetic fields and reduce cross-coupling from power coils. In such designs, a pair of planar figure-8 coils is utilized in which the electromotive force (EMF) induced from the power carrier in one loop opposes the same in the other loop. Therefore, in a perfectly aligned condition, the power carrier interference becomes negligible (Jow and Ghovanloo, 2010).

7.5.4 Pulse-Based Data Transmission

The majority of modulation techniques that have been used in near-field inductive links and discussed earlier modify a sinusoidal carrier signal based on the data to be transferred across the link. Even though modulating a carrier signal provides a robust mean to transfer data, generation of the carrier signal at a power level that ensures sufficient signal-to-noise ratio (SNR) at the Rx involves consuming a considerable amount of power at the Tx, which is scarce on the IMD side. Therefore, carrier-based modulation techniques are more suitable for the downlink. Because of the significant electromagnetic field absorption in the tissue, which exponentially increases with the carrier frequency, high bandwidth must be achieved at the lowest possible carrier frequencies. This requirement rules out the majority of commercially available wide-band wireless protocols, such as Bluetooth or WiFi, which operate well in the air at 2.4 GHz but not in the tissue. On the other hand, there are specific standards, such as Medical Implant Communication Service (MICS), operating in the 402–405-MHz band, which can only offer a limited bandwidth (300 kHz).

One solution, recently proposed by Inanlou and Ghovanloo (2011), is to substitute the carrier signal with a series of sharp and narrow pulses, which require much less power to generate them. The timing and amplitude of these pulses have been carefully selected to reduce the intersymbol interference (ISI) on the Rx side and make it easier to detect and recover the serial data bit stream. This method, called *pulse harmonic modulation*, takes advantage of the residual oscillation in high-Q LC tanks. To transmit each bit 1, the PHM transmitter generates a sharp pulse at the onset of the bit period to initiate an oscillatory response in the Rx high-Q LC tank, as shown in Figure 7.17. A second pulse is then generated with specific amplitude, $P < 1$, and delay, t_d, with respect to the initial pulse that suppresses the residual oscillation across the Rx LC tank well before the end of the bit period. No pulses are transmitted

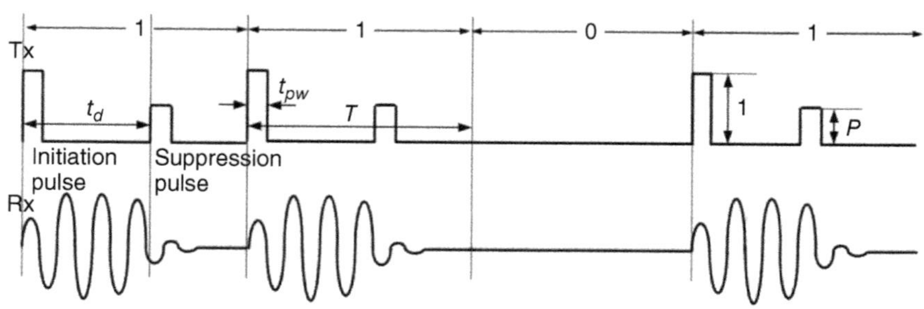

Figure 7.17 PHM conceptual waveforms including their key parameters.

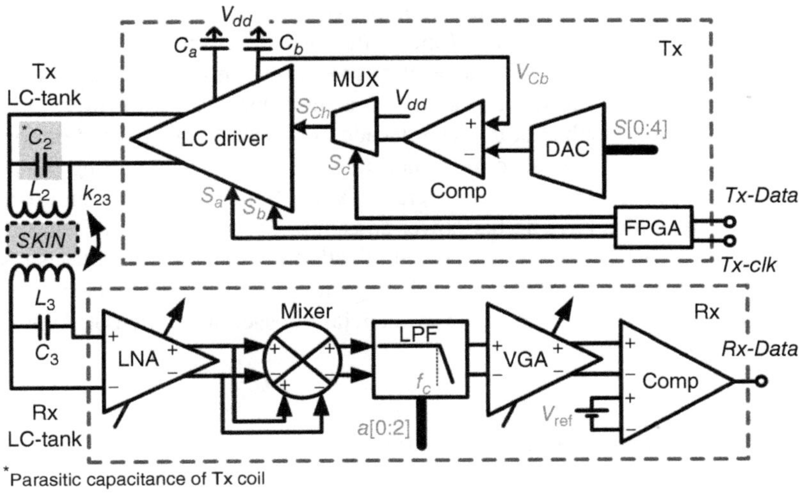

Figure 7.18 Block diagram of PHM-based transceiver.

in this scheme for bit 0. This method allows for reaching high data rates in excess of 10 Mbps without reducing the inductive link Q factor, thus significantly improving the transmission range and selectivity of the data link in rejecting out-of-band interferences such as the power carrier, without consuming too much power in the IMD.

Figure 7.18 shows the block diagram of a PHM transceiver. On the Tx side, C_b is charged up to a voltage set by a digital-to-analog converter (DAC) and C_a is charged up to the supply voltage, V_{dd}. Then C_a is discharged into the primary data coil followed by C_b via an LC driver circuit according to a specific timing that is dictated by an field-programmable gate array (FPGA), which accepts the serial data bit stream (Tx-Data). In the Rx block, which operates based on a noncoherent energy detection (ncED) scheme, the received oscillatory signal is amplified, squared, and low-pass filtered. Finally, a comparator recovers the serial data bit stream. This transceiver has achieved a data rate of 10.2 Mbps with bit error rate of 6.3×10^{-8} at 1 cm coupling distance.

It should be noted that there are also other pulse-based near-field data transmission methods, developed for chip-to-chip communication and body area networks (Miura et al., 2006; Yoo et al., 2009; Lee et al., 2010). However, they require an inductive link with a low Q to achieve wide bandwidth, which is not suitable for the IMD applications, where higher transmission distance and better selectivity for noise and interference rejection are necessary (Inanlou and Ghovanloo, 2011).

7.6 BROADER APPLICATIONS

Inductive coupling as a viable technique to power up devices and also establish a wireless data communication link is widely used to remove direct electrical contact between the energy source and the target device. Such applications include RFID transponders and cochlear implants (one of the most successful IMD applications) with power consumptions in the range of microwatts to milliwatts. The use of this technique to wirelessly transfer energy across a short distance is, however, expected to see an explosive growth over the next decade in a much broader range of applications from advanced IMDs, such as retinal implants and BCIs to cutting the power cord in charging mobile electronics, operating small home appliances, and recharging electric cars, which have higher power consumptions in the order of hundreds of milliwatts to kilowatts. Yet another new application for inductive data transmission is known as near-field communication (NFC), which will be added to the next generation of smartphones mainly to perform financial transactions in lieu of credit cards (Leavitt, 2010).

7.7 FUTURE RESEARCH DIRECTIONS

PTE is a key merit in inductive links for reducing heat dissipation, exposure to electromagnetic fields, size of the main energy source, and interference with nearby electronics. Multicoil links in the simple form of three and four coils have been proposed to increase the efficiency. However, they have not been studied as well as the conventional two-coil links. Several research groups have considered using metamaterials to further increase the PTE by focusing the Tx magnetic field on the Rx and, therefore, increasing the mutual coupling (Choi and Seo, 2010; Urzhumov and Smith, 2011; Wang et al., 2011). However, these efforts are still at early stages and their possible applications in IMDs have not been explored. Moreover, designing efficient coils, inductive links, and antennas inside and around the body, while considering the complex implant and body environment, still needs additional work.

As the number of electrodes in neural recording and stimulation devices increases, there is a tremendous need for high-bandwidth short-rage data links with low power per bit data transferred. Using multiple carriers for power and data can potentially be a feasible candidate for this purpose. However, utilization of such links for actual IMDs while considering their small size and other practical constraints, such as the effects of misalignment, has not been studied yet. Pulse-based data transmission, proposed recently for IMDs, can significantly reduce the Tx power consumption. However, it could still be susceptible to the power carrier interference and coil misalignments. Therefore, proposing new techniques for pulse-based inductive data links that are robust enough against the power carrier interference and motion artifacts are of interest.

7.8 CONCLUSION

In this chapter, fundamental principles, design constraints, and practical considerations of wireless data and power transmission by means of inductive coupling have been described. Fundamental near-field equations governing self- and mutual inductance between magnetically coupled coils have been reviewed. Also important geometric and circuit parameters that can affect the coupling coefficient and PTE of an inductive link on both Tx and Rx sides have been specified. The inductive link PTE was formulated utilizing reflected load theory. It can be concluded that the mutual coupling between the coils, k_{23}, has the most significant effect on the PTE followed by the primary and secondary coil quality factors, Q_3 and Q_2, as well as the load resistance (R_L), source resistance (R_s), and carrier frequency (f_0). In order to achieve the optimal loading condition and reduce the negative effect of R_s on Q_2, three- and four-coil inductive links have been proposed. An optimization procedure for certain types of links that utilize PSCs, which are suitable for IMD and RFID applications, was presented.

Safety is paramount in the design of every medical device, especially those that are meant to be implantable. In the design of transcutaneous power transmission links, the intensity of the magnetic field and its frequency of operation are the key factors that need to be chosen based on the electromagnetic safety standard guidelines, expressed in terms of the SAR of the power induced in the human body.

Three major carrier modulation techniques for data transmission, ASK, FSK, and PSK, have been discussed and compared. It can be concluded that ASK provides the simplest solution for forward data transmission and would be an appropriate choice when low-data-rate and high-PTE link with a single carrier is needed. FSK provides high data rates and a robust link at the expense of more complexity and reduced PTE. PSK can offer the highest bandwidth; however, the synchronization issues might result in a high bit error rate (BER) or high sensitivity to interference and artifacts. Conflicting requirements for high PTE and high bandwidth through the same inductive link have led designers to multicarrier inductive links, in which two or three separate carrier signals are used for power transmission, uplink, and downlink. Pulse-based near-field data transmission has recently become popular due to its low power consumption, robustness, and high bandwidth, which are important in the IMD and FRID applications.

IMDs have a relatively short yet outstanding history of success in the treatment of a range of ailments and disabilities, such as heart disease, some neurological disorders, and deafness. Recent advancements in microelectronics, low-power circuits, and communications have made it possible to extend the range of IMD applications into challenging areas such as blindness, epilepsy, stroke, and paralysis. Nevertheless, today's aging population drives a wide range demand for even more advanced and efficacious medical treatments using smaller, safer, more functional, and more cost effective IMDs, which are in turn in need of more power and larger data volumes.

REFERENCES

Abrial A, Bouvier J, Renaudin M, Senn P, Vivet P. 2001. A new contactless smart card IC using an on-chip antenna and an asynchronous microcontroller. *JSSC*, 36:1101–1107.

Allan R. 2003. Medtronic sets the pace with implantable electronics. *Elect. Des.*, 51:52–56.

Arabi K, Sawan MA. 1999. Electronic design of a multichannel programmable implant for neuromuscular electrical stimulation. *IEEE Trans. Rehab. Eng.*, 7:204–214.

Baker MW, Sarpeshkar R. 2007. Feedback analysis and design of RF power links for low-power bionic systems. *TBCAS*, 1:28–38.

Bawa G, Ghovanloo M. 2008. An active high power conversion efficiency rectifier with built-in dual-mode back telemetry in standard CMOS technology. *TBCAS*, 2:184–192.

Boyer S, Sawan M, Abdel-Gawad M, Robin S, Alhilali MM. 2000. Implantable selective stimulator to improve bladder voiding: Design and chronic experiment in dogs. *IEEE Trans. Rehab. Eng.*, 8:789–797.

Catrysse M, Hermans B, Puers R. 2004. An inductive power system with integrated bi-directional data-transmission. *Sensors Actuators* 115:221–229.

Chae MS, Yang Z, Yuce MR, Hoang L, Liu W. 2009. A 128-channel 6 mW wireless neural recording IC with spike feature extraction and UWB transmitter. *TNSRE*, 17:312–321.

Chen K, Yang Z, Hoang L, Weiland J, Humayun M, Liu W. 2010. An integrated 256-channel epiretinal prosthesis. *JSSC*, 45:1946–1956.

Choi J, Seo C. 2010. High-efficiency wireless energy transmission using magnetic resonance based on negative refractive index metamaterial. *Prog. Electromagn. Res.*, 106:33–47.

Fernandez JM, Borras JA. 2001. Contactless battery charger with wireless control link. US patent 6, 184,651.

Fernandes R, Diniz B, Ribeiro R, Humayun M. 2012. Artificial vision through neuronal stimulation. *Neurosci. Lett.*, 519:122–128.

Finkenzeller K. 2003. *RFID-Handbook*, 2nd ed. Hoboken, NJ: Wiley.

Fisher R. 2011. Direct brain stimulation is an effective therapy for epilepsy. *Neurology* 77:1220–1221.

Galbraith DG, Soma M, White RL. 1987. A wide-band efficient inductive transdermal power and data link with coupling insensitive gain. *TBME*, 34:265–275.

Gao H, Walker R, Nuyujukian P, Makinwa K, Shenoy K., Murmann B, Meng TH. 2012. HermesE: A 96-channel full data rate direct neural interface in 0.13μm CMOS. *JSSC*, 47:1043–1055.

Ghovanloo M, Atluri S. 2007. A wideband power-efficient inductive wireless link for implantable microelectronic devices using multiple carriers. *TCASI*, 54:2211–2221.

Ghovanloo M, Najafi K. 2004. High data rate frequency shift keying demodulation for wireless biomedical implants. *TCASI*, 51:2374–2383.

Ghovanloo M, Najafi K. 2004. A modular 32-site wireless neural stimulation microsystem. *JSSC*, 39:2457–2466.

Grover FW. 1946. *Inductance Calculations Working Formulas and Tables*. New York: D. Van Nostrand Company.

Gosalia K, Lazzi G, Humayun M. 2004a. Investigation of a microwave data telemetry link for a retinal prosthesis. *TMTT*, 52:1925–1933.

Gosalia K, Weiland J, Humayun M, Lazzi G. 2004b. Thermal elevation in the human eye and head due to the operation of a retinal prosthesis. *TBME*, 51:1469–1477.

Hamam RE, Karalis A, Joannopoulos JD, Soljacic M. 2009. Efficient weakly-radiative wireless energy transfer: An EIT-like approach. *Ann. Phys.*, 324:1783–1795.

Harrison RR. 2007. Designing efficient inductive power links for implantable devices. *ISCAS*, 2080–2083.

Hawley ES, Hargreaves EL, Kubie JL, Rivard B, Muller RU. 2002. Telemetry system for reliable recording of action potentials from freely moving rats. *Hippocampus*, 12:505–513.

Inanlou F, Ghovanloo M. 2011. Wideband near-field data transmission using pulse harmonic modulation. *TCASI*, 58:186–195.

Inanlou F, Kiani M, Ghovanloo M. 2011. A 10.2 Mbps pulse harmonic modulation based transceiver for implantable medical devices. *JSSC*, 46:1296–1306.

Institute of Electrical and Electronics Engineers (IEEE). 1999. IEEE standard for safety levels with respect to human exposure to radio frequency electromagnetic fields, 3 kHz to 300 GHz.

Jow U, Ghovanloo M. 2007. Design and optimization of printed spiral coils for efficient transcutaneous inductive power transmission. *TBCAS*, 1:193–202.

Jow U, Ghovanloo M. 2009. Modeling and optimization of printed spiral coils in air, saline, and muscle tissue environments. *TBCAS*, 3:339–347.

Jow U, Ghovanloo M. 2010. Optimization of data coils in a multiband wireless link for neuroprosthetic implantable devices. *TBCAS*, 4:301–310.

Kaiser U, Steinhaugen W. 1995. A low-power transponder IC for high-performance identification systems. *JSSC*, 30:306–310.

Kazimierczuk MK, Czarkowski D. 1995. *Resonant Power Converters*. New York: Wiley-Interscience.

Kazimierczuk MK, Puczko K. 1987. Exact analysis of class E tuned power amplifier at any Q and switch duty cycle. *TCASI*, 34:149–159.

Karalis A, Joannopoulos J, Soljacic M. 2007. Efficient wireless non-radiative mid-range energy transfer. *Ann. Phys.*, 323:34–48.

Kendir GA, Guoxing W, Sivaprakasam M, Bashirullah R, Humayun MS, Weiland JD. 2005. An optimal design methodology for inductive power link with class-E amplifier. *TCASI*, 52:857–866.

Kiani M, Ghovanloo M. 2010. An RFID-based closed-loop wireless power transmission system for biomedical applications. *TCASII*, 57:260–264.

Kiani M, Jow U, Ghovanloo M. 2011. Design and optimization of a 3-coil inductive link for efficient wireless power transmission. *TBCAS*, 5:579–591.

Kiani M, Ghovanloo M. 2012. The circuit theory behind coupled-mode magnetic resonance-based wireless power transmission. *TCASI*, 59.

Kuiken TA, Miller LA, Lipschutz RD, Lock BA, Stubblefield K, Marasco PD, Zhou P, Dumanian GA. 2007. Targeted reinnervation for enhanced prosthetic arm function in a woman with a proximal amputation: a case study. *Lancet*, 369:371–380.

Kurs A, Karalis A, Moffatt R, Joannopoulos JD, Fisher P, Soljacic M. 2007. Wireless power transfer via strongly coupled magnetic resonances. *Science*, 317:83–86.

Leavitt N. Payment applications make E-commerce mobile. *IEEE Comput. Soc.*, 43:19–22.

Lee HM, Ghovanloo M. 2011. An integrated power-efficient active rectifier with offset-controlled high speed comparators for inductively powered applications. *TCASI*, 58:1749–1760.

Lee J, Rhew H, Kipke DR, Flynn M. 2010a. A 64 channel programmable closed-loop neurostimulator with 8 channel neural amplifier and logarithmic ADC. *JSSC*, 45: 1935–1945.

Lee S, Song K, Yoo J, Yoo HJ. 2010b. A low-energy inductive coupling transceiver with cm-range 50-Mbps data communication in mobile device applications. *JSSC*, 45:2366–2374.

Lee SB, Lee HM, Kiani M, Jow U, Ghovanloo M. 2010c. An inductively-powered scalable 32-channel wireless neural recording system-on-a-chip for neuroscience applications. *TBCAS*, 4:360–371.

Lee TH. 2004. *The Design of CMOS Radio-Frequency Integrated Circuits*, 2nd ed. New York: Cambridge University Press.

Lin JC. 1986. Computer methods for field intensity predictions. In *CRC Handbook of Biological Effects of Electromagnetic Fields, pp.* 273–313.

Liu W, Vichienchom K, Clements M, DeMarco SC, Hughes C, McGucken, E, Humayun MS, De Juan E, Weiland JD, Greenberg R. 2000. A neuro-stimulus chip with telemetry unit for retinal prosthetic device. *JSSC*, 35:1487–1497.

Mandal S, Sarpeshkar R. 2008. Power-efficient impedance-modulation wireless data links for biomedical implants. *TBCAS*, 2:301–315.

Marschner C, Rehfuss S, Peters D, Bolte H, Laur R. 2002. A novel circuit concept for PSK-demodulation in passive telemetric systems. *Microelectron. J.*, 33:69–75.

Mathieson K, Loudin J, Goetz G, Huie P, Wang L, Kamins T, Galambos L, Smith R, Harris J, Sher A, Palanker D. 2012. Photovoltaic retinal prosthesis with high pixel density. *Nature Photonics*, 6:391–397.

Miura N, Mizoguchi D, Inoue M, Sakurai T, Kuroda T. 2006. A 195-Gb/s 1.2-inductive inter-chip wireless superconnect with transmitter power control scheme for 3-D-stacked system in a package. *JSSC*, 41:23–33.

Mohseni P, Najafi K. 2004. A fully integrated neural recording amplifier with dc input stabilization. *TBME*, 51:832–837.

Morrel M. 2011. Responsive cortical stimulation for the treatment of medically intractable partial epilepsy. *Neurology*, 77:1295–1304.

Neihart NM, Harrison RR. 2005. Micropower circuits for bidirectional wireless telemetry in neural recording applications. *TBME*, 52:1950–1959.

Nicolelis MAL. 2003. Brain-machine interfaces to restore function and probe neural circuits. *Nature*, 4:417–422.

Obeid I, Morizio JC, Moxon KA, Nicolelis MAL, Wolf PD. 2003. Two multichannel integrated circuits for neural recording and signal processing. *TBME*, 50:255–258.

Raab FH, Sokal NO. 1978. Transistor power losses in the Class E tuned power amplifier. *JSSC*, 13:912–914.

Raker P, Connell L, Collins T, Russell D. 2001. Secure contactless smartcard ASIC with DPA protection. *JSSC*, 36:559–565.

RamRakhyani AK, Mirabbasi S, Chiao M. 2011. Design and optimization of resonance-based efficient wireless power delivery systems for biomedical implants. *TBCAS*, 5:48–63.

Rauschecker JP, Shannon RV. 2002. Sending sound to the brain. *Science*, 295:1025–1029.

Rizk M, Bossetti CA, Jochum TA, Callender SH, Nicolelis MAL, Turner DA, Wolf PD. 2009. A fully Implantable 96-channel neural data acquisition system. *Neural Eng.*, 6.

Sadiku MNO. 2007. *Elements of Electromagnetics*, 4th ed. Oxford University Press.

Sawan M, Hu Y, Coulombe J. 2005. Wireless smart implants dedicated to multichannel monitoring and microstimulation. *IEEE Cir. Sys. Mag.*, 5:21–39.

Schwartz A, Cui T, Weber D, Moran D. 2006. Brain-controlled interfaces: Movement restoration with neural prosthetics. *Neuron*, 52:205–220.

Shire DB, Kelly SK, Jinghua C, Doyle P, Gingerich MD, Cogan SF, Drohan WA, Mendoza O, Theogarajan L, Wyatt. JL, Rizzo JF. 2009. Development and implantation of a minimally invasive wireless subretinal neurostimulator. *TBME*, 56:2502–2511.

Simard G, Sawan M, Massicotte D. 2010. High-speed OQPSK and efficient power transfer through inductive link for biomedical implants. *TBCAS*, 4:192–200.

Smith B, Tang Z, Johnson MW, Pourmehdi S, Gazdik MM, Buckett JR, Peckham PH. 1998. An externally powered, multichannel, implantable stimulator-telemeter for control of paralyzed muscle. *TBME*, 45:463–475.

Sokal NO, Sokal AD. 1975. Class-E—A new class of high-efficiency tuned single-ended switching power amplifiers. *JSSC*, 10:168–176.

Soma M, Galbraith DG, White RL. 1987. Radio-frequency coils in implantable devices: misalignment analysis and design procedure. *TBME*, 34:276–282.

Suaning GJ, Lovell NH. 2001. CMOS neuro-stimulation ASIC with 100 channels, scalable output, and bidirectional radio-freq. telemetry. *TBME*, 48:248–260.

Sullivan CR. 1999. Optimal choice for number of strands in a Litz-wire transformer winding. *IEEE Trans. Power Electron.*, 14:283–291.

Tang Z, Smith B, Schild JH, Peckham PH. 1995. Data transmission from an implantable biotelemeter by load-shift keying using circuit configuration modulator. *TBME*, 42:524–528.

Terasawa Y, Uehara A, Yonezawa E, Saitoh T, Shodo K, Ozawa M, Tano Y, Ohta J. 2008. A visual prosthesis with 100 electrodes featuring wireless signals and wireless power transmission. *IEICE Electron Express*, 5:574–580.

Terman FE. 1943. *Radio Engineers Handbook*. New York: McGraw-Hill.

Theogarajan L. 2012. Strategies for restoring vision to the blind: Current and emerging technologies. *Neurosci. Lett.*, 519:129–133.

Tourkhani F, Viarouge P. 2001. Accurate analytical model of winding losses in round Litz wire windings. *IEEE Trans. Magn.*, 37:538–543.

Troyk PR, DeMichele GA. 2003. Inductively-coupled power and data link for neural prostheses using a class-E oscillator and FSK modulation. *EMBC*, 3376–3379.

Troyk PR, Schwan MAK. 1992. Closed-loop class E transcutaneous power and data link for MicroImplants. *TBME*, 39:589–599.

Urzhumov Y, Smith DR. 2011. Metamaterial-enhanced coupling between magnetic dipoles for efficient wireless power transfer. *Phys. Rev.*, 83.

Venkatraman S, Elkabany K, Long JD, Yao Y, Carmena JM. 2009. A system for neural recording and closed-loop intracortical microstimulation in awake rodents. *TBME*, 56:15–22.

Wang B, Teo KH, Nishino T, Yerazunis W, Barnwell J, Zhang J. 2011. Experiments on wireless power transfer with metamaterials. *Appl. Phys. Lett.*, 98.

Weiland J, Humayun M. 2008. Visual prosthesis. *Proc. IEEE.*, 96:1076–1084.

Yin M, Ghovanloo M. 2011. A low-noise clockless simultaneous 32-channel wireless neural recording system with adjustable resolution. *Analog Integr. Circ. Signal Proc.*, 66:417–431.

Yoo J, Lee S, Yoo HJ. 2009. A 1.12 pJ/b inductive transceiver with a fault tolerant network switch for multi-layer wearable body area network applications. *JSSC*, 44:2999–3010.

Zeng F, Rebscher S, Harrison W, Sun X, Feng H. 2008. Cochlear implants: System design, integration, and evaluation. *Rev. Biomed. Eng.*, 1:115–142.

Zhou L, Donaldson N. 2003. A fast passive data transmission method for eng telemetry. *Neuromodulation*, 6:116–121.

Zhou D, Greenbaum E. 2009. *Implantable Neural Prostheses 1*. New York, NY: Springer.

Zhou M, Yuce MR, Liu W. 2008. A non-coherent DPSK data receiver with interference cancellation for dual-band transcutaneous telemetries. *JSSC*, 43:2003–2012.

Ziaie B, Nardin MD, Coghlan AR, Najafi K. 1997. A single-channel implantable microstimulator for functional neuromuscular stimulation. *TBME*, 44:909–920.

Ziaie B, Rose SC, Nardin MD, Najafi K. 2001. A self-oscillating detuning-insensitive class-E transmitter for implantable microsystems. *TBME*, 48:397–400.

Zierhofer CM, Hochmair ES. 1990. High-efficiency coupling-insensitive transcutaneous power and data transmission via an inductive link. *TBME*, 37:716–722.

Zierhofer CM, Hochmair ES. 1996. Geometric approach for coupling enhancement of magnetically coupled coils. *TBME*, 43:708–714.

8

ANTENNAS AND RF COMMUNICATION

ASIMINA KIOURTI AND KONSTANTINA S. NIKITA

School of Electrical and Computer Engineering, National Technical University of Athens, Athens, Greece

8.1 INTRODUCTION

Wireless biomedical telemetry systems performing in the RF bands attract significant scientific interest due to their promising applications in the field of patient care inside hospital, home, or even outside environments. The major drawback of historically reported biomedical telemetry systems was the wired communication between the medical devices and an exterior monitoring/control equipment, which significantly limited patient comfort and convenience. Therefore, biomedical telemetry devices with wireless communication functionalities appear as a highly promising option toward improving the patients' quality of life and providing medical systems with constant availability, context awareness, reconfigurability, and unobrusiveness (Jovanov et al., 2003). Nowadays, RF biomedical telemetry systems are further facilitated by the continuously increasing advancements in wireless communications, standards, and components.

Wireless medical devices and, thus, antennas integrated into them can be divided into three categories, according to their placement on or inside the patient's body: (1) on-body, (2) implantable, and (3) ingestible antennas.

Handbook of Biomedical Telemetry, First Edition. Edited by Konstantina S. Nikita.

1. *On-body antennas* are intended to be placed on the human body or be worn as part of an individual's garment. In the first case, antennas are designed and meant to be integrated on medical devices that are placed on the patient's body. In the second case, antennas are designed to form part of the patient's clothing, thus, attempting to enhance the functionalities of everyday clothing. Optimal integration of a wearable antenna into clothing can be achieved by making the antenna itself out of a conductive and smart textile material. Example on-body medical devices include temperature monitors, accelerometers, pulse oximeters, electromyogram sensors, and the like.
2. *Implantable antennas* are intended for integration into medical devices that will be implanted inside the human body by means of a surgical operation. Epidermal as well as deep implantation of the medical device are included in this category. Millions of people worldwide depend upon implantable medical devices to support and improve the quality of their lives. Wireless medical implants are already in use for a wide variety of applications such as temperature monitors, pacemakers, functional electrical stimulators, cochlear and retinal implants, and the like. As technology continues to evolve, their use is expected to rapidly increase from an already large base.
3. *Ingestible antennas* are intended for integration into medical devices that have the form of a capsule to be swallowed and ingested by the patient. Wireless capsule endoscopy receives considerable attention for diagnosing diseases of the gastrointestinal (GI) tract. The purpose is for the capsule to be swallowed and wirelessly transmit images and video outside the body. Ideally, wireless transmission will be performed in real time, while the capsule travels along the GI tract.

   Regardless of their positioning, the utmost aim of on-body, implantable, and ingestible devices is to improve the patients' quality of life by providing them with convenient and continuous monitoring systems. Ever growing miniaturization of antennas along with the recent advancements in electronics components and information and communication technologies (ICTs) promise the development of a wide variety of miniature medical devices that can be worn by, implanted into, or ingested by a patient without causing any sort of discomfort.

Before the widespread use of on-body, implantable, and ingestible antennas, however, there are a number of technical challenges that need to be addressed, mainly in terms of design and channel modeling. Antennas are essential parts of RF biomedical telemetry systems, with their complexity highly depending on the surrounding tissue environment (proximity to the human body), and the electronics, communications, and medical requirements. For example, such antennas are required to be small, lightweight, robust, conformal to the body surface, and biocompatible, yet maintaining an improved radiation and patient safety performance. An overview of these challenges is outlined below for on-body, implantable, and ingestible antennas, along with solutions proposed so far in the literature.

8.2 BACKGROUND INFORMATION

An antenna can be defined as the transitional structure between free space and a guiding device (coaxial line or waveguide), which is used to transport electromagnetic energy from the transmitting source to the antenna or from the antenna to the receiver. In addition to transmitting or receiving energy, an antenna in an advanced wireless system is usually required to optimize or accentuate the radiation energy in some directions and suppress it in others. Thus, the antenna must also serve as a directional device in addition to a probing device.

To describe the performance of an antenna, definitions of various parameters are necessary (Balanis, 2002):

- An *antenna radiation pattern* is defined as a mathematical function or a graphical representation of the radiation properties of the antenna as a function of space coordinates. Radiation properties may include power flux, density, radiation intensity, and the like. In most cases, the radiation pattern is determined in the far-field region and is represented as a function of the directional coordinates.
- The space surrounding an antenna is usually sub-divided into three regions. The *reactive near-field region* is the portion of the near field where the reactive field predominates and is commonly taken to exist at a distance $R < 0.62\sqrt{D^3/\lambda}$, where λ is the wavelength and D is the largest dimension of the antenna. The *radiating near-field (Fresnel) region* is the region wherein radiation fields predominate and wherein the angular field distribution depends upon the distance from the antenna. The outer boundary is taken to be $R < 2D^2/\lambda$. Finally, the *far-field (Fraunhofer) region* is defined as the region where the angular field distribution is essentially independent of the distance from the antenna and is commonly taken to exist at distances greater than $2D^2/\lambda$ from the antenna.
- *Directivity* of an antenna is defined as the ratio of the radiation intensity in a given direction from the antenna to the radiation intensity averaged over all directions. On the other hand, the *gain* of an antenna is defined as the ratio of the intensity, in a given direction, to the radiation intensity that would be obtained if the power accepted by the antenna was radiated isotropically. Therefore, an antenna's directivity is a component of its gain.
- *Input impedance* is defined as the impedance presented by an antenna at its terminals or the ratio of the voltage to current at a pair of terminals or the ratio of the appropriate components of the electric to magnetic fields at a point.
- The *reflection coefficient* of an antenna is defined as the ratio of the reflected waves' amplitude to the amplitude of the incident wave. The reflection coefficient is zero if the transmission line impedance is the complex conjugate of the antenna impedance.
- Finally, the *bandwidth* of an antenna is defined as the range of frequencies, on either side of a center frequency, within which the performance of the antenna (e.g., input impedance, reflection coefficient, etc.) is within an acceptable value of that at the center frequency.

8.3 ON-BODY ANTENNAS

8.3.1 Antenna Design

There exist several challenges regarding the design of on-body antennas, as summarized below. Research on on-body antennas attracts growing interest in both academia and industry, and many efforts have been reported in the literature toward the design of on-body antennas for communication with exterior antennas or even other on-body antennas. It is important to highlight that even though these challenges are more or less common for most on-body antennas, precise requirements on the antenna design should, in each case, be determined by the designer given the medical application in hand.

Selection of Operation Frequency In the United States, the Federal Communications Commission (FCC) has allocated the bands of 608–614, 1395–1400, and 1427–1432 MHz for wireless medical telemetry service (WMTS), as well as the bands of 902–928 and 2400.0–2483.5 MHz for industrial, scientific, and medical (ISM) applications. In Europe, the frequency bands of 433.1–434.8 and 868.0–868.6 MHz are used for ISM applications, and they are defined by the Electronic Communications Committee (ECC). Finally, the Ultrawide Band (UWB) of 3.1–10.6 GHz, which has been authorized by the FCC, receives considerable attention because of the wide bandwidth offered.

Type and Material of the Antenna Microstrip and loop designs are generally applied for on-body antennas because of their conformability and light weight. However, for several on-body links (i.e., links formed between on-body antennas placed on the patient's body), and for many body postures, quarter-wavelength ($\lambda/4$) monopole antennas placed on a small ground plane have been shown to perform even better. The main reason is that the monopole antenna exhibits an omnidirectional radiation pattern, which is highly preferable in cases where the geometry and the characteristics of the wireless link are unknown. On the other hand, in certain medical application and posture scenarios, directive planar inverted-F antenna (PIFA) geometries placed so that the direction of maximum radiation is oriented toward the direction of the receiving antenna have been shown to achieve reduced loss in the wireless link as compared to that achieved by monopole antennas.

Miniature coplanar waveguide (CPW)–fed tapered slot antennas (TSA), as well as planar inverted cone antennas (PICA) have also been reported for UWB biomedical telemetry (Alomainy et al., 2009) (Figure 8.1). Both antennas were proved to meet the UWB requirements with regards to impedance, radiation bandwidth, and impulse responses (pulse fidelity) with average fidelity of 86 and 88% for TSA and PICA, respectively. Unlike the conventional wide-band CPW-fed antenna (Chen, 2005), the TSA employed two diverging tapered slots to provide better impedance matching between 3.1 and 10.6 GHz. The TSA antenna did not show any strong resonance within the band and hence exhibited less group delay variation. Furthermore, its radiation was found to be more directive, thus making it resilient to variations in the tissue and surrounding indoor/outdoor environment.

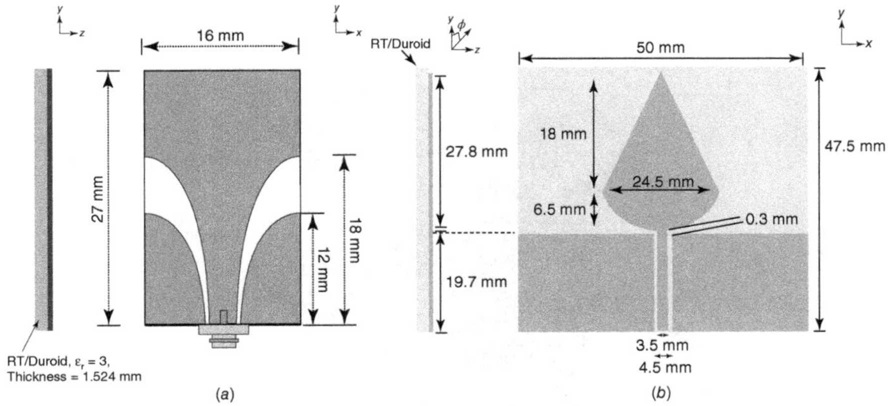

Figure 8.1 Geometry of (*a*) TSA and (*b*) PICA proposed for on-body medical telemetry applications (Alomainy et al., 2009).

In terms of material, the aim should be for the antenna structure to be easy to attach to the body or clothing, without the selected material restricting its possible placements onto the body. Practically, there is no limitation for the material of antennas that are intended for integration into on-body medical devices. However, despite having the potential to be relatively inexpensive, such antennas may be inflexible and high profile. For this purpose, textile antennas with the ability to be attached into everyday clothing are studied. Conductive textile materials, known as electrotextiles (e-textiles), which can function as electronics while physically behaving as textiles, enable the fabrication of textile antennas (Klemm et al., 2004). Both electrical properties (e.g., conductivity) and mechanical properties (e.g., flexibility) are crucial in the fabrication of e-textiles. The methods most commonly used to integrate conductivity into textiles include stitching, weaving, knitting, and printing.

For example, a UWB antenna made from textile materials has been presented (Osman, et al., 2011). The antenna was characterized as a "fully textile" antenna because textiles were used for both its substrate and its conducting parts. The antenna exhibited a wide, 17 GHz of bandwidth, omnidirectional patterns and adequate gain and efficiency values, and was found to meet the requirements of robustness, low power consumption, and compactness. Experimental investigations were carried out, and measurement results were compared with simulations, indicating good agreement.

A weakness of most planar textile antennas proposed in the literature is that they require a coaxial cable to be connected to the transceiver (Tronquo et al., 2006). Such a feeding technique is rigid, and, thus, relatively disturbing to the patient wearing the antenna. To meet this requirement and overcome such difficulties with feeding structures, the coaxial cable feed can be replaced by a microstrip feed line, which couples its power into the antenna through an aperture in the ground plane. The first aperture-coupled patch antenna (ACPA) made of textile materials has been presented in Hertleer et al. (2007) (Figure 8.2). The result was a highly efficient, fully flexible

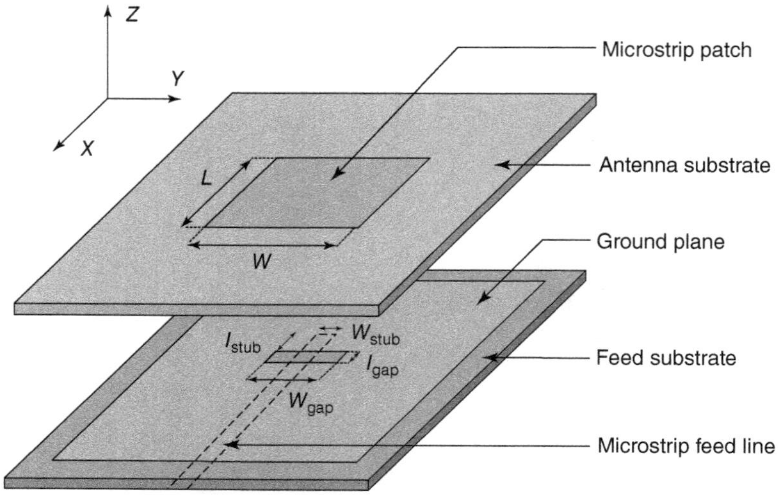

Figure 8.2 Geometry of an ACPA proposed for integration into wearable textile systems (Hertleer et al., 2007).

antenna, that was easy to integrate into garments for wireless medical telemetry in the 2.45 GHz ISM band.

Influence of Human Body Performance of on-body antennas is considerably affected by their close proximity to human tissues. Commonly reported issues include antenna detuning (shift of the resonance frequency and change in the input impedance of the antenna), distortion of the radiation pattern, and degradation of the radiation efficiency (Scanlon and Evans, 2001; Okoniewski and Stuchly, 1996; Wong and Lin, 2005). To get a deeper understanding of the way that on-body antennas interact with the human tissues, it becomes necessary to investigate the performance of such antennas when they are mounted close to the body.

For example, experimental investigations inside a reverberation chamber for five compact on-body antennas operating at 2.45 GHz and worn by nine individual test subjects have shown standard deviations of the antenna radiation efficiency of less than 0.6 dB and resonance frequency shifts of less than 1% (Conway et al., 2008). Such variabilities were found to significantly depend on body tissue coupling, which is, in turn, related to the antenna geometry and radiation characteristics. Even though the test subjects were asked to remain stationary, the measured results included all natural body movements. Further measurements carried out in tissue-emulating phantoms rather than in live subjects demonstrated the suitability of phantoms for assessing the performance of on-body antenna configurations.

For medical applications, antennas need to be immune from frequency detuning. For this purpose, wide-band designs need to be targeted for on-body antennas. At this point, it is important to highlight that, unlike its narrow-band counterpart, the design of a UWB antenna is determined not only by its reflection coefficient characteristics

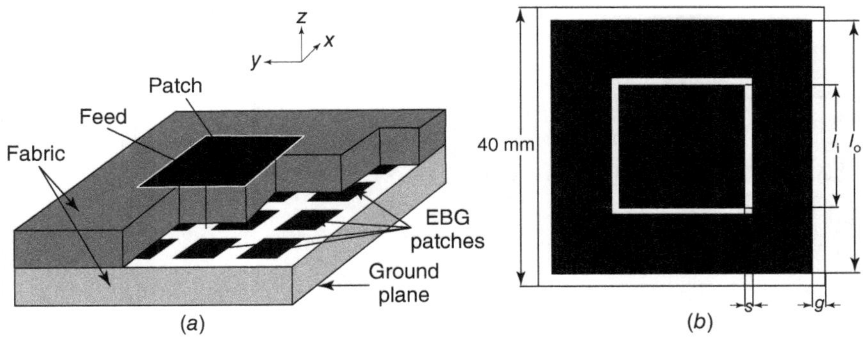

Figure 8.3 Geometry of (*a*) wearable electromagnetic band gap antenna (WEBGA) (Salonen et al., 2004a) and (*b*) dual-band coplanar antenna on EBG substrate (Zhu and Langley, 2009) proposed for overcoming the radiation into the human body.

but also by its ability to preserve the pulse shape. As a result, antennas for UWB systems are required to have very broad impedance bandwidth, as well as stable and constant channel transfer response and high efficiency.

Furthermore, understanding the influence of the human body on the antenna radiation pattern becomes vital in determining its actual radiation performance (Scanlon and Evans, 2001; Chen, 2007). The technique of designing on-body antennas on electromagnetic band gap (EBG) substrates has been reported for overcoming the radiation absorption inside the human body and achieving higher values of antenna gain. Example antennas are shown in Figure 8.3 (Salonen et al., 2004a; Zhu and Langley, 2009). Another reported solution is the design of highly directive antennas that are not affected from the reflections of the human body, such as the miniature TSA UWB antenna proposed in Alomainy et al. (2009) (Figure 8.1*a*).

Multi-band Operation On-body antennas most commonly focus on single-frequency operation. However, there exists an increased demand on multi-frequency or, equivalently, multi-functional on-body antennas. In other words, the antenna is required to operate at different frequencies in order to perform more than one task at the same time, thus accounting for new technologies and services available to the patient/user.

For example, a dual-band wearable textile antenna has been reported for simultaneous operation at 1900 [Global System for Mobile Communications (GSM)] and 2400 MHz [wireless local area network (WLAN) and ISM] (Salonen, et al., 2004b). The antenna had a U-shaped slot structure mounted on the surface of fleece fabric, which enabled the antenna to be flexible, low profile, and lightweight (Figure 8.4*a*). In this study, the antenna was fabricated and measured, with good agreement being observed between measurements and simulations (Figure 8.4*b*).

Effect of Surrounding Electronics Components Design of on-body antennas is most commonly performed by ignoring the structure of the medical device into which the antenna is to be integrated. This involves the surrounding circuitry of the medical

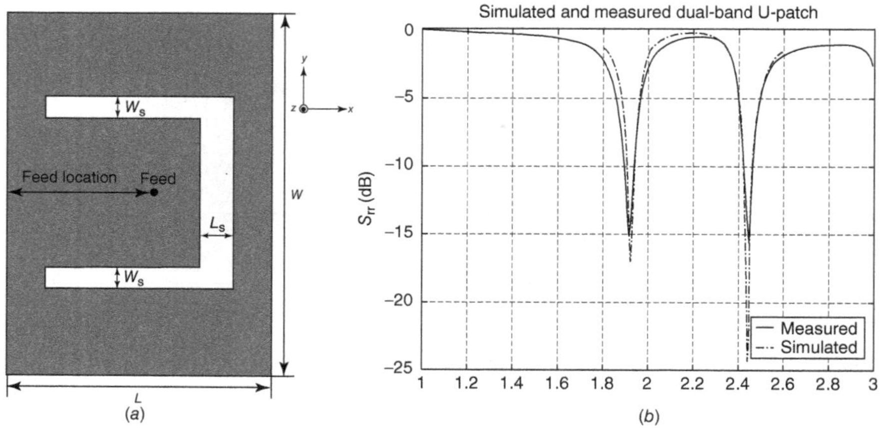

Figure 8.4 (*a*) Geometry of a dual-band wearable textile antenna and (*b*) simulated and measured reflection coefficient frequency response (Salonen et al., 2004b).

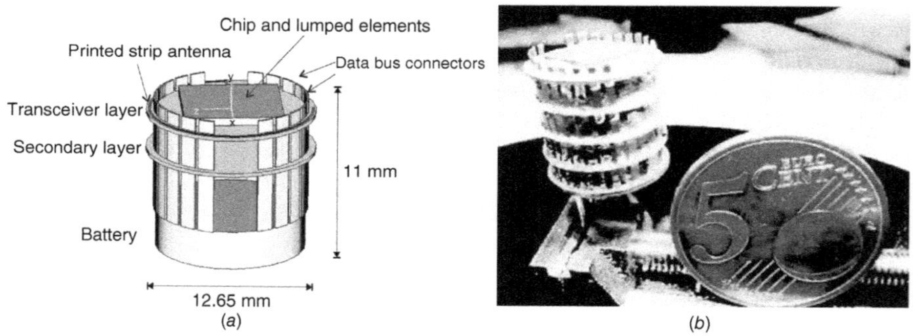

Figure 8.5 (*a*) Schematic of the sensor used in numerical analyses and (*b*) picture of the fabricated sensor used in experimental investigations to assess the effect of the surrounding components on the performance of an on-body antenna (Alomainy et al., 2007a).

device, including integrated circuits (ICs) and lumped elements. However, antenna operation and performance are expected to vary according to those parameters that govern the operation of the overall medical device.

To address this issue, a study has been performed that illustrated the importance of including the details of the medical device in determining and analyzing the performance of an on-body antenna (Alomainy et al., 2007a). The overall sensor structure was designed for operation in the 2.4-GHz band, and was compact enough to address the miniaturization requirements of medical devices. A schematic of the sensor structure applied in the numerical analysis, as well as a picture of the fabricated prototype sensor are shown in Figure 8.5. Table 8.1 compares the antenna gain, radiation, and total efficiency of a printed monopole antenna designed in free space to that of a sensor-matched printed monopole antenna. As compared to a conventional printed

TABLE 8.1 Comparison of Stand-Alone and Sensor-Matched Printed Monopole Antennas (Alomainy et al., 2007a)

Antenna	Gain (dBi)	Radiation Efficiency (%)	Total Efficiency (%)
Stand-alone monopole	−1.2	48	2
Sensor-matched monopole	1.6	77	75

antenna, an antenna that had been optimized for a specific sensor design demonstrated an increase in gain by 2.8 dB, an improvement in radiation efficiency by 29%, and an enhancement in the sensor coverage area by 25%. Limitations related to the placement of the components around the antenna were found to be highly challenging in determining and improving its performance in terms of impedance matching, gain, efficiency, and front-to-back ratio of the radiated energy.

Antenna Diversity The requirements of high data rate and reliable data transmission between on-body medical devices and exterior equipment or between on-body medical devices themselves necessitate the use of multiple antennas. Antenna diversity is a technique where two or more signals from several independent and, thus, uncorrelated diversity branches are combined in different ways to form a unique signal, known as the diversity combined signal. This technique can be accomplished in a number of ways. It may involve the use of different antennas (space diversity), different radiation patterns (pattern diversity), or different polarizations (polarization diversity). Space diversity is achieved by using more than one antenna at the transmitter or receiver side and is by far the most popular technique. For example, a space diversity monopole antenna with variable spacing d is shown in Figure 8.6a (Khan et al., 2009). Pattern diversity is achieved by using different radiation patterns in the same or separate antenna. Finally, polarization diversity is achieved by using a single antenna with multiple polarizations or separate antennas with different polarizations.

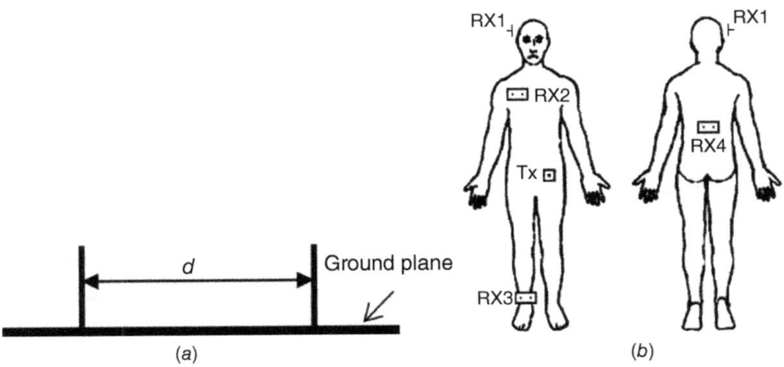

Figure 8.6 (*a*) Space diversity monopole antenna with variable spacing (Khan et al., 2009), and (*b*) analyzed positions for diversity antennas (Serra et al., 2007).

The aim of antenna diversity is to overcome channel fading and provide an efficient link in terms of power. It should be noted that, in the case of on-body channels, fading may occur due to the relative movement of the body parts, shadowing, polarization mismatch, and scattering due to the body and surrounding environment. Use of diversity removes the deep fades, thus decreasing the rate of crossing and the duration of the lower fade levels for the resultant combined signal.

The improvement offered by the diversity technique is most commonly quantified in terms of the diversity gain (DG). Diversity gain is defined as the improvement in the signal-to-noise ratio (SNR) or the bit error rate (BER) over a single antenna with no diversity at some level of outage probability, and strongly depends on the correlation and power imbalance among the signals of the diversity branches. However, second-order statistics can also be used to quantify the results of applying the diversity technique.

Preliminary diversity measurements with monopole antennas at 2.45 GHz have shown that antenna diversity can offer significant improvement for the on-body channels, that is, the channels formed between medical devices placed onto a patient's body. For example, in Serra et al., (2007) a monopole antenna was used as a transmitter (Tx), and two monopoles on a common ground plane were used as a receiver (Rx). Antenna locations under study are shown in Figure 8.6*b*. Measurements were carried out in an anechoic chamber in order to solely assess the scattering and fading effects of the body itself. Table 8.2 summarizes measured diversity gain results at 99% reliability levels. Different combining techniques [selection combining (SC), equal-gain combining (EGC), and maximum ratio combining (MRC)] have been considered for comparison purposes. In Khan et al. (2009), three different types of 2.45-GHz antennas were considered for five body channels, and measurements were taken in anechoic and indoor environments. High diversity gains achieved suggested that multipath is noticeable in the on-body propagation channels. In this study, the PIFA was shown to be the best choice of antenna, exhibiting reasonable path gains and high diversity gains.

Diversity analysis for on-body channels at 5.8 and 10 GHz has also been performed, by using different types of antennas for space and pattern diversity (Khan and Hall, 2009). Same as for the 2.4-GHz case, the PIFA was proved to be the best choice of antenna for the deployment of the diversity technique. The 5.8-GHz frequency was shown to be the best choice in terms of cost (hardware cost and complexity at 10 GHz are higher than those at 5.8 GHz) and path loss (path loss at 5.8 GHz was found to be

TABLE 8.2 Diversity Gain for Antenna Placements Indicated in Figure 8.6*b*

	Standing				Sitting		Jogging	
DG	Rx3	Rx1	Rx2	Rx4	Rx2	Rx4	Rx3	Rx1
SC	2.57	4.61	2.17	5.86	3.10	0.79	5.14	8.57
EGC	3.69	5.53	3.16	6.65	3.29	2.44	6.13	9.62
MRC	4.28	6.00	3.89	7.12	4.22	2.90	6.46	10.28

Source: Serra et al. (2007).

less than that at 10 GHz). Positioning and orientation of the diversity antennas were found to introduce minor discrepancies.

8.3.2 Channel Modeling

Two communication channels can be considered for on-body medical devices:

(a) *On-body channels* refer to the wireless communication between on-body medical devices. Such channels involve transmission paths on the body as well as paths scattering off the local environment of the body (indoors or outdoors).

(b) *Off-body channels* refer to the wireless communication between an on-body medical device and an exterior/remote device and deal with electromagnetic wave propagation around the body.

For example, in the case of a biomedical telemetry application where on-body medical devices communicate with a wearable device, both on- and off-body propagation channels need to be considered. In this case, the wearable device may act as a controller that relays data between the on-body devices and remote stations.

Radiowave propagation in on-body and off-body channels is much more complicated as compared to propagation between two antennas placed in free space, for example, between a cellular phone and a cellular base station. The first reason is the dynamic nature of the human body. During normal activity, movements of the human body can be significant or, in certain cases (e.g., sports), extreme. Even when standing or sitting, the human body is subject to several small movements. As a result, the characteristics of the wireless link and, hence, system performance change rapidly. Furthermore, the human body is a complex problem to be solved numerically, as attributed to variations in anatomy and dielectric properties between individuals, as well as varying dielectric properties of tissues with frequency. In terms of numerical simulations, the body is a rather large problem, which can be considered to be at the edge of what can currently be managed on single-PC-based solvers and requires the use of cluster computing. The need for accurate simulation of the body's postures and movements is also highly challenging.

In order to enable the design of reliable and robust communication links for on-body medical devices, it becomes necessary to develop deterministic and generic channel models. The aim is to provide a clear picture of the propagation channel and its behavior with regard to different body dynamics and exterior environments. Given that there already exist models for indoor and outdoor propagation, it becomes crucial to combine the on-body propagation models with existing environmental models or statistics in order to get the full picture. In any case, during channel characterization, there are difficulties in de-embedding the antenna characteristics from those of the propagation path or, equivalently, separating the performance of the antenna from the channel characteristics.

There have been a number of research studies that aim to characterize and analyze the on- and off-body channels. Results for various transmitter and receiver antenna combinations over various on- and off-body paths have been presented at 2.45 GHz as well as in the UWB, as outlined below.

Studies at 2.45 GHz Measurements of path loss for several on-body antenna locations at 2.45 GHz have been carried out in Hall et al. (2007). Transmitter or receiver locations are shown in Figure 8.7*a*, while measurements were performed both in an anechoic chamber and in other surroundings. Monopoles over a small ground plane were chosen because of their omnidirectional radiation pattern, which rendered the results independent of antenna orientation. Variability in path loss due to different antenna placement and changes in posture was found to be as much as 50 dB. Example path gain variations are shown in Figure 8.7*b*. In Figure 8.7*c*, path gain values are plotted as a function of the separation between the antennas measured along the shortest geodesic path around the body. The path gain of antennas in free space, as given by the Friis formula, is shown in solid line. Data points for line-of-sight (LOS) propagation

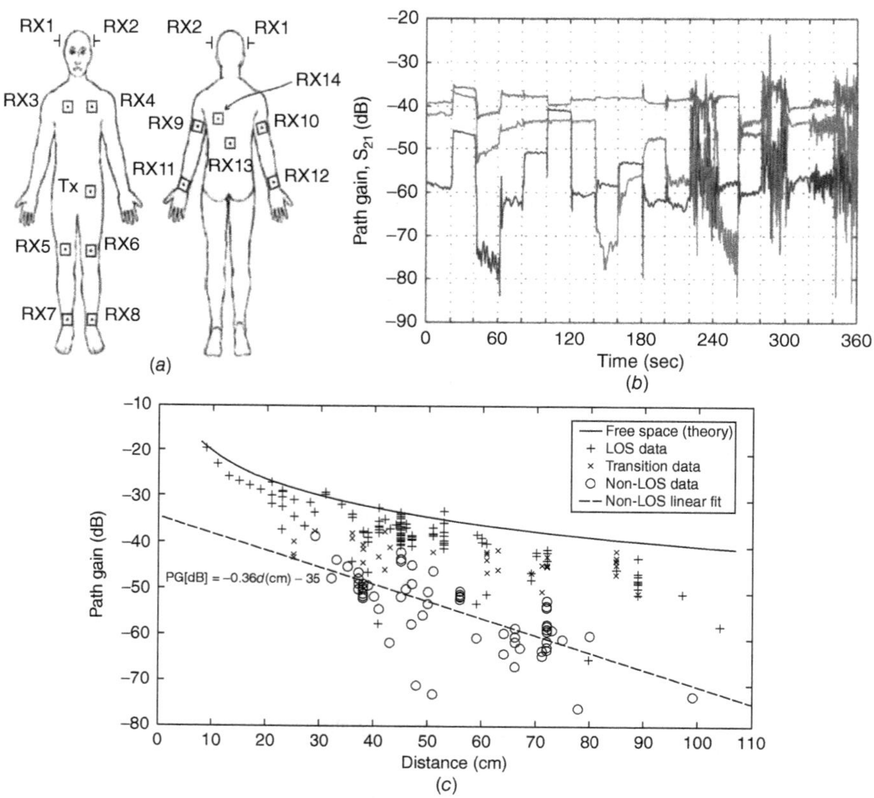

Figure 8.7 (*a*) Possible transceiver locations on the body, (*b*) measured path gain variations for Rx1 (blue), Rx3 (green), and Rx12 (red) and, (c) path gain as a function of the separation between the antennas measured along the shortest geodesic path around the body. (*See insert for color representation of b.*)

scenarios are denoted by +. Most of the LOS data followed the free-space curve with a mean difference between the measured and theoretical values of − 5.1 dB. Thus, the path gain in LOS scenarios was said to be given by

$$G_P\ (\text{dB}) = -5.33 - 20\log_{10} d \tag{8.1}$$

where d is the distance between the antennas in centimeters.

Non-line-of-sight (NLOS) data points are denoted by ○, and were also shown to follow a trend corresponding to an exponential attenuation, according to

$$G_P\ (\text{dB}) = -0.36d - 35 \tag{8.2}$$

Propagation scenarios that could not be identified as LOS or NLOS are shown by × marks. Some of them seemed to follow the same trend as the LOS data points, while others were clustered together with the NLOS data points.

Given the high number of communication paths and their variability associated with different postures, no single channel model was shown to be satisfactory. A classification of body postures based on their spatial and temporal characteristics was rather suggested for future research.

The coupling performance of patch antennas placed in close proximity to a lossy medium that simulated body tissues has further been studied (Conway and Scanlon, 2009). The performance of higher mode microstrip patch antennas (HMMPA) was evaluated both numerically and experimentally and further compared with that of a microstrip patch excited at its fundamental mode (MPA-F) and a fundamental microstrip patch antenna with the addition of a shortening wall (MPA-S). Significant channel fading was recorded during normal activity, as attributed to the dynamic nature of the human body and the surrounding multipath.

Statistical analyses for on-body radio propagation channels have also been performed. Channel measurements using two microstrip patch antennas for a range of body positions and postures were presented and statistically analyzed in Alomainy et al. (2007b). Attenuation attributed to factors such as the body, head, and clothing was estimated as 19.2, 13.0, and 1.7 dB, whereas the measured cumulative distribution function of the data was found to fit to the log-normal distribution. An effort toward gathering and statistically analyzing a large amount of data in practical environments (inside and outside an office block, in the home, while driving) has been performed in Hu et al. (2007). The measured variations in path loss appeared to be nonstationary, thus leading to multimodal probability distributions, which could not be represented by a simple theoretical formula. However, an approximate fit to one of standard probability distributions was found to be achieved by limiting the number of data samples.

Studies in UWB Given the frequency dependency of the dielectric properties of human tissues, characterization of the UWB on-body channels appears to be more challenging than that of the narrow-band channels. The conventional and empirical models available for many narrow-band and wide-band systems are insufficient to describe the UWB channel.

An initial characterization of the UWB on-body channel was presented in Zasowski et al. (2003). A predefined set of nodes with multi-hopping networking was considered in order to determine the energy and power requirements. However, realistic behaviors of the human body, such as different body postures and movements, were not analyzed.

A detailed description of the measurement procedure performed to obtain sufficient path loss data for acceptable characterization of UWB on-body channels has appeared in Alomainy et al. (2005). The performance of horn-shaped self-complementary (HSCA) and PICA antennas was analyzed. Results showed that the hybrid use of different types of UWB antennas could effectively improve channel behavior. As compared to the PICA, the HSCA exhibited reduced values of the mean RMS delay spread for cases where surface waves were dominant in the wave traveling along the human body. Path loss analysis demonstrated NLOS behavior as attributed to attenuation and shadowing by the human body.

Free-space and body-mounted characterization of UWB CPW-fed TSAs and inverted cone antennas (PICAs) was performed in Sani et al. (2010). Results demonstrated that the Nakagami model provided the best fit to the root mean square (RMS) delay spread. If a more omnidirectional antenna (PICA) was used, the goodness of the statistical model became lower, whereas the time delay behavior was degraded. Using a less sensitive receiver, the statistical model was proved to be more deterministic.

Investigations on the effect of body movements on the UWB radio channel have also been performed. An experimental investigation to derive suitable radio propagation models for UWB channels of on-body medical devices was carried out in Abbasi, et al. (2010). Measurements were performed for 35 radio links, considering both stationary and pseudodynamic movements of the human body, and radio channel parameters were extracted and statistically analyzed to provide a radio propagation model with the inclusion of pseudodynamic body motion effects. For the static and pseudodynamic motion scenarios, less than 25 and 30% of the on-body links demonstrated bad link quality, respectively. The upper and lower parts of the body were shown to provide relatively stable radio channels. Another measurement campaign with the subject performing movements of different nature in order to cover a wide range of scenarios and ensure sufficient data collection for statistical analysis was presented in Abbasi et al. (2011a). Four links were targeted (belt to head, belt to chest, belt to wrist, and belt to ankle), and measurements were performed inside an anechoic chamber and indoor environments. The normal distribution was found to provide the best fitting for the path loss, while the RMS delays were better modeled with a log-normal distribution.

Finally, investigations on potential power-efficient UWB orthogonal frequency division multiplexing-based channels (UWB-OFDM) for on-body medical devices have been carried out (Abbasi et al., 2011b). The BER performance of the system under study showed that the lower part of the body for the on-body channels and the lower part of the trunk for the off-body channels provided relatively stable radio channels with a good-quality link.

In terms of modeling, the ray tracing (RT) technique and the finite-difference time-domain (FDTD) method have been widely applied to propagation modeling for UWB systems. For example, a simulation model for UWB indoor radio channels using RT has been proposed in Attiya (2004). The FDTD method has been applied to characterize the UWB radio channel for the chest-to-waist link (Wang and Wang, 2009) and simulate pulse propagation around the torso at 2–6 GHz (Fort, 2005). A deterministic on-body channel model using a subband FDTD method was suggested in Zhao et al. (2006). Numerical results were compared to those from a hybrid uniform geometrical theory of diffraction (UTD)/RT model, demonstrating the ability of the proposed method to model materials with any frequency dependence and deal with complicated on-body radio channels.

8.4 IMPLANTABLE ANTENNAS

8.4.1 Antenna Design

A key and critical component of wireless implantable medical devices is the integrated implantable antenna that enables its bidirectional communication with exterior monitoring/control equipment. Design of implantable antennas is highly intriguing, with the main challenges being outlined below (Kiourti and Nikita, 2012a).

Selection of Operation Frequency A few years ago no globally accepted frequency band had been dedicated for biomedical telemetry of implantable medical devices. The situation changed with the International Telecommunications Union—Radiocommunications Recommendation SA.1346 (ITU-R, 1998), which outlined the use of the 402–405 MHz frequency band for Medical Implant Communications Systems (MICS). The 433.1–434.8 MHz, 868–868.6 MHz, 902.8–928 MHz, and 2400–2500 MHz ISM bands are also suggested for biomedical telemetry of medical implants. However, focus is on the MICS band because of its advantages to be available worldwide and feasible with low-power and low-cost circuits, reliably support high data rate transmissions, fall within a relatively low noise portion of the spectrum, lend itself to small antenna designs, and propagate acceptably through human tissue.

Type and Material of Antenna Patch designs are mostly preferred for implantable antennas because they are highly flexible in design and conformability. In a realistic scenario, the implantable patch antenna will be mounted on the existing hardware of the medical implant, which will also serve as its ground plane.

Implantable antennas must be biocompatible in order to preserve patient safety and prevent rejection of the implant. Furthermore, human tissues are conductive and would short-circuit the implantable antenna if they were allowed to be in direct contact with its metallization. To preserve biocompatibility, while at the same time separating the metal radiator from human tissue, the most widely used approach is to cover the structure by a superstrate dielectric layer (Figure 8.8a). Commonly used biocompatible materials include Teflon (permittivity, $\varepsilon_r = 2.1$, dielectric loss tangent,

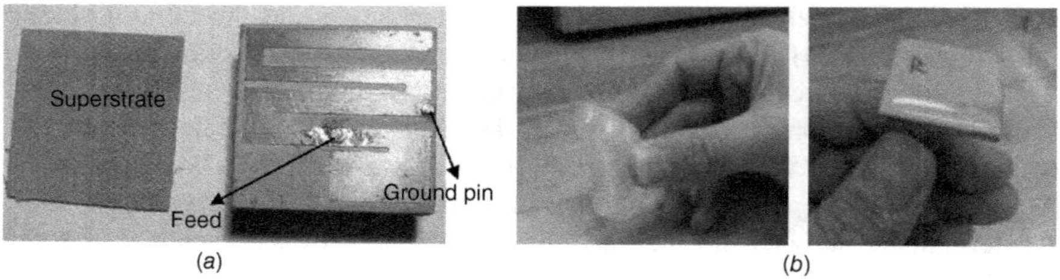

Figure 8.8 Biocompatibility issues for implantable antennas: (*a*) addition of a superstrate (Karacolak et al., 2009) and (*b*) thin-layer encapsulation (Karacolak et al., 2010).

tan $\delta = 0.001$), MACOR ($\varepsilon_r = 6.1$, tan $\delta = 0.005$), and ceramic alumina ($\varepsilon_r = 9.4$, tan $\delta = 0.006$) (Soontornpipit et al., 2004). Insulating the implantable antenna with a thin layer of low-loss biocompatible coating is another reported approach (Figure 8.8*b*). Materials proposed for biocompatible encapsulation include zirconia ($\varepsilon_r = 29$, tan $\delta \approx 0$) (Skrivervik, 2011), polyether ether betone (PEEK) ($\varepsilon_r = 3.2$, tan $\delta = 0.01$) (Abadia et al., 2009), and Silastic MDX–4210 Biomedical–Grade Base Elastomer ($\varepsilon_r = 3.3$, tan $\delta \approx 0$) (Karacolak et al., 2010).

Miniaturization Dimensions of the traditional half-wavelength ($\lambda/2$) or quarter-wavelength ($\lambda/4$) antennas at the frequency bands allocated for medical implants, and especially at the low-frequency MICS band, make them useless for implantable applications. Therefore, miniaturization becomes one of the greatest challenges in implantable antenna design.

Human tissue in which implantable antennas are intended to operate exhibits relatively high permittivity or, equivalently, reduced wave propagation velocity, which, in turn, work to advantageously miniaturize the physical size of the antenna. Use of patch designs for implantable antennas allows for several additional miniaturization techniques, including:

1. *Use of High–Permittivity Dielectric Materials* High-permittivity dielectrics are selected for implantable patch antennas [e.g. ceramic alumina ($\varepsilon_r = 9.4$) (Kiourti et al., 2011) or Rogers 3210 ($\varepsilon_r = 10.2$) (Kiourti and Nikita)] because they shorten the effective wavelength and result in lower resonance frequencies, thus assisting in antenna miniaturization.
2. *Lengthening of Current Flow Path on Patch Surface* Longer effective current flow paths excited on the radiating patch can bring the resonance frequency to lower values and achieve a more compact size for the implantable antenna. Meandered (Kiourti and Nikita, 2011), spiral (Kiourti and Nikita, 2011), waffle-type (Soontornpipit et al., 2005) and hook-slotted (Liu et al., 2008) shaped patches have been suggested for this purpose.
3. *Addition of Shorting Pins* Inserting a shorting pin between the ground and patch planes increases the effective size of the antenna, and, in turn, reduces the required physical dimensions, given a specific operation frequency scenario.

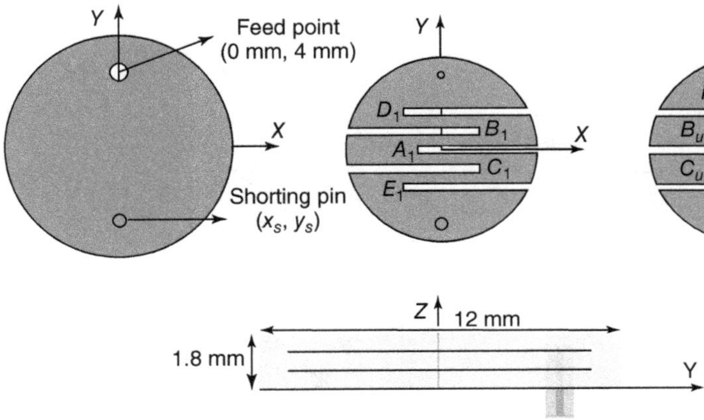

Figure 8.9 Example of a miniature skin-implantable MICS antenna (Kiourti and Nikita, 2012b).

The technique works in much the same way a ground plane doubles the height of a monopole antenna, that is, it typically produces a PIFA with the same resonance performance as a double-sized antenna without the shorting pin (Soontornpipit et al., 2004).

4. *Patch Stacking* Vertically stacking two radiating patches reduces antenna size by increasing (nearly doubling) the length of the current flow path (Kiourti and Nikita, 2012b).
   An example of a skin-implantable antenna that operates in the MICS band and employs all aforementioned techniques to reduce size is shown in Figure 8.9 (Kiourti and Nikita, 2012b). The model is parameterized to include variable-length meanders [denoted by the x coordinate (x_{ij}) of the points marked as $\{ij, i = A\text{–}F, j = 1,u\}$] and a variable positioned shorting pin (x_s, y_s), and the two-step design methodology of Figure 8.10 can be applied to optimize its resonance characteristics for any implantation scenario, in an accurate and fast way (Kiourti and Nikita, 2012b, 2012c). The methodology consists of an initial approximate design inside a tissue-simulating cube, and further quasi-Newton optimization inside a canonical model of the intended implantation site.

Patient Safety Considerations Patient safety issues limit the maximum allowable power incident to the implantable antenna. The specific absorption rate (SAR) (rate of energy deposited per unit mass of tissue) is generally accepted as the most appropriate dosimetric measure, and compliance with international guidelines is assessed. For example, the Institute of Electrical and Electronics Engineers (IEEE) C95.1–1999 standard restricts the SAR averaged over any 1 g of tissue in the shape of a cube to less than 1.6 W/kg (SAR 1 g, max $\leq$ 1.6 W/kg) (IEEE, 1999). The International Commission on Non-Ionizing Radiation Protection (ICNIRP) basic restrictions limit

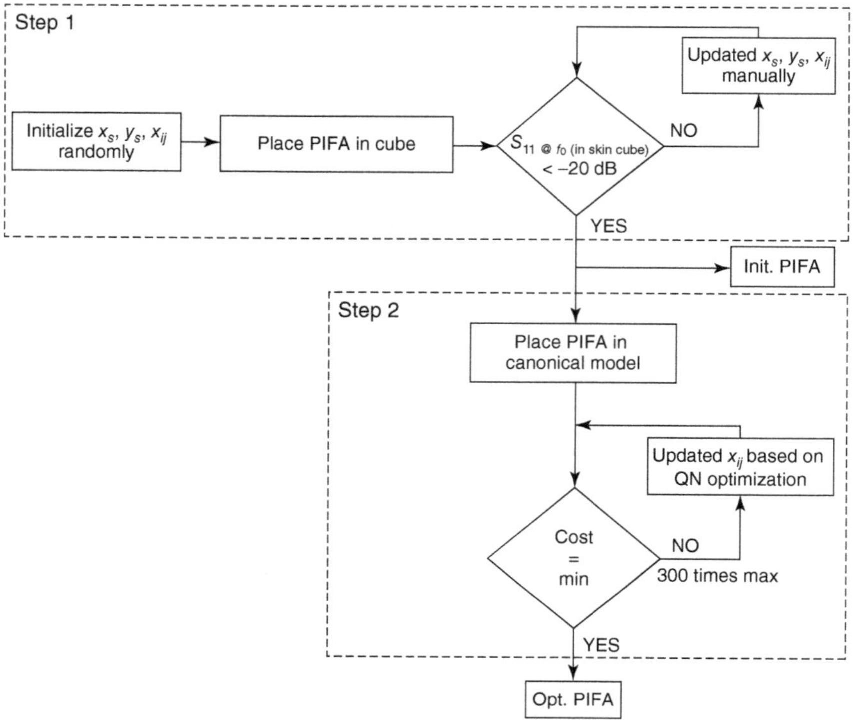

Figure 8.10 Proposed two-step methodology for implantable antenna design (Kiourti and Nikita, 2012b).

the SAR averaged over 10 g of contiguous tissue to less than 2 W/kg (ICNIRP, 1998). To harmonize with the ICNIRP guidelines, the IEEE C95.1–2005 standard restricts the SAR averaged over any 10 g of tissue in the shape of a cube to less than 2 W/kg (SAR 10 g, max ≤ 2 W/kg) (IEEE, 2005).

The power absorbed by the human body by an incident electromagnetic field is given by

$$P_{\text{abs}} = \frac{1}{2}\int \sigma |E|^2 dV \tag{8.3}$$

where σ is the conductivity of the human tissues, and $|E|$ is the intensity of the electric field inside the body. Absorbed power is, thus, related to the electric field, so that maximum SAR values are recorded in the areas where maximum electric field intensities occur.

Based on the deduction that peak averaged SAR values are generated from the high near field, advanced implantable patch antennas can be designed aiming at lower electric field intensities. An attempt to understand the radiation mechanism of an implantable antenna toward appropriately modifying its design for reducing the spatial-averaged SAR in human tissue has been presented in Kim and Rahmat-Samii (2006). Replacing the uniform-width spiral radiator of an implantable MICS PIFA with a non-uniform-width radiator was found to decrease the electric field intensity and, in turn, SAR.

It is worth noting that, contrary to wearable antennas, the high local SAR values achieved for implantable antennas need to be considered more carefully due to higher local energy deposition from the RF source. High values of the local electric field induced in human tissues might even cause some nonthermal effects.

Multi-band Operation If operated continuously, the transceiver of the implantable medical device will consume significant energy and reduce the lifetime of the device. Even though methods for recharging the battery are available (e.g., via an inductive loop approach (Valdastri et al., 2004)), it would be highly advantageous to use the biotelemetry link only when necessary. For this purpose, a transceiver with dual-band operation may be used. The system uses two frequency bands, one for "wake-up," and one for transmission. The transceiver stays in "sleep mode" with a low power consumption until a wake–up signal is sensed in the 2.45 GHz ISM band. In the normal mode, the implantable medical device (IMD) is fully powered and exchanges data in the MICS band. Following the data transfer, the IMD transceiver returns back to the sleep mode.

For example, a dual-band (MICS and ISM) implantable antenna has been proposed in the literature for continuous glucose monitoring (Karacolak et al., 2008). A meandered antenna configuration was considered for optimizing the antenna surface area, and particle swarm optimization was applied to achieve the desired resonance characteristics (Figure 8.11*a*). An advanced antenna design was suggested using a π–shaped

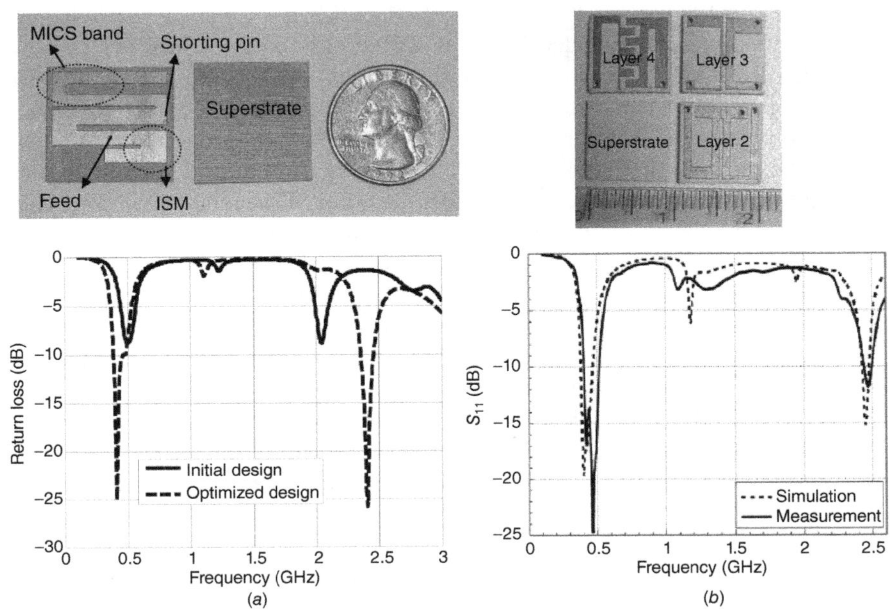

Figure 8.11 Geometry and reflection coefficient frequency response of (*a*) dual-band (Karacolak et al., 2008) and (*b*) triple-band (Huang et al., 2011) implantable antennas.

radiator with stacked and spiral structure, to support triple-band operation with data telemetry (402 MHz), wireless power transmission (433 MHz), and wake-up controller (2450 MHz) (Huang et al., 2011) (Figure 8.11*b*).

Influence of Human Body Several strategies have been proposed for implantable antenna design, which are mainly dictated by the fact that antennas are intended to operate inside biological tissue rather than in free space. The antenna should, therefore, be designed inside free space and further refined for tissue implantation or designed directly inside an environment surrounded by human tissue.

Rucker et al. (2007) designed a patch antenna operating in the MICS band in free space and further implanted it inside the skin tissue of an anatomical head model. A resonance frequency detuning was observed as attributed to the capacitive loading effect of the surrounding tissues, and a varactor diode with tuning capability was inserted to refine resonance.

In Abadia et al. (2009), a MICS antenna was designed in free space aiming at a high gain value (higher than − 20 dB) to account for subsequent body absorption losses. The antenna was optimized in free space to minimize size and further be covered by a biocompatible layer and placed inside tissue material. Design modifications were performed to account for the frequency shift induced by the presence of encapsulation and human tissue.

When designing implantable antennas directly inside tissue material, the simplest and fastest option is to use a single-layer tissue model of the intended implantation tissue. Following this methodology, antennas are designed for a "generic" tissue implantation scenario. Simplified tissue models in the shape of a cube (Kiourti and Nikita, 2011, 2012b, 2012d; Kiourti et al., 2011; 2013), a rectangular parallelepiped (Liu et al., 2008), and a cylinder (Liu et al., 2009) have been used for this purpose. Design is performed by selecting the dielectric material, and subsequently optimizing all antenna design parameters to refine tuning at the desired operation frequency. Another option is to design the antenna for a specific implantation site. A multilayer tissue model is selected in this case, with either finite or infinite dimensions. For example, implantable antennas intended for trunk (Karacolak et al., 2008) and chest (Kim and Rahmat-Samii, 2004) implantation have been directly designed inside three-layer planar tissue models consisting of skin, fat, and muscle tissues.

An advanced two-step design methodology was proposed for implantable antennas that emphasizes design speed-up and optimized resonance performance inside specific implantation sites (Kiourti and Nikita, 2012b). This involves approximate antenna design inside a simplified tissue model (a cube filled with the intended tissue material) and further quasi-Newton optimization inside a canonical model of the intended implantation site. Despite being optimized inside a canonical tissue model, the designed antennas were shown to exhibit insignificant resonance discrepancies inside detailed anatomical tissue models (Kiourti and Nikita, 2013).

Experimental Investigations Prototype fabrication of implantable antennas meets all classical difficulties of miniature antennas. For example, additional glue layers used to affix all components together strongly affect antenna performance by shifting

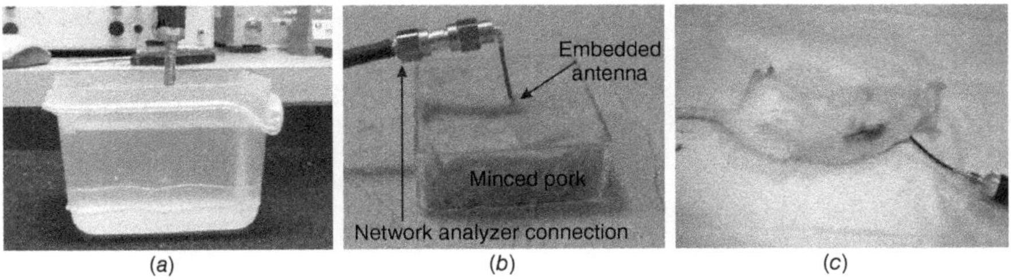

Figure 8.12 Experimental investigations for implantable antennas: (*a*) in vitro testing (Kiourti and Nikita, 2012b), (*b*) testing inside animal tissue samples (Huang et al., 2011), and (*c*) in vivo testing (Karacolak et al., 2010). (*See insert for color representation of b and c.*)

its resonance frequency and degrading its matching characteristics. Furthermore, the coaxial cable feed used to connect the antenna with the network analyzer may give rise to radiating currents on the outer part of the cable, thus deteriorating measurements. Based on the above, design parameters of the numerical antenna model must be slightly adjusted to take prototype fabrication considerations into account.

Testing inside phantoms is relatively easy and practical to implement. The fabricated prototype is immersed inside a tissue phantom (i.e., a container filled with a liquid or gel material that mimics the electrical properties of biological tissue) and measured (Figure 8.12*a*). Reflection coefficient measurements are performed by immersing the prototype antenna inside a tissue-emulating phantom and connecting it to a network analyzer through a coaxial cable.

Use of animal tissue samples provides an easy approach to mimicking the frequency dependency characteristic of the electrical properties of tissues (Figure 8.12*b*). This can prove highly advantageous when carrying out measurements for multiband implantable antennas. In the literature, an implantable dual-band patch antenna that resonates at 380 and 440 MHz has been tested inside test tissue obtained by grinding the front leg of a pig (Lee et al., 2009). A dual-band skin-implantable patch antenna operating in the MICS and 2450 MHz ISM bands has also been tested in real animal skin (Karacolak et al., 2009). Skin samples were extracted from the dorsal area of three donor rats to cover the designed antenna, and measurements were performed within 30 min of euthanization. Finally, a triple-band implantable patch antenna has been tested inside a minced front leg of a pig (Huang et al., 2011).

In vivo testing inside living animals is highly challenging too (Figure 8.12*c*). An in vivo testing protocol needs to be developed before the experimental investigations, which will deal with the choice and number of animals, presurgical preparation, anesthesia, surgical procedure, measurements, and postsurgical treatment. In vivo studies reported in the literature are very limited. The return loss frequency response of a skin-implantable antenna has been measured using rats as model animals (Karacolak et al., 2010). The antenna was implanted by means of a surgical operation inside the dorsal midline of three rats, and euthanasia was applied after the measurements (approximately 13–15 min after the surgery). Canine studies for trans-scalp evaluation of a scalp-implantable antenna at 2450 MHz have also been presented (Kawoos et al., 2008). Canine models were selected to ensure a large head size, and an intracranial pressure monitoring device with an integrated PIFA was fixed to the skull. The monitor was tested while the dog was still under anesthesia. After the measurements, the animal was allowed to emerge from anesthesia and taken to the recovery area.

8.4.2 Channel Modeling

Modeling the biomedical telemetry channel for implantable antennas is a highly challenging task. Unlike free-space propagation, the various tissues and organs within the body have their own unique electrical characteristics, the lossy medium absorbs electromagnetic (EM) energy, and the presence of objects causes EM field refraction, diffraction, reflection, and absorption.

Transmission Coefficient Studies Even though most studies for implantable antennas are limited to reflection coefficient, radiation pattern, and safety investigations, transmission coefficient (S_{21}) and numerical and experimental results have also been presented:

- Warty et al., (2008) implanted 2.45 GHz PIFA inside a gel scalp phantom for intracranial pressure monitoring, and a linearly polarized 2.45 GHz chip antenna was used as the receiving (probing) antenna. Transmission coefficient measurements were reported for the first time. The drift of the transmission coefficient values over time was evaluated, and the effective radiated power (ERP) was computed from the transmission measurements. An increase in S_{21} of 2.2–2.4 dB was observed over a period of two days. Transmission measurements performed using both S_{21} and the received power measurement (for an intracranial pressure device mimic) yielded a maximum ERP of approximately 2 mW per 1 W of power delivered to the antenna. Values of the received power (P_r) and ERP per 1 W delivered to the PIFA are shown in Figure 8.13*a* as a function of distance.
- Weiss et al. (2009) presented simulations and measurements for a 433 MHz mock telemetry link in a 6 cm-thick porcine thigh. Electrically, short linear wire antennas were used at both the embedded transmitter and external receiver, whereas the receiver was considered to be placed on the outer surface of the thigh. Simulated and measured S_{21} values in the porcine experiments are shown in Figure 8.13*b* and are found to vary in the −45 to −55 dB range for different transmitter-to-receiver distances. Simulations for a 10 cm-radius cylindrical human thigh and a 3 cm-long receiver antenna demonstrated a coupling coefficient of −63 dB. When a 6 cm-long external antenna was used as the receiver, S_{21} increased to −54 dB, a 9 dB improvement. The lowest coupling was computed for muscular patients. To accommodate such a patient with a thigh radius of 21 cm, a 30 dB link margin was found to be required as compared to the average male.

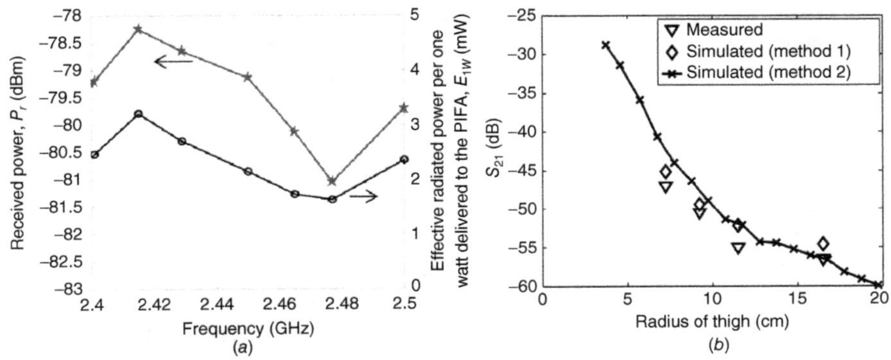

Figure 8.13 (*a*) Values of the received power (P_r) and ERP per 1 W delivered to the PIFA as a function of distance from a receiving chip antenna (Warty et al., 2008) and (*b*) simulated and measured S_{21} values in the porcine experiments of Weiss et al. (2009).

Out-of-Body Channel Modeling Link budget investigations have also been carried out for implantable antennas. Characteristic studies are summarized below.

- A link budget calculation was proposed for the communication between a left-ventricular wireless implant and an off-body base station, as an example for MICS applications (Sani et al., 2009). In far-field communication in free space, the received power at the base station was calculated using the Friis formula:

$$P_r = P_t(1 - |S_{11}|^2)G_t\left(\frac{\lambda}{4\pi d}\right)^2 G_r(1 - |S_{22}|^2) \tag{8.4}$$

where $|S_{11}|$ and $|S_{22}|$ are the reflection coefficients of the transmitting and the receiving antennas, P_t is the transmitter's available power, G_t is the gain of the implantable transmitting antenna, G_r is the gain of the exterior receiving antenna, λ is the free-space wavelength, and d is the distance between the transmitter and the receiver.

In the indoor environment, due to reflections, diffraction, and scattering of electromagnetic waves, the transmitted signal reaches the receiver via more than one path. The most common path loss (PL) model, that is, the log-distance model, was suggested, so that

$$\mathrm{PL}(d)\ (\mathrm{dB}) = 10 \cdot n \log\left(\frac{d}{d_0}\right) + \mathrm{PL}(d_0) \tag{8.5}$$

where n indicates how fast the received power decays with distance, d_0 is a reference distance, chosen to be 1 m, and $\mathrm{PL}(d_0)$ is the PL value at the distance d_0, which is assumed to be equal to the free-space loss.

The received power at the exterior base station is then calculated as

$$P_r\ (\mathrm{dBm}) = P_t\ (\mathrm{dBm}) + G_t(\mathrm{dB}) - \mathrm{PL}(\mathrm{dB}) + G_r\ (\mathrm{dB}) \tag{8.6}$$

Furthermore, it was demonstrated, that due to the presence of human tissues, the antenna radiation from wireless implants tended to be directive, and, therefore, the SNR in the communication link was strongly dependent on the orientation of the human subject with respect to the base station.

- The wave propagation between an off-body reader and an ultra-high-frequency (UHF) (868 MHz) subcutaneous tag antenna has also been investigated (Sani et al., 2010). Both LOS and NLOS scenarios were studied, and Equations (8.2)–(8.4) were applied.

  Results demonstrated that, due to the body losses, the electrically small size of the antenna, and the directional radiation pattern, a passive tag solution allows a very limited communication range. If the tag is powered (active tag) with a limited power (−20 dBm), a maximum communication range of 10 m was calculated for propagation within a room.
- Propagation losses of body-implanted antennas have been studied at the ISM bands of 433, 915, 2450, and 5800 MHz (Gemio, 2010). A single- and a three-layer body models were used, and the antenna pair gain was analyzed at different depths of the implantable antenna. Free-space wireless links were assumed between half-wavelength resonant dipole antennas, and Equation (8.4) was applied to investigate the link budget.

  Simulation results proved that, when far-field conditions are fulfilled, body losses can be included in the gain of the implanted antenna, and a free-space propagation model can be assumed to evaluate the system link budget. The single-layer model was found to be a conservative approach that tended to underestimate the antenna pair gain as compared to the three-layer model. Changes in frequency revealed a trade-off between antenna dimensions and system performance. When the frequency was doubled, antenna dimensions were reduced by half, and propagation losses were increased by 6 dB, whereas the antenna pair gain was further degraded due to the higher loss of body tissues. Similar trade-offs were observed when the antenna position was changed from one layer to another.
- Finally, Xia et al. (2009) carried out a link budget calculation between a 2.45 GHz H-shaped cavity slot antenna implanted in a human arm and a receiving monopole antenna placed 4 m away inside a sickroom. The availability of communication was decided by the C/N_0 ratio, that is, if the link's C/N_0 exceeds the required C/N_0, then wireless communication is possible. Table 8.3 shows the parameters used to calculate the link C/N_0 and the required C/N_0. The link C/N_0 and the required C/N_0 used were calculated as follows.

$$\text{Link}\ \frac{C}{N_0} = P_t - L_{\text{feed}} + G_t - L_f - L_a + G_r - L_{\text{feed}} - N_0 \qquad (\text{dB/Hz}) \tag{8.7}$$

$$\text{Required}\ \frac{C}{N_0} = \frac{E_b}{N_0} + 10\log_{10}(B_r) - G_c + G_d \qquad (\text{dB/Hz}) \tag{8.8}$$

TABLE 8.3 Parameters Used to Calculate Link and Required Signal-to-Noise Ratio

Transmitter		Propagation	
Frequency f (GHz)	2.45	Distance d (m)	4
Transmitter power P_t (W)	1.25×10^7	Free-space loss L_f (dB)	52.27
Feeding loss L_{feed} (dB)	1.0	Air propagation loss L_a (dB)	0.5
Transmitter antenna gain G_t (dBi)	– 26.5		
Receiver		**Signal Quality**	
Receiver antenna gain G_r (dBi)	2.15	Bit rate B_r (kbps)	7
Feeding loss L_{feed} (dB)	1	Bit error rate	1×10^{-5}
Ambient temperature T_0 (K)	293	E_b/N_0 (ideal PSK) (dB)	9.6
Receiver noise factor NF (dB)	3.5	Coding gain G_c (dB)	0
Boltzmann constant k	1.38×10^{-23}	Fixing deterioration G_d (dB)	2.5
Noise power density N_0 (dB/Hz)	– 199.95		

Source: Xia et al. (2009).

$$L_f = 10 \log_{10}\left(\frac{4\pi d}{\lambda}\right)^2 \quad \text{(dB)} \tag{8.9}$$

$$N_0 = 10 \log_{10}(k) + 10 \log_{10}(T_i) \quad \text{(dB/Hz)} \tag{8.10}$$

$$T_i = T_0(\text{NF} - 1)\ \text{(K)} \tag{8.11}$$

In the scenario under study, the link was computed to be higher than that required by approximately 0.3 dB.

Inside-the-Body Channel Modeling In estimating the propagation loss for wireless transmission inside the human body, changing the loss coefficient of the free-space formula will only change the rate of decrease in power and will not help in estimating the total loss in the form of absorption.

A propagation loss model for homogeneous tissue bodies (PMBA) was reported in Gupta et al. (2003) and verified for the frequency range 900 MHz to 3 GHz, both numerically and experimentally. The study considered an elemental oscillating electric dipole in a lossy medium of conductivity σ, permittivity ε, permeability μ, complex propagation constant γ, complex intrinsic impedance $\eta = \gamma/(\sigma + j\omega\varepsilon)$ at frequency ω, as shown in Figure 8.14*a*. The dipole consisted of a short conducting wire of length dl, terminated in two small conductive spheres or disks, whereas the current I was assumed to be uniform and vary sinusoidally with time. The power absorbed in the near field of the lossy tissue was said to be obtained by computing the average SAR over the entire tissue mass in the near field, that is, from the surface of the antenna ($R = r$) to the end of the near-field region ($R = d_0$, where d_0 is the point where the far field starts), as

$$P_{\text{NF}} = \sigma\mu\omega \frac{|\eta|}{|\gamma|} \frac{I^2 dl^2}{6\pi}(A + B + C) \tag{8.12}$$

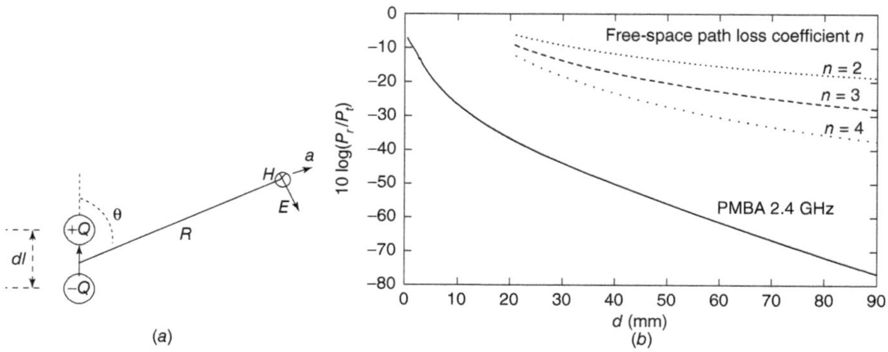

Figure 8.14 Development of propagation model for in-body transmission: (*a*) a Hertzian dipole and (*b*) comparison of the proposed model with the free-space loss formula (Gupta et al., 2003).

where

$$A = e^{-2\alpha r}\left[\frac{|\gamma|^2}{2\alpha} + \frac{d_0 - r}{4r^2} + \frac{|\gamma|(d_0 - r)}{2r}\right] \tag{8.13}$$

$$B = e^{-2\alpha d_0}\left[\frac{-|\gamma|^2}{2\alpha} + \frac{d_0 - r}{4d_0^2} + \frac{|\gamma|(d_0 - r)}{2d_0}\right] \tag{8.14}$$

$$C = e^{-\alpha(d_0+r)}\left[\frac{2\left(d_0 - r\right)}{(d_0 + r)^2} + \frac{2|\gamma|(d_0 - r)}{(d_0 + r)}\right] \tag{8.15}$$

The total power absorbed in the far field of the lossy tissue between the transmitting and receiving antennas was said to be obtained by computing the average SAR over the entire tissue mass in the far field from distance d_0 to d as

$$P_{\mathrm{FF}} = \sigma|\eta|^2|\gamma|^2\frac{I^2 dl^2}{12\pi\alpha}(e^{-2\alpha d_0} - e^{-2ad}) \tag{8.16}$$

The ERP is obtained by subtracting the loss in the near field (P_{NF}) and the far field (P_{FF}) from the transmitted power P_t (i.e., $(P_t - P_{\mathrm{loss}})G_t$), where

$$P_{\mathrm{loss}} = P_{\mathrm{NF}} + P_{\mathrm{FF}} \tag{8.17}$$

The power received by the receiving antenna in the near field was approximated by

$$P_r = \frac{16\delta(P_t - P_{\mathrm{NF}})}{\pi L^2}A_e \tag{8.18}$$

where L is the largest dimension of the antenna, and δ is the aperture efficiency (approximated as $\delta = A_e/A$, with A_e being the effective aperture, and A the physical area of the antenna).

The power received by the receiving antenna in the far field was calculated as

$$P_r = \frac{(P_t - P_{\mathrm{NF}} - P_{\mathrm{FF}})\lambda^2}{(4\pi d)^2} G_t G_r \tag{8.19}$$

where G_t and G_r are the gains of the transmitting and receiving antennas.

Figure 8.14*b* compares PMBA with the free-space propagation model $P_r = P_t G_t G_r \left[\lambda/4\pi d\right]^n$ with loss coefficients $n = 2, 3, 4$. It is observed that power received in PMBA decreases more rapidly than free-space loss. Compared to free space, there is an additional 30–35 dB of attenuation at small distances in the far field.

It is worth noting that even though the derived formulas are applicable only in the case of small dipole implantable antennas, the power loss formula for similar applications with other antennas could be derived in a similar manner.

8.5 INGESTIBLE ANTENNAS

The conventional methods used for diagnosing disorders of the human gastrointestinal (GI) tract cause significant patient discomfort. As a result, there is considerable ongoing work in developing ingestible antennas that can be integrated into a capsule and swallowed for examination of the entire digestive tract.

8.5.1 Antenna Design

Selection of Operation Frequency Selection of operation frequency for ingestible antennas has received significant attention from the scientific community. The reason is that selection of frequency is accompanied by a number of competing effects. For example, antenna efficiency can improve with frequency. On the other hand, higher frequencies may cause increased radiation absorption because of the high water content of body tissues. This in turn may deteriorate the performance of the wireless link, requiring increased levels of supply powers and posing questions regarding patient safety.

The effects of the human body on the performance of an ingestible antenna inside the frequency range of 150 MHz–1.2 GHz have been investigated in Chirwa et al. (2003a). Maximum radiation was found to occur between 450 and 900 MHz, whereas radiation intensity outside the body had a Gaussian form relationship with frequency. Although the frequency range under study covered the ISM bands of 434 and 915 MHz, these bands are not well developed for digital video transmission. Chirwa et al. (2003b) found a peak in the power transmitted by an ingestible medical device at approximately 650 MHz, and adequate communication performance was demonstrated between 600 MHz and 1 GHz.

On the other hand, video transmission in the 2.45 GHz band is better developed for WLAN and Bluetooth applications in terms of technology, antennas, camera modules, and other RF components. Furthermore, ingestible medical devices operating at this frequency can be directly connected to the WLAN or Bluetooth networks for real-time monitoring and remote control. Finally, higher transmission frequencies allow the use of smaller antennas and electronic components, which is a prerequisite for ingestible medical devices. Therefore, the 2.45 GHz band appears to be a promising solution. For example, an IC design for wireless capsule endoscopy at 2.4 GHz has been proposed in Xie et al. (2004).

By taking into account that the whole ingestible telemetry system has about 80–110 dB of room for loss in system efficiency and body tissue attenuation, it becomes obvious that ingestible medical devices operating at 5.8 GHz can also be applied. Attenuation achieved by such devices is expected to be within the desired attenuation range; however, great care must be taken to design an efficient telemetry system without (or minimum) additional losses.

Type and Material of Antenna Different from implantable antennas, the ingestible antenna needs to be omnidirectional and exhibit circular polarization in order to transmit signals independent of its position and orientation. Since the capsule device travels along the entire GI tract, its exact position and orientation are generally unknown. Therefore, an isotropic radiation pattern is required for ingestible antennas.

Given the above considerations, normal mode helical antennas are most commonly employed for such applications (Figure 8.15*a*). Furthermore, as wireless capsule endoscope systems transmit real-time and high-resolution data, antennas with a miniature size but wide bandwidth are required. For example, a wideband spiral antenna for ingestible capsule endoscope systems at 500 MHz was presented (Lee et al., 2011) (Figure 8.15*b*). A thick-arm spiral structure was applied to achieve a fractional bandwidth of 21% and an isotropic radiation pattern.

Numerical modeling of ingestible antennas is also intriguing. For example, although the FDTD method is able to model anatomically detailed human body

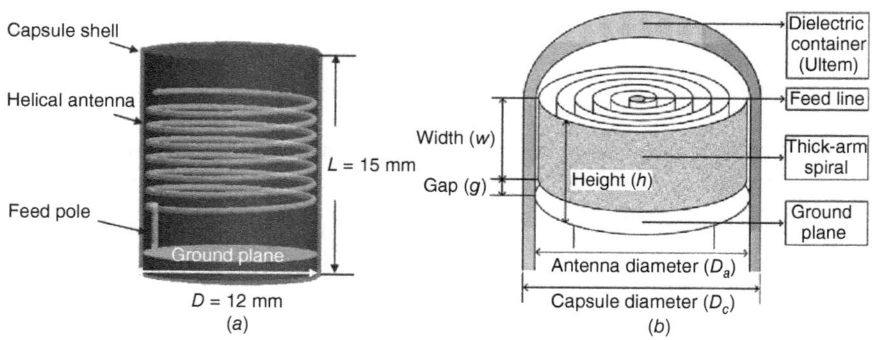

Figure 8.15 Example ingestible antennas: (*a*) helical (Xu et al., 2009b), and (*b*) spiral (Lee et al., 2011) geometries.

structures, there are still significant difficulties in modeling small antennas. Furthermore, it is important to highlight that employing a one-cell (delta gap) excitation in FDTD simulations, where the excitation feed gap corresponds to one entire spatial interval of the FDTD lattice, is not effective. In such cases, the size of the feeding gap is expected to be comparable with that of the antenna to be modeled, thus resulting in erroneous numerical results.

Same as for implantable antennas, ingestible antennas must be biocompatible so that they do not harm the patient and durable so that the body fluids do not harm the antenna. These requirements entail packaging of the ingestible antennas inside a shell. Capsule casings and circuitry have been found to have a negligible effect on the performance of the ingestible antenna. Therefore, taking the effects of casing into account while modeling the antenna is not necessarily compulsory.

Patient Safety Considerations Safety performance of ingestible antennas operating at several frequency bands attracts significant scientific interest. Usually, low-frequency wireless devices cause less significant biological effects than higher-frequency devices as attributed to decreased levels of tissue absorption. Design of an ingestible antenna at high frequencies needs to be carefully considered in order to guarantee compliance with safety guidelines.

The SAR and temperature rise performance of ingestible antennas has been analyzed at frequencies from 430 MHz to 3 GHz (Xu et al., 2008b). Simulations were carried out inside a high-fidelity human body model consisting of 34 types of tissues, and a helical ingestible antenna was considered inside a plastic capsule with a diameter and length of 12 and 5 mm, respectively. Results showed that high values of SAR and temperature rise were localized at the area near the location of the ingestible device. Figure 8.16 shows the near-field intensities both inside and outside the human body model at frequencies of 430 MHz, 800 MHz, 1.2 GHz, 2.4 GHz, and 3 GHz. The ingestible antenna was safe and could be used in ingestible medical devices at input power levels of less than 25 mW in order to conform with safety regulations.

Simulation results for a helical ingestible antenna at 2.4 GHz, carried out in 21 scenarios (3 orientations at 7 source positions), showed that all temperature rises, peaks of the SAR, and averaged SAR values were under the safety limits when the delivered power was no more than 26.16 mW (Xu et al., 2008a). In other words, the frequency of 2.4 GHz was proved to be able to be employed safely and efficiently in the ingestible wireless device, as long as its input power is no more than 26.16 mW.

Xu et al. (2009b) assessed the compliance of ingestible medical devices with international safety guidelines at 430, 800, 1200, and 2400 MHz. A high-fidelity female body with 34 types of tissues and a high-fidelity male body with 23 types of tissue, and scaled into the same height with the female body, were studied. Among all simulated scenarios, the maxima of the 1 g and 10 g averaged SAR were computed as 3.71 and 1.37 W/kg at an input power of 25 mW, respectively. The ingestible antenna was found to be safe to be applied in the medical device at an input power less than 36 and 11 mW according to the ICNIRP and IEEE safety standards, respectively. As far

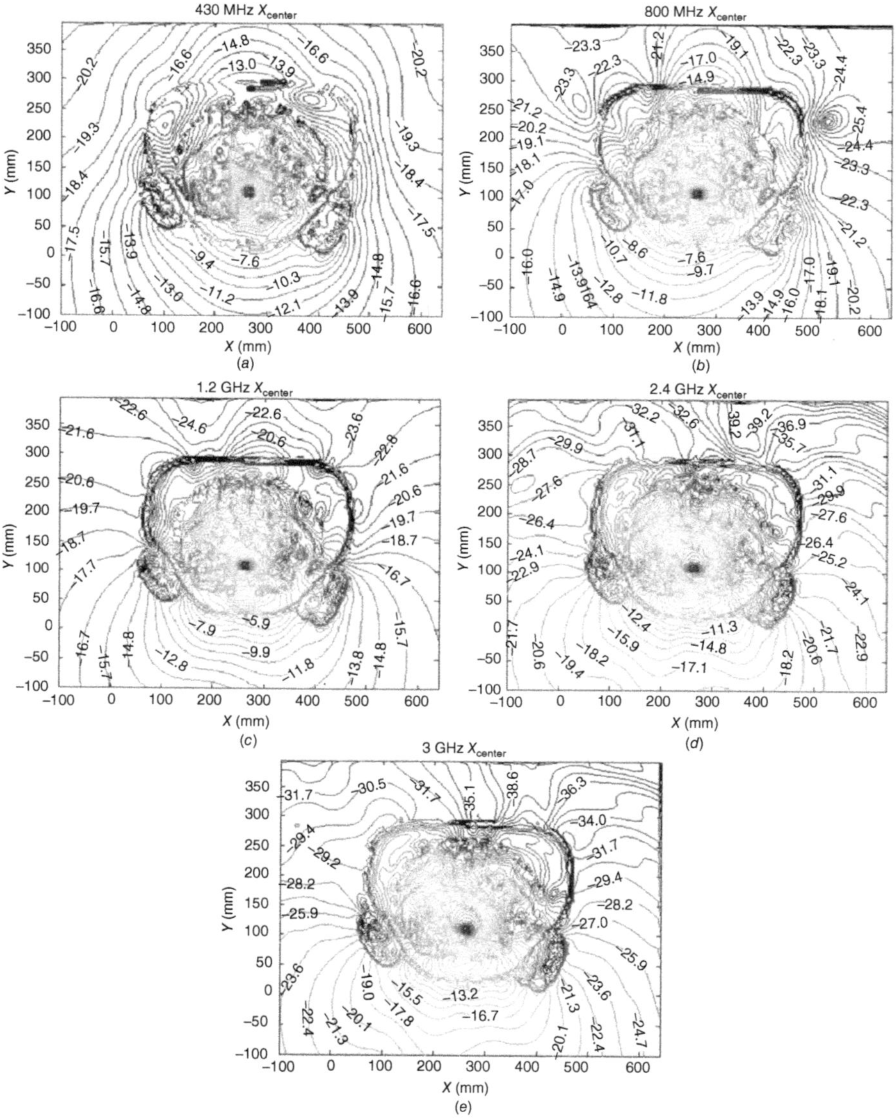

Figure 8.16 Near fields for ingestible antenna at (*a*) 430 MHz, (*b*) 800 MHz, (*c*) 1.2 GHz, (*d*) 2.4 GHz, and (*e*) 3 GHz (Xu et al., 2008b). (*See insert for color representation of the figure.*)

as the thermal effect is concerned, it was shown that even in the cases of maximum SAR values, temperature rise maxima remained low.

Finally, in Xu et al. (2009a), the safety performance of ingestible antennas was evaluated in two (male, female) realistic human body models whose dielectric parameters (permittivity and conductivity) were varied from the original by ±10% and ±20%. Three operation frequencies were considered, that is, 430, 800, and 1200 MHz. SAR values were found to increase with the increase in conductivities of human body tissues, and usually decreased with the increase in permittivities. From the equation of SAR calculation, one might have assumed that the SAR value would increase with the increment of dielectric values. However, SAR values are not only influenced by the dielectric properties of the medium but also by impedance matching, which is, in turn, influenced by the dielectric properties. A variation by up to 20% in conductivities and permittivities alone or simultaneously always caused a SAR variation by less than 10, 20, and 30% at 430, 800, and 1200 MHz, respectively. The ingestible antenna was shown to be safe used at input power levels less than 12.6, 9.3, and 8.4 mW, according to the IEEE safety standards at 430, 800, and 1200 MHz, respectively. SAR depositions in the female body model were larger than those in the male body model.

Position and Orientation Considerations Performance of ingestible antennas for various locations inside the body and antenna orientations has also been investigated.

Chirwa et al. (2003a) performed FDTD numerical investigations considering a male human subject and an ingestible monofilar helix antenna with a diameter of 8 mm, length of 4 mm, and pitch of 1 mm (four turns). Frequencies were in the range 150 MHz–1.2 GHz. Three antenna orientations were investigated (vertical, transverse, and longitudinal), and simulations were performed at many locations inside the GI tract, as shown in Figure 8.17*a*. Near- and far-field results showed that maximum radiation was experienced on the anterior side of the body, while no direct correlation was found between the near and far fields. For the majority of locations, vertically polarized radiation was more greatly attenuated than horizontal radiation. Anatomy around the gut region highly affected the electromagnetic field results, while radiation characteristics differed significantly based on the distance between the ingestible antenna and the outer surface of the body.

In another study, the same monofilar helix antenna was used, and simulations were carried out for five locations in the GI: near the skin, top of the small intestine, behind the stomach, right extremity of the small intestine next to the colon, a central location, and the bottom of the small intestine (Chirwa et al., 2003b). Results showed that the exact location of the ingestible antenna had a strong effect on radiation. The radiated field was predominantly strongest in the same horizontal transverse plane as the source and in the direction that was on the line from the source to the nearest body surface. Therefore, the optimum location for a receiver antenna would be an anterior location slightly to the left of the abdomen (as the small intestine does not extend so much to the right due to the presence of the colon).

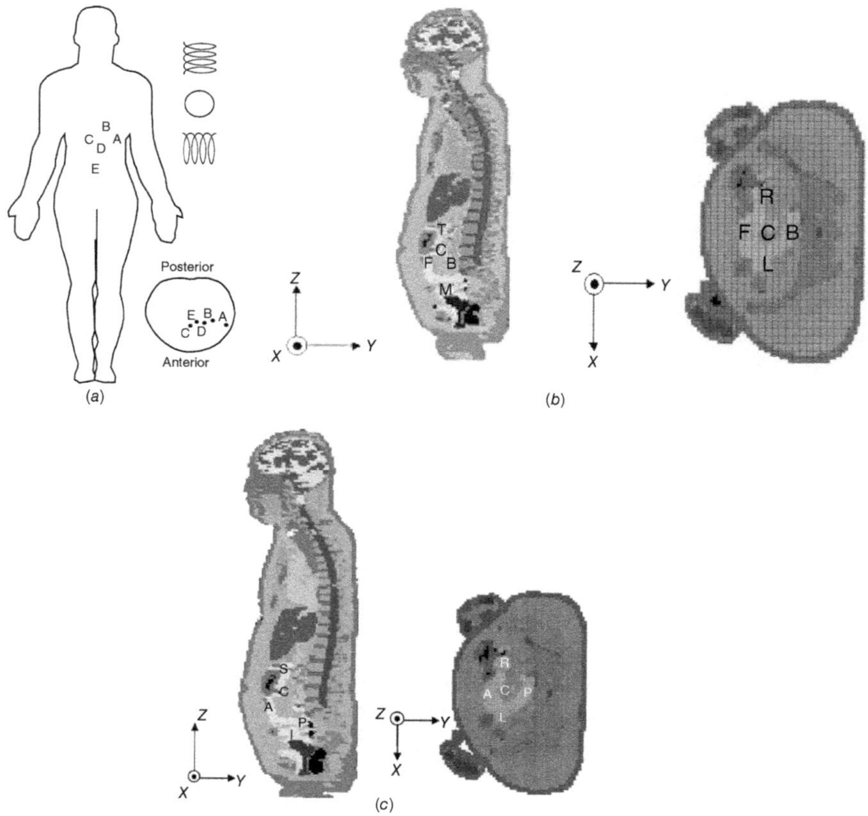

Figure 8.17 Location and orientation of ingestible antennas in (*a*) Chirwa et al. (2003a), (*b*) Xu et al. (2008a, 2008b), and (*c*) (Xu et al., 2009b).

Numerical analyses have further been carried out for three orientations of an ingestible antenna at frequencies from 430 MHz to 3 GHz, and at seven positions inside a human body model (Figure 8.17*b*) (Xu et al., 2008*b*). A helical antenna was assumed that was integrated inside a plastic shell with a thickness of 1 mm, a diameter of 12 mm, and a length of 15 mm. It was shown that the orientation and position of the ingestible antenna, which gave maximum radiation efficiency, was frequency dependent. Furthermore, when the ingestible antenna was located at the frontmost and backmost positions of the small intestine, the electric intensity outside the human body was maximum and minimum, respectively. Position was found to influence the radiation efficiency of the antenna more than its orientation.

Simulations carried out for a 2.4-GHz ingested source at 3 orientations and 7 positions (Figure 8.17*b*) demonstrated that port impedances, radiation efficiencies, and link performance for the antenna varied greatly with location and orientation

(Xu et al., 2008a). Similarly, Xu et al. (2009b), assessed ingestible antennas for 21 scenarios (3 orientations, 7 positions) at 430, 800, 1200, and 2400 MHz (Figure 8.17*c*). Radiation characteristics and radiation absorption were found to be frequency, position, and orientation dependent. The study demonstrated that if the motion of the ingestible antenna inside the GI tract could be controlled, then the quality of communication would be improved by 3 dB at least. The electric field intensity in the anterior of the human body was higher than that in the posterior.

8.5.2 Channel Modeling

Only a limited number of research studies have been performed regarding the modeling of the channel between an ingestible antenna integrated into an ingestible medical device and an exterior antenna integrated into an exterior monitoring/control device. For example, an attempt has been made to estimate the attenuation of the human body trunk for ingestible antennas in the frequency range of 100 MHz–6 GHz (Chan et al., 2005). A simplified rectangular ($55 \times 24 \times 16$ cm^3) phantom was used to emulate the human trunk, which was filled with distilled water, saline solution, and porcine tissue alternately (Figure 8.18*a*). Patch antennas matched at 300 MHz, 900 MHz, 1.5 GHz, 3.0 GHz, and 4.8 GHz were used as radiating sources inside the model, whereas a broadband yagi antenna and a broadband horn antenna were used as the receiving antennas for the frequency ranges of 100 MHz–1.3 GHz and 1.5 GHz–6 GHz, respectively. The received power considering free space, the tank filled with distilled water, the tank filled with saline solution, and the tank filled with porcine tissue were denoted as $P_{r,\text{air}}$, $P_{r,\text{water}}$, $P_{r,\text{saline}}$, and $P_{r,\text{tissue}}$, respectively. With these values recorded, attenuation of different water and saline solutions was calculated as

$$\text{Attenuation in water (dB)} = P_{r,\text{air}}\ (\text{dBm}) - P_{r,\text{water}}\ (\text{dBm}) \tag{8.20}$$

$$\text{Attenuation in saline (dB)} = P_{r,\text{air}}\ (\text{dBm}) - P_{r,\text{saline}}\ (\text{dBm}) \tag{8.21}$$

$$\text{Attenuation in porcine tissue (dB)} = P_{r,\text{air}}\ (\text{dBm}) - P_{r,\text{tissue}}\ (\text{dBm}) \tag{8.22}$$

Averaged attenuation values were plotted as shown in Figure 8.18*b*. For comparison purposes, an ideal curve of the attenuation exhibited for a plane wave traveling through 6 cm of water (Debye equations) is also plotted. All curves indicated a similar trend of attenuation. Since the ideal curve considered the attenuation of water in only one dimension, experimental data appear as reasonable and reliable. Saline solution exhibited higher attenuation than water (by 5–15 dB), as attributed to its higher conductivity. Attenuation within porcine tissue laid between the water and saline curves.

Link budget studies have also been performed. Rajagopalan and Sahmat-Samii (2010) developed a link budget for ingestible antennas based on frequency, SAR, and

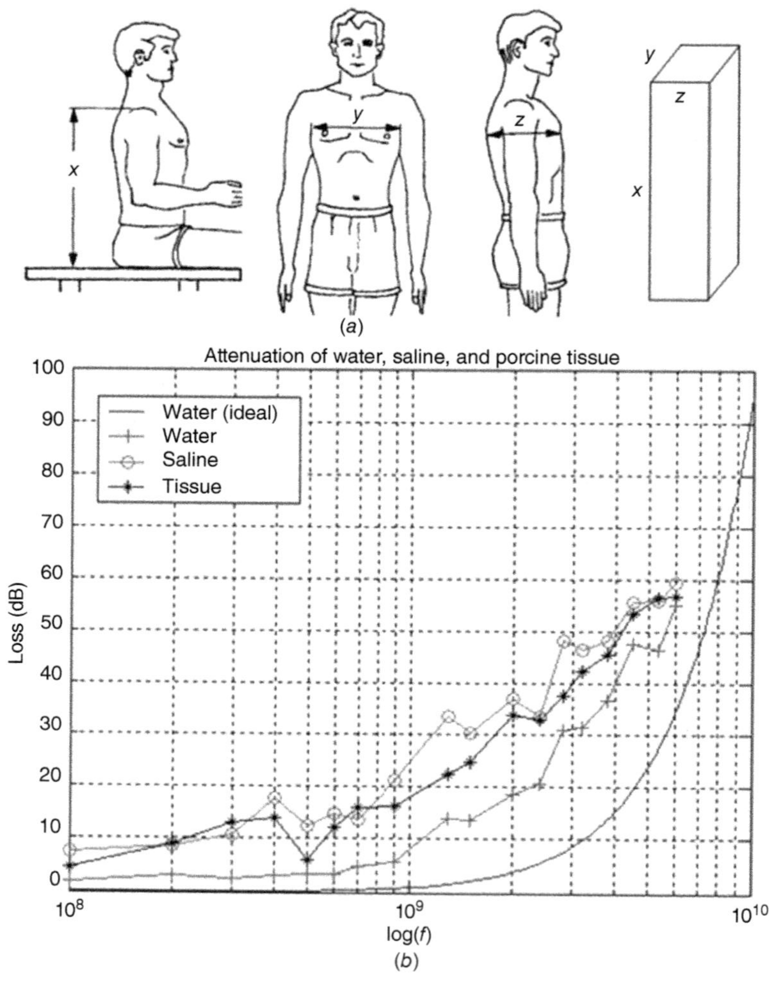

Figure 8.18 Estimation of attenuation of the human body trunk for ingestible antennas: (*a*) simplified phantom of the human body trunk and (*b*) attenuation of water, saline solution, and porcine tissue (Chan et al., 2005).

power limitations, and it was applied to assess the performance of a specific antenna system. The link budget analysis in terms of transmission, propagation, and reception is shown in Table 8.4.

The antenna along with the body were assumed to act as a transmitter, and the SNR was calculated as

$$\frac{C_1}{N_0} = P_t + G_t + L_f + G_r - N_0 = 72.45 \text{ dB/Hz} \tag{8.23}$$

TABLE 8.4 Link Budget Evaluation for Ingestible Antenna

Transmission		Propagation		Reception	
Frequency f (GHz)	1.4	Distance d (m)	2	Receiver gain G_r (dBi)	2.15
Transmitter power P_t (dB)	−57	Free-space loss L_f (dB)	41.38	Noise power density N_0 (dB/Hz)	− 202.17
Transmitter Gain G_t (dBi)	−33.5				

Source: Rajagopalan (2010).

Using basic communication theory for the signal quality, the link was found to require a SNR of

$$\frac{C_2}{N_0} = \frac{E_b}{N_0} + 10\log_{10} B_r - G_c + G_d = 72.10\ \text{dB/Hz} \tag{8.24}$$

Since $C_1/N_0 > C_2/N_0$, a successful link was shown to be established between the transmitter and receiver in the scenario under investigation.

Following the same approach, three different links were compared (free-space link, in-body link where a simple dipole was used as the transmitter, and in-body link where an ingestible capsule antenna was used as the transmitter), and it was shown that, when linearly polarized antennas (dipoles) were used, the capsule link was the most robust with respect to the orientation of the transmitting antenna. A comparison of the maximum available power at the receiver from the three links for rotating transmitting antennas (Figure 8.19*a*) is shown in Figure 8.19*b*.

Lee et al. (2011) measured the received power level using a wideband spiral ingestible antenna, a circular polarized receiver antenna at a distance R, and a pig under general anesthesia. Measurements of the received power were performed inside the stomach and intestine of the pig, which are typical digestive organs. Results are shown in Figure 8.20*a*. The width (w) and gap (g) of the spiral antenna are defined in Figure 8.15*b*. When the transmitted power was 0 dBm, the received powers in the stomach and intestine were measured as −51.06 and −47.19 dBm, respectively.

If the values of the transmitted power (P_t), the received power (P_r), and the gain of the receiving antenna (G_r) are known, then the gain of the transmitting antennas (G_t) was said to be calculated with the Friis transmission equation inside a lossy medium as

$$\frac{P_r}{P_t} = \left(\frac{\lambda}{4\pi R}\right)^2 G_t G_r (e^{-\alpha R})^2 \tag{8.25}$$

where the attenuation constant is given as

$$\alpha = \text{Re}(\gamma) = \text{Re}\left(j\omega\sqrt{\mu\varepsilon}\sqrt{1 - j\frac{\sigma}{\omega\varepsilon}}\right) \tag{8.26}$$

and ε and σ are the permittivity and conductivity values of the medium at an angular frequency of ω, respectively.

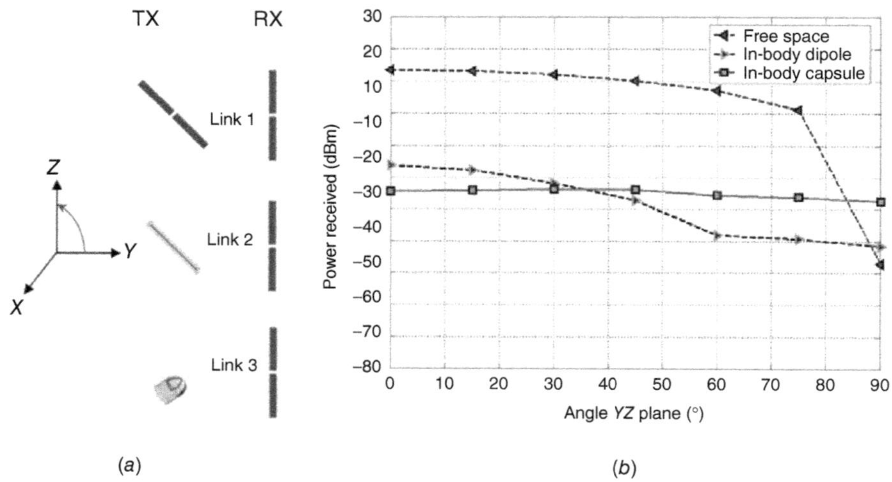

Figure 8.19 Link budget analysis for ingestible linearly polarized antennas: (*a*) orientation of antennas for three links and (*b*) comparison of the maximum power available at the receiver from the three links when the transmitting antennas rotate in the *yz* plane (Rajagopalan and Sahmat-Samii, 2010).

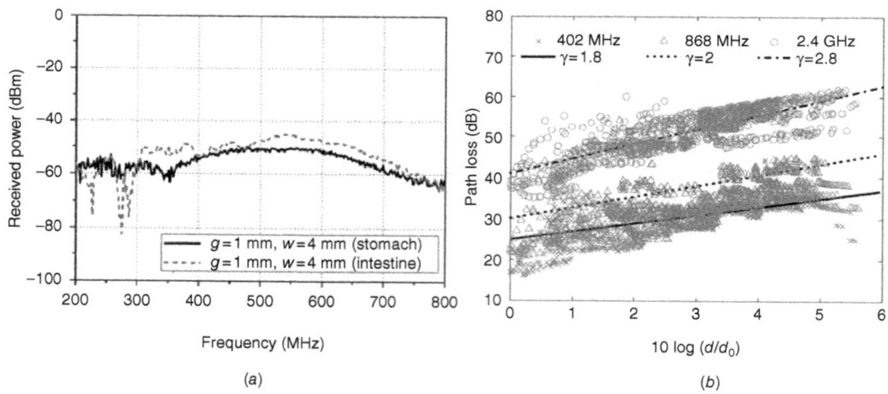

Figure 8.20 (*a*) Measured received power of a spiral ingestible antenna inside a pig (Lee et al., 2011) and (*b*) measured and empirically modeled path loss for biomedical telemetry links between ingestible antennas and receiver antennas placed in parallel to the human body (Alomainy and Hao, 2009).

Finally, biomedical telemetry radio channels from ingestible antennas at 402, 868, and 2400 MHz were numerically modeled and experimentally tested in Alomainy and Hao (2009). The aim of the study was to demonstrate a detailed analysis of wave propagation in different human tissues. The average path loss for a distance *d* between

TABLE 8.5 Exponent Values of Modeled Path Loss for Ingestible Antenna

Frequency (MHz)	Measured γ	Simulated γ
402	1.90	1.85
868	2.00	1.90
2400	2.80	2.60

Source: Alomainy and Hao (2009).

the transmitter and the receiver was expressed as

$$\mathrm{PL}(d) \propto \left(\frac{d}{d_0}\right)^{\gamma} \tag{8.27}$$

where γ is the path loss exponent that indicates the rate at which the path loss increases with distance, and d_0 is a reference distance (set to 10 cm in the study).

The path loss observed at any given point was said to deviate from this average value due to variations in the environment, according to a log-normal distribution. Therefore, the average path loss was represented as

$$\mathrm{PL}_{\mathrm{dB}}(d) = \mathrm{PL}_{\mathrm{dB}}(d_0) + 10\gamma \log\left(\frac{d}{d_0}\right) + X_{\sigma} \tag{8.28}$$

where X_{σ} is a zero-mean Gaussian distributed random variable with standard deviation σ.

The measurement setup included a human phantom of approximately 1.7 m in height, and an average width of 0.35 m. Animal organs (sheep liver, heart, and lungs) were used to represent human tissues. A 4 cm-long monopole without the ground plane was used as both the transmitter and the receiver, and near-field scanning was used to obtain path loss data at different distances from the body.

The exponent values of the modeled path loss based on numerical and experimental data applying empirical linear power law and least-square fitting techniques are shown in Table 8.5.

The measurement-derived exponent values were highly comparable to the numerically derived ones. Deviations may be attributed to differences in loss factors and conductivities between simulated and animal tissues. Variation of the measured path loss with distance away from the body surface is shown in Figure 8.20*b*, along with the empirically modeled path loss.

8.6 CONCLUSION AND FUTURE RESEARCH DIRECTIONS

Research performed so far in the field of design and channel modeling for on-body, implantable, and ingestible antennas is by no means exhaustive. Many more research directions need to be considered, and further optimization of the antenna design and channel modeling procedures may be possible.

In the field of on-body antennas, further research is required on textile antennas and e-textile materials themselves before they can actually be used for reliable transmission of medical data. Treating textile antennas in the way we treat our daily garments is very demanding: washable packaging of the electronics, durable interconnections, and long-term behavior remain major challenges to be addressed. Furthermore, future designers will need to focus on multifunction and multifrequency on-body antennas.

Effort should also be directed toward ensuring that the antennas resonate and operate properly in the close vicinity of the human body, while exhibiting improved efficiency and reduced power absorption by human tissues. Metamaterials may offer the potential to reduce surface currents, thus decreasing the antenna-to-body coupling and improving the link performance of the antennas. The field of antenna diversity, that is, the use of multiple antennas at the transmitter and receiver side, will also need to be explored in an attempt to improve the speed and reliability of the on- and off-body channels.

In the field of implantable antennas, it becomes necessary to assess and quantify the trade-off between size and performance. Electrically small antennas present poor radiation performance and relatively narrow bandwidth. Design of multiband antennas, which wake up the implantable medical device when there is a need for information exchange, is also significant for conserving energy and extending the lifetime of the device. Use of efficient and accurate simulation tools and tissue models can further be considered as a key issue for the design and performance analysis of implantable antennas. The highest challenge, however, lies in experimental testing and, especially, measurements within living animals, in which careful consideration is required for developing the optimal testing protocol. Regarding the development of propagation models inside the human body, limited research has been done so far, and channel modeling has only been conducted for particular scenarios with specific antennas and orientation.

In the field of ingestible antennas, further scientific research efforts are required in determining the dependence of radiation characteristics on antenna position and orientation. Analyses presented so far provide advanced and useful information for ingestible antenna design. Nevertheless, there are still uncertainties that need to be taken into account, such as the placement of the arms, which may influence the near field, or the existence of clothing, which may affect the temperature rise. Studies regarding the wireless communication link give an insight into electromagnetic wave propagation and radiation properties of ingestible antennas; however, results are mostly given as guidelines since different human subjects exhibit different postures, body mass index, and other physiological data.

REFERENCES

Abadia K, Merli F, Zurcher JF, Mosig JR, Skrivervik AK. 2009. 3D Spiral small antenna design and realization for biomedical telemetry in the MICS band. *Radioengineering*, 18(4):359–367.

Abbasi QH, Sani A, Alomainy A, Hao Y. 2010. On-body radio channel characterization and system-level modeling for multiband OFDM ultra-wideband body-centric wireless network. *IEEE Trans. Microw. Theory Tech.*, 58(12):3485–3492.

Abbasi QH, Sani A, Alomainy A, Hao Y. 2011a. Experimental characterization and statistical analysis of the pseudo-dynamic ultrawideband on-body radio channel. *IEEE Antennas Wireless Propag. Lett.*, 10:748–751.

Abbasi QH, Alomainy A, Hao Y. 2011b. Characterization of MB-OFDM-based ultrawideband systems for body-centric wireless communications. *IEEE Antennas Wireless Propag. Lett.*, 10:1401–1404.

Alomainy A, Hao Y, Parini CG, Hall PS. 2005. Comparison between two different antennas for UWB on-body propagation measurements. *IEEE Antennas Wireless Propag. Lett.*, 4:31–34.

Alomainy A, Hao Y, Pasveer F. 2007a. Numerical and experimental evaluation of a compact sensor antenna for healthcare devices. *IEEE Trans. Biomed. Circ. Syst.*, 1(4):242–249.

Alomainy A, Hao Y, Owadally A, Parini CG, Nechayev Y, Constantinou CC, Hall PS. 2007b. Statistical analysis and performance evaluation for on-body radio propagation with microstrip patch antennas. *IEEE Trans. Antennas Propag.*, 55(1):245–248.

Alomainy A, Hao Y. 2009. Modeling and characterization of biotelemetric radio channel from ingested implants considering organ contents. *IEEE Trans. Antennas Propag.*, 57:999–1005.

Alomainy A, Sani A, Rahman A, Santas JG, Hao Y. 2009. Transient characteristics of wearable antennas and radio propagation channels for ultrawideband body-centric wireless communications. *IEEE Trans. Antennas Propag.*, 57(4):875–884.

Attiya AM, Safaai-Jazi A. 2004. Simulation of ultra-wideband indoor propagation. *Microw. Opt. Technol. Lett.*, 42(2):103–108.

Balanis CA. 2002. *Antenna Theory: Analysis and Design*, 2nd ed. New York: Wiley.

Chan Y, Meng MH, Wu KL, Wang X. 2005. Experimental study of radiation efficiency from an ingested source inside a human body model. *IEEE Eng. Med. Biol. Soc.*, 7754–7757.

Chen ZN. 2005. Novel bi-arm rolled monopole for UWB applications. *IEEE Trans. Antennas Propag.*, 53(2):672–677.

Chen ZN. 2007. *Antennas for Portable Devices*. New York: Wiley.

Chirwa LC, Hammond PA, Roy S, Cumming DRS. 2003a. Electromagnetic radiation from ingested sources in the human intestine between 150 MHz and 1.2 GHz. *IEEE Trans. Biomed. Eng.*, 50:484–492.

Chirwa LC, Hammond PA, Roy S, Cumming DRS. 2003b. Radiation from ingested wireless devices in biomedical telemetry bands. *IEEE Electron. Lett.*, 39(2):178–179.

Conway GA, Scanlon WG. 2009. Antennas for over-body-surface communication at 2.45 GHz. *IEEE Trans. Antennas Propag.*, 57(4):844–855.

Conway GA, Scanlon WG, Orlenius C, Walker C. 2008. In situ measurement of UHF wearable antenna radiation efficiency using a reverberation chamber. *IEEE Antennas Wireless Propag. Lett.*, 7:271–274.

Fort A, Desset C, Ryckaert J, De Doncker P, Van Biesen L, Donnay S. 2005. Ultra wide-band body area channel model. *IEEE Int Conf. Commun.*, 4:2840–2844.

Gemio J, Parron J, Soler J. 2010. Human body effects on implantable antennas for ISM bands applications: models comparison and propagation losses study. *Prog. Electrom. Res.*, 110:437–452.

Gupta SKS, Lalwani S, Prakash Y, Elsharawy E, Schwiebert L. 2003. Towards a propagation model for wireless biomedical applications. *IEEE Int. Conf. Commun.*, 1993–1997.

Hall PS, Hao Y, Nechayev YI, Alomainy A, Constantinou CC, Parini C, Kamarudin MR, Salim TZ, Hee DTM, Dubrovka R, Owadally AS, Song W, Serra A, Nepa P, Gallo M, Bozzetti M. 2007. Antennas and propagation for on-body communication systems. *IEEE Antennas Propag. Mag.*, 49(3):41–58.

Hertleer C, Tronquo A, Rogier H, Vallozzi L, Van Langenhove L. 2007. Aperture-coupled patch antenna for integration into wearable textile systems. *IEEE Antennas Wireless Propag. Lett.*, 6:392–395.

Hu ZH, Nechayev YI, Hall PS, Constantinou CC, Hao Y. 2007. Measurements and statistical analysis of on-body channel fading at 2.45 GHz. *IEEE Antennas Wireless Propag. Lett.*, 6:612–615.

Huang FJ, Lee CM, Chang CL, Chen LK, Yo TC, Luo CH. 2011. Rectenna application of miniaturized implantable antenna design for triple-band biotelemetry communication. *IEEE Trans. Antennas Propag.*, 59(7):2646–2653.

Institute of Electrical and Electronics Engineers (IEEE). 1999. IEEE Standard for Safety Levels with Respect to Human Exposure to Radiofrequency Electromagnetic Fields, 3 kHz to 300 GHz, IEEE Standard C95.1-1999.

Institute of Electrical and Electronics Engineers (IEEE). 2005. IEEE Standard for Safety Levels with Respect to Human Exposure to Radiofrequency Electromagnetic Fields, 3 kHz to 300 GHz, IEEE Standard C95.1-2005.

International Commission on Non-Ionizing Radiation Protection (ICNIRP). 1998. Guidelines for limiting exposure to time–varying electric, magnetic, and electromagnetic fields (up to 300 GHz). *Health Phys.*, 74:494–522.

International Telecommunications Union—Radiocommunications (ITU-R). 1998. Recommendation ITU-R SA.1346.

Jovanov E, O'Donnell-Lords A, Raskovic D, Cox P, Adhami R, Andrasik F. 2003. Stress monitoring using a distributed wireless intelligent sensor system. *IEEE Eng. Med. Biol. Mag.*, 22(3):49–55.

Karacolak T, Cooper R, Topsakal E. 2009. Electrical properties of rat skin and design of implantable antennas for medical wireless telemetry. *IEEE Trans. Antennas Propag.*, 57(9):2806–2812.

Karacolak T, Cooper R, Butler J, Fisher S, Topsakal E. 2010. In vivo verification of implantable antennas using rats as model animals. *IEEE Antennas Wireless Propag. Lett.*, 9:334–337.

Karacolak T, Hood AZ, Topsakal E. 2008. Design of a dual-band implantable antenna and development of skin mimicking gels for continuous glucose monitoring. *IEEE Trans. Microw. Theory Techn.*, 56(4):1001–1008.

Kawoos U, Tofighi MR, Warty R, Kralick FA, Rosen A. 2008. In-vitro and in-vivo trans-scalp evaluation of an intracranial pressure implant at 2.4 GHz. *IEEE Trans. Microw. Theory Tech.*, 56(10):2356–2365.

Khan I, Hall PS. 2009. Multiple antenna reception at 5.8 and 10 GHz for body-centric wireless communication channels. *IEEE Trans. Antennas Propag.*, 57(1):248–255.

Khan I, Hall PS, Serra AA, Guraliuc AR, Nepa P. 2009. Diversity performance analysis for on-body communication channels at 2.45 GHz. *IEEE Trans. Antennas Propag.*, 57(4):956–963.

Kim J, Rahmat-Samii Y. 2004. Implanted antennas inside a human body: Simulations, designs, and characterizations. *IEEE Trans. Microw. Theory Techn.*, 52(8):1934–1943.

Kim J, Rahmat-Samii Y. 2006. SAR reduction of implanted planar inverted F antennas with non–uniform width radiator. *IEEE Int. Symp. Antennas Propag.*, 1091–1094

Kiourti A, Christopoulou M, Nikita KS. 2011. Performance of a novel miniature antenna implanted in the human head for wireless biotelemetry. *IEEE Int. Symp. Antennas Propag.*, Spokane, Washington, pp. 392–395.

Kiourti A, Nikita KS. 2011. Meandered versus spiral novel miniature PIFAs implanted in the human head: Tuning and performance. 2nd ICST Int. Conf. Wireless Mobile Commun. Healthcare, Kos Island, Greece, pp. 80–87.

Kiourti A, Nikita KS. 2012a. A review of implantable patch antennas for biomedical telemetry: Challenges and Solutions. *IEEE Antennas Propag. Mag.*, 54(3):210–228.

Kiourti A, Nikita KS. 2012b. Miniature scalp-implantable antennas for telemetry in the MICS and ISM bands: Design, safety considerations and link budget analysis. *IEEE Trans. Antennas Propag.*, 60(6):3568–3575.

Kiourti A, Nikita KS. 2012c. Accelerated design of optimized implantable antennas for medical telemetry. *IEEE Antennas Wireless Propag. Lett.*, 11:1655–1658.

Kiourti A, Nikita KS. 2012d. Recent advances in implantable antennas for medical telemetry. *IEEE Antennas Propag. Mag.*, 54(6):190–199.

Kiourti A, Nikita KS. 2013. Design of implantable antennas for medical telemetry: Dependence upon operation frequency, tissue anatomy and implantation site. *Int. J. Monit. Surv. Technol.*, 1(1):16–33.

Kiourti A, Psathas KA, Costa JR, Fernandes CA, Nikita KS. 2013. Dual-band implantable antennas for medical telemetry: A fast design methodology and validation for intra-cranial pressure monitoring. *Prog. Electrom. Res.*, 141:161–183.

Klemm M, Locher I, Tröster G. 2004. A novel circularly polarized textile antenna for wearable applications. 7th Europ Microw Week, Amsterdam, The Netherlands, pp. 137–140.

Lee CM, Yo TC, Huang FJ, Luo CH. 2009. Bandwidth enhancement of planar inverted–F antenna for implantable biotelemetry. *Microw. Opt. Technol. Lett.*, 51(3):749–752.

Lee SH, Lee J, Yoon YJ, Park S, Cheon C, Kim K, Nam S. 2011. A wideband spiral antenna for ingestible capsule endoscope systems: Experimental results in a human phantom and a pig. *IEEE Trans. Biomed. Eng.*, 58(6):1734–1741.

Liu WC, Chen SH, Wu CM. 2008. Implantable broadband circular stacked PIFA antenna for biotelemetry communication. *J. Electromagn. Waves Appl.*, 22:1791–1800.

Liu WC, Chen SH, Wu CM. 2009. Bandwidth enhancement and size reduction of an implantable PIFA antenna for biotelemetry devices. *Microw. Opt. Technol. Lett.*, 51(3):755–757.

Okoniewski M, Stuchly MA. 1996. A study of the handset antenna and human body interaction. *IEEE Trans. Microw. Theory Tech.*, 44(10):1855–1864.

Osman MAR, Rahim MKA, Samsuri NA, Salim HAM, Ali MF. 2011. Embroidered fully textile wearable antenna for medical monitoring applications. *Prog. Electrom. Res.*, 117:321–337.

Rajagopalan H, Rahmat-Samii Y. 2010. Link budget analysis and characterization for ingestible capsule antenna. *Int. Workshop Antenna Technol.*, pp. 1–4.

Rucker D, Al-Alawi A, Adada R, Al-Rizzo HM. 2007. A miniaturized tunable microstrip antenna for wireless communications with implanted medical devices. ICST 2nd Int Conf on Body Area Networks, Brussels, Belgium, pp. 1–4.

Salonen PO, Yang F, Rahmat-Samii Y, Kivikoski M. 2004a. WEBGA-Wearable electromagnetic band-gap antenna. *IEEE Antennas Propag. Int. Symp.*, 1:451–454.

Salonen P, Rahmat-Samii Y, Hurme H, Kivikoski M. 2004b. Dual band wearable textile antenna. IEEE Antennas Propag. *Soc. Int. Symp.*, pp. 463–466.

Sani A, Alomainy A, Hao Y. 2009. Numerical characterization and link budget evaluation of wireless implants considering different digital human phantoms. *IEEE Trans. Microw. Theory Tech.*, 57(10):2605–2613.

Sani A, Rajab M, Foster R, Hao Y. 2010. Antennas and propagation of implanted RFIDs for pervasive healthcare applications. *Proc. IEEE*, 98(9):1648–1655.

Scanlon WG, Evans NE. 2001. Numerical analysis of bodyworn UHF antenna systems. *IEE Electron Commun. Eng. J.*, 13(2):53–64.

Serra AA, Nepa P, Manara G, Hall PS. 2007. Diversity measurements for on-body communication systems. *IEEE Antenna Wireless Propag. Lett.*, 6(1):361–363.

Skrivervik AK, Merli F. 2011. Design strategies for implantable antennas. Antennas Propag Conf, pp. 1–5.

Soontornpipit P, Furse CM, Chung YC. 2004. Design of implantable microstrip antenna for communication with medical implants. *IEEE Trans. Microw. Theory Tech.*, 52:1944–1951.

Soontornpipit P, Furse CM, Chung YC. 2005. Miniaturized biocompatible microstrip antenna using genetic algorithm. *IEEE Trans. Antennas Propag.*, 53(6):1939–1945.

Tronquo A, Rogier H, Hertleer C, Van Langenhove L. 2006. A robust planar textile antenna for wireless body lans operating in the 2.45-GHz ISM band. *Inst. Elect. Eng. Electron. Lett.*, 42(3):142–143.

Valdastri P, Menciassi A, Arena A, Caccamo C, Dario P. 2004. An implantable telemetry platform system for in vivo monitoring of physiological parameters. *IEEE Trans. Inf. Technol. Biomed.*, 8(3):271–278.

Wang Q, Wang J. 2009. Performance of on-body chest-to-waist UWB communication link. *IEEE Microw. Wireless Compon. Lett.*, 19(2):119–121.

Warty R, Tofighi MR, Kawoos U, Rosen A. 2008. Characterization of implantable antennas for intracranial pressure monitoring: Reflection by and transmission through a scalp phantom. *IEEE Trans. Microw. Theory Tech.*, 56(10):2366–2376.

Weiss MD, Smith JL, Bach J. 2009. RF coupling in a 433 MHz biotelemetry system for an artificial hip. *IEEE Antennas Wireless Propag. Lett.*, 8:916–919.

Wong KL, Lin CI. 2005. Characteristics of a 2.4-GHz compact shorted patch antenna in close proximity to a lossy medium. *Microw. Opt. Technol. Lett.*, 45(6):480–483.

Xia W, Saito K, Takahashi M, Ito K. 2009. Performances of an implanted cavity slot antenna embedded in the human arm. *IEEE Trans. Antennas Propag.*, 57(4):894–899.

Xie X, Li G, Chen XK, Li XW, Chi BY, Han SG. 2004. A novel low power IC design for bi-directional digital wireless endoscopy capsule system. IEEE Int. Workshop Biomed. Circuit. Syst., pp. S1.8-5–1.8-8.

Xu L, Max QH, Ren HL. 2008a. Electromagnetic radiation from ingested sources in the human intestine at the frequency of 2.4 GHz. Progress Electrom. Res. Symp., pp. 893–897.

Xu L, Maz MQH, Ren H, Chan Y. 2008b. Radiation characteristics of ingested wireless device at frequencies from 430 MHz to 3 GHz. IEEE Conf. Eng. Med. Biol. Soc., pp. 1250–1253.

Xu L, Meng MQH, Chan Y. 2009a. Effects of dielectric parameters of human body on radiation characteristics of ingestible wireless device at operating frequency of 430 MHz. *IEEE Trans. Biomed. Eng.*, 56:2083–2094.

Xu L, Meng MQH, Ren H, Chan Y. 2009b. Radiation characteristics of ingestible wireless devices in human intestine following radio frequency exposure at 430, 800, 1200, and 2400 MHz. *IEEE Trans Antennas Propag.*, 57:2418–2428.

Zasowski T, Althaus F, Stager M, Wittneben A, Troster G. 2003. UWB for noninvasive wireless body area networks: Channel measurements and results. IEEE Conf. UWB Syst. Technol., Reston, VA, pp. 285–289.

Zhao Y, Hao Y, Alomainy A, Parini C. 2006. UWB on-body radio channel modeling using ray theory and sub-band FDTD method. *IEEE Trans. Microw. Theory Tech.*, 1–9.

Zhu S, Langley R. 2009. Dual-band wearable textile antenna on an EBG substrate. *IEEE Trans. Antennas Propag.*, 57(4):926–935.

9

INTRABODY COMMUNICATION

Laura M. Roa, Javier Reina-Tosina, Amparo Callejón-Leblic, David Naranjo, and Miguel Á. Estudillo-Valderrama

University of Seville, Seville, Spain

9.1 INTRODUCTION

Recent advances in Information and Communication Technologies (ICTs) have become a key ally in the development of Personal Area Networks (PANs) for remote and ubiquitous monitoring, together with the search for minimally invasive wearable systems embedded within cordless networks capable of gathering data of the user's surroundings. Current PANs normally use short-range radio frequency (RF) solutions based on widespread commercial standards such as Bluetooth (2012) and Zigbee Alliance (2012). However, these products still have limitations when taking into account their energy consumption and the increasing saturation present in the free band of frequencies where they operate [Industrial, Scientific, and Medical (ISM) radio bands], which cause interferences along with a greater vulnerability to attacks on the privacy of the transmitted data. Other options have been proposed to solve these inconveniences (ANT, 2012; 6LowPan, 2012), among which new low-power standards [Bluetooth Low Energy (BLE), 2012] take an important role, although they are not as massively used as the previously mentioned.

A different approach is becoming popular nowadays, although it was originally proposed by Zimmerman in 1996. It is based on signal transmission through the human body, called intrabody communication (IBC), and has led to the first definition

Handbook of Biomedical Telemetry, First Edition. Edited by Konstantina S. Nikita.

of Body Area Networks (BANs). Zimmerman's original architecture comprised a transmitter placed on the body surface that modulates an electric field, which is conducted by means of small currents (50 pA) through the body toward the receiver, which demodulates the received signal, and thus the encoded information is recovered. Besides that, the external environment provides a return path for the transmitted signal via capacitive coupling. It is worth mentioning that low carrier frequencies associated with low-power waves can be used, which favors a lower consumption of the terminals and less interference with nearby devices. Thus, Zimmerman's approach to data communication in BANs paved the way for the design of systems that could minimize RF-based solution disadvantages.

In general terms, a BAN communication is based on a transmitter that perturbs an electric potential, thus disturbing its immediate external environment, and a receiver that detects the disturbance. In electrical terms, the transmitter is capacitively coupled to the receiver. In order not to produce an excessively radiative signal, the voltages to be handled are small enough to fulfill the maximum permitted levels and ensure safety (Deventer et al., 2007). For this type of coupling, power decreases as the cube of the distance; however, for BAN applications, this property is considered more an advantage than a problem. This is because the distances involved in the human body are small enough so that the signal is received correctly at the receiving end, and, in turn, the signal is hardly perceptible outside; hence, the system is robust against external attacks, making them difficult to access and thus improving data privacy.

The IBC propagation channel is mainly established through the human body, but depending on the application and the coupling type, there are different signal pathways. Hall and Hao (2006) distinguish three types of body-centric communications types, called *off body*, *on body* and *in body*, depending on the degree of interaction of the human body with the surrounding space. For example, galvanic coupling should always be used when sensors are implanted inside the human body (in body), and, therefore, the signal pathway is the human body itself. However, when all sensors are deployed on the surface of the skin, we refer to body surface sensor networks (on-body and off-body types), and both IBC galvanic and capacitive coupling schemes can be used. In the former case, the signal propagates differentially between a pair of transmitter electrodes and a pair of receiver electrodes, all of them attached to the skin. Thus, the signal pathway is only formed through the human body. In the case of capacitive coupling, the signal travels through two different pathways, one of them being the human body itself, and the second being the capacitive return path through the air toward the external ground. On the other hand, in the case of body surface sensor networks, IBC is based on near-field coupling mechanisms, and the signal is confined in the body surface, mainly through the skin, which acts as a signal guide that quasi-statically couples the signal. For more details about IBC coupling strategies refer to Section 9.2.

IBC applications vary, as the communication channel is seamless to the context, but with remarked advantages compared to the available commercial RF standards in the field of BANs. A brief review of the latest market applications and academic publications shows that IBC can be applied in very different fields of people's daily lives. On the one hand, those related to consumer electronics, entertainment, and leisure activities can be highlighted (Hyoung et al., 2006), with some companies already taking advantage of the potentiality of the IBC technology (KDDI, 2012; Kado and Shinagawa, 2010). In a second place, there have been significant advances in the field of space and military applications (Vasilopoulos, 2010), personal security (Fujikawa and Nishigaki, 2011; Oguma et al., 2011), and disaster management (Bessis et al., 2010). Finally, the main and original application of BAN is the remote monitoring of biomedical signals through the management of on-body and in-body physiological parameters measured by sensors either attached to the user's body (Warren and Natarajan, 2011) or implanted (Black, 2011). There are many cases taking into account the relevance that BAN technologies acquire in the health care domain and the rapid outcomes that IBC prototypes achieve (for more details, see Table 9.1). Some relevant examples of IBC at present are eased by the deployment of sensors in direct contact with human skin [like electromyograms (EMGs) or electrocardiograms (ECG)] for the welfare enhancement of disabled people (Liolios et al., 2010) or the estimation of muscular fatigue (Lucev et al., 2010c) through EMG that also facilitates the brain–computer interface design for affective applications (Pun et al., 2011b).

The last decade represents an important boost for the research in IBC, by means of modeling approaches and techniques that have taken advantage of the rapid development of microelectromechanical systems (MEMS) and nanotechnologies for the implementation of several laboratory prototypes with better performance. Thus, there are a relevant number of papers that describe the electronic approaches followed and the experimental results obtained in terms of communications performance (data rate,

TABLE 9.1 Summary of Relevant Intrabody Communication Systems

Reference	Amplitude	Coupling Type	Carrier Frequency	Modulation Scheme	Data Rate (bps)
Zimmerman,1995	30 V	Capacitive	330 kHz	OOK	2.4 k
Handa et al., 1997	20 μA	Galvanic	70 kHz	PWM	0.9 k
Fukumoto and Tonomura, 1997	21 V	Capacitive	90 kHz	FM	9.6 k
Partridge et al., 2001	22 V	Capacitive	160 kHz	FSK	38.4 k
Hachisuka et al., 2003	1 V	Capacitive	10.7 MHz	FSK	9.6 k
NTTRedTacton, 2005	25 V	Capacitive	10 MHz	OOK	10 M
Ruiz and Shinamoto, 2005	—	Capacitive	200–600 MHz	MSK	—
Wegmueller, 2007a	1 mA	Galvanic	256 kHz	BPSK	64 k
Lucev et al., 2009	1.15 V	Capacitive	1 MHz	FM	64 k
Lin et al., 2011	0.5 V	Capacitive	200 MHz	OOK	2 M

OOK: on-off keying
PWM: pulse width modulation
FM:frequency modulation
FSK:frequency-shift keying
MSK: minimum-shift keying
BPSK: binary phase-shift keying
Source: Adapted from Wegmueller (2007a) and Lucev et al. (2010b).

return path loss, attenuation, etc.). These approaches are intimately related to the coupling strategies followed, which determine the different modulation schemes as well as the disposal of electrodes to be used, which are key issues due to the different results obtained. For instance, Tang and Bashirullah (2011) reports results of attenuation on experimental measurements performed on a human cadaver obtaining a range of −10 to −40 dB in the frequency range of 1–20 MHz. In Lucev et al. (2011), a characterization of the IBC channel for the frequency range from 100 kHz to 100 MHz is presented, obtaining a transmission gain increase of 20 dB/decade up to 1 MHz independently of the transmitter and receiver configuration. Similarly, Zedong et al. (2012) suggest the 24- to 30-MHz frequency range as the optimal for IBC communications, which was obtained through experimentation in a shielded chamber. On the other hand, Attard and Zammit (2012) determined a frequency band between 280 and 500 MHz for capacitive body-coupled communications. Okamoto et al. (2010) developed an interesting transcutaneous communication system able to bidirectionally transmit 115 kbps from the left wrist to the right forearm without communication error. As can be seen, there is an important variability of the performances obtained, which demands research efforts focused on different techniques and approximations to measure channel propagation characteristics (Yang et al., 2010, 2011).

Table 9.1 shows the main features of some of the most relevant IBC prototypes reported in the literature depending on the type of coupling, operating frequencies, data rate, and consumption. The maximum rate reported for IBC has been 10 Mbps, thanks to the introduction of high-impedance electro-optical sensors. Unfortunately, these sensors have the disadvantage of a more complex electronic design, resulting in low capacity for integration, greater power consumption, and size.

Despite the great performance obtained by IBC systems, there are still important remaining advances to be made, which have been indirectly mentioned in this section. From a theoretical viewpoint, although IBC prototypes have significantly improved in performance, bit rate, and consumption, there is no common methodology yet to establish the design specifications of IBC systems, although some recent exceptions can be made (Bae et al., 2012). Undoubtedly, this is due to the fact that the electromagnetic (EM) mechanisms that govern the transmission through biological tissues still remain unknown. From a review of the IBC literature, the lack of bioelectrical models to explain the nature of the communication principles that allow defining simple expressions for the design of the devices becomes obvious. This chapter aims at clarifying the existing gap between measurements, models, and technology development, which do not evolve together as they should. From a technological viewpoint, the main conclusions derived are that IBC systems must consider present and future trends in the development of micro - and nano-MEMS, which can ease more unobtrusive and compact solutions, like the proposed IBC system on a chip in Yan et al. (2011), while facing the need to compromise electronic complexity with performance in order to obtain realistic and sustainable solutions able to be translated to the health care domain. Finally, and for the sake of a successful commercial use of IBC solutions, there are other considerations to be made regarding the features and quality of service (QoS) of the communications, like physical layer security issues (Zhu et al., 2012).

The chapter starts with an introduction to the concept of IBC, the main differences with other similar technologies, and its potential applications, remarking on the main achievements made and the challenges to be faced. Section 9.2 analyzes the principal transmission methods employed in IBC communications, which influence the modeling approaches as well as the technology to be used. Section 9.3 is devoted to the dielectric properties of the human body and the skin, which play a fundamental role in the propagation of IBC signals. The most significant outcomes in terms of the experimental characterization of the IBC channel are analyzed in Section 9.4. Section 9.5 addresses the modeling of the IBC channel, as a first step to the study of its propagation behavior (Section 9.6). Finally, the chapter finishes with Section 9.7 containing the summary and concluding remarks.

9.2 INTRABODY COMMUNICATION TRANSMISSION METHODS

In communications throughout the body, human tissues establish the signal path between a transmitter and a receiver (Wegmueller, 2007a). According to the signal coupling method used for transmission in the body environment, the following modalities can be distinguished: galvanic coupling and capacitive coupling (Estudillo et al., 2010).

9.2.1 Galvanic Coupling

An electric current is injected into the body through a pair of electrodes attached to the skin (M. S. Wegmueller et al., 2006a; Gao et al., 2009) (see Figure 9.1). Most of the current density flows through the direct path between the electrode pairs; however, other weaker currents are induced inside the conductive tissues of the human body. As we move away from the source electrodes, current density inside the human body decreases; hence, only a weak current flows between the electrode pair at the receiver end. This weak current generates a low voltage drop that can be detected by an instrumentation amplifier (Wegmueller, 2007a; Wegmueller et al., 2009). These devices usually have very high input impedances, and therefore interelectrode voltage generates a negligible current at the device input. In this way, the measured signal

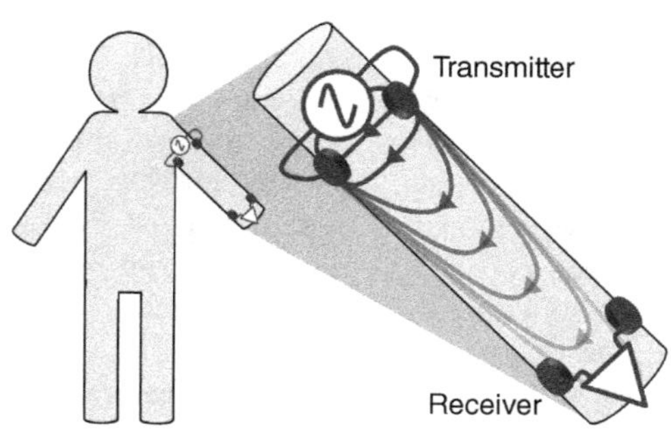

Figure 9.1 Galvanic coupling transmission method.

corresponds to that existing between the electrodes, as interelectrode voltage can also be neglected.

As an electric current is injected into the body, galvanic-type IBC devices must strictly comply with safety recommendations and standards, avoiding the occurrence of any interference with pacemakers, defibrillators, and other biomedical devices. Moreover, in order to prevent any adverse health effects due to exposition to electromagnetic fields (EMFs), or interference with physiological signals, EM emissions must be kept well within safety levels established by international regulation [International Commission on Non-Ionizing Protection (ICNIRP), 1998, 2009]. Usually, a transmission frequency above 10 kHz is selected so as not to interfere with the subject's biological signals. Furthermore, injection current levels in the range of 1–2 mA are typically used (Wegmueller, 2007a). Experimental results show that galvanic coupling is an appropriate method for data transmission between devices located on the breast or at short ranges along the extremities in the frequency range of 10 kHz–1 MHz. Besides that, as the injected current is very weak, this modality is much more energy efficient than standard wireless RF transceivers (Wegmueller et al., 2007b).

Electrodes play a major role in IBC systems. For the galvanic coupling modality, the electrodes provide an electrochemical interface between the IBC transceiver subsystem and the human body, both for the transmitter and receiver sides. Despite their key role, the dispersion in their impedance values, noise, fluctuations, and adherence problems make them exhibit variable properties, and the inability to tackle these effects lead to serious problems (Roa et al., 2012).

Generally, and in order to couple the largest amount of signal in the human body for a better transmission, it is desirable that the impedance of the electrodes have a small value so as not to lose power in the contact with the skin. In order to reduce this impedance, a gel electrolyte is often added to the interface, typically with Cl^- ions (Roa et al., 2012). This skin–electrode interface has a frequency behavior with three distinguishable regions. At very low frequencies, the behavior is resistive (faradic region). If frequency is increased, there is a range in which the impedance decreases (Warburg region) because the fringing capacitances start to come into play (at about 80 kHz). Finally, the last zone is characterized by having a flat response with frequency, and the impedance is about two orders of magnitude lower than that of the faradic region. One of the most employed type of electrodes in the literature is that composed of Ag–AgCl (Hachisuka et al., 2003; Wegmueller et al., 2006b; Song et al., 2007), with an electrolyte containing Cl^-, due to their good noise performance.

On the other hand, as the impedance of the electrode–skin interface is inversely proportional to the area of the electrodes, the increment of the electrode area can lead to lower impedances, providing better results in terms of communication performance. However, large-area electrodes are not suitable for the development of nonintrusive IBC devices. It has been experimentally proven that attenuation decreases with the size of the transmission electrodes, while the size of the receiving electrodes has a little contribution in terms of attenuation, as a consequence of the high input impedance of the reception amplifiers.

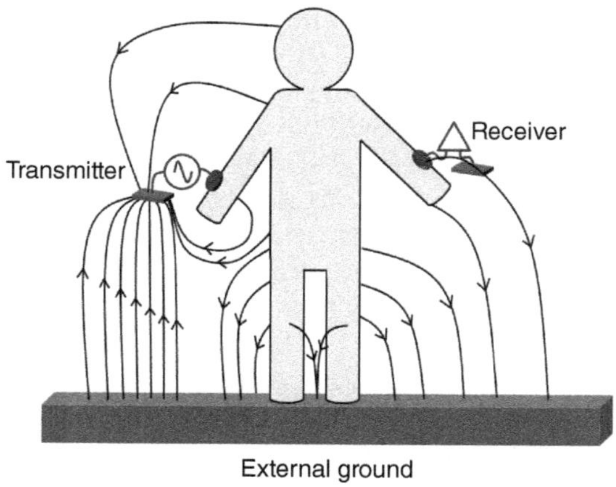

Figure 9.2 Capacitive coupling transmission method.

9.2.2 Capacitive Coupling

In the case of capacitive coupling (Zimmerman, 1995; Partridge et al., 2001), electric fields are induced between the signal electrode and the ground electrode of the transmitting device (see Figure 9.2). Part of this field passes through the human body in order to return to the device through the air or aided in part by the return path of the external ground (capacitive return path). Finally, a small portion flows through the signal and ground electrodes of the receiver device (capacitive forward path). In this configuration, the human body behaves as a conductor, while the electrical coupling of the electrodes with the human body, the air, and the external ground can be modeled as capacitances, hence the name of this modality. The capacitances of the links through the air are extremely small, at the level of femptofarads (Zimmerman, 1995), thus the weak signals in transmission systems throughout the body are mainly due to poor return paths and not to the low conductivity of the human body. Since most of the signal is confined into the human body owing to its behavior as an electric conductor, the capacitive coupling techniques reduce power consumption and increase communication security compared to conventional RF methods.

As in the galvanic coupling case, in capacitive coupling the electrodes also play a key role in the signal coupling within the human body. Parameters such as the characteristics of the electrodes, the polarization of the material, the area, the radius of curvature, or the position of the electrodes on the human body can influence the amount of signal coupled (Hachisuka K. et al., 2005, 2006; Northrop, 2004; Neuman 2000; Wegmueller et al., 2007b). However, the ground electrode is the one that causes the larger variability in this modality. In the electrostatic coupling, the grounds of the transmitter and receiver devices are not connected to any surface and generate a return path by capacitive coupling through the air (Fukumoto and Tonomura 1997; Matsushita et al., 2000) or by capacitive coupling through the air between the grounds of the transmitter–receiver devices and the external ground plane (Zimmerman, 1995;

Matsushita et al., 2000; Partridge et al., 2001). In these cases, it is very common to use a large ground plane in the circuit board of the device (Cho et al., 2007). Since the human body behaves as a low-pass filter due to the capacitive return path, as the transmission frequency increases, the impedance of the capacitive return path decreases, and as a result the impedance of the body cannot be ignored. In order to secure the link between the transmitter and receiver, it may be necessary to enlarge the ground plane of the devices to ensure the external coupling (Cho et al., 2007).

Fujii et al. (2006, 2007) concluded that the use of the ground electrode in the transmitter strengthens the electric field when the device is placed on the arm. However, the existence of a ground electrode in the receiver decreases the level of the received signal. On the other hand, the transmission is improved if the transmitter has an electrode orthogonal to the direction of propagation to the receiver. Hwang et al. (2007) found that signal losses are reduced when the ground electrode is directly attached to the human body, but the current density nearby the electrode also increases, as the current flows directly from the signal electrode to ground electrode, thereby increasing power consumption. For this reason, in the case of small distances, it may be desirable not to connect the ground electrode to the body aiming to minimize power consumption. In the case of longer distances, higher than 1 m, it may be advisable to connect the ground electrode to the body in order to reduce the signal losses.

9.3 DIELECTRIC PROPERTIES OF HUMAN BODY

The context for which IBC communications are defined is closely related to the physical channel where they are developed: the human body. This matter imposes limitations and design conditions for the transceiver devices due to the interaction between the human body and EMFs (Wegmueller et al., 2007b; Estudillo et al., 2010).

In a biological environment, there is an inseparable relationship between electricity and chemistry. Biological tissues behave as conductors of electric current depending on their composition. In an aqueous medium, the electric current is carried by the ions dissolved therein. As the conductivity is directly proportional to the content of ions (electrolytes) present in the fluids, electrolyte solutions in the interior of cells and extracellular fluids provide good conductivity (Roa et al., 2012).

On the other hand, in any physical media, in addition to conductivity, there is another property related to the flow of electric current: permittivity (Gabriel et al., 1996; Roa et al., 2012). This is a physical property that is related to the ability of the material to store electric field energy. In general, dielectric materials present a reduction of the permittivity with frequency due to the absorption of electromagnetic energy. This absorption can be associated with resonance effects, arising from the rotation or vibration of atoms, ions, or molecules, which normally occur in the proximity of their characteristic absorption frequencies. However, the major energy losses are caused by relaxation effects associated with the molecular dipoles of the medium. At low frequencies, the field changes slowly enough to allow dipoles to reach equilibrium before the field changes its polarity. For frequencies in which the orientations of the dipoles cannot follow the applied field due to the viscosity of the medium, absorption of field energy, dissipated as heat, and a time delay in the response of the medium take place.

In biological tissues, dispersive effects are the result of the interaction of EM fields with the molecular and cellular components. The main dielectric characteristics of tissues as a function of frequency are well known and were already shown by Foster and Schwan in 1989, who showed that the dielectric spectrum of a tissue is characterized by four regions in which the permittivity decreases as frequency increases (three major dispersions: α, β, and γ at low, medium; and high frequencies; and a lower dispersion that is usually referred to as δ dispersion) (Gabriel et al., 1996) (see Figure 9.3):

- α Dispersion: from 10 Hz to 10 kHz; mainly due to ion diffusion effects in the cell membrane.
- β Dispersion: between 10 kHz and 30 MHz; it is associated with the polarization of cell membranes but also a consequence of the polarizations of proteins and other organic macromolecules.
- γ Dispersion: over the gigahertz range; it is primarily due to dipolar polarization of water molecules.

Each of these dispersions or relaxations is the manifestation of a polarization mechanism characterized by a time constant (τ), which, in a first-order approximation, gives the following expression for the relative complex permittivity as a function of the angular frequency ($\omega = 2\pi f$):

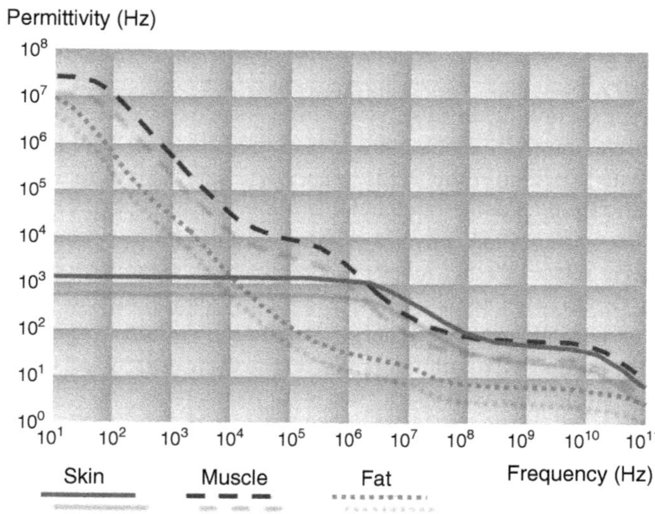

Figure 9.3 Permittivity and dispersions of the skin, muscle, and fat as a function of frequency.

$$\varepsilon_r^*(\omega) = \varepsilon_r'(\omega) - j\varepsilon_r''(\omega) = \varepsilon_\infty + \frac{\varepsilon_s - \varepsilon_\infty}{1 + j\omega\tau} \tag{9.1}$$

This is the well-known Debye expression, where $\varepsilon_r'(\omega)$ and $\varepsilon_r''(\omega)$ stand for the real and imaginary parts, respectively, ε_∞ is the permittivity at frequencies where $\omega\tau \gg 1$, ε_s is the permittivity at frequencies where $\omega\tau \ll 1$, and $j^2 = -1$. The magnitude of the dispersion is described as $\Delta\varepsilon = \varepsilon_s - \varepsilon_\infty$. In 1985, Hurt proposed modeling the dielectric spectrum of tissues as a sum of five dispersions plus a conductivity term (Gabriel et al., 1996):

$$\varepsilon_r^*(\omega) = \varepsilon_\infty + \sum_{k=1}^{5} \frac{\Delta\varepsilon_k}{1 + j\omega\tau_k} + \frac{\sigma_s}{j\omega\varepsilon_0} \tag{9.2}$$

where σ_s is the static ionic conductivity of the tissue and ε_0 is the permittivity of free space.

However, the complexity of both the structure and composition of biological materials is such that each dispersion region can be expanded by introducing distribution parameters, as defined in the Cole–Cole model for dispersions, introducing the parameter α $(0 < \alpha \leq 1)$,

$$\varepsilon_r^*(\omega) = \varepsilon_\infty + \sum_{k=1}^{5} \frac{\Delta\varepsilon_k}{1 + (j\omega\tau_k)^\alpha} + \frac{\sigma_s}{j\omega\varepsilon_0} \tag{9.3}$$

According to the literature and previous experiments, Gabriel et al. (1996) provided values to the equation parameters, which modeled the complex relative permittivity for different tissues (blood, bone, fat, brain, kidney, spleen, heart, liver, lung, muscle, skin) in the frequency range of 10 Hz–20 GHz. These parametric equations of the electric permittivity and conductivity as a function of the frequency have been widely used by many authors to establish the dielectric properties of the human body (Gabriel et al., 1996). Table 9.2 summarizes the relevant parameters of the Cole–Cole models for the most relevant tissues in the context of IBC.

As previously mentioned, the β dispersion is primarily caused by cell membranes. Its behavior is comparable to that of a capacitance, which occurs due to the existence of regions of high conductivity (intracellular and extracellular media) separated by low conductivity regions (cell membranes). In this case, the regions of high conductivity behave as the plates of a capacitor.

At a frequency of 0 Hz, the "biological plates" can be charged, but there will not be current conduction through the membrane (Roa et al., 2012). At very low frequencies, there will be a minimum conduction through the cells due to the high impedance of the cell membrane (see Figure 9.4). Current will flow in this case through the extracellular fluid surrounding the cells. As the frequency increases, capacitances of the membrane are charged and discharged, reducing the impedance of the membrane and allowing a small current penetration into the intracellular compartment. Thus, the impedance of the medium will decrease as frequency increases because the amount of conductive volume increases. At high frequencies, above 5 MHz, the rate of charging

TABLE 9.2 **Parameters of Cole–Cole Equation**

	ε_∞	$\Delta\varepsilon_1$	τ_1 (ps)	α_1	$\Delta\varepsilon_2$	τ_2 (ns)	α_2	$\Delta\varepsilon_3$	τ_3 (μs)	α_3	$\Delta\varepsilon_4$	τ_4 (ms)	α_4	σ
Skin	4.0	39.0	7.96	0.10	280	79.58	0.00	3.0×10^4	1.59	0.16	3.0×10^4	1.592	0.20	0.0004
Muscle	4.0	50.0	7.23	0.10	7000	353.68	0.10	1.2×10^6	318.31	0.10	2.5×10^7	2.274	0.00	0.2000
Fat	2.5	3	7.96	0.2	15	15.92	0.1	3.3×10^4	159.15	0.05	1.0×10^7	7.958	0.01	0.01

Source: Adapted from Gabriel et al. (1996).

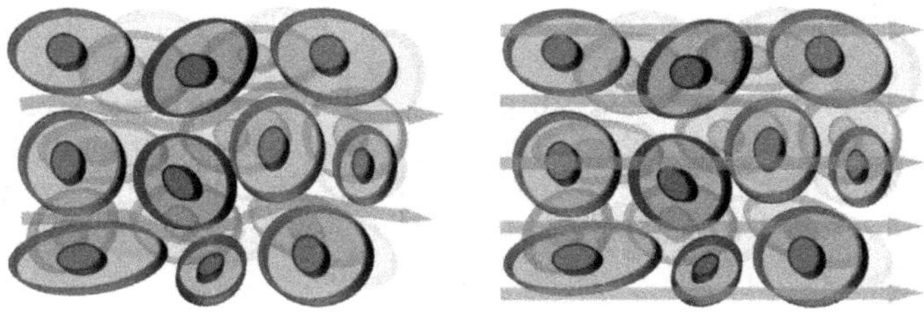

Figure 9.4 Behavior of an electric current in a biological tissue at low and high frequency.

and discharging of the membrane is such that the capacitive effect may be considered negligible (the capacitance of the membrane behaves as a short circuit) and current flows directly through the extracellular and intracellular compartments as a function of the relative conductivity of these mediums. The IBC method by galvanic coupling described in Section 9.2.1 can be largely explained by this current flow through the biological media.

9.3.1 Electrophysiological Properties of Skin

In the IBC context, skin is the mayor path traversed by EMFs on the human body; hence its electrophysiological properties have a special influence on the parameters of attenuation and distortion of an EM signal.

Human skin consists of three layers: the epidermis, the dermis, and the subcutaneous fat (Healthsystem, 2012). The epidermis contains some other layers, the outer of which is referred to as the stratum corneum (SC), which is made of dead, flat skin cells called keratinocytes (Armijo and Camacho, 1998). This is the body's first line of defense that interfaces with a hostile environment, providing a protective barrier. From an electrical standpoint, the SC is the layer that most influences the skin impedance behavior at low frequencies (Yamamoto and Yamamoto, 1976).

Experimental studies specifically show that the skin mainly presents two dispersive regions (Gabriel et al., 1996); the first of these includes the low-frequency region, up to 1 MHz, where the total skin impedance is practically determined by the impedance of the SC (Tronstad et al., 2010), which is composed of keratinized cells that do not contain cytoplasm and whose electrical behavior is similar to a dielectric material. In this first dispersive region, conductivity is determined by electric current paths through the sweat glands. The frequency response of the conductance G and susceptance B of the skin up to 1 MHz was reported in Grimmes and Martinsen (2000) (Grimnes and Martinsen, 2000), and, in particular, it was found that the SC represents only 10 % of the total skin impedance around 100 kHz. The second dispersive region ranges from 1 MHz to 1 GHz, approximately. In this band, the skin impedance is determined by the deeper layers; therefore, the dielectric properties of the skin in this region are given by the membrane of the living inner cells.

Thus, permittivity is influenced by the lipid bilayer and conductivity by the ionic channels that cross it.

Considering only two regions of dispersion as the dominant processes in the skin, it is common to represent the dielectric properties of the skin, with accuracy and effectiveness, through a Debye model with two dispersions and a conductive loss term (Callejón et al., 2012c; Gabriel et al., 1996; Fujii et al., 2010). Therefore, the complex relative permittivity of the skin can be expressed as

$$\varepsilon_r^*(\omega) = \varepsilon_\infty + \frac{\sigma_s}{j\omega\varepsilon_0} + \sum_{k=1}^{2} \frac{\Delta\varepsilon_k}{1 + j\omega\tau_k} \tag{9.4}$$

and the complex conductivity can be found by means of (9.1) through the expression

$$\sigma^*(\omega) = \sigma'(\omega) + j\sigma''(\omega) = j\omega\varepsilon_0\varepsilon_r^*(\omega) \tag{9.5}$$

where σ' and σ'' stand for the real and imaginary parts of $\sigma^*(\omega)$. Typical values for these parameters are reported in Gabriel et al. (1996).

For the sake of a compact representation of the skin frequency response, a reduced circuital structure can be used to represent the transcutaneous skin admittance, Y (Callejón et al., 2012c): a shunt two-component circuit with one conductance, G, and one susceptance, B, (i.e., $Y = G + jB$) such that Y retains the same frequency characteristics of the two-pole Debye model. Although several authors have modeled skin transcutaneous admittance emphasizing the electrical differences of each layer and the hydration effect (Edelberg, 1968; Fowles, 1986), the two-component circuit is the simplest and most effective equivalent structural circuit mimicking the admittance found in tissue measurements (Grimnes and Martinsen, 2000). Thus, in order for this simplified circuit to reproduce the two-pole Debye model frequency behavior, both G and B must be frequency dependent. Thus, a simple but accurate basic electrical transcutaneous model is depicted in Figure 9.5, which can be regarded as a starting point for IBC modeling.

In order to find the frequency responses of G and B, and assuming homogeneous electrical properties of the skin (Yamamoto and Yamamoto, 1976), equations (9.4) and (9.5) can be used, together with the values of permittivity and conductivity found by means of equations (9.6) and (9.7), with parameters reported for dry skin in Gabriel et al. (1996):

$$G(\omega) = K\sigma'(\omega) = K\omega\varepsilon_0\varepsilon_r''(\omega) \tag{9.6}$$

$$B(\omega) = \omega\frac{\varepsilon_r'(\omega)\varepsilon_0}{\sigma'(\omega)} G(\omega) = K\omega\varepsilon_0\varepsilon_r'(\omega) \tag{9.7}$$

where K is a constant dependent on the measurement electrode, its geometry, and active area. Without loss of generality, a unit value can be chosen for the sake of simplification (Grimnes and Martinsen, 2000). The values found for conductance G and susceptance B are represented by means of a Wessel diagram and shown in

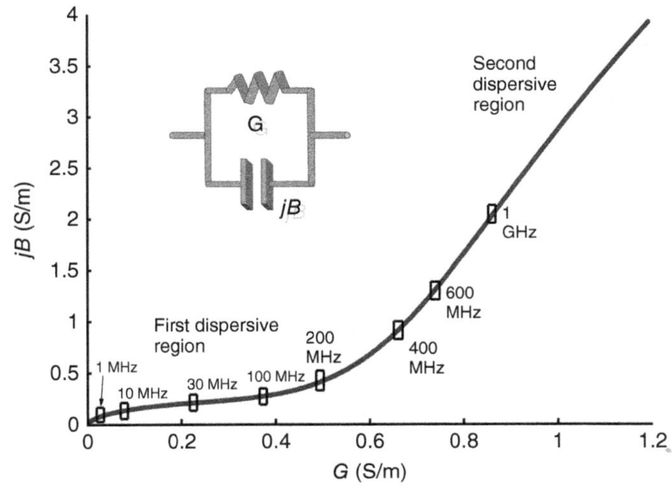

Figure 9.5 Wessel diagram of skin admittance for the complex relative permittivity given by equation (9.4).

Figure 9.5. While the Debye model (9.4) has two finite poles and one at zero, by virtue of equations (9.6) and (9.7) the latter is moved to infinity in the admittance locus. On the other hand, it can be noticed that the skin admittance Y presents two dispersive regions, as other authors have investigated (Grimnes and Martinsen, 2000). The frequency interval where G and B are depicted ranges from 1 Hz to 1 GHz, which is the frequency band for which Gabriel et al. (1996) reported their experimental values of the skin dielectric parameters.

9.4 EXPERIMENTAL CHARACTERIZATION OF IBC CHANNEL

Section 9.1 concluded that there is neither consensus on the optimum frequency ranges nor design methodologies for IBC prototyping due to the variability of the performance obtained by different authors, among other reasons. Actually, this performance is seriously influenced by the coupling modalities, the particular electronic issues affecting circuit design, and especially the influence of the environment around the subject. Therefore, the experimental characterization of the IBC channel can assist in gaining knowledge on the human body as a transmission medium. These experimental data can be useful in order to analyze key issues underlying IBC performance, such as the optimum frequency range, thus assisting in the design of IBC devices. At the same time, experimental results allow the validation of theoretical models, contributing to knowledge generation about electromagnetic interactions with living tissues.

The characterization of the IBC channel has commonly been carried out by means of the evaluation of path loss as a function of frequency, which is commonly accepted to establish frequency ranges for optimum IBC performance. Diverse measurement setups have been reported in the literature (Cho et al., 2007; Li et al., 2010; Liu et al., 2010), but an agreement on how to obtain comprehensive IBC measurements is yet to be reached. For instance, a diverse range of circuit structures has been applied in order to tackle different issues such as grounding (Lucev et al., 2011; Xu et al., 2011), differential signaling (Song et al., 2011; Wegmueller, 2007a), electrode–impedance matching (Hwang et al., 2007), and so forth. Indeed, the disparity between authors' outcomes is due to the fact that results are strongly dependent on the specific experimental conditions, thereby making a comparison between them unfeasible. In this sense, grounding has proved to be a key issue that must be carefully addressed since realistic galvanic and capacitive coupling schemes have not always been successfully achieved, as an efficient grounding strategy has not consistently been applied. In fact, if the electrodes are directly connected to a measurement device, such as an oscilloscope or a spectrum analyzer, and their internal ground has not been properly isolated, these could distort the experimental IBC channel characterization, leading to a measurement error. In summary, a grounding strategy has to be applied to ensure that actual capacitive and galvanic couplings schemes are achieved.

The most common ground strategy reported in the literature is based on the use of baluns, which are electrical transformers that provide proper isolation of the internal grounds of the measurement devices. On the other hand, the influence of the type of electrodes on IBC performance must also be carefully studied, since different results have been reported in the literature when using diverse types of electrodes, as has been already mentioned in Sections 9.2.1 and 9.2.2. Another key issue is the frequency range under which IBC performance is optimum: In the case of galvanic coupling, its range has been limited to units of megahertz (Callejón et al., 2012b; Pun et al., 2011a; Wegmueller, 2007a) because it is based on the application of currents into the human body, whose bioelectrical properties are related to low-frequency ranges. In the case of capacitive coupling, higher frequency ranges, up to hundreds of megahertz, have usually been chosen, due to the fact that capacitive coupling is based on the application of voltages that couple to the human body and to the external ground, thereby allowing higher frequencies to be used (Callejón et al., 2012b; Xu et al., 2011, 2012). In order to mitigate discrepancies over different authors' outcomes and to ensure the validity and rigor of the measurement methods, harmonized setups have to be used. In this context, two harmonized setups for both galvanic and capacitive coupling, addressing important issues such as the aforementioned frequency range, grounding strategy, and types of electrode, have been proposed in Callejón et al. (2012a). Below, the technical characteristics of both experimental setups are explained in detail and have also been summarized in Table 9.3.

9.4.1 Experimental Setup for Galvanic Coupling

The galvanic coupling measurement setup proposed in Callejón et al. (2012a), and shown in Figure 9.6, consists of a GFG-8015G function generator of GW Instek to provide the signal, an MSO6032A digital oscilloscope of Agilent Technologies Inc. ($R_{\text{input}} = 1\ \text{M}\Omega$) to acquire it, a pair of PT4 balun transformers of Oxford Electrical

TABLE 9.3 Galvanic and Capacitive Coupling Measurement Setup Parameters

	Galvanic Coupling	Capacitive Coupling
Signal amplitude/level	5 V	0 dBm
Frequency range	10–2000 kHz	1–100 MHz
Channel length	5, 10, 15, 20, 25 cm	15, 30, 45, 125 cm
Type of electrodes	I, II, III, IV	I, II, III, IV

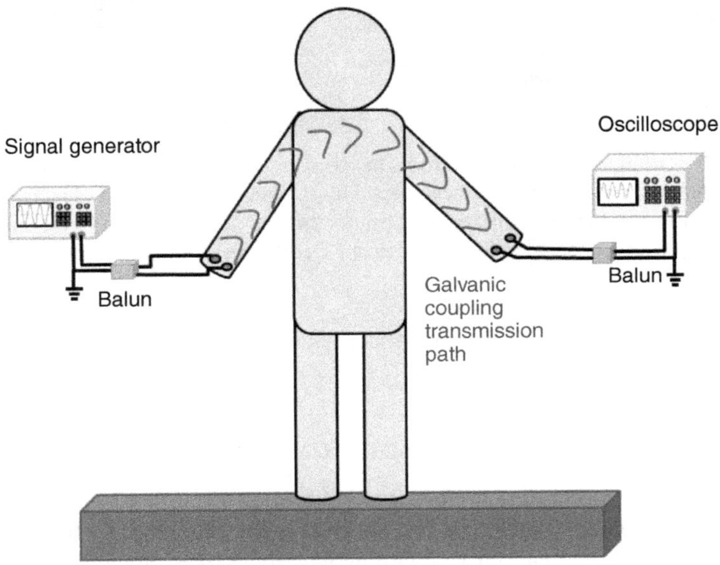

Figure 9.6 Galvanic coupling measurement setup.

Products, and a number of galvanic coupling electrodes. The baluns are used in order to remove the effect of the internal ground of both the signal generator and the oscilloscope, allowing a realistic IBC galvanic coupling transmission path to be obtained. Two transmitting electrodes are attached to the skin near the wrist and two receiving electrodes are moved along the forearm. At the same time, diverse types of electrodes, which are shown in Figure 9.7, are used with the aim of analyzing their influence on path loss results: commercial round pregelled silver/silverchloride. Swaromed ECG electrodes (0.5 cm radius), square aluminum electrodes, and square copper electrodes with different active areas. On the other hand, in order to study the effect of an increase in the channel length, several distances of 5, 10, and 15 cm are fixed. Regarding the input signal, sinusoidal signals with a peak-to-peak voltage of 5 V are applied. Finally, a frequency range from 10 kHz up to 2 MHz is selected due to evidence that as the frequency increases, some other nondeterministic effects,

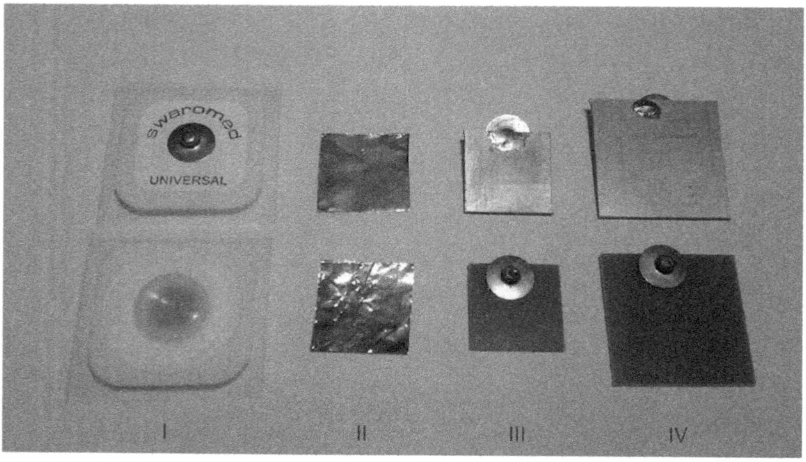

Figure 9.7 Different types of electrodes used: type I (commercial round pregelled silver/silverchloride Swaromed ECG electrodes with 0.5 cm radius); type II (square aluminum electrodes with area 2 cm × 2 cm); type III (square copper electrodes with area 2 cm × 2 cm); and type IV (square copper electrodes with area 3 cm × 3 cm).

such as radiation from cables and electrodes, lead to a nonnegligible scattering effect (Koutitas, 2010; Xu et al., 2012).

9.4.2 Experimental Setup for Capacitive Coupling

The capacitive coupling measurement setup, as shown in Figure 9.8, consists of a FSL18 spectrum analyzer of Rohde and Schwarz, a pair of FTB-1-1+ baluns of Mini-Circuits Inc., and some capacitive coupling electrodes. The signal is launched and received by the spectrum analyzer using the internal tracking generator. Again, the baluns are used to isolate the spectrum analyzer's internal ground, thereby obtaining a realistic capacitive coupling transmission path. Square copper electrodes with an area of 3 cm × 3 cm are used. A signal-transmitting electrode is attached to the skin near the wrist and a signal-receiving electrode is moved along the forearm at distances of 15, 30, 45, and 125 cm. On the other hand, the ground electrodes remain floating in the air. In order to mitigate the human body antenna effect (Cho et al., 2007), the frequency ranges from 1-100 MHz. The signal level is set to 0 dBm. Finally, it should be noted that the insertion losses of the baluns and coaxial cables are calibrated and normalized in order for them to be removed.

9.4.3 Experimental Results for Galvanic Coupling

The authors carried out experimental tests with the proposed galvanic setup over a period of several days, using type A electrodes and a channel length of 10 cm. Despite the influence of the environmental conditions, electrode-to-skin interfacing

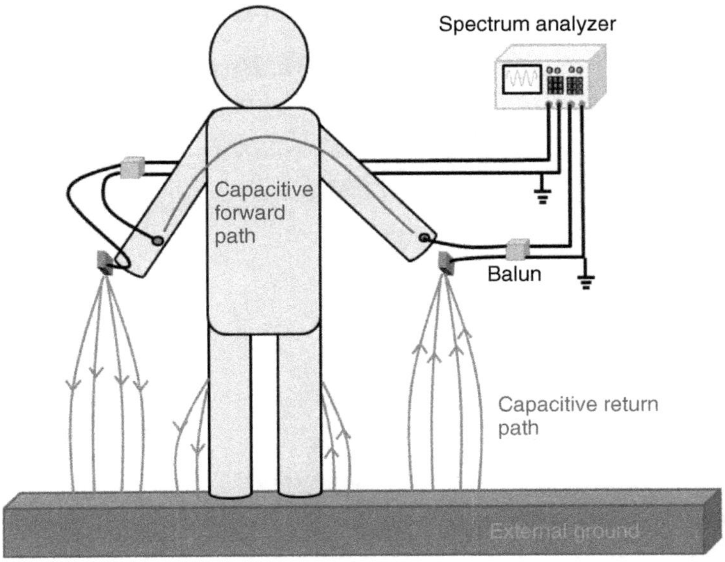

Figure 9.8 Capacitive coupling measurement setup.

issues, and particular skin properties, measurement variability showed a deviation of only ±4 dB (see Figure 9.9), which fairly compares to other authors outcomes. This confirms the reliability of the measurement setup Figure 9.9 also reveals a large attenuation found at low frequencies, up to tens of kilohertz. Higher level signals

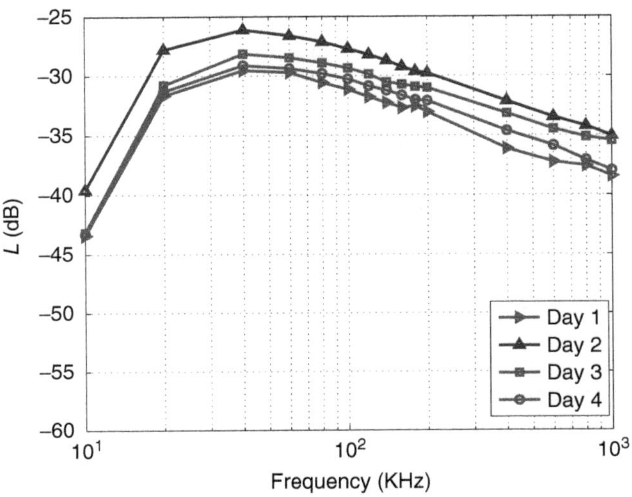

Figure 9.9 Path loss experimental results for IBC galvanic coupling over four different days, considering a channel length of 10 cm.

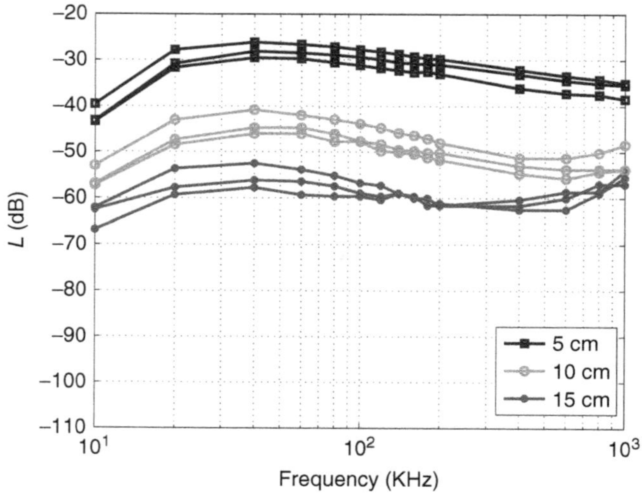

Figure 9.10 Path loss experimental results for IBC galvanic coupling, for three different channel lengths of 5, 10, and 15 cm and over a period of several days.

begin to be received within the range of 10–200 kHz, with path loss values between −35 and −25 dB, and a peak at 40 kHz for each daily measurement, thus suggesting that this frequency could be an optimum candidate for IBC galvanic coupling. Beyond 200 kHz, and up to 1 MHz, the path loss decreases and subsequently remains approximately constant with a value of −40 dB. Regarding the channel length, l, several distances were also studied with the aim of analyzing the effect of increasing the distance between the electrodes. These results are shown in Figure 9.10, and it is seen that an increase in channel length leads to a proportional increase in attenuation. In fact, a magnitude increase in attenuation of more than 10 dB is caused by an increase of only 5 cm in the channel length. These results reveal that channel length is a key parameter in IBC galvanic coupling performance. In fact, for a distance of 20 cm, the received signal was below the noise floor (−85 dBm).

On the other hand, a comparative study using different types of electrodes, shown in Figure 9.7, was conducted. For the sake of simplification, the channel length was set to 10 cm and a single frequency of 40 kHz, at which the maximum signal could be detected, was studied. A bar graph showing the attenuation found for each type of electrode under the same conditions is presented in Figure 9.11. The results confirm that the electrode material does not have too much influence on the IBC performance whenever a conductor is used (Hachisuka et al., 2005). However, commercial AgCl electrodes seem to have better performance even when their active area is the smallest. This could be explained by the fact that they are pregelled, which enhances both conductive properties and adherence to the skin. In conclusion, a good attachment must be guaranteed in order to receive higher level signals under IBC galvanic coupling.

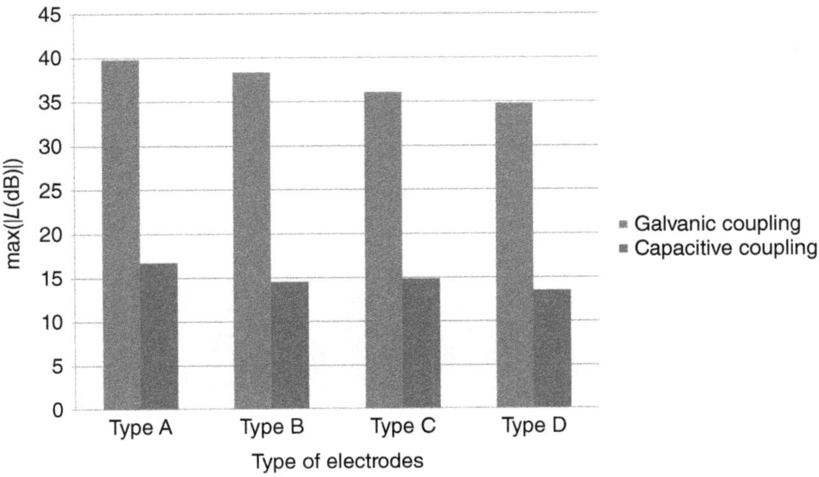

Figure 9.11 Maximum results of path loss magnitude (at 40 kHz) using different types of electrodes for both galvanic and capacitive coupling.

9.4.4 Experimental Results for Capacitive Coupling

The experimental results obtained for capacitive coupling over two consecutive days, shown in Figure 9.12, reveal that this technique presents a bandpass profile within the range 1–100 MHz. Beyond this upper limit, measurements are highly uncorrelated, due to unpredictable effects such as radiation from cables and electrodes, external

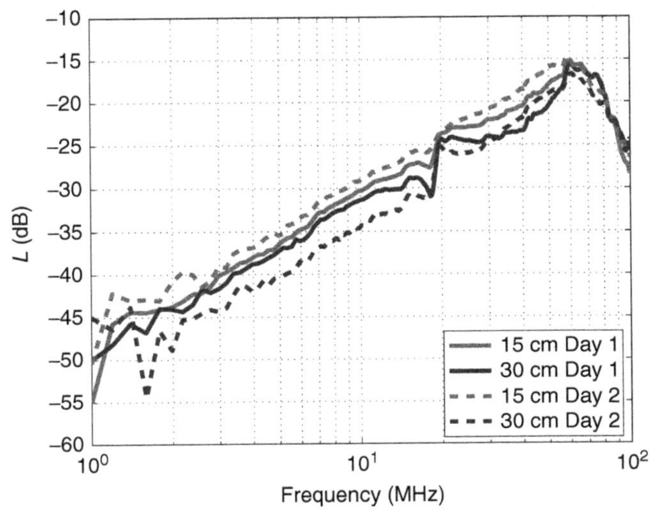

Figure 9.12 Path loss experimental results for IBC capacitive coupling, for two different channel lengths of 15 and 30 cm and over two consecutive days.

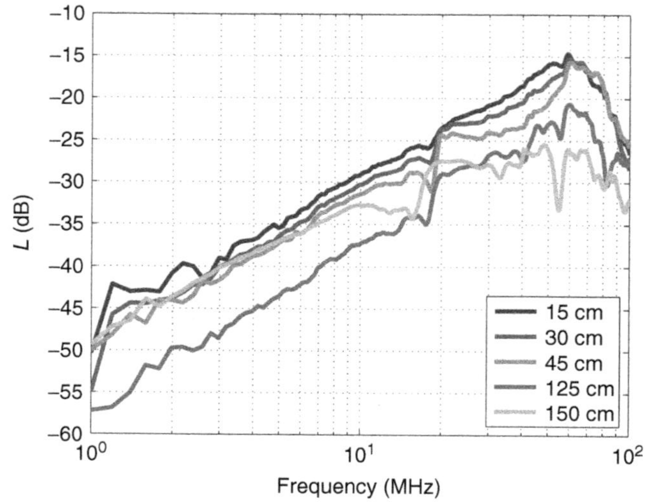

Figure 9.13 Path loss experimental results for IBC capacitive coupling for several channel lengths. (*See insert for color representation of the figure.*)

interference sources, multipath propagation and human body antenna effect, among others. Therefore, a frequency range between 1 and 100 MHz is suggested for capacitive coupling, based on the repeatability concerns and optimum attenuation results. At the same time, as frequency increases, the capacitive return path, which is established through the air toward the external ground plane, becomes dominant. This means that the signal pathway is extremely influenced by the external environmental conditions. The results obtained for several channel lengths up to 150 cm are shown in Figure 9.13. The signal reaches its peak at approximately 60 MHz for different days and channel lengths. Furthermore, capacitive coupling is not as much influenced by the channel length in this case as it is in the case of galvanic coupling, since a distance as large as 150 cm leads to a decrease in signal level of only 10–15 dB, whereas in the case of galvanic coupling, a decrease of the same order is caused by increasing the channel length by only 5 cm.

These differences observed in behavior patterns evidence that the two types of coupling behave quite differently, and that a comparative study can be useful to determine their key parameters. Furthermore, measurements show marked resonance effects when longer distances, such as 150 cm, are considered. This demonstrates the fact that capacitive coupling is strongly influenced by the interferences that appear in the capacitive return path through the external ground, whose effect is more pronounced as the distance between the electrodes increases (Callejón et al., 2012b). Thus, capacitive coupling is not as dependent on skin properties as galvanic coupling because the electrode–skin impedance contact does not influence the IBC performance. In fact, the electrodes can be closer to the skin, but not directly attached to it.

As in the case of galvanic coupling, a comparative study to analyze the effect of electrodes made with diverse materials and active areas was carried out. In this case, a channel length of 30 cm and a frequency of 60 MHz were used. The results are quite similar for all the types and also confirm that the electrode material does not bear too much influence on IBC performance whenever a conductor is used. However, it should be noticed that at such high frequencies, the dominant signal pathway is established through the external ground, and indeed, using electrodes with different active areas does not lead to major changes into the path loss.

9.5 INTRODUCTION TO IBC MODELS

As already mentioned in Sections 9.1 and 9.4, there is still a lack of criteria for the design of optimal electronic designs for IBC systems. This is due, in part, to the high variability of results in terms of IBC performance for reported prototypes. While this disadvantage can be mitigated through the use of harmonized measurement setups, it also reveals the need of modeling research, aimed at gaining a better understanding of the signal propagation with the skin–human body medium. Unlike empirical approaches to IBC design, based on a purely experimental optimization of the key design parameters and sensor placement, models offer simple expressions that can aid this process, simulation tools that can provide realistic performance, and the chance to get feedback from measurements for the adjustment of model parameters.

Since the inception of IBC as a promising communication method for BANs, three major modeling approaches have been used so far: empirical models obtained from measurements derived with phantoms or realistic setups, circuit-level models, and electromagnetic models. Considering that the issues concerning experiments have been covered in Section 9.4, this section is focused on circuit-level and electromagnetic approaches to IBC modeling.

9.5.1 Circuit-Level Approaches

With the purpose of gaining insight into the characteristics of the human body as a transmission medium, several attempts to model the human body channel by means of circuit-level approaches have been accomplished in the literature. A simple lumped-parameter circuit modeling the human body as a single node was proposed by Zimmerman (1996). This node is coupled to the transmitter, the receiver; and the external ground through capacitive effects. Such simplification is valid as long as the body impedances are negligible compared with the rest of the impedances that form the signal pathway. In the scheme proposed by Zimmerman, a transmitter generates an oscillatory voltage between the signal electrode, attached to the human body, and the ground electrode, which points away from the body. The receiver electrodes are coupled to the transmitter electrodes by means of capacitive impedances that model all existing couplings within the system. Thus, the signal pathway is established between both the receiver and transmitter signal electrode, through the node modeling the human body. The signal path is subsequently closed through the capacitive effects that join both the transmitter and the receiver ground electrodes with the external ground plane.

Thus, the signal quality is determined by the characteristics of the environment. One way to increase its quality is by locating both the transmitter and receiver ground electrodes near the external ground, for example, in the subject's shoes. This model presents a 400-kHz bandwidth, with a central frequency of 330 kHz. Some resistances were added to Zimmerman's model in Partridge et al. (2001) in order to emulate the signal losses through the capacitive return path. Hachisuka et al. (2005) modeled the IBC channel as an electric circuit formed by 6 impedances between 2 transmitter and 2 receiver electrodes (4-port circuit). They used this model to simulate the IBC transmission through the arm, and they found that the gain was considerably higher when both the transmitter and the receiver ground electrodes were not directly attached to the skin. The gain values obtained were shown to be approximately constant with frequency (−10 dB), decreasing drastically from 1 GHz. In order to improve the simulation results, the electrode coupling impedances were included into the model in (Wegmueller, 2007a), therefore obtaining a 4-port circuit with 10 impedances, which emulates the galvanic coupling configuration. This kind of structure was subsequently used by Song et al. (2007) while introducing some new resistive effects associated to the internal elements of the IBC devices in order to simulate the attenuation signal through different signal pathways over the human body (arm and torso), both modeled as multilayer cylinders. Variable attenuation values between 16 and 24 dB were obtained, within a frequency range of 100 kHz and 5 MHz.

Due to the progressive increase in work frequency, lumped-parameter models become more imprecise, and, consequently, distributed-parameter circuits are a natural choice. Cho et al. (2007) proposed a distributed circuit modeling the human trunk and arms, formed by the cascade of basic *RC* cells, which includes a capacitive element that emulates the coupling with the external ground. Model simulations matched well with experimental results, showing that the human body behaves as a high-pass filter up to 4 MHz. It should be noted that bioelectric properties of the skin must be addressed due to the biological context under which this type of communication technique takes place. In this way, some electrophysiological properties were gathered in (Wegmueller, 2006a), where a Cole–Cole model is applied in order to emulate the attenuation through the skin and muscle up to 1 MHz, with the aim of using galvanic coupling to interconnect implanted devices inside the human body. Song et al. (2011) modeled the entire human body as a set of cylinders, each formed by a multilayer tissue. Depending on the area of the cylinder, an impedance is defined, allowing the galvanic coupling path to be emulated through a 4-terminal circuit containing 11 impedances. The galvanic coupling transfer function is calculated in order to simulate different transmission paths through the body, considering diverse parts, such as arm, torso, and leg.

Another example of distributed-circuit approach is the transmission line model proposed in Callejón et al. (2012c). It is formed from the cascade of basic electrical cells along a longitudinal axis lengthwise to the signal path. The objective of this kind of circuital structure is to study the influence of a cross-sectional admittance on the propagation characteristics of a particular signal. This admittance simulates specific physiological behavior that is repeated periodically in space. In this way, the effects of the transverse admittance of the skin on the propagation characteristics of a signal traveling through the skin can be analyzed. According to Section 9.3, this transcutaneous admittance can be represented as a shunt circuit composed of a conductance $G(\omega)$ that emulates the conductive pathways of the skin, which are mainly the sweat glands and the ionic channels of the cell membrane (Grimnes and Martinsen, 2000), and a susceptance $B(\omega)$ that accounts for the keratinized cells of the SC and the lipid bilayer, which are, respectively, more or less negligible depending on the frequency range (Tronstad et al., 2010). Subsequently, a resistor R_u and a resistor R_i (Ω/m) are added to model the signal propagation through the surface of the SC and the deeper layers, respectively. This would be the ideal electrical current path if the signal did not penetrate transversely inside the skin. This model, shown in Figure 9.14, allows some insight into the effect of the dielectric properties in the signal propagation. It can be seen that the basic cell in the frame can correspond to the equivalent electrical circuit model of a lossy transmission line without the inductive element. Notice also that the upper and inner lines are, respectively, associated to the SC and the deeper layers of the skin. In addition, the bottom line is assumed to be an internal reference with respect to which the voltage at the upper side of the skin is measured. The resistor R_i can be neglected since the SC presents a high resistivity compared to that of the deeper layers. Therefore, only R_u is considered in the analytic expression of the model. The natural response for this voltage drop is given by the homogeneous parabolic equation in the frequency domain (Collin, 1992):

$$\frac{d^2V(x,\omega)}{dx^2} - R_u\ [G(\omega) + jB(\omega)]V(x,\omega) = 0 \tag{9.8}$$

where

$$V(x,\ \omega) = V^+e^{-\gamma(\omega)x} +\ V^-e^{\gamma(\omega)x} \tag{9.9}$$

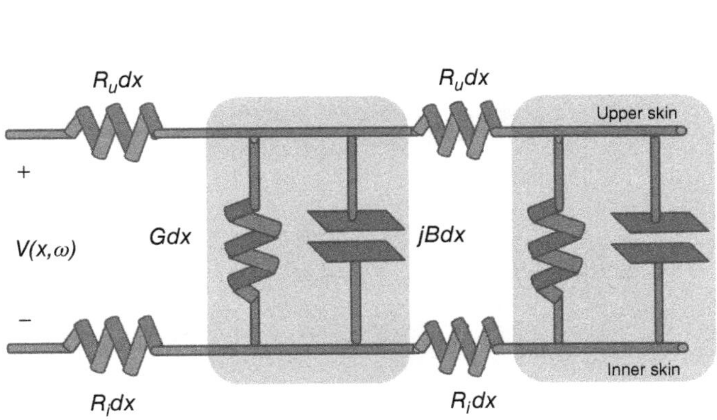

Figure 9.14 Skin propagation model based on a distributed-parameter circuit, by the cascade of basic *GB* cells modeling the transcutaneous admittance of the skin. This basic cell is composed by the transcutaneous admittance [represented by the shunt conductance *Gdx* (S) and susceptance *Bdx* (S)], connected in series with the resistor $R_u dx$ and $R_i dx$ (Ω), that model the electrical current path through the upper and inner surface of the skin, respectively.

where V^+ and V^- represent the complex amplitudes of the voltage waves that travel through the skin in directions $+x$ and $-x$ referred to the $x = 0$ plane, respectively, and γ is the propagation constant, whose expression can be obtained by substituting equation (9.8) in (9.9),

$$\gamma(\omega) = \sqrt{R_u[G(\omega) + jB(\omega)]} = \alpha(\omega) + j\beta(\omega) \tag{9.10}$$

where α is the attenuation constant (Np/m) and β is the phase constant (rad/m). In addition, the characteristic impedance of the line is

$$Z_0(\omega) = \sqrt{\frac{R_u}{G(\omega) + jB(\omega)}} = R_0(\omega) + jX_0(\omega) \tag{9.11}$$

where R_0 and X_0 stand for the real and imaginary parts of the characteristic impedance, respectively. It should be noted that the propagation characteristics of a signal traveling through the skin can be obtained by means of equation (9.10), which is related to the circuit elements that model the transcutaneous admittance. Thus, a relation between signal propagation and electrophysiological properties of the skin is established.

Once the skin communication channel is modeled as a transmission line characterized by a propagation constant γ and a characteristic impedance Z_0, special considerations have to be taken into account about the effect of the electrodes, by means of their equivalent impedance Z_e. The complete scheme is represented in Figure 9.15, where two different reference voltage symbols are introduced in order to apply the model to diverse coupling mechanisms preserving the same circuit structure. The impedance mismatch caused by the electrode–skin impedance can be represented by a reflection coefficient Γ

$$\Gamma(x,\ \omega) = \Gamma_l(\omega)e^{2\gamma(\omega)(x-l)} \tag{9.12a}$$

$$\Gamma_l(\omega) = \frac{z_l(\omega) - z_0(\omega)}{z_l(\omega) + z_0(\omega)} \tag{9.12b}$$

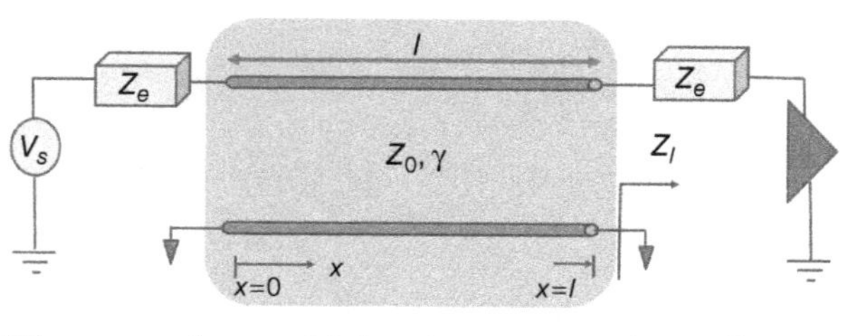

Figure 9.15 Skin propagation model based on a transmission line with characteristic impedance Z_0. The transmitter and receiver electrodes are modeled by Z_e impedances. The distance between electrodes is l. V_s is the transmitter generator voltage source and the triangle at the reception side represents the detection circuit.

where l is the distance between electrodes and Z_l is the impedance seen in $x = l$. In this case, Z_l corresponds to the electrode impedance Z_e. The detection circuit, represented in Figure 9.15 as a triangle, is assumed to exhibit a low input impedance, in such a way that the load impedance at the end of the skin channel is given by Z_e. Therefore, the voltage can be now expressed as

$$V(x,\omega) = V^+ e^{-\gamma(\omega)x}[1 + \Gamma(x,\omega)] = V^+ e^{\alpha'(\omega)x} e^{-j\beta'(\omega)x} \tag{9.13}$$

The average power delivered at a distance x by a transmission line with the complex-valued characteristic impedance of equation (9.11) is given by

$$\langle P(x,\omega)\rangle = \frac{|V^+|^2}{2|Z_0(\omega)|^2} e^{-2\alpha(\omega)x}\left[\left(1 - |\Gamma(x,\omega)|^2\right) R_0(\omega)\right] + 2\mathrm{Im}\{\Gamma(x,\omega)\}X_0(\omega) \tag{9.14}$$

However, for the sake of a simpler representation for the attenuation, the ratio of voltages can be used instead of the power loss derived from equation (9.14). Thus, the path loss of the IBC system takes the form

$$L\ (\mathrm{dB}) = 20\ \log_{10} \frac{1 + \Gamma_l(\omega)e^{-2\gamma(\omega)l}}{[1 + \Gamma_l(\omega)]e^{-\gamma(\omega)l}} \tag{9.15}$$

Once a formal expression of the propagation constant of the skin is formulated, some modifications are introduced with the aim of adapting the model to both IBC galvanic and capacitive coupling techniques. In this way, the configuration of the electrodes and the effect of the external ground are the main parameters that define each coupling technique, respectively. Both configurations, along with their respective models, are schematically shown in Figure 9.16. New considerations can be taken into account, for example, signal propagation is now defined through the 2D longitudinal surface of the skin, which is composed of two parallel longitudinal axes, each of which models the signal pathway between a pair of transmitter and receiver electrodes. It must be noticed that these two signal pathways are the same in the case of galvanic coupling, because of its differential operation, but different in the case of capacitive coupling, since the return signal pathway is accomplished through the external ground. In order to model this characteristic, a distributed capacitive element C_{ad} is introduced to emulate the capacitive effect between each point on the surface of the skin and the external ground plane (Callejón et al., 2011). In addition, a discrete capacitance, C_a, is used to represent the capacitive coupling between both transmitter and receiver devices and the external ground plane. Now, $Y_{\mathrm{skin}}(\omega)$ represents the skin admittance, defined again as a *GB* shunt circuit. In addition, the impedance $Z_{\mathrm{skin}}(\omega)$ corresponds to a resistive characteristic $R(\omega)$ that emulates the signal propagation between the basic cells repeated along the propagation axis modeling the signal pathway through the skin (Callejón et al., 2012b).

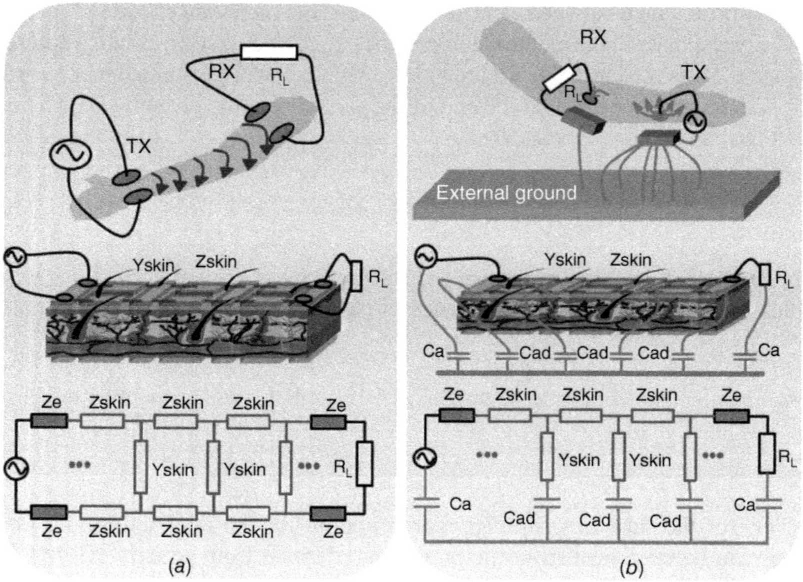

Figure 9.16 (*a*) Galvanic coupling IBC technique and equivalent circuital model. (*b*) Capacitive coupling IBC technique and equivalent circuital model. (*See insert for color representation of the figure.*)

The propagation constant $\gamma(\omega)$ for both coupling schemes can be found through

$$\gamma(\omega) = \sqrt{Z_{\text{skin}}(\omega) Y'_{\text{skin}}(\omega)} \tag{9.16}$$

where $Y'_{\text{skin}}(\omega)$ depends on the coupling type. In agreement with the configurations depicted in Figure 9.16 , two cases can be distinguished:

- For galvanic coupling,

$$Y'_{\text{skin}}(\omega) = 2Y_{\text{skin}}(\omega) = 2\left[G(\omega) + jB(\omega)\right] \tag{9.17}$$

  where the constant factor is equal to 2 due to the differential configuration of galvanic coupling. Notice that a virtual ground plane appears in the middle of the distributed circuit, thus dividing $B(\omega)$ into $2B(\omega)$-value susceptances and $G(\omega)$ into $2G(\omega)$-value conductances. It can be seen that in this case the propagation mechanisms mainly depend on both the electrode and skin properties since the signal path is primarily accomplished through the skin.
- For capacitive coupling, $Y'_{\text{skin}}(\omega)$ is given by the series association of $Y_{\text{skin}}(\omega)$ and C_{ad},

$$Y'_{\text{skin}}(\omega) = \left(\frac{1}{Y_{\text{skin}}(\omega)} + \frac{1}{j\omega C_{ad}}\right)^{-1} \tag{9.18}$$

In this case, it can be noticed that the propagation mechanisms depend not only on the skin channel but also on the capacitive return path toward the external ground, modeled through a coupling capacitance, C_{ad}, which highlights the fact that capacitive coupling has two different signal paths, which are, respectively, more or less dominant depending on the frequency range and the external conditions.

Finally, a propagation constant gathering the electrophysiological properties of the skin, modeled by means of $R(\omega)$, $G(\omega)$, and $B(\omega)$, as well as the coupling type, by means of two different expressions of $Y'_{\text{skin}}(\omega)$, is obtained. In addition, the expression for $\gamma(\omega)$ will be different according to the coupling type. Regarding data used for model simulations, $G(\omega)$ and $B(\omega)$ can be obtained from equations (9.6) and (9.7) and $R(\omega)$ is set as the inverse of $G(\omega)$. The magnitude of the electrode impedance, Z_e, is reported in Hachisuka et al. (2005), for different materials, including AgCl and copper. At the same time, the value of the coupling capacitances for the case of capacitive coupling have been studied in Cho et al. (2007), where a value of 5 pF was reported for C_a. Subsequently, C_{ad} is estimated as the quotient between the discrete capacitance C_a and the distance between the electrodes and the floor. It must be noticed that this is a simplification, as, in fact, the value of C_a depends strongly on the particular environmental conditions under which the experiments are carried out. In fact, its value not only increases with the presence of interfering devices but also varies with frequency. Some attempts to combine the coupling with the environment as well as the effect of the external interferences in the IBC propagation channel have been proposed in the literature (Xu et al., 2012), however, an exhaustive analysis toward the understanding of the capacitive coupling forward path is still needed.

In conclusion, the main advantage of circuit-level models is their simplicity, allowing an accurate approach toward IBC channel characterization as well as the design of efficient and secure IBC links and prototypes. Unlike complex propagation models and electromagnetic numerical simulations with high computational costs, circuit-level approaches represent a trade-off between simplicity and precision. However, as frequency increases, some other nonpredictable effects, such as radiation from cables and electrodes, the human body antenna effect, multipath components, and the like, which cannot be gathered by a circuit-level approach, and which in fact depend on the distance and the position of the electrode, become more dominant, thereby limiting their validity range to low frequencies.

9.5.2 Electromagnetic Models

Electromagnetic modeling applied to the IBC context comprises two parallel lines of research. The first is based on solving the problem with computational techniques of Maxwell's equations by using different methods (Kost, 2011), like the finite-difference time-domain (FDTD) method (Fujii et al., 2007), the finite-element method (FEM) (Takahashi et al., 2012), or the method of moments (MoM) (Eid and Wallace, 2011). The other line, more incipient, is based on the theoretical modeling of IBC EM propagation through formulas and simple geometries able to obtain closed-form expressions that provide information about the main propagation mechanisms that occur in the human body. Both research lines have their advantages and disadvantages.

Due to the high complexity presented by the human body, although being a more direct approach to the problem, the simulations by numerical techniques need to generate a large amount of data to ensure a certain accuracy, which increases the computational cost and the simulation time. Hence, after using an electromagnetic model based on these numerical techniques, many authors present a circuital simplification of these expressions for a simpler operation and in order to aid the design. In addition to this, these models are difficult to reproduce, while not accompanied by a set of formulas or expressions to infer the results presented by their authors. Due to their development from scratch, these models usually lead to different results that depend on the software and the simulation conditions employed. Because of these disadvantages, more and more authors prefer theoretical models of signal propagation through the human body. However, the main lack associated with theoretical EM procedures is the fact that the geometries presenting an analytical solution are simple ones such as the cylinder and the plane, which can serve as a first approximation to understand the mechanisms of propagation, but provide results that differ significantly from those obtained in experiments. In any case, these issues stimulate deeper research that may provide a comprehensive vision of the inner mechanisms related to IBC communications. An overview of the most remarkable advances with the two approaches is presented below.

9.5.3 Computational Models

One of the key parameters in IBC transmission is the attenuation the signal undergoes through the human body and its possible dependence on the characteristics of the subject's environment and his/her posture (Igou et al., 2011). Thus, EM simulation may serve as a vehicle to study the influence of the geometry and the dielectric properties of human tissues by using FE approaches.

One of the first papers to address these issues was Fujii et al. (2003), where the authors only considered the communications along the arms and computed a parallelepiped structure with the dielectric characteristics of muscle tissue. On the other hand, the attenuation and current distribution inside the human body by using a 3D FEM was analyzed in (Wegmueller, 2007a) through a superposition of tissue layers considered in the geometry of the 3D human body arm model, including a representation of human joints. A similar model was developed in Lucev et al. (2010a), with the objective of studying the capacitive coupling by calculating the distribution of current density at three frequencies: 100 kHz, 1 MHz, and 10 MHz. The study revealed that most of the current density was confined into the muscle tissue, and that this effect increased with frequency, skin moisture, and the size of the signal excitation electrodes.

Other authors take into account not only the arm but also model the influence of the trunk and extremities. Li et al. (2010) proposed an FEM model of the human body where the trunk was modeled as a parallelepiped, and arms and limbs were considered to have cylindrical geometry consisting of several layers of tissue, as in previous cases. The model was validated with experimental measurements, which show a high-pass behaviour for the channel due to the capacitive return path. Moreover, Fujii et al. (2007) studied the distribution of the electric field within a high-resolution model of the human body of two Japanese adults of both sexes and found that most of the electric field was confined into the lower arm and there was no significant difference between genders. Song et al. (2011) also used cylindrical models of up to five layers of fabric to simulate the galvanic coupling with different signal paths. Finally, a more realistic model, called NIH Visible Man Model (National Library of Medicine, 2012) and based on magnetic resonance images and reconstructed in 3D, was used by Reusens et al. (2009) in order to study the formation and efficient design of topology channel wireless PAN applications. First, 583 measurements were made and subsequently treated by statistical methods in order to characterize the signal attenuation for different body parts (arm, back, torso, and leg) as a previous step to the design of a very low consumption PAN communication scheme based on a multihop technology.

9.5.4 Theoretical Models of EM Propagation

The main proposals for the theoretical modeling of IBC communications are based on the definition of basic geometries, such as the cylinder, for modeling the arm and the trunk, and the infinite plane to generally model a body region formed by the superposition of several layers of tissues.

Gao et al. (2008) and Pun et al. (2009) studied the current flow through the arm from a model based on a cylindrical geometry of 30 cm long and 5 cm diameter. In order to solve Maxwell's equations, a cross section of the cylinder was considered formed by two homocentric environments with different conductivity properties, which represent the muscle and skin, respectively. Pun et al. (2010), went a step further by presenting a three-dimensional multilayered cylindrical model. They considered again the quasi-static approximation to convert Maxwell's equations to Laplace's. Gain results provided by the simulations and experimental measurements showed the same trend at low frequencies but a mismatch in magnitude of at least 20 dB.

Gupta and Abhayapala (2008) presented an analytical model of the channel for PAN networks based on the study of electromagnetic propagation. Specifically, they proposed the dyadic Green function for the study of a model body with cylindrical geometry. The study considers four different cases depending on whether the transmitter/receiver is on the human body or implanted in it. On the other hand, in Ma and Zhang (2011), the use of Mathieu functions instead of Fourier transform is proposed, along the axes, in a cross section and the impedance boundary condition on surface, for the characterization of the propagation channel for the band of 915 MHz up to 2.40 GHz.

Other approaches have relied on methods such as high-frequency asymptotic representation (Sasamori and Takahashi, 2009) in which the arm was modeled as a dielectric cylinder whose resolution was approached by the method of the eigenfunctions and boundary condition based on impedance (Ma and Zhang, 2009), considering an ideal dipole radiating at a frequency of 2.45 GHz. Lea et al. (2009) proposed a simplified model based on an infinite plane formed by the superposition of several layers: air, skin, fat, and muscle. The frequencies of the study were in the range of the GHz band, which is the most used in the ISM band to implement most BAN applications.

Pun et al. (2011a) proposed a multilayer mathematical model for human limb using inhomogeneous properties of human tissue and quasi-static approximations to decouple the Maxwell equations, concluding that most of the power signal was confined in the limb for frequencies below 1 MHz. Chen et al. (2012) published results of the transfer function using the quasi-static assumption and obtained similar outcomes.

Finally, Eid and Wallace (2011) developed a theoretical study to analyze the surface transmission that afterwards is implemented by MoM. In this context, Bae et al. (2012) present a comprehensive approach to theoretical analysis of the communication surface modes from a triple point of view related to the operating frequencies and channel distances employed: the quasi-static near-field coupling part, the reactive induction-field radiation part, and the surface wave far-field propagation component.

9.6 IBC PROPAGATION CHANNEL

An introduction to the modeling of IBC has been presented in Section 9.9.5 with the aim of obtaining an overview about the main modeling efforts presented in the literature. In general, circuit-level and computational EM solutions have been the most common approaches presented by authors. Models can be used to evaluate some key parameters of the communication channel, such as path loss, dispersion, and impact on digital communications. We will address these issues by considering the models proposed by the authors in Callejón et al. (2012b, 2012c) and others mentioned in the previous section. Some experimental results of IBC path loss reported in the literature have been used to identify model parameters and validation purposes. Subsequently, the proposed models have also been validated by means of the experimental results for attenuation introduced in Section 9.9.3, differentiating the particular considerations between both galvanic and capacitive coupling. Finally, other parameters such as dispersion, velocity of propagation, and impact of modulation schemes are analyzed in this section as a proof of usefulness of the models.

9.6.1 Path Loss

The model proposed in Callejón et al. (2012b), which is shown in Figure 9.14, can be validated through the comparison between model simulations and some experimental results reported in the literature. Specifically, in order to obtain the value of R_u, which is the only unknown parameter in the model, a basic process identifying the parameters is conducted. First, the value for which the theoretical results were closer to an initial set of experimental results reported in the literature (Lin et al., 2011; Ruiz and Shimamoto, 2006a) has been searched for. Subsequently, this value has been used to compare the model simulation with another set of path loss experimental results different from the previous one, thus showing the validity of the model. In fact, the value of R_u that best matched these reported results, is around 150 ± 50 Ω/m. In order to illustrate the validity of the model when using different frequency ranges and electrode types, the data of two authors that worked in the same frequency range (0–1000 MHz) (Liu et al., 2010; Song et al., 2007) but with different electrodes material have been chosen. In addition, the results of a third author that worked in a lower frequency range (Xu et al., 2009), about tens of megahertz, have been also considered. Two different channel lengths of 15 and 20 cm have been compared at the same time. The model simulations along with the reported results are shown in Figures 9.17 and 9.18. A satisfactory agreement between the path loss results predicted by the model and the experimental data reported in the literature can be seen, notwithstanding that there is a large variability in the order of magnitude of the attenuation results reported in the literature, as a consequence of the different test setups and measurement methods. Moreover, skin admittance varies considerably between people and environmental conditions. Changes in hydration mechanisms due to sweat gland activity and temperature can be manifested in large variations of skin admittance (Tronstad et al., 2010). Thus, the tuning capability of the model by means of the parameter R_u is showed. It should be noted that such a parameter not only retains a physiological significance but is also related to the effect of the unavoidable experimental conditions (type of coupling, noise, interferences, etc.).

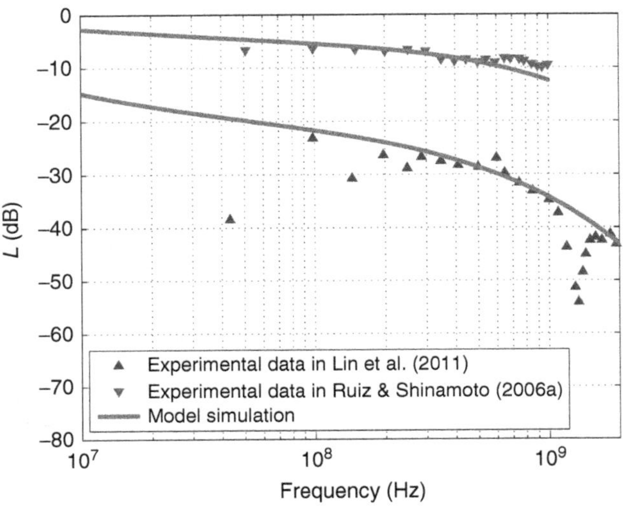

Figure 9.17 IBC path loss (dB) versus frequency (Hz). Model simulations (solid line) along with experimental results reported in the literature. Experimental data in (Lin et al., 2011) were obtained with AgCl electrodes and a channel length of 15 cm, whereas experimental data in (Ruiz and Shinamoto, 2006a) were obtained by using copper electrodes at a distance of 20 cm.

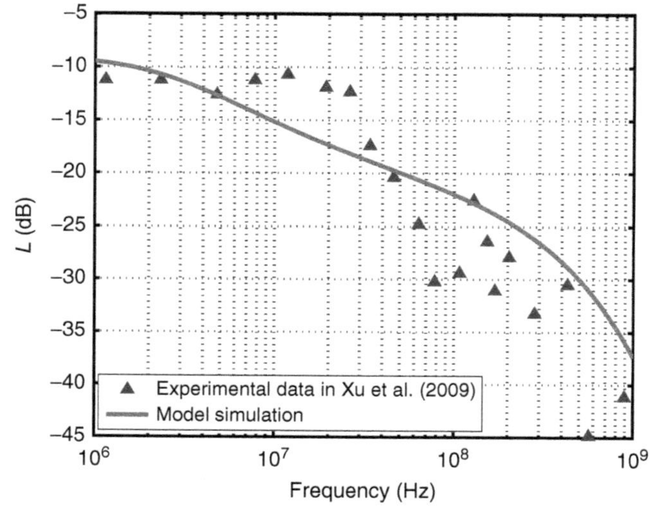

Figure 9.18 IBC path loss (dB) versus frequency (Hz). Model simulations (solid line) along with experimental results reported in the literature. Experimental data in Xu et al. (2009) were obtained with copper electrodes and a channel length of 15 cm.

Regarding the type of coupling, simulations of IBC path loss with the model described in Callejón et al. (2012c), taking into account the specific considerations of each type of coupling, have been compared with the experimental results reported for both coupling techniques in Section 9.4. In the case of galvanic coupling, the experimental results that were considered were path loss measurements using type A electrodes, over a period of several days and with three different channel lengths of 5, 10, and 15 cm. Experimental results at longer distances could not be obtained since the received signal was below the noise floor for channel lengths above 20 cm. Model simulations and experimental results, which are depicted in Figure 9.19, show a satisfactory agreement in the range 50–200 kHz. It also shows that above 200 kHz, other effects, such as signal penetration across other tissues, which in turn could lead to an increase in attenuation, or radiation from cables and electrodes, could affect the measurement process. In addition, skin properties are not constant and in fact are influenced by external conditions.

Regarding capacitive coupling, measurements using type C electrodes and considering three channel lengths of 15, 30, and 45 cm were used in order to validate the model, now gathering the environmental coupling through the air by means of C_a and C_{ad} capacitances. The results and model simulations are presented in Figure 9.20. As stated in Section 9.4, this technique presents a bandpass profile within the range of 1–100 MHz. Notice that the model is able to emulate this attenuation characteristic up to approximately 80 MHz, from which the model tends toward a constant value, rather than decreasing, up to 100 MHz. This discrepancy at higher frequencies could be due to the fact that C_a and C_{ad} have been set as constant parameters, while actually a more complex frequency dependence exists. Some efforts to characterize this

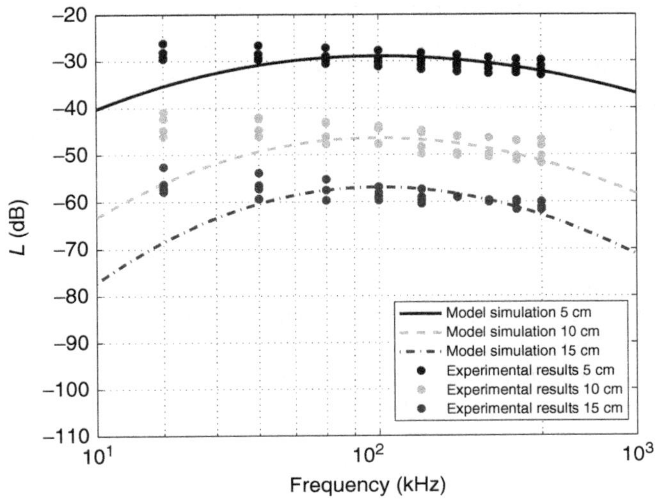

Figure 9.19 IBC galvanic coupling path loss (dB) versus frequency (kHz) for three different channel lengths: 5, 10, and 15 cm, each of them repeated over several days. Model simulations (lines) along with experimental results (marks) are shown within the 50- to 300-kHz range.

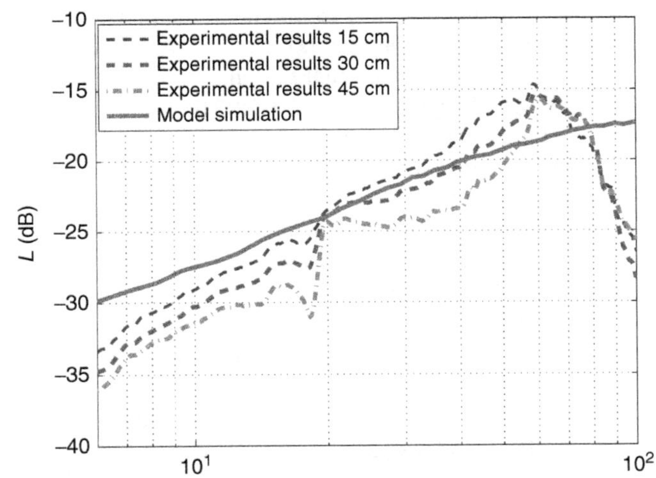

Figure 9.20 IBC capacitive coupling path loss (dB) versus frequency (MHz) for three different channel lengths of 15, 30, and 45 cm. Model simulation (solid line) along with experimental results (dotted line) are shown within the 6- to 100-MHz range.

external coupling have been reported in the IBC literature (Xu et al., 2012). However, more comprehensive research is required in order to understand the capacitive coupling return path. For longer distances, such as 150 cm, other phenomena such as radiation and resonance effects become dominant. Notice that despite their simplicity, models reveal that capacitive coupling is not primarily dependent on skin properties but also on the capacitive return path.

9.6.2 Dispersion

Dielectric dispersion constitutes the dependence of the permittivity of a dielectric material on frequency. Concerning skin and other human tissue, this dependence was already reported in Gabriel et al. (1996). This dispersion is quantitatively characterized by the phase constant β' in equation (9.13), from which the phase velocity of a wave that propagates through the skin can be easily derived as

$$v_f(\omega) = \frac{\omega}{\beta'} \tag{9.19}$$

As a consequence of the dependence of the phase velocity on frequency, modulated pulses are spread when they propagate through a dispersive medium, thus degrading the signal. In case of digital communications, this phenomenon limits the width of the baseband transmission pulses and, therefore, the maximum data rate. For this reason, dispersion, as well as attenuation, becomes a key parameter involving IBC performance. Once the propagation constant γ is obtained, the phase constant β' can be derived through equation (9.13), obtaining its characteristic versus frequency, which is shown in Figure 9.21. It can be seen that the phase constant increases with frequency, in agreement with Vaughan et al. (2009). The normalized phase velocity has been obtained and presented in Figure 9.22. It can also be seen that it increases with frequency, being negligible at low frequencies. Specifically, considering copper electrodes and a channel length of 10 cm at 1 GHz, a value of 2.6986×10^8 m/sec, which is similar to that reported in Van Roy et al. (2007), is found. Therefore, the model predicts that skin begins to behave as a dispersive medium as the frequency increases, becoming significant from tens of megahertz, which in fact means that capacitive coupling is influenced by the dispersive behavior of the skin more than galvanic coupling, as the former works with higher frequencies, within this order of magnitude.

In order to evaluate both attenuation and dispersion mechanisms at the same time, a Gaussian-shaped RF burst and a square-pulsed RF sequence propagating through the skin have been simulated with the model proposed in Callejón et al. (2012b). This way, both Gaussian and square pulses of width τ have been, respectively, considered as voltage signals at $x = 0$. To obtain the received signal at $x = l$, the inverse Fourier transform has been calculated. Two different pulse widths of 1 μs and 10 ns and three channel lengths of 10, 20, and 30 cm between copper electrodes have been studied. It was observed that the received signals were the same Gaussian and square pulses but attenuated and dispersed. The simulations for both the Gaussian and square pulses are shown in Figures 9.23 and 9.24. For the sake of clarity, both

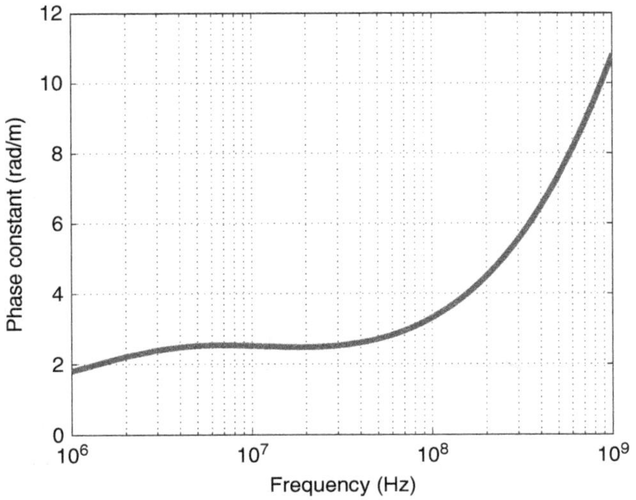

Figure 9.21 Modeled phase constant (rad/m) versus frequency (Hz).

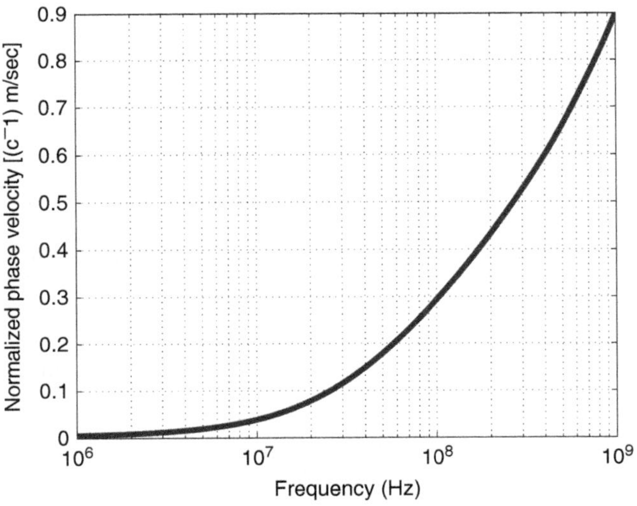

Figure 9.22 Normalized phase velocity (c^{-1}m/sec) versus frequency.

transmitted and received pulses have been represented at the same time reference, in spite of being delayed with respect to each other. The delays found for 1 μs and 10-nsec-width pulses for a channel length of 20 cm were 20 μsec and 50 nsec, respectively. Therefore, it can be stated that pulses become wider as distance increases, as other authors remark (Cho et al., 2007; Vaughan et al., 2009). Regarding the delay

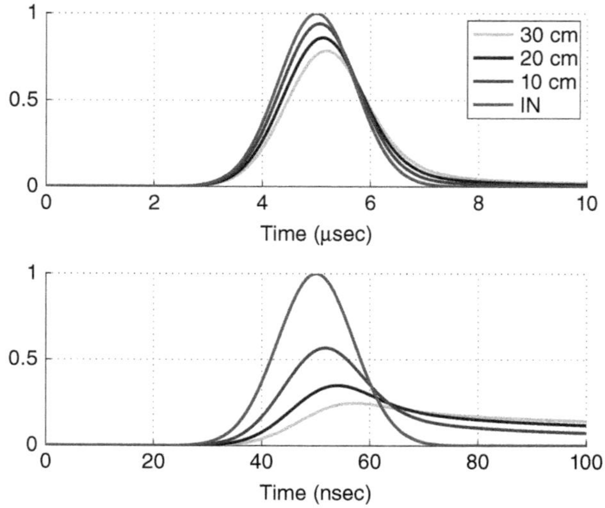

Figure 9.23 Attenuation and dispersion of 1-μsec- and 10-nsec-wide Gaussian pulses that propagate through the skin several channel lengths of 10, 20, and 30 cm. IN refers to the input pulse.

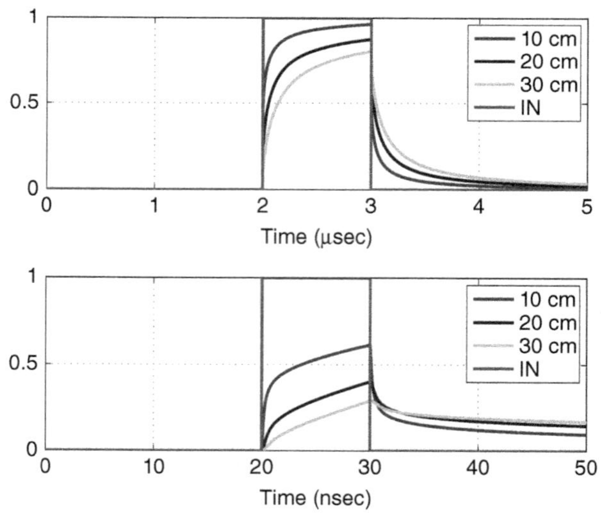

Figure 9.24 Attenuation and dispersion of 1-μsec- and 10-sec-wide square pulses that propagate through the skin several channel lengths of 10, 20, and 30 cm. IN refers to the input pulse.

spread, the closed-form expression obtained for the propagation constant γ allows a quantitative analysis of spreading/delay caused by the skin communication channel. Specifically, a dispersion coefficient of about 5.2 nsec/(cm · rad · MHz) has been estimated by simulation. In conclusion, this analysis shows that the width of a pulse is a major design parameter in the communication performance when the transmission medium is dispersive. In much the same way, wider pulses are not as much affected as shorter ones by the attenuation and the dispersion of the skin communication channel. Unfortunately, the data rate achieved by using wider pulses is lower, making a trade-off necessary.

9.6.3 Modulation Schemes

The impact of the IBC channel on digital communications can be analyzed by using simulation techniques relying on appropriate channel models. This way, it is possible to evaluate the performance of the communications system under different modulation schemes, several data rates, or different channel lengths. Despite the fact that most biomedical monitoring applications require a low data rate according to Paksuniemi et al. (2005), new IBC broad-band systems and applications, which require higher data rates and bandwidths, have been proposed (Fujikawa and Nishigaki, 2011; Pun et al., 2011b; Ruiz and Shimamoto, 2006b; Shinamoto et al., 2007; Song and Yoo, 2011). In this section, we follow a simulation approach to trade-off the data rate with bit error rate (BER), mediated by the IBC channel model.

As in previous sections, the band-limited channel of skin, given by equation (9.13), is tested using Matlab to filter the modulated signal with the IBC channel model. Symbols have been shaped as root-raised cosine pulses in order to minimize the intersymbol interference (ISI). Bit error rate has been calculated as a function of bit rate and channel length for different modulation schemes. Some of the results obtained for each type of modulation scheme are detailed below.

BPSK The constellation and the eye diagram obtained for a BPSK modulation scheme, considering a bit rate of 2 Mbps and a channel length of 30 cm, are shown in Figure 9.25, which can be fairly compared to results reported in Ruiz and Shimamoto (2006a). In addition, a graph with BER results versus data rate and several channel lengths is depicted in Figure 9.26.

It can be seen that a maximum bit rate of 10 Mbps (corresponding to a BER of 10^{-1}) is obtained for all channel lengths; hence, it can be inferred that BPSK is more suitable for longer distances, even up to 30 cm, for which higher bit rates with BER below 10^{-1}, can be achieved. In conclusion, BPSK is sufficiently robust with respect to the dispersion introduced by skin. Obviously, when the channel length increases, the maximum data rate achieved for a required BER decreases.

QPSK and 8PSK The constellation and the eye diagram obtained for a QPSK modulation scheme, considering a bit rate of 2 Mbps and a channel length of 30 cm, are in Figure 9.25. In addition, a comparative analysis of BER for BSPK, QPSK, and 8PSK schemes, considering a channel length of 25 cm, is summarized in Figure 9.27

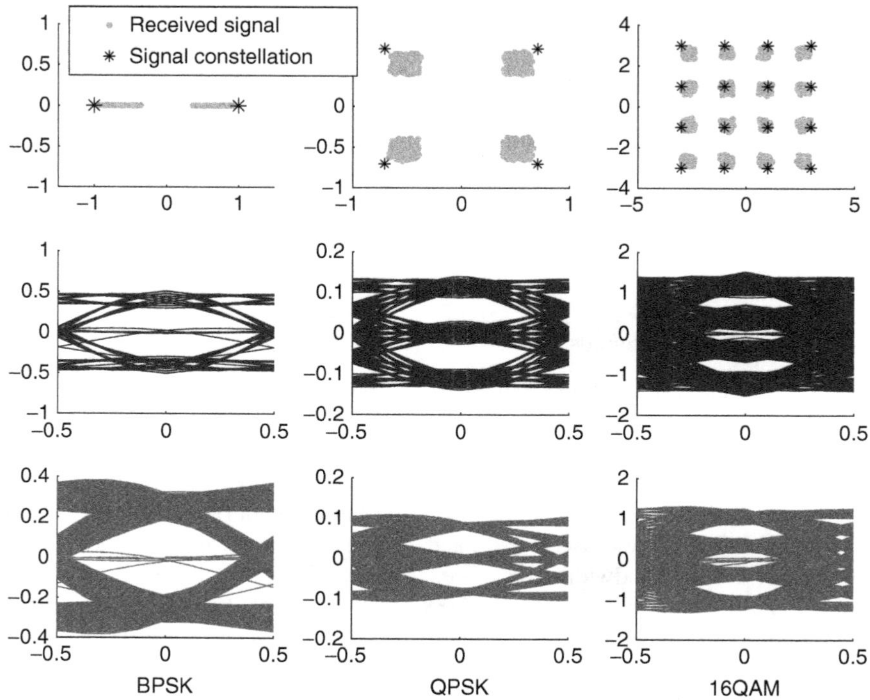

Figure 9.25 Constellation and eye diagrams for BPSK, QPSK, and 16QAM. A channel length of 30 cm is considered in all cases. Regarding bit rate, 2 Mbps is considered for BPSK and QPSK, whereas a bit rate of 1 Mbps is considered for 16QAM. Above, the eye diagram of the transmitted signal, and below, the eye diagram of the received signal after propagation through the skin. In all cases, a root-raised cosine pulse is considered for the shaping filter.

16QAM In this case, a bit rate of 1 Mbps has been chosen to generate both the constellation and eye diagram of 16QAM, which are shown in Figure 9.25. It is obvious that 16QAM is not as robust as the other schemes with respect to channel dispersion since lower bit rates with higher BERs are obtained compared to the rest of the modulation schemes studied. For this reason, the BER has been limited to the range between 10^{-4} and 10^{-1}, and the corresponding data rate values for this range were searched for several channel lengths. In this way, for instance, for channel lengths of 50, 40, 35, and 30 cm, the maximum bit rates achieved for a BER of 10^{-1} were 2, 3, 4, and 6 Mbps, respectively. In conclusion, if QAM is compared with BPSK for all channel lengths considered, QAM always obtains lower data rates for the same BER than BPSK.

Comparison of Modulation Schemes Simulation results show that BPSK and QPSK provide the best performance, with BER values below 10^{-3} for a bit rate of

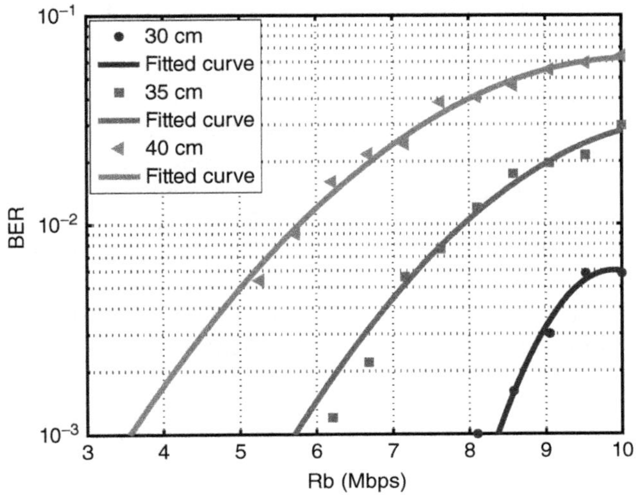

Figure 9.26 BPSK results for BER as a function of bit rate R_b (Mbps) and three different channel lengths of 35, 40, and 50 cm. Simulation results (marks) are presented along with its fitted curve (line).

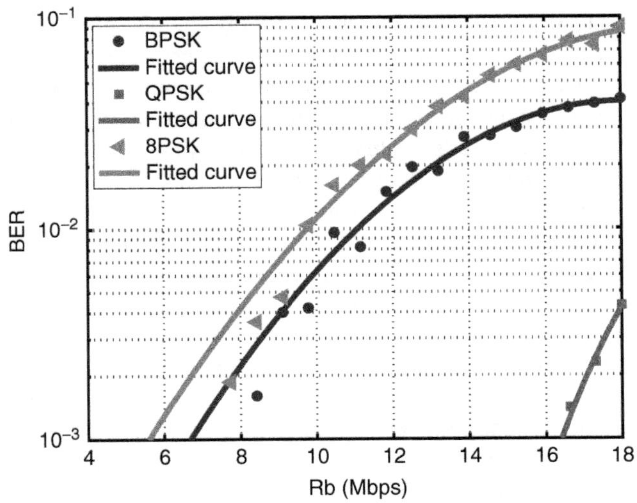

Figure 9.27 Comparative study for BPSK, QPSK, and 8PSK. BER as a function of bit rate R_b (Mbps) for a channel length of 25 cm. Simulation results (marks) are presented along with the fitted curve (line).

8 Mbps and a channel length of 30 cm. QPSK has marginally better results than BPSK for bit rates up to 20 Mbps, after which the results of both modulation schemes begin to match. The worst results come from 16QAM, with BER values below 10^{-3} only for bit rates lower than 3 Mbps, taking into account the same channel length of 30 cm. 8PSK provides results midway between BPSK and 16QAM. BPSK, 8PSK, and QPSK results for bit rates between 5 and 20 Mbps and channel lengths of 25 and 30 cm have been extracted. QPSK provides higher bit rates, followed by BPSK, about 20 Mbps, with BER values lower 10^{-2} for a distance of 25 cm. 8PSK also provides the same BER but with bit rates below 15 Mbps. The increase in distance causes the results to become worse in all modulation schemes, and, in fact, for a distance of 30 cm the maximum bit rate achieved by both BPSK and 8PSK drops below 10 Mbps for the same BER of 10^{-2}, whereas the bit rate for QPSK remains above 15 Mbps in the range of BER of 10^{-3} up to 10^{-2}.

9.7 CONCLUSION

This chapter has focused on the study of IBC as a promising technique based on the use of the human body as a transmission medium for electrical signals, providing a channel to interconnect sensor devices in BANs. Nowadays, interdevice communication is established through RF wireless standards such as Bluetooth and Zigbee. However, it should be noted that such standards have serious problems regarding power limitations, which in turn pose a number of technical challenges for BAN if more ambitious features regarding battery autonomy are to be achieved. In fact, in the case of biomedical monitoring applications, where a pervasive monitoring of biosignals is sought, such limitations are critical since continuously having to exchange batteries represents a major handicap in their usefulness, subsequently leading to a detrimental effect on the user's quality of life. This is especially true in the case of surgical operations. This way, IBC presents several advantages over RF standards: being based on the near-field coupling mechanism, the EMF is confined along the surface of the human body, ensuring that much energy is not radiated into the air, and thereby reducing consumption and interferences with other external devices. Furthermore, IBC possesses some positive features regarding bandwidth and immunity, which are also some of the most critical aspects within BAN. These advantages have led researchers to improve IBC prototypes regarding data rate, electronic complexity, and consumption. At the same time, two main techniques have been discussed: galvanic and capacitive coupling, and although a wide nomenclature for each can be seen in the literature, their main features have been established. In the case of galvanic coupling, two electrodes at the transmission side and another two electrodes at the reception side configure the differential characteristic of this form of coupling. For capacitive coupling, only one signal electrode is placed at both the transmission and the reception sides, which can either be directly attached to the skin or positioned near it. The signal path is then closed through the air by means of capacitive coupling between the external ground plane and the ground electrodes, which point away from the body.

Another issue related to IBC technology is the experimental characterization of the human body as a transmission medium. Many experimental approaches have been proposed in the literature, showing the impact of topics such as grounding, differential signaling, electrode–impedance matching, and the like, which require clarification. An approach to harmonized experimental setups for both galvanic and capacitive coupling has been proposed, and it has been concluded that a consistent ground strategy must be considered in order to obtain realistic IBC galvanic and capacitive couplings. Regarding channel length, galvanic coupling exhibits lower ranges, given the fact that very low level signals are obtained for channel lengths longer than 20 cm. In contrast, capacitive coupling shows higher ranges, distances as long as 150 cm can be easily covered, and considerable attenuation changes with respect to those obtained for shorter distances such as 30 cm have not been evidenced. However, in this case, as the channel length increases, more pronounced resonance phenomena affect the attenuation results, due to the influence of the capacitive return path. As for the frequency range, galvanic coupling shows better path loss results for low frequencies, up to units of megahertz, with a maximum peak being found at 40 kHz, whereas in the case of capacitive coupling, a bandpass profile is seen up to 100 MHz. In this case, the maximum is located within the range of 60–70 MHz.

The complexity of IBC measurements and the discrepancies over diverse data sets reported by different authors highlight the need for models that help to understand the main propagation mechanisms through the human body and the skin and, therefore, to assist in the design of IBC prototypes. Models provide a set of accurate expressions that allow simulations to be carried out as a preceeding step toward experimentation and design. These simulations permit IBC designers not only to understand the experimental results but also to confirm their reliability. Computational EM models can be solved by using numerical techniques and specialized software, which also has the ability to calculate the distribution of the electromagnetic energy throughout the body. Unfortunately, reproducing these models is not an easy task due to their complexity and the large amount of data generated. Circuit-level models, however, are easy to reproduce and accurate enough to predict experimental results, thus assisting the design of IBC systems. The main disadvantage is a limited frequency range together with their inability to gather EM phenomena such as radiation at higher frequencies. The models described throughout this chapter are an example of how useful they can be to get knowledge about IBC performance: a simple skin transmission line model permits a propagation constant to be obtained as a function of both the electrophysiological properties of the skin and the coupling technique, and from this, some other key issues such as path loss, dispersion, and modulation schemes can subsequently be studied. Moreover, since experimental results are seen to vary from person to person, models shall be personalized with the aim of addressing different subjects' bioelectric properties and anthropometrical characteristics.

The main idea of this chapter is that IBC is a promising and efficient approach to interconnect wireless devices located both inside and outside the human body, in the context of BAN and PAN networks. Over the last few years, biomedical sensors on micro- and nanolevels have been greatly experimented and, in fact, will constitute the biomedical monitoring networks of the future. In the area of communications further research should be conducted in order to solve some of the technical challenges and to discover new possibilities, with the objective always being to offer an efficient and secure communication system for the user.

ACKNOWLEDGMENTS

This work has been supported in part by the Consejería de Economía, Innovación, Ciencia y Empleo, Government of Andalucía (Spain), under Grants P08-TIC-04069 and P10-TIC-6214, and in part by the Fondo de Investigaciones Sanitarias, Instituto de Salud Carlos III (Spain), under Grant PI082023. The authors are grateful to G. Barbarov, E.C. Wiegers, and D. Plant for their helpful comments and support in this work.

REFERENCES

ANT. 2012. Available: http://www.thisisant.com; last visited: June 2012.

Armijo M, Camacho F. 1998. *Handbook of Dermatology* (in Spanish). Madrid: Aula Médica. pp. 3–33.

Attard S, Zammit S. 2012. An empirical investigation of the capacitive body coupled communications channel for body area networks. Proc IEEE Topical Conf. Biomedical Wireless Technologies, Networks, and Sensing Systems (BioWireleSS), pp. 85–88.

Bae J, Cho H, Song K, Lee J, Yoo H-J. 2012. The signal transmission mechanism on the surface of human body for body channel communication. *IEEE Trans. Microw. Theory Tech.*, 60(3):582–593.

Bessis N, Asimakopoulou E, French T, Norrington P, Xhafa F. 2010. The big picture, from grids and clouds to crowds: A data collective computational intelligence case proposal for managing disasters. Proc. Int. Conf. P2P, Parallel, Grid, Cloud and Internet Computing (3PGCIC), pp. 351–356.

Black RD. 2011. Recent advances in translational work on implantable sensors. *IEEE Sensors J.*, 11(12):3171–3182.

Bluetooth Low Energy (BLE). 2012. Available: http://www.bluetooth.com/Pages/Low-Energy.aspx; last visited: June 2012.

Callejón MA, Naranjo D, Reina-Tosina J, Roa LM. 2012a. A first approach to the harmonization of intrabody communications measurements. In Long M (ed.): 2012 *World Congress on Medical Physics and Biomedical Engineering*. Beijing, China: IFMBE Proceedings, pp. 704–707.

Callejón MA, Naranjo D, Reina-Tosina J, Roa LM. 2012b. Distributed circuit modeling of galvanic and capacitive coupling for intrabody communication. *IEEE Trans. Biomed. Eng.*, PP(99):1–7 (early access).

Callejón MA, Roa LM, Reina-Tosina J, Naranjo D. 2011. A distributed-parameter approach to model galvanic and capacitive coupling for intrabody communications. Proc. 2nd Int. ICST Conf. Wireless Mobile Commun. and Healthcare (MobiHealth 2011), pp 1–8.

Callejón MA, Roa LM, Reina-Tosina J, Naranjo D. 2012c. Study of attenuation and dispersion through the skin in intrabody communications systems. *IEEE Trans. Inf. Technol. Biomed.*, 16(1):159–165.

Chen XM, Pun SH, Gao YM, Mak PU, Vai MI, Du M. 2012. Study on transfer function of intra-body communication based on quasi-static electric field modeling. Proc. IEEE-EMBS Int. Conf. Biomedical and Health Informatics (BHI), pp. 388–391.

Cho N, et al. 2007. The human body characteristics as a signal transmission medium for intra-body communication. *IEEE Trans. Microw. Theory Tech.*, 55(5):1080–1086.

Collin, RE, 1992. *Foundations for Microwave Engineering*. New York: McGraw-Hill.

Deventer DS, Van E, Repacholi M. 2007. EMF standards for human health. In C. Press (ed.). *Handbook of Biological Effects of Electromagnetic Fields: Biological and Medical Aspects of Electromagnetic Fields*. Boca Ratón: CRC Press Taylor & Francis Group

Edelberg R. 1968. Biopotentials from the skin surface: The hydration effect. *Ann. NY Acad. Sci.*, 148:252–262.

Eid AM, Wallace JW. 2011. Accurate nodeling of body area network channels using surface-based method of moments. *IEEE Trans. Antennas Propag.*, 59(8):3022–3030.

Estudillo MA, Naranjo D, Roa LM, Reina LJ. 2010. Intrabody communications as an alternative proposal for biomedical wearable systems. In Cávado, Ave (eds). *Handbook of Research on Developments in e-Health and Telemedicine: Technological and Social Perspectives*. Portugal: IGI Global.

Fowles DC. 1986. The eccrine system and electrodermal activity. *Psychophysiology*. New York: Guilford Press, pp. 51–96.

Fujii K, Ito K, Tajima S. 2003. A study on the receiving signal level in relation with the location of electrodes for wearable devices using human body as a transmission channel. Proc. 2003 IEEE Int. Symp. Antennas and Propagation Society, Vol. 3, pp. 1071–1074.

Fujikawa M, Nishigaki, M. 2011. A study of prevention for social engineering attacks using real/fake organization's uniforms: Application of radio and intra-body communication technologies. Proc. 6th IEEE Int. Conf. Availability, Reliability and Security, pp. 597–602.

Fujii K, Takahashi M, Ito K. 2006. Study on the electromagnetic field distributions of realistic Japanese adult male and female models with a wearable device using the human body as a transmission channel. Proc. 2006 IEEE Int. Symp. Antennas and Propagation Society, pp 2121–2124.

Fujii K, Takahashi M, Ito K. 2007. Electric field distributions of wearable devices using the human body as a transmission channel. *IEEE Trans. Antennas Propag.*, 55(7):2080–2087.

Fujii M, et al. 2010. Exploration of whole human body and UWB radiation interaction by efficient and accurate two-debye-pole tissue models. *IEEE Trans. Antennas Propag.*, 58(2):515–524.

Fukumoto M, Tonomura Y. 1997. Body coupled FingeRing: Wireless wearable keyboard. Proc. CHI Conf. Human Factors in Computing Systems, pp. 147–154.

Gabriel S, Lau RW, Gabriel C. 1996. The dielectric properties of biological tissues: III. Parametric models for the dielectric spectrum of tissues. *Phys. Med. Biol.* 41(11):2271–2293.

Gao Y, Hang-Pun S, Du M. I-Vai M, Un-Mak P. 2008. A preliminary two dimensional model for intra-body communication of body sensor networks. Proc. Int. Conf. Intelligent Sensors, Sensor Networks and Information Processing (ISSNIP), pp. 273–278.

Gao YM, Pun SH, Du M, Vai MI, Mak PU. 2009. Quasi-static field modeling and validation for intra-body communication. Proc. 3rd Int. Conf. Bioinformatics and Biomedical Eng.

Grimnes S, Martinsen G. 2000. *Bioimpedance and Bioelectricity Basics*. San Diego, CA: Academic Press.

Gupta A, Abhayapala T. 2008. Body area networks: Radio channel modelling and propagation characteristics. Proc. Australian Communications Theory Workshop (AusCTW), pp. 58–63.

Hachisuka K, et al. 2003. Development of wearable intra-body communication devices. *Sensors Actuators A: Phys.*, 105:109–115.

Hachisuka K, et al. 2005. Intra-body data transmission for the personal area network. *Microsyst. Technol.*, 11(8–10):1020–1027.

Hachisuka K, et al. 2006. Simplified circuit modeling and fabrication of intrabody communication devices. *Sensors Actuators A: Phys.*, 130–131:322–330.

Hall PS, Hao Y. 2006. *Antennas and Propagation for Body-Centric Wireless Communications*. Boston: Artech House.

Handa T, Shoji S, Ike S, Takeda S, Sekiguchi T. 1997: A very low-power consumption wireless ECG monitoring system using body as a signal transmission medium. Proc. 1997 Int. Conf. Solid-State Sensors and Actuators, Chicago, pp. 1003–1006.

Healthsystem. 2012. Available: http://www.healthsystem.virginia.edu; last visited: June 2012.

Hwang JH, et al. 2007. Effect of load impedance on the signal loss of human body communication. Proc. 2007 IEEE Int. Symp. Antennas and Propagation, pp. 3217–3220.

Hyoung CH, Sung JB, Hwang JH, Kim JK, Park DG, Kang SW. 2006. A novel system for intra-body communication: Touch-and-play. Proc. 2006 IEEE Int. Symp. Circuits and Systems, pp. 21–24.

Igou S, Taki M, Furuya A, Shibata N, Shinagawa M, Hattori M. 2011. Dependence of transmission characteristics of intra-body communication system on the body posture and surrounding environment. Proc. 30th URSI General Assembly and Scientific Symposium, pp. 1–4.

International Commission on Non-Ionizing Radiation Protection (ICNIRP). 1998. Guidelines for limiting exposure to time-varying electric, magnetic, and electromagnetic fields (up to 300 GHz). *Health Phys.*, 74(4):494–522.

International Commission on Non-Ionizing Radiation Protection (ICNIRP). 2009. ICNIRP statement on the guidelines for limiting exposure to time-varying electric, magnetic and electromagetic fields (up to 300 GHz). *Health Phys.*, 97(3):257–258.

Kado Y, Shinagawa. 2010. RedTacton near-body electric-field communications technology and its applications. NTT Microsystem Integration Laboratories, pp. 1–6.

KDDI. Designing Studio. 2012. Available: http://www.kds.kddi.com/; last visited: June 2012.

Kost A. 2011. Different methods for computational electromagnetics: Their characteristics and typical practical applications. Computational methods for the innovative design of electrical devices. *Stud. Comput. Intell.*, 327:261–275.

Koutitas G. 2010. Multiple human effects in body area networks. *IEEE Antennas Wireless Propag. Lett.*, 9:938–941.

Lea A, Ping H, Ollikainen J, Vaughan RG. 2009. Propagation between on-body antennas. *IEEE Trans. Antennas Propag.* 57(11):3619–3627.

Li CY, Wang CC, Rieger R. 2010. On the capacitively coupled transmission channel for body network application. Proc. Int. Symp. Next-Generation Electronics (ISNE), pp. 5–8.

Lin Y-T, et al. 2011. A 0.5-V biomedical system-on-a-chip for intrabody communication system. *IEEE Trans. Ind. Electron.*, 58(2):690–699.

Liolios C, Doukas C, Forulas G, Maglogiannis. 2010. An overview of body sensor networks in enabling pervasive healthcare and assistive environments. Proc. 3rd Int. Conf. Pervasive Technologies Related to Assistive Environments (PETRA).

Liu Y, et al. 2010. Measurement system for propagation characteristics of intra-body communication. Proc. Sixth Int. Conf. Wireless Communications Networking and Mobile Computing, pp 1–4.

Lucev Z, Korican A, Cifrek M. 2010a. A finite element method study of the current density distribution in a capacitive intrabody communication system. 12th Mediterranean Conference on Medical and Biological Engineering and Computing. *IFMBE Proc.*, 29(3):422–425.

Lucev Z, Krois I, Cifrek M. 2009. A multichannel wireless EMG measurement system based on intrabody communication. Proc. 19 IMEKO World Congress. Fundamental and Applied Metrology, Lisbon, Portugal, pp. 1711–1715.

Lucev Z, Krois I, Cifrek M. 2010b. Intrabody communication in biotelemetry. Wearable and autonomous biomedical devices and systems for smart environment. *Lect. Notes Electr. Eng.*, 75:351–368.

Lucev Z, Krois I, Cifrek M. 2010c. Application of wireless intrabody communication system to muscle fatigue monitoring. Proc. IEEE Instrumentation and Measurement Technology Conf. (I2MTC), pp. 1624–1627.

Lucev Z, Krois I, Cifrek M. 2011. A capacitive intrabody communication channel from 100 kHz to 100 MHz. Proc. IEEE Instrumentation and Measurement Technology Conf., pp. 1–4.

Ma D, Zhang WX. 2009. Analytic propagation model for body area network channel based on impedance boundary condition. Proc. 3rd European Conf. Antennas and Propagation (EuCAP), pp. 974–978.

Ma D, Zhang WX. 2011. Analytic propagation model for wireless body-area networks. *IEEE Trans. Antennas Propag.*, 59(12):4749–4756.

Matsushita N, et al. 2000. Wearable key: Device for personalizing nearby environment. Proc. 4th Int. Symp. Wearable Computing, pp. 119–126.

National Library of Medicine. 2012. Available: http://www.nlm.nih.gov/research/visible/visible_human.html; last visited: June 2012.

Neuman MR. 2000. Biopotential electrodes. In Joseph D. Bronzino (ed.) *The Biomedical Engineering Handbook*, 2nd ed. Boca Raton: Florida.

Northrop RB. 2004. Sources and properties of biomedical signals. In *Analysis and Application of Analog Electronic Circuits to Biomedical Instrumentation*, pp. Boca Raton: Florida.1–21.

NTTRedTacton. 2005. Available: http://www.ntt.co.jp/news/news05e/0502/050218.html; last visited: March 2014.

Oguma H, Nobata N, Nawa K, Mizota T, Shinagawa M. 2011. Passive keyless entry system for long term operation. Proc. IEEE Int. Symp. on a World of Wireless, Mobile and Multimedia Networks (WoWMoM), pp. 1–3.

Okamoto E, Sato Y, Seino K, Kiyono T, Kato Y, Mitamura Y. 2010. Basic study of a transcutaneous information transmission system using intra-body communication. *J. Artif. Organs* 13(2):117–120.

Paksuniemi M, et al. 2005. Wireless sensor and data transmission needs and technologies for patient monitoring in the operating room and intensive care unit. Proc. 27th Annual Int. Conf. IEEE Engineering in Medicine and Biology Society, pp. 5182–5185.

Partridge K, et al. 2001. Empirical measurements of intrabody communication performance under varied physical configurations. Proc. 14th Annual ACM Symp. User Interface Software and Technology, pp. 183–190.

Pun SH, Gao Y, Mak P, Du M, Vai M. 2009. Modeling for intra-body communication with bone effect. Proc. 2009 Annual Int. Conf. IEEE Engineering in Medicine and Biology Society, pp. 693–696.

Pun SH, Gao Y, Mou P, Mak P, Mai M, Du M. 2010. Multilayer limb quasi-static electromagnetic modeling with experiments for galvanic coupling type IBC. Proc. 2010 Annual Int. Conf. IEEE Engineering in Medicine and Biology Society, pp. 378–381.

Pun SH, et al. 2011a. Quasi-static modeling of human limb for intra-body communications with experiments. *IEEE Trans. Inf. Technol. Biomed.*, 15(6):870–876.

Pun SH, et al. 2011b. Galvanic intrabody communication for affective acquiring and computing. *IEEE Trans. Affective Comp.*, PP(99):1 (early access).

Reusens E, Joseph W, Latre B, Braem B, Vermeeren G, Tanghe E, Martens L, Moerman I, Blondia. 2009. Characterization of on-body communication channel and energy efficient topology design for wireless body area networks. *IEEE Trans. Inf. Technol. Biomed.*, 13(6):933–945.

Roa LM, Naranjo D, Reina J, Oliva JS, Lara A, Milán JA, Estudillo MA. 2012. Applications of bioimpedance to end stage renal disease (ESRD). *Modeling and Control of Dialysis Systems*. Heidelberg: Springer.

Ruiz JA, Shimamoto S. 2005. A study on the transmission characteristics of the human body towards broadband intra-body communications. Proc. 9th International Symposium on Consumer Electronics (ISCE 2005), pp. 99–104.

Ruiz JA, Shimamoto S. 2006a. Experimental evaluation of body channel response and digital modulation schemes for intra-body communications. Proc. IEEE Int. Conf. Communications, Vol. 2, pp. 349–354.

Ruiz JA, Shimamoto S. 2006b. Novel communication services based on human body and environment interaction: Applications inside trains and applications for handicapped people. Proc. IEEE Wireless Communications and Networking Conf., Vol. 4, pp. 2240–2245.

Sasamori T, Takahashi M, Uno T. 2009. Transmission mechanism of wearable device for on-body wireless communications. *IEEE Trans. Antennas Propag.*, 57(4):936–942.

Shinamoto S, Alsehab AM, Kobayashi N, Dovchinbazar D, Ruiz, JA. 2007. Future applications of body area communications. Proc. 6th Int. Conf. Information, Communications and Signal Processing, pp. 1–5.

Song S, Cho N, Yoo H. 2007. A 0. 2-mW 2-Mb/s digital transceiver based on wideband signaling for human body. *IEEE J. Solid-State Circuits*, 42(9):2021–2033.

Song S, Yoo H. 2011. Body channel communication for energy-efficient BAN. In Yoo H-J, Hoof C (eds). *Bio-Medical CMOS ICs. Integrated Circuits and Systems, Part 2*, Boston: Springer, pp. 303–336.

Song S, et al. 2011. The simulation method of the galvanic coupling intrabody communication with different signal transmission paths. *IEEE Trans. Instrum. Meas.*, 60(4):1257–1266.

Takahashi M, Shimoda R, Kusunoki T, Yanagawa T, Kobase T, Nagai R, Shimasaki H, Kado Y. 2012. Characterization of MHz-band near-field coupling communication using finite element electromagnetic simulation: Body-channel communication system for human-area networking. Proc. 6th European Conf. Antennas and Propagation, pp. 1127–1131.

Tang C-M, Bashirullah R. 2011. Channel characterization for galvanic coupled in vivo biomedical devices. Proc. IEEE Int. Symp. Circuits and Systems, pp. 921–924.

Tronstad C, et al. 2010. A study on electrode gels for skin conductance measurements. *Physiol. Meas.*, 31(10):1395–1410.

Van Roy S, et al. 2007. On-body propagation velocity estimation using ultra-wideband frequency-domain spatial correlation analyses. *Electron. Lett.*, 43(25):1405.

Vasilopoulos K. 2010. Implementation of networking-by-touch to small unit, network-enabled operations. University: Naval Postgraduate School, Monterey, CA.

Vaughan, RG, et al. 2009. Theory of propagation for direct on-body wireless sensor communications. Proc. 2nd IET Seminar on Antennas and Propagation for Body-Centric Wireless Communications, pp. 1–5.

Warren S, Natarajan B. 2011. Wireless communication technologies for wearable systems. In Annalisa Bonfiglio and Danilo De Rossi (eds.) *Wearable Monitoring Systems*, New York: Springer US pp. 51–80.

Wegmueller MS, Huclova S, Froehlich J, Oberle M, Felber N, Kuster N, Fichtner W. 2009. Galvanic coupling enabling wireless implant communications. *IEEE Trans. Instrum. Meas.*, 58(8):2618–2625.

Wegmueller MS. 2007a. Intra-body communication for biomedical sensor networks. Ph.D. Thesis Diss. ETH No. 17323, ETH Zürich, Suisse.

Wegmueller MS, et al. 2006a. From dielectrical properties of human tissue to intra-body communications. 2006 World Congress on Medical Physics and Biomedical Engineering, IFMBE Proceedings, Vol. 14 (7), pp. 613–617.

Wegmueller MS, et al. 2007b. An attempt to model the human body as a communication channel. *IEEE Trans. Biomed. Eng.* 54(10):1851–1857.

Wegmueller S, Oberle M, Felber N, Kuster N, Fichtner W. 2006b. Galvanical coupling for data transmission through the human body. Proc. IEEE Instrumentation and Measurement Technology Conf., pp. 24–27.

Xu R, et al. 2009. Circuit-coupled FEM analysis of the electric-field type intra-body communication channel. Proc. 2009 IEEE Biomedical Circuits and Systems Conf. (BioCAS 2009), November 26–28, 2009, Beijing, China, pp. 221–224.

Xu R, et al. 2011. Electric-field intrabody communication channel modeling with finite-element method. *IEEE Trans. Biomed. Eng.*, 58(3):705–12.

Xu R, et al. 2012. Environment coupling and interference on the electric-field intrabody communication channel. *IEEE Trans. Biomed. Eng.*, 59(7):2051–2059.

Yamamoto T, Yamamoto Y. 1976. Electrical properties of epidermal stratum corneum. *Med. Biol. Eng. Comput.*, 14(2):151–158.

Yan L, Bae J, Lee S, Roh T, Song K, Yoo H-J. 2011. A 3.9 mW 25-electrode reconfigured sensor for wearable cardiac monitoring system. *IEEE J. Solid-State Circuits*, 46(1):353–364.

Yang L, Jingming K, Zunwen H, Jun F. 2010. Measurement system for propagation characteristics of intra-body communication. *Proc. 6th Int. Conf. on Wireless* Communications Networking and Mobile Computing (WiCOM), pp. 1–4.

Yang L, Jingming K, Zunwen H, Xiaobo X, Ling J. 2011. Measurement for channel characteristics of intra-body communication. Proc. Cross Strait Quad-Regional Radio Science and Wireless Technology Conf. (CSQRWC), Vol. 2, pp. 1138–1140.

Zedong N, Tengfei L, Wenchen W, Feng G, Lei W. 2012. Experimental characterization of human body communication in shield chamber. Proc. IEEE-EMBS Int. Conf. Biomedical and Health Informatics (BHI), pp. 759–762.

Zhu H, Ng WC, Shan H, Yuan J. 2012. A physical layer security analysis on the electric-field intra-body communication. Proc. Int. Conf. Computing, Networking and Communications (ICNC), pp. 14–17.

Zigbee Alliance. 2012. Available: http://www.zigbee.org; last visited: June 2012.

Zimmerman TG, 1995. Personal Area Networks (PAN): Near-field intrabody communication. MIT Media Laboratory, Cambridge.

Zimmerman TG, 1996. Personal Area Networks: Near-field intrabody communication. *IBM Syst. J.*, 35(3,4):609–617.

6LowPan. 2012. Available: http://6lowpan.net/; last visited: June 2012.

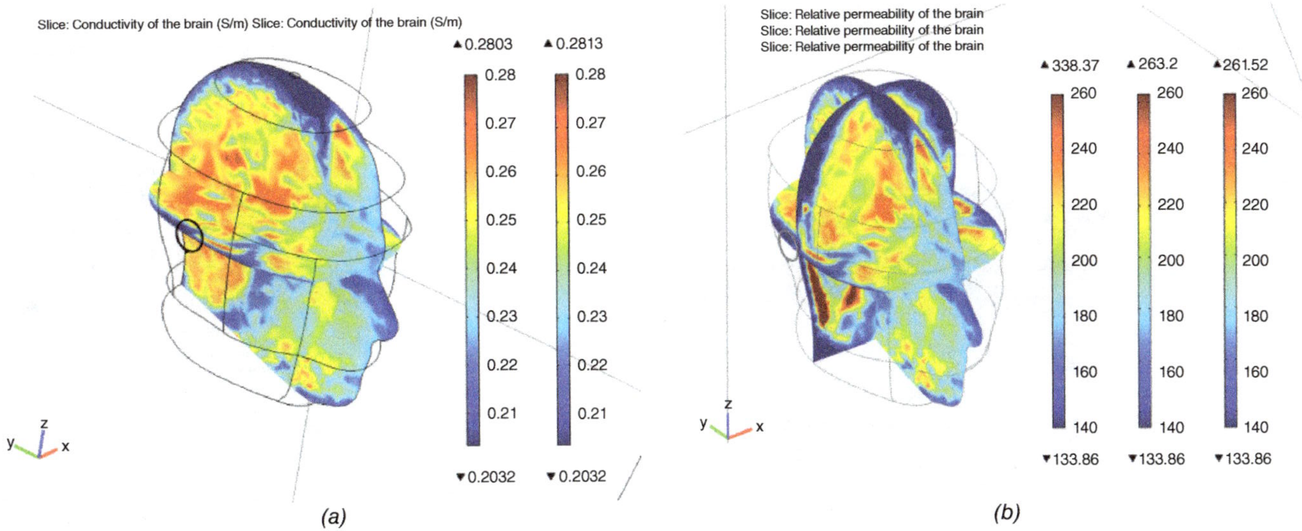

Figure 2.13 Model of human head at 13.56 MHz in COMSOL simulations: (*a*) conductivity of brain tissue; (*b*) relative permeability.

Handbook of Biomedical Telemetry, First Edition. Edited by Konstantina S. Nikita.

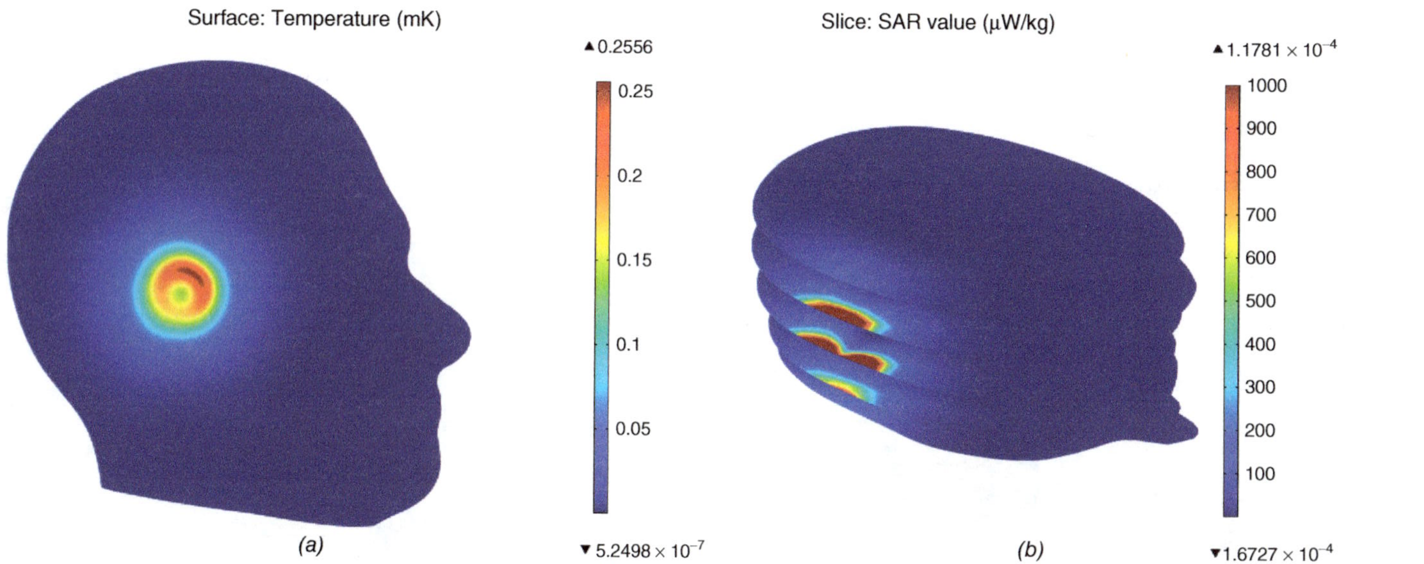

Figure 2.14 Result of simulation at 13.56 MHz with current of 200 mA through coil: (*a*) surface temperature increase due to electromagnetic heating of head; (*b*) SAR in human head.

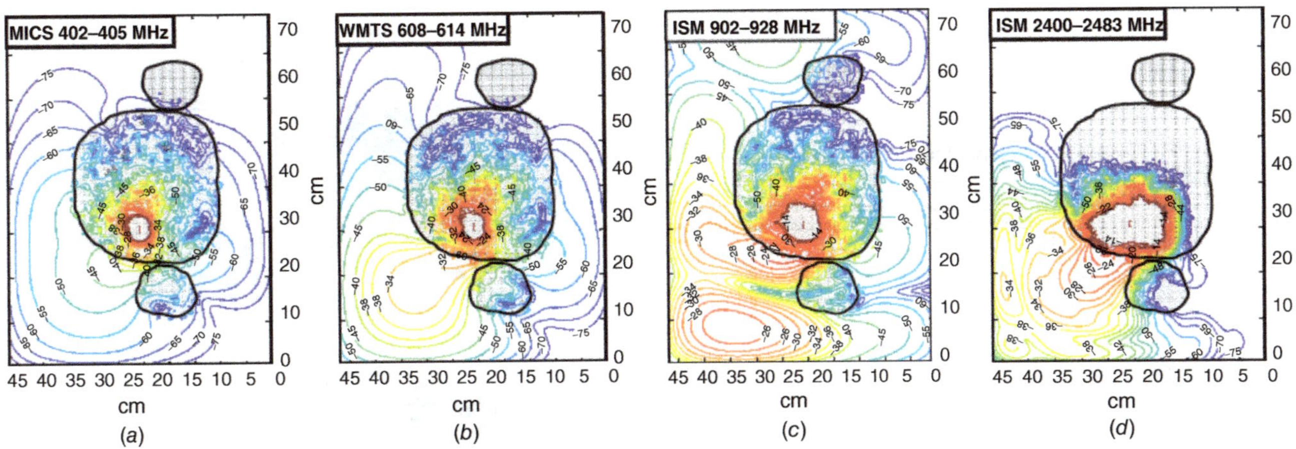

Figure 2.15 Normalized contours of radiated fields from capsule antenna inside human body torso (cross sections shown) operating at (*a*) MICS 402–405 MHz, (*b*) WMTS 608–614 MHz, (*c*) ISM 902–928 MHz, and (*d*) ISM 2.4–2.483 GHz FCC frequency bands (Yu et al., 2009).

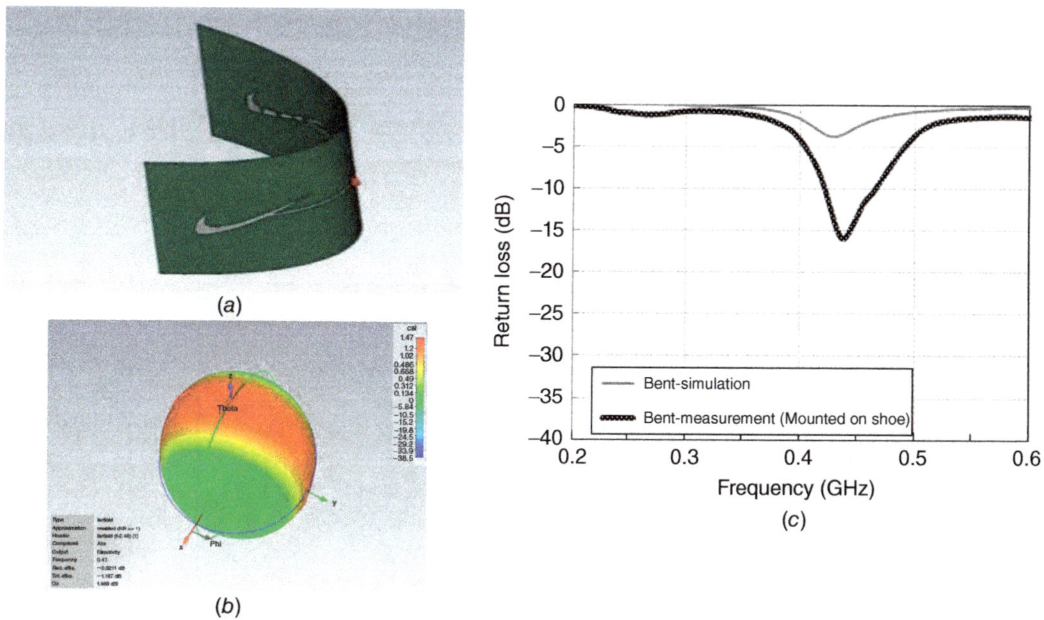

Figure 5.4 (*a*) Nike logo antenna without body effect, (*b*) Nike logo antenna simulated radiation pattern, and (*c*) Nike logo antenna simulated and measured return loss.

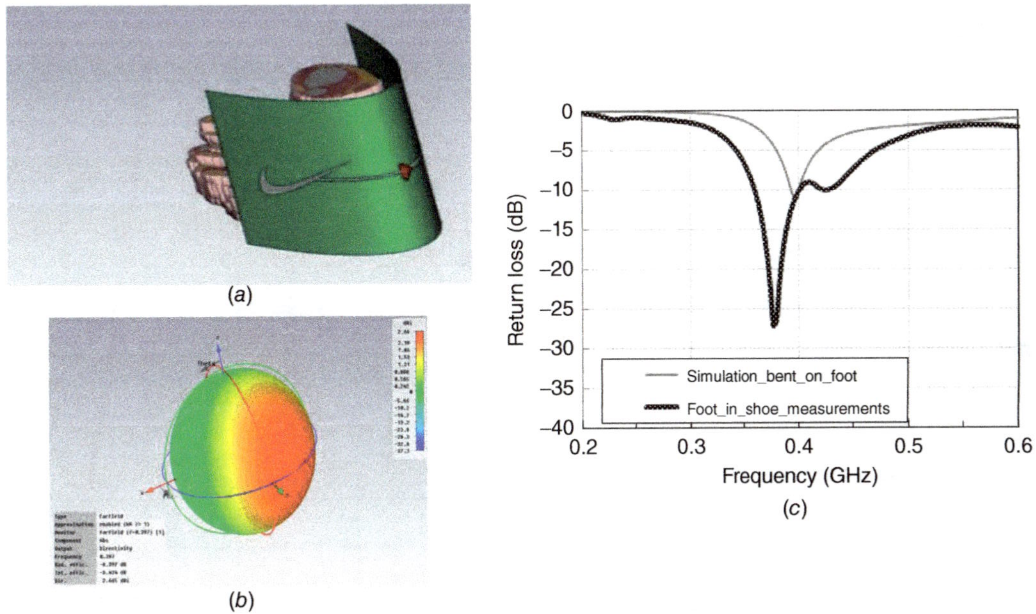

Figure 5.5 (*a*) Nike logo antenna with body effect, (*b*) Nike logo antenna simulated radiation pattern in presence of human tissue, and (*c*) Nike logo antenna simulated and measured return loss in presence of human tissue.

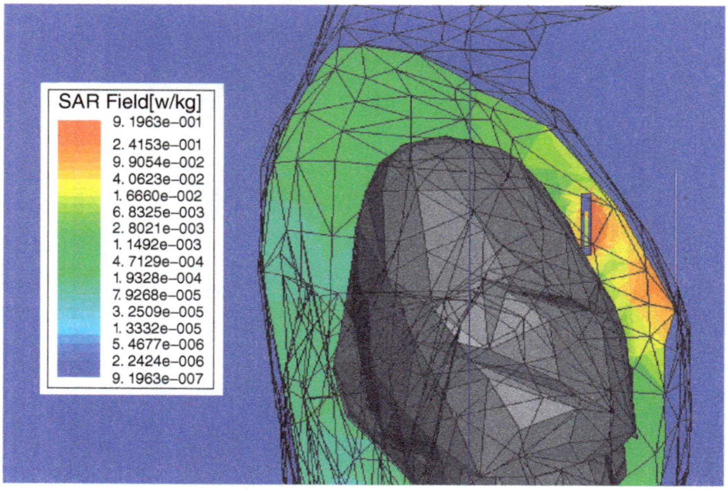

Figure 5.17 CSCMR in human body: SAR distribution.

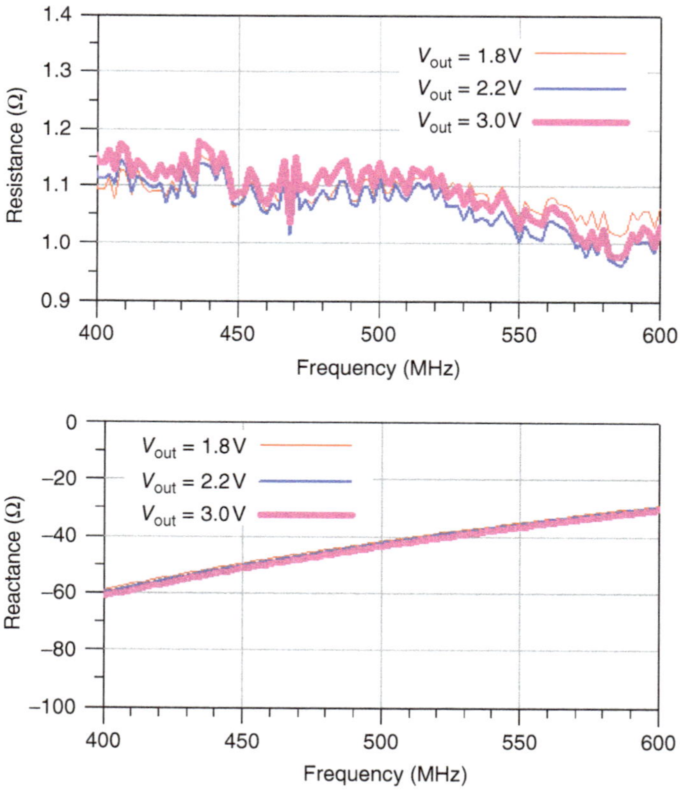

Figure 5.21 Measured impedance of RF−DC converter circuit.

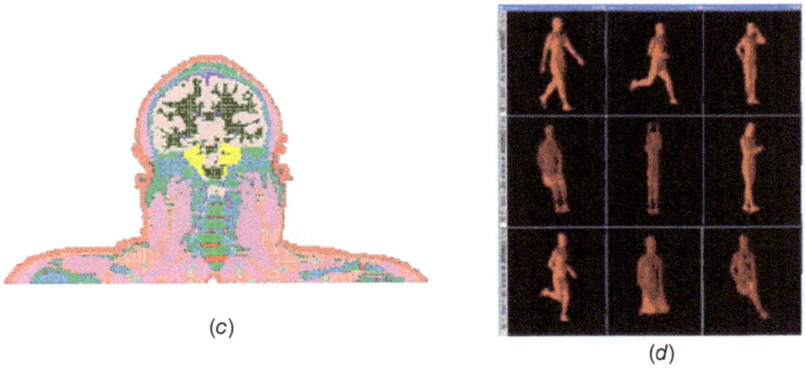

Figure 6.7 Anatomical phantom models used to model the human head with shoulders (Chen et al., 2009), and (*d*) body (Wang et al., 2009) in recent biomedical telemetry studies.

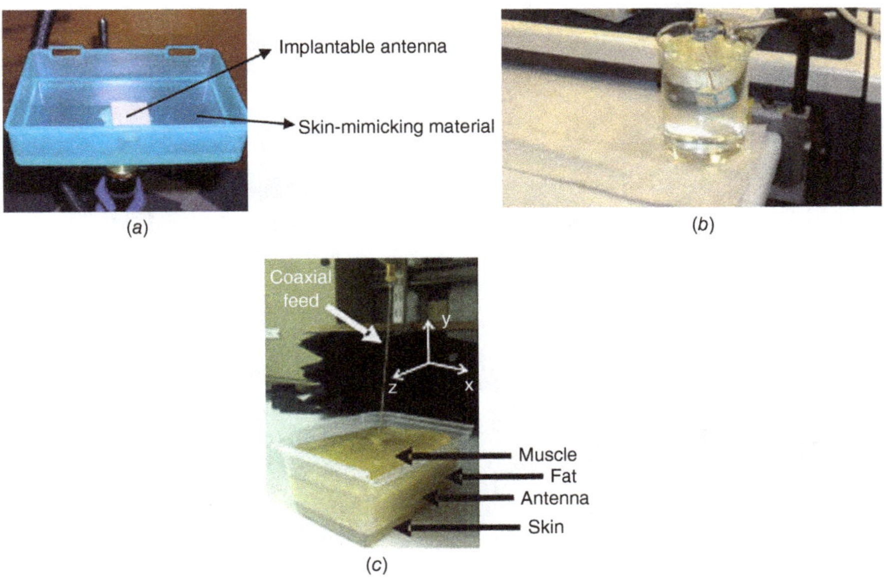

Figure 6.9 Canonical physical phantoms proposed in (*a*) Karacolak et al. (2009), (*b*) Abadia et al. (2009), and (*c*) Sani et al. (2010).

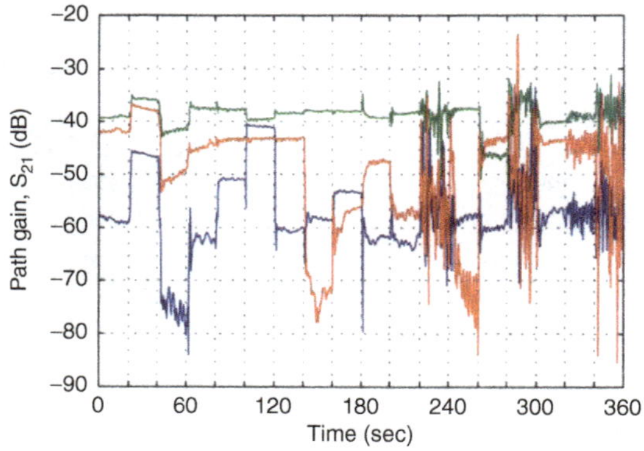

Figure 8.7 Measured path gain variations for Rx1 (blue), Rx3 (green), and Rx12 (red).

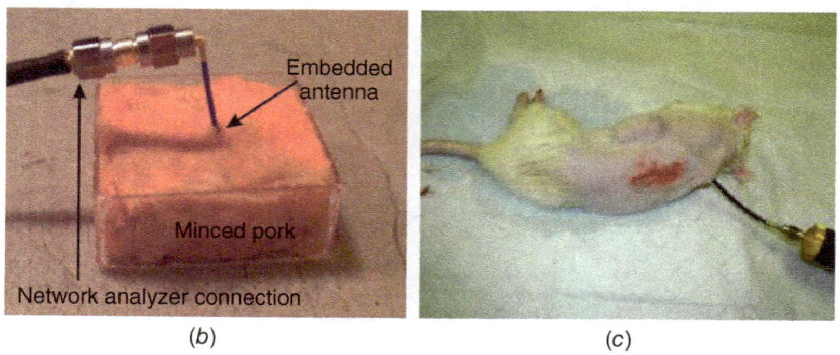

Figure 8.12 Experimental investigations for implantable antennas: testing inside animal tissue samples (Huang et al., 2011), and (*c*) in vivo testing (Karacolak et al., 2010).

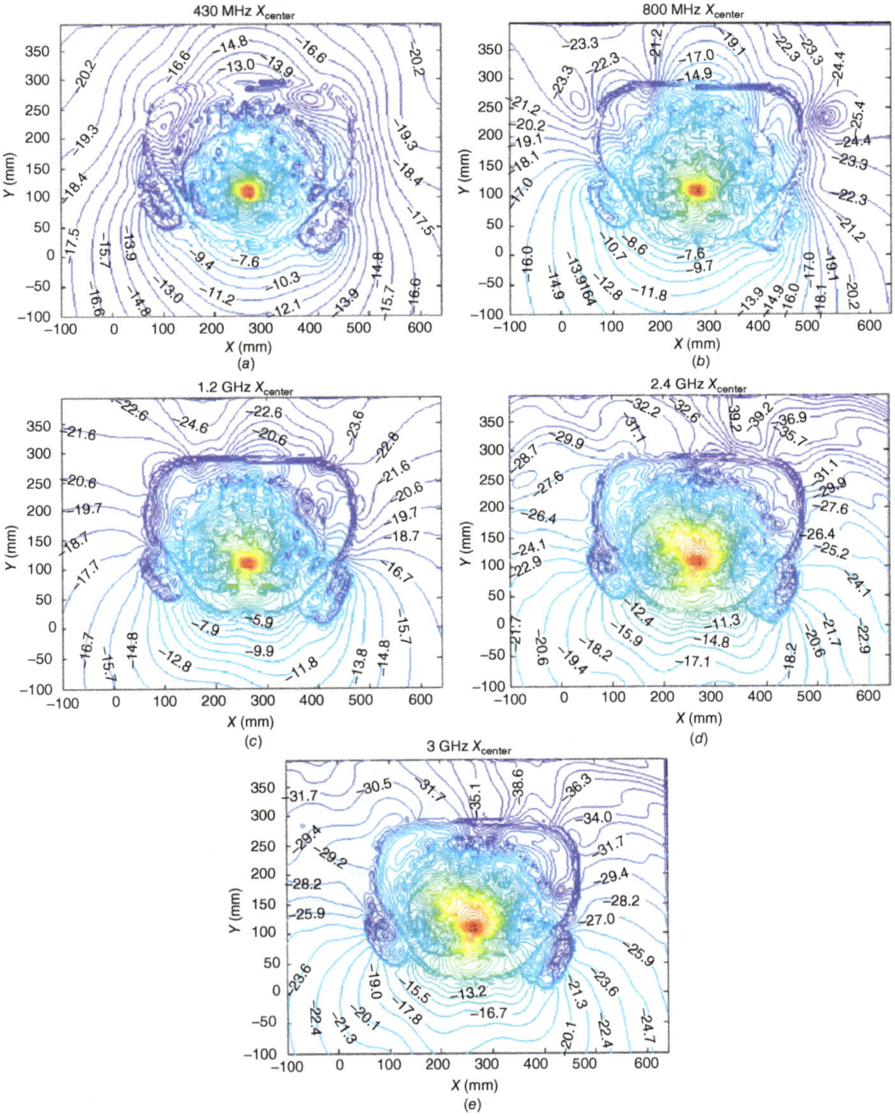

Figure 8.16 Near fields for ingestible antenna at (*a*) 430 MHz, (*b*) 800 MHz, (*c*) 1.2 GHz, (*d*) 2.4 GHz, and (*e*) 3 GHz (Xu et al., 2008b).

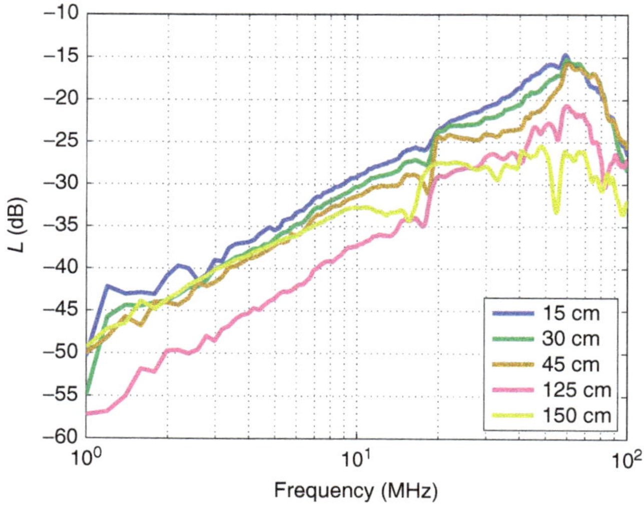

Figure 9.13 Path loss experimental results for IBC capacitive coupling for several channel lengths.

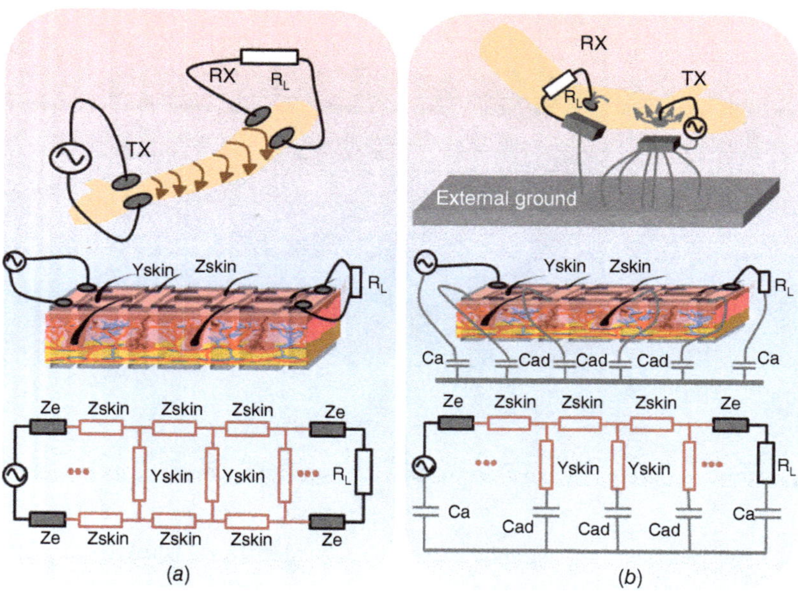

Figure 9.16 (*a*) Galvanic coupling IBC technique and equivalent circuital model. (*b*) Capacitive coupling IBC technique and equivalent circuital model.

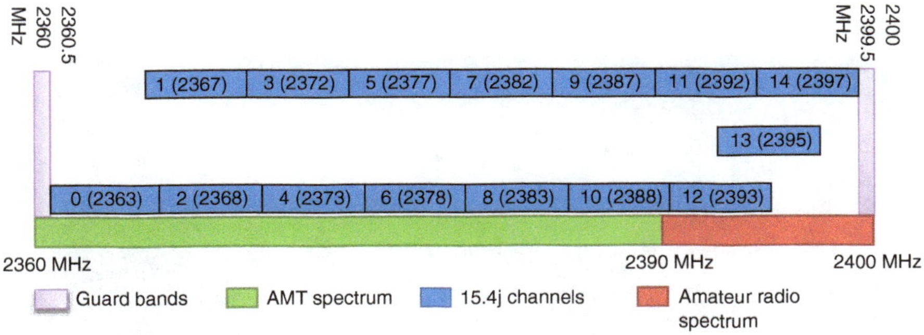

Figure 11.15 Proposed 15.4j channelization.

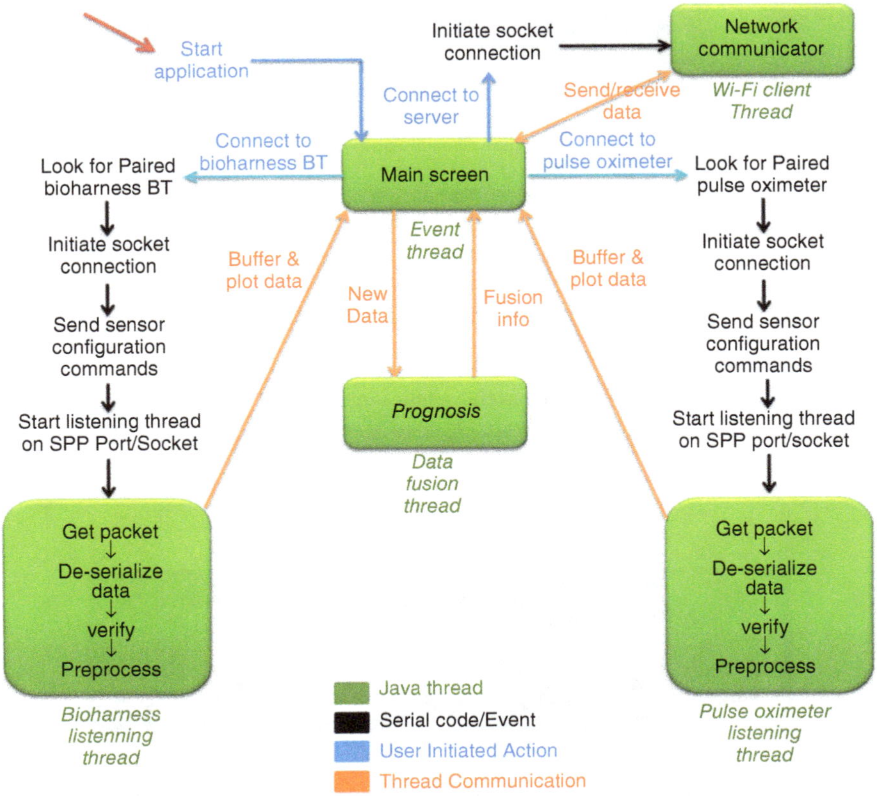

Figure 13.8 J2ME software architecture.

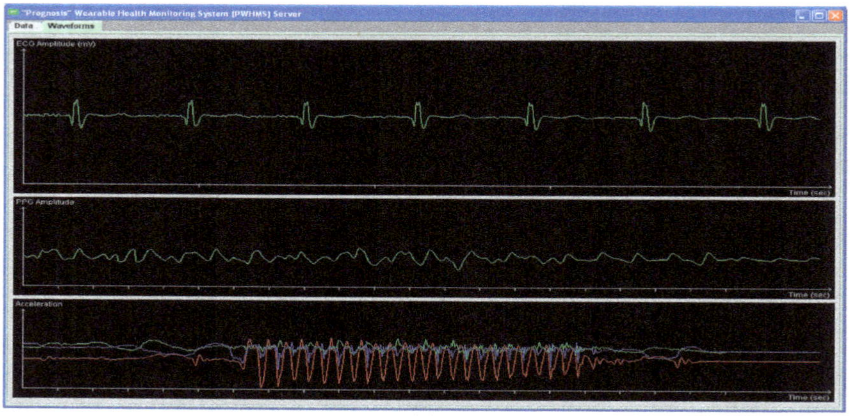

(*a*)

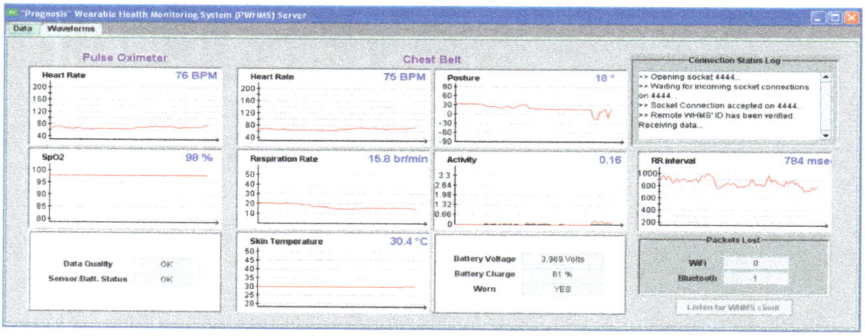

(*b*)

Figure 13.9 Interface of remote monitoring application: (*a*) recorded waveforms and (*b*) data trends.

(a) (b) (c) (d)

Figure 13.19 (*a*) Acquired iris image, (*b*) edge image, (*c*) circular mask, and (*d*) localization of iris.

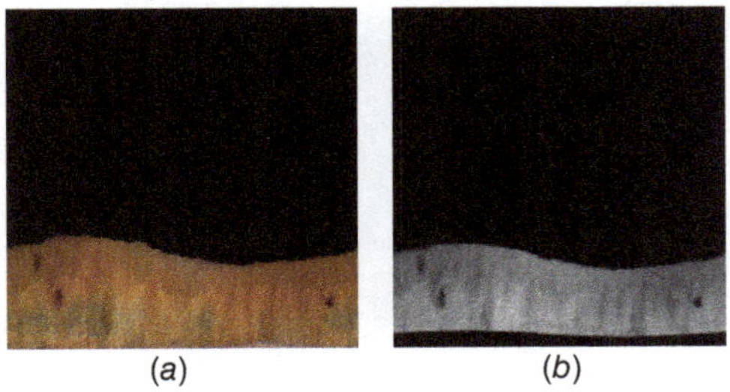

Figure 13.20 (*a*) Log-polar transformation after iris localization. (*b*) Segmentation (based on region labeling) of logpolar image using k-means clustering (R-color space with $k = 51$).

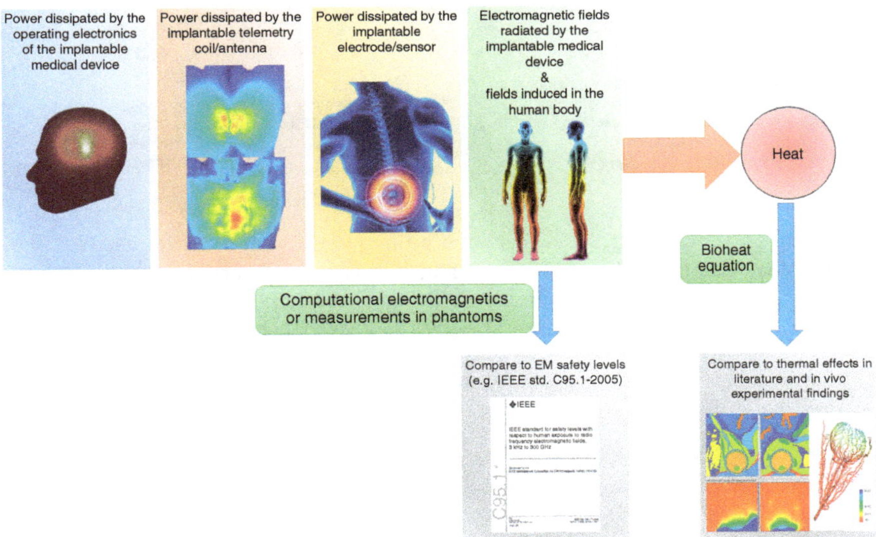

Figure 15.2 Schematic of potential causes of temperature increase in human body associated with operation of invasive devices.

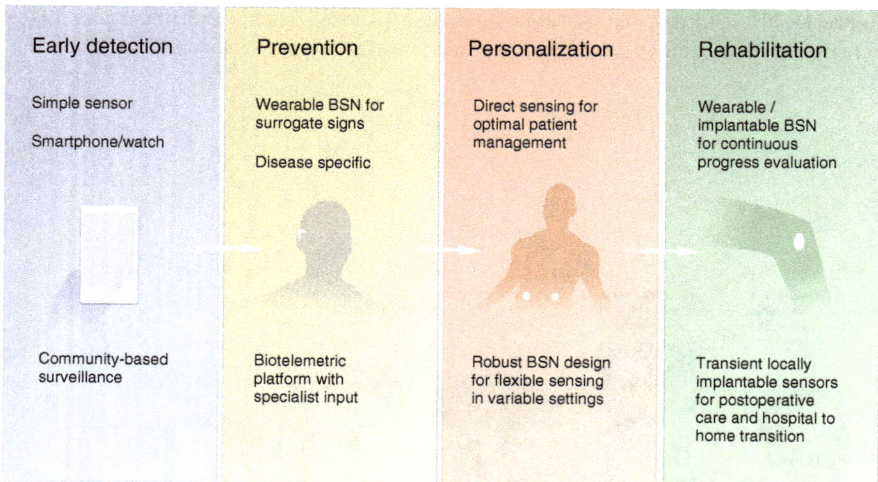

Figure 16.3 Examples of BSN in healthcare pathway. In all user scenarios, the BSN must have a positive benefit–demand ratio. A healthy user requires appropriately pervasive monitoring, preferably through a device already carried. Implementation of a BSN with higher user demand is more acceptable in high-risk subjects as their potential benefit is likely to be greater.

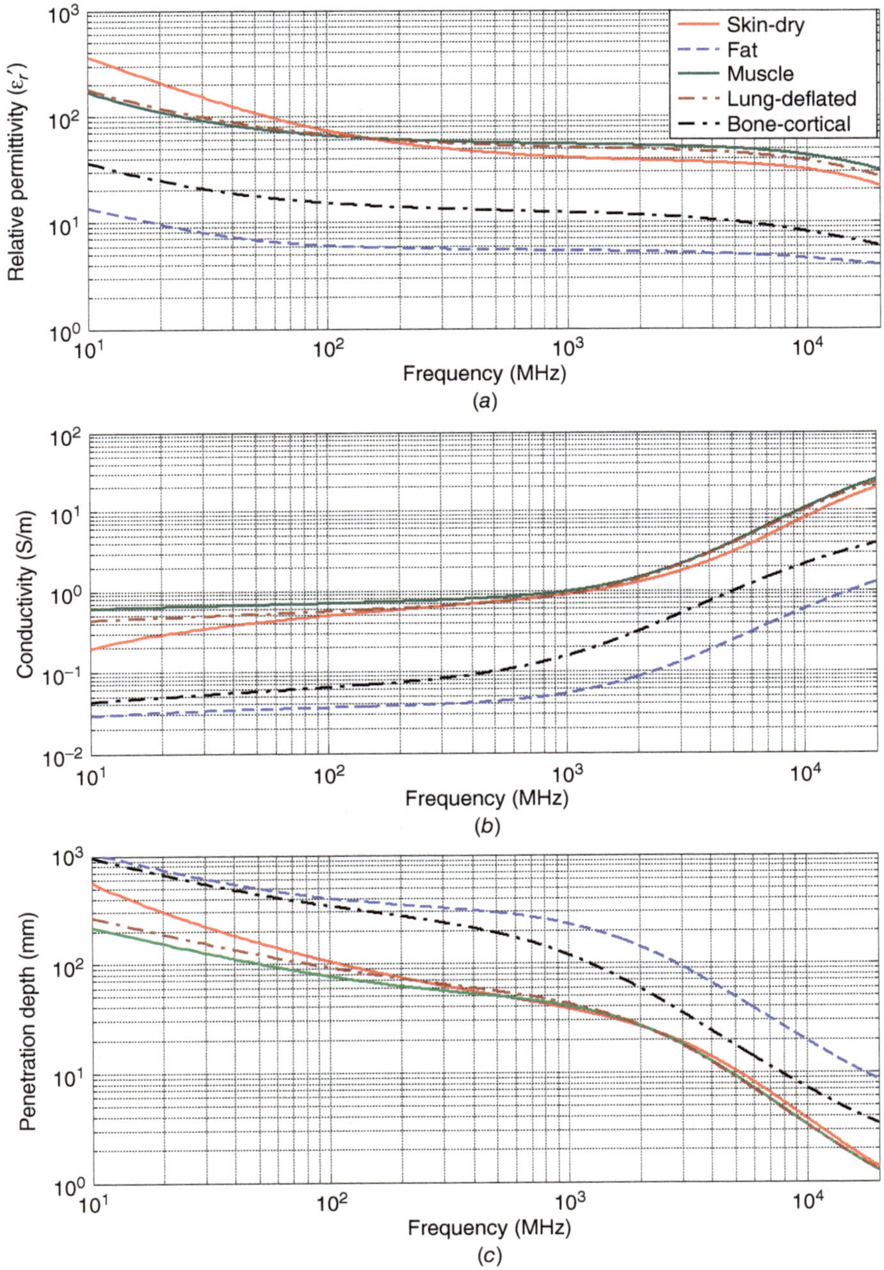

Figure 18.1 Electromagnetic properties of various tissues: skin, fat, muscle, and lung (deflated): (*a*) relative permittivity (real value), (*b*) conductivity, and (*c*) penetration depth in millimeters.

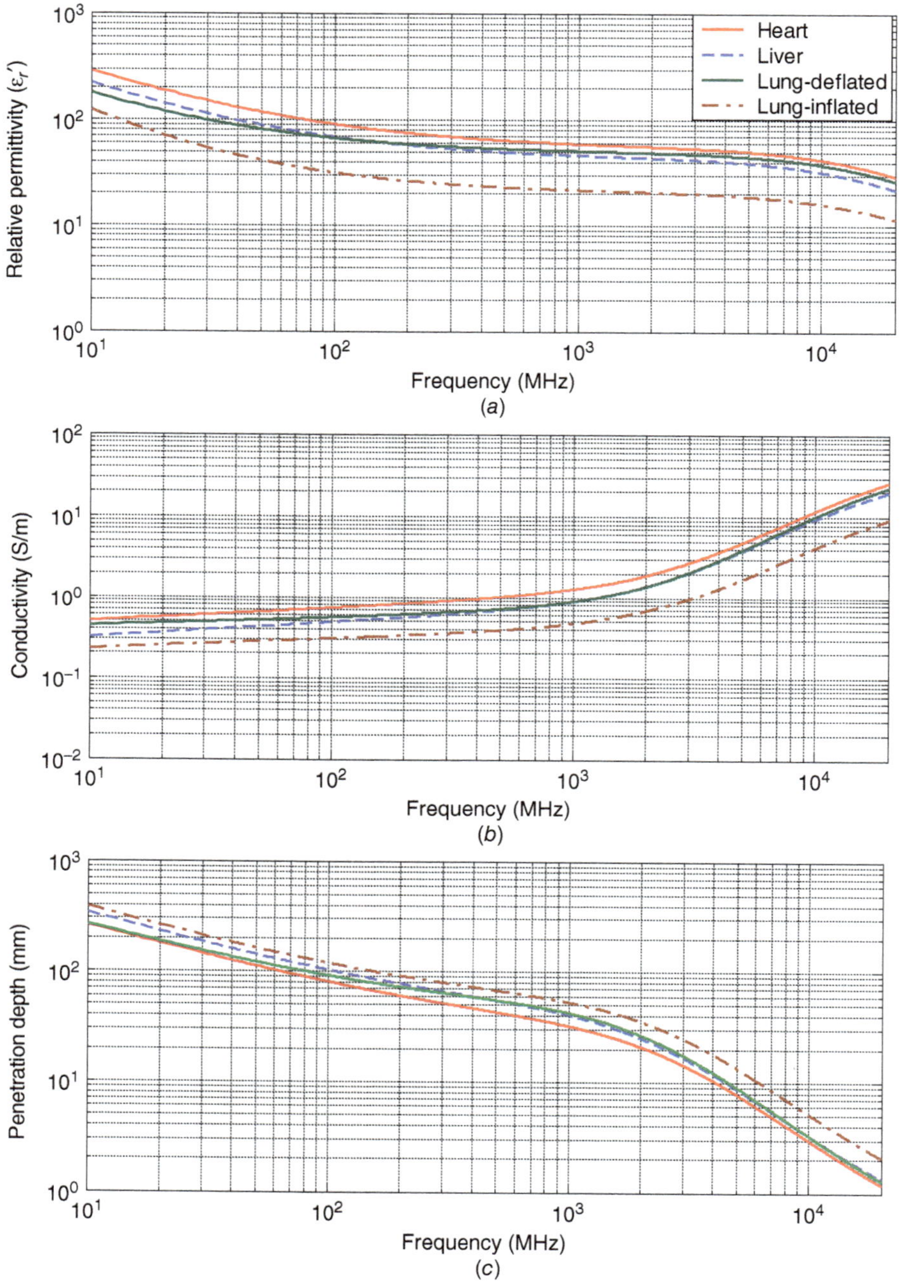

Figure 18.2 Electromagnetic properties of various internal organs: heart, liver, lung (deflated), and lung (inflated): (*a*) relative permittivity (real value); (*b*) conductivity, and (*c*) penetration depth in millimeters.

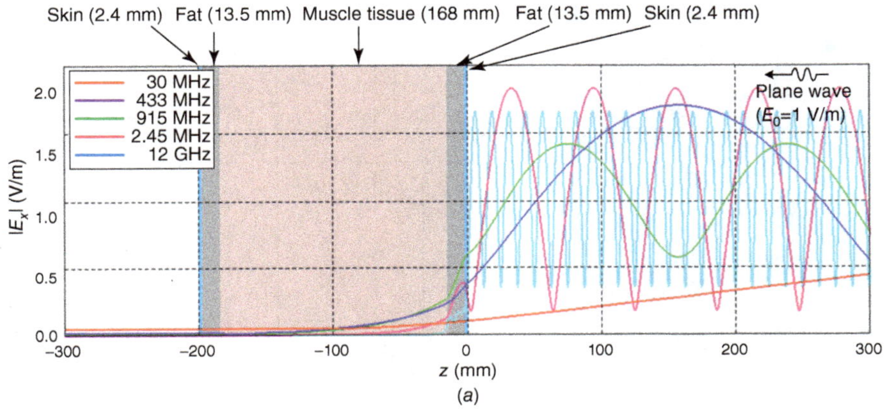

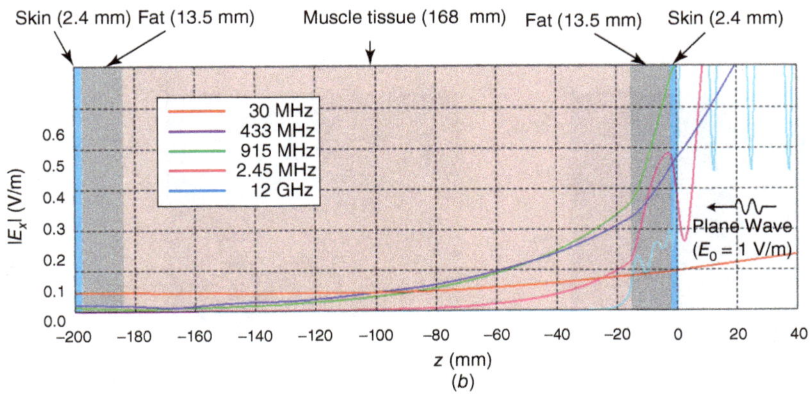

Figure 18.4 Electric field magnitude due to plane wave incidence on the layered tissue model depicted in Figure 18.3 (rotated clockwise by 90°). The electric field, generated using FEKO is sampled along the direction of propagation, normal to the tissue surface: (*a*) holistic view and (*b*) zoomed-in view focusing on human tissue layers.

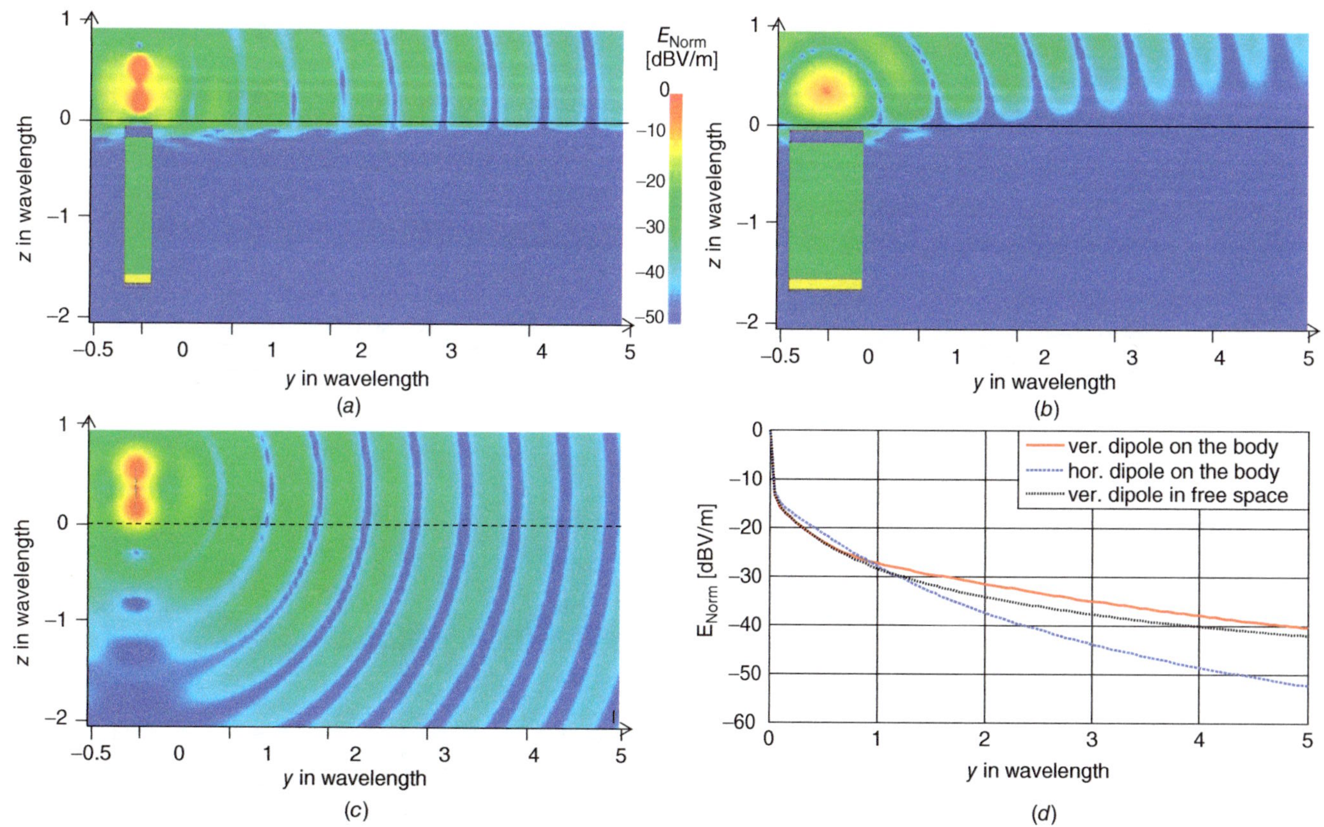

Figure 18.5 Dipole radiation near the infinite layered human tissue model as depicted in Figure 18.3: (*a*) vertical resonant dipole fields, (*b*) horizontal resonant dipole fields, (*c*) vertical dipole fields in free space, and (*d*) normalized field magnitude along a line ($z = 0.4\lambda$) parallel to the body surface.

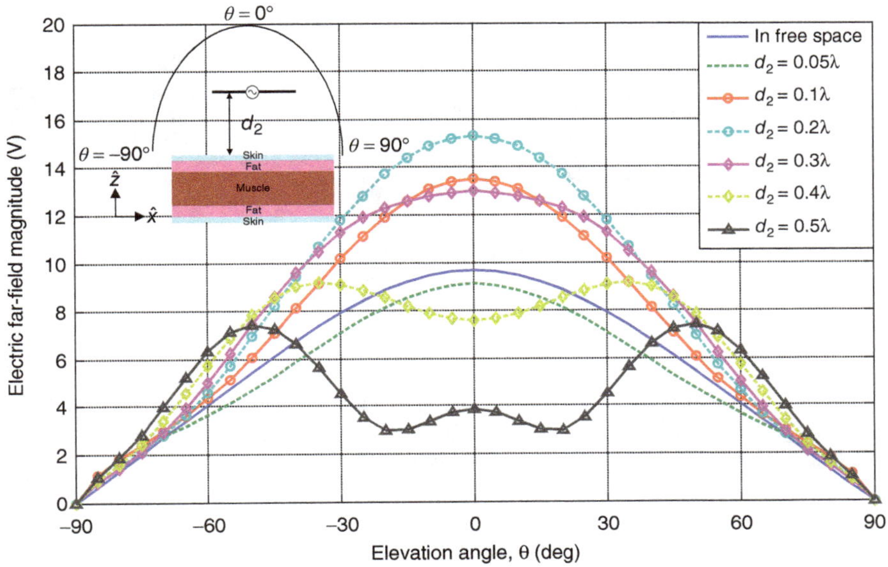

Figure 18.6 Elevation angle far-field pattern for different distances between antenna and human body.

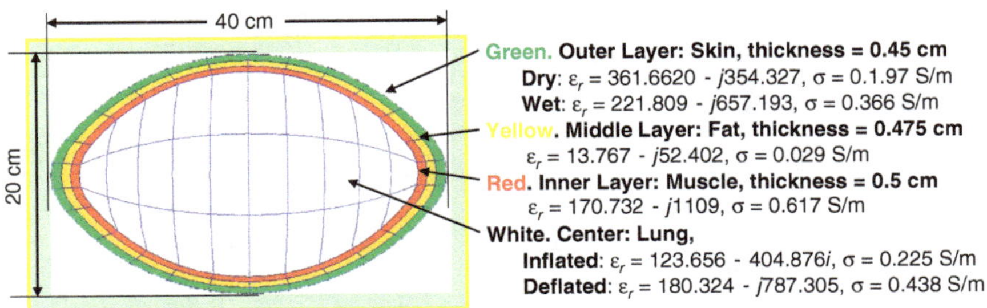

Figure 18.11 Depiction of four-layer torso model and corresponding tissue values for investigating field penetration at 10 MHz. Dielectric constant and conductivity values are from Salman et al. (2012).

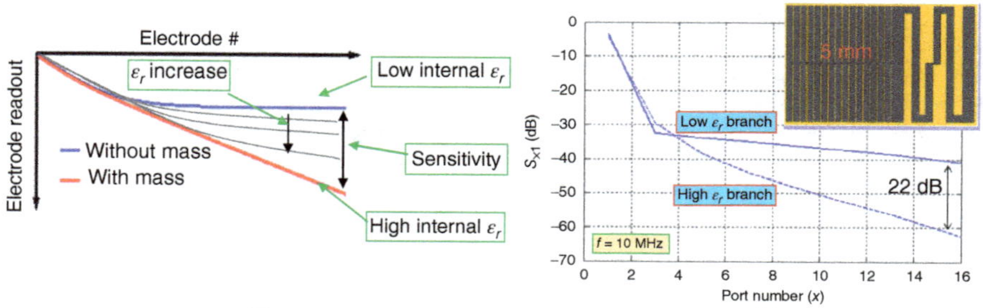

Figure 18.13 (*a*) Graphical representation of the voltage readout changes due to changes in the dielectric of the innermost layer in the four-layer model. The outer layers are assumed to be unchanged. (*b*) Lung sensor sensitivity simulation results when the lung (mass) is inflated or deflated (no mass) using the sensor design given in Salman et al., (2012).

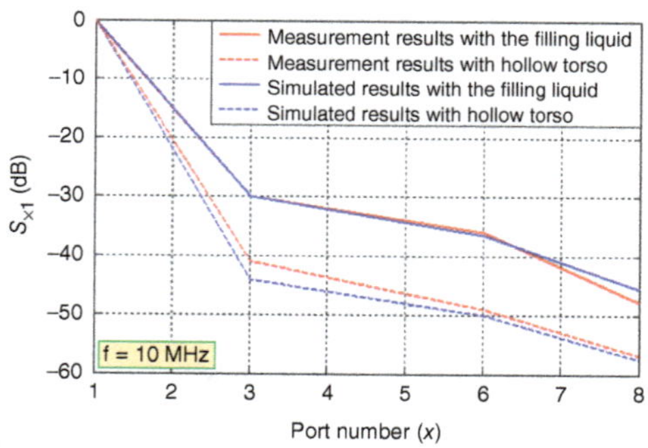

Figure 18.14 Graphic representation of electrode outputs (S_{x1}) showing agreement between simulations and experiment.

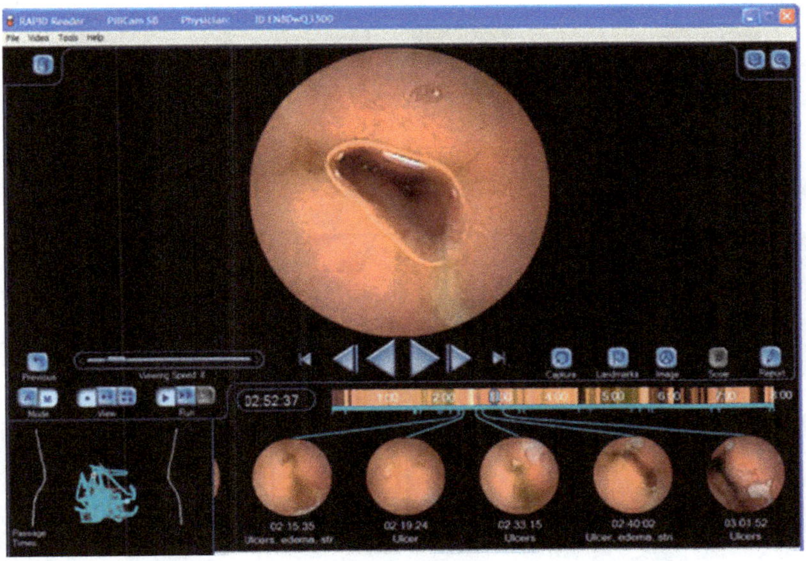

Figure 20.2 RAPID Reader software. (Picture Courtesy of Given Imaging Inc.).

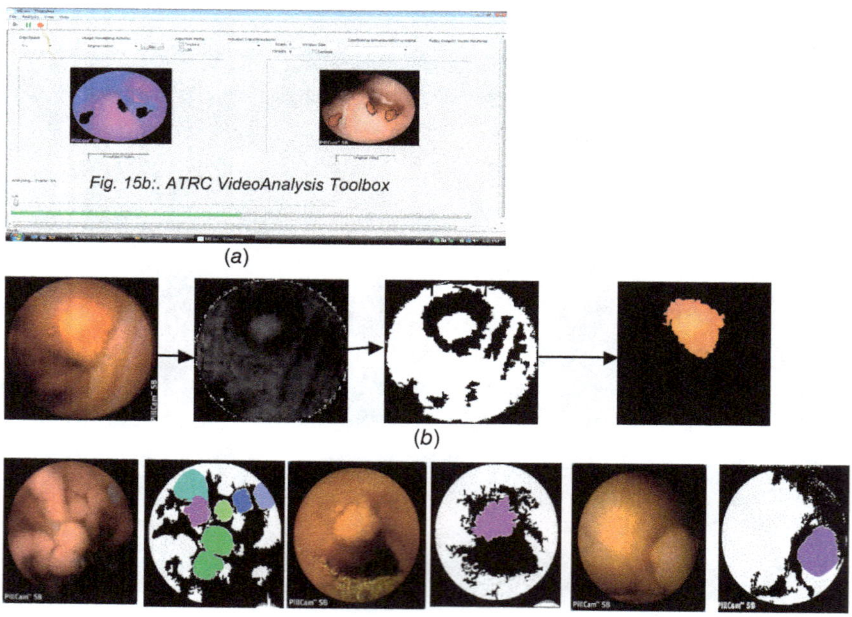

Figure 20.3 Modules for detecting (*a*) bleeding, (*b*) polyps, and (*c*) ulcers in WCE frames.

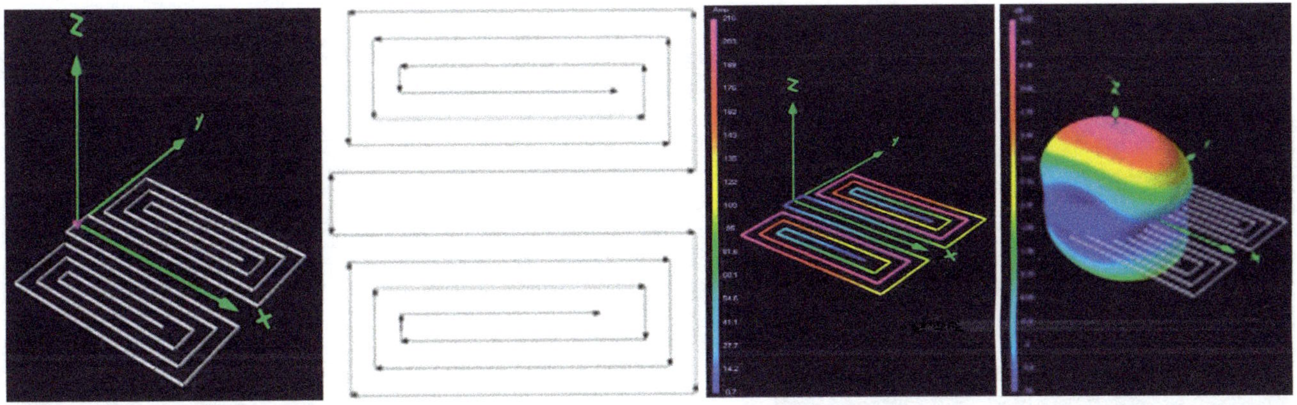

Figure 21.5 Meander antenna structure (2D spiral), current vector alignment, and radiation pattern (Soora, 2005; Soora et al., 2008).

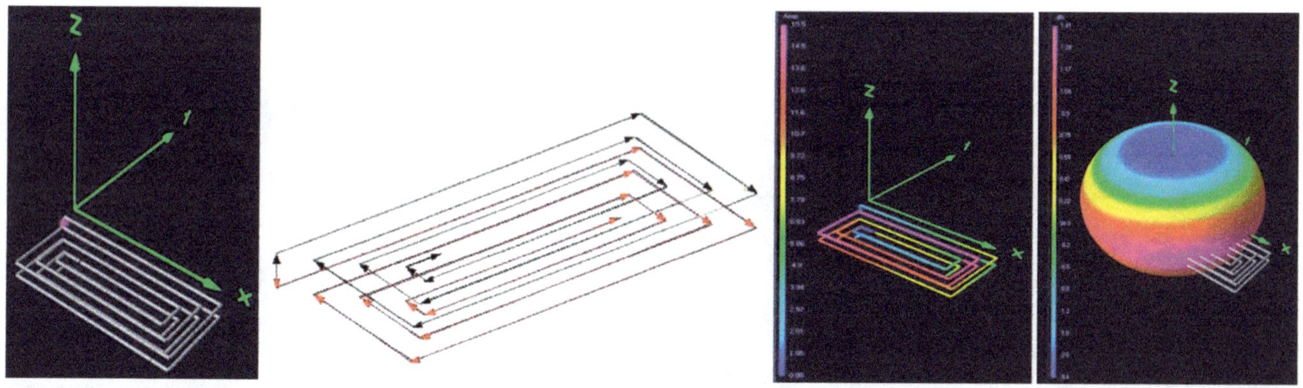

Figure 21.6 Meander antenna structure (3D spiral, folded), current vector alignment, and radiation pattern (Soora, 2005; Soora et al., 2008).

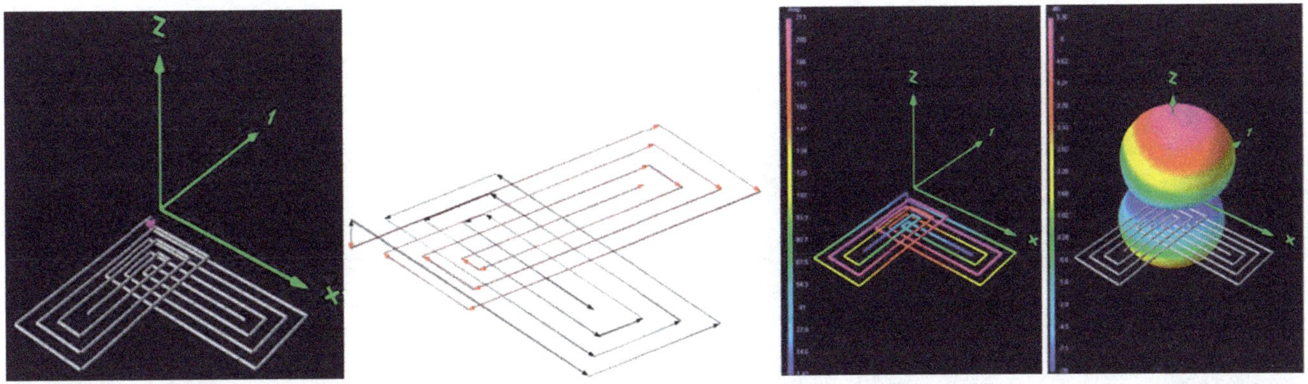

Figure 21.7 Meander antenna structure (3D rotated), current vector alignment, and radiation pattern (Soora, 2005; Soora et al., 2008).

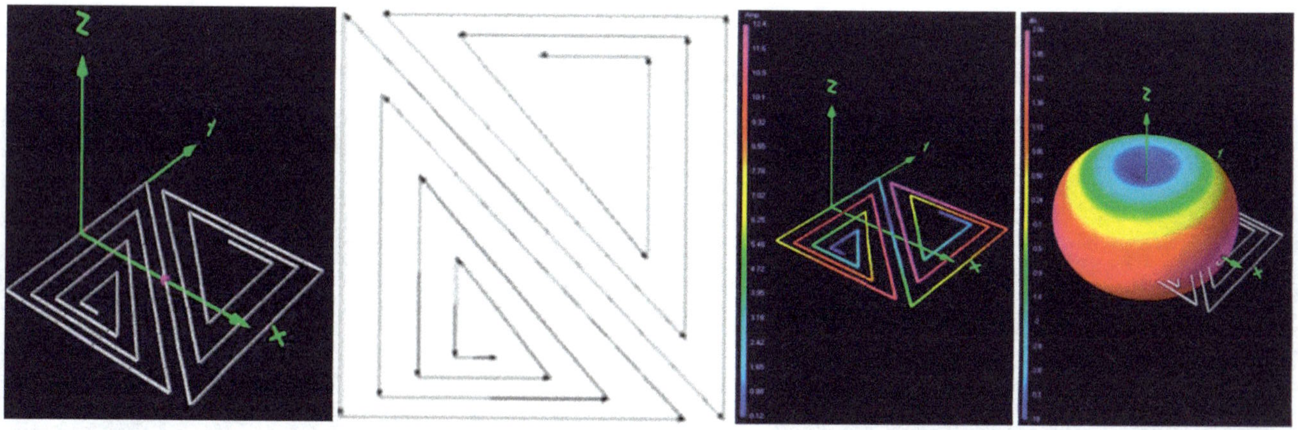

Figure 21.8 Meander antenna structure (2D triangular spiral), current vector alignment, and radiation pattern (Soora, 2005; Soora, et al., 2008).

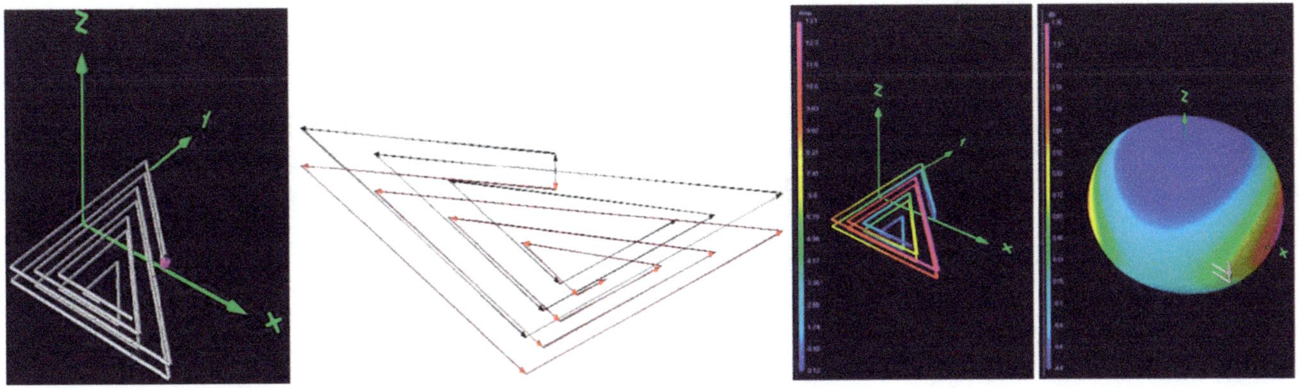

Figure 21.9 Meander antenna structure (3D triangular spiral, folded), current vector alignment, and radiation pattern (Soora, 2005; Soora et al., 2008).

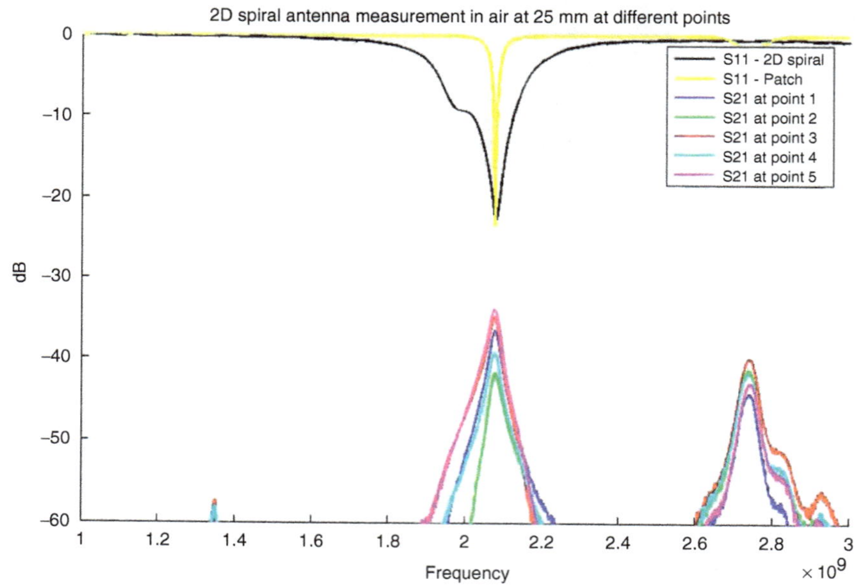

Figure 21.16 2D spiral meander antenna performance (coupling) at test points (Soora, 2005).

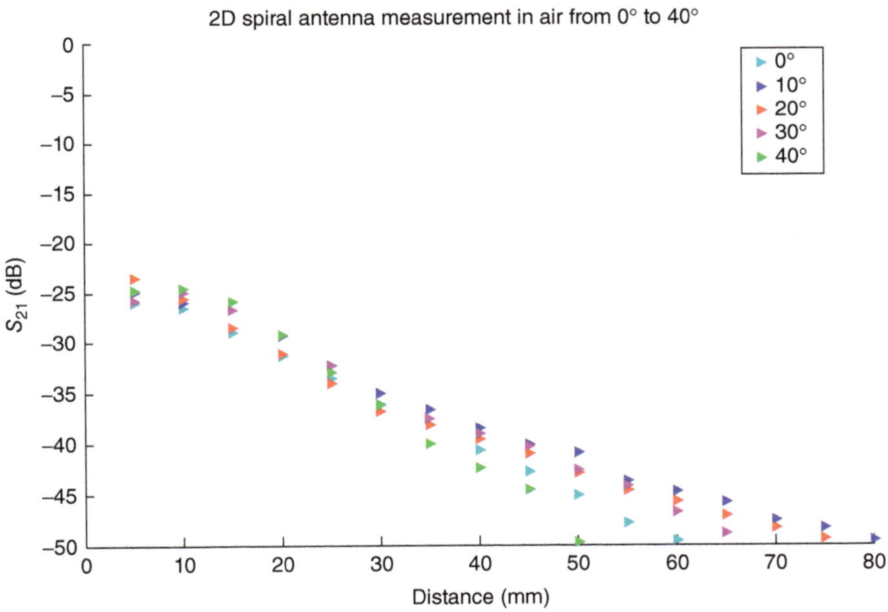

Figure 21.17 2D spiral meander antenna coupling as function of eye rotation (Soora, 2005).

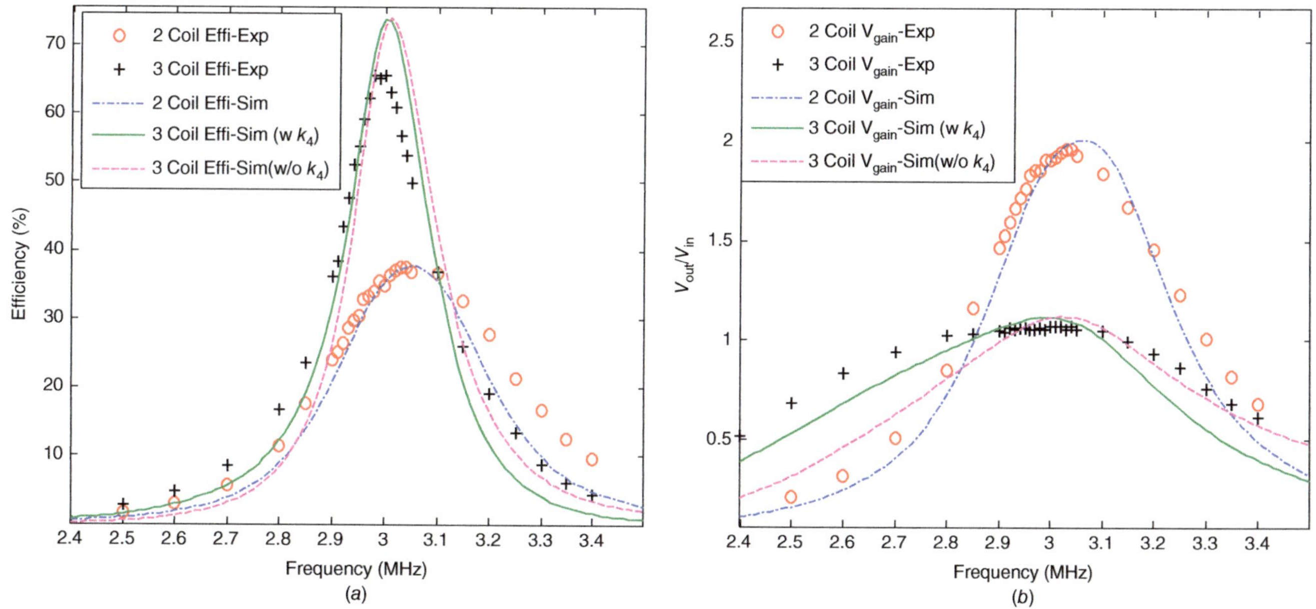

Figure 21.20 Simulated (Sim) and experimental (Exp) (*a*) power transfer efficiency and (*b*) voltage gain of two-coil system and its three-coil equivalent system with respect to frequency for design 2 in Table 21.6. Simulation is done with (w k_4) and without (w/o k_4) the consideration of coupling between driver and load coils k_4. [(From RamRakhyani and Lazzi (2012).]

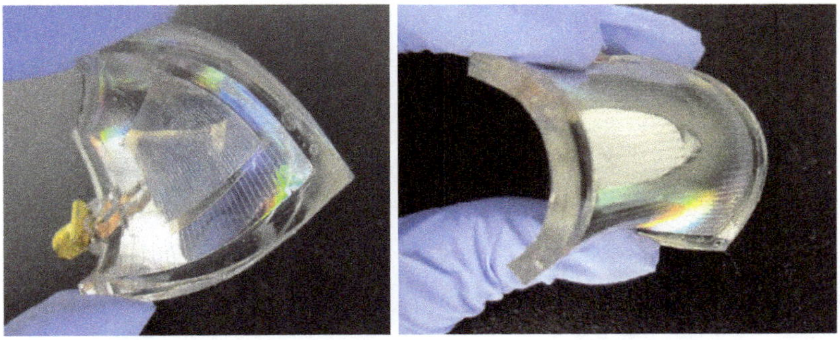

Figure 21.21 Photographs of multilayer, flexible microstrip patch antenna composed of liquid metal encased in elastomer.

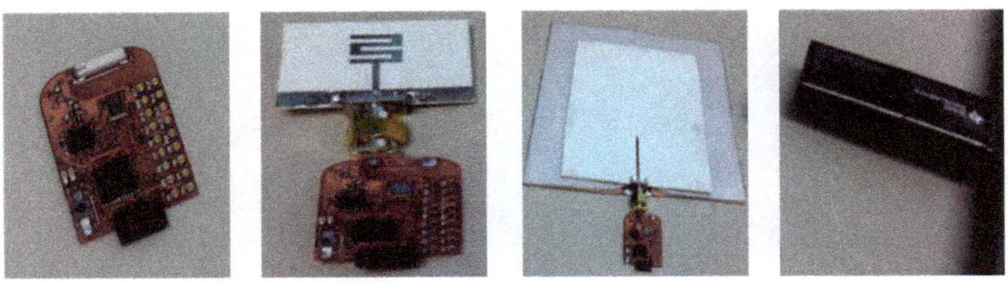

(*a*) Chip antenna[3] (*b*) Inkjet-printed meandered monopole (*c*) Monopole with EBG ground (*d*) Receiver

Figure 23.15 Communication range measurement setup.

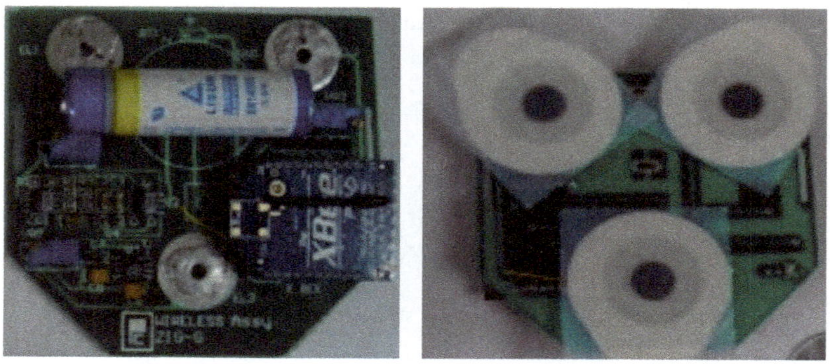

Figure 23.44 Wireless EKG transmitter.

10

OPTICAL BIOTELEMETRY

KOICHI SHIMIZU
Graduate School of Information Science and Technology, Hokkaido University, Sapporo, Japan

10.1 INTRODUCTION

The invention of laser and optical fiber has brought great progress in optical technologies. Particularly, the rapid progress in optical communication technology has promoted the advancement of many engineering techniques. However, the application of optical communication methodology to the field of biotelemetry has not been common. In most telemetry applications, the radio wave has been used as the communication medium (e.g., Mackay, 1993). We noticed the significant potential of the optical technology for biotelemetry and have introduced the optical techniques to various areas of biotelemetry (Shimizu, 2005).

In biotelemetry the measured signal is transmitted to a remote place by means of the transmission medium for communication. A radio wave is a typical example of the medium. Since it has been used widely, people naturally think of it as radio telemetry when they say biotelemetry. The use of different media for data transmission has been attempted, that is, light and ultrasound. The feasibility and the usefulness of optical biotelemetry have been suggested early (Shimizu, 1980; Kimmich, 1982). However, few studies have followed them. It is one of the reasons that optical technology has not matured to support the development of optical biotelemetry. For example, a high-power transmitter and a sensitive detector could not be made portable. Over

Handbook of Biomedical Telemetry, First Edition. Edited by Konstantina S. Nikita.

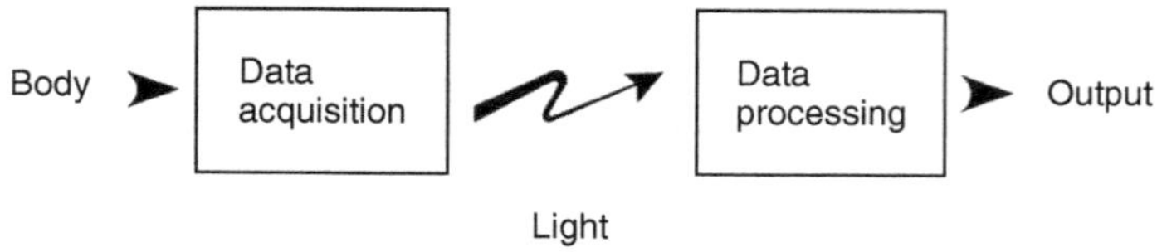

Figure 10.1 Basic principle of optical biotelemetry.

the last few decades optical technology has advanced greatly, and now scientists have a reliable laser diode and photodiode at hand.

Figure 10.1 shows the fundamental principle of optical biotelemetry. Based on the telemetry principle, the data acquisition part is separated from the data processing and display parts by means of optical communication. By this separation the following merits of biotelemetry are expected:

1. The data acquisition part can be made very small and light. This makes the measuring procedure possible with minimal perturbation on the object or the subject. The ambulatory measurement is a typical example.
2. The signal can be measured in an environment which is not easily accessible, such as inside the body or deep in water and space.
3. Many subjects can be monitored for a long period of time in the centralized system, such as an intensive care unit (ICU).

Moreover, by using the light the following advantages over conventional radio telemetry are expected.

(i) The wide-band signal transmission can be relatively easily achieved. This corresponds to the possibility of information transmission with a good frequency characteristic or a high-speed temporal response. In another view, it indicates the possibility of high-speed large-capacity transmission of information.
(ii) The EMI can be controlled much easier than with radio telemetry. The signal light of the optical telemetry does not interfere with other instruments. The shielding and securing safety can be done relatively easily.
(iii) The legal restriction to use light for communication is not as strict as the radio wave.

There are of course some disadvantages with optical biotelemetry: for example, the vulnerability to optical noises and the inability to reach a position out of the line of sight. However, they are not serious and can be overcome by technical measures which are discussed in the following sections.

The significance of the above advantages in biotelemetry are as follows. The first advantage implies the improvement of the telemetry function itself over radio telemetry. It increases the practical value of the telemetry technique. With the wide-band transmission channel, we can use various intelligent techniques in the data transmission. The significance of this advantage over radio telemetry is increasing with the progress of intelligent communication techniques. The second advantage is important for the serious problems of EMI in a hospital. In a clinical environment, there are many sources of electromagnetic noise, such as an electric scalpel (radio knife, acusector), microwave hyperthermia, and many devices with built-in microcomputers. Thus, the telemetry system should not receive interference from them. Moreover, in a clinical environment, delicate equipment, such as life support systems and centralized patient monitoring systems, have increased. The trouble and malfunction of these systems result in serious problems. Therefore, there are strict requirements to prevent a telemetry system from interfering with these systems. The third advantage becomes an important merit in the practical use of the telemetry system in a clinical environment and in field works.

As mentioned above, optical biotelemetry is a new technique with abundant potential. However, the basic technology has not been sufficiently established compared to radio telemetry. This chapter deals with the outline of the basic technologies necessary in order to use optical biotelemetry in practice. The common topics to radio telemetry such as the topics of electrodes and batteries are excluded. As concrete examples, some of the optical biotelemetry systems we have developed are introduced.

10.2 OPTICAL TECHNOLOGY FOR OPTICAL BIOTELEMETRY

Optical biotelemetry is an integrated technology which consists of basic technologies such as optics, communication, biosignal measurement, and information processing. In the following sections, these basic technologies necessary for optical biotelemetry are outlined.

10.2.1 Selection of Wavelength

To design an optical biotelemetry system, the wavelength of optical communication has to be selected. Any wavelength of light, or the ultraviolet–visible–infrared range, can be used for optical telemetry in principle. However, with ultraviolet light, the adverse effect on our health becomes a problem for a prolonged use. As for visible light, there may be a psychological effect on the subject. Further, in this wavelength, there are many sources of optical noise such as indoor lighting. With infrared light particularly of 700–1200 nm wavelength, such problems are few. Moreover, in this wavelength, the optical absorption of body tissue is relatively low, and we can expect high optical transmission through our body. Figure 10.2 shows the absorption spectra of the major constituents of a human body. It shows the relatively low absorption of light in the near-infrared range, or 700–1200 nm wavelength. In the ranges less than 700 nm and more than 1200 nm, the absorption of hemoglobin and water are predominant, respectively. Furthermore, this wavelength range has been used in optical fiber communication, and there are many kinds of sources and detectors available.

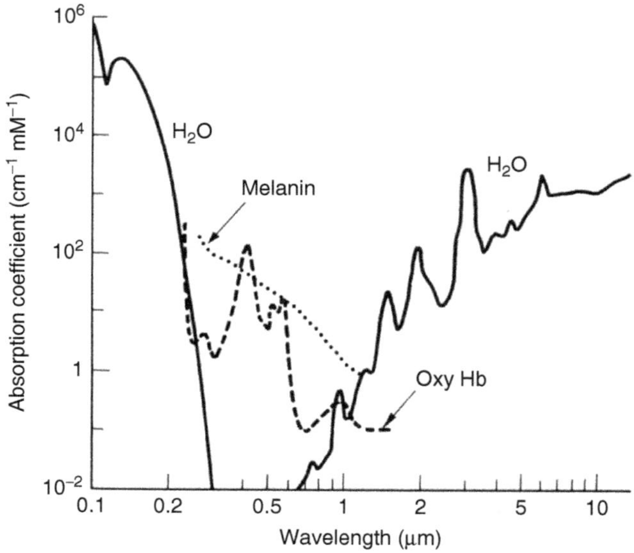

Figure 10.2 Absorption spectra of major constituents of human body.

10.2.2 Light Source

A basic part of the optical signal transmission consists of a light source and a photoreceiver. As a light source, a light-emitting diode (LED) and a laser diode (LD) are used, generally. Semiconductor elements are small, light weight, and operable with low voltage and with low power consumption. With each element, direct modulation is possible as well. If we compare them, each has different merits in practical use. The LED is generally inexpensive and easy to handle. The LD is superior in monochromaticity, coherency, and response speed. These merits of the LD become important in optical communication through an optical fiber and in the case of high-frequency modulation (more than several hundred megahertz). However, these features are not necessarily required for the transmission of general biosignals such as an ECG (frequency range of zero to several hundred hertz). In such a case, LEDs are often used.

When a large amount of optical power is required,such as in the optical communication using indirect scattered light, many elements of the light source are used. To drive many elements simultaneously, the elements can be connected in series or in parallel at the last stage of a driver circuit. They are shown in Figures 10.3*a* and 10.3*b*. Since a current-source drive is a standard method with LED, the series connection is preferable to make the light emission of each LED uniform. Moreover, the composite capacitance of LED becomes small in the series connection. It is advantageous when high-speed operation such as a pulse-wave drive is required. However, in the series connection the number of elements (n) cannot exceed a certain limit. This is because the forward voltage V_F times n cannot exceed the power-supply voltage V_{CC}. As a

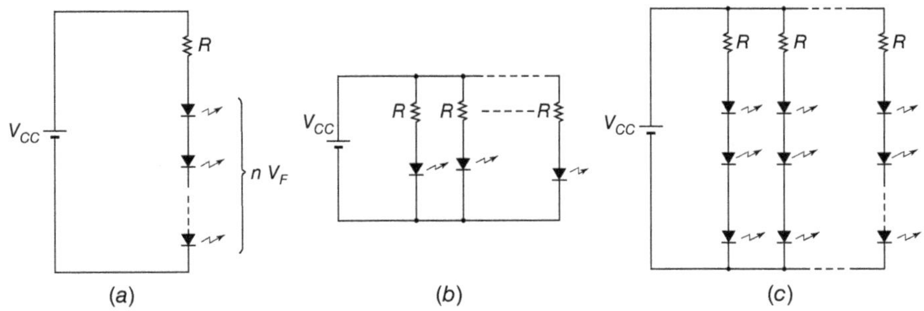

Figure 10.3 Driving of multiple light-emitting elements: (*a*) series, (*b*) parallel, (*c*) mixed; V_{cc} = source voltage, n = number of elements, V_F = forward voltage.

result, we often need to divide the series connection into some parallel connections of short series elements as shown in Figure 10.3*c*.

10.2.3 Light-Detecting Elements

A light detector forms the front end of the receiver of data transmission. As the detector, a photodiode (PD), a photo transistor (PTr), an avalanche photodiode (APD), or a photomultiplier (PMT) is commonly used. To detect a low-level light at a photon counting level, the detector with an internal gain such as an APD and PMT is used. The PTr is used as a photo-receiving part of such a composite element as a photocoupler and a photointerrupter. In optical biotelemetry, the PD is often used since it is easy to handle. Particularly a PIN photodiode can meet the requirement of high-speed operation (typically from some tens of megahertz to some gigahertz) relatively easily. Therefore, it is suitable for a wide-band digital data transmission system.

In optical telemetry, it is necessary to amplify the electric signal from the PD and to process the signal in a wave-shaping circuit and a demodulation circuit. When high-sensitivity and high-speed operation is required, the performance of the whole receiver is often dominated by the first part of the receiver circuits rather than by the detector element itself. To solve this problem, special devices have been developed. They include the PD with built-in operational amplifiers as a preamplifier and the OPIC (optical IC), an IC with a photodiode and signal processing circuit integrated on a single chip of IC.

10.2.4 Measures for Optical Noises

In practical applications of biotelemetry, the measurements are particularly vulnerable to the foreign noise coming into the communication channel. Since optical telemetry uses light as the carrier of communication, it shows extremely high resistance to electromagnetic noise. However, it is certainly vulnerable to the effects of optical noise. To deal with this problem, two kinds of filters, an optical filter, and an electric filter, are used. The noise source with the largest effect in a room is a fluorescent

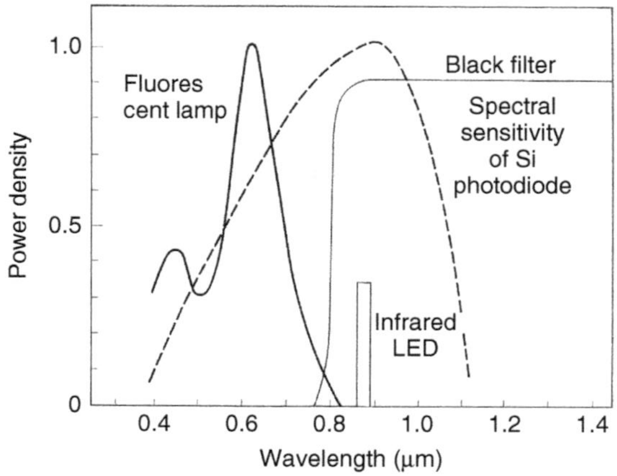

Figure 10.4 Spectral characteristics of optical elements.

lamp. As shown in Figure 10.4, the major emission spectrum of a fluorescent lamp lies between 400 and 700 nm. Therefore we can greatly suppress its effect using the optical filter, which cuts the wavelength range shorter than a visible light. The filter is often called a black filter.

As for sunlight and the light of an incandescent lamp, we can eliminate a considerable part of the effect using an electronic filter. After the photoelectric conversion at the photodetector, the frequency component of these lights is well defined. They are either very low frequency near DC or some harmonics of the power line frequency (50 or 60 Hz). Thus, it is not difficult to eliminate them by an electronic filter. However, we have to be careful not to saturate the photodetector in any case.

10.3 COMMUNICATION TECHNOLOGY FOR OPTICAL TELEMETRY

Figure 10.5 shows a basic flow of the signal transmission in optical biotelemetry. A special consideration to optical biotelemetry is necessary in the part from the modulator to the demodulator. In this section, the communication technology concerning this part is described.

10.3.1 Analog/Digital Transmission

Most of the biomedical information can be measured as an analog signal. A conventional medical telemeter such as an ECG telemeter transmits analog information. In analog transmission the electronic circuit is generally simple. It is important to make the telemeter small in size and low in power consumption.

With the progress of digital technology and components, digital techniques have often been used in biotelemetry. It may sound meaningless to convert the analog biological signal into a digital one and then back into an analog signal at the receiving

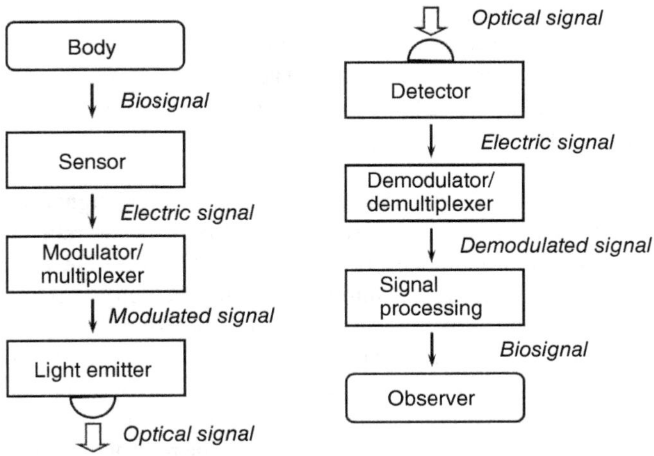

Figure 10.5 Signal flow in optical biotelemetry.

end. However, the advantage of the digital signal transmission is particularly significant in the intelligent data transmission, as mentioned in the following Sections.

10.3.2 Modulation Method

In conventional radio telemetry, a continuous-wave modulation (AM, FM, etc.) and a pulse wave modulation (PPM, PIM, PCM, etc.) were used (Shimizu, 1991). Commonly, in optical biotelemetry, we first modulate the electric signal using these modulation methods and then modulate the light intensity by the modulated electric signal. Thus, we express the modulation method as FM/IM (frequency modulation/intensity modulation) or PPM/IM (pulse position modulation/intensity modulation) in optical telemetry.

To choose an appropriate modulation method, the characteristics of the biological signal, the multiplexing method, and other practical factors have to be taken into account. In many of our applications, the PPM/IM and PIM/IM (pulse interval modulation/intensity modulation) have been used. With pulse modulation, we can drive LEDs in higher luminance. With PPM and PIM, we can make the power consumption of the transmitter low by reducing the duty ratio of the light pulses.

10.3.3 Toward Intelligent Transmission

When we use light as a transmission medium, high-speed data communication becomes possible. Its great advantage is that we can use various techniques of intelligent communication. We can call this telemetry with intelligent communication, the telemetry of the next generation. It is compared with the conventional telemetry in which the measured biological signal is transmitted in the modulation signal of one-to-one correspondence.

As for the intelligent communication, the progress of the portable terminal which collects and displays a biological signal is remarkable. However, this chapter deals with the intelligent signal transmission itself. There are two levels of intelligent communication, that is, the level of a biological signal and that of a carrier signal. In the former case, the biological signal is processed using information processing techniques.

For example, the various redundancies such as noises and signal repetitions are eliminated in the transmission side (Tsuda et al., 1993). In the receiving side, the necessary features of the biological signal are extracted from the transmitted signal. In the latter case, the communication part of the telemetry is made intelligent. It includes the error control techniques in data transmission. Typical examples of the error control techniques are the ARQ (automatic repeat request) and FEC (forward error correction). In both cases, some redundancy to the signal to be transmitted is added. Then, at the receiving side the error caused in the transmission process is detected. With the ARQ, the system keeps requesting the repeated transmission of the data until no error is detected at the receiving side. With the FEC, the error is automatically corrected using the special code embedded (inscribed) in the transmitted signal (Shimizu, 1991).

In such error control processes, some duration of the data processing time is required for the error correction. Generally, the longer is the time required, the more complicated is the error control, and thus the reliability of data transmission is increased. If this processing time exceeds the temporal capacity of the transmission system, the real-time operation is hindered. In biotelemetry, the real-time operation is the most crucial. Therefore it is necessary to devise the error control method to secure high reliability while meeting the demand for the real-time operation. The balance between these contradicting requirements has been taken into account, for example, by limiting the number of repeated retransmissions in ARQ or by using a simple logic FEC such as a majority rule (Murakami et al., 1994).

It should be noted that the difficulty of this problem has been gradually reducing due to the remarkable progress in the data processing algorithm and the data transmission speed.

10.3.4 Multiplexing Method

In optical biotelemetry, multichannel data transmission is often required due to its large transmission capacity. To answer this, a multiplexing technique is necessary. The multiplexing techniques in optical communication are categorized largely into space division multiplexing (SDM), wavelength division multiplexing (WDM), time division multiplexing (TDM), frequency division multiplexing (FDM), and code division multiplexing (CDM). Each has the following different characteristics and is chosen according to the purpose and the usage.

SDM uses the spatial difference in the light propagation path. It is effective for the telemetry using direct light transmission or optical fiber transmission. With the indirect light transmission, the light path has to be separated by shading material, and the constraints in practical use are sometimes severe.

With WDM, the channel is separated by the difference of the wavelength of the transmission light. It can be realized relatively easily using light sources with different wavelengths and optical filters. It can be used in combination with other multiplexing techniques. Generally, it is difficult to have a large number of multiplexed channels due to the practical restrictions such as the monochromaticity of light source, the trade-off between the filter bandwidth and the signal-to-noise (SN) ratio of an optical signal, and so on. With the indirect light transmission, the effective wavelength separability of the interference filter cannot be fully utilized, since the orientation of the light incidence to the optical detector is not constant.

With TDM, multiple channels are allocated on the time axis of the optical signal. With the recent rapid increase in the data transmission speed, the multiplexing of many channels has become possible. To perform efficient multiplexing in telemetry, the synchronization between multiple transmitters and receivers is necessary. For wireless synchronization, a two-way transmission is necessary and the receiving function is required for the apparatus equipped on a subject as well as the transmitting function. In biotelemetry, there is a fundamental demand to make the apparatus equipped on a subject as small as possible. Thus, the requirements of the synchronization often make it difficult to satisfy this demand.

With FDM, multiple channels are differentiated by the difference in the frequency of intensity modulation of carrier light. This can be done relatively easily. As mentioned above, a pulse modulation method is often used in optical biotelemetry. In the pulse modulation, the number of multiplexed channels is limited due to the higher harmonics of the pulse shape. With indirect light transmission, the effect of multipath transmission becomes nonnegligible when the modulation frequency becomes large. Therefore, the bandwidth of signal transmission is limited.

With CDM, we encode the electric signal that modulates the light source and differentiate multiple channels by the code. The spread-spectrum technique is a typical example. This technique is suitable to the optical communication with which a broad-band communication is possible. This technique has much useful potential such as strength against foreign noises and multipath propagation. However, at present, some problems in practical use remain, such as the complexity of the circuits in both a transmitter and a receiver.

10.4 PROPAGATION OF OPTICAL SIGNAL

For optical biotelemetry, it is necessary to understand the propagation of the optical signal in various conditions. This chapter summarizes the optical propagation in transcutaneous telemetry and ambulatory telemetry.

10.4.1 Optical Characteristics of Body Surface Tissue

In transcutaneous signal transmission, the absorption and the scattering characteristics of body tissue are important. The optical absorption characteristics of biological tissue largely depend on the wavelength of light. Figure 10.6 shows the light attenuation in body surface tissues. The exponential decay is approximated to be linear. This

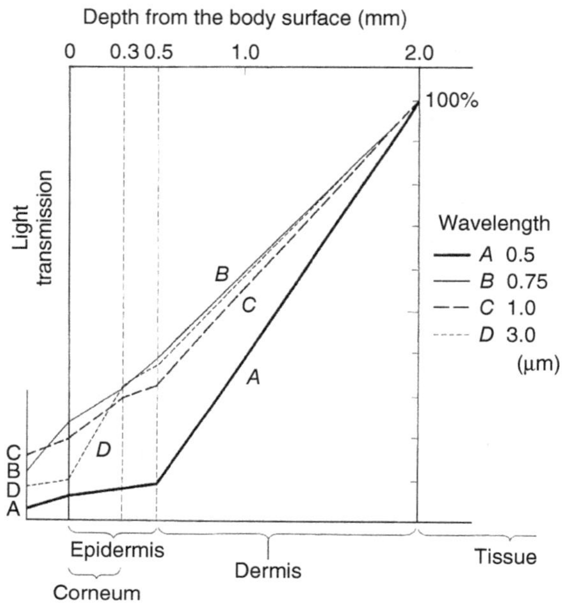

Figure 10.6 Attenuation of light in human surface tissue.

shows how the optical signal from the hypodermic (2 mm depth) light source is attenuated as the light propagates to the skin surface (Kudo et al., 1988). With near-infrared light, 10–20% light transmission is expected through the 2-mm skin tissue.

An example of the measured scattering characteristics of the skin is shown in Figure 10.7. This is the case of a mouse, where the effect of relatively thick white fur is obvious (Kudo et al., 1988). Even with the complex emission-pattern of LED in the air, the angular pattern of light transmitted through the skin becomes fairly uniform in all directions. This is due to the strong scattering at the body surface tissue, and is also expected with human skin with less hair. This orientational uniformity of signal transmission is a significant merit in biotelemetry in which the subject moves around freely. Other optical properties of mammalian tissue can be found elsewhere (Tuchin, 2000; Vo-Dinh, 2003).

10.4.2 Distribution of Indirect Light in a Room

In a closed space like a room, ambulatory telemetry is possible using the indirect light scattered from the ceiling, the wall and the floor, and so on. To understand the behavior of the indirectly scattered light, the spatial distribution of received light intensity is investigated in a computer simulation (Kudo et al., 1987).

Figure 10.8*a* shows the conditions of the simulation. A light emitter with a hemispherical emission pattern toward the ceiling is placed 1.5 m high (a shoulder height of a human adult) above each of the floor mesh points. The light intensity received by

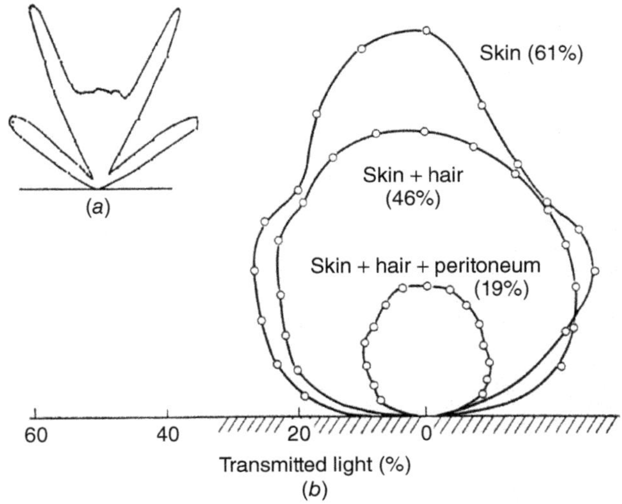

Figure 10.7 Scattering property of animal skin: (*a*) radiation pattern of LED in air (wavelength 890 nm); (*b*) radiation pattern of transmitted light through mouse skin (values are maximum transmission rates).

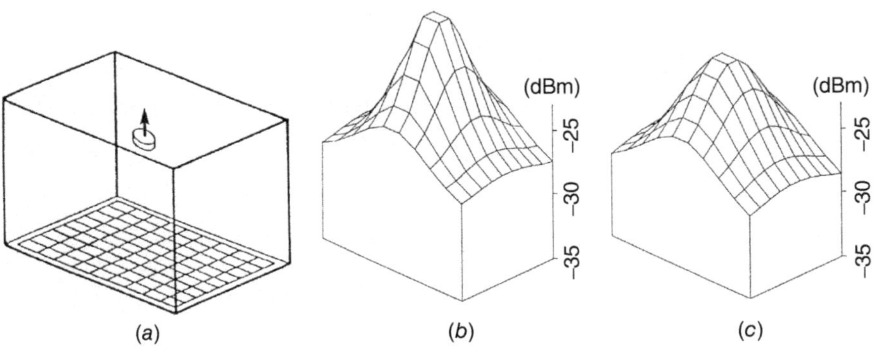

Figure 10.8 Conditions and results of simulation of indirectly scattered light. (*a*) Dimension of room 6.4 m width, 4.2 m long, 2.7 m high; reflection coefficient of inner surface of room 0.7 (Lambertian surface), optical power of transmitter 1 W; receiving area of receiver 10 × 10 mm^2; directivities of transmitter and receiver are both cosine. (*b*) Intensity distribution of signal transmission with receiver facing down toward floor (max −17.0 dBm, min −27.5 dBm). (*c*) Intensity distribution of signal transmission with receiver (hung 0.6 m down from ceiling) facing up toward ceiling (max −20.4 dBm, min −28.6 dBm).

the detector which is placed at the center of the ceiling is calculated according to the position of the emitter. The positions of the emitter are set at each of the floor mesh points. The light receiver is hung 0.6 m down from the center of the ceiling and the detecting face is oriented toward the ceiling or the floor. Figures 10.8*b*, *c* and 10.9*b*, *c* show the spatial distribution of the received light or the intensity of the transmitted signal.

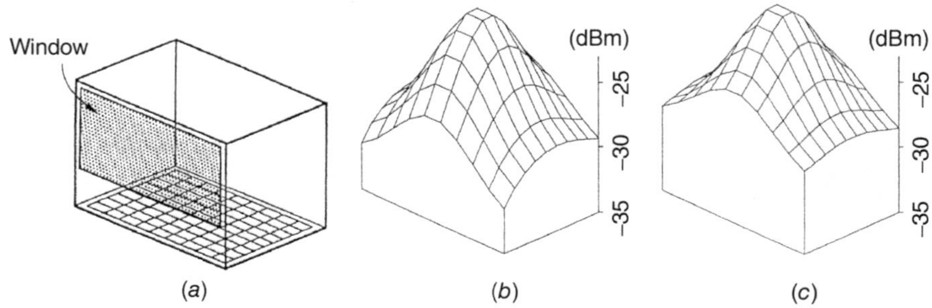

Figure 10.9 Change of signal transmission caused by window: (*a*) position of window, (*b*) with window (max −20.6 dBm, min −31.5 dBm, and (*c*) without window (max −20.4 dBm, min −28.6 dBm).

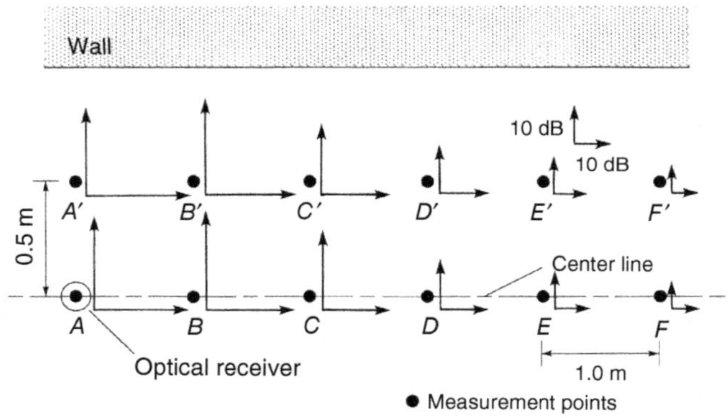

Figure 10.10 Measured distribution of signal transmission in corridor: length and orientation of vectors indicate C/N of received signal and direction of subject's face, respectively.

Figures 10.8*b*, *c* show the results with different orientations of the light detector. In both cases, a sufficient amount of light for telemetry was received in all parts of the room, even in a corner area, due to the scattering effect. When the transmitter approaches the central position of the floor, the received signal increases. In the case of Figure 10.8*c*, the light-detecting face of the receiver is pointed up toward the ceiling and the variation of the received signal becomes relatively small. In ambulatory telemetry, the small variation is often more preferable than the high peak intensity of the received signal. That is, stable telemetry can be expected with the light detector hung down from the ceiling with its receiving face pointing up toward the ceiling.

In the telemetry using indirect light transmission, the room is filled up with the signal light by the scattering from the ceilings and the walls. This is the basic principle of ambulatory telemetry. Thus, the signal strength is reduced considerably when there is an open window or a dark curtain on the wall. To examine this effect, a simulation was conducted for the case when there is a large open window (the wall with zero reflection coefficient) covering most of one of the walls. Figure 10.9 shows the result. The signal strength decreases when the transmitter approaches the window. However, the decrease is small (2–3 dB) and there is little problem in practice. This example typically shows that the distribution of the indirectly scattered light is relatively stable against the change in environmental physical conditions.

The validity of this simulation result was confirmed in the real conditions and measurements, as well (Kudo et al., 1987). An example of the measured result is shown in Figure 10.10. This is the spatial distribution of the received signal in the corridor in which two sides are considered as open space. (width 2.1 m × height 2.7 m × length 30.0 m) This shows the signal strength received at the ceiling at the position indicated in the figure. The subject equipped with a light emitter (1 W optical power) on both shoulders stands at different points ($A'–F'$, $A–F$) facing against a wall or facing along the center line of the corridor. The vectors in the figure indicate the orientation of the subject and the magnitude of the received optical signal in the C/N (ratio between the strengths of carrier and noise). This result shows that we can expect more than 10 dB C/N in entire area by installing light detectors every 8–10 m. Through these analyses it was confirmed that ambulatory biotelemetry can be realized using indirect light transmission in a regular room with little effect from the subject's position and movement.

10.4.3 Optical Signal Propagation in Open Space

In the ambulatory measurement using indirect light transmission, the optical signal filled the closed space, generally. Therefore, it is considered to be effective in a patient's room, an examination room, and a cabin of a vehicle, but not in an open space such as outdoors. Two techniques can be used to overcome this problem.

The first is the use of direct light transmission with a tracking function. Using a beam of light, data can be transmitted by tracking either the transmitting side or the receiving side. Various systems that track a light spot automatically have been commercialized, and it is not difficult to follow the speed of human movement. However, when direct light transmission is used, the light propagation path has to be secure.

In another technique, the indirect light transmission available in an open space is used. For instance, the ambulatory measurement becomes possible to some extent using ground reflection effectively or using many receivers equipped with a diffuser or a reflector. An example of the unit of reflectors and a photodetector is shown in Figure 10.11. Eight to 16 units of this structure are combined around the z axis in the positions of rotational symmetry. With this kind of receiver, ambulatory measurement can be achieved with a light emitter of 1 W optical power within 10 m distance from the receiving unit in an open space.

10.5 MULTIPLEXING IN OPTICAL TELEMETRY

As mentioned in Section 10.4, there are various multiplexing techniques for optical telemetry. Every technique has problems to be solved in the practical use for optical telemetry. We have also developed the multiplexing technique suitable for

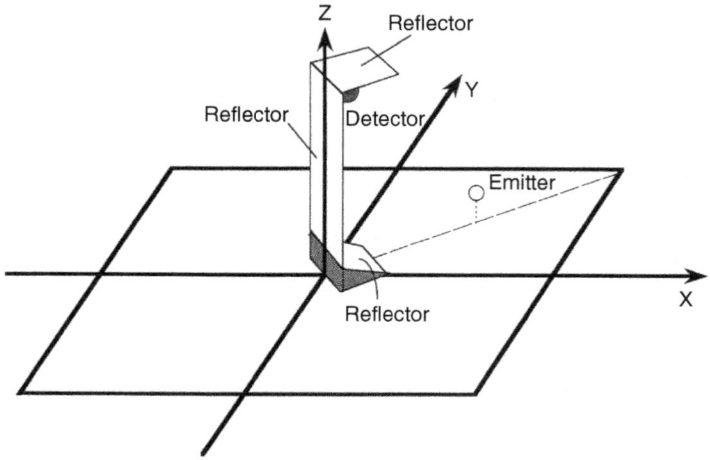

Figure 10.11 Example of reflector structure for indirect light transmission in open space.

biotelemetry using indirect light transmission (Matsuda et al, 1991). Here, two representative techniques are introduced.

10.5.1 Pulse-Burst Method

In optical data transmission in free space, the pulse modulation technique is more advantageous than the continuous-wave (CW) modulation technique from the viewpoints of optical transmission power and noise immunity. Pulse width modulation (PWM) is a typical pulse modulation technique. PWM can be used in a relatively simple manner, and the demodulation is easy as well. However, the power consumption of PWM is generally high. So, PWM can be modified as the PPM by representing the positions of the rise and fall of the PWM pulse by two short pulses. The first and second pulses indicate the temporal positions of synchronization and modulation, respectively. In the proposed pulse-burst method, the above-mentioned short pulse is modified into a train of narrow pulses. This train of pulses consists of a few pulses to several tens of pulses. In multiplexing, each channel is distinguished by the frequency of the burst pulses. Figure 10.12*a* shows the principle of this technique, and Figure 10.12*b* shows the frequency spectra of the pulse-burst signal with 500 kHz repetition of 1 μsec width pulses. As the number of pulses increases, the spectral peak becomes narrow. In this case, we can expect several channels multiplexing with eight pulses in a single burst. The effectiveness of this pulse-burst technique in the multiplexing of several channels has been confirmed in experiments (Matsuda et al., 1991).

10.5.2 Spread-Spectrum Method

For biotelemetry using indirect light transmission, the application of the spread-spectrum (SS) method has been proposed as a promising multiplexing technique.

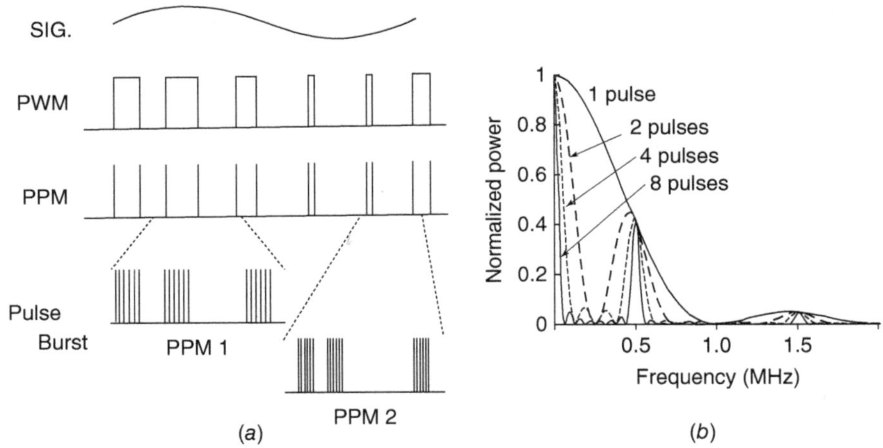

Figure 10.12 Pulse-burst multiplexing technique: (*a*) principle; (*b*) power spectra.

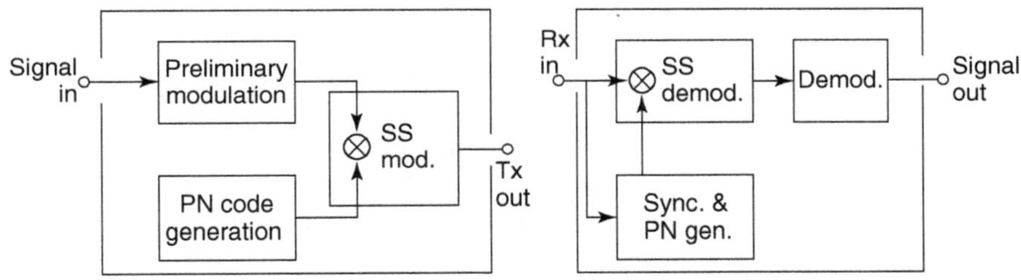

Figure 10.13 Telemetry system using spread-spectrum technique.

With the SS method the signal to be transmitted is modulated into a much wider bandwidth than that required for the original signal.

Figure 10.13 shows the principle of the SS modulation system we have developed for optical biotelemetry. In this system, we modulate the signal to be transmitted into a digital signal as the first modulation. Then, as the second modulation, we modulate the digital signal by multiplying it with a pseudonoise signal generated in the transmitter. Since the pseudonoise signal has much wider bandwidth than the original signal, the spectrum of the modulated signal is spread widely. In the receiving side, we take the correlation between the received signal and the same pseudonoise signal as the one in the transmitting side. As a result, we obtain the digital signal of the first modulation. By demodulating the digital signal, the original signal can be recovered. Through these processes, the effects of noise, interference, and multipath propagation are greatly suppressed. With this SS method, we can multiplex more than several tens of channels relatively easily by using different kinds of pseudonoise. On the other hand, a wide-band transmission is required to utilize this technique effectively. Since the wide width of the transmission band is one of the typical merits of optical biotelemetry, it is suitable for the SS method. The detail of this technique and the results of experiments can be found in the literature (Shimizu et al., 1998).

10.6 APPLICATIONS OF OPTICAL TELEMETRY

We have applied the methods and techniques previously mentioned to realize the optical biotelemetry. Let us see some of them as concrete examples of optical biotelemetry.

10.6.1 Transcutaneous Biotelemetry

One of the techniques used to take the internal information out of the body is transcutaneous biotelemetry. The radio capsule is a well-known examples. With this technique, pH electrodes or temperature sensors are encapsulated in a swallowable small capsule which transmits the pH or temperature information of the digestive tract from the inside to the outside of the body by a radio wave. Capsule endoscopy has been widely used to examine small intestine. The image of the internal surface of the digestive tract is transmitted by a radio wave (e.g., Munoz-Navas, 2009; Rammohan, and Naidu 2011). The principle of optical telemetry can also be applied for this purpose.

We place a light-emitting element inside the body and receive the light propagated through the body tissue. Figure 10.14 shows this principle. The feasibility of this principle has been confirmed through various fundamental studies (Kudo et al., 1988). Then, a telemetry system was manufactured, and it was applied to an animal experiment (Kudo et al., 1988).

Figure 10.15 illustrates the animal experiment. The electrodes, the transmitter, and the light emitters were implanted inside the body of a rat. The light signal diffusely propagated through the skin and was captured by the optical receiver placed outside the animal cage. In this way, the internal ECG can be measured remotely without restraining the animal. There is no need to attach any element on the body surface of the animal. This makes the animal experiment much easier than using many attachments on the body.

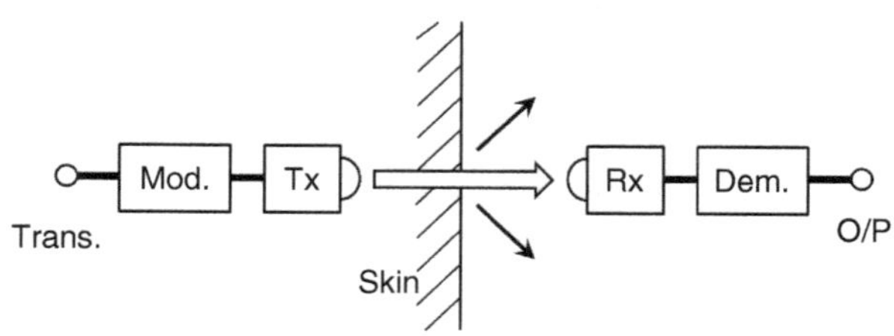

Figure 10.14 Transcutaneous optical telemetry: Trans = transducer, Mod = modulator, Tx = transmitter, Rx = receiver, Dem = demodulator, O/P = output.

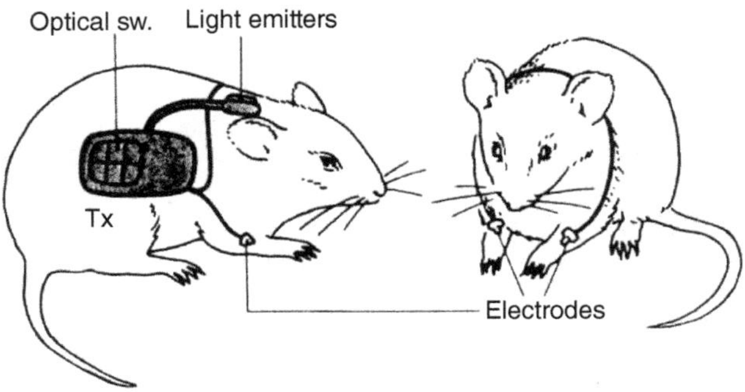

Figure 10.15 Animal experiment of transcutaneous optical telemetry (all equipment was beneath the skin) : optical. sw. = optical switch.

The size of the implanted transmitter is $40 \times 25 \times 15$ mm^3 and the weight is 16.6 g. The transmitter operates for about 100 h in continuous use. We installed a power switch which can be driven by the illumination of near-infrared light pulses from outside the body. Therefore, the battery life was extended considerably by turning on the power only when it was necessary. The light emitters were two LEDs with optical power of 70 mW in total. The transmission range was 3 m in free space. If there was a scattering object such as a wall or a human body within about 1 m, the telemetry was possible even when the direct propagation path was intercepted.

10.6.2 Optical Body Area Network

The usefulness of a wearable computer has been widely recognized and many reports are found on the body area network (BAN) (e.g., Hanson et al., 2009). In most of the studies the signal carrier for the BAN is an electrical signal such as an electric field, electric current, or electromagnetic wave (e.g., Cao et al., 2009). Many attempts have been made to use the electric BAN in medical practice (e.g., Khan et al., 2010). However, in the use of electric signals for BAN, there are restrictions of signal bandwidth, problems of electromagnetic interference, and difficulties in controling information leakage by electromagnetic radiation.

To overcome these problems, we propose the use of light for the signal carrier (Shimizu, 2005). There are few studies reported on the optical signal transmission through the human body for a BAN and also few practical applications of the optical BAN for medicine. When light enters our body, it is diffusely scattered. We analyzed the characteristics of light transmission thoough body tissue, and utilizing this scattered light for data transmission, we attempted a detection technique to examine the feasibility of the optical BAN in practical applications.

Figure 10.16 illustrates one of the examples of the optical BAN. When we illuminate near-infrared (NIR) light on the wrist area, the light propagates widely and reaches the fingers. Therefore, the BAN from/to a wristwatch-type device to/from fingertips can be used. For example, the individual with the wristwatch can be indentified

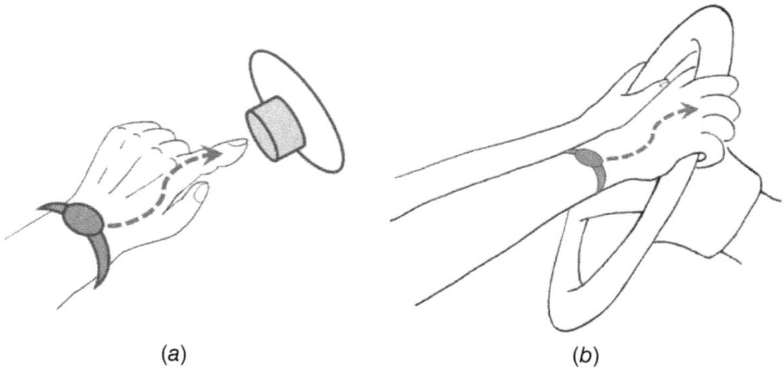

Figure 10.16 Examples of application of optical BAN: (*a*) activation of restricted switch; (*b*) information transmission in a car.

by just touching the doorknob without any action of personal authentication such as finger print imaging or iris scanning. There are many other applications possible with this optical BAN from the wrist to the fingertips.

In the BAN using diffusely scattered light, the propagation path of light is spread in a wide area of the body tissue. The temporal shape of the optical pulse is broadened in the propagation through the body. This pulse broadening restricts the data transmission rate. To evaluate the transmission rate, the optical responses of a human arm and a human hand were measured. Short pulses of laser light (Ti–Sapphire, wavelength 786 nm, pulse width 20 psec full width at half maximum (FWHM), optical power 200 mW) were illuminated on one side of the body. The transmitted light through the body was received at the other side of the body with an optical fiber. The light was led to a streak camera (temporal resolution <15 psec, repetition rate 2 MHz) and the broadened pulse shape was measured. The 20-psec pulses were broadened to 780 and 980 psec in the hand and arm, respectively (Saito et al., 2008). From this result, we can expect signal transmission in the order of 100 MHz. With signal processing techniques such as the linear equalizing process, we could transmit an 800-MHz square-wave signal in our experiment (Saito et al., 2008).

The attenuation of an optical signal in our body is much higher than that of the electric signal. This makes the control of information leakage through unnecessary radiation easier than the electric signal. However, if the range is too small, it is not practically useful. Thus, the data transmission range was analyzed using a model phantom and a human body. Figure 10.17 shows the result of the measurement. It shows that the optical data transmission from the wrist is possible to all the fingers. Among the wavelengths examined, the highest performance was obtained with the 800-nm wavelength.

In optical BAN, the signal attenuation due to the strong scattering in the body is severe, and the data transmission is vulnerable to optical and electrical noise. To deal with this problem, we introduced the space diversity (SD) technique. In this technique, we place some detectors on the body surface at separate positions and apply

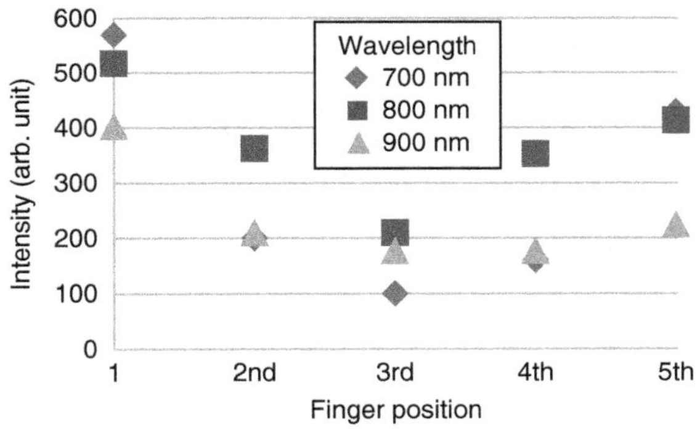

Figure 10.17 Received optical signal intensity at different fingers and wavelength: light source at wrist.

an operation to the outputs of some detectors. As the SD operation, the following three methods were used:

1. Selection Method: The instantaneous signal intensities from all the detectors are compared, and the largest signal is selected.
2. Equal-Gain Combining Method: The signal intensities from all the detectors were summed with the same weights.
3. Maximum Ratio Combining Method: The signal intensities from all the detectors were summed with the weight of each signal intensity.

As expected, the SD performance was different for different noise types. The results suggested that the maximum ratio combining method had superior noise suppression ability compared to the other methods.

10.6.3 Noncontact Measurement of Body Surface Displacement

The displacement of the body surface can be measured without any direct contact by using light. Figure 10.18 illustrates the principle of this technique. This example shows that we can use light not only for the signal carrier but also for the sensor to obtain a biological signal (Shimizu et al., 1989). When we illuminate the measurement point on the body surface with a thin parallel beam of light, a scattering light spot appears on the body surface. With a lens, the scattered light is focused on a solid-state position-sensitive device (PSD) as a light spot. Since the body surface displacement in the direction of the light beam changes the position of the light spot on the PSD, we can measure the displacement as the current change of the PSD. We used a laser diode (1 mW, 780 nm) as a light source and a one-dimensional PSD (1×12 mm^2 effective receiving area) as a light receiver. The sensitivity and the accuracy were 200 mV/mm and 0.01 mm (frequency bandwidth 0.1 – 20 Hz), respectively.

This principle was applied to the respiratory and cardiac monitoring of neonates (Shimizu et al., 1989). Figure 10.19 shows the concept of this technique. Figure 10.20

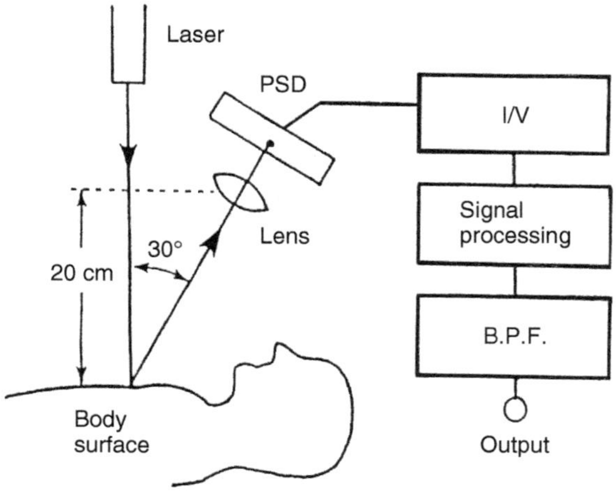

Figure 10.18 Surface displacement measurement with light.

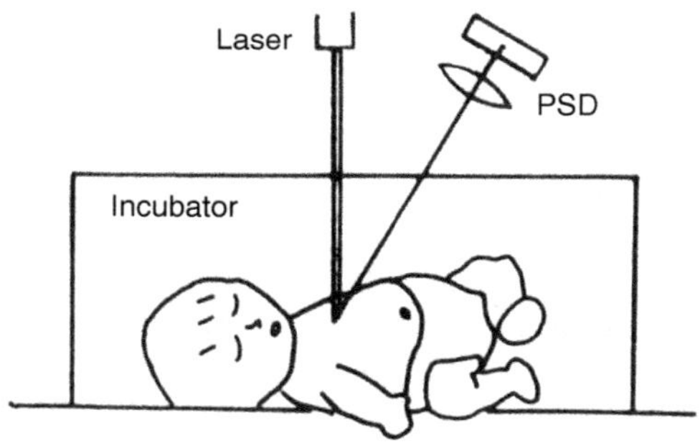

Figure 10.19 Respiration monitoring of neonate with light.

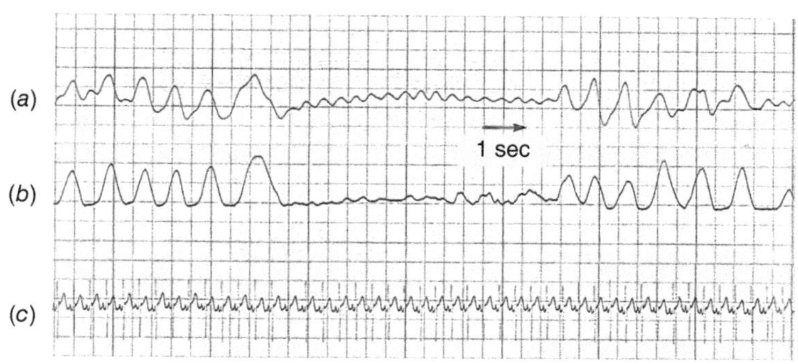

Figure 10.20 Measured results of chest movement of neonate: (*a*) optical technique; (*b*) impedance measurement; (*c*) ECG.

shows the result of the measurement obtained with the neonate lying on the back. It shows the waveform obtained by the proposed technique (Figure 10.20*a*, 0.1 – 3 Hz bandwidth), the respiration waveform obtained by an impedance measurement technique (Figure 10.20*b*), and the ECG (Figure 10.20*c*) for comparison. Figures. 10.20*a*, *b* show the temporary apnea between normal respirations. Even with simple bandpass filtering, the waveform obtained by the optical technique shows a good agreement with that obtained in the conventional impedance measurement. In Figure 10.20*a* we can observe the fluctuation in the apnea period. In comparison with Figure 10.20*c*, it was found to be the cardiac movement of the chest wall. This shows that we can monitor the vital signs of neonates with a beam of light and without any electrodes and cables (Shimizu et al., 1989).

There have been different applications of this technique, or the optical noncontact measurement of body surface movement. They include the measurements of a mechano-cardiogram and the pulse propagation velocity (Shimizu, 1990). It has been shown that this technique is also useful in detecting blood vessel constriction due to physical and mental stress without any physical contact on the body (Shimizu, 1990).

10.6.4 Ambulatory Telemetry

One of the most important purposes of biotelemetry is to measure physiological signals without any constraint on the natural activity of the subject. The ECG telemeter which has been used in clinical routines is a typical example. In hospitals and medical institutions, various kinds of telemeters have been used, such as integrated monitoring in an ICU and coronary care unit (CCU) and daily cardiac monitoring of hospitalized patients. In almost all the cases, a radio wave has been used as the transmission medium in these kinds of telemetry. EMI problems become more serious than ever due to the deterioration of the electromagnetic environment in hospitals and the enforcement of legal restrictions.

Many of these problems can be solved by using light as a transmission medium for the biological information. Using indirect light which has been reflected and scattered at a ceiling, a floor, and walls, a stable signal can be transmitted by light without the problem of interception of a direct path. Figure 10.21 shows the principle of this technique. The biological or physiological signals obtained from a subject are modulated for transmission. If multichannel data transmission is required, they are multiplexed into a single channel of data. The modulated signal is converted into a light signal at a transmitter. The light signal is emitted, reflected, and scattered at the inner surface of a room and eventually fills the room. In this way, the signal transmission can be maintained by the indirect light transmission even if the direct light path between the transmitter and the receiver is intercepted.

Figure 10.22 shows the outline of the system to realize this technique. This system consists of a transmitter to collect and to transmit biological signals and a receiver to receive and to display/record the signals. To examine the effectiveness of this technique, an optical telemetry system was manufactured. Table 10.1 shows major characteristics of the system. The effectiveness and the usefulness of this technique have been confirmed in real conditions through measuments with a

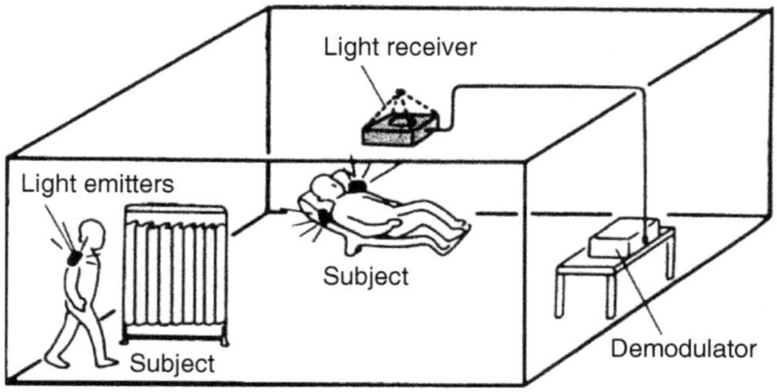

Figure 10.21 Ambulatory telemetry using indirect light transmission.

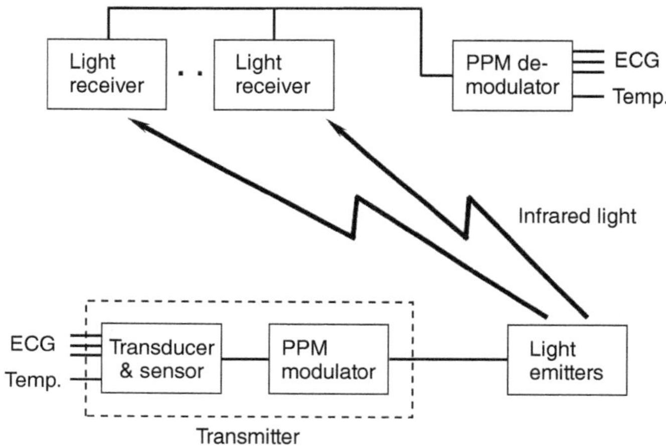

Figure 10.22 Telemetry system using indirect light transmission.

subject in various moving conditions. Figures 10.23 and 10.24 show the positions of the ECG electrodes and the ECG waveforms obtained with the developed system in the exercise of the subject. The optical telemetry system using indirect light transmission developed in this study has also been used for the biological data acquisition of a payload specialist of a space shuttle in the space lab. The waveforms of ECG, electro-oculogram (EOG), skin potential reflex (SPR), and respiration were monitored in the spacelab successfully and downloaded to the National Aeronautics and Space Administration (NASA) control center on Earth (Shimizu, 1995).

10.6.5 Multichannel Biotelemetry

Using the biotelemetry technique, biological signals and medical data can be transmitted to a remote place. The ECG transmission from isolated islands and

TABLE 10.1 Performance of Optical Biotelemetry System

Transmitter	Light emitters (set of two)
ECG: 3 ch. standard limb leads or precordial unipolar leads	Light emitter: 80 IR LEDs (940 nm)
Signal bandwidth: 0.05–100 Hz	Total optical output (when pulse on): 1 W
Non-linearity: less than 4%	Current drain: 20 mA
Total gain: 46 dB	Dimension: L 90 × W 80 × H 18 mm^3
Temperature: 1 ch. thermistor	Weight: 85g
Temperature range: 0–42°C	
Time constant: 0.42 deg/sec	Light receiver
Modulation: 4 ch. PWM/PPM/IM	Photo detector PIN Si photodiode (effective receiving area 100 mm^2)
Pulse frequency: 2.6 kHz	Optical filter: long-pass filter (cutoff wavelength 800 nm)
Pulse interval	Dimension: L 100 × W 120 × H 50 mm^3
ECG ch.: 0.3 msec	Weight: 450g
Temperature ch: 0.6 msec	
Pulse width: 2 μsec	ch: channel

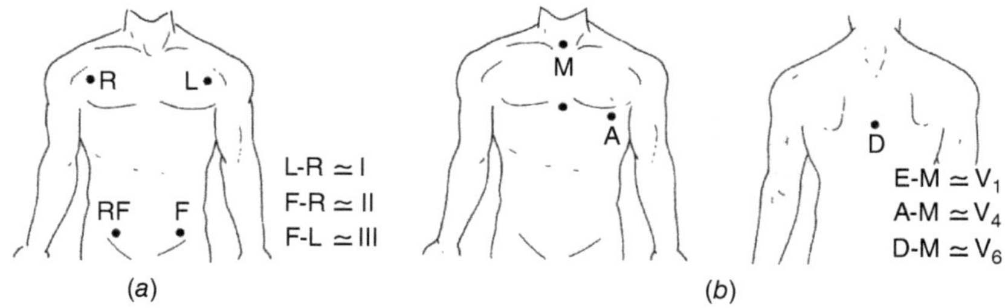

Figure 10.23 Arrangement of ECG electrodes for ambulatory optical telemetry: (*a*) analogous bipolar standard limb leads; (*b*) analogous unipolar precordial leads.

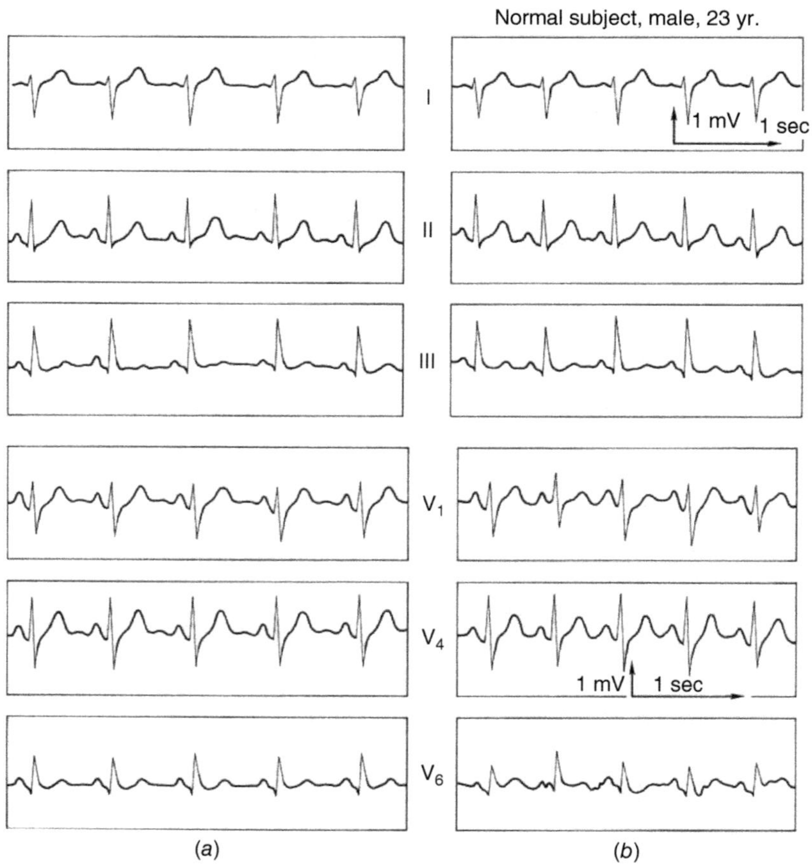

Figure 10.24 ECG waveforms obtained from subject in exercise: (*a*) walking; (*b*) jogging.

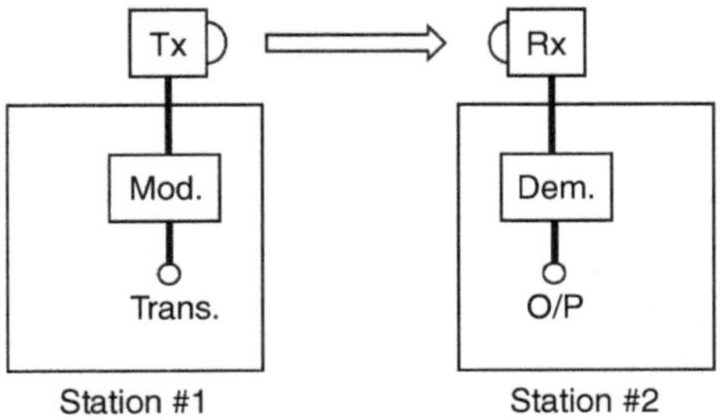

Figure 10.25 Multichannel data transmission to remote place by optical telemetry: Trans = transducer, Mod = modulator, Tx = transmitter, Rx = receiver, Dem = demodulator, O/P = output.

remote rural areas using a public telephone network is a well-known example. Using mobile communication techniques, mobile telemedicine can be realized (Shimizu, 1999, 2006). With the optical technique, large-capacity high-speed data transmission is attained.

Figure 10.25 shows the principle. Different kinds of biological signals and medical data are multiplexed in a primary station. They are transmitted through an optical communication channel. The optical signal is received at the secondary station and demultiplexed in order to reproduce the original signals and data. This technique makes it possible to not only separate the data collection site and the data analyzing site but also transmit many different kinds of data to a remote place simultaneously in real time. In this way, we can significantly increase the mobility and the flexibility of each station. This is particularly important in the case of animal or human experiments and clinical practices. In these cases, it is generally difficult to bring large data collection/processing equipment to the measurement site. Since we can send multichannel data, including color moving images and audio signals, in real time, this technique provides a powerful tool for telemedicine and homecare medicine.

Figure 10.26 illustrates the outline of the system manufactured to show the possibility of this technique. From the primary station to the secondary station, the biological signals (color video image, audio signal, ECG, EEG, and body temperature) are modulated by the PPM/IM method and transmitted through open air. In the return path, the signal for ARQ is transmitted back from the secondary station to the primary station. Since the capacity of the information through this return path is very small, such as the existence of channel disconnection, four kinds of signals were enough. They were differentiated by the frequency in an FM/IM modulation.

An experimental system was developed and the effectiveness of this technique was examined (Odagiri et al., 1985, 1988, 1989). In the experiment, a color moving image, an audio signal, and three channels of biological signals (ECG, EEG, body temperature) were transmitted simultaneously to show the condition of a patient. In this experiment, the signals were transmitted by light through open air. The transmission distance was up to 1 km. Using a laser with higher optical power, larger transmission distance is expected. If the mobility of the primary station is not required, optical fiber

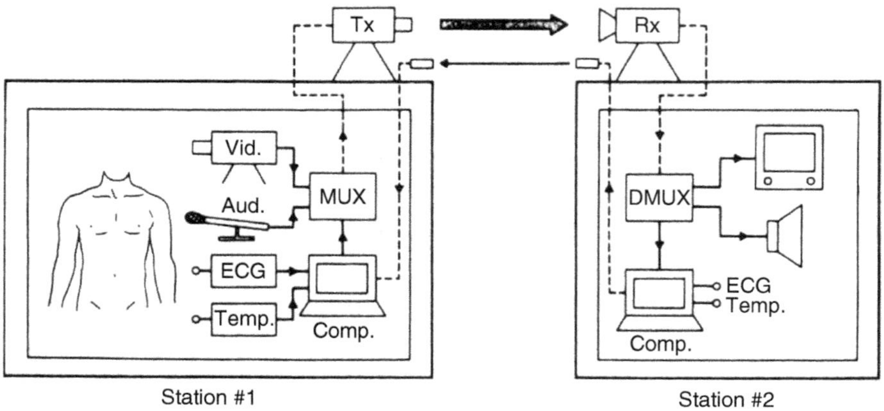

Figure 10.26 System for multichannel data transmission to remote place by optical telemetry: Vid = video signal, Aud = audio signal, MUX = multiplexer, Tx = transmitter, Rx = receiver, DMUX = demultiplexer, Comp = computer, broken line = optical fiber.

communication can be used. Then, the transmission distance can be much larger than the open-air communication.

10.6.6 Data Transmission between Medical Equipment

With the recent increase of various electrical equipment in medical institutions, there has been a strong demand for them to be integrated into a small number of systems. Particularly in an operating room and an ICU, centralized control and data processing are required among many monitoring, imaging, and automatically controlled equipments. However, such integration inevitably increases the number of electric wires connecting the equipment. This not only hinders the operation but also increases the risk of accidents, such as leakage of electric current. The wireless transmission of data and control signals is one of the solutions. However, the use of a radio wave in a hospital is not preferable due to the problem of EMI. The optical biotelemetry technique can be apllied to answer this demand.

For multiplexing, the SS method has been used. It is robust against various noises and can provide many channels for multiplexing. To examine the feasibility of this technique, a central part of the transmission system has been developed (Shimizu et al., 1998). Figure 10.27 shows the outline of this system. The digital data to be sent are fed from computer A through an RS-232C interface. At the transmitter, the data are processed in the preliminary modulation circuit to be suitable for the next SS modulation. Then, the data are multiplied with the pseudonoise (PN) code generated at the transmitter. For the preliminary modulation and the PN code, we used the ASK (amplitude shift keying) and the Gold series with 127 chips period, respectively. To make the data transmission rate at 9600 bps, the clock frequency of a chip was set at 7.315 MHz. The SS modulated signal is emitted from the LED array which consists of 60 LEDs (nominal optical power 30 mW each, wavelength 850 nm). The indirect

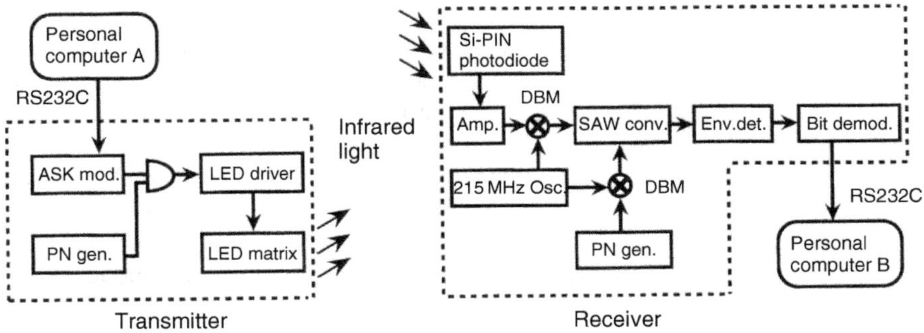

Figure 10.27 System for optical data transmission between medical equipments using SS technique.

light scattered by a ceiling, a floor, and walls reaches the Si–PIN photodiode and is transformed into an electric signal. The received signal is amplified and its frequency is shifted to 215 MHz in an up-conversion circuit using a double balanced mixer (DBM). This is the operating frequency of a surface acoustic wave (SAW) convolver. At the receiver, the time-reversed Gold series which is the same as that used at the transmitter is generated. This PN code is also frequency shifted to 215 MHz and fed to the SAW device. In the SAW device, the received signal is convolved with the PN code, and the correlation data are obtained in real time. After this SS demodulation, the signal is demodulated according to the preliminary modulation. In the developed system, the signal is processed in the envelope detection and the bit-demodulation circuits. The bit data are fed to computer B through an RS-232C interface and an original signal is reproduced by the computer.

An experimental system which consists of two transmitters and one receiver has been manufactured. In the analysis of transmission characteristics, the practical feasibility of the proposed technique has been verified (Toya and Shimizu, 2000).

10.7 CONCLUSION

A concept of optical biotelemetry has been proposed and summarized. Conventionally, a radio wave is commonly used for biotelemetry. However, optical biotelemetry is possible in various applications as shown in this chapter. Due to the recent progress in optical and electronic technologies, the introduction of the optical technique can offer new useful features. We do not intend to replace radio telemetry by optical biotelemetry. They should be used to support each other. For example, in a closed space such as in a building, optical telemetry has many useful merits. However, in an open space such as outdoors, radio telemetry has much greater transmission range.

Until recently, the applications of biotelemetry have been limited in a relatively small area, such as ECG telemetry and animal tracking. However, the demand for biotelemetry has been steadily expanding. There has been a notable change in our society. Cardiac and cerebral disorders are increasing rapidly as the elderly population increases. The necessity of home care and telemedicine is higher than ever. The importance of rehabilitation medicine and sports medicine has been widely recognized. Health management and self-diagnosis have become common among the general public. The demand for providing emergency medical support to moving vehicles such as ships, aircrafts and space ships has been increasing as well (Shimizu, 1999, 2006). In these social trends, biotelemetry has been expected to play an important role as a key technology. Optical biotelemetry is only one of the techniques of such biotelemetry, but it has promising potential to open a new possibility to answer the above demands.

REFERENCES

Cao H, Leung V, Chow C, Chan H. 2009. Enabling technologies for wireless body area networks: a survey and outlook. *IEEE Comm. Mag.*, 47:84–93.

Hanson MA, Powell HC, Barth AT, Ringgenberg K, Calhoun BH, Aylor JH, Lach J. 2009. Body area sensor networks: Challenges and opportunities. *Computer*, 42:58–65.

Khan JY, Yuce MR, Bulger G, Harding B. 2010. Wireless body area network (WBAN) design techniques and performance evaluation. *J. Med. Syst.*, 36:1441–1457.

Kimmich HP. 1982. Biotelemetry based on optical transmission. *Biotelemet. Patient Monitor.*, 9:129–143.

Kudo N, Shimizu K, Matsumoto G. 1987. Optical biotelemetry using indirect light transmission. *Biotelemetry*, 9:55–58.

Kudo N, Shimizu K, Matsumoto G. 1988. Fundamental study on transcutaneous biotelemetry using diffused light. *Frontiers Med. Biol. En.*, 1:19–28.

Mackay RS. 1993. *Bio-Medical Telemetry*. New York: IEEE Press.

Matsuda S, Shimizu K, Yamamoto K, Mikami T. 1991. Multiplexing technique for the optical telemetry using indirect light transmission. *Biotelemetry*, 11:54–57.

Munoz-Navas M. 2009. Capsule endoscopy. *World J. Gastroenterol.*, 15:1584–1586.

Murakami H, Shimizu K, Yamamoto K, Mikami T, Hoshimiya N, Kondo K. 1994. Telemedicine using mobile satellite communication. *IEEE Trans. Biomed. Eng.*, 41:488–497.

Odagiri H, Tsuda S, Shimizu K, Matsumoto G. 1985. Free-space optical data link for biological multi-dimensional information. *Tech. Rep. IEICE*, 85:21–28.

Odagiri H, Shimizu K, Matsumoto G. 1988. Free-space optical data link for multi-dimensional biological information. *Memoirs Hokkaido Instit. Tech.*, 16:1–8.

Odagiri H, Shimizu K, Matsumoto G. 1989. Free-space optical data link for multi-dimensional biological information II. *Memoirs Hokkaido Inst. Tech.*, 17:1–9.

Rammohan A, Naidu RM. 2011. Capsule endoscopy: New technology, old complication. *J. Surg. Tech. Case Rep.*, 3:91–93.

Saito T, Kakita S, Kato Y, Shimizu K. 2008. Feasibility of transcutaneous data transmission using scattered light. *Tech. Rep. IEICE*, 107:57–60.

Shimizu K. 1980. Optical biotelemetry. *Jpn. J. Med. Electron. Biol. Eng.*, 18:492–498.

Shimizu K, Kobayashi K, Matsumoto G. 1989. Respiratory and cardiac monitoring of neonate by non-contact optical technique. *Biotelemetry*, 10:626–632.

Shimizu K. 1990. Measurement of body surface displacement by light. *BME*, 4:33–43.

Shimizu K. 1991. Fundamentals for biotelemetry. *Biotelemetry*, 11:179–184.

Shimizu K. 1995. Optical biotelemetry. *Jpn. J. Phys. Fitness Sports Med.*, 44 :317–324.

Shimizu K, Toya N, Yamamoto K. 1998. Application of spread spectrum technique for multichannel transmission in optical telemetry. *Biotelemetry*, 14:97–102.

Shimizu K. 1999. Telemedicine by mobile communication. *IEEE Eng. Med. Biol.*, 18:32–44.

Shimizu K. 2005. Optical biotelemetry. In Lin JC (ed.). *Advances in Electromagnetic Fields in Living Systems*, Vol. 4. New York: Springer, pp. 131–154.

Shimizu K. 2006. Application of mobile communication techniques for emergency telemedicine. In Istepanian RSH, Laxminarayan S, Pattichis CS (eds). *M-Health, Emerging Mobile Health Systems*. New York: Springer, pp. 417–433.

Toya N, Shimizu K. 2000. Fundamental analysis of multichannel spread spectrum data transmission using indirect scattered light. *IEICE Trans.*, J83-B:780–788.

Tsuda S, Shimizu K, Matsumoto G. 1993. Efficient application of coding technique for data compression of ECG. *IEICE Trans. Inf. Syst.*, E76-D:1425–1433.

Tuchin V. 2000. *Tissue Optics*. Bellingham: SPIE Press.

Vo-Dinh T (ed.). 2003. *Biomedical Photonics Handbook*. Boca Raton, FL: CRC Press.

11

BIOSENSOR COMMUNICATION TECHNOLOGY AND STANDARDS

LARS SCHMITT, JAVIER ESPINA, THOMAS FALCK, AND DONG WANG

Philips Research High Tech Campus, Eindhoven, the Netherlands

11.1 INTRODUCTION

Over the last decade health care has been dramatically evolving—50% of the developed world is projected to be chronically ill by 2050. Thereby the demand for care is growing and there will be not enough caregivers to cope with the aging population. In addition, people have higher expectations about care and demand more voice and choice in their health care. However, we do not take good care of ourselves to prevent diseases: Obesity, heart disease, and cancer are global health issues that are worsened by the way we live, for example, poor diet, physical inactivity, and smoking. Since rising health care costs are unsustainable, there is a growing need for finding new ways to deliver health care.

Economists at the Centers for Medicare and Medicaid Services (CMS) estimate that health care spending in the United States will grow by 5.8% per year for the period 2010–2020, 1.1% more than the expected growth of the gross domestic product (GDP). By 2020, health care costs are projected to reach $4.64 trillion, representing then almost 20% of the GDP (Keehan et al., 2011).

Figure 11.1 shows the top 10 medical expenses in the United States (WebMD, 2012). Chronic diseases such as heart conditions, cancer, chronic obstructive pulmonary disease (COPD) and asthma, hypertension, and diabetes make up 75% the

Handbook of Biomedical Telemetry, First Edition. Edited by Konstantina S. Nikita.

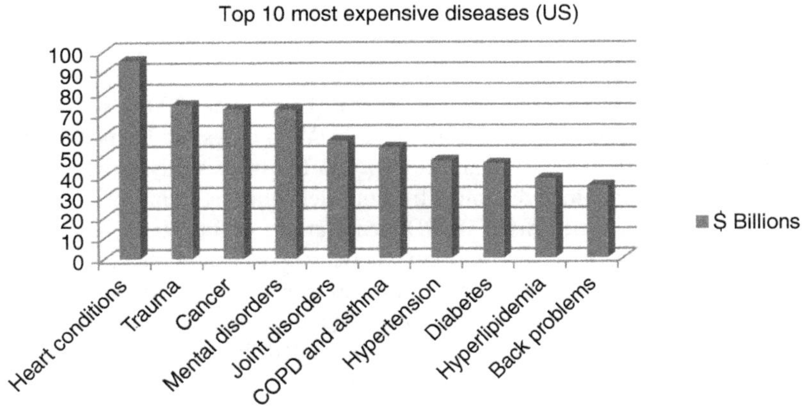

Figure 11.1 Top 10 most expensive diseases in the United States.

lion's share of the overall costs. For many of these conditions obesity is a risk factor and a healthier lifestyle could help prevent them. Remote patient management is a new paradigm in health care where patients are empowered to take more responsibility and control of their health and to be more proactive in general. Medical practitioners are encouraged to draw up a disease management plan in close partnership with the patients, which in turn improves their compliance. By continuously monitoring and analyzing vital signs and contextual information of patients in their own home environment, deterioration of their condition can be detected at an early stage, allowing for treatment before critical situations arise.

Figure 11.2 compares the traditional care model, where the patient's state of health is subject to major changes, including emergency cases due to only irregular health

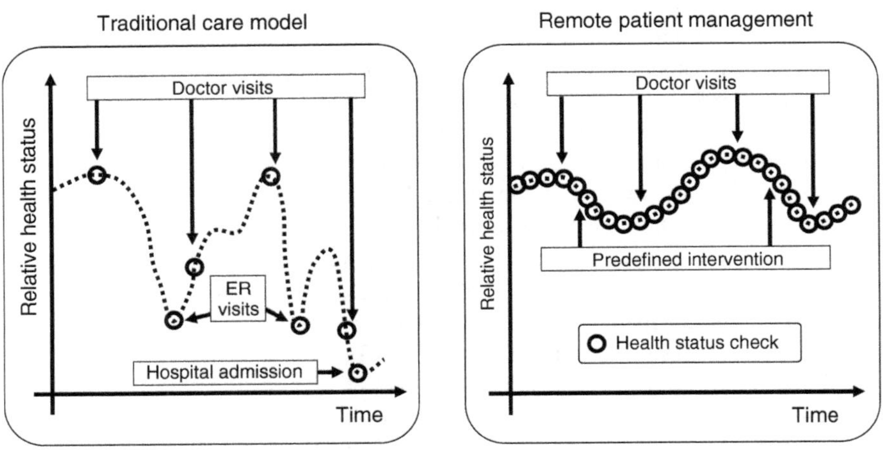

Figure 11.2 Traditional care model versus remote patient management.

checks, with the concept of remote patient management that helps extreme falls to be avoided and thereby sustains the state of health on a higher level due to close-meshed patient monitoring. The aim is to reduce costly emergency visits and to prevent unnecessary (re-)hospitalizations of patients in general (Falck et al., 2006).

The U.K.'s Department of Health initiated the Whole System Demonstrator program in May 2008 (Steventon et al., 2012). The two-year study is believed to be the world's largest randomized control trial of telehealth involving more than 3000 patients with a clinical diagnosis of COPD, heart failure, or diabetes. The participants were split into two groups, a control group receiving usual care without telehealth and an intervention group with telehealth. It aimed to provide evidence that the adoption of remote patient management reduces the financial burden of institutional care for the chronically ill. The findings of the study included that telehealth can deliver a 15% reduction in accident and emergency department visits, a 20% decline in emergency admissions, a 14% reduction in bed days, and an 8% saving in tariff costs. Most strikingly, they also demonstrate a 45% drop in mortality rates, which from the patient's perspective is a compelling reason to opt for remote patient management. In summary, the results indicate that telehealth can improve both the quality of life for individual patients and their families and efficient use of resources in the health care system.

11.2 BIOSENSOR APPLICATION SCENARIOS

11.2.1 Reference Use Case

A typical use case of a remote patient management system that drives transformation of care outside traditional settings by empowering and motivating patients to actively participate in managing their health and connecting them well with their caregivers is depicted in Figure 11.3 (Simons et al., 2006).

It consists of the following elements:

- Wireless measurement devices to monitor vital signs such as electrocardiography (ECG), blood pressure, oxygen saturation, glucose, and weight. This category comprises both body-worn biosensors [like an SpO_2 wrist watch shown in Figure 11.4 (Philips, 2012a)] and stationary devices (such as a weighing scale). Also more unobtrusive ways of monitoring vital signs and activities of daily living will become feasible in the future. For example, cameras integrated in the environment can be used to measure heart and respiration rate (Philips, 2012b).
- One or more base stations can collect and aggregate the observations from all wireless measurement devices in the patient's home and transmit the gathered data to a remote medical data center via the Internet. These access points can be either stationary (e.g., dedicated embedded TeleStation devices or off-the-shelf PCs) or mobile (e.g., smartphones or tablet PCs).
- The medical data center in the back end stores the collected data from all subscribed patients in a secure way, makes it available for review to the responsible care managers, and analyzes them to provide clinical decision support.

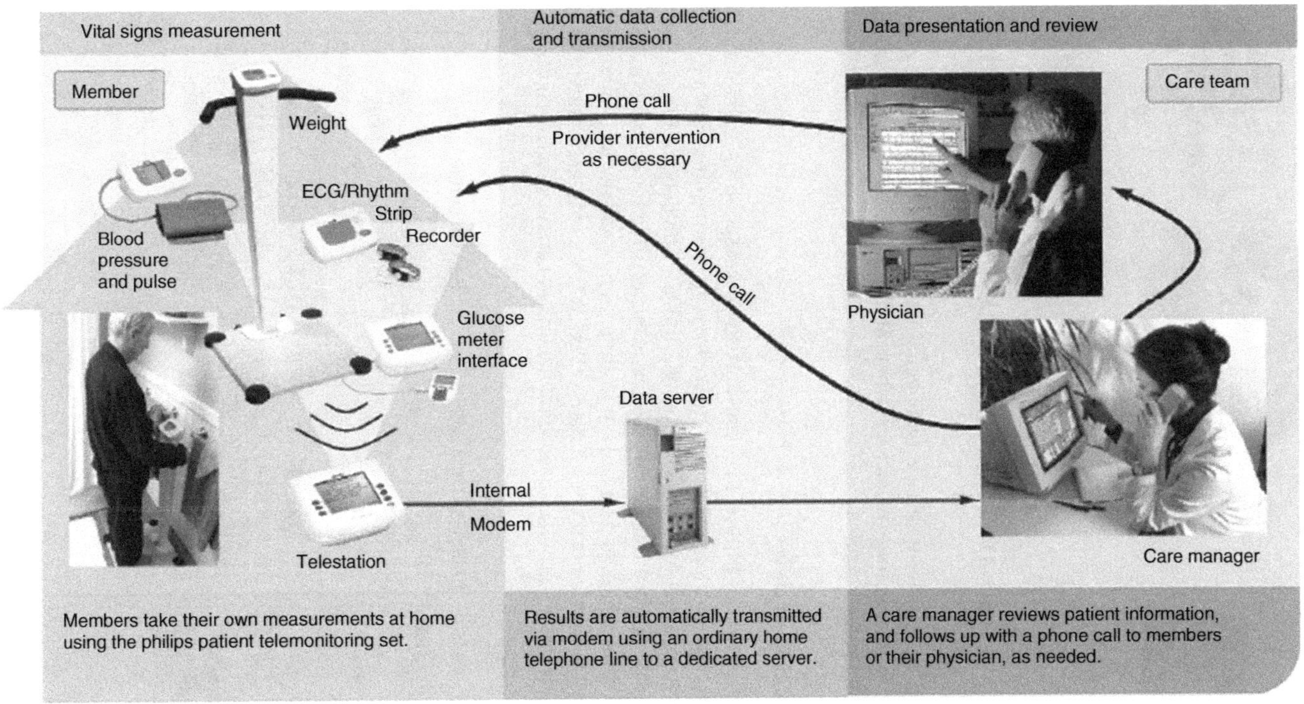

Figure 11.3 Use case of Philips telehealth solution.

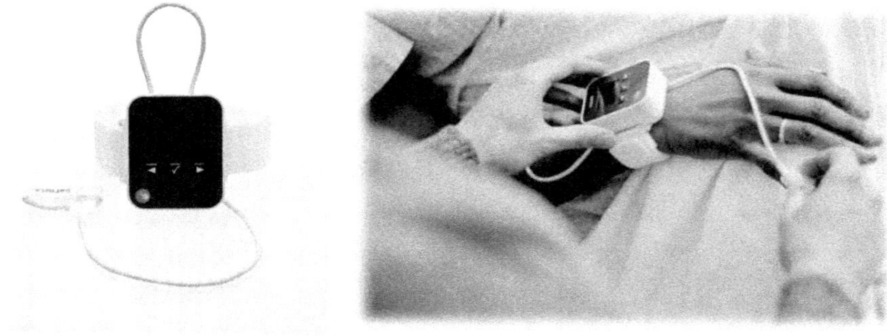

Figure 11.4 Cableless SpO_2 measurement device.

- The care team consists of care managers in a remote health service center that involve corresponding physicians when abnormal measurements or signs of deterioration are detected.

11.2.2 System Overview

Based on that use case, Figure 11.5 presents the corresponding architectural view of a remote patient management system identifying the required device types and the interfaces to connect them (Schmitt et al., 2007a). The general idea is that at the patient's home an application hosting device is responsible for aggregation and processing of the sensed patient data and controlling all medical devices in the home. Its counterpart at the service center is the so-called WAN (wide-area network) device. These back-end servers provide disease management services based on the gathered

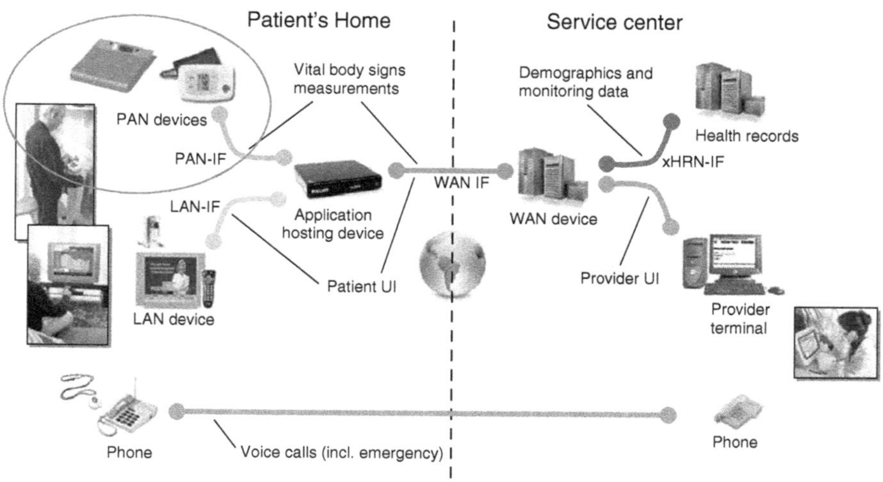

Figure 11.5 Reference architecture of remote patient management systems.

data from the application hosting devices of all subscribed patients. The demographics and monitoring data are stored in a health record system and presented at a terminal of a health care service provider.

In order to create a telehealth ecosystem, interoperability is important and can be achieved by means of well-defined and standardized interfaces between the various device classes. The Continua Health Alliance (2012), with currently more than 240 member companies, is working on selecting connectivity standards and defining complementary guidelines for ensuring interoperability (Wartena et al., 2009). The PAN (personal or peripheral area network) interface describes how biosensors (e.g., for measuring vital signs such as ECG, blood pressure, and activity) and actuators (such as an insulin pump) connect to an application hosting device via wired (such as USB) or short-range wireless communication (such as Bluetooth). To extend the distance between medical devices and the application hosting device, instead of the PAN interface, the LAN (local area network) interface can be used utilizing communication technologies such as Ethernet, ZigBee, or WiFi. The WAN interface defines how application hosting devices connect to one or more WAN devices which are often under control of specialized service providers focusing on certain diseases such as diabetes or obesity. Finally, the xHRN (Electronic/Personal Health Records Network) interface enables interoperability between a WAN device and a health records database operated by a traditional health care service provider or new initiatives like Microsoft HealthVault.

11.3 BIOSENSOR COMMUNICATION TECHNOLOGIES

This section describes how biosensors can communicate with other components of a remote patient management system.

11.3.1 Frequency Spectrum Regulations

The use of the frequency spectrum for radio communication is regulated, and often managed, by authorized national administrations or agencies. The most known is the U.S. Federal Communication Commission (FCC, 2012a). Moreover, international organizations coordinate regulations and standards at a supranational level. Worldwide frequency spectrum regulations are defined by the Radio Regulations of the International Telecommunication Union (ITU, 2012). Regionwide regulations are defined by organizations such as the Electronic Communications Committee (ECC) of the European Conference of Postal and Telecommunications Administrations (CEPT), which brings together the national administrations of 48 countries of the European continent.

International frequency spectrum regulations are the most relevant to be considered, since they can enable a widespread and economically viable use of biosensor communication systems. In practical terms, the main unlicensed spectrum bands internationally available for application-generic low-power radios are ultrahigh-frequency (UHF) radio (e.g., 315, 433, 868–928 MHz), the worldwide

2.4-GHz industrial–scientific–medical (ISM) band, the worldwide 5.8-GHz ISM band, and, for the ultrawide band (UWB), the 3.1–10.6-GHz band. Some biosensors can or do make use of such application-generic bands, albeit without exclusivity of use and hence without protection from other types of communication systems.

The UHF radio band is a set of frequency spectrum bands with a largely geography dependent availability, which is detrimental to globally deployable systems. Moreover, some regions impose strict duty cycle restrictions, such as in the 868-MHz band in Europe (ERC/REC 70-03, 2012). Nevertheless, low-power systems that only require the transmission of small amounts of data may benefit from the relatively good propagation characteristics of the band, in comparison with higher frequency bands.

The worldwide 2.4-GHz ISM band is a commonly utilized frequency spectrum band. Systems operating in the 2.4-GHz ISM band—unlike in the UHF band—enjoy fewer restrictions regarding duty cycle or channel spacing. Furthermore, this band offers a significant portion of frequency spectrum: from 2400 to 2483.5 MHz in the United States and Europe and from 2471 to 2497 MHz in Japan. All these characteristics have made the 2.4-GHz ISM band the most popular among low-power radio systems. A proof of its popularity is the adoption of the band by largely widespread technologies such as Bluetooth, WiFi [based on the IEEE 802.11b/g/n standards (IEEE 802.11, 2012)], and ZigBee [based on the IEEE 802.15.4 (2011) standard. Many proposed biosensor networks are using such technologies, as discussed in Section 11.3.2.

The worldwide 5.8-GHz ISM band has, in general, similar characteristics to the 2.4-GHz band. Its most common users are WiFi systems based on the IEEE 802.11a (1999) standard. However, the poor propagation characteristics in proximity of the human body make the band unsuitable for many medical applications.

UWB systems operate at very low radiated power density levels by employing very narrow pulses resulting in very large bandwidths. A number of countries allow UWB operation in parts or the totality of the 3.1–10.6-GHz spectrum band. UWB technology promises to enable high transmission data rates, low power operation, and radar and ranging capabilities. Despite the slow progress in standards development, the cost of initial implementations, and performance significantly lower than initially expected, there are several reasons for the limited success of UWB radios so far (Takahashi, 2009). The use of UWB technologies and the UWB spectrum remains an option for future biosensor communication systems.

Beyond the internationally available spectrum for application-generic communication, some medical-specific spectrum bands are available.

Implantable medical devices using inductive communication technologies operate worldwide on spectrum bands in the lower part of the frequency spectrum, typically ranging from a few kilohertz to less than 40 MHz. Most noninductive medical implants, such as cardiac pacemakers and defibrillators, are allowed to operate worldwide without individual spectrum license within the 401–406-MHz spectrum range. Such operations are known as Medical Implant Communications Service (MICS) in the United States (FCC, 1999) and Japan and Ultra Low Power Active Medical Implants (ULP-AMI) in Europe (ERC/REC 70-03, 2012;) (ETSI EN 301 839,

2009). It is worth noting that the original MICS regulation in the United States has been superseded by the Medical Device Radiocommunications Service (MedRadio) (FCC, 2012b), which includes the core MICS spectrum at 402–405 MHz and adds additional spectrum bands that are also usable for on-body diagnostic and therapeutic devices. MedRadio includes 24 MHz of spectrum for the so-called medical micropower networks (MMNs).

There are no truly worldwide available bands for nonimplantable biosensors—and in general medical devices. However, two regional spectrum band regulations are relevant enough to be mentioned herewith: medical body area networks (MBANs) and Wireless Medical Telemetry Service (WMTS). Both spectrum bands serve the purpose of enabling wireless diagnosis and treatment of patients, mostly via vital sign monitoring. They do not require individual licenses for operation.

An MBAN consists of one or more on-body wireless biosensors and/or medical actuator devices that can communicate with a monitoring device placed on/around (up to 10 m from) the human body. Such monitoring devices display and process vital sign parameters from MBAN devices and may also forward them (e.g., to a central nurse station) by using wired or wireless technologies other than MBANs. The MBAN frequency spectrum can only be used for short-range communications. The FCC published its MBAN regulation in May 2012, allowing the deployment of MBANs in the 2360–2400-MHz spectrum range (FCC, 2012c), in both hospitals and patient homes. At the time the present book was written, ETSI and ECC were in the process of regulating and standardizing MBANs in Europe (ETSI TR 101 557, 2012). ECC considered some spectrum bands within the 2300–2500-MHz range for future MBAN operation.

Similarly to MBAN, WMTS permits bidirectional data communications related to medical care, with the exception of voice and video. Yet WMTS enables longer range biosensor communications (known as telemetry) at the expense of higher power consumption and with the restriction of hospital use only. WMTS is available in the United States, operating in the spectrum bands 608–614, 1395–1400, and 1427–1432 MHz (FCC, 2012d). Similar WMTS regulations are active in Japan, where WMTS operates in the 420–430 and 440–450-MHz ranges. In Australia the spectrum available for biomedical telemetry transmitters is the 520–668-MHz range under an application-specific subcategory of so-called low interference potential devices (LIPDs).

In conclusion, the nature and usage of biosensors often determine the spectrum band choice and the applicable regulations. Such choice is, in turn, one of the factors that may influence the selection of the wireless technology and/or standard for enabling biosensor communication.

Figure 11.6 gives an overview of the frequency spectrum regulations discussed above.

11.3.2 Wireless Technologies and Standards

This section describes some examples of recent standard-based wireless technologies that are suitable for biosensor communication.

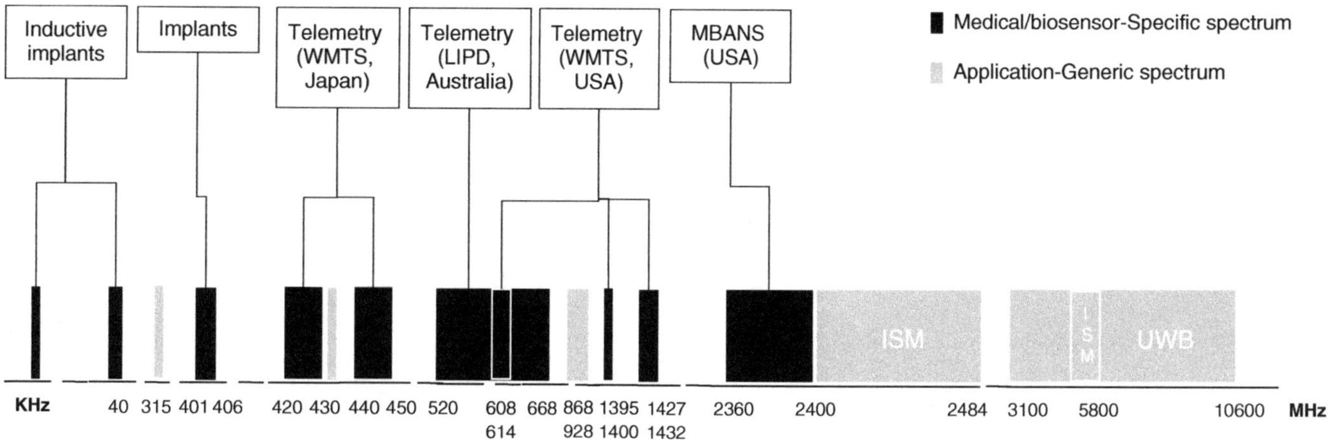

Figure 11.6 Spectrum regulation overview.

IEEE 802.15.4 The IEEE 802.15.4 (2011) standard specifies the physical and media access control layers for low-rate wireless personal area networks (LR-WPANs) and is the basis for the ZigBee (2012), ISA100.11a (Kinney, 2007), WirelessHART (HART, 2008), and MiWi (Microchip, 2010) specifications, which further extend the standard by developing the upper layers. The standard was released in 2003, amended in 2006, and enables communication of data at a maximum rate of 250 kbps.

IEEE 802.15.4 WPANs can form either star or peer-to-peer network topologies. The latter may be used to form arbitrary patterns such as cluster trees. Star topologies have one PAN coordinator, whereas peer-to-peer topologies can have one or more. The role of such a coordinator is to act as the primary controller of the network by controlling the association of nodes as well as initiating, terminating, or routing the communication. To allow for very low cost low-complexity devices, IEEE 802.15.4 defines reduced-function devices (RFDs) and full-function devices (FFDs). RFDs implement a subset of the IEEE 802.15.4-defined primitives and cannot act as coordinator. FFDs have a full implementation of IEEE 802.15.4 and can adopt any role in the WPAN.

Table 11.1 summarizes the properties of the IEEE 802.15.4 physical layer, which consists of four options, known as PHYs. The 2450-MHz DSSS PHY is available worldwide. The rest of the PHYs are defined for operation in the UHF frequency spectrum range around 868 MHz in Europe and 915 MHz in the United States and Canada.

The transmission power of IEEE 802.15.4 implementations depends, among other things, on the respective local spectrum regulations. A maximum transmission power between 0 and 5 dBm is usually employed.

The IEEE 802.15.4 medium-access control (MAC) layer is a simple protocol designed to cater for low-cost low-power devices but yet flexible enough to enable the delivery of periodic data, intermittent data, and repetitive low-latency data (e.g., real-time ECG streaming). It features an acknowledged protocol for reliable delivery of data and is able to efficiently support very low duty cycle operations.

The MAC layer optionally supports a superframe structure defined by the coordinator. The superframe is bounded by network beacons sent periodically by the coordinator of a beacon-enabled network to synchronize the associated devices, identify the WPAN, and describe the structure of the superframes. The superframe can have an active and an inactive portion, during which the coordinator does not interact with its WPAN and may enter a low-power mode. For low-latency applications and/or applications requiring fixed data rates—such as ECG monitoring—the coordinator may dedicate portions of the active superframe to that application. These portions are called guaranteed time slots (GTSs) and are allocated in the contention-free period (CFP). In the rest of the active superframe, known as the contention access period (CAP), devices use slotted CSMA-CA (carrier sense multiple-access collision avoidance) to access the channel. In non-beacon-enabled networks, in which there is no superframe structure or GTS, the channel access mechanism used is always unslotted CSMA-CA. Figure 11.7 shows a representative example of a superframe structure with GTS.

TABLE 11.1 Overview of IEEE 802.15.4 Physical Layer

PHY	Modulation	Frequency Band (MHz)	Number of Channels	Chip Rate (kchips/s)	Symbol Rate (ksps)	Data Rate (kbps)	Support
868/915 MHz BPSK PHY (DSSS)	BPSK	868–868.6	1	300	20	20	Mandatory
		902–928	10	600	40	40	
868/915 MHz ASK PHY (PSSS)	ASK	868–868.6	1	400	12.5	250	Optional
		902–928	10	1600	50	250	
868/915 MHz O-QPSK PHY (DSSS)	O-QPSK	868–868.6	1	400	25	100	Optional
		902–928	10	1000	62.5	250	
2450 MHz PHY (DSSS)	O-QPSK	2400–2483.5	16	2000	62.5	250	Mandatory

BPSK: Binary Phase-Shift Keying
ASK: Amplitude-Shift Keying
O-QPSK: Offset Quadrature Phase-Shift Keying
DSSS: Direct Sequence Spread Spectrum
PSSS: Parallel Sequence Spread Spectrum

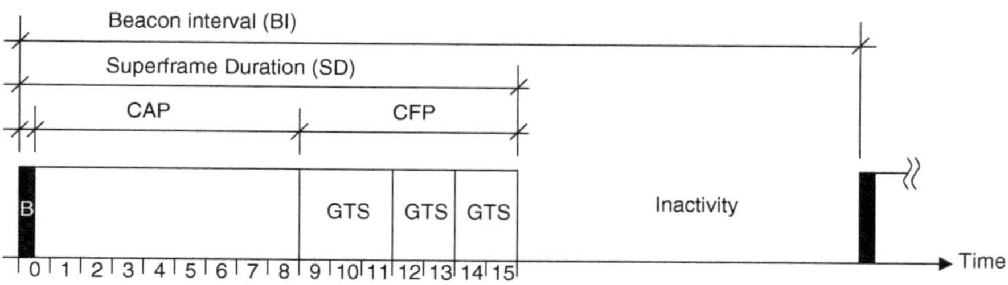

Figure 11.7 IEEE 802.15.4 superframe structure.

IEEE 802.15.4 also supports security mechanisms. Symmetric cryptography can be used to protect the transmitted data and restrict it to a group of devices or just a point-to-point link. This ensures the confidentiality and authenticity of the data. Furthermore, the IEEE 802.15.4 MAC computes freshness checks between successive receptions, which ensures that presumably old frames or data no longer considered valid are not forwarded to higher layers.

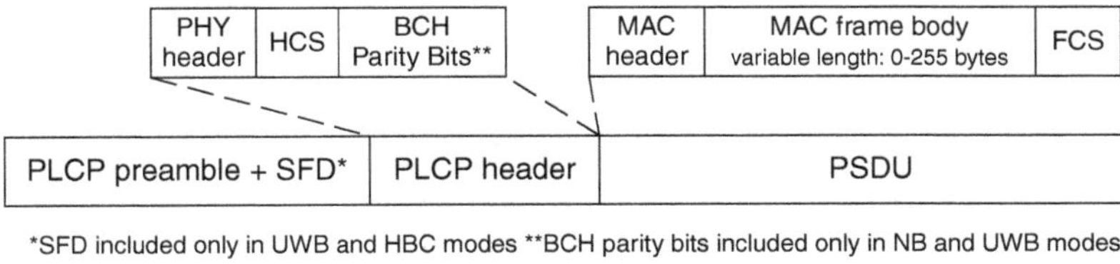

Figure 11.8 IEEE 802.15.6 PPDU format.

IEEE 802.15.6 In February 2012, the IEEE published the IEEE 802.15.6 standard (referred to as 15.6) for a short-range wireless communication inside, on, or around a human body, with support of the combination of high reliability, quality of service, low power, scalable data rates (up to 10 Mbps), and interference mitigation (IEEE 802.15.6, 2012). The 15.6 standard specifies three PHYs and a unified MAC and security solution to address the broad requirements of medical as well as nonmedical (e.g., entertainment and well-being) body area network (BAN) applications.

Overview of IEEE 802.15.6 PHY The current 15.6 standard defines three PHYs, the narrow-band (NB) PHY, UWB PHY, and human body communication (HBC) PHY for different frequency bands available in different regions. A common physical layer protocol data unit (PPDU) format is defined for all the three PHY schemes, as shown in Figure 11.8.

A physical layer convergence protocol (PLCP) preamble is attached at the beginning of each PPDU packet in order to aid the receiver in packet detection, timing synchronization, and frequency offset recovery. In the UWB and HBC schemes, an extra start frame delimiter (SFD) field is included after the PLCP preamble to assist fine timing synchronization and start-of-frame detection at the receiver. The second component of the PPDU is the PLCP header, which conveys the PHY parameters needed by the receiver to decode the physical layer service data unit (PSDU). A header check sequence (HCS) is calculated over the PHY header with a cyclic redundancy check (CRC) code to assist the receiver to detect the PHY header decoding error. In the NB and UWB PHYs, the PHY header and the HCS are encoded with a BCH code to improve robustness of the PLCP header. The PLCP header is transmitted at the given header data rate in an operating frequency band. The last component of the PPDU is the PSDU, which is a concatenation of the MAC header, the MAC frame body, and the frame check sequence (FCS).

The NB PHY is specifically optimized for medical applications and provides scalable data rates to allow for smooth rate-range trade-off and low-complexity implementation. For each frequency band supported by the NB PHY, at least 10 channels are defined.

A NB PHY PSDU construction block diagram is shown in Figure 11.9 and the major system parameters are listed in Table 11.2. A systematic (63, 51) BCH encoder is used to protect the PSDU since it allows for low-complexity implementation with reasonable coding gain. The encoded bits are then spread by being repeated K times, where K is the spreading factor. By changing K from 1 to 4, multiple data rates are

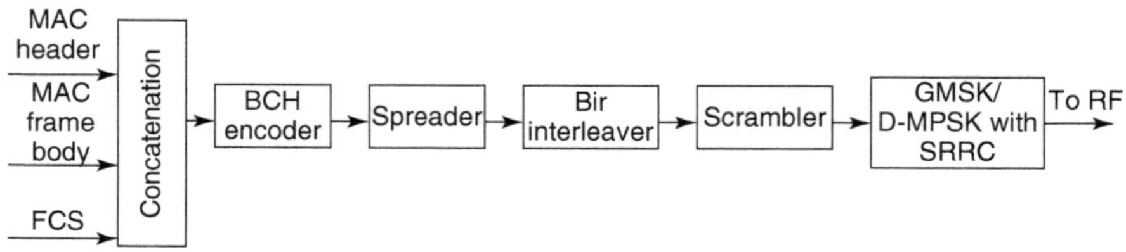

Figure 11.9 IEEE 802.15.6 NB PHY PSDU construction block diagram.

achieved with different spreading gains while the symbol rate keeps constant. The constant symbol rate simplifies the NB transceiver design by allowing only a single set of fixed analog filters to be used for all the data rate modes. A simple block bit interleaver is employed after the scrambler to mitigate error propagation effect and reserve the spreading gain. The interleaved bits are then scrambled by a pseudorandom bit sequence to make the spectrum of the RF signal flat. At last, the scrambled bits are modulated to RF signal. The Gaussian minimum shift keying (GMSK) modulation is used for the 420–450-MHz band (Japan WMTS band) while the rotated differential M-ary phase shift keying (D-MPSK, $M = 2, 4, 8$) modulation scheme is used for the other bands. After the rotated D-MPSK modulation, a square-root raised cosine (SRRC) pulse shaping filter is used to compact the RF spectrum. With such design, the NB PHY can provide sufficient link margin for a typical communication range of 3 m under realistic BAN channels with a transmission power of − 10 dBm.

In the UWB PHY, two different types of UWB technologies, impulse radio UWB (IR-UWB) and frequency modulation UWB (FM-UWB), are specified with IR-UWB as mandatory and FM-UWB as optional. The UWB channelization defines three physical channels in the low band (3–5 GHz) and eight physical channels in the high band (6–10 GHz) with channel bandwidth of 499.2 MHz. For each physical channel, four logical channels are created with unique preamble sequences. In total, eight preamble sequences are defined using 63-bit Kasami sequences, half of which should be used for an odd number of physical channels and the other half should be used for an even number of physical channels. Such arrangement can mitigate cochannel and adjacent-channel interference effect. The PSDU construction procedure is illustrated in Figure 11.10.

For general medical and nonmedical applications (i.e., default mode), the (63, 51) BCH code used in the NB PHY is adopted to protect PSDU, while a (126, 63) 10-correcting BCH code is used for high-priority medical applications [i.e., high-quality-of-service (QoS) mode] to provide better protection. To further improve link robustness, a type II hybrid automatic repeat request (ARQ) mechanism

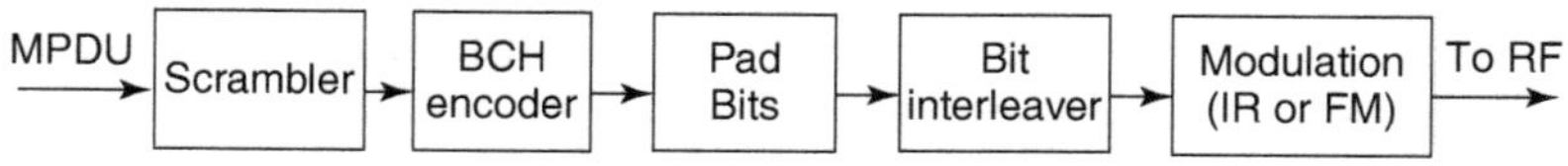

Figure 11.10 IEEE 802.15.6 UWB PHY PSDU construction block diagram.

TABLE 11.2 Narrow-Band PHY System Parameters

Band (MHz)	Number of Channels	Modulation	Symbol Rate (ksps)	BCH Code Rate (*k*/*n*)	Spreading Factor (*K*)	Pulse Shape	Data Rate (kbps)	Receiver Sensitivity (dBm)	Support
402–405	10	π/2-DBPSK	187.5	51/63	2	SRRC	75.9	–95	Mandatory
		π/2-DBPSK	187.5	51/63	1	SRRC	151.8	–92	Mandatory
		π/4-DQPSK	187.5	51/63	1	SRRC	303.6	–89	Mandatory
		π/8-D8PSK	187.5	51/63	1	SRRC	455.4	–83	Optional
420–450	12	GMSK	187.5	51/63	2	Gaussian BT = 0.5	75.9	–90	Mandatory
		GMSK	187.5	51/63	1	Gaussian BT = 0.5	151.8	–87	Mandatory
		GMSK	187.5	1/1	1	Gaussian BT = 0.5	187.5	–84	Optional
863–870	14	π/2-DBPSK	250	51/63	2	SRRC	101.2	–94	Mandatory
902–928	60		250	51/63	1	SRRC	202.4	–91	Mandatory
950–958	16	π/4-DQPSK	250	51/63	1	SRRC	404.8	–87	Mandatory
		π/8-D8PSK	250	51/63	1	SRRC	607.1	–82	Optional
2360–2400	39	π/2-DBPSK	600	51/63	4	SRRC	121.4	–92	Mandatory
		π/2-DBPSK	600	51/63	2	SRRC	242.9	–90	Mandatory
		π/2-DBPSK	600	51/63	1	SRRC	485.7	–87	Mandatory
2400–2483.5	79	π/4-DQPSK	600	51/63	1	SRRC	971.4	–83	Optional

is supported in the high-QoS mode. In the IR-UWB option, time hopping is adopted to randomize impulse positions for coexistence. The time hopping sequence is specifically designed to provide a guard interval between two consecutive hopping intervals only when necessary. Compared to allocating a fixed guard interval in each hopping interval, this method has much less overhead. The modulation schemes supported in the IR-UWB option include on–off keying (OOK) with either a single pulse option or a burst pulse option and DPSK with spreading. There is no mandatory pulse shape for IR-UWB; however, the used shape has to meet a certain correlation criterion for interoperability. The FM-UWB option exploits high modulation index of FM to generate an UWB signal. The bits of the PPDU are first modulated with continuous-phase binary frequency shift keying (CP-2FSK) and then the resulting narrow-band CP-2FSK signal is modulated with wide-band frequency modulation to generate a constant-amplitude UWB signal. The FM index of FM-UWB is 131.28 and such a high index can provide a processing gain similar to the one achieved in a spread-spectrum system. In summary, the UWB mode is designed to provide a large scope of implementation opportunities for high-reliability, low-complexity, and ultralow-power operations.

HBC is a technique that uses the human body as the transmission medium for communications via electric fields instead of electromagnetic waves. A HBC transceiver uses an electrode capacitively coupled to the human body to generate a weak yet detectable electric field for transmission of information. The HBC PHY in 15.6 operates in the frequency band centered at 21 MHz with a bandwidth of 5.25 MHz. The supported data rate ranges from 164 kbps to 1.3125 Mbps. The transmitter block diagram is shown in Figure 11.11.

The PSDU data are scrambled and then grouped into 4-bit vectors. Then each 4-bit vector is mapped into a 16-bit Walsh codeword. At last, the output bits are spread with a frequency shift code (FSC), which is generated by repeating [0 1]. The final chip rate is fixed to 42 Mcps regardless of data rate.

Overview of IEEE 802.15.6 MAC The network topology supported by the 15.6 standard is the one-hop star topology with optional features defined to support a two-hop star network extension. In a 15.6 BAN, there is one and only one hub, which is responsible for creating/maintaining the BAN and coordinating medium access. A hub operates in one of the three access modes, the beacon mode with beacon periods (superframes), the non–beacon mode with superframes, and the non–beacon mode without superframes. In the first two modes, the time axis is divided into superframes of equal length and each superframe is composed of allocation slots of equal length.

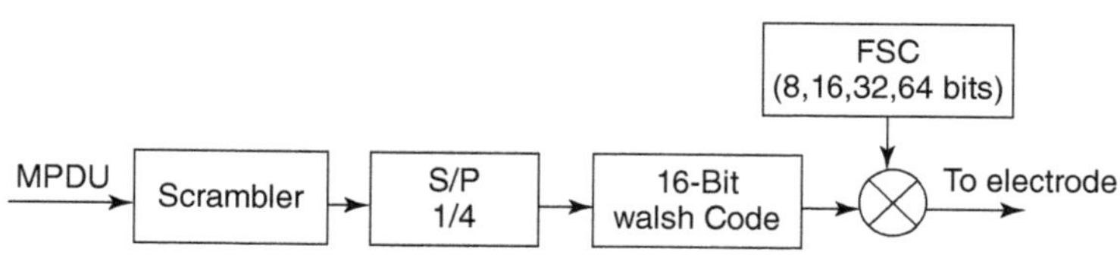

Figure 11.11 IEEE 802.15.6 HBC PHY PSDU construction block diagram.

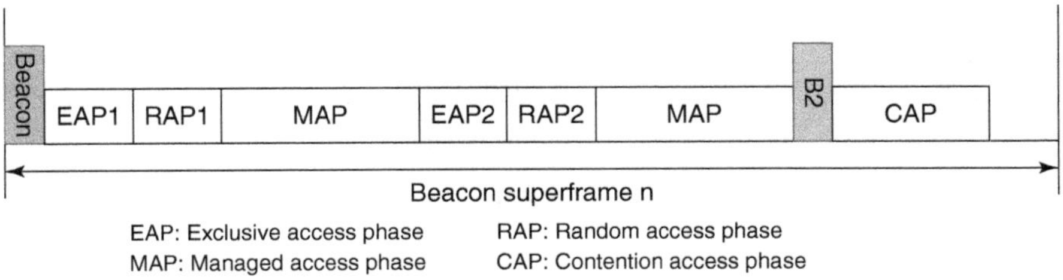

Figure 11.12 Access phases in an active superframe for beacon mode.

The hub establishes a time reference base, which is followed by nodes, and maintains the superframe structure.

1. *Beacon Mode with Superframes* In this mode, the hub transmits a beacon in each active superframe to provide the time reference base and the superframe, structure information to nodes and organizes applicable access phases in each active superframe, as illustrated in Figure 11.12. The hub may set to zero the length of any of the access phases but does not have RAP1 end before the guaranteed earliest time as communicated to the connected nodes in the node connection establishment procedure. To provide a nonzero-length CAP, the hub needs to transmit a preceding B2 frame. EAP1, EAP2, RAP1, RAP2, and CAP are random-access phases in which a node may obtain and initiate frame transactions using CSMA/CA or slotted Aloha based random access. EAPs are only used for the hub or a node to send data-type frames of the highest user priority UP (i.e., medical emergency event reports) while RAPs and CAPs are usually for regular data traffic. Only nodes may obtain contended allocation intervals in RAP1, RAP2, and CAP to send management- or data-type frames. MAP can only be used by the hub to provide contention-free time allocation intervals. Specifically, the hub may arrange scheduled uplink, downlink, or bilink allocation intervals, unscheduled bilink allocation intervals, and improvised polled and posted intervals in the MAP phase.
2. *Non–Beacon Mode with Superframes* In this mode, the whole period of each active superframe is dedicated to a MAP phase for the hub to arrange scheduled, unscheduled, and improvised allocation intervals, as illustrated in Figure 11.13.
3. *Non–Beacon Mode without Superframes* In this mode, a hub may provide unscheduled bilink time allocation intervals comprising type II polled and/or posted allocations (i.e., allocation the length of which is specified in terms of the number of frames granted for transmission). A node may employ CSMA/CA-based random access to obtain a contended allocation.

The 15.6 standard provides three access types: (1) scheduled access (contention free), which schedules reoccurring 1-periodic (i.e. available every superframe) or m-periodic (i.e., available every multiple superframes) allocation intervals during node connection establishment procedure and is supported only in the beacon mode; (2) improvised access and unscheduled access, which uses unscheduled

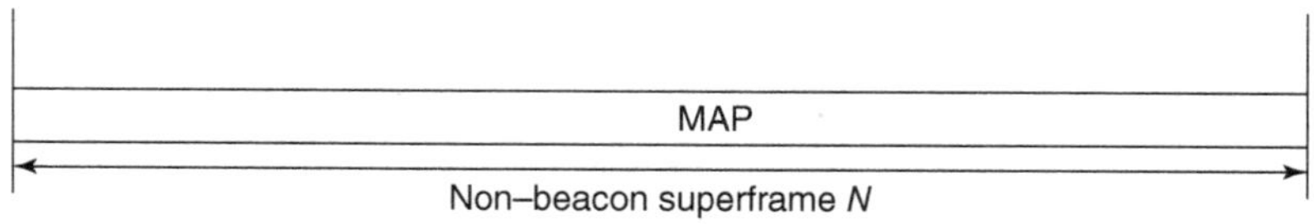

Figure 11.13 Access phases in an active superframe for non–beacon mode with superframes.

polling/posting for time interval allocation and is supported in both beacon and non–beacon modes; and (3) random access, which uses CSMA/CA in the NB PHY or Slotted Aloha in the UWB PHY. Scheduled access provides guaranteed periodic access and allows nodes to wake up at a prenegotiated, scheduled time interval to perform frame transactions. Therefore, it is suitable for medical data streaming applications (e.g., ECG waveform data streaming). Improvised access facilitates impromptu exchange for medical alarms, variable-rate medical, configuration, and network management data. Random access is usually used for episodic medical applications, such as home and fitness monitoring and medical event/alert reporting, and prioritized access is provided to accommodate different QoS requirements of medical applications.

The 15.6 standard also defines several optional MAC features to facilitate BAN coexistence and interference mitigation. Those features include (1) adaptive channel hopping (for the NB mode operating outside the MICS band or the FM-UWB PHY), which can adaptively select the operating channel set and guarantee the minimum channel separation to achieve frequency diversity; (2) active superframe interleaving, which may coordinate the schedules of active superframes of multiple overlapping 15.6 BANs for coexistence via inter-BAN communications; and (3) beacon shifting, which allows a hub to transmit its beacon in a random time hopping manner to mitigate potential repeated beacon collisions and scheduled allocation conflicts between overlapped and adjacent BANs operating on the same channel.

Moreover, a built-in security mechanism is provided in the 15.6 MAC to protect data privacy/integrity, which is critical in medical applications. For unicast secured communication, the node and the hub need to generate necessary security keys based on the security hierarchy illustrated in Figure 11.14.

First, an unauthenticated or authenticated association is needed for a hub and a node to activate a preshared master key (MK) or establish a new MK. With the activated MK, a pairwise temporal key (PTK) is created once per session, where "session" refers to a time span in which a temporal key remains valid. The PTK is used to generate security check sum for message authentication and reply defense and to

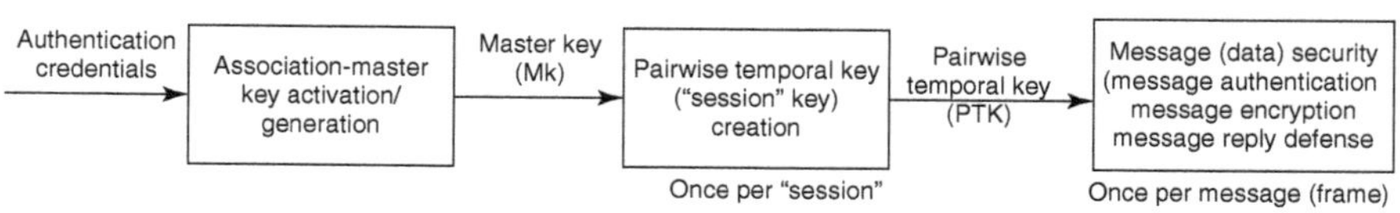

Figure 11.14 Security hierarchy.

perform encryption. Using a PTK, rather than the MK, to secure communications in a session is for security reasons, for example, compromise of a PTK does not break the MK. For multicast secured communications, the hub distributes a group temporal key (GTK) to the corresponding multicast group in a unicast manner.

IEEE 802.15.4j Standardization IEEE 802.15.4j is an ongoing standardization activity. The scope is to define necessary PHY/MAC modifications for the IEEE 802.15.4 standard, making it able to operate in the 2360–2400-MHz band in compliance with the FCC MBAN rules (FCC, 2012c). Two considerations have been taken into account in the TG 4j standardization: (1) to keep the channelization scheme flexible to accommodate harmonized coexistence with in-band primary/MBAN services and (2) to provide MAC support to enable MBAN low-power implementations. In the current draft specification (Evans, 2012), the channelization scheme defines 15 overlapping channels with a 0.5-MHz guard band at each band edge, as shown in Figure 11.15. A channel bitmap may be used by a hub device to dynamically control the MBAN access to the channels in the 2360–2390-MHz band. By configuring such a parameter, a health care facility can enforce that all its MBAN devices only have access to the part of the MBAN spectrum authorized by the MBAN coordinator so as to comply with the FCC regulation and achieve harmonized coexistence with the primary services, including aeronautical mobile telemetry (AMT) and amateur radio services. Moreover, a health care facility can configure this parameter to enforce the use of nonoverlapping channels in order to avoid interference caused by using overlapping channels. With 15 candidate channels defined, the number of available nonoverlapping channels is maximized in the case that part of the MBAN spectrum is occupied by primary users. For in-hospital deployment, up to 7 nonoverlapping channels are provided. For out-of-hospital deployment, 3 overlapping channels are defined in the 2390–2400-MHz band to provide flexibility to avoid colocated primary services (e.g., amateur radio) and other MBAN systems. A channel with a central frequency of 2395 MHz allows the use of 20 mW transmission power to achieve a large communication range with compliance to the strict FCC out-of-band emission (OOBE) limit.

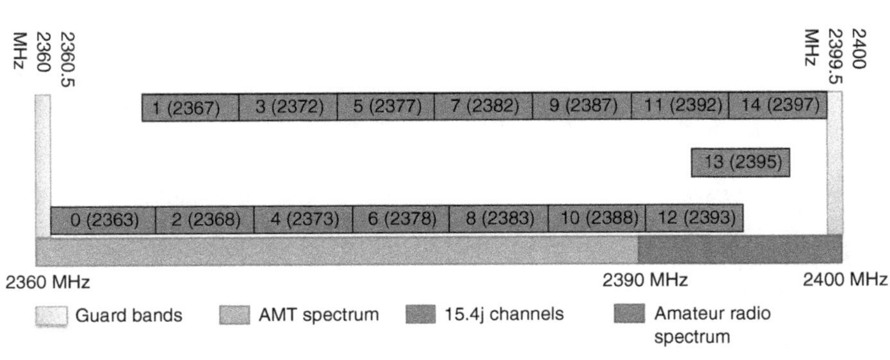

Figure 11.15 Proposed 15.4j channelization. (*See insert for color representation of the figure.*)

In the MAC layer, a new channel switch command is proposed to accommodate the need to dynamically vacate a portion of the MBAN spectrum when the MBAN coordinator requests to do so in order to protect primary services. Compared to the coordinator realignment command defined in 802.15.4, the new command includes a time stamp to define the time instance at which the MBAN should switch to a new channel. This new feature allows 802.15.4j users to rechannelize different types of MBAN services with different priorities to avoid collisions and ensure QoS. In addition, there are several proposed MAC features, such as Coordinator Switch and Association Proxy, which aim to simplify MBAN device pairing and save the power of biosensor devices.

11.3.3 Health Data Interoperability Standards

According to the IEEE, interoperability is defined as "the ability of two or more systems or components to exchange information and to use the information that has been exchanged" (IEEE, 1990). This problem has to be solved on three principal levels: At the lower layers, a standardized transport technology enabling basic connectivity has to be developed. In the upper layers, application profiles have to be developed which define what capabilities of the transport technology have to be used to best support the requirements of the particular domain, here the biomedical telemetry domain. Finally, at the application level common data models and formats need to be developed which represent an abstract unique mapping of the real-world entities. In the previous sections an introduction to several recent wireless transport standards suitable for connected devices in biomedical telemetry has been given. This has shown that a significant number of problems on the lower layers have been solved already and mature standards are available. However, more work at levels closer to the application is needed.

A solution to this last gap of application layer interoperability has been developed within the family of International Standards Organization (ISO)/IEEE 11073 personal health device (PHD) communication standards (Schmitt et al., 2007b). These define a common framework for making an abstract model for personal health data available in transport-independent transfer syntax to establish logical connections between personal health sensors or devices (e.g., ECG sensors, heart rate meters, thermometers, or blood pressure monitors) and in-home compute engines (e.g., cell phones, personal health appliances, or set-top boxes). The scope of personal health device data as defined in the ISO/IEEE PHD series of standards includes the biomedical telemetry data domain as a subset. However, beyond this set of physiological data the scope comprises also the domain of aging independently applications, where data from sensors external to the body are measured. Examples include data from sensors for detecting motion, smoke, gas, water, usage of objects, and so on.

Besides general requirements like patient and user safety of medical devices, minimal user interaction, and unambiguous association between devices and sensors, the protocol suite is optimized to personal health device usage requirements. Personal health devices are often battery powered, which mandates the need for very low computational complexity and low power consumption. For wireless devices the latter requirement suggests not only minimizing transmit power but also reducing transmission time by minimizing protocol overhead.

As a result, the encoding and parsing of protocol data units using ISO/IEEE 11073 PHD are very efficient due to the concept of precoded messages (message templates can be filled in memory in which only the actual updated values must be copied). A guiding principle for the development of ISO/IEEE 11073 PHD has also been to place the greater computational burden on aggregators (aka *managers* in the ISO/IEEE 11073 domain), which have typically richer capabilities in terms of memory and CPU power, compared to the sensor/device side (aka *agents* in the ISO/IEEE 11073 domain). Managers often use larger batteries or are even mains powered.

Another distinct feature is a mechanism to reduce message overhead for very simple personal health devices having a static configuration. By negotiating virtually all static context information once in a so-called configuration phase, only the dynamic information is transmitted in a device's measurement report. The term static here refers to information that does not change from measurement to measurement and which is usually exploited for parsing (e.g., attributes denoting the unit of a measurement value or attributes describing the type of measurement). This mechanism reduces message overhead significantly and hence results in less time spent for transmission and thus a reduction of transmit power consumption, improving battery life.

The architecture of ISO/IEEE 11073 PHD is based on an object-oriented system management paradigm, defined in ISO/IEEE 11073-20601 (2010), which represents a generic optimized exchange protocol. An object-oriented data model, the domain information model (DIM), is used to specify objects, attributes, and data types for describing personal health data, such as measurements, device settings, and units of measurement. The standardized nomenclature comprises a set of numeric codes that identify every item that is communicated between systems. A service model defines access to the data objects of the devices. A communication model manages the connection state machine for message exchange.

The base protocol ISO/IEEE 11073-20601 provides a generic framework and feature set, whereas interoperability for a particular device type is guaranteed by properly constraining this generic feature set to the needs of a particular device type and by adding device-specific nomenclature terms. This is specified in so-called device specialization documents for various personal health devices as part of the ISO/IEEE 11073-104zz series (see Table 11.3). The list of devices covered by the 104zz series has grown over the past years and is expected to further expand in the future.

The ISO/IEEE 11073 PHD standards have been adopted by the Continua Health Alliance and are leveraged by its design guidelines to enable a consistent, interoperable data layer concept across the different Continua interfaces. In the following sections the concepts of ISO/IEEE 11073 PHD are described in more detail.

ISO/IEEE 11073 PHD DIM Nomenclature A key aspect of the DIM is that all entities are referenced using unique nomenclature codes found in ISO/IEEE 11073-10101 (2004). By using a consistent nomenclature, interoperability is

TABLE 11.3 ISO/IEEE 11073 Personal Health Device Communication Series of Standards

Part	Title	Status
00103	Health Informatics—Personal Health Device Communication—Technical Report—Overview	D
10404	Health Informatics—Personal Health Device Communication—Device Specialization—Pulse Oximeter	P
10406	— Basic ECG (1 to 3-lead)	P
10407	— Blood Pressure Monitor	P
10408	— Thermometer	P
10413	— Respiration Rate Monitor	D
10415	— Weighing Scale	P
10417	— Glucose Meter	P
10418	— International Normalized Ratio (INR) Monitor	P
10419	— Insulin Pump	D
10420	— Body Composition Analyzer	P
10421	— Peak Expiratory Flow Monitor	P
10441	— Cardiovascular Fitness and Activity Monitor	P
10442	— Strength Fitness Equipment	P
10471	— Independent Living Activity Hub	P
10472	— Medication Monitor	P
20601	Health Informatics—Personal Health Device Communication—Application profile—Optimized Exchange Protocol	P

Note: Status: P = published, D = in draft.

enhanced as all implementations maintain the same semantic meaning for the numeric codes. The usage of codes rather than strings supports international implementations and allows for compact data payloads.

The nomenclature is defined as a set of context-dependent partitions. The nomenclature code in each context-dependent partition is defined by a 16-bit code that supports up to 65,536 independent terms per partition. The partitions are referenced by a 16-bit partition code. When the partition of the nomenclature code is defined through context, it is possible to use only the 16-bit term code. If the context is not defined or a context-independent term code is required, this situation is specified by a 32-bit code constructed from the 16-bit partition code together with the 16-bit term code. Partitions that are defined in ISO/IEEE 11073-10101 refer to, for example, physiological identifiers, dimensions, body sites, events, personal health device disease management, personal health device health and fitness, personal health device aging independently, and so on.

ISO/IEEE 11073 PHD DIM Object Classes The DIM defines several classes for modeling an agent. An agent device is described as a set of objects that represent the data sources, with each object having one or more attributes. The attributes describe measurement data and device status information that is communicated to a manager as well as elements that control behavior. There are three categories of attributes: static, dynamic, and observational. Static attributes do not change during the lifetime of an association between agent and manager. In contrast, dynamic and observational attributes do typically change during the lifetime of an association. Observational attributes represent measurement values and time stamps. The static and dynamic attributes provide context information for the observational attributes; that is, when a set of observational attributes is received by the manager, these values are combined with all the related static and dynamic attributes in order to enable the manager to correctly interpret the values of the observational attributes.

Figure 11.16 represents the potential objects of an agent along with class relationships using Unified Modeling Language (ISO/IEEE 11073-20601, 2010).

MDS Class The top-level object of an agent is an instantiation of the medical device system (MDS) class. The MDS object attributes represent the identification and status of an agent. Among others, the attributes of the MDS object describe the agent's device type,[1] globally unique device identifier, manufacturer model number, serial number, date and time information, power source, battery level, remaining battery time, and regulatory and certification information.

Metric Class This is an abstract common base class for all objects representing measurements, status, and context data. The metric class is not instantiated but contains common and shared attributes that are inherited by the numeric, real-time sample array (RT-SA), and enumeration objects. Among others, the attributes of the metric object describe the measurement's type (e.g., pulse rate), body location, unit code, time stamp, observation period, and status.

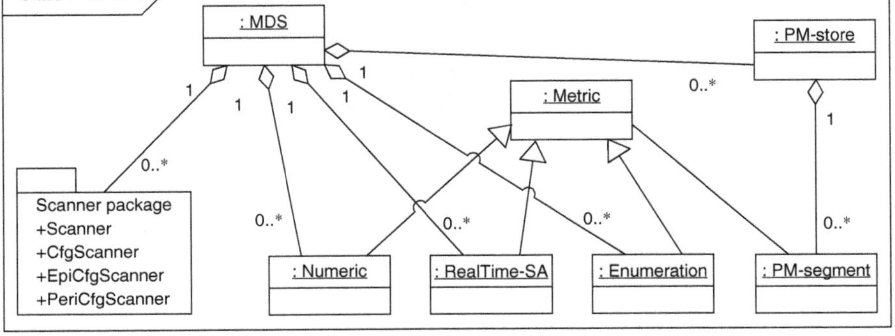

Figure 11.16 ISO/IEEE 11073 personal health device—domain information model.

[1]If the agent is compliant to one or more of the ISO/IEEE 11073-104zz device specializations, this is indicated in the type attribute or a corresponding list of type values.

Numeric Class This class is derived from the metric base class and inherits all its attributes. An instantiation of the numeric class represents a numerical measurement, as for example blood pressure, heart rate, or weight. The attributes of the numeric object describe the measurement's observed value and accuracy.

RT-SA Class This class is derived from the metric base class and inherits all its attributes. The RT-SA class instantiations represent waveform measurements, as for example electrocardiographic or plethysmographic waveforms. The attributes of the RT-SA class describe the observed values of the sample array, the sample period, the sample array size, the size of the samples (e.g., 16-bit), and scale and range information for mapping the scaled values to absolute values.

Enumeration Class This class is derived from the metric base class and inherits all its attributes. An instantiation of the enumeration class represents status information and/or annotation information, as for example information classifying a cardiac event or information on the device status and its potential malfunction.

Scanner Classes The scanner package comprises a set of scanner classes. A scanner is a packaging and reporting mechanism that enables collected sets of observations from one or more metric-derived objects to be packed together in a very efficient manner. Furthermore, a scanner allows the manager to control the flow of data. There are different scanner classes for episodic and periodic measurements. Regarding the former, an episode is defined by the application, with the time between episodes being unknown. In contrast, for periodic scanners the data are reported within fixed time intervals. As an example, heart rate measurements can be periodic (e.g., average heart rate reported every second) or episodic (e.g., beat-to-beat heart rate measurements).

PM-Store Class Objects of the persistent metric (PM) store class represent the permanent storage mechanisms on an agent for metric data that are accessed by the manager at a later time. The data are stored in a variable number of PM segment objects. An example for the use of PM store objects is an overnight study, where a pulse oximetry agent may store many minutes or hours of oximetry data while being unable to retain a communication link with a manager. After the long-term data acquisition is complete, the manager retrieves the data from the agent.

PM-Segment Class Objects of the PM segment class contain persistent metrics associated with a PM store. An instance of the PM segment class represents a persistently stored episode of measurement data. There can be many PM segments associated with a PM store object.

ISO/IEEE 11073 PHD Service Model The service model provides services to access the information objects defined in the DIM. These services are mapped to messages that are exchanged between the agent and manager. They are used for data reporting and by data access functions provided by the agent. The following object access services are supported:

EVENT REPORT Service This service is used by the agent to send measurement data to the manager. It is available for the MDS object, scanner objects, as well as PM store objects. For the MDS object this service is also used to send the agent's configuration information to the manager.

GET Service This service is used by the manager to retrieve object attribute values. It is available for the MDS object and PM store objects.

SET Service This service is used by the manager to set object attribute values to the agent. The scanner objects support the SET service. In particular, the operational state attribute of the scanner object can be set to enabled/disabled. This provides the manager with control over the data flow generated by the scanner.

ACTION Service This service is used by the manager to invoke actions supported by the agent. ACTION services are supported by the MDS object and PM store objects.

The MDS object supports ACTION services for the manager setting the time of the agent's clock as well as for the manager requesting the transmission of measurement data. In the latter case, the measurement data are transmitted by the agent as part of the response to the manager's request or via the event report service of the MDS object.

The PM store object supports three action services for the manager. One allows the manager to delete the data currently stored in one or more selected PM segments. Another one allows the manager to retrieve PM segment object attributes of one or more selected PM segments. Finally, another action service allows the manager to start the transfer of stored data of a specified PM segment. This latter action is executed by the agent via the event report service supported by the PM store object.

ISO/IEEE 11073 PHD Communication Model ISO/IEEE 11073-20601 defines a connection-oriented protocol supporting a topology of one or more agents communicating over point-to-point connections to a single manager. For each point-to-point connection, the dynamic system behavior is defined by a connection state machine. The connection state machine defines the states and substates an agent and manager pair passes through and the detailed entry, exit, and error conditions for the respective states. A simplified view of the agent's state machine is illustrated in Figure 11.17.[2]

In order to start using the object access services described in the previous section, the agent and manager need to enter an association, first. This is achieved by exchanging so-called association messages, which can be considered as connection-management-related command messages.

Association Procedure Starting form the unassociated state, the agent sends an association request message to the manager, thus transitioning (①) to the associating state. An association request message contains basic information for the manager

[2]For the full state machine, refer to ISO/IEEE 11073-20601, which also specifies a corresponding state machine for the manager.

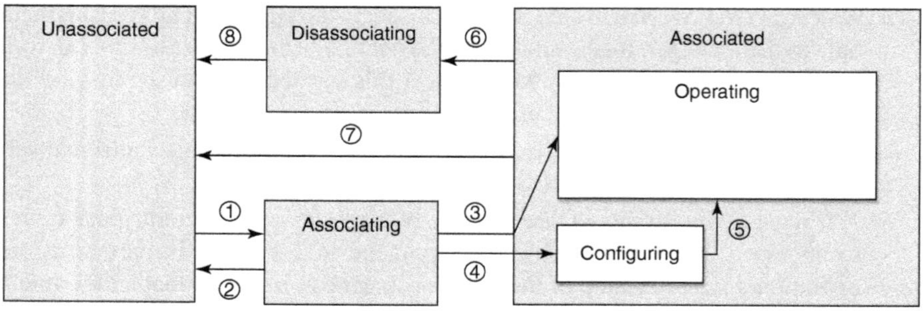

Figure 11.17 ISO/IEEE 11073 personal health device—agent state machine (simplified view).

to assess whether it is feasible to enter an association with the agent. The association request message includes information related to the version of the association protocol used, the data protocol and its version to be used after association has been reached, the rules for encoding[3] the messages into binary messages for transmission, the version of the nomenclature used, some basic device capabilities, the globally unique identification of the agent device, and the identification of the agent's configuration.

If basic interoperability is not possible, the manager replies with an association response message, indicating that the agent's request is rejected. In this case, the agent transitions (②) back to the unassociated state. If the agent's request is accepted, the agent transitions to the associated state. Being associated the agent either directly transitions (③) to the Operating state or it transitions (④) to the configuring state. The former is the case if the configuration is known a priori to the manager, either because the agent has previously connected using the same configuration or because the agent is using a standard configuration[4] as it is defined in some of the device specialization documents (ISO/IEEE 11073-104zz). The latter is the case if the configuration of the agent is unknown to the manager.

Being associated, the agent can always decide to release the association by sending an association release request message to the manager and transition (⑥) into the disassociating state. Once the agent's request is acknowledged by receiving the manager's association release response, the agent transitions (⑧) into the unassociated state. The agent may directly transition (⑦) from the associated into the unassociated state by sending an association abort message to the manger.

Configuring Procedure The set of objects, except the MDS object, and attributes that exists in an agent denotes the agent device configuration. The device configuration is

[3]ISO/IEEE 11073-20601 mandates to support the medical device encoding rules (MDER); see ISO/IEEE 11073-20101 (2004) for a definition of MDER.

[4]Note in this case the value of the configuration identifier is from within a particular range, reserved for standard configurations.

associated with an identifier, which is locally unique, in case the agent owns multiple device configurations.

During the lifetime of an association, the configuration of an agent remains fixed in the sense that the set of objects cannot be changed. However, new attributes can be added to an object or attribute values can be changed. In case another configuration is to be used on an agent, the association is released and established again with the desired new configuration.

The identifier of the configuration that the agent wants to use is sent to the manager as part of the association request. The manager indicates via the association request response message whether or not the configuration of the agent is known to the manager. In case it is not known, the agent transitions to the configuring state and transfers its configuration to the manager using a configuration event report. This report describes all objects that are part of the configuration and contains for each of them the list of present static and dynamic attributes.

If the manager indicates via the configuration report response message that it accepts the configuration of the agent, the agent transitions (⑤) to the operating state. Otherwise, the agent can try a different configuration. If neither of the agent's configurations is acceptable to the manager, then the agent requests to release the association. Agent and manager do not interoperate in this case (see Table 11.4).

TABLE 11.4 Summary of Styles for Transferring Data from Agent to Manager

Reporting Object	Reporting Scope	Service for Data Transfer	Service for Triggering Data Transfer by Manager	Manager Controls Data Flow	Confirmed Mode	Unconfirmed Mode	Variable Format	Fixed Format	Grouped Format
MDS	Metric-derived objects, MDS object	Event report, MDS-Dynamic-Data-Update-*	Initiated by agent	—	•	•	•	•	—
MDS	Metric-derived objects, MDS object	Event report, MDS-Dynamic-Data-Update-*	Action, MDS-Data-Request	•	•	•	•	•	—
Scanner	Metric-derived objects	Event report, Buf-Scan-Report-*, Event report, Unbuf-Scan-Report-*	Set	•	•	•	•	•	•
PM-Store	PM-Segment objects	Event report, Segment-Data-Event	Action, Trig-Segment-Data-Xfer	•	•	—	—	—	•[a]

[a]Though termed differently, the format of the Segment-Data-Event messages is comparable to the grouped format of the scanner event report services.

Operating Procedure During the operating procedure the actual communication of health data and status information occurs.

The concept of device configurations makes the transmission of observational attribute values very efficient during the operating state, because the related static and dynamic attributes relevant for the correct interpretation of the observational attribute values are already known to the manager. They have been reported initially as part of the configuration report and do not have to be included with each transmission of observational attributes. However, if the value of a dynamic attribute changes, the updated value needs to be conveyed to the manager prior to the time when the value would be needed for interpreting a reported observation.

In order to actually transmit measurement data (observational attributes) from the manager, the services of the MDS, PM store, or scanner objects can be used.

(a) MEASUREMENT DATA TRANSFER USING MDS OBJECT

Agent Initiated When the agent has small amounts of episodic data to be transferred to the manager, it can use the MDS-Dynamic-Data-Update-*[5] event report service of the MDS object. The agent includes a spontaneous measurement into the message that corresponds to the MDS-Dynamic-Data-Update-* service and transmits it on its own initiative, that is, without being requested by the manager. The agent can even include multiple measurements that have been temporarily stored on the agent into a single message. The limit for the number of measurements to be included in a single MDS-Dynamic-Data-Update-* service message is 25. In general, if storage of more than 25 measurements is required on the agent, it is recommended to use the PM store mechanism for persistently stored metrics as described below.

Manager Initiated Another style of transmitting measurement data via the MDS object is to use the MDS-Data-Request action service. This is an optional service, useful for applications involving streaming measurement data (e.g., waveform data) as it allows the manager to control the flow of data. By sending a MDS-Data-Request service message, the manager can start or stop requesting data from the agent. The manager may choose to request all data available at the agent, data available according to a particular object class, or data available according to particular objects. There is no requirement that the agent starts performing measurements for collecting data based on such a request by the manager; rather the request aims at retrieving the data currently available at the agent.

Regarding the temporal scope of the request, three modes can be used: single response, time period, and no time limit. After having confirmed the request from the manager, the agent transfers available data to the manager by

[5]The asterisk is a placeholder for various variants of this service. There are variants for fixed and variable format messages as well as variants suitable for multiperson applications, with a user identifier included in the data update messages. For confirmed messages, the manager sends a response to acknowledge the reception of the sent data.

using the MDS-Dynamic-Data-Update-* event report service until either the time period has expired (time period mode) or the manager sends another MDS-Data-Request indicating to stop the sending of data (time period mode and no time limit mode). For the single response mode, the data sent by the agent is included in the response message to the manager's request, that is, the Dynamic-Data-Update-* service is not used in this case.

(b) MEASUREMENT DATA TRANSFER USING SCANNER OBJECTS Measurement data transfer via scanner objects is done by using the event report services of the scanner objects, that is, the Unbuf-Scan-Report-* service for periodic scanners and the Buf-Scan-Report-* service for episodic scanners. For periodic scanners, Buf-Scan-Report-* service messages are sent periodically with a fixed time interval between two messages. More than one measurement from the same object can be included in each message. In contrast, for an episodic scanner, no more than one measurement per object is included in an Unbuf-Scan-Report-* service message, as the message is transmitted when the measurement episode is completed.

Commonly, multiple scanners are implemented if different measurement reporting requirements are to be accomplished. For example, a pulse oximeter might use a periodic scanner with a reporting interval of 50 msec for the RT-SA object representing the plethysmogram, a periodic scanner with a reporting interval of 1sec for the numeric representing the saturated oxygen level and any enumeration objects reporting on events associated with the value, and an episodic scanner for beat-to-beat metric objects (numeric or enumeration).

Similar to the manager-initiated MDS event reporting, as described in (a), the data transfer for scanners is controlled by the manager. The manager needs to set the operational state attribute of the scanner object to enable data transfer via Buf-/Unbuf-Scan-Event-Report* messages by the scanner. This is done by utilizing the SET service of the scanner object. Similarly, the manager can stop the scanner sending Buf-/Unbuf-Scan-Event-Report* messages by using the SET service for setting the scanner object's operational state attribute to disabled.

Another property of scanner-based event reporting is that the Buf-/Unbuf-Scan-Event-Report* service messages allow an optimized data format, which is more compact compared to the data format when using MDS object event report services in particular when data from multiple objects are reported. This and the property of manager-controlled data flow make measurement data transfer via scanners the preferred choice of applications involving multiple metric objects and streaming measurement data.

(c) MEASUREMENT DATA TRANSFER USING PM STORE OBJECT Measurement data by transfer via a PM store object is done by using the Segment-Data-Event event report service of the PM store object. This service enables sending the data stored in one of the PM segments associated to the PM store object.

The manager triggers the data transfer of a PM segment by using the Trig-Segment-Data-Xfer action service of the PM store object. By means of a

Trig-Segment-Data-Xfer response message, the agent indicates whether the manager's request can be honored. If yes, the agent sends Segment-Data-Event messages until all the data stored in the PM segment is transferred. The Segment-Data-Event messages are confirmed messages. The manager can use a certain bit field in the confirmation message to abort the transfer of the PM segment.

Formatting Styles for Measurement Data According to the object-oriented DIM, the data to be transferred from the agent to the manager are represented as attribute values. Figure 11.18 illustrates the three different styles for arranging attributes within messages that correspond to the different event report services: variable format, fixed format, and grouped format. There are dedicated services for each of the three styles. For example, within the set of MDS-Dynamic-Data-Update-* event report services, the MDS-Dynamic-Data-Update-Var service represents the service that uses the variable format.

The variable format, as illustrated in Figure 11.18*a*, explicitly defines each reported attribute by including the object handle, which identifies the object the attribute is part of, the attribute identifier, the length of the attribute value in terms of bytes, and the attribute value. This format is self-contained with regard to the manager being able to determine to which attributes and objects the data values correspond. As a consequence the variable format provides most flexibility

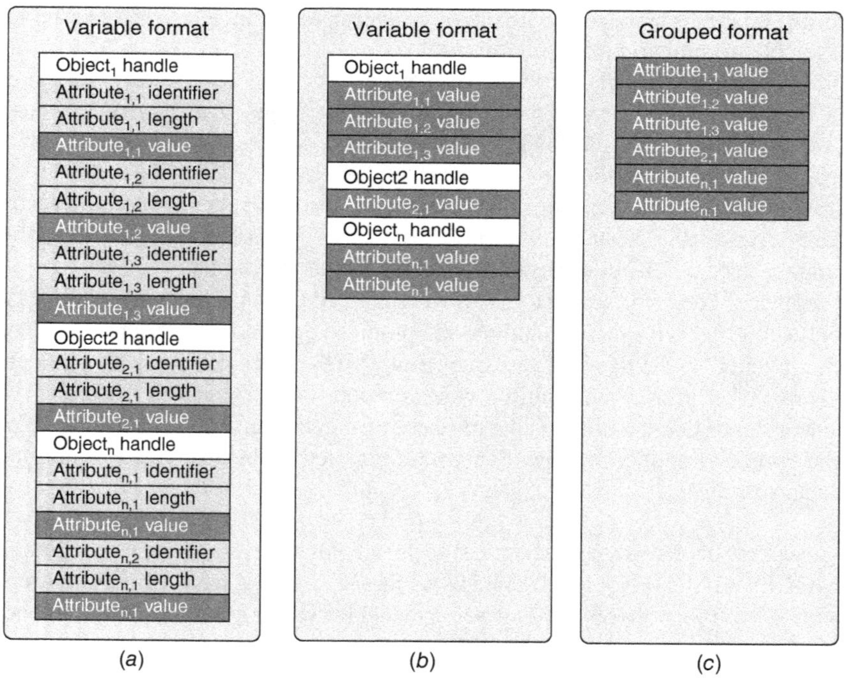

Figure 11.18 Formatting styles for event report services.

for including a different set of attributes per message. The disadvantage is a considerable message overhead for applications where the same set of attributes is reported repeatedly.

For the fixed-format style, as illustrated in Figure 11.18*b*, the messaging is optimized by defining for each object a specific list of attributes to communicate and the order in which they appear in the message. This list is an attribute of the metric object. It holds the attribute identifiers and the length of the attribute values in the order they appear in the fixed-format event report messages. The list is transmitted to the manager before fixed-format messages are transmitted. On receipt of a fixed-format event report message, the manager uses the previously transmitted list to know how to extract the data. By means of the fixed-format messaging style, the overhead of sending attribute identification and length in each event report is avoided.

The grouped-format style, as illustrated in Figure 11.18*c*, is further optimized by defining a single list of attributes of one or more different objects. The notion is similar to that of the fixed-format style. The attributes are listed in the order they appear in the grouped-format message. This list holds the attribute identifiers, the length of the attribute values, as well as the object handles to identify the corresponding objects. This style is available for scanner objects only and the list is an attribute of the scanner object. The list is transmitted to the manager before grouped-format messages are transmitted. On receipt of a grouped-format event report message, the manager uses the previously transmitted list to know how to extract the data. By means of the grouped-format messaging style, the overhead of sending the scanned object's handles, their attribute identification, and data lengths in each event report is avoided.

Example: IEEE 11073-10406 Basic ECG (1–3 Lead) In this section an example of a particular device specialization is considered which serves to illustrate the concepts described in the previous sections. The IEEE 11073-10406 Basic ECG (1–3 Lead) standard (IEEE 11073-10406, 2011) specifies a device specialization for basic ECG devices used for monitoring purposes. Monitoring ECG devices are distinguished from diagnostic ECG equipment with respect to including support for wearable ECG devices, limiting the number of leads supported by the equipment to three and not requiring the capability of annotating or analyzing the detected electrical activity to determine known cardiac phenomena. In the context of the standard a basic ECG is used for the purpose of acquiring and recording one to three channel (leads) electrocardiographic waveforms or analyzing the acquired signals to measure heart rate.

In general, an ECG device with associated lead wires and electrodes measures the electrical activity of the heart. More precisely, it measures the electrical potential differences between electrodes placed on the person's body, reflecting the sum of the electrical activities of muscle fibers. These electrical activities are related to the myocardial muscle but also include artifacts caused by electrical activities of other muscle fibers after movement. The electrical potential is measured by means of a

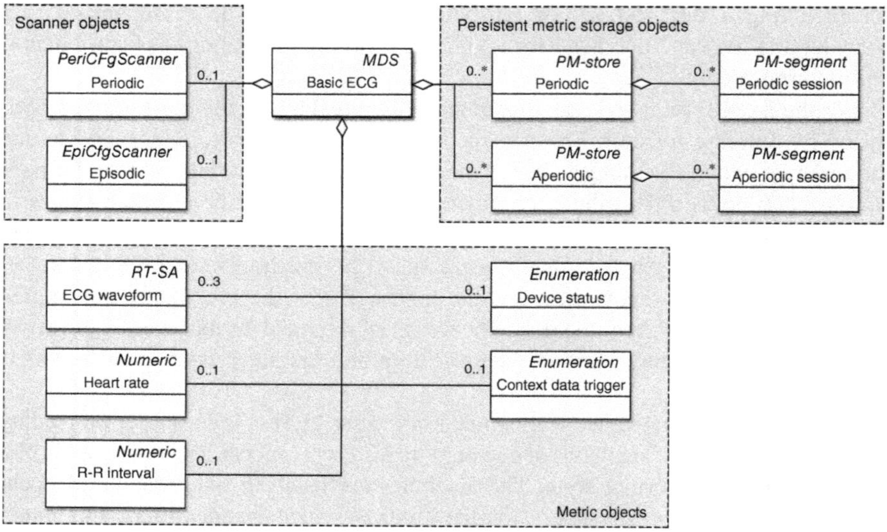

Figure 11.19 IEEE 11073-10406 Basic ECG (1—3 Lead)—domain information model.

system of conducting wires attached to a person using electrodes. The electrodes are placed on specific locations on the surface of the person's body.

Domain Information Model The domain information model of a basic ECG as defined in IEEE 11073-10406 (2011) is depicted in Figure 11.19. In the following, the different objects are described:

- *ECG Waveform RT-SA Object* ECG waveforms represent a continuous stream of measured electrical potential differences within a certain time period and are typically used for cardiac rhythm monitoring. ECG waveforms are modeled as a series of samples, with each waveform (lead) represented as a separate RT-SA object. Up to three leads are supported. As per the specification in IEEE 11073-10406, ECG waveforms are transmitted by means of the scanner event report service. Regarding electrode placement, the ECG lead types that are supported by the standard are illustrated in Figure 11.20.
- *R–R Interval Numeric* Figure 11.21 illustrates the basic ECG signal and specific waveforms that may occur, for example, during atrial depolarization (P wave), ventricular depolarization (QRS complex), and ventricular repolarization (ST-T wave). The R–R interval (interbeat interval) is an instantaneous measurement defined as the time between the maximums of two consecutive R waves (early ventricular depolarization), and it is typically indicated in milliseconds.
- *Heart Rate Numeric* Heart rate is defined as the number of heartbeats per unit of time, typically normalized to beats per minute. A heart rate measurement

Lead type
Unspecified lead
Lead I
Lead II
Lead III
Augmented voltage right (aVR)
Augmented voltage left (aVL)
Augmented voltage foot (aVF)
Lead V1
Lead V2
Lead V3
Lead V4
Lead V5
Lead V6

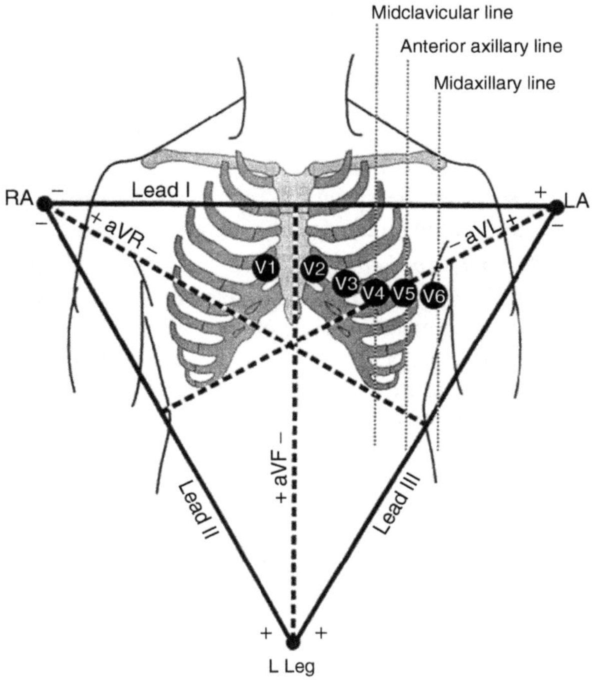

Figure 11.20 Supported ECG lead types.

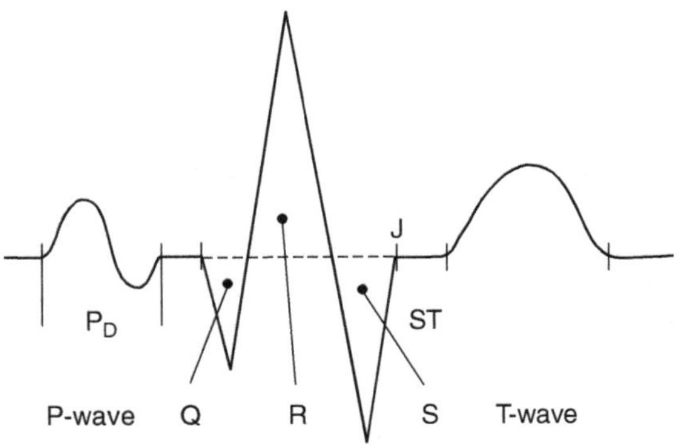

Figure 11.21 Basic form of ECG signal.

based on a single beat interval is called the instantaneous heart rate, defined as the inverse of a single R–R value. The instantaneous heart rate is often fluctuating and is therefore filtered (averaged), reducing higher frequency content for the purposes of display.

- *Device Status Enumeration* The device status enumeration object informs the user of device-related events, such as lead wire loss or lead signal loss.

- *Context Data Trigger Enumeration* The context data trigger enumeration object can be used to inform the user for the reason for transmitted segments of ECG waveform, heart rate, or R–R interval measurement data, such as a user pressing a button or an automatic event due to a data analysis mechanism in the agent (e.g., in case of detection of a cardiac event). In the existing version of the standard, the data trigger context attribute is defined rather general with respect to the specificity of detected cardiac events, not providing detailed information on the specific cause of the event.
- *Periodic/Aperiodic PM Store* In the context of personal health devices, basic ECG agent devices are portable or mobile devices. These devices may be used to collect measurements or observations at a time when out of the network and agent/manager associations cannot be established. It is also common that a given set of measurements or observations made by basic ECG agents may need to be uploaded to more than one manager, for example, in the home and at a medical facility.
  In order to account for these use cases, IEEE 11073-10406 supports PM store-based data transfer. The defined PM store model utilizes PM store objects for the persistent storage of periodic data, as for example, ECG waveform measurements or periodic heart rate measurements, and for aperiodic data, related for example to device status enumerations.
- *Periodic/Episodic Scanner* Periodic scanners are supported to enable efficient streaming of ECG waveform data. Episodic scanners account for nonperiodic data transmission employing, for example, the streaming of R–R interval measurements.

Standard Configuration IEEE 11073-10406 defines a standard configuration for simple heart rate devices, such as heart rate breast straps used in the fitness domain. It is identified by means of a configuration identifier (0 x 258), which is unique within the range of 11073 PHD standard configurations. The DIM for this standard configuration contains a single heart rate numeric object. The corresponding object instance diagram is illustrated in Figure 11.22, showing also the specified attribute values for the instantiation of the heart rate numeric object. Standardized MDC

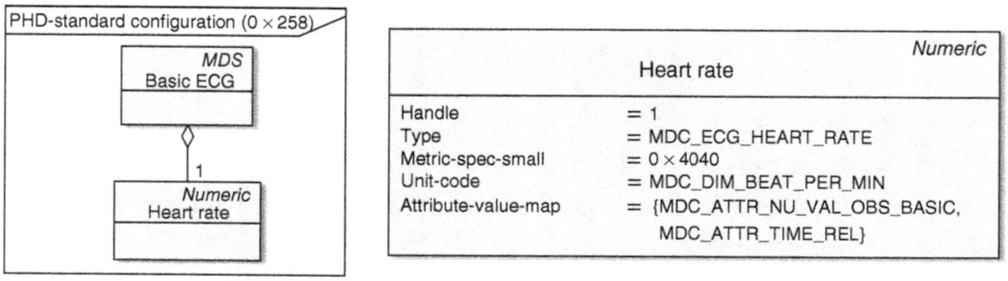

Figure 11.22 Basic ECG standard configuration for simple heart rate devices.

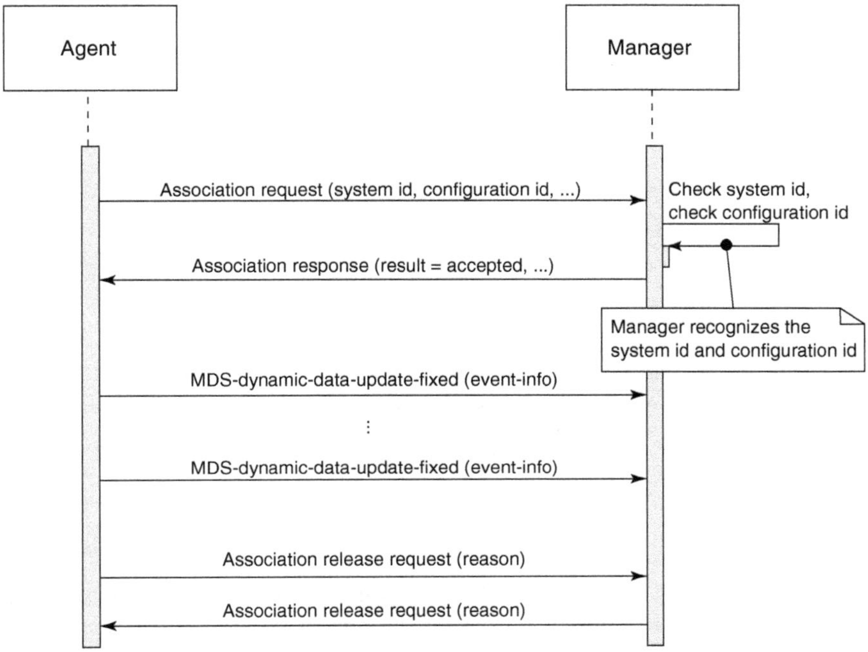

Figure 11.23 Sequence chart example for unconfirmed event reporting.

codes are used to uniquely define the type of measurement, the unit of measurement, and the list of attribute types included in the Attribute-Value-Map attribute for fixed-format messaging.

Message Sequence Example In Figure 11.23 a sequence diagram of the messaging procedure is shown for the example of an agent implementing the standard configuration (0 x 258) of the Basic ECG (1–3 lead) device specialization. According to the standard configuration the agent is capable of performing heart rate measurements.

In this example, the manager checks its internal data base and recognizes the standard configuration used by the agent. It sends an association response to the agent's association request with the result "accepted.'' As a consequence of this, both devices immediately enter the operating state. Subsequently, the agent transmits a series of measurements using the fixed format version of the MDS event report service (MDS-Dynamic-Data-Update-Fixed). In this example, the data messages are sent in the unconfirmed mode, with the manager not acknowledging the receipt of data messages. Finally, the user ends the measurement session, for example, by pushing a proper button on the device or just by not using the device for a time period with duration longer than a specified one. As a consequence, the agent disassociates from the manager by sending an association release request. The manager responds with an association release response.

0×E7 0×00	APDU CHOICE Type (PrstApdu)
0×00 0×38	CHOICE.length = 36
0×00 0×36	OCTET STRING.LENGTH = 34
0×00 0×04	invoke-id = 4
0×01 0×00	CHOICE(Remote Operation Invoke | Unconfirmed Event Report)
0×00 0×30	CHOICE.length = 28
0×00 0×00	obj-handle = 0 (MDS object)
0×00 0×00 0×00 0×00	event-time = 0
0×0D 0×1D	event-type = MDC_NOTI_SACN_REPORT_FIXED
0×00 0×26	event-info.length = 18
0×F0 0×00	ScanReportInfoFixed.data-req-id = 0xF000
0×00 0×01	ScanReportInfoFixed.scan-report-no = 1
0×00 0×01	ScanReportInfoFixed.obs-scan-fixed.count = 1
0×00 0×1E	ScanReportInfoFixed.obs-scan-fixed.length = 10
0×00 0×01	ScanReportInfoFixed.obs-scan-fixed.value[0].obj-handle = 1
0×00 0×06	ScanReportInfoFixed.obs-scan-fixed.value[0]. obs-val-data.length = 6
0×00 0×79	Basic-Nu-Observed-Value = 121 bpm
0×00 0×00 0×E2 0×90	Relative-Time-Stamp = 7.25 s

Figure 11.24 Encoded message example.

Protocol Data Unit Example Finally, an example of an MDER-encoded data message is shown in Figure 11.24. The message in binary format represents a MDS-Dynamic-Data-Update-Fixed unconfirmed event report message with length 40 bytes sent by a Basic ECG (1–3 lead) agent. The last 6 bytes contain the actual heart rate measurement and relative time stamp information. Due to the fact that during the operating state the configuration of the agent is known to the manager, static context information such as unit of measurement, type of measurement, and numerical format of the measurement value does not need to be included in the event report messages.

Note that the highlighted bytes represent the actually varying parts of the message. This is in accordance with the principle of canned messages, which allows the use of predefined message templates. Only the fixed-location, varying parts need to be modified before sending. This reduces the parsing efforts at the agent side significantly.

11.4 CONCLUSION

To address the paradigm shift in health care delivery toward utilizing telehealth services, payers, care providers, device manufacturers, software and middleware vendors, information technology suppliers, and silicon device manufacturers have organized efforts via industry alliances such as the Continua Health Alliance and other standards development organizations to solve the challenge of developing simple and interoperable personal health care devices.

ISO/IEEE 11073 PHD offers high-level, semantic interoperability between sensors/actuators and aggregators, which are connected to back-end care provider information management systems. Being portable to different transport technologies, the ISO/IEEE 11073 PHD protocols have entered dedicated transport profile standards such that interoperability of the overall communication stack is guaranteed. One example is the ZigBee Health Care profile (ZigBee Alliance, 2010), which contains specifications for using ISO/IEEE 11073 PHD over the ZigBee/IEEE 802.15.4 transport. ZigBee Health Care has been adopted into the Continua interoperability guidelines, which enable the development of interoperable end-to-end products and solutions.

IEEE 802.15.6 and IEEE 802.15.4j are the latest examples of standard-based wireless transport technologies that are particularly suitable for biosensor communication, with the potential to leverage the new regulated MBAN spectrum opened by the FCC for medical applications.

There is now high potential to establish ecosystems for achieving widespread interoperability and interconnectedness required to support a robust health monitoring infrastructure at home and inside institutional settings as well. This will allow patients using these personal health care devices to live their lives with greater comfort and autonomy.

REFERENCES

Continua Health Alliance. 2012. Available: http://www.continuaalliance.org/index.html.

ERC/REC 70-03. 2012. ERC Recommendation 70-03 Relating to the Use of Short Range Devices (SRD). CEPT. Available: http://www.erodocdb.dk/docs/doc98/official/pdf/rec7003e.pdf.

ETSI EN 301 839. 2009. Electromagnetic Compatibility and Radio Spectrum Matters (ERM); Short Range Devices (SRD); Ultra Low Power Active Medical Implants (ULP-AMI) and Peripherals (ULP-AMI-P) Operating in the Frequency Range 402 MHz to 405 MHz. ETSI. Available: http://www.etsi.org/deliver/etsi_en/301800_301899/30183901/01.03.01_60/en_30183901v010301p.pdf.

ETSI TR 101 557. 2012. System Reference Document (SRdoc); Medical Body Area Network Systems (MBANSs). Available: http://www.etsi.org/deliver/etsi_tr/101500_101599/101557/01.01.01_60/tr_101557v010101p.pdf.

Evans D. 2012. IEEE P802.15-11-0836-06-0004j,TG 4j Amendment Draft, March 2012.

Falck T, et al. 2006. BASUMA—the sixth sense for chronically ill patients. In Proc. 3rd Intl. Workshop on Wearable and Implantable Body Sensor Networks, Cambridge, MA, April 2006.

Federal Communication Commission (FCC). 1999. 47 CFR 95.601-95.673 Subpart E. Available: http://wireless.fcc.gov/services/index.htm?job=operations&id=medical_implant.

Federal Communications Commission (FCC). 2012a. Available: http://www.fcc.gov.

Federal Communication Commission (FCC). 2012b. Medical Device Radiocommunications Service (MedRadio). Available: http://www.fcc.gov/encyclopedia/medical-device-radiocommunications-service-medradio.

Federal Communication Commission (FCC). 2012c. FCC 12-54. Amendment of the Commission's Rules to Provide Spectrum for the Operation of Medical Body Area Networks. First Report and Order and Further Notice of Proposed Rulemaking. Available: http://hraunfoss.fcc.gov/edocs_public/attachmatch/FCC-12-54A1.pdf.

Federal Communication Commission (FCC). 2012d. Wireless Medical Telemetry Service (WMTS). Available: http://www.fcc.gov/encyclopedia/wireless-medical-telemetry-service-wmts.

HART. 2008. WirelessHART Device Specification, HCF_SPEC-290.

IEEE. 1990. IEEE Standard Computer Dictionary.

IEEE 802.11. 2012. IEEE Standard for Information Technology—Telecommunications and Information Exchange between Systems Local and Metropolitan Area Networks—Specific Requirements Part 11: Wireless LAN Medium Access Control (MAC) and Physical Layer (PHY) Specifications. IEEE, March 2012.

IEEE 802.11a. 1999. IEEE Standard for Telecommunications and Information Exchange between Systems—LAN/MAN Specific Requirements—Part 11: Wireless Medium Access Control (MAC) and Physical Layer (PHY) Specifications: High Speed Physical Layer in the 5 GHz band. IEEE, January 1999.

IEEE 802.15.4. 2011. IEEE Standard for Information Technology—Telecommunications and Information Exchange between Systems—Local and Metropolitan Area Networks—Specific Requirements Part 15.4: Wireless Medium Access Control (MAC) and Physical Layer (PHY) Specifications for Low-Rate Wireless Personal Area Networks (WPANs); September 2011.

IEEE 802.15.6. 2012. IEEE Standard for Local and Metropolitan Area Networks—Part 15.6: Wireless Body Area Networks, IEEE, February 2012.

IEEE 11073-10406. 2011. Health Informatics—Personal Health Device Communication—Part 10406: Device Specialization—Basic Electrocardiograph (ECG) (1- to 3-lead ECG).

ISO/IEEE 11073-10101. 2004. Health Informatics—Point-of-Care Medical Device Communication—Part 10101: Nomenclature.

ISO/IEEE 11073-20101. 2004. Health Informatics—Point-of-Care Medical Device Communication—Part 20101: Application Profiles—Base Standard.

ISO/IEEE 11073-20601. 2010. Health Informatics—Personal Health Device Communication—Application Profile—Optimized Exchange Protocol.

International Telecommunication Union (ITU). 2012. Radio Regulations. Available: http://www.itu.int/pub/R-REG-RR.

Keehan SP, et al. 2011. National Health Spending Projections through 2020: Economic Recovery and Reform Drive Faster Spending Growth. *Health Affairs*, 30(8): 1594–16052011.

Kenney P. 2007. Available: https://www.isa.org/store/products/product-detail/?productId=121023.

Microchip. 2010. MiWi Wireless Networking Protocol Stack, AN1006.

Philips. 2012a. Available at: http://www.healthcare.philips.com/main/products/patient_monitoring/products/intellivue_cableless/.

Philips. 2012b. Available: http://www.vitalsignscamera.com/.

Schmitt L, Falck T, Wartena F, Simons D. 2007a. Towards plug-and-play interoperability for wireless personal telehealth systems. In Proc. 4th Intl. Workshop on Wearable and Implantable Body Sensor Networks, Aachen, Germany, March 2007.

Schmitt L, Falck T, Wartena F, Simons D. 2007b. Novel ISO/IEEE 11073 Standards for Personal Telehealth Systems Interoperability. 2007 Joint Workshop on High Confidence Medical Devices, Software, and Systems and Medical Device Plug-and-Play Interoperability, pp. 146–148.

Simons D, et al. 2006. Remote patient monitoring solutions. In Spekowius G, Wendler T (eds). *Advances in Healthcare Technologies*. Springer.

Steventon A, et al. 2012. Effect of telehealth on use of secondary care and mortality: findings from the Whole System Demonstrator cluster randomised trial. *BMJ*, 344:e3874.

Takahashi D. 2009. Tzero Technologies shuts down; that's the end of ultrawideband. Available: http://venturebeat.com/2009/02/12/tzero-technologies-shuts-down-thats-the-end-of-ultrawideband/.

Wartena F, Muskens J, Schmitt L. 2009. Continua: The impact of a personal telehealth ecosystem. In *Proc. Intl. Conference on ehealth, Telemedicine, and Social medicine*.

WebMD. 2012. Available: http://www.webmd.com/healthy-aging/medical-cost-disability-11/slideshow.

ZigBee Alliance, 2010. Health Care Profile Specification, version 1.0.

ZigBee Alliance, 2012. Available: http://www.zigbee.org/Specifications/ZigBee/Overview.aspx.

12

CONTEXT-AWARE SENSING AND MULTISENSOR FUSION

STEFAN HEY
KIT Karlsruhe Institute of Technology, Karlsruhe, Germany

12.1 INTRODUCTION

The assessment of context information and the analysis of information from multiple sensors, including environmental sensors or context sensors, is one of the most challenging tasks in mobile biomedical applications. Beside this, various other sensing applications in different domains are possible due to the availability of multiple sensors that can be networked using the wireless technologies. During the last years, there are a lot of activities in the development of such systems for environmental and habitat monitoring, including applications for ambient assisted living, tracking, travel and transportation, disaster management, and many others.

The deployment of such multisensor systems allows us to obtain fine-grain real-time information about the physical world as well as to analyze this information together.

12.2 CONTEXT-AWARE SENSING

The use of mobile and telemedical monitoring systems for the surveillance of patients suffering from chronic diseases is a very promising method to avoid negative health consequences. Some of the main advantages are the avoidance of hospitalization as a consequence of an improved therapy control, for patients with chronic heart failure,

Handbook of Biomedical Telemetry, First Edition. Edited by Konstantina S. Nikita.

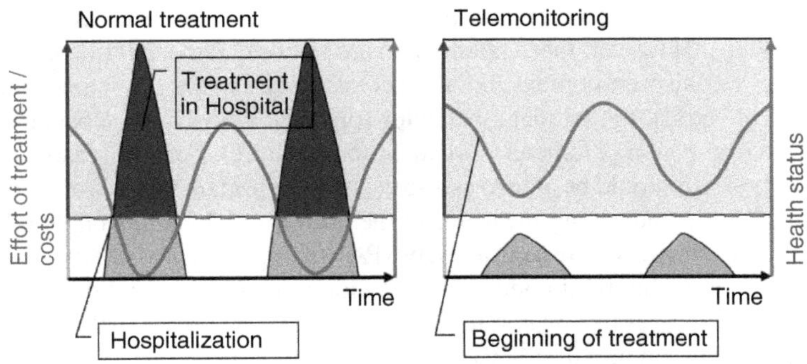

Figure 12.1 Treatment and health status in telemedical applications.

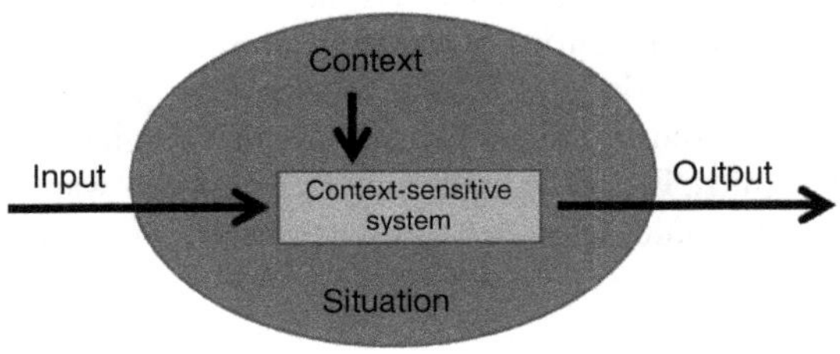

Figure 12.2 Context-sensitive system.

for example, the reduction of cost for the whole health care system and the increase of health status and quality of life of patients (see Figure 12.1).

In normal treatment, the intensity of treatment and the related cost are increasing rapidly when a patient is going to the hospital because of an increasing deterioration in health status. In case of telemedical monitoring of chronic diseases, a decrease in health status could be detected much earlier and hospitalization could be avoided.

A context-sensitive system is defined as a system that gets input information (e.g., from sensors) and information about the context of a specific situation to generate the output signal (see Figure 12.2).

Context sensitivity and context-sensitive behavior allow the adaption of the properties of a system according to the user's needs. In contrast to conventional systems that react always in the same way to a user interaction, context-sensitive systems can react in different ways according to the patient's situation. Possible context information that could be taken into account are, for example, the location of a person (at home, at work, in bed, in vacation, etc.), physical parameters from the environment (like noise, temperature, light, air pressure, weather, etc.), the actual situation (alone or together with other persons like friends, family, colleagues, chief, etc.), or behavior of a person (eating, drinking, smoking, doing sports, etc.).

The importance of context monitoring and context recognition could be explained if we take into account one of the most widely used application in telemedicine. In mobile monitoring of cardiovascular diseases, for example, parameters such as the electrocardiogram (ECG) and blood pressure are normally used as indicators of the health status. However, these parameters are not only dependent on the patient's health status but also on context like the actual situation (e.g., mood, emotions, or psychological stress), the patient's behavior (physical activity vs. sedentary behavior), and environmental parameters (weather, temperature). For that reason, a mobile telemedical system should be able to assess the most important factor from the context that influences the underlying measure. In the above-mentioned example of cardiovascular monitoring, the physical activity (PA) of a person and the psychological stress are the most important context information with a strong and direct influence to the primary measures like heart rate (HR), heart rate variability (HRV), pulse transit time (PTT), and blood pressure (BP). In order to get a reliable result while analyzing the physiological signals, these parameters (physical activity, stress) of a person have to be taken into account as additional context information.

The context-aware analysis of the ECG has the goal of reducing wrong abnormality detections like tachycardia during physical exercises. In order to reach this goal, the context-sensitive monitoring system has to assess the physical activity of a person and adapt the threshold in the analysis of the ECG signal in order to improve the validity of the results. This is of great importance if you want to realize a warning or alarming function in case of an emergency based on the analysis of the vital signals.

12.2.1 Classification of Context-Sensitive Systems

Up to now there is no common definition of a context-sensitive system. One of the first attempts to classify context sensitivity was done by Schilit and Theimer (1994). According to this classification, the system could be described by the way context information is acquired (manually or automatically) and the functionality influenced by the contextual information (see Figure 12.3). In the first case context information is used for additional information for a user or for the system; in the second case context information is the basis for actions that are triggered either by the user or automatically by the system.

The range of definitions and classifications of context-sensitive systems varies from more general approaches, where a context-sensitive system is defined by "the ability of computing devices to detect and sense, interpret and respond to aspects of a user's local environment and the computing devices themselves,"(Pascoe et al. 1998) to more specific ones, where context-sensitive systems are characterized by

	Manual	Automatic
Information	Proximate selection & contextual information	Automatic contextual reconfiguration
Command	Contextual commands	Context triggered actions

Figure 12.3 Schilit's taxonomy.

a dynamic change or adaption of their behavior based on the information about the context of the system and the behavior of the user.

12.2.2 Sensor Technologies

The realization of context-aware systems in biomedical applications normally requires a combination of body-worn sensors and environmental sensors. Parameters that have to be assessed for context recognition are temperature, light, sound or noise, body movements and location (indoor and outdoor), interaction with other persons or things (mobile telecommunication devices, household devices, etc.), and many others. If you want to assess a holistic picture of the context, you have to measure as many relevant signals as possible. All these sensors have to be integrated in a multisensor system, and in case of body sensors, they should be small and unobtrusive in order to keep the mobility and compliance of the person. The connection between sensors could be realized by wired or wireless interfaces. Other nonfunctional requirements of the sensors that have to be fulfilled are the wearability, unobtrusiveness, price, size, weight, energy consumption, robustness, calibration, accuracy, and reliability.

Beside the selection of the sensors, the placement on the body or in the environment plays an important role. The placement has to be defined according to the specific application of the system. If you want to detect the movement of a person during a whole day in his everyday life, body-worn sensors are required. But you have to take into account that a placement of an acceleration sensor on the wrist could lead to misinterpretation of the signal. In general, it is possible to place the different sensors all together in one place or distributed over the body. A central adjustment leads to difficulties in context recognition if compared with a system with distributed sensor nodes. On the other hand, a system with a number of different sensors is much more difficult to use and could lead to lower compliance.

12.2.3 Preprocessing

In order to get the desired information about context from sensor signals, different preprocessing methods are applied to the sensor signals to calculate a set of features representing this information. These techniques can be divided in three main areas: algorithms that calculate features of the signal in the time domain, analysis of the frequency parameters of the signal (frequency domain), or the so-called discrete representation domain (Figure 12.4).

Time Domain Some basic information about sensor signal can be calculated by simple mathematical and statistical functions applied to the signal in the time domain. A basic characterization of a signal could be done by calculating the maximum and minimum of a given parameter in a specified time interval or by calculating the range of values. The mean value of a signal in a given time interval is often used to easily filter noise or artifacts from the signal (e.g., caused by movements of the body)

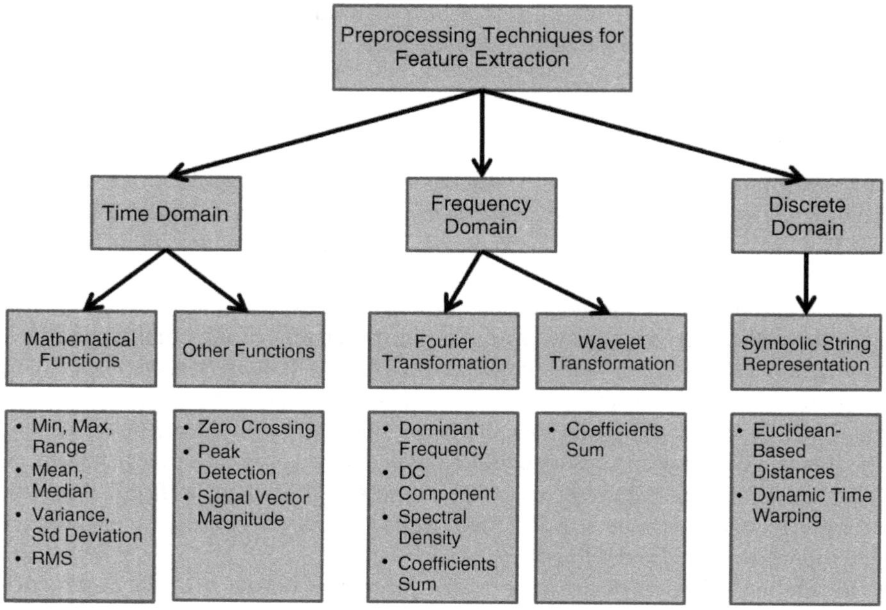

Figure 12.4 Classification of preprocessing techniques for feature extraction in sensor signals.

(Ottenbacher and Heuer, 2009; Kirst et al., 2009a). Most of these mathematical functions can be calculated with low computational effort and are suitable for online analysis of sensor signals also in mobile applications with limited performance (Kirst et al., 2009b).

But there are also more complex signal preprocessing algorithms in time-domain analysis like peak detection [e.g., for R peak detection (Hamilton, 2002) in the analysis of ECG signals or heart rate calculation from photoplethysmogram] or detection of zero crossing that is often used in step detection algorithms for acceleration sensor signals (Tansella et al., 2011).

Frequency Domain To analyze the characteristics of a signal with respect to frequency, the original signal has to be transformed in frequency domain. In signal analysis, the most widely used transformations are Fourier transformation and wavelet transformation (von Wagner et al., 2004). Depending on the transformation method, different techniques for feature calculation are used.

One of the most widely used methods in cardiology is the analysis of HRV (Malik, 1996), which could serve as a risk marker (Sghir et al., 2012) for sudden cardiac arrest. To calculate the power spectrum density (PSD) of short-term HRV, the use autoregressive (AR) modeling is an attractive method instead of fast fourier transform (FFT), which is not particularly suitable (Schäfer et al., 2005).

Discrete Domain Symbolic string representation is used to map an interval of values to defined strings to describe the content of the signal with a limited alphabet. This could lead to a substantial compression of the signal but also to a loss of information (Lin et al., 2003). There are different methods to do this mapping; one of the newest ones is the so-called symbolic aggregate approximation (SAX) (Lin et al., 2007) with piecewise approximation. In monitoring the physical activity of a person, this method could be used to map the acceleration signals to different activity levels.

After the transformation of the signal to the discrete alphabet, matching and distance evaluation techniques are used to classify the signal by identifying known patterns. One example is dynamic time warping (DTW) (Sakoe and Chiba, 1978), which is often used in speech recognition algorithms.

12.3 MULTISENSOR FUSION

Multisensor fusion is the most widely used methodology for context recognition in a variety of applications (Wu et al., 2002). The architecture of multisensor systems was initially inspired by the human cognitive process, where the brain fuses the information from various sensory organs and links these signals to a cognitive understanding of the environment. From the beginning of multisensor systems, they have outgrown the disciplines of robotics and machine learning and can be found in a wide range of applications like medical diagnosis or ambient assisted living (Jatoba et al., 2008; Kunze et al., 2002).

Multisensor fusion implies the combination of output variables from different sensors to achieve information with higher quality. In this context, the definition of "higher quality" depends on the application of the multisensor system. You can increase the quality of the information by increasing the number of the same sensors (e.g., wider visual range in video signals, more channels in ECG or EEG monitoring), by increasing the different modalities of measurement (e.g., acquisition of physical activity and electrocardiogram) or by increasing the measurement range (e.g., combination of infrared sensors, light sensors, and UV sensors).

In general, there are some basic requirements for sensor systems that are also valid for multisensor systems. They have to fulfill the specifications for accuracy, reliability, security, completeness of the required data, as well as some nonfunctional requirements like weight, dimensions, housing, and price. In telemedical systems, some additional requirements have to be taken into account. The sensors have to be biocompatible and also certified as medical products. Special attention has to be given on safety and security aspects as well as on the usability of the sensor system.

Through multisensor fusion, it becomes easier to fulfill the technical requirements. But if you want to increase the accuracy of a sensor system by combining different sensor information, the errors of the sensors have to be independent. Having various types of sensors has in addition the potential to increase the information provided by these sensors. If a system consists of sensors based on different measurement principles, the possibility of detecting a certain event could also be increased. A further advantage of using multiple sensors is the possibility of compensating the failure of a single sensor in order to increase the reliability of the whole system.

Last but not least, sensor fusion techniques could be used to reduce the costs of an expensive measurement system by replacing it by several cheaper sensors.

The above-mentioned advantages of multisensor fusion techniques do have some disadvantages that are caused by the usage of multiple sensors and the connection of the output data.

One of the main problems in multisensor applications is the growth of communication in distributed systems. Depending on the topology of sensor nets and the amount of data generated from each sensor, data transfer rate could lead to a capacity overload of the communication system. But there is also an increase of the communication's complexity due to the higher number of communication partners that require specialized communication protocols with increased overhead for transmitting and processing the data (Sohraby et al., 2007; Dargie and Poellabauer, 2010; Kumar et al., 2012; Gnawali and Welsh, 2011).

Beside the growing requirements in communication, a higher number of sensors is connected with a higher complexity of the overall system. For this reason, the probability of errors in the specification and implementation of the systems increases equivalently. But the higher complexity of a system is also related with higher prerequisite to the user's skills and in case of body-worn sensor systems with higher obtrusiveness and reduced acceptance by the user.

Multisensor systems with separated signal acquisition units and information processing units have to be synchronized before fusing and analyzing the data. The assessment of the different sensors normally differs in sampling time points as well as in sampling frequency. In addition to this, a desynchronization of signals with the same sampling parameters could take place because of instabilities in the clock of the sampling processor, caused by external sources like temperature. To overcome these problems, many efforts for developing synchronization algorithms and methods have been made in recent years (Sivrikaya and Yener, 2004; Chen et al., 2010, Ringwald and Römer, 2007).

All the advantages and disadvantages are summarized in Table 12.1.

TABLE 12.1 Advantages and Disadvantages of Multisensor Systems

Advantages	Disadvantages
• Higher accuracy	• Communication overhead
• Wider range	• Complexity
• More modalities	• Usability
• Redundancy	• Synchronization
• Reliability	• Cost
• Security	
• Costs	

If it is not possible to replace an expensive measurement system by a combination of several cheaper sensors, the cost of a multisensor system can be higher than that of the original single-sensor system.

The fusion of sensor data can be classified in four different types according to their functionality.

- *Concurrent fusion*: To increase the accuracy of a sensor, different sensors of the same type and with the same measurement signal could be combined.
- *Cooperative fusion*: In many sensor fusion applications, one single sensor is not able to deliver the desired information. If you want to differentiate if a person is walking up or down a hill or to get an accurate estimation of the energy expenditure, you can combine the output signal of a classification method for acceleration signals with the data of an air pressure sensor to measure changes in altitude.
- *Complementary fusion*: This fusion is done by independent sensors in order to increase the integrity of the data.
- *Hybrid fusion*: In real systems, different combinations or mixed forms of the fusion types described above are used.

12.3.1 Fusion Architecture and Different Levels of Sensor Data Fusion

The fusion of multisensor data could be done in different levels of the signal analysis process (see Figure 12.5). There are three main methods that could be differentiated.

Data fusion describes a method to combine different raw signals. This type of fusion is also called sensor-level fusion.

In case of feature fusion, the combination of different sensor information is done after preprocessing and feature extraction. The result of this fusion process is the input information for classification.

Decision fusion is done after all the other signal processing steps are completed.

Besides the three main fusion levels, mixed forms of information fusion are possible (hybrid fusion). In hybrid fusion, preprocessed and analyzed data is calculated in parallel with direct sensor data.

Data Fusion Data fusion is the fusion of sensor data from a set of heterogeneous or homogeneous sensors, soft sensors, and history values of sensor data.

This method could be used, for example, in the compensation of movement artifacts of physiological signals like ECG based on the information of an acceleration sensor. Feature extraction and classification is done after the fusion of the signals. In case of data fusion, the amount of data for fusion is not significantly decreased in comparison to raw data.

Feature Fusion If data fusion is done after some preprocessing of the raw signal, this fusion is called central level fusion. In general, a preprocessing unit is located in front of the fusion process module. A widely used method for preprocessing is the

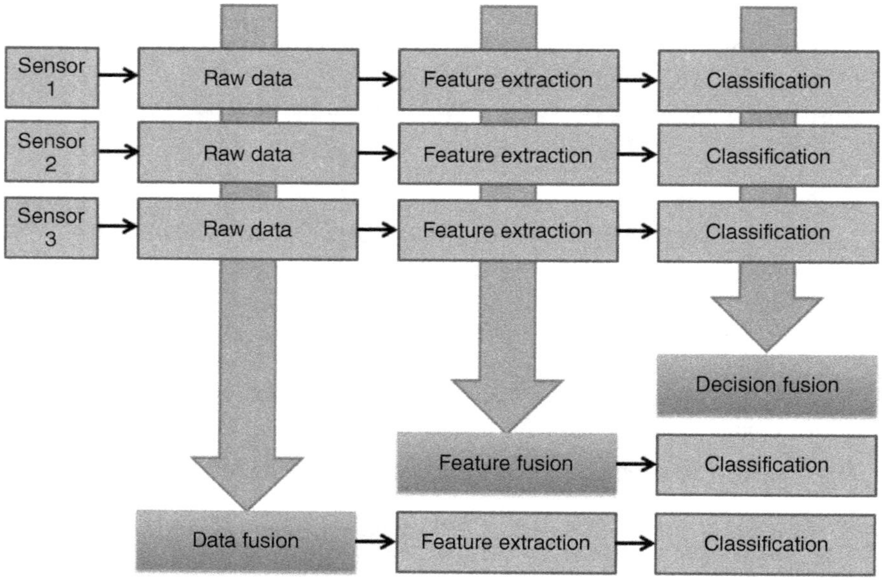

Figure 12.5 Different levels of sensor fusion.

extraction of specific features of the signal. One of the main advantages of preprocessing in general and especially of feature extraction is the significant reduction of the amount of data that have to be processed by the fusion method.

Feature Extraction Feature extraction is one of the central signal processing methods on multisensor fusion applications. The aim of feature extraction is to facilitate the identification of pattern by reducing the amount of data of a continuous signal into a vector of discrete values. The features to be extracted should optimally describe each different class that has to be identified by pattern recognition. In real applications, the feature vectors built up a feature space with uncertain borders between the different patterns.

A classification of features could be done by dividing them in basic features and complex features. Basic features are independent from the signal and the application and can be applied to every signal. Some common examples for basic features are the amplitude, the signal energy, or the standard deviation of a signal. Complex features depend on the application and the modality of the signal.

If you want to extract features for pattern recognition, these features should fulfill a number of requirements. An optimal feature is a measure of an observable and physically measurable parameter that is easy to understand and can be interpreted clearly. Besides this, the feature has to be easy to obtain and to calculate.

Feature Selection The goal of feature selection is to minimize the number of features used for classification and to optimize the classification result. The determination of the optimal number of features used for classification is one of the major issues of feature selection. If the number of features is too big, this can lead to an overfitting and a large number of features increases the complexity of calculation. This could cause problems in mobile applications with limited resources.

To select the optimal number and subset of features, optimization algorithms are normally used. They select the optimal features by calculating the relevance and redundancy of the feature. The relevance of a feature has to be as high as possible; this means that the feature has high data content about the different pattern, and the redundancy has to be low.

There are multiple methods in terms of mathematically obtaining output values through multisensor data, also combined with other data input. Each of these models has an intake of certain explanatory variables representing the features and results in a dependent variable representing output value.

The most simple regression technique is linear regression, entirely calculated by vector, respectively, matrix multiplication. The linear relation is usually determined by minimizing the least squares of the distance from the data values to the resulting linear function.

The first nonlinear, quite similar but more complex approach is support vector regression (SVR) based on support vector machines (SVMs). SVM is a classification technique applied when data points in a vector space are not linearly separable (Anastasopoulou et al., 2012). It is difficult to satisfy reliability because an error always remains, caused by outliers, measurement instability, or bad synchronization (Hsu et al., 2003; Cox, 1972).

In SVR, the data is transformed to a higher dimension. A regression resulting in a function in this higher dimension is performed. Finally, the result is retransformed into the original dimension, resulting in a nonlinear function. The transformation function—into a higher dimension—does not have to be on a linear base. There are four types of kernel functions used: linear, polynomial, radial basis, and sigmoid.

The nonlinear multivariate regression spline model (MARS) has similarities with the linear one. It consists of the sum of weighted basis functions, resulting in a piecewise linear regression function. The basic functions are segments of linear functions calculated in a way that they are connected at each end. Training proceeds by a forward and backward pass. The forward pass starts with an initial term, adding basis functions in pairs that minimize the sum of squares. The backward pass aims to avoid overtraining, meaning to generalize the model better. Basis functions are discarded one by one also in order to minimize complexity.

Artificial neural networks (ANNs) are models inspired by biological neural networks as found in the human brain, consisting of synapses, neurons, and atone. The ANN models consist of neurons, weights, and functions. The input over a specifically designed network form passes from one layer to the next. The sum of weighted income values arriving at one neuron causes it to trigger when a certain threshold is exceeded. The values of weights and thresholds must be trained, which is done with forward and backward training. The quality of the network and run time depends on the predetermined design, defined by the number of layers and number of neurons in the hidden layers.

12.3.2 Decision-Level Fusion

The last type of sensor data fusion uses the output values of different classification modules and is therefore called decision-level fusion. In decision-level fusion, besides the sensor information other sources like a priori knowledge about the environment or context and subjective information from user input are often used to enhance the quality of analysis. In the analysis of physical activity of a person, the output of a classificator that can differentiate between different types of movement (e.g., standing, walking, jogging) can be combined with input from the user to specify the sport (e.g., football, basketball, etc.). Fusion techniques at the decision level vary from simple ones like decision trees to neural networks and modular neural networks.

12.4 EXAMPLE APPLICATION: STRESS MEASUREMENT

The consequences of stress on the health of a person as well as the impact on the health system attracted more and more the public attention in recent years. But there is a lack of instruments that are able to measure stress in everyday life. For a holistic assessment of stress, a number of different parameters from different areas are necessary. These parameters can be derived from a model for the phenomenon "stress" that helps us understand the complex correlations between psychological, physiological, and behavioral levels. According to this model, a telemedical system that allows a comprehensive assessment of stress-related parameters consists of a device for psycho-physiological measurement and a wirelessly coupled Smartphone used as an e-diary and e-questionnaire to assess subjective, behavioral, and environmental parameters.

The integrated stress model that contains aspects from different disciplines was developed by Ice and James (2007); see Figure 12.6. According to this model, stress can be defined as a process that induces emotional, behavioral, and physiological reactions. These reactions depend on the personal, biological, and cultural context of the individual and affect the physical and mental health of that person.

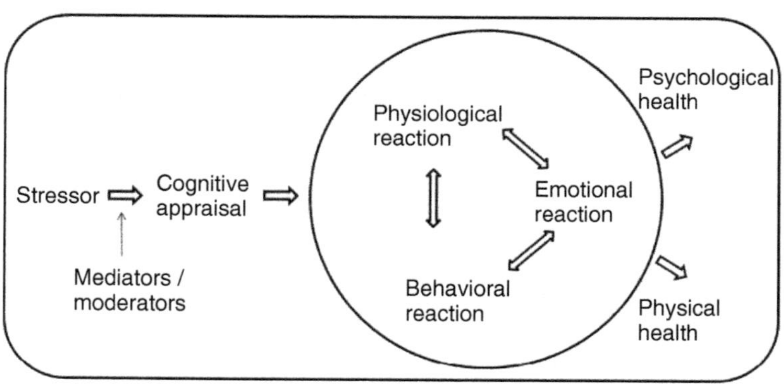

Figure 12.6 Stress model based on Ice and James (2007).

The physiological reactions are caused by an activation of the sympathetic nervous system and by the release of hormones (adrenalin, noradrenalin, testosterone, and cortisol). These hormones are released as a consequence of the activation of the hypothalamic–pituitary–adrenal axis (HPA or HTPA axis), the neuroendocrine system that controls reactions to stress and regulates many body processes, that is, mood and emotions, the immune system, as well as many others. In addition, we can observe behavioral and emotional reactions (like changes in physical activity or time spent sleeping), which lead to negative effects in physical and psychological health (e.g., heart attack, depression).

The different parts of this stress process need different measurement methods. To get a broad and valid assessment of the whole process, you have to acquire as many parameters as possible and take them into account for the analysis of the stress level. In recent years, many hardware and software solutions for multiparametric assessment of physiological signals and experience sampling have been developed. But none of them is able to provide an integrated platform for the assessment of all the parameters defined above.

From the sensor signals, physiological features and contextual information are extracted in real time and can be transmitted to a Smartphone via Bluetooth connection. Online communication is necessary for interactive assessment where subjective data are collected according to changes or events in physiological signals. The mobile data collecting software is designed to run on Smartphones. Smartphones are not only a very good platform for mobile data acquisition but also provide multimedia features such as digital photo or video capture.

12.5 CONCLUSION AND FUTURE RESEARCH DIRECTIONS

In future, focus will be on fusion multisensor and multimodal data. Information from totally different sources like physiological signals, medical image information, and genetic information, subjective information from users as well as context information will be integrated and fused. In addition to the different modalities of this information, these signals have a totally different time behavior and have to be processed online or offline. Therefore, stored data like genetic information or medical images can be fused to a personalized model that will be fused with the more dynamic signals during online analysis.

REFERENCES

Anastasopoulou P, Härtel S, Tubic M, Hey S. 2012. Using support vector regression for assessing human energy expenditure using a triaxial accelerometer and a barometer. Proceedings of the 3rd International Conference on Wireless Mobile Communication and Healthcare, Paris, France.

Chen S, Brantley JS, Kim T, Lach J. 2010. Characterizing and minimizing synchronization and calibration errors in inertial body sensor networks. BodyNets'10, Corfu, Greece.

Cox D. 1972. Estimation of walking energy expenditure by using support vector regression. *J. Roy. Stat. Soc.* Series B, 3526–3529.

Dargie W, Poellabauer C. 2010. *Fundamentals of Wireless Sensor Networks: Theory and Practice.* Hoboken, NJ: Wiley.

Ebner-Priemer UW, Kubiak T. 2007. Psychological and psychophysiological ambulatory monitoring. *Eur. J. Psychol. Assess*, 23:214–226.

Figo D, Diniz PC, Ferreira DR, Cardoso JMP. 2010. Preprocessing techniques for context recognition from accelerometer data. *Personal Ubiquitous Comput.*, 14(7):645–662.

Gnawali O, Welsh M. 2011. Sensor networks architectures and protocols. In *Emerging Wireless Technologies and the Future Mobile Internet*, Chapter 5, Cambridge University Press.

Hamilton P. 2002. Open Source ECG Analysis (OSEA). Comput. Cardiol., 101–104.

Happel B, Murre J. 1994. The design and evolution of modular neural network architectures. *Neural Netw.* 7:985–1004.

Hastie T, Tibshirani R, Friedman J. 2009. *The Elements of Statistical Learning.* Springer.

Hey S, Gharbi A, von Haaren B, Walter K, König N, Löffler S. 2009. Continuous non-invasive pulse transit time measurement for psycho-physiological stress monitoring. Proceeding of the International Conference on eHealth, Telemedicine, and Social Medicine eTELEMED, 2009, pp. 113–116.

Hey S, Sghir H. 2011. Psycho-physiological stress monitoring using mobile and continous pulse transit time measurement. Proceeding of the International Conference on eHealth, Telemedicine, and Social Medicine eTELEMED, 2011.

Hsu C, Chang C, Lin C. 2003. A practical guide to support vector classification. *Bioinformatics*, 1(1):1–16.

Hull R, Neaves P, Bedford-Roberts J. 1997. Towards situated computing. 1st International Symposium on Wearable Computers, pp. 146–153.

Ice GH, James GD. 2007. *Measuring Stress in Humans: A Practical Guide for the Field.* Cambridge Studies in Biological and Evolutionary Anthropology. Cambridge University Press, Cambridge.

Jatobá LC, Grossmann U, Kunze C, Ottenbacher J, Stork W. 2008. Context-aware mobile health monitoring: evaluation of different pattern recognition methods for classification of physical activity. Conf. Proc. IEEE Eng. Med. Biol. Soc., pp. 5250–5253.

Kirst M, Lamparth S, Fuhrhop S, Ottenbacher J, Kunze C. 2009a. Transferring an artifact handling system into a mobile long-term ECG device with dry electrodes. IFMBE Proceedings, World Congress on Medical Physics and Biomedical Engineering.

Kirst M, Ottenbacher J, Kunze C, Stork W. 2009b. Improvement of automatic ECG analysis for telemonitoring systems based on context signals. Proceedings of the Doctoral Consortium on Biomedical Engineering Systems and Technologies BIOSTEC 2009.

Kumar P, Singh MP, Trair US. 2012. A review of routing protocols in wireless sensor network. *IJERT*, Vol. 1.

Kunze C, Grossmann U, Stork W, Müller-Glaser KD. 2002. Application of ubiquitous computing in personal health monitoring systems. *Biomed Tech (Berl).*, 47(Suppl 1 Pt 1):360–362.

Lin J, Keogh E, Lonardi S, Chiu B. 2003. A symbolic representation of time series, with implications for streaming algorithms. In Proc. of the 8th ACM SIGMOD Workshop on Research Issues in Data Mining and Knowledge Discovery.

Lin J, Keogh E, Wei L, Lonardi S. 2007. Experiencing SAX: A novel symbolic representation of time series. *Data Mining Knowledge Discov.*, 15(2):107–144.

Malik M. 1996. Heart rate variability—Standards of measurement, physiological interpretation, and clinical use. *Eur. Heart J.*, 17:354–381.

Ottenbacher J, Heuer S. 2009. Motion artefacts in capacitively coupled ECG electrodes. IFMBE Proceedings, World Congress on Medical Physics and Biomedical Engineering.

Quinlan JR. 1993. *C4.5: Programs for Machine Learning*. Morgan Kaufmann.

Pascoe J, Ryan NS, Morse DR. 1998. Human-Computer-Giraffe Interaction - HCI in the field. Proceedings of the Workshop on Human Computer Interaction with Mobile Devices, Glasgow, Scotland.

Ringwald M, Römer K. 2007. Practical Time Synchronization for Bluetooth Scatternets. BROADNETS 2007, Raleigh, NC.

Sakoe H, Chiba S. 1978. Dynamic programming algorithm optimization for spoken word recognition. *IEEE Trans. Acoust., Speech, Signal Process.*, 26(1):43–49.

Schäfer D, von Wagner G, Hey S, Müller-Glaser K.-D, Meffert B. 2005. Comparative study of different methods and constraints in heart rate variability spectrum estimation. *Biomed. Tech.* 50(Suppl 1, P 1):143–144.

Schilit B, Adams N, Want R. 1994. Context-aware computing applications. 1st International Workshop on Mobile Computing Systems and Applications. pp. 85–90.

Schilit B, Theimer M. 1994. Disseminating active map information to mobile hosts. *IEEE Netw.*, 8(5):22–32.

Sghir H, Löffler S, Ottenbacher J, Stumpp J, Hey S. 2012. Investigation of different HRV parameters during an extended Trier social stress test. 7th Conference of the European Study Group on Cardiovascular Oscillations.

Sivrikaya F, Yener B. 2004. Time synchronization in sensor networks: A survey. *IEEE Netw.*, 4(18):45–50.

Sohraby K, Minoli D, Znati T. 2007. *Wireless Sensor Networks: Technology, Protocols, and Applications*. Hoboken, NJ: Wiley, pp. 203–209.

Stumpp J, Anastasopoulou P, Hey S. 2010. Platform for ambulatory assessment of psycho-physiological signals and online data capture. 7th International Conference on Methods and Techniques in Behavioral Research "Measuring Behavior 2010".

Tansella M, Stumpp J, Hey S. 2011. A ubiquitous system for ambulatory assessment of physical activity in everyday life. 2nd International Congress on Ambulatory monitoring of physical activity and movement ICAMPAM.

von Wagner G, Sommermeyer D, Hey S, Stork W, Müller-Glaser K.-D. 2004. Wavelet-based QRS detection and classification using a neuro-fuzzy approach. *Biomed. Tech.*, 49(2, pt. 1):316–319.

Wu H, Siegel M, Ablay S. 2002. Sensor fusion for context understanding. IEEE Instrumentation and Measurement Technology Conference.

13

SECURITY AND PRIVACY IN BIOMEDICAL TELEMETRY: MOBILE HEALTH PLATFORM FOR SECURE INFORMATION EXCHANGE

NIKOLAOS BOURBAKIS, ALEXANDROS PANTELOPOULOS, AND RAGHUDEEP KANNAVARA

Wright State University, Assistive Technologies Research Center, Dayton, Ohio

13.1 INTRODUCTION

During the past years there has been an abundance of research efforts in the area of wearable systems for health monitoring purposes (Pantenopoulos, 2010; Pantenopoulos and Bourbakis, 2010a). Mainly motivated by increasing health care costs, there has been a trend toward facilitating out-of- hospital patient monitoring using wearable and wireless technology (Tröster, 2005; Akay et al., 2012; Istepanain and Zhang, 2012). By employing on-body sensors that record physiological signals of a user such as the electrocardiogram, the body temperature, the oxygen saturation, the systolic and diastolic blood pressure, and the respiration rate, wearable health monitoring systems (WHMS) can provide real-time updates about a user's physiological status. Data are usually communicated to some type of network managing device, such as a personal digital assistant (PDA) or a custom microcontroller board, which can in turn preprocess and then transmit the captured measurements to a remote station or hospital via a wireless network such as WiFi and General Packet Radio Service (GPRS). As a result, these data can be made available to a remote supervising location in real time.

Handbook of Biomedical Telemetry, First Edition. Edited by Konstantina S. Nikita.

Three important questions that arise now are: (1) Under what conditions are these data recorded and where? (2) What do those data actually mean? (3) How are these health data protected during information exchange? Regarding the first question, the issue is whether the recorded data are actually usable and not corrupted by noise. As physiological sensor technology is far from being robust to measurement artifacts, which can be caused by the user's movement, or by bad contact between the sensor and the skin or by sensor fault, there is a requirement for a means to deduce whether and to what extent the recorded data are to be considered as reliable and thus if they should be further analyzed by signal processing algorithms and decision-making schemes.

The second question raises the issue of data processing, feature extraction, and interpretation. More concretely, pertaining to ECG, the signal needs to be processed in order to perform denoising (Ercelebi, 2004), detect QRS complexes (Kohler et al., 2002), and finally to classify the recorded beats and rhythms as normal or abnormal (Chazal et al., 2004). In addition to that, other signals such as the triaxial acceleration captured from on-body accelerometers can be analyzed to quantify the amount and type of motion the user is performing. Finally, the recorded data and the corresponding extracted features can be all fused together to estimate the progression of the user's health condition (Pantelopoulos and Bourbakis, 2009, 2010c).

The third question is related with the protection scheme that has to be developed to secure the private health information during exchange between the human user and the medical center or remote physician (Canim et al., 2012; Constantinescu et al., 2012). There are many methods in the research field, but there is no "silver bullet" that will offer the optimum protection under the lowest cost. For this particular reason we present here two methodologies that synergistically offer a good solution to this issue. The first methodology (called SCAN) offers lossless compression, strong encryption, and robust hiding in one package (Bourbakis and Dollas, 2003). The second methodology provides a strong authentication and authorization method based on three biometrics (iris, fingerprint, voice) (Kannavara et al., 2011). The efficient collaboration of these two methods offers the necessary protection of the health data during communication.

In this chapter we will present a prototype wearable platform based on commercially available Bluetooth-enabled wearable sensors and a BlackBerry Smartphone. We will describe how recorded physiological data can be analyzed continuously via appropriate methods based on DWT in order to detect signal portions of the ECG and PPG corrupted by noise and to denoise the recorded signals. Much of our efforts in these areas constitute fusion of approaches proposed by other researchers, which will be referenced accordingly. Finally, we briefly describe the secure way of protecting health information during exchange with a remote center.

13.2 DIGITAL SECURITY

Security has always been an important concern in computer systems with the advent of the Internet, sharing of information and services between distant points never being easier before. However, this very easy and global connectivity, being part of the Internet infrastructure itself, allowed many security threats to occur. For instance, one of the first and most infamous attacks was the Morris worm. It was released in 1988 and, through the Internet, it managed to infect thousands of computer systems, including military and medical research systems, having an economic impact of millions of dollars.

Threats to security in computer systems can be classified into four broad categories: interception, interruption, modification, and fabrication. *Interception* threat means that an unauthorized party can gain access to valuable information or resources. In an *interruption* incident, an asset of the system is destroyed or becomes unavailable or unusable. An example would be an attacker destroying a hardware device, deleting a system file, or swamping a server with a deluge of requests. *Modification* refers to gaining illegal access to an asset and performing changes to it. For example, an attacker could alter the content of an electronic message or a data file. Finally, by *fabrication* threat is meant that an unauthorized party inserts a spurious object into the system, for instance, the insertion of fake records into a database system or spurious messages into a network.

Thus, security can be successfully used to protect three different but critical areas: (i) the host computer (processor), (ii) the information (or data), and (iii) the network. The same critical components are used in mobile or telemedicine, where processors handle medical information and networks are used to communicate and exchange health information with medical centers. In particular, personal health data (biosignals) are collected by wearable health monitoring devices from human subjects (e.g., the elderly). These biosignals (especially pathological ones) with the human subject's identification data compose a private health information record to be sent to a medical center for a decision to be made. Thus, the transmission of such valuable health records has to be secure by employing strong data security schemes to protect them. In a medical center, only the appropriate physicians should have access to these health records by using their biometrics as authentication–authorization mechanisms to decrypt these records and make a decision regarding the health status of that particular human subject. Thus, below we briefly describe secure biometrics processors, information security, and network security in order to connect these techniques in an integrated scheme that offers a secure health information exchange. Figure 13.1 briefly shows the general idea.

13.2.1 Host Computer Security

To this end, much research effort has been made to secure host computers and guard the coveted confidentiality, integrity, and availability of their information,

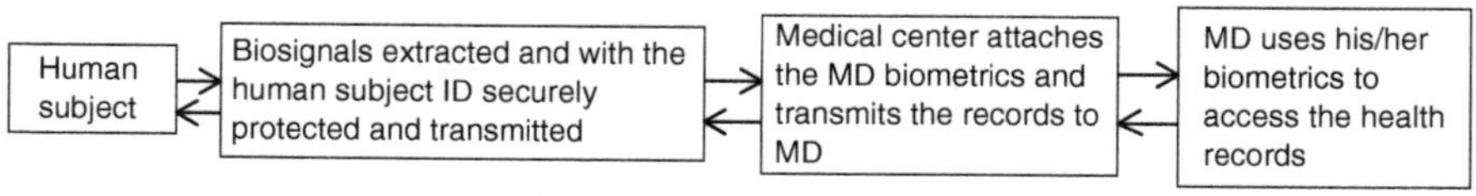

Figure 13.1 General scheme for health information exchange.

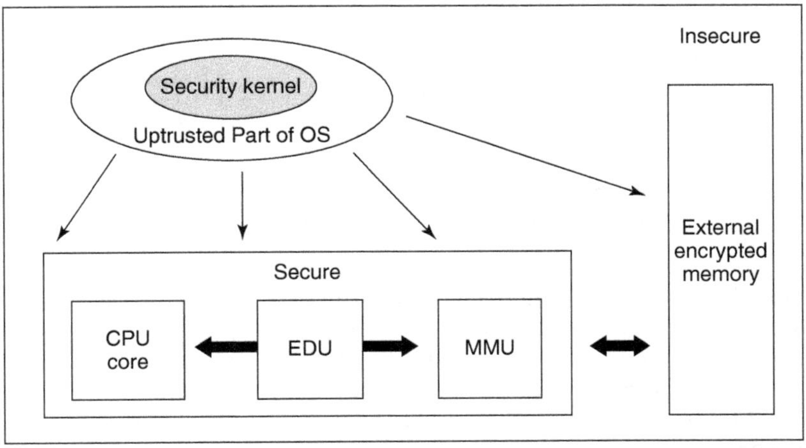

Figure 13.2 Secure computing model.

operations, and resources. Thus, techniques such as access control rules, encryption of data, and/or transactions and administration tools have been developed. These general methods are able to counterattack the aforementioned threats that utilize a class of attacks (Kannavara and Bourbakis, 2009, 2013)

The secure computing model shown in Figure 13.2 is a single-processor system with external memory and peripherals. The processor is implemented on a single-chip integrated circuit (IC) and is assumed to be trusted and protected from physical attacks, that is, its internal state cannot be tampered with or observed directly by physical means. The processor contains secret information such as a public key that identifies it and allows it to communicate securely with the outside world. The processor is protected by a tamper-sensing scheme (Kannavara and Bourbakis, 2009).

The trusted computing base (TCB) consists of the processor chip and optionally some core parts of the operating system such as the security kernel. The processor is used in a multitasking environment, which uses virtual memory and runs mutually mistrusting processes. External memory and peripherals are assumed to be untrusted; they may be observed and tampered with at will by an adversary. Additionally, data values stored in off-chip memory are encrypted. The encryption and decryption of data values is done by a hardware cryptoengine placed either between the cache and the external memory or between the cache and the central processing unit (CPU) core.

13.2.2 Information Security

If the cryptoengine or encryption–decryption unit (EDU) is placed between the cache and the external main memory as seen in Figure 13.2, encryption and decryption is done blockwise, pagewise, or segmentwise. This is because a cache miss causes

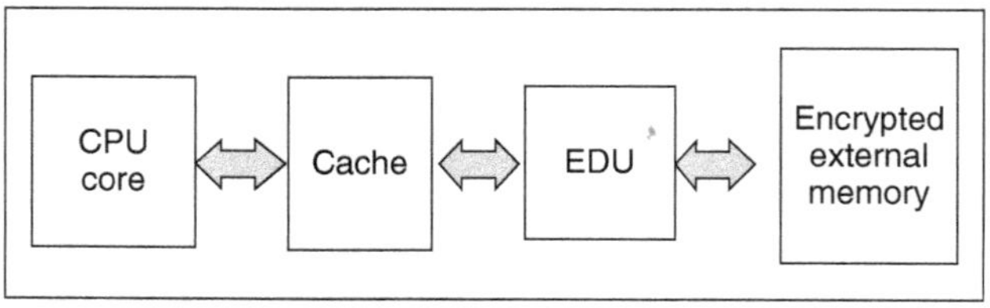

Figure 13.3 EDU between cache and external memory.

flushing of the cache and filling the cache with new data. Since data movement between the cache and the external main memory occurs pagewise or segmentwise, encryption and decryption is implemented by employing a block cipher. In this case the cache contains unencrypted or plain text instructions.

If the EDU is placed between the cache and the CPU core as seen in Figure 13.3, encryption and decryption is done instructionwise. Any branching in the instruction execution sequence requires the instruction to be fetched from the branched location in the cache. Therefore, the encryption or decryption technique has to accommodate the discontinuities in program execution. A simple stream cipher may not be able to achieve the required encryption because of the jumps in program execution due to branching. Therefore, each and every instruction may be encrypted as a separate entity, that is, instructionwise encryption is implemented. In this case the cache contains encrypted instructions, which are decrypted just before execution in the pipelined execution model.

The TCB needs to ensure the integrity of memory accesses before it performs any operation or stores data into the external memory space and, thus, memory integrity verification is required. Memory behaves like valid memory if the value the processor loads from a particular address is the most recent value stored to that address (Kannvara and Bourbakis, 2013). Also the programs must be well written in order not to leak secrets via memory access patterns.

Information Security Encryption Protecting data or information can be mainly expressed as information security (Bourbakis and Dollas, 2003; Yang et al., 2004, Maniccam and Bourbakis, 2004), while information security mainly refers to the way of encrypting the information from undesirable attacks. Thus, encryption can be expressed in different ways. Depending on the type of plaintext, data encryption systems can be classified as text encryption, audio encryption, image encryption, and video encryption. In order to have a generic cryptosystem that can encrypt different types of data, some encryption standards have been developed. Among them, DES (data encryption standard), RSA, AES, IDEA, TEA, SEED, and SCAN are elaborately designed and widely adopted. In general, these encryption standards consider all forms of multimedia data—such as text, image, and video—as bit stream and encrypt them without any differentiation. Later in this chapter we briefly describe the SCAN scheme as an example.

SCAN encryption scheme was introduced in 1986 (Bourbakis, 1986) as a generic strategy for 2D data accessing/processing and later (Alexopoulos, 1988; Bourbakis) was implemented as an encryption-hiding methodology. The SCAN encryption scheme is based on a (nxk)! permutation scheme of the data associated with a strong confusion function, which converts the data histogram to a flat one. The SCAN scheme alone offers a robust protection of the data but carries the permutation weak points. The combination of the SCAN scheme with its confusion function provides a better and stronger protection.

13.2.3 Network Security

Security in networks can be provided in different layers of the protocol stack (Reddy et al., 2010; Papadimitratos, 2002; Goth, 2012). Network security services generally fall into two categories: (1) data security that provides authentication, confidentiality, integrity, and nonrepudiation; and (2) access control that protects networked resources—hosts, servers, and other networking devices such as bridges, routers—from unauthorized use. There are security protocols operating at different layers of the protocol stack providing protection of user data of various degrees. However, securing the networking infrastructure itself is essentially orthogonal to securing the data transfer over the infrastructure. An adversary can purposefully launch an attack against the infrastructure instead of between any two securely communicating end points, which in fact presents far more profound ramifications. For example, by interfering with the routing protocol, data may be redirected over paths controlled by the attacker, or worse, the use of incorrect connectivity information (e.g., an attack against a single element of the routing infrastructure) can result in failure to have data delivered to a large number of destinations, causing a denial of service for a large portion of the user population. Network routing infrastructure presents a single obvious vulnerability since routers have to acquire network link-state information by periodically exchanging information with their neighboring routers. Injected false routing information would propagate throughout the network and deceive many routers. Therefore, an attack of this type would potentially allow the adversary to control over the network, for example, intercept data, obstruct the data flows by dropping or redirection if user traffic is encrypted, or even disable the network operation altogether. On the other hand, with the proliferation and increasing usage of networked, embedded, portable, and wearable devices, in addition to protecting against attacks from malignant software, we also have to be concerned with physical attacks that corrupt data, discover private data, or violate copy protection, as well as combinations of physical and software attacks.

In addition, the wide use of mobile computing has made wireless networking (like WiFi) a very popular communication medium with which to work. Portable computers, mobile smart phones, and a variety of other devices take advantage of online networking. This networking "necessity" is exposed to serious security attacks from individuals or groups. Researchers and practitioners have developed protection techniques from encryption to trustworthy mechanisms and firewalls, but the network is still under attack. For the general public and average users there are several tools to scan and monitor a number of applications. These tools are available and can "assist" a user to find out many things about a WiFi network that is within signal strength. Some of these tools are Cain, NetStumbler, Kismet, and WireShark, and others. These tools offer network analysis and monitoring for security and some degree of optimizations. They can also monitor an area and test the security of the existing networks within range.

Some of the techniques used to defend network attacks are based on recognition of trustworthy sources. For instance, several network users have developed certain natural language (NL) phrases for communication (except the password protection). Other techniques use human-only recognizable pictures with hidden words requesting that the user recognize and correctly answer the hidden message. These techniques do not provide strong secure machine-to-machine communication since the recent computer vision and pattern recognition algorithms are so sophisticated that can break most of these messages. Some other researchers proposed secure processors with cryptoauthentication capabilities (Kannavara et al., 2011).

13.2.4 Biometrics

By knowing the capabilities of the host information network exploiting tools, researchers have started to use biology-oriented mechanisms or properties for protection from undesirable attacks (Kannavara et al., 2011). Thus, they have introduced human biometrics for security. Biometrics is the study of methods for uniquely identifying or authenticating humans based on intrinsic physical or behavioral traits. *Identification* means characteristics are selected from a database to produce a list of possible or likely matches. *Authentication* means that when a person makes a claim that he or she is that specific person, then that specific person's characteristics are checked to see if they match.

Biometrics are those characteristics that are both measurable and distinctive, thus they can be used in order to label a specific person and distinguish him from others. Physical traits are the ones that relate to the body and the physical structure of the person in general. The most commonly used include fingerprint, iris recognition, retina, palm print, face recognition, and DNA. Moreover, behavioral traits, such as, for example, voice, gait, and typing rhythm, refer to characteristics that relate to a person's behavior and can be measured somehow in order to be stored in a database.

The functionality of the biometrics selected to be used for a specific application depends on some characteristics that they should have. First of all, collectability refers to the ease with which they can be measured and collected not only in order to be stored at the central database but also every time that an identification process is required. Besides that, the trait should be characterized by stability. Thus, traits that remain almost invariant over time are the ones selected. The uniqueness of the characteristic is also of great importance in order for the system to be reliable and avoid mistaking a person with another one. Finally, issues regarding the concerns about the use of the biometric identifier should be considered, for example, collecting fingerprints is more acceptable than collecting DNA.

The two important operations in a biometric system are enrollment and test. During *enrollment* the biometric of the individual is stored in a database. The signal captured is processed in order to collect the required information that distinguishes it from those of other subjects, and then it is transformed to an appropriate form. Afterward, it is represented appropriately according to the application for which it will be used and stored in the database. During *test* the biometric information of the individual is detected and compared with that stored in the database. The procedure related to the signal capture is exactly the same as the one followed during enrollment, except from the fact that the biometric information is not stored in the database but is compared with the information already stored there in order to verify the authentication of the person examined. The scheme in Figure 13.4 graphically provides the steps needed for biometrics authentication authorization for biometric information that are transformed to images.

Various biometric techniques are used to protect information exchange using traits mentioned above. Later in this chapter we present three different biometric schemes for this purpose.

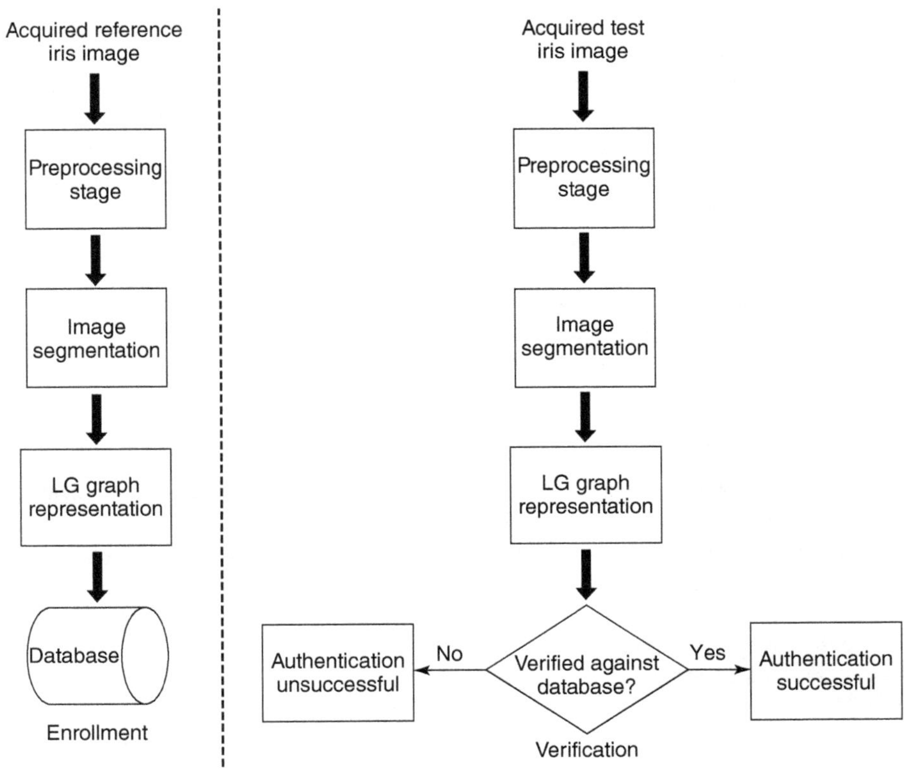

Figure 13.4 Sequence of steps used for biometrics authentication.

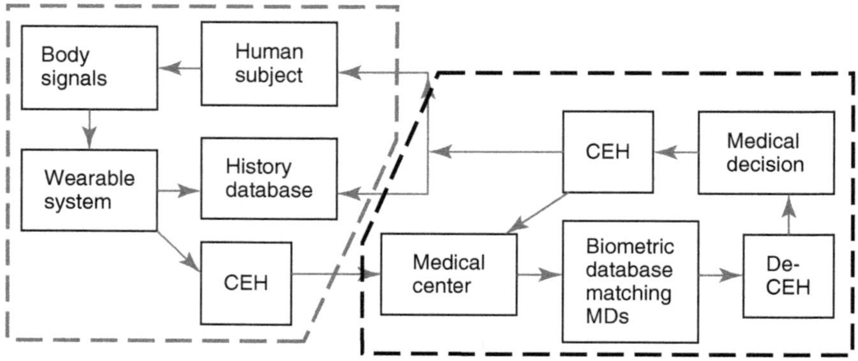

Figure 13.5 Overall mobile health architecture (CEH = compression, encryption, and hiding).

13.3 WEARABLE HEALTH MONITORING SYSTEMS (WHMS) PLATFORM

The overall WHMS mobile health architectural scheme is depicted in Figure 13.5. This scheme actually consists of two main parts. The first part represents the selection of the body signals using the wearable system, the filtering of these signals (important symptoms) according to medical health rules for the individual using the history database, the compression–encryption of these signals for their protection using the SCAN language, and the hiding of the medical doctor (MD) biometrics into the compressed–encrypted signals for authorization.

The second part represents the medical center and the physician's actions for authentication of the biometrics in order to allow him/her to decrypt–decompress the body signals (symptoms) and make a decision regarding the patient health condition. The MD response is then compressed–encrypted using the SCAN language and sent to the human subject for further action.

13.3.1 System Setup

The WHMS prototype (displayed in Figure 13.6) is set up around the BlackBerry Bold 9000 PDA/Smartphone, which has been chosen due to its powerful processing capabilities (624-MHz processor), J2ME programmability, multithreading support, and Bluetooth piconet support.

The set of sensors employed in our system consists of the Bioharness BT chest belt by Zephyr, which is capable of measuring one-lead ECG, heart rate, respiration rate, skin temperature, posture, and activity (via a triaxial accelerometer), and a wireless wrist-mounted pulse oximeter with a finger clip sensor from Nonin Medical, which can capture the PPG waveform and also provide the levels of pulse rate and oxygen saturation. Figure 13.7 shows two screen shots of real-time displayed data on the Smartphone.

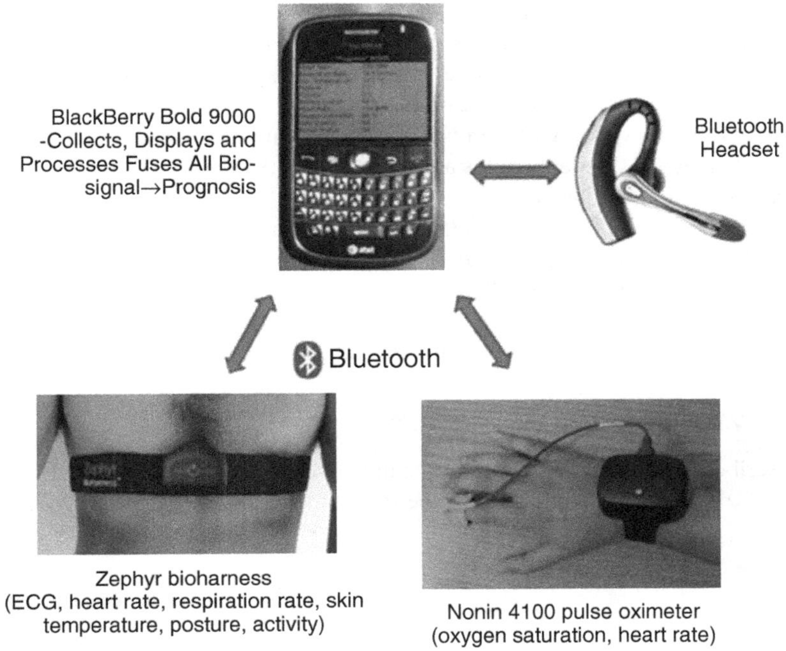

Figure 13.6 Components of the current WHMS prototype.

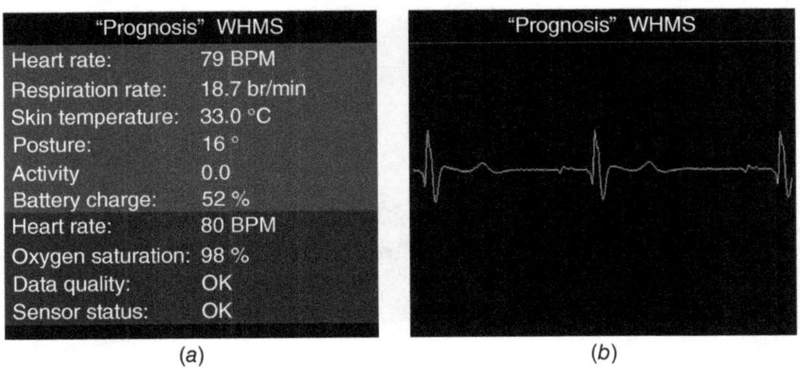

Figure 13.7 Visual display of symptoms.

The communication between the cell phone and the wireless sensors is implemented by forming a Bluetooth personal area network (PAN). In this formation the PAN's master (e.g., the BlackBerry Smartphone) assigns dedicated time slots to the peripherals and data is received in a protocol-inherent round-robin manner via Serial Port Profile (SPP), for example, virtual serial port connections. For every sensor there is a dedicated thread on the phone, which is responsible for (i) receiving a new data packet, (ii) deserializing the received data, (iii) performing basic signal processing and feature extraction as we will see in the following section (although these steps are implemented on the server's side, e.g., on a workstation to which data are sent via WiFi), (iv) displaying and plotting the data, and (v) handing over the new data to the main thread, which will then process the measurements based on the Prognosis formal language (Pantenopoulos and Bourbakis, 2010c) and will finally derive an estimation of the user's health.

In case a health risk is detected or there is a significant abnormality in the measured biosignals, the system may choose to ask the user for additional feedback (via a simple questionnaire) or to notify a medical center, a supervising physician, or even a close contact by means of the 3G cellular network or a WiFi connection. Figure 13.8 depicts the schematic of the overall J2ME software architecture.

13.3.2 Voice Interaction

As described previously and as it is documented in Pantenopoulos and Bourbakis (2010b, 2010c), we wish for the user to be able to interact with WHMS in a productive

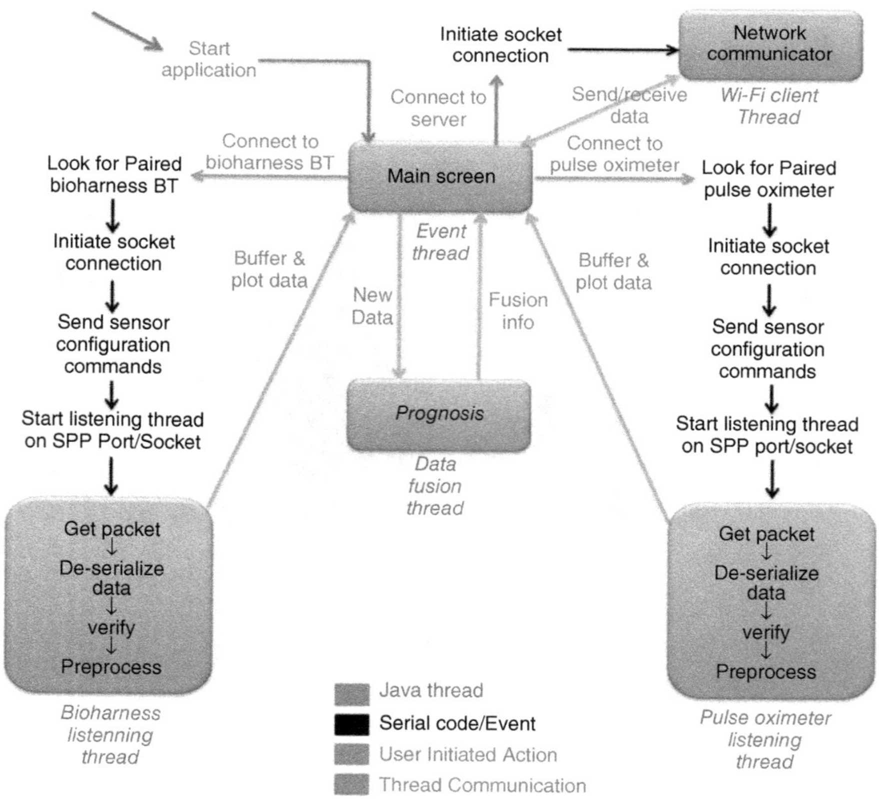

Figure 13.8 J2ME software architecture. (*See insert for color representation of the figure.*)

way. The main motive behind this approach is that such a capability would enable the capturing of additional nonmeasurable symptoms such dizziness, headache, chest pain, coughing, and the like. As a result, the capturing of such incidents can provide a more comprehensive picture of the user's overall status and context, which may be proven useful in the decision-making process.

To facilitate this feature in the current prototype, a speech recognition mechanism needs to be employed. However, as the cell phone is already charged with the burden of handling multiple network connections (Bluetooth and WiFi) and that of preprocessing the data, it would be unreasonable given its processing capabilities to require an embedded speech recognition scheme. Furthermore, constrained resources would severely hinder the real-time performance (both in terms of accuracy and speed) and the recognition capabilities in terms of vocabulary size of such an approach.

To address the previous issue, the approach that has been taken is similar to that of several commercial applications for voice recognition on cell phones, such as the application (Vlingo). More specifically, a voice recording can be initiated on the phone by means of pressing a side function key of the phone. Once the recording is complete, the recorded speech can be transmitted over the WiFi link to a remote server where a powerful speech recognition engine can be utilized to perform reliable and robust speech recognition. Such an engine is provided by Sphninx- 4 (CMU), a speech recognizer written in Java from researchers in Carnegie Mellon University that is utilized in this research effort.

13.3.3 Remote Monitoring Application

Data that is gathered from the WHMS platform will be transmitted to a remote location, where a supervising physician or care provider should be able to access them in real time. In response to this demand, we have developed a Java-based desktop application that is capable of receiving the data from the PDA via a socket connection over Wi-Fi or the cellular network. The application offers the capabilities of viewing remotely recorded data in real time, as well as accessing the patient's recent measurement history. In addition to that, this application is responsible for receiving the speech recordings taken on the phone and invoking the appropriate Sphinx- 4 application programming interfaces (APIs) to perform speech recognition. A log of recorded symptoms is also saved. The interface of the remote monitoring application (which has also been implemented in Java) can be seen in Figure 13.9. Figure 13.9*a* depicts a graphical user interface (GUI), which displays the trends of measured data such as heart rate, respiration rate, oxygen saturation, and activity level, as well as information regarding the quality of the received data and regarding the connection status between the workstation and the BlackBerry Smartphone. In addition, a panel in Figure 13.9*b* displays the progression of the instantaneous R-to-R interval of the ECG, which is computed by employing the well-known algorithm described in the pioneering work of Pan and Tomkins (1985). Finally, Figure 13.9*a* depicts real time graphs of measured ECG, PPG, and triaxial acceleration signals.

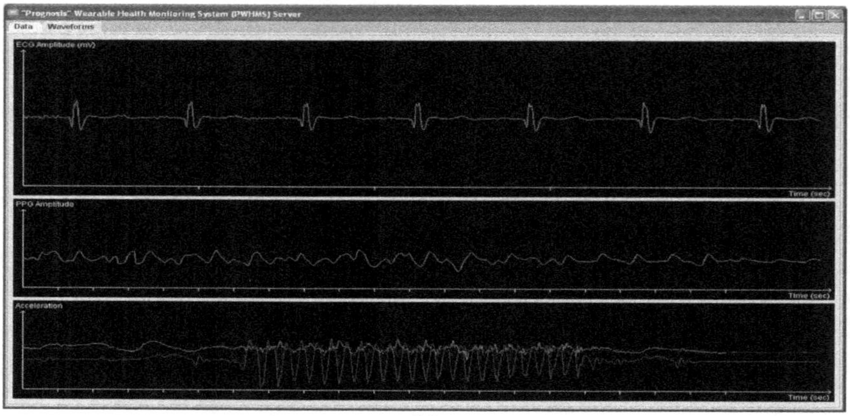

(a)

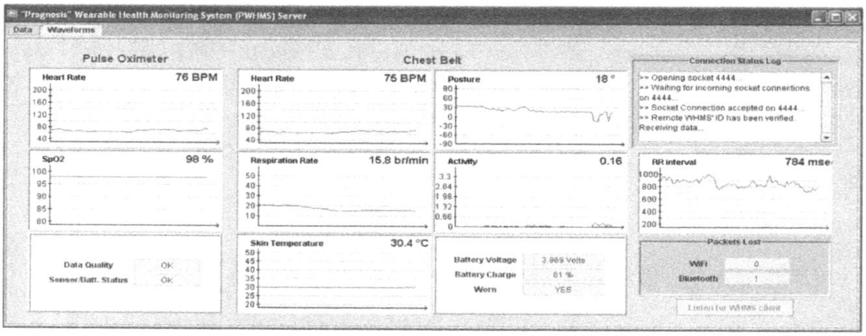

(b)

Figure 13.9 Interface of remote monitoring application: (*a*) recorded waveforms and (*b*) data trends. (*See insert for color representation of the figure.*)

13.4 PROCESSING OF PHYSIOLOGICAL DATA

As mentioned in the introduction, the purpose of this chapter is (a) to illustrate the operation and setup of the wearable platform, (b) to describe a scheme for accurately detecting severely corrupted data segments of measured signals and for denoising the measured data, and (c) to offer a secure information exchange scheme via encryption and biometrics.

The approach that has been adopted for ECG denoising is based on the well-known method of thresholding the coefficients obtained via DWT decomposition. This approach has been very popular among researchers in the field (Addison, 2005; Donoho, 1995) and has been proven particularly effective. Here we will present an enhanced denoising method that combines conventional wavelet-based and wavelet-packet-based denoising. Before doing that, we will briefly provide some background on the DWT and on wavelet packets.

13.4.1 DWT and Wavelet Packets

The wavelet transform has emerged as the most powerful tool for the analysis of transient nonstationary signals. There have been numerous research efforts that employ the WT in signal processing of ECG signals (Addison, 2005). Its abilities to (a) employ a variable window in order to realize high temporal resolution in high frequencies and high spectrum resolution in low frequencies as well as (b) the large set of available wavelet functions that can be utilized to detect various feature morphologies have been proven very useful in the analysis of electrocardiograms.

The continuous wavelet transform is defined as the inner product of the function in question and a set of dilated and translated wavelet functions $\psi(t)$:

$$\mathrm{WF}\,(s,\tau) = \int f(t)\frac{1}{\sqrt{s}}\psi^*\left(\frac{t-\tau}{s}\right)dt \tag{13.1}$$

where s is the scale or dilation and τ is the translation or location. Instead of computing the inherently redundant wavelet transform for an infinite amount of scales and translations, s and τ can be chosen to be discrete. A very common choice is the dyadic grid:

$$\psi_{i,k}(t) = 2^{-1/2}\psi(2^{-i}t - k) \tag{13.2}$$

The above form expresses the DWT and in this case the wavelet functions (if properly selected) constitute an orthonormal basis. The DWT is implemented via a pair of conjugate mirror filters (Mallat, 1998, Chapter VII), a low-pass FIR filter h (associated with a scaling function φ) and a high-pass far-infrared (FIR) filter g (associated with the wavelet function ψ), which result in what is usually referred to as subband coding. In this case, a signal that is decomposed up to a scale n is initially filtered by h and g in order to produce approximation coefficients a_i and detail coefficients d_i. Approximation coefficients in every scale can be further decomposed via the same scheme, in order to produce the a_i and d_i coefficients of the next scale. If the discrete signal is of length $N = 2^M$, then this scheme allows for up to M levels of decomposition (coefficients are subsampled in each scale to avoid redundancy). From the above discussion it is clear that the DWT acts as an octave band filter whereby by adding higher octave bands we add more detail or resolution to the signal. Hence comes the name "multiresolution analysis."

Wavelet packets extend the idea of the DWT by basically decomposing both approximation and detail coefficients in every scale to finer components (the DWT decomposes only the a_i in every scale). Beginning with the scaling and the wavelet function and the filters h and g, one can iteratively generate appropriate analyzing functions that can be used to produce a full binary wavelet packet decomposition tree. The produced binary tree is usually pruned by applying some criterion in order to select an optimal decomposition scheme. The applied criterion is usually the Shannon entropy (Wood and Asada, 2006) given by

$$E(x) = -\sum_t x^2(t)\,\log[x^2(t)] \tag{13.3}$$

The idea behind this approach is to look at each parent node and its two children and quantify the information that is gained by splitting the parent node. If the sum of the entropies of the two children is smaller than the entropy of the parent node, then the split is performed as there is obviously a gain in terms of compression of information.

13.4.2 Detecting Unusable ECG Data Portions

An important issue in ambulatory ECG recordings is that of complete signal corruption of the signal due to measurement artifacts or sensor malfunction. The DWT can be employed in this case in order to analyze the energy content of the recorded signal at different scales, and, if values are detected over or under empirically set thresholds, then we may deduce that the signal is corrupted or that there is a low-energy unusable section (in case the wearable sensor has been removed or misplaced). A similar approach utilizing FIR and infinite impulse response (IIR) filters is described in Redmond et al. (2008).

Here we want to describe the detection of noise- corrupted ECG segments using a DWT based approach. In Figure 13.10 we can see part of an ECG recording taken with the WHMS platform and processed offline using MATLAB. It is evident that there is a part of the signal, starting at about 28 sec, that is severely corrupted by noise (in this case from abrupt chest movements that significantly affected the contact surface between the ECG sensors and the subject's chest). In Figure 13.11 we can see four levels of approximation coefficients (obtained using the Coiflet wavelet of order $N = 5$) of the same signal that is depicted in Figure 13.10. We notice that the noise we are dealing with here has significant low-frequency contents, which of course interfere with the corresponding QRS complexes, which at the first half of the signal, where the ECG is almost noise free, have been smoothed down. This is particularly obvious at the finest decomposition scale. As a result, the approximation coefficients at level 4 offer a means by which we could discriminate between normal ECG and the noise type in question. A way to take advantage of this feature is to compute the amount of energy being carried in a window of length N of DWT coefficients by

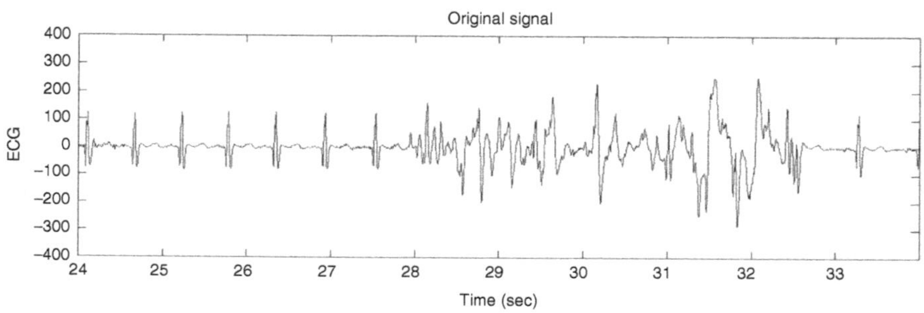

Figure 13.10 An ECG segment, part of which is severely corrupted by noise caused by movement.

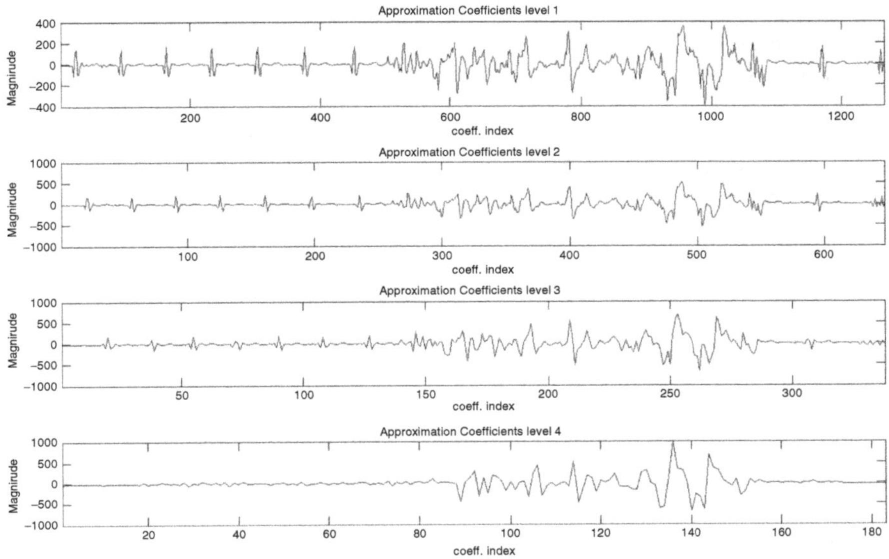

Figure 13.11 Four levels of approximation (using Coif5 wavelet) coefficients of the ECG shown in Figure 13.10.

calculating:

$$E_{a4} = \sum_{i=0}^{N-1} |a_{4,i}|^2 \tag{13.4}$$

For a more smooth power estimate over time, it is more favorable to first square the approximation coefficients and then apply a low-pass filter, as proposed in Redmond et al. (2008), to smoothly average the power estimate. This is done by employing a variable-length Hamming window, whereby the length of the window is adjusted according to the instantaneous heart rate estimate. The length of the Hamming window is computed as

$$L = \frac{60}{\text{HR}} \frac{F_s}{2^n} + b \tag{13.5}$$

where F_s is the sampling frequency of the ECG (250 Hz in the case of the bioharness), HR is the instantaneous heart rate estimate (which can be taken from the estimate given by the bioharness or by an averaged HR estimate from recent RR intervals), n is the decomposition level (level 4 in this case) and, finally, b is a bias to account for the heart rate variability as a result of which successive RR intervals may show significant variation (here $b = 10$, chosen empirically).

Before showing how the resulting energy estimate can be utilized to detect corrupted ECG segments, we should briefly refer to other types of signal corruption in the ambulatory ECG. Of particular interest is the case whereby the ECG becomes corrupted due to permanent displacement of the wearable sensor or even due to the user taking it off. Such a case is depicted in Figure 13.12, where an artifact is initially created due to sensor displacement and then due to the user taking of the chest belt. We wish to be able to instantly detect such a scenario as well.

By employing the energy estimation scheme described before, we are able to detect the described cases where severe "nonreversible" artifacts are created in the ECG. This is schematically illustrated in Figure 13.13. To detect events such as the one depicted in Figure 13.10, we can employ a heuristically set threshold (shown

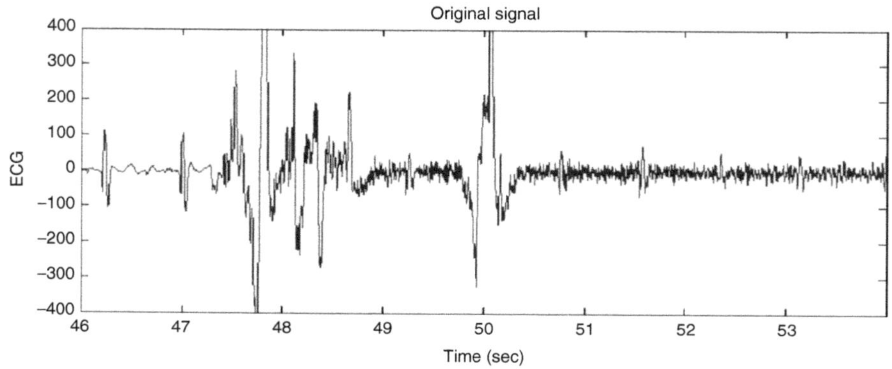

Figure 13.12 An ECG segment, part of which is severely corrupted by noise caused by sensor displacement.

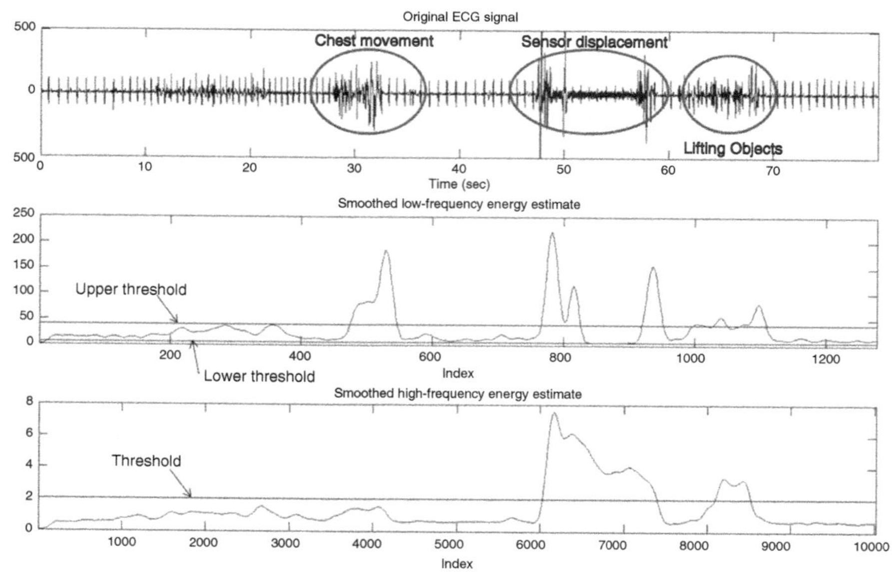

Figure 13.13 Detection of corrupted ECG segments with appropriate thresholding on the energy estimates.

as the upper black line in Figure 13.13), above which we can deduce that the ECG has been corrupted. Similarly, for low-energy portions such as in the case shown in Figure 13.11, we can employ a lower threshold on the smoothed energy estimate, under which samples are discarded as well. To improve the detection of corrupted ECG segments, we can utilize the high-frequency detail coefficients (d_1) to recognize segments of abnormally high concentration of high-frequency components (keeping in mind that the ECG transmitted by the bioharness chest band has been low-pass filtered to remove high frequencies).

Putting everything together we get the power estimates shown in Figure 13.13. By selecting heuristically appropriate thresholds, we can detect the severely distorted signal portions. In the 80-sec ECG segment shown in Figure 13.13, there are three types of occasions whereby noise is introduced in a nonreversible manner in the ECG. By empirically estimating two thresholds for the energy estimate shown in the second graph of Figure 13.13 (computed from the fourth level of approximation coefficients) and another threshold for the lower graph (taken from the energy estimate of d_1), we can recognize the parts of the signal that need to be discarded. This thresholding operation creates two separate masks and by combining them we identify the samples to be discarded.

13.4.3 Approach on ECG Denoising

The DWT has been widely employed in the denoising of the ECG (Addison, 2005). The conventional wavelet denoising scheme, as proposed by Donoho (1995), consists of employing DWT decomposition up to scale 4 and then applying the soft-thresholding technique to the detail coefficients at all levels in order to suppress high-frequency ripple noise. Finally, the denoising algorithm reconstructs the signal using the inverse DWT with the modified detail coefficients. More details on the algorithm by Donoho (1995).

A similar approach is to perform optimal wavelet packet decomposition and then perform thresholding on the resulting coefficients before reversing the whole process to reconstruct the denoised signal estimate. As discussed above, the criterion for optimal wavelet packet decomposition is usually considered some entropy criterion, like the one given from equation (13.3). However, such a criterion is appropriate for compression applications, whereby a decomposition of the signal is searched, which minimizes the overall entropy.

A different approach could perhaps be to try to identify an admissible tree (from the full binary tree produced at the beginning), which provides a decomposition that maximizes the information content of the signal. Motivated by this idea, we applied a modified criterion for merging nodes in the wavelet packet tree. Namely, for every node, based on its coefficients, we compute a normalized histogram, and the frequency counts of the resulting groups of coefficients are taken as probability estimates, based on which a new entropy modEntropy criterion is computed. Now if the resulting combined modEntropy of two children are found to be bigger than the modEntropy of the parent node, the two nodes are not merged together, in order to keep maximal information content. In the modEntropy case, nodes whose coefficients are more "spread" and less concentrated produce a higher metric, indicating the presence of noise. From our experiments with this modified entropy metric, we found that the resulting decomposition trees are significantly different (and longer) from the ones produced from the optimality criterion, which in many cases tends to resemble the conventional wavelet decomposition tree (decomposing only the approximation coefficients). In our case higher frequencies are analyzed in finer scales than the lower frequencies.

We experimented with the modified wavelet packet decomposition tree and found that the resulting denoising scheme tends to overestimate the low-frequency noise and, thus, suppresses actual low-frequency components of the ECG. However, we noticed that the wavelet denoising scheme tended to underestimate the low-frequency noise and left unwanted low variations in the filtered signal. By combining the denoised signals from both approaches, we were able to address the issues of both schemes and to produce superior results than each method separately. To illustrate this, we can refer to Figure 13.14 where a signal portion that has been recorded with the wearable platform, while the user is standing still, has been intentionally corrupted by additive white Gaussian noise (AWGN) which results in an ECG with SNR = 0 dB, where the signal-to-noise ratio has been computed as

$$\mathrm{SNR} = 10\log_{10}\left(\frac{\sum x^2[n]}{\sum (x[n]-y[n])^2}\right) \tag{13.6}$$

where $x(n)$ is the original sequence and $y(n)$ is the noisy or the denoised signal. The decomposition was performed using the Coif5 wavelet, and soft thresholding was applied with the scale-dependent threshold proposed by Donoho, which we do not need to reproduce here. The decomposition was performed until level 4. The efficiency of the method is illustrated by visually inspecting the results shown in Figure 13.14 as well by the fact that the proposed scheme resulted in an SNR of 6.48 dB and a normalized mean square error (MSE) of 0.22, while the conventional wavelet denoising produced SNR = 2.04 dB and normalized MSE = 0.63.

The main disadvantage of this method remains the fact that it is considerably more time consuming than conventional wavelet denoising, but in addition to that it requires the production and then the pruning of the wavelet packet tree. As a result it is not particularly attractive for real-time applications. However, it can still be considered as a promising approach for offline processing of ECG records.

13.5 SECURE INFORMATION EXCHANGE

In the digital information era, the vast majority of information exchanges take place via e-mail and mobile communication. Through time, this type of communication has generated painful experiences and valuable lessons of security and privacy protection from undesirable attacks and abuses. In response to these lessons, researchers and practitioners have developed software tools mainly for protecting customer's privacy and network security as well. Nevertheless, every day we read about new kinds of

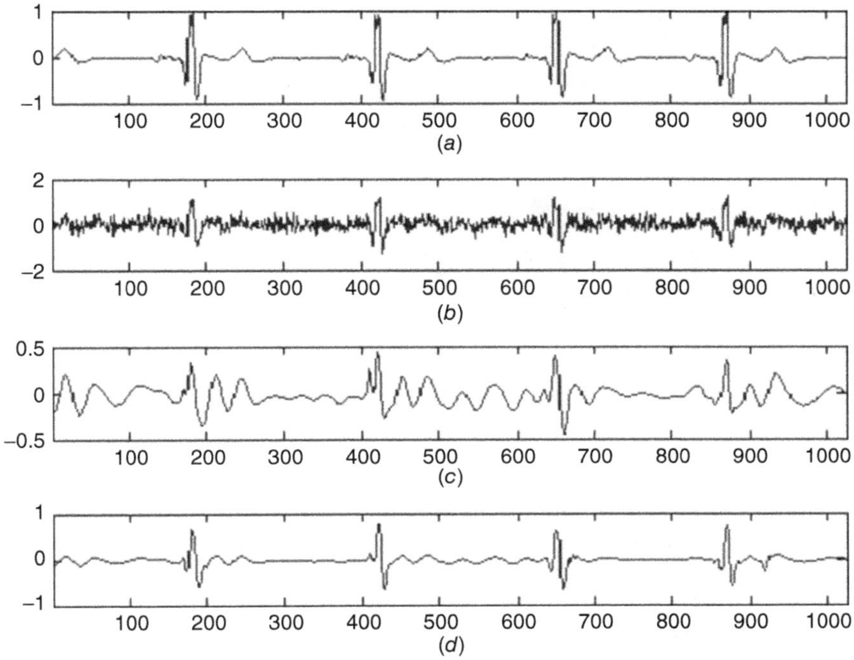

Figure 13.14 (*a*) Original ECG, (*b*) noisy, (*c*) wavelet denoising, and (*d*) combined wavelet and modified wavelet- packet denoising.

attacks or abuses, for instance, e-theft, on networks and private PCs, mobile communicators, and networks remain vulnerable to these illegal actions, and as a result additional security (constraints) have been imposed on networks for dealing with these issues. In addition, extra protection is required especially on medical records not only to secure their safety and the lossless of their nature for liability purpose but also to protect them (encryption and authentication) in a robust way resistant to undesirable abuses and attacks. Here such a protection is provided by a robust information security scheme composed by two main parts: the CEH based on the SCAN language and the authentication based on biometrics (see Figure 13.4).

13.5.1 CEH Scheme

The CEH scheme is based on the SCAN methodology, which can perform a lossless compression, encryption, and hiding in the same package. SCAN is applied on digital signals, images, and text and a secure version of the health information extracted from a human subject.

Scan Methodology The SCAN is a formal language based on 2D spatial accessing methodologies, which can represent and generate a large number of scanning paths easily. SCAN language is defined by a grammar and has a set of basic

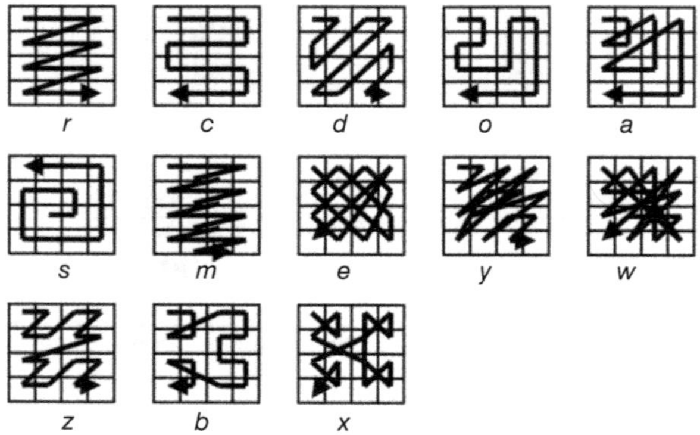

Figure 13.15 Basic SCAN patterns.

scan patterns, a set of transformations, and a set of rules to compose simple scan patterns to obtain complex scan patterns. The basic scan patterns are shown in Figure 13.15. The rules for building complex scan patterns from simple scan patterns are specified by the production rules of the grammar of each specific language. The reader is referred to, Bourbakis and Dollas (2003),Yang and Bourbakis (2004), and Maniccam and Bourbakis (2004) for detailed descriptions of syntax and semantics of various SCAN languages and their applications. The SCAN language has applications in information compression, information encryption, and information hiding. The SCAN encryption method involves permutation and substitution.

We use the SCAN methodology to generate scan patterns to permute the values, and the methodology replaces the values using a substitution rule. The complex SCAN patterns form the encryption keys. The substitution rule adds confusion and diffusion properties. The permutation and substitution operations are applied in an intertwined and iterative manner.

Lossless Image Compression The lossless image compression algorithm consists of four main steps. These steps are (1) scanning and prediction (2) scan-path encoding, (3) context modeling, and (4) arithmetic coding. The image is first partitioned into blocks of size $2k \times 2k$ bytes, $k \geq 2$. Each block is then scanned with various scan paths, and pixel values are predicted using different predictors along the scan paths, that is, for each scan path, the sum of absolute values of prediction errors (the difference in pixel intensity values) and the number of bits needed to encode the scan path are computed. For each block, the scan path that minimizes the prediction errors and encoding bits is chosen as the best scan path of the block. Recursive hardware to find the best scan path is presented in Bourbakis (1986) and Alexopoulos (1988).

Image Encryption and Decryption Image encryption using the SCAN encryption methodology and its reconfigurable architecture is discussed in this section. SCAN encryption methodology offers a symmetric private key encryption. The image encryption method is based on permutation of the pixels of the image and replacement of the pixel values. The permutation is done by scan patterns generated by the SCAN methodology. The pixel values are replaced using a simple substitution rule, which adds confusion and diffusion properties. The permutation and substitution operations are applied in intertwined manner and iteratively producing an iterated product cipher.

Information Hiding Information hiding in images (or image steganography) seeks to identify the complex regions of the cover image and embed the secret data into those regions. The bits from the secret data are embedded into a variable number of least significant bits of the pixels in complex regions depending on their complexity. The bits from the secret data are embedded into complex regions in random order determined by the secret SCAN key chosen by the user. The embedding algorithm consists of four main steps. These steps are (1) complexity identification of cover image, (2) SCAN rearrangement of cover image and complexity matrix, (3) bit embedding of secret data into rearranged cover image, and (4) reverse SCAN rearrangement of rearranged embedded image. It should be noted that this technique has sever deficiencies resulting in its easy detection.

CEH Protection Scheme The CEH information protection scheme is depicted in Figures 13.15, 13.16, and 13.17. Figure 13.16 shows the results of the compression–encryption–hiding of an image by using the SCAN methodology. Figure 13.17 graphically shows the steps for converting 1D signals and natural language text data into 2D images and their integration into a single 2D image, on which the CEH is applied to secure and protect their transmission to medical center.

13.5.2 Authentication–Authorization Scheme

The authentication–authorization scheme is based on the local–global graph approach, which can efficiently represent biometrics (Kannavara et al., 2011).

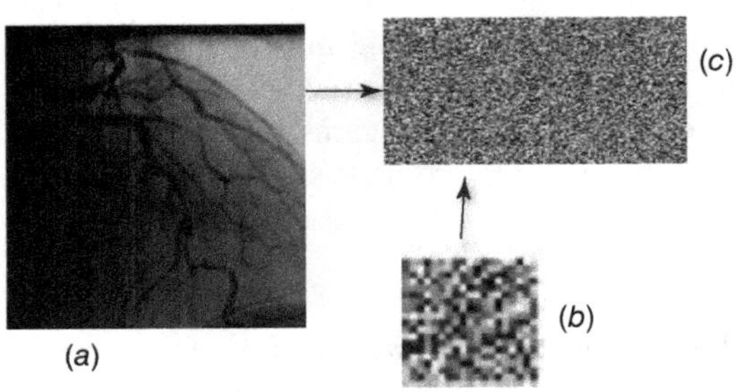

Figure 13.16 It shows (*a*) an original image, (*b*) a message (text or signal) converted into a 2D image, and (*c*) the compressed and encrypted image containing the message.

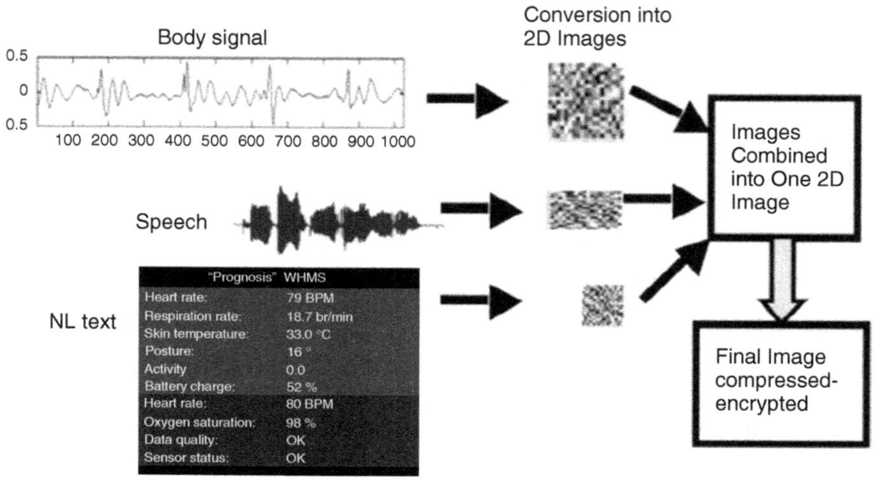

Figure 13.17 Scheme that integrates pieces of health information (biosignals, voice, text, etc.) extracted from the human user and the CEH of these pieces of information for secure transmission.

Local–Global Graph Method *The local–global (LG) graph* method provides local and global pieces of information (features, attributes, relationships, etc.) into a graph. The LG graph is a more accurate representation of an object using a nonlinear graph-matching function. The LG graph is used after an image segmentation to improve the object recognition accuracy without increasing computation complexity. Thus, here the LG graph is capable of describing with adjustable accuracy and robustness the features contained in an image. The main components of the LG graph are: (i) the local graph that represents the information related with shape and size and (ii) the skeleton graph that provides information about the internal shape of each segmented region. The global graph represents the relationships among the segmented regions for the entire image. The nodes of the global graph include the color, the texture, and the overall connectivity of the region—segments that compose the entire image (Figure 13.18). For example, if the shape of a region changes, the local graph and the skeleton graph record the changes attached them to the LG graph. In addition, the centroid may change and the global graph will register that change as well. The process used here for objects recognition is based on the synthesis of segmented adjacent regions, using the LG graph, and association (comparison) of the integrated region models available in a database. The method can search objects in one given image based upon the provided object model database.

LG Graph-Based Iris Biometric Authentication Iris recognition is a method of biometric authentication that uses pattern recognition techniques based on high-resolution images of the iris. Iris-scan-based security systems offer robust and reliable authentication as compared to other biometric methodologies. Here, we

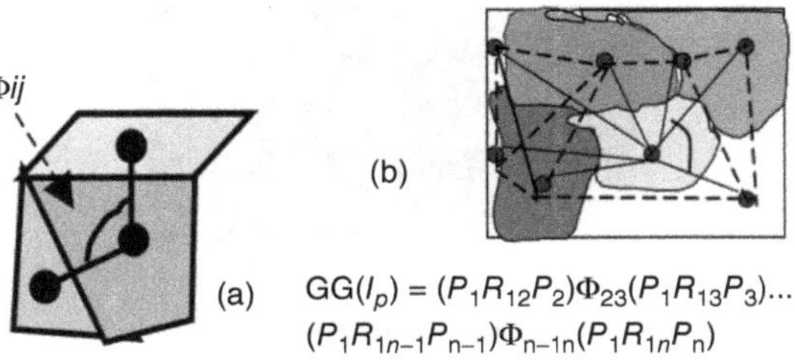

Figure 13.18 (*a*) Shows the connectivity among three neighboring regions and their centroids and the LG graph. (*b*) The LG graph of a synthetic image consisted of seven segmented regions.

present a biometric authentication system based on iris scan, that is, the identity claimed by the subject is verified by comparing his (or her) iris biometric with the one stored at a database. A unique LG graph methodology to extract, store, and compare iris patterns for the purpose of authentication is presented. Local features from the iris image are first extracted and then combined to form a global graph that is stored in the system database. Authenticity is established by comparing the global graph of the presented iris image with the global graph of the stored reference image. A comparison is performed, by calculating the distance between the two LG graphs, to establish or not the authenticity. Thus, we present the LG graph methodology as an incremental methodology for efficiently representing iris patterns via their features and the features' relationships.

Preprocessing Iris Image Preprocessing the acquired iris image is necessary before the application of LG graph methodology in order to extract the iris features. Preprocessing consists of four major steps: (1) iris localization, see Figure 13.19; (2) image smoothing; (3) log-polar transformation; and (4) image segmentation, see Figure 13.20.

A database of LG graph representations of the reference (enrolled) iris images is created, and the LG graph representation of the presented test image is compared against this database. Based on a distance metric calculated between the reference and test LG graphs, the authenticity of the subject is established.

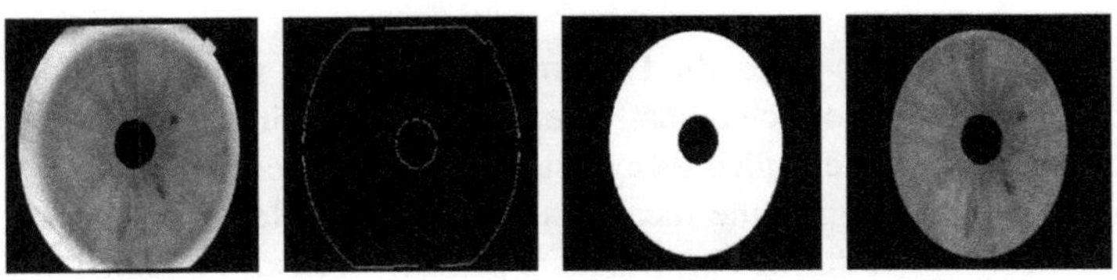

Figure 13.19 (*a*) Acquired iris image, (*b*) edge image, (*c*) circular mask, and (*d*) localization of iris. (*See insert for color representation of the figure.*)

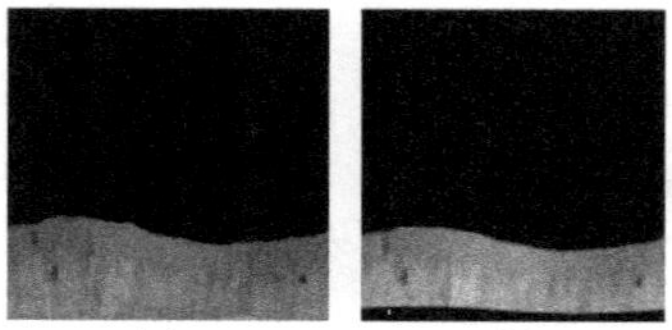

Figure 13.20 (*a*) Log-polar transformation after iris localization. (*b*) Segmentation (based on region labeling) of logpolar image using k-means clustering (R-color space with $k = 51$). (*See insert for color representation of the figure.*)

LG Graph-Based Voice Biometric Authentication Prior to comparing the reference and test speech signals, some kind of time alignment has to be performed between those two. This alignment is implemented by preprocessing the reference and the test speech signals. Preprocessing the speech signals eliminates the parts of the speech signal that do not convey any information (i.e., silence), before or after the utterance of the speech. Regions of the signal that do not convey any information are eliminated using the method of standard deviation. The standard deviation of the samples of speech signal is calculated using a moving window. If the window size is too small, the window will not contain enough signal information to be able to measure the desired features, if the window size is too large, the feature values will be averaged over too much, and will lose precision. In our experiments, we found that the length of samples retained varied with the size of the window. By trial and error, the optimum window size was found to range from 75 to 125 samples. The speech signal was padded with the mean value of the windowed signal at the beginning and end points of the signal in order to handle boundary problems. Mean padding was used instead of zero padding because zero padding introduces abrupt variations in the signal that are not suitable for the standard deviation method used to eliminate sections of no interest in the speech signal.

The calculated values of the standard deviation are stored in an array. The mean value of the stored standard deviation values is calculated, and the standard deviation values in the array that are less than the calculated mean are forced to zero. Resampling of the presented speech signal is done based on the new standard deviation array. Samples of the speech signal that correspond to zero standard deviation are eliminated. Figure 13.21 shows the speech sample and its preprocessed version, respectively.

Once the alignment is done, the LG graphs of test and reference signals can be extracted and compared to each other. For each pair, consisting of a test and reference signal, a distance measure, called local distance, is computed. The global distance is then computed by averaging the local distances over the length of the signal. The reference speech signal during enrollment is preprocessed as explained in the previous section. After preprocessing, the reference signal is represented as an LG graph and stored in the system database. When a test speech signal is presented, it is preprocessed using the same technique and represented as an LG graph. The LG graph

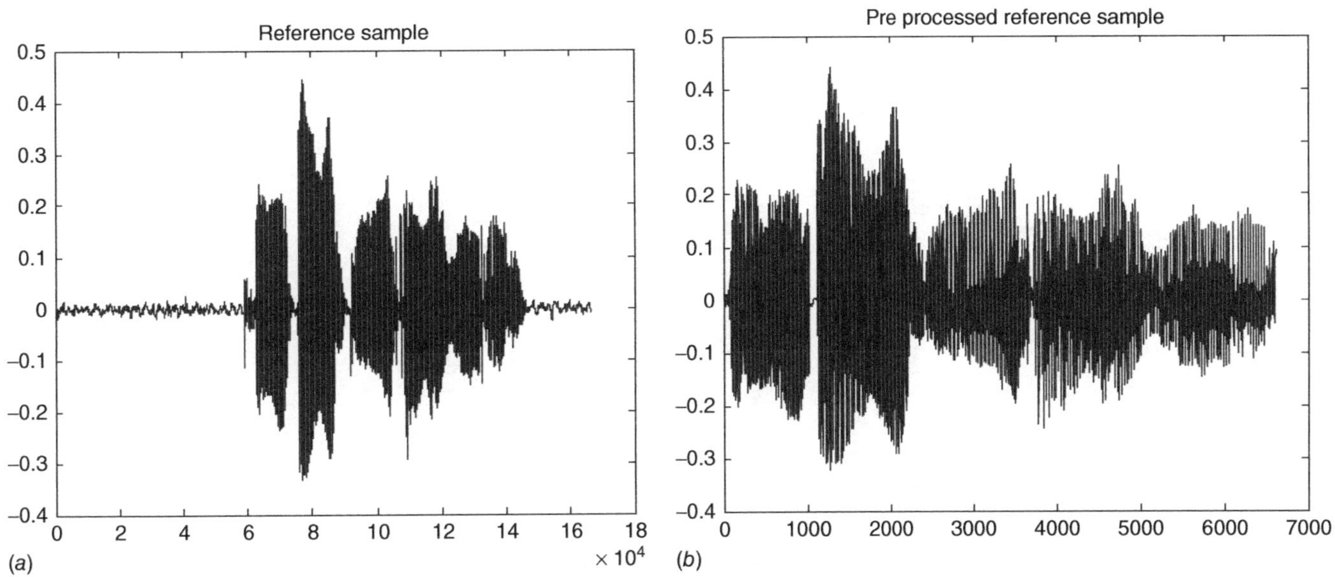

Figure 13.21 (*a*) Speech sample and (*b*) preprocessed speech sample.

embeds both local information (the shape of signal, i.e., parameters of a triangle, stored within the local graph at each node) and global information (the topology of the signal). Individual features of interest are extracted from the given signal and these features are mapped to other extracted features of the signal by finding relations between them. Thus, the LG graph describes the individual features of the signal in relation to the entire signal, thus offering a robust methodology for signal description and comparison. A database of LG graphs can be used to compare and match the LG graphs of test and reference speech signals in order to verify their authenticity.

Maxima–Minima Detection The first step is to extract the local features by detecting the maxima and minima points of the signal. This is done by finding the slope of the straight-line segments, which form the signal, and identifying points for the signal changes. A maximum is defined as the point on the curve where the tangent changes from positive to negative. A minimum is defined as the point on a curve where the tangent changes from negative to positive. Figure 13.22 shows the maxima and minima detection in a section of the presented speech signal. Similarly, maxima and minima points are identified for the entire speech signal.

LG Graph Representation of Signal Once all the maxima and minima in the signal have been identified, the next step is to construct triangles between two minima and one maximum (or two maxima and one minimum) in the given signal. Triangles above the zero x axis are formed by two minima and one maximum, while triangles below the zero x axis are formed by two maxima and one minimum. This enables us to describe the signal in the form of triangles of varying sizes. Figure 13.23 shows the representation of a given signal in the form of triangles. We also see that between the two minima and one maximum (or two maxima and one minimum), which form an approximate triangle, there are points where the slope changes. At these points the slope either continues to increase toward the maximum (if the point lies on the part of signal that is approaching a maximum from a minimum) or continues to decrease toward a minimum (if the point lies on the part of the signal that is approaching a minimum from a maximum). These points, along with the maxima and minima are termed

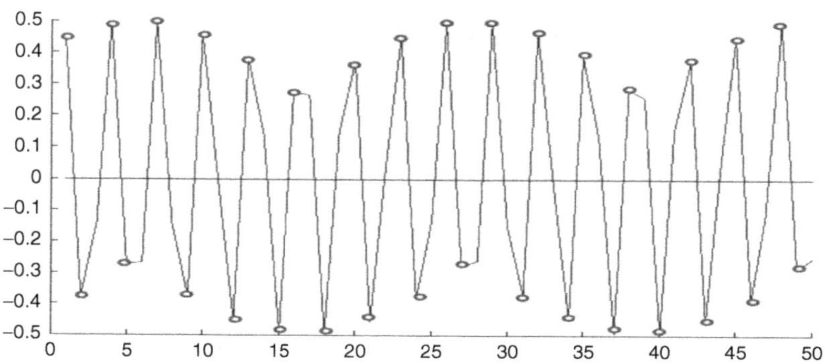

Figure 13.22 Maxima–minima detection in a section of the speech signal.

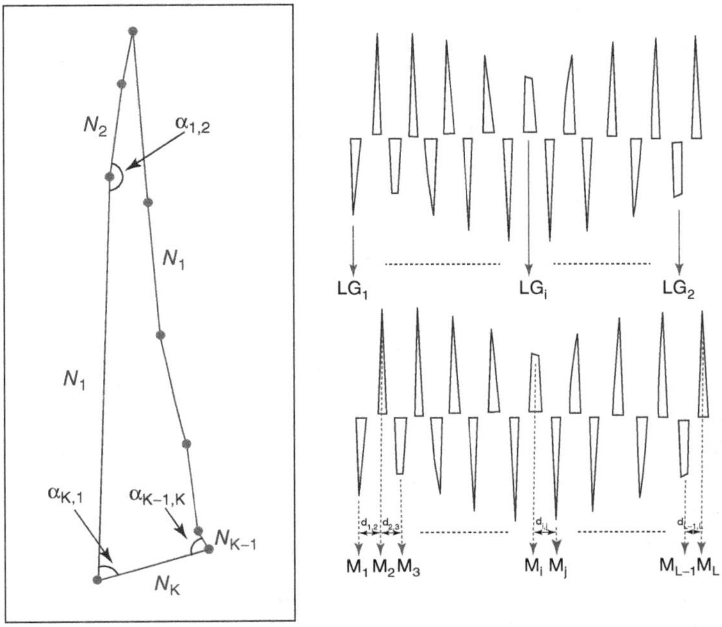

Figure 13.23 Nodes in an approximated triangle from a speech sample; formation of local graphs in a speech sample; formation of global graphs in a speech sample.

nodes. Precisely, points where the slope of the signal changes are termed nodes, and these nodes occur between the two minima and one maximum that form a triangle.

Thus, we see that the triangles form approximations of the signal at those points. To make this approximation precisely match the given signal, we extract features from the nodes that occur in the signal approximated by the triangles. The features that are extracted from these nodes are explained further. A node given by N_i is represented by

$$N_i = \{\mathrm{SP}_i,\ \mathrm{EP}_i,\ \phi_i,\ L_i\}$$

where N_i stands for the ith node, SP_i stands for the starting point of a segment joining node$_i$, EP_i stands for the ending point of a segment joining node$_i$, ϕ_i stands for the angle between node$_i$ and the horizontal axis (slope), and L_i stands for the length of the segment joining node$_i$.

The relationship between two nodes is given by α_{ij}, where α_{ij} stands for $\{*N_i, \theta_{ij}, *N_j\}$, where N_i and N_j stand for the two consecutive nodes that occur in the signal and θ_{ij} stands for the angle "connecting" nodes i and j.

Figure 13.23 shows the occurrence of the nodes in the signal in the approximated triangles. From the features extracted as described before, we construct the local graph, which is as shown next:

$$\mathrm{LG}_i = N_1\alpha_{1,2}N_2\alpha_{2,3}N_3\alpha_{3,4}\ \cdots\ N_{K-1}\alpha_{K-1,K}N_K\alpha_{K,1}$$

where LG_i stands for the local–global graph at the ith node.

From the local graphs obtained from the previous step, we generate a global map, which is maintained as a database or can be matched with the database to establish the authenticity of the speech signal.

LG Graph-Based Fingerprint Biometric Authentication Fingerprint biometrics is by far the most widely deployed identification and authentication technique. Most of the fingerprint-based biometric authentication schemes either rely on minutiae information or on ridge information to establish authenticity of the subject. Invariance to elastic distortion during fingerprint matching is an open problem that has not been completely addressed by these methodologies. In this direction, we offer a unique LG graph methodology for authenticating the presented fingerprint image. The LG graph methodology seeks to combine the minutiae information with the ridge feature information to enhance the robustness and reliability of the fingerprint matching process. Features from the individual ridges in the preprocessed fingerprint image are first extracted to represent the local graph. The individual ridges are also referred to as regions and used interchangeably in this chapter. The local information extracted from these regions includes region starting, region ending, ridge bifurcation, and region centroid. Euclidean distances between the individual regions' centroids are calculated and represented as a distance matrix to form the global graph. Thus, a combination of the local information combined with the global distance matrix yields the LG graph representation of the fingerprint image. This LG graph representation of the reference or enrolled fingerprint image is stored in a database. For subject verification, an LG graph is extracted from the presented test fingerprint image and compared with the corresponding LG graph in the database. The comparison is achieved by calculating the absolute difference between the two graphs and using a threshold, the authenticity of the subject is then established.

The database used in our experiments was obtained online from Fingerprint Verification Competition (FVC, 2000). The fingerprints were synthetically generated by using SFinGe. The image size is 240×320 pixels and 8-bit depth. The parameters of the synthetic generator were tuned to emulate a low-cost sensor with a small acquisition area.

Preprocessing Fingerprint Image The LG graph methodology relies on successful extraction of the individual ridges. Therefore, preprocessing the fingerprint image is critical prior to extraction of the LG graph from the fingerprint image. The important steps in preprocessing include image binarization, skeletonization, and elimination of spurious regions. The acquired fingerprint image has a bimodal histogram as seen in Figure 13.24. Hence it is easy to set up a hard threshold to binarize and generate a negative of the acquired image.

Figure 13.25*a* shows the acquired fingerprint image and Figure 13.25*b* shows the binarized negative of the acquired fingerprint image. The next step is to skeletonize the binary image and eliminate spurious regions. This is achieved by implementing an "OPEN" operation on the binary image followed by a thinning operation. "OPEN" operation reduces false minutiae while making sure that previously unconnected regions remain unconnected. The result of skeletonization operation

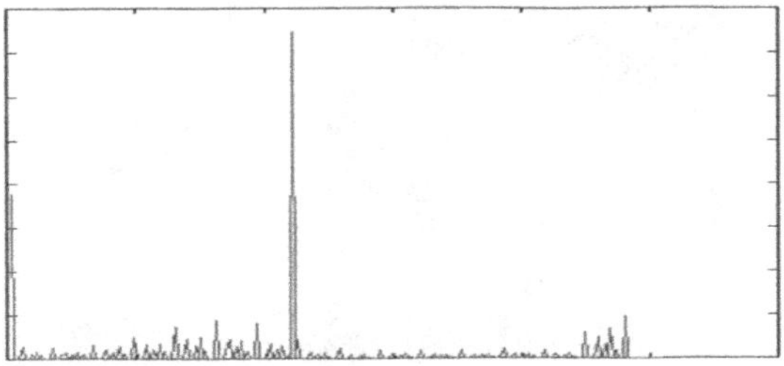

Figure 13.24 Bimodal histogram of synthetic fingerprint image.

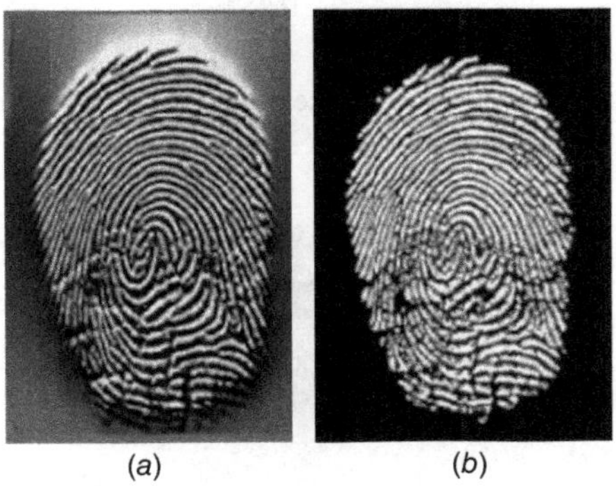

Figure 13.25 (*a*) Sample synthetic fingerprint image and (*b*) binarized version of the fingerprint image to obtain its negative.

on Figure 13.25*b* is shown in Figure 13.26*a*. We then apply connected component labeling algorithm with 8-neighbor connectivity to isolate different regions and remove regions that have pixel count less than a predetermined threshold, thus eliminating spurious regions. This resultant image is as shown in Figure 13.26*b*. After this preprocessing of the fingerprint image is completed, the LG graph is then extracted from the preprocessed image.

The ridge starting and ending points are discovered using the hit-or-miss structuring elements. These structuring elements traverse the fingerprint image in a raster scan and identify the ridge starting and ending points, based on matching the patterns as described in (Kannavara et al., 2011).

Individual regions (or ridges) can be isolated by using the connected component labeling technique with the 8-neighbor connectivity, as explained before. A region's centroid is calculated by averaging the region's pixel positions. The region (or ridge) centroid calculated by averaging its pixel positions may not necessarily lie within the region and may be located outside the region. Therefore, the calculated centroid is projected back into the region (or ridge) by choosing a point on the ridge having the least Euclidean distance from the calculated centroid. The point on the ridge with the least Euclidean distance from the calculated centroid is chosen as the ridge (or region) centroid. Figure 13.27 shows an isolated region (ridge), its corresponding

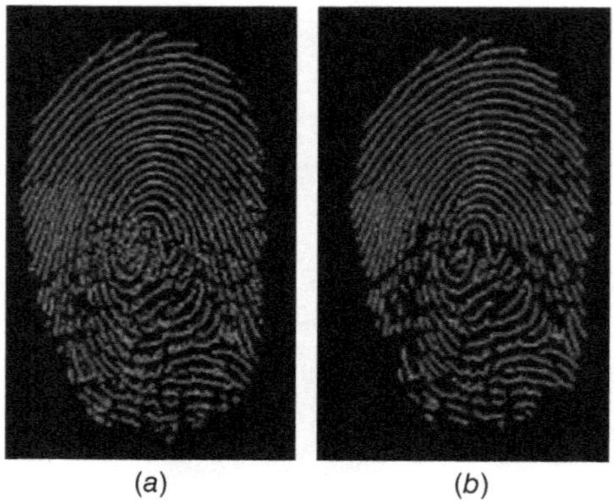

Figure 13.26 (*a*) Results after thinning process has been applied on fingerprint image and (*b*) elimination of spurious segments.

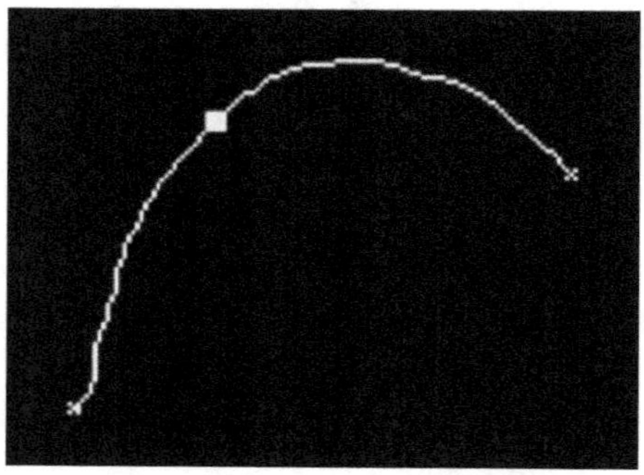

Figure 13.27 Isolation of a single ridge, its centroid, its end, and start points.

region centroid, and its starting and ending points. Figure 13.28 shows the centroids of all the different region ridges.

A local distance matrix is computed based on the Euclidean distances between the region's centroid and its corresponding starting, ending, and region bifurcation points. The magnitude of the region is obtained by counting the number of pixels in the region. The regions are then ranked based on their magnitude in descending order and a Euclidean distance matrix is calculated based on the location of the regions' centroids. The distance matrix represents the relative distances between different regions on a global scale. Figure 13.28 also shows the global graph that connects all the regions' centroids. The relative magnitude can be obtained by dividing all the regions' magnitude values by the magnitude value of the biggest region.

Thus, in the LG graph methodology we effectively combine the local minutiae features with the corresponding ridge description and map them on a global scale to enable a robust representation of the fingerprint image for the purpose of authentication.

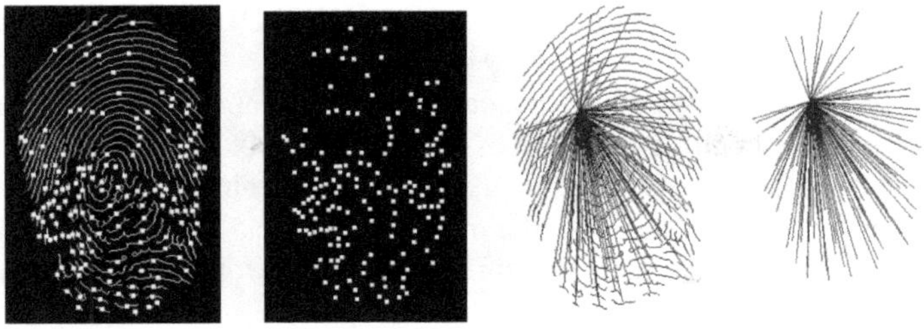

Figure 13.28 Centroids of different regions (centroids highlighted as squares), connecting the centroids to form a global graph.

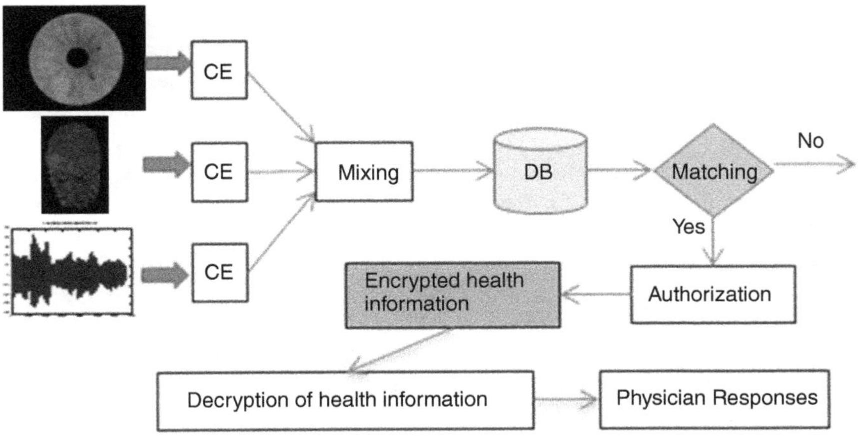

Figure 13.29 Graphically shows the three biometrics and the steps of process for a person to access the patient personal information and make a decision, if needed.

Authentication–Authorization Architecture The authentication–authorization scheme is presented in Figure 13.29. In this case the doctor or the appropriate person in a medical center uses his/her biometrics (iris, fingerprint, and voice password) to get access (after matching in the biometrics database) and decrypt the private information of a human subject under surveillance in a smart home environment.

This is a strong authentication scheme since it is difficult for an unauthorized person to "steal" all these biometrics and the rotated password in order to decrypt the private information of a patient. As graphically indicated in Figure 13.29, these three biometrics are combined and matched against a secure database in a medical center.

After a successful matching, the personal health information is automatically decrypted, and the authorized person (like a physician) makes a decision and automatically his/her medical instructions are securely compressed–encrypted (CE) and transmitted to the human subject for further actions. Note that most of the Smartphones of the near future will offer these capabilities.

13.6 CONCLUSION AND FUTURE RESEARCH DIRECTIONS

In this chapter we presented a mobile health generic wearable platform that continuously monitors the health conditions of a human subject by extracting biosignals, denoising them for further processing (like extracting health symptoms), and finally securing them for the exchange with the medical center via a robust protection (compression, encryption, hiding) and authentication scheme for accessing private health information.

The discussion and the results presented in all sections illustrate the feasibility and the efficiency of employing methodologies for a real implementation of such a system prototype. In particular, the DWT-based methods for detecting portions of ECG recordings corrupted by noise and for denoising the usable signal segments were analyzed. The proposed modified denoising technique combining conventional wavelet and modified wavelet-packet denoising produced promising results, at the cost of significantly more computation, restricting its applicability to offline processing.

A similar approach as the one described for detecting severe signal artifacts in the ECG can be applied in the case of the PPG sensor. Photoplethysmography is notorious for being particularly susceptible to motion-induced corruption (Wood and Asada, 2006). As a result, a corruption detection scheme could be facilitated for the case of the PPG after the DWT decomposition of the received signal is carefully studied for various artifact cases.

We would also like to point out that the presented wearable platform along with described preprocessing steps constitute the front end of a larger project as it has been described in our research efforts (Pantenopoulos and Bourbakis, 2009, 2010b, 2010c). Estimation of noise levels and denoising of usable data portions is the first step after data have been collected. The estimated noise level from all signals along with estimated activity level from the acceleration signals can be fused together to produce signal quality estimators. After this step, the ECG needs to be processed in order to detect the QRS complexes (Kohler et al., 2002) and then detected beats and rhythms need to be classified as normal or abnormal (6). This extracted higher level information is then in turn fused with other detected symptoms from the rest of the biosignal measurements to produce estimates of the user's health condition (Pantenopoulos and Bourbakis, 2010c).

Detected events (which can also be noise events and not only health events) can trigger verbal interaction with the user in order to further evaluate his status and context. At the same time these measured data that are transmitted to a remote server can be utilized to adjust user-defined models and to evaluate long-term health trends. In this loop a supervising physician can play an active role by judging the collected data and adjusting system parameters. The authors hope that such a model of health care for people at risk can potentially increase the quality of life of this population as well as possibly lead to better understanding and management of several common health disorders.

In addition, the secure information exchange scheme is working well by protecting health data. The SCAN methodology has been tested on various cases and works efficiently by offering a robust protection via compression–encryption–hiding. In particular, the denoised biosignals are converted into 2D matrices and altogether are combined to produce a 2D matrix. This matrix is then compressed, encrypted, and additional messages (like voice, biometrics, etc.) are hidden in it. This matrix is transmitted to the medical center and the authentication schemes after matching allow the decryption of the health data and the further process. Also the authentication scheme was tested against database sets and the results were very high.

Some future directions of the research areas presented in this chapter are:

1. Robust and real-time detection of pathological conditions with immediate communication with the medical center. The reason that we report this direction is that very few wearable systems are capable of performing in real time and most of them in limited scenarios.
2. Robust human user interaction with the wearable system (device) via speech recognition and synthesis. The reasoning behind this is that people of a certain age (elderly) and people with disabilities are not capable of verbal communication with good quality. Thus, the new WHMS must be able to do better than the typical speech recognition systems.
3. Implanted sensors. The main problems of the elderly using WHMS units is that they forget to use it, refuse to use it, or do not know how to use it. Thus, implanted sensors will eliminate the worries, and the problem will be transferred to the technology and not to the human with limited capabilities. Of course, we have to mention here that the Food and Drug Administration approval is a major obstacle for such devices. In addition, there is a serious educational effort to convince the elderly users to accept the implanted sensors.
4. Another important issue here is the security versus privacy. The elderly believe that the government will steal their health data or that the commercial sector is violating their privacy. Thus, education on these issues is a necessary effort. It seems like the refusal of the old fashion physicians to accept the computers as data management and storage devices.
5. New category of miniaturized sensors with potential to detect and monitor pathological conditions very early. Sensors that will work at biomarker levels.
6. Finally, new educational courses involving physicians, scientists, and engineers in a synergistic collaboration.

ACKNOWLEDGMENT

This work is partially supported by AIIS Inc.

REFERENCES

Addison PS. 2005. Wavelet transforms and the ECG: A review. *J. Physiol. Measure.*, 26:R155–R199.

Akay M, Exarchos T, Fotiadis D, Nikita K. 2012. Emerging technologies for patient-specific healthcare. *IEEE Trans. IT Biomed.*, 16(2):185, 189.

Alexopoulos C. 1988. A language for sequential data accessing. *University of Patras, Dept. of Computer Engr.*, Patras, Greece.

AMCC. 2005. Secure embedded processors enabling high-performance network protection, 5:26–5.

Ayat M, Shamsollahi MB, Mozaffari B, Kharabian S. 2009. ECG denoising using modulus maxima of wavelet transform. In Proceedings of the 31st Intl. IEEE EMBS, Minneapolis, MN, September 2–6, pp. 416–419.

Bonh D, Franklin M. 2001. Identity-based encryption from Weil pairing. In *Advances in Cryptology*, CRYPTO- 01, Lecture Notes in CS, Vol. 2139. Berlin: Springer, pp. 213, 229.

Bourbakis N. 1986. SCAN—A language for effective accessing of a 2-D array. IEEE Workshop on LFA, Aug., Singapore, pp. 52–58.

Bourbakis N. 2002. Emulating human visual perception for measuring differences in images using an SPN graph approach. *IEEE Trans SMC*, 32(2):191, 201.

Bourbakis N, Dollas A. 2003. *SCAN based multimedia information on demand*. IEEE Multimedia Mag., Sept. pp. 79–87

Canim M, Kantarcioglu M, Malin B. 2012. Secure mnagement of biomedical data with cryptographic hardware. *IEEE T-IT Biomedicine*, 16(2):166, 175.

Chazal PD, Dwyer MO, Reilly RB. 2004. Automatic classification of heartbeats using ECG morphology and heartbeat interval features. *IEEE Trans. Biomed. Eng.*, 51(July):1196, 1206.

CMU Sphinx Group. 0000 Available: http://cmusphinx.sourceforge.net/sphinx4/.

Cocks C. 2001. *An identity based encryption scheme based on quadratic residues*. In Honary B. (ed). *Cryptography &Coding, Lecture Notes in CS*, Vol. 2260. Berlin: Springer. Available: http://www.cert.org/advisories.

Constantinescu L, Kim J, Feng D. 2012. SparkMed: A framework for dynamic integration of multimedia medical data into distributed m-health systems. *IEEE Trans. IT Biomed.*, 16(1):40–52.

Deng H, Agrawal DP. 2004. TIDS: Threshold and identity-based security scheme for wireless ad hoc networks. *Ad Hoc Netw.*, 2:291–307.

Deng H, Wei L. Agrawal DP. 2002. Routing security in wireless ad-hoc networks. *IEEE Commun. Mag.*, 40:70–75.

Donoho DL. 1995. De-noising by soft-thresholding. *IEEE Trans. Inf. Theory*, 41:612–627.

Ercelebi E. 2004. Electrocardiogram signals de-noising using lifting-based discrete wavelet transform. *J. Comput. Biol. Med.*, 34(Sept.):479–493.

Fingerprint Verification Competition (FVC). 2000. Available: http://bias.csr.unibo.it/fvc2000/databases.asp.

Goth G. 2012. *Mobile security issues come to the forefront*. IEEE CS, May/June, pp. 7–9.

Hauser R, Przygenda T. Tsudik G. 1999. Lowering security overhead in link state routing, *Comput. Netw.*, 31:885–894.

Istepanain R, Zhang Y-T. 2012. 4G- Health the long term evolution of m- health. *IEEE Trans IT Biomed.*, 16(1):1–5.

Johnson DB, Maltz DA. 2001. Dynamic source routing in ad hoc networks. In Perkins CE (ed). *Ad Hoc Networking*. Reading, MA: Addison- Wesley, pp. 139, 172.

Kannavara R, Bourbakis N. 2009. Surveying secure processors. *IEEE IT Potential*, 28(1):28–34.

Kannavara R, Bourbakis N. 2013. SP-SCAN: A secure processor and its image crypto capabilities. *J. on Microprocessors and Microsystems* (submitted).

Kannavara R, Mertoguno S, Bourbakis N. 2011. A secure processor with biometric capabilities, *SPIE J. Electron. Imaging*, 20:2.

Kohler BU, Hennig C, Orglmeister R. 2002. The principles of software QRS detection. *IEEE EMBS Mag.*, 21:42–57.

Krawczyk H, Bellare M, Canetti R. 1997. HMAC: Keyed hashing for message authentication. *RFC*, 2104.

Li S, Liu G, Lin Z. 2009. Comparisons of wavelet packet, lifting wavelet and stationary wavelet transform for de-noising ECG. In Proceedings of 2nd IEEE ICCSIT, Beijing, China, August 8, 11, pp. 491–494.

Mallat S. 1998. *A Wavelet Tour of Signal Processing*. San Diego: Academic Press.

Maniccam S, Bourbakis N. 2004. Image and video encryption using SCAN patterns. *PR Soc. Pattern Recog. J.*, 37(4):725–738.

Nonin Medical. Available: www.nonin.com.

Pan J, Tompkins W. 1985. A real-time QRS detection algorithm. *IEEE Trans. Biomed. Eng.*, 32:230, 236.

Pantelopoulos A. 2010. Prognosis a wearable system for health monitoring of people at risk. Ph.D. Dissertation, ATRC, CSE-WSU.

Pantelopoulos A, Bourbakis NG. 2009. *A health prognosis wearable system with learning capabilities using NNs*. In Proceedings of the 21st Intl. IEEE ICTAI, Newark, NJ, November 2–5, pp. 243, 247.

Pantelopoulos A, Bourbakis NG. 2010a. A survey on wearable sensor-based systems for health monitoring and prognosis. *IEEE Trans. Syst., Man*, Cybern. C, Appl. Rev., 40(Jan.): 1, 12.

Pantelopoulos A, Bourbakis NG. 2010b. Design of the new prognosis wearable system- prototype for health- monitoring of people at risk. In *Advances in Biomedical Sensing, Measurements, Instrumentation and Systems*. Berlin: Springer, pp. 29–42.

Pantelopoulos A, Bourbakis NG. 2010c. Prognosis—A wearable health monitoring system for people at risk: Methodology and modeling. *IEEE Trans. Inf. Tech. Biomed.*, 14(3):613–621.

Papadimitratos P, Haas ZJ. 2002. *Securing the Internet routing infrastructure*. IEEE Commun. Mag., October.

Reddy SV, et. al. 2010. Wireless hacking—A WiFI hack by cracking WEP. *IEEE Second Int. Conf. on Education Tech. and Computer*, Vol. 1, pp. 189–193.

Redmond SJ, Lovell NH, Basilakis J, Celler BG. 2008. *ECG quality measures in telecare monitoring*. In Proceedings of the 30th Ann. Intl. IEEE EMBC, Vancouver, Canada, Aug. 20–24, pp. 2869, 2872.

Shamir A. 1979. How to share a secret. *Com. ACM*, 22(11):612–613.

Tikkanen PE. 1999. Nonlinear wavelet and wavelet packet de-noising of electrocardiogram signal. *Biol. Cybern.*, 80(4):259, 267.

Tröster G. 2005. The agenda of wearable healthcare. *IMIA Yearbook of Med. Informatics*, pp.125, 138.

Vlingo. Available: www.vlingo.com.

Wood LB, Asada HH. 2006. *Noise cancellation model validation for reduced motion artifact wearable PPG sensors using MEMS accelerometers*. In Proceedings of the 28th IEEE EMBC, New York City, Aug. 30–Sept 3, pp. 3525, 2528.

Yang M, Li S, Bourbakis N. 2004. A survey on image and video encryption. *IEEE Potentials*, Aug/Sept: 28–34.

Zephyr™. Available: www.zephyr- technology.com.

14

CONNECTION BETWEEN BIOMEDICAL TELEMETRY AND TELEMEDICINE

EMMANOUIL G. SPANAKIS, VANGELIS SAKKALIS, KOSTAS MARIAS AND MANOLIS TSIKNAKIS

Computational Medicine Laboratory, Institute of Computer Science, Foundation for Research and Technology - Hellas (FORTH), Heraklion, Crete, Greece;
Biomedical Informatics Laboratory, Department of Informatics Engineering, TEI Crete, Heraklion, Crete, Greece

14.1 INTRODUCTION

Recent advances in information and communications technology (ICT) enable the acquisition, transmission, and interpretation of different biosignals, from fixed or mobile locations, at an acceptable cost. This can support better prevention and well-being and provide valuable and prompt diagnostic tools in various application domains, ranging from home care to emergency care, or situations in which a second or a specialist opinion is required before taking a clinical decision (Breen et al., 2010). Important trends in health care include citizen mobility and the consequent move toward shared or integrated care in which the single doctor–patient relationship has changed to one in which an individual's health care is the responsibility of a team of professionals in a geographically extended health care system. In this new scenario, the possibility of consulting and collecting clinical information

Handbook of Biomedical Telemetry, First Edition. Edited by Konstantina S. Nikita.

from different points is a common need for citizens and physicians. Moreover, the increase in life expectancy has produced an older population that may need continuous assistance, especially in cases of serious or chronically ill people that wish to live independently.

In this perspective, with focus to the provision of high quality of care, assuring life independency, the significant new research challenge is the integration of multiple biomedical sensor streams to extract local and global health state indicator variables that can be queried and monitored by an unobtrusive system. Using biomedical telemetry, ubiquitous computing, social user interfaces, and wireless communication technologies, intelligent e-health spaces can be created, accelerating the extensive deployment of sensor technology in tomorrow's telemedicine services. The vision for future biomedical telemetry and telemedicine informatics is to offer seamless e-health services to people through an environment capable of recognizing and unobtrusively responding to their needs by allowing intelligent communication and interaction among users and systems.

14.2 BIOMEDICAL INSTRUMENTATION

A medical instrument is any appliance, apparatus, material, device, or other, including necessary software for its proper application, used for the purpose of diagnosis, prevention, monitoring, treatment, or alleviation of a disease and which does not achieve its principal intended action in or on the human body by pharmacological, immunological, or metabolic means, although they may be assisting in its function. We can distinguish medical instruments and devices into three different categories:

General Control Devices Simple devices with minimal potential for harming the user's body

Special Control Devices Noninvasive devices, including x-ray, functional magnetic resonance imaging (f-MRI), wheelchairs, infusion pumps, electronic stethoscopes, spirometers, electrocardiographic (ECG) and electroencephalographic (EEG) devices, SpO_2 sensors, blood pressure sensors, surgical drapes, needles and suture material, and more

General and Special Control Devices With premarket approval or scientific review that ensure the safety and effectiveness (most of them are life-sustaining or life-supporting devices), such as replacement heart valves, silicone gel-filled breast implants, implanted cerebral stimulators, ventilators, bed side monitors, and many more

Biomedical instrumentation covers this area of research focusing on selecting and properly using instruments for measuring medical variables and indicators for medicine or biomedical research. Biomedical telemetry involves devices designed and connected together in a scientifically appropriate manner to sense (or capture) biomedical signals and process them for human display and/or further processing for control and therapy using intelligent electronic health information and communication systems.

14.3 BIOMEDICAL TELEMETRY AND TELEMEDICINE: RELATED WORK

It is evident that the delivery of health care and health care organizations themselves are transformed by advances in biomedical telemetry technologies. These advances are giving rise to a range of reforms in the way in which health care services are delivered. Preventive medical care will be emphasized for individual health management, data will constantly be transmitted to the hospital through built-in sensor and monitoring systems, for example, in the patient's watch, accessories, or other items worn daily, and results will be fed back to the patient. Mobile and wireless technologies promise to transform the health care industry, supporting the acquisition, integration, analysis, and storage of clinical data in all its forms; information tools provide access to the latest findings while communication tools enable collaboration among many different organizations and health professionals. Patients and health professionals are becoming increasingly mobile and the need for support in managing their health is growing. As a result, wireless connectivity is intensive for ubiquitous radio communications (Timmons et al., 2004).

Webster (2006) presented various types of biotelemetry systems from the early 1950s to the late 1990s. The authors summarize the advances and reviews of the biotelemetry-related technologies in this area. A classic reference book in biotelemetry by MacKay (1993) represents an excellent starting point and includes some ingenious techniques used by early investigators to remotely acquire physiological information.

The use of wireless communications in a health care environment raises two crucial issues. First, radio frequency (RF) transmission can cause electromagnetic interference (EMI) of biomedical devices, which could as a result critically malfunction. Second, the different types of e-health applications require different quality of service (QoS). Phond et al. (2010) introduce an innovative wireless access scheme called EMI-aware prioritized wireless access. This protocol focuses on guaranteeing the safety of biomedical devices from harmful interference by selecting the transmission power of the wireless devices based on the EMI constraints. The landmark paper (Akyildiz et al., 2002) describes major concepts of sensor networks which have been made viable through the convergence of microelectromechanical systems technology, wireless communications, and digital electronics. This paper presents an extensive review of factors influencing the design, implementation and deployment of sensor networks and related applications. The design of a complete communication platform (from sensors to remote applications) by Fernanda et al. (2006) describes the basis for the construction of a nano- and microsensor-based sleep management system. The innovation of the work presented by the authors is the introduction of a framework and system for hypovigilance monitoring and management for cross-sector applications based on micro- and nanosensors.

Galarraga et al. (2007) present a survey for the ISO/IEEE 11073 family of standards, trying to identify the changes that these hardstands are undergoing in order to meet communication demands for interpolatable personal health monitoring systems. The article identifies the need for standardized, open, and robust solutions allowing the development and deployment of cost-effective and interoperable telemedicine systems, with the International Organization for Standardization/Institute for Electrical and Electronic Engineers (ISO/IEEE) 11073 family of standards being the most appropriate international standards for telemedicine applications.

As the population ages and the risk of chronic disease increases, the cost of health care rises. Technology for mobile telemetry is expected to contribute to the dual objective of cost reduction and improvement of the efficiency of care. In achieving these goals, several technical challenges must be successfully addressed, including sufficient system lifetime, high signal fidelity, and adequate security. Jurik et al. (2009) present the design, implementation, and evaluation of a mobile biotelemetric system that addresses the efficiency of treatment and other remote medical monitoring and technical challenges. The system differs primarily for its ability to provide flexible, robust, and precise end-to-end communication and to enable trade-offs in metrics such as power and transmission frequency. A good example of a mobile application is presented in Nguyen et al. (2009), where a mobile waist-mounted device is used for monitoring a subject's acceleration signal, subsequently used for detecting fall events in real time with high accuracy and triggering appropriate emergency messages. Brunelli et al. (2006) presented the design and implementation of a biofeedback system for rehabilitation based on a dedicated, distributed, wireless body sensor network. This work introduces the concept of a closed-loop system, deploying multiple body-mounted sensors, multisensorial signal fusion, and intelligent action through the provision of audio biofeedback using personal computer devices and headphones.

In parallel, implanted sensors can provide precise measurements and continuous patient monitoring as described in the approach of an artificial retina prosthesis and cortical implant in Schwiebert et al. (2002), where a complete system for restoring vision to visually impaired persons from the signals generated by an external camera to an array of sensors that electrically stimulate the retina via a wireless interface was presented. The authors aim to create a smart sensor implant to restore vision to persons with diseased retinas or suffering from other damage to the visual system that is believed to have tremendous potential for improving the quality of the life for people. Other examples focus on the management of chronic diseases, like the work presented in using monitoring sensors and close look monitoring systems for and this paper we describe the current status of diabetes management and insulin administration and the future perspectives and aspects to support long term management of diabetes based on wearable continuous blood glucose monitoring sensors, (Spanakis, Chiarugi 2011) and blood pressure monitoring (Fassbender et al. 2008).

Cerny et al. (2008) focus on the analysis and design of sensing electrodes by the use of unconventional materials for long-term bioelectric activity measurement with advanced capabilities in the transmission of signal, time–impedance stability, and low half-cell voltage. The principle of the electrodes is conducting polymers which can be built up on a textile or other nonprimary conducting materials. These types of sensors could be used, among others, in mobile human telemetry systems. Lastly Khoor et al. (2001) explored Bluetooth for collecting short- and long-term digitized ECGs together with relevant clinical data for patient management. A wireless communication protocol was developed for short-range radio data transmission able to send compressed records though cellular networks. Panescu (2006) provides a review of microelectromechanical systems (MEMSs), a technology developed to create miniature sensors and actuators, and fabrication technologies, materials, and power supplies in order to better understand their design. The paper discusses some of the applications of MEMSs in medicine and biology, such as pressure sensors, accelerometers, human retinal prosthesis, tactile sensor skin, and MEMS-based renal replacement systems.

14.4 THEORY AND APPLICATIONS OF BIOMEDICAL TELEMETRY

Biomedical telemetry refers to the use of ICT for the transmission of biological or physiological data using any kind of biomedical instruments and devices from a remote location to a location that has the capability to interpret the data and affect clinical decision making to provide effective diagnostic, therapeutic, and prosthetic tools in physiological research and pathological intervention. Biomedical telemetry systems, used when direct observation is impossible, are able to acquire a wide spectrum of environmental, physiological, and behavioral data. Biomedical telemetry offers wireless, restraint-free, simultaneous, real-time, long-term data gathering of physiological variables in conscious, unrestrained humans. Biomedical telemetry systems consist of various medical and other devices to create a medical sensor network that can be used in various application domains, such as:

- *Diagnostic Applications* Diagnostic biomedical telemetry systems are used to gather biomedical, physiological, or histological data from the sensors within or around the body in order to identify and diagnose any kind of pathologies (i.e., eye implants measure intraocular pressure to diagnose low tension glaucoma, endoscopic wireless camera pill, designed to be swallowed, captures images from the digestive track, etc.).
- *Therapeutic Applications* Therapeutic biotelemetry systems are designed to alleviate certain symptoms and help in the treatment of a disease (i.e., drug delivery systems, glucose transponder to capture glucose fluctuations, ePatch, ECG holders, etc.)
- *Rehabilitative Applications* Rehabilitative biotelemetry systems used to substitute a lost function, such as vision, hearing, or motor activity (i.e., neuromuscular microstimulator stimulates paralyzed muscle groups in paraplegic and quadriplegic patients, visual prosthetic device stimulates ganglion cells in retina in order to restore vision to people, fall detection sensors, etc.).

14.5 INTEGRATION OF BIOMEDICAL TELEMETRY WITH TELEMEDICINE

Telemedicine (also referred to as *telehealth* or *e-health*) is defined as the delivery of health care and sharing of medical data over distances using communication technologies. Today's telemedicine systems are supported by state-of-the-art technologies like interactive video, high-resolution monitors, biomedical telemetry data acquisition systems, powerful computer systems, expert clinical decision support or recommendation systems, electronic patient record systems high-speed computer networks, and telecommunications systems such as fiber optics, wireless networks, satellites, and cellular telephony (Meystre, 2005; Bashshur, 2002).

Telemedicine services include but are not limited to:

(i) *Specialist referral health services* typically involving a specialist or a group of specialists assisting or consulting a remote distant doctor or other medical and paramedical personnel to finalize rendering a medical diagnosis (Traganitis et al., 2001; Chiarugi et al., 2003). Surveys and research have shown a rapid increase in the number of specialty and subspecialty areas that have successfully used telemedicine (Tsiknakis and Spanakis, 2010; Sakkalis et al., 2003).

(ii) *Patient teleconsultations* using telecommunication methods to acquire and transmit medical data, including audio, video, and still live images between a patient and a health professional (i.e., using Web-based communications, wireless or wired computer networks, and the Internet) (Spanakis et al., 2005).

(iii) *Remote patient monitoring* using smart medical devices to collect and send data to a remote monitoring station for collection, interpretation, and evaluation, usually from a patient's home using special vital sign monitors able to acquire the patient's vital signs (blood pressure, temperature, heartbeat, blood glucose, ECG, and many more biomedical indicators) or even in the case of emergency care and ambulances (Tsiknakis and Spanakis, 2010).

(iv) *Medical education services* that provide continuing medical education for health professionals and special medical education for targeted groups in remote locations.

(v) *Health 2.0 services,* referring to the use of a specific set of Web tools (blogs, podcasts, tagging, search, wikis, etc.) by actors in health care, including doctors, patients, and scientists, using principles of open source and generation of content by users and the power of networks in order to personalize health care, collaborate, and promote health education (Eysenbach, 2008; Hughes et al., 2008). Health 2.0 (as well as the closely related concept of Medicine 2.0) is evolving fast as the technology landscape evolves. However, there are already signs of Health 3.0 emerging, which is defined as delivery of health care which leverages the use of elements of Semantic Web such as location awareness, the emerging Internet of Things and embedded sensors.

A survey of the telemedicine applications and biotelemetry systems to be used in e-health is presented by Übeyli (2010). Developing a biotelemetry system and the principal operation of such a system are presented, and its components and the telemetry types are explained. Matusitz and Breen (2007) discuss the use of distant communication technologies within the context of clinical health care and the effects it has on health communication.

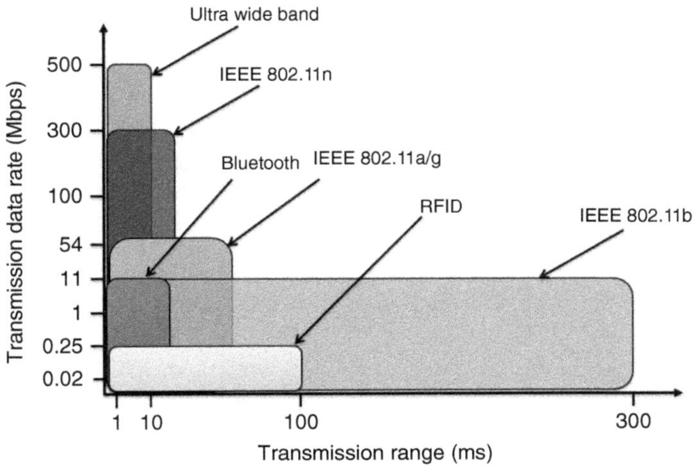

Figure 14.1 Distribution of various communication protocols (data rate vs. range).

14.6 WIRELESS COMMUNICATION PROTOCOLS AND STANDARDS

The most common delivery mechanisms for telemedicine are network applications and services designed to interconnect hospitals, emergency units, medical clinics, and other medical facilities and health community centers in rural and urban areas. The communication mechanisms used may include any kind of Internet connections or dedicated high-speed communication channels, both wired and wireless. Focusing on wireless technologies, Figure 14.1 presents the most prominent wireless technologies depicting the order of magnitude of their transmission data rate versus their connectivity range. It is obvious that for all the spectrum of applications in telemedicine it is necessary to incorporate many technological standards to take advantage of their unique characteristics (Table 14.1).

14.7 CROSS-LAYER DESIGN OF WIRELESS BIOMEDICAL TELEMETRY AND TELEMEDICINE HEALTH NETWORKS

14.7.1 Electromagnetic Spectrum

The electromagnetic spectrum is the range of all possible frequencies that characterizes the distribution of electromagnetic radiation emitted or absorbed by an object whether it is a wireless receiver or transmitter, a biomedical instrument (implanted or not), an electrical circuit, or any other object. It extends from low frequencies used for advanced radio communication to gamma radiation at the short wavelength (high frequency). The electromagnetic spectrum (Figure 14.2) in wireless communications is of paramount importance since it characterizes the common medium of all devices,

TABLE 14.1 Comparison of Various Wireless Technologies

	ZWave	ZigBee	Bluetooth	Bluetooth Low Energy
Primary usage/ applications	Domestic	Domestic industrial	Domestic industrial	Domestic medical
Topology	Mesh	Star/mesh/tree	Star	Star
Interoperability	No	No	Yes	Yes
Transmit frequency, MHz	900	900/2400	2400	2400
Throughput	40 kbps	40/250 kbps	1–2 Mbits (max. 24 Mbits)	1000 kbps
Current consumption (Tx), mA	35	25	30	15
Price	$4	$3		$1
Number of suppliers	1	<10	10+	10+
Range	30 m	30/10 m	10 m (1–100 m)	10 m
Frequency hopping and number of channels (ch)	No (1 ch)	No (16 ch)	Yes	Yes (40 ch)
Encryption	Yes	Yes	Yes	Yes
IEEE standard	No	802.15.4	802.15.1	802.15.1
Max. number of nodes in a net	232	65.000	Piconet/Scatternet	Unlimited
Max. number of node hops	4	32	N/A	N/A
Modulation type	BFSK	QPSK	GFSK / AFH	GFSK

BFSK: Binary Frequency-shift keying
QPSK: Quadrature Phase-shift keying
GFSK: Gaussian Frequency-shift keying
AFH: Audio Frequency-shift keying

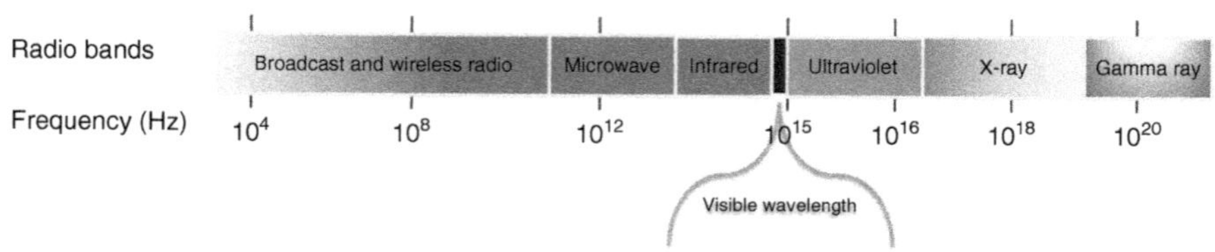

Figure 14.2 Electromagnetic spectrum range and usage (http://www.kollewin.com/EX/09-15-03/SURA_Electromagnetic_Spectrum_Full_Chart.jpg).

instruments, applications, and services offered over them. Each frequency has different physical properties, which are utilized in order to achieve greater effectiveness of the specific purpose for which it is used. The frequencies of the electromagnetic spectrum are treated as a public resource, so their usage is regulated by national

TABLE 14.2 ISM Band Frequencies

Frequency Range	Center Frequency	Availability
6.765–6.795 MHz	6.780 MHz	Subject to local regulations
13.553–13.567 MHz	13.560 MHz	
26.957–27.283 MHz	27.120 MHz	
40.66–40.70 MHz	40.68 MHz	
433.05–434.79 MHz	433.92 MHz	
902–928 MHz	915 MHz	Subject to local regulations
2.400–2.500 GHz	2.450 GHz	
5.725–5.875 GHz	5.800 GHz	
24–24.25 GHz	24.125 GHz	
61–61.5 GHz	61.25 GHz	Subject to local regulations
122–123 GHz	122.5 GHz	Subject to local regulations
244–246 GHz	245 GHz	Subject to local regulations

Source: http://en.wikipedia.org/wiki/ISM_band; http://www.itu.int/ITU-R/terrestrial/faq/index.html#g013.

or international organizations [International Telecommunication Union (ITU)[1]]. The usage of certain frequencies can be restricted (e.g., for military purposes, medical purposes) and regulated by the guidelines of national or international authorities or can be publicly available [e.g., the industrial, scientific, and medical (ISM) band]. The ISM band is commonly used by most communication technologies for telemedicine applications. The ISM band is actually a set of bands which cover multiple frequencies (Table 14.2).

14.7.2 Interference Management for Biomedical Telemetry Communication Networks

As we know from physics, the superposition of different waves results in a new wave pattern. This means that in the telecommunications domain the transmission in the same frequencies of two or more transmitters causes interferences to the signal as it travels from its source to the receiver. Interference is a central phenomenon in wireless communication when multiple uncoordinated links share a common communication medium. Most state-of-the-art wireless systems deal with interference in one of two ways, even though both of them might be suboptimal:

Orthogonalize the communication links in time or frequency so that they do not interfere with each other at all

Allow the communication links to share the same degrees of freedom but treat each other's interference as adding to the noise floor

The first approach entails an a priori loss of degrees of freedom in both links, no matter how weak the potential interference is. The second approach treats interference

[1] http://www.itu.int/.

as pure noise while it actually carries information and has structure that can potentially be exploited in mitigating its effect. These considerations lead to the natural question of what is the best performance one can achieve without making any a priori assumptions on how the common resource is shared.

Throughput enhancement and energy saving are two contradictory goals in an interference wireless networking environment where multiple links share a common communication medium. The problem of finding the performance trade-offs between throughput and energy efficiency becomes a difficult challenge in the design of joint rate and power adaptation. Spanakis et al. (2009) examined this topic for a simplified wireless telecommunication system where networking nodes use a specific modulation scheme with a specific BER (bit error rate) and a constant bandwidth assuming the absence of any error control coding or any multiuser decoding scheme or any cooperation between transmitters and/or receivers. This model reflects the situation of many practical low-power, low-complexity communication networks such as sensor networks, suitable to work in networking biotelemetry system, and it enables us to make observations, with practical significance, for the operation of these networks. The results show the achievable rate region with respect to the network topology, the degree of interference, and the individual maximum power constraints of each transmitter. This work also defined the necessary conditions that maximize the system's total aggregate rate and provide conditions under which simultaneous link operation outperforms time-sharing operation (Spanakis et al. 2014).

14.8 TELECOMMUNICATION NETWORKS IN HEALTH CARE FOR BIOMEDICAL TELEMETRY

Telemedicine and e-health systems are created as a synergy of mobile medical computing, medical sensor, and communication technologies. Possible medical applications include wearable systems and personal health care monitoring, emergency medical care and mass casualty event, home monitoring, computer-assisted rehabilitation and therapy, and social networking of relatives and peers of chronically ill patients. A proper integration of biomedical telemetry in telemedicine would allow physicians to diagnose, monitor, and treat patients remotely without compromising standards of care.

Wireless networks allow seamless system integration of various sensors. The advances in this domain are closely related if not correlated with recent advances in pervasive and ubiquitous computing systems for health care. The broad research areas that address the development in emerging mobile and network technologies will also support the complex systems that are related to matching current health care processes and infrastructures against the relevant cost–benefit business models. These include relevant business issues, such as interaction, visualization, planning and management, creating, monitoring, collaboration, information extraction, and education/training. This will allow the integration of ubiquitous computing and biomedical signal telemetry in future e-health research creating a comprehensive and successful telemedicine environment that can establish a holistic health care framework. The goals are:

(a) To adapt current health care infrastructures and existing medical (legacy) technologies to support future reliable, resilient, and cost-effective personalized health care services
(b) To address issues beyond classical health care disciplines and expand our understanding of the disease formation and correlating issues, such as genomics and the changing environments
(c) To increase the understanding of the effect of emerging health care technologies on human life and the fair access of health care delivery and services to the global population

14.8.1 Body Area and Personal Area Networks

In the communications field the terms LAN (local area network) and WAN (wide-area network) are used to denote a network able to cover an area such as a house, a housing complex, a hospital, an institution, or a city; the terms BAN (body area network) and PAN (personal area network) are used to denote a network of a smaller scale, analogous to humans, their dimensions, and their personal area, that is, a few meters or less. These terms are used to denote the proximity of interest (depending on the application) to a specific person and not the actual capability of their underlying technology, which in some cases might be capable of delivering a wider coverage. So, the same network and the same technology could be characterized by different terms depending on its application and not its maximum coverage allowed by the technology, which may vary. The term BAN usually refers to networks with limited coverage which cover only a person's body or refers to networks with applications having to do only with that specific person and not his or her surroundings. The term PAN usually refers to wider networks, in contrast to a BAN, or applications which use it involve not only that specific person but also his or her surroundings, for example, other computers in proximity, environmental sensors, or devices of other individuals in proximity.

The rapid growth of sensors, low-power integrated circuits, wireless communications, and relevant technologies has enabled the extensive development of applications in the fields of PAN and BAN. In order to provide smart, intuitive, and unrestricting applications to users, the preferred communication technologies for PAN and BAN are wireless communications, since they allow the user to be on the move, to formulate dynamic networks by adding and removing components at will, and to integrate seamlessly without requiring high expertise from users. A typical network of this type could include or require vital signal monitoring sensors and devices, motion detectors (such as accelerometers) or other kind of equipment to track and locate individual and environmental sensors, and communication technology (usually wireless) for the communication between those network components or for the communication and transmission of gathered information to other, back-end systems. In this context wireless sensor network (WSN) technologies are considered one of the key research areas of computer science and health care applications (Darwish and Hassanien, 2011).

14.8.2 Medical Device Connectivity

It is evident, as described in Section 14.7, that no single technology is perfectly balanced to support all of the required characteristics. Selecting the appropriate protocol depends heavily on the specific use case and the required trade-off between data rate, power consumption, range, and other factors, such as mobility, connectivity, stability, storage capabilities, routing protocols, data formats, and streaming data transfers. Today, on the one hand many standards are available, while on the other hand a diverse range of applications exist. Despite this diversity it is our view that four primary parameters must guide the selection of the most appropriate wireless medical technology:

(a) Range over which the device(s) needs to operate
(b) Amount of data that needs to be transferred
(c) Throughout speed needed to fulfill accurate estimations, the frequency of these transfers—how often data need to be sent
(d) Power available—typically whether the devices are battery operated or main-supply powered

In the following we review and present the main technical characteristics of key communication technologies and protocols available today, namely Bluetooth and Bluetooth low energy (BLE), ZigBee (IEEE 802.15.4), universal serial bus (USB), wireless universal serial bus (WUSB), ZWave, RFID (radio frequency identification), Infrared Data Association (IrDA), WiFi (IEEE 802.11a/b/g/n), light fidelity (LiFi), and near-field communication (NFC).

Bluetooth is an open wireless technology standard for exchanging data over short distances from fixed and mobile devices by using short-length radio waves and creating PANs. Bluetooth was initially created by telecom vendor Ericsson in 1994 and today is managed by the Bluetooth Special Interest Group. It can, in its current specification, connect several devices, forming small short-range networks.

Bluetooth low energy wireless technology (initially called Wibree) is a feature of the Bluetooth version 4.0 specification, special at new applications for wireless devices. Its chief advantage is enabling very low battery use within a short range (10 m). This facilitates a smaller form factor of devices and a wide range of applications in the health care, fitness, and home entertainment industries. BLE is designed with two implementation alternatives: single-mode and dual-mode. Small, highly integrated and compact devices like watches and sports sensors based on a single-mode BLE implementation will benefit from the low-power consumption advantages. In dual-mode implementations BLE functionality is integrated into classic Bluetooth circuits. The architecture will share classic Bluetooth technology radio and antenna, enhancing classic Bluetooth chips with the new low-energy stack, and will be backward compatible with previous implementations of the protocol.

ZigBee (IEEE 802.15.4) is a WPAN standard for a suite of high-level communication protocols using small, low-power digital devices with short-range radios. Typically used for industrial automation and domestic light control applications, ZigBee is governed by the ZigBee Alliance. ZigBee operates in the ISM bands around 900 MHz and 2.4 GHz and is intended to be a simpler, low-cost alternative to standard Bluetooth. As such, the ZigBee chip vendors typically offer integrated low-power consumption solutions with ZigBee radio, microprocessor unit, and flash in one application-specific integrated circuit (ASIC).

The *universal serial bus* is a wired point-to-point connection technology that is capable of high throughput (480 Mbps for USB 2.0 and 4800 Mbps for USB 3.0). The USB signals are transmitted through a twisted-pair data (channels) cable. A USB connection is always between a host [or a hub, e.g., personal data assistant (PDA)] at the connector end and a device (or hub's upstream port, e.g., a biosensor's transport gate) at the other end.

Wireless universal serial bus is a logical evolution of USB that combines speed and security of wired technology with the ease-of-use feature of wireless technology. WUSB intends to support robust high-speed wireless connectivity by utilizing the ultrawide-band (UWB) radio platform developed by the WiMedia Alliance. WUSB is the first high-speed wireless personal interconnection technology that meets the needs of multimedia consumer electronics (CE), PC peripherals, and mobile devices. It preserves the functionality of wired USB while also unwiring the cable connection and providing enhanced support for streaming media CE devices and peripherals. Its performance is targeted at 480 Mbps at 3 m and 110 Mbps at 10 m.

ZWave is a proprietary low-power wireless technology designed specifically for home automation and remote control applications. It is based on an innovative mesh networking technology that routes 2Way command signals from one ZWave device to another around any obstacles or radio dead spots that might occur throughout the home. ZWave is based on a low-bandwidth half-duplex protocol designed for reliable wireless communication in a low-cost control network. The protocol's main purpose is to communicate short control messages in a reliable manner from a control unit to one or more nodes in the network. All devices based on the ZWave standard are regulated by the ZWave Alliance.

Radio frequency identification is a communication technology that enables point-to-point wireless communication, typically used for identifying the whereabouts of humans, animals, or goods but lately also serving the purpose of payment registration and even sensor applications. A typical RFID tag consists of a microchip attached to a radio antenna mounted on a substrate. To retrieve the data stored on an RFID tag, you need a reader, a device that has one or more antennas that emit radio waves and receive signals back from the tag.

The *Infrared Data Association* is an industry-based group of companies that have developed the specifications for a communication protocol standard for short-range data exchange over infrared light for uses such as PANs. The IrDA protocol was developed as a communication standard especially suited for low-cost, short-range, cross-platform and point-to-point communications at a range of speeds.

A *wireless fidelity* network is used to connect computers to each other and to the Internet. Current WiFi networks operate in the ISM bands (2.4 and 5 GHz), offer speeds up to 54 Mbps, and support QoS with managed levels for data, voice, as well as video applications. Their medium access is based on the carrier sense multiple access/collision avoidance (CSMA/CA) principle. The IEEE802.11 standard specifies the physical and medium-access control (MAC) layers of a wireless LAN (WLAN). The technology is developed and standardized further in a number of working groups. Below we give a summary of the currently available substandards:

- *IEEE 802.11b* This standard, ratified in 1999, specifies WLAN systems operating in the frequency range of 2.4–2.4835 GHz with a data rate of up to 11 Mbps that is comparable to a fixed Ethernet. It uses direct-sequence spread-spectrum/complementary code keying (DSSS/CCK) technology. Although 802.11b products have had a significant share in the WLAN market, the problem of interference within the 2.4-GHz ISM band from other devices, including medical equipment and household appliances, led to the creation of the 802.11a standard operating in the 5-GHz range.
- *IEEE 802.11a* This standard enables the transmission of voice and data with high data rates (up to 54 Mbps). The devices operate in frequency bands of 5.15–5.35 GHz and 5.47–5.725 GHz. These frequency ranges are also used by other applications such as satellites, amateur radio, and radar, which have priority over data communications. The orthogonal frequency division multiplexing (OFDM) transmission technique is used to minimize signal delay spread that is caused by multipath propagation, especially in indoor environments. The maximum range is about 50 m. However, as 802.11a is not compatible with 802.11b and since the 5-GHz spectrum is not license free globally, an enhanced version of 802.11b resulted in the 802.11g standard ratified in 2003.
- *IEEE 802.11g* Operating in the unregulated 2.4-GHz range, this standard supports compatibility with 802.11b devices. It defines two optional modulation schemes: packet binary convolution code (PBCC) that supports 22 and 33 Mbps data rates and OFDM that offers 54 Mbps data rate. The disadvantages of IEEE 802.11b concerning nonoverlapping networks and noise sensitivity are also valid for devices built using this standard. It offers a range of 50–100 m.
- *IEEE 802.11n* IEEE 802.11n is based on previous 802.11 standards and adds multiple-input multiple-output (MIMO) antenna technology and 40-MHz channels to the PHY layer and frame aggregation to the MAC layer. It can operate in both 2.4- and 5-GHz frequency bands. These enhancements offer significant interference mitigation as well as increase in the maximum raw data rate from 54 to 600 Mbps.

Other IEEE 802.11 working group standards, such as 802.11h/j/k/s and so on, are extensions or offshoots of WiFi technology where each serves a specific purpose and there are also other region-specific standardization activities such as HIPERLAN standards developed by ETSI, which are similar to the IEEE 802.11 standards. WiFi networks support two modes: ad hoc and infrastructure. The ad hoc mode allows stations to instantly form a WLAN, allowing all stations to communicate directly with each other in a peer-to-peer fashion.

LiFi is the optical version of WiFi and the newly launched Li-Fi Consortium says that the technology will provide a secure, reliable, and ultrahigh-speed wireless communication interface that relies on gigaspeed technology, an optical mobile technology with several LiFi environmental features and services. It will use existing communication technology or technical topics regarding optical wireless communication in order to increase available services in the consumer area, industry, medical area, or logistics.

Lastly, *Near-field communication* is a short-range high-frequency wireless communication technology which enables the exchange of data between devices over about 10–20 cm. The technology is an extension of the ISO/IEC 14443 proximity card standard and is compatible with existing smartcards and readers and other NFC devices. The ISO/IEC 14443 international standard defines the transmission protocols for communicating with identification cards and proximity cards. Similar to ISO/IEC 14443, NFC communicates via magnetic field induction, where two loop antennas are located within each other's near field. It operates within the unlicensed ISM radio frequency band of 13.56 MHz with a bandwidth of 14 kHz. The working distance with compact standard antennas is usually around 10–20 cm. The supported data rates are between 100 and 800 kbps. NFC primarily targets the mobile phone industry to provide new applications and usage schemes.

14.8.3 Biomedical Telemetry Monitoring Devices for Telemedicine

Advances in material science and signal processing enable the design of smart medical sensors for real-time data recording and processing of multi-physiological signals (Darwish and Hassanien, 2011). The rapid development in microelectronics and digital wireless technology makes it possible to realize wireless medical sensors with networking capability to facilitate the joint processing of spatially and temporally collected physiological information from different parts of the body and the external communication for mobile health care. Examples of medical devices using these technologies include weighing scales, lifestyle devices (e.g., pedometers), fitness products, and the range of sensors required for assisted living (typically devices sending simple state information). If the major requirement is to transfer large amounts of data, the choice is likely to be WiFi, Bluetooth, or ZigBee. A defining reason for the choice of wireless can also be the topology of the connection. If it is a mobile device which is more likely to connect to a mobile phone or a tablet PC, then Bluetooth and ZigBee are the obvious choices. Following our review of technologies, in the subsequent section we present state-of-the-art biomedical telemetry devices for patient monitoring that are commercially available and approved to be used in telemedical applications.

Blood Glucose Meters Achieving the goal of tight glycemic targets requires extensive nursing efforts, including frequent glucose measurements, training to handle control algorithms or guidelines with intuitive decision taking, and most importantly additional responsibility to prevent hypoglycemic episodes. Some examples of devices that could be effectively used for blood glucose monitoring are the Bayer Contour USB meter, which allows the patient to measure the glucose levels and share information with a diabetes management system, and the MyGlucoHealth meter, the first U.S. FDA (Food and Drug Administration)–approved integrated Bluetooth diabetic care product.

Weighing Scales Digital weighing scales are designed to measure, store, and transmit, via a communication protocol (usually Bluetooth) to an application hosting device or an electronic patient record system, weight data of a patient. Examples include BC-590, which is designed to wirelessly transmit multiple readings to a personal computer, LifeStar Scale for precise and fast weight measurements, and Cardiocom TELESCALE, which takes part of the industry telemedicine standard for CHF (congestive heart failure) patient home monitoring management. In this category, lastly we refer to the Alere DayLink monitor, a biometric measurement device that records a patient's weight and where the patient can answer to preprogrammed questions regarding symptoms of the chronic condition or comorbidities that may be monitored.

ECG Meters Electrocardiography gives a transthoracic-over-time collected interpretation of the electrical activity of the heart. The data are captured by detecting and amplifying the tiny electrical changes that are caused when the heart muscle depolarizes during each heartbeat. Adequately, the ECG is measured on the skin surface. The in-hospital setting usually requires bed monitoring, but there are some mobile ECG meters that can be used: for example, Zenicor-ECG, a simple device for both patients and carergivers, and LifeStar ECG, a lightweight and easy-to-use personal one-lead ECG event monitor which is intended for monitoring symptoms that may suggest abnormal heart rhythms.

Biodevices is a company developing biomedical products such as VitalJacket, an intelligent wearable T-shirt textile garment accompanied with a small electronic pocket box that allows continuous monitoring of a person's heart activity, ECG, and heart rate in medical and other applications. It can either store data on a Secure Digital (SD) memory card for posterior analysis in a PC or visualize it online using a Smartphone or PDA. The ePatch sensor (Figure 14.3) is made for recording measured data using a sensor fitted in a small container patched on the subject's body. Its highlights include the possibility to stream or store data locally, include embedded software and algorithms, support real-time synchronization among sensors, support flexible power management, and communicate over the ZigBee radio communication. It is also easy to clean and to snap on electrodes.

Blood Pressure Sensors A blood pressure meter is a device used to measure blood pressure, comprising an inflatable cuff to restrict blood flow and a mercury or mechanical (i.e., later analog-to-digital) manometer to measure the pressure. State-of-the art devices include the LifeStar BP Pro wireless blood pressure and pulse rate monitor, an advanced device for measuring and assessing hypertension, and Omron's HEM-790IT, an advance diagnostic device including morning hypertension and irregular heartbeat with the specific software.

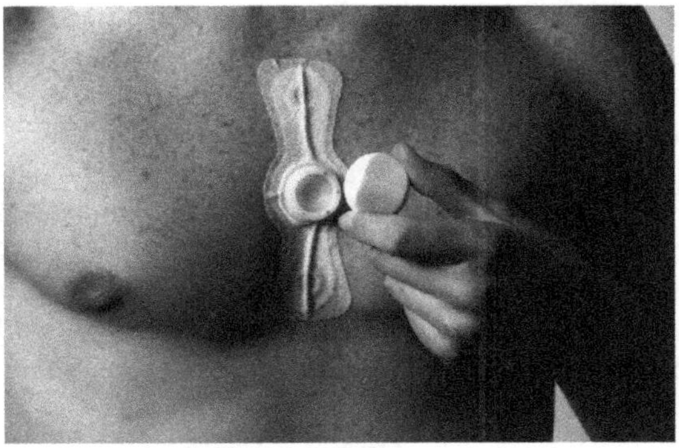

Figure 14.3 ePatch sensor.

Oximeter Pulse Meters LifeStar Oxy Pro is an easy-to-use oximeter for measuring both oxygen saturation levels in the blood and the pulse rate. It displays the medical data on the LifeStar PDA where oxygen saturation and pulse rate measurements are transmitted automatically via Bluetooth. The Alive Pulse Oximeter is a wearable medical device that uses wireless technology. It reads oxygen saturation data from a sensor on the finger or earlobe and transmits the data via Bluetooth to a mobile phone, PDA, laptop PC, or other Bluetooth-enabled device that can act as an application host device (AHD). This oximeter is well-suited for remote respiratory monitoring, for example, for remotely monitoring patients receiving home oxygen therapy (AlivePulseOximeter, 2010). PM-50 is a miniature, lightweight device designed exclusively for spot-check monitoring of SpO_2 and pulse rate. It weighs less than 200 g and is easy to carry and to store. Deploying this pulse oximeter gives efficiency when performing intermittent monitoring of a wide range of patients. The Nonin Avant 4000 system with Bluetooth wireless technology provides freedom from wired connections while ensuring reliable and secure transfer of patients. This lightweight wrist-worn patient module wirelessly sends data to a small table-top display and thus improves patient mobility and reduces bedside clutter. The highlights with this product are that it uses Bluetooth, is light and comfortable, has high power (33.5 hr memory and 10 m signal radius), is easy to use, has customizable alarms, is secure by data encryptions, and has long battery life.

Combo Devices MDKeeper is an effortless, uninterrupted vital signs monitor by which the elderly or chronically ill can maintain a normal active lifestyle without being confined to their homes or making frequent visits to their doctor. MDKeeper integrates unique, powerful biosensors into a lightweight wearable device that can monitor multiple vital signs (such as pulse rate, one-lead ECG, and blood oxygen saturation level) without discomforting users. By using its proprietary integrated expert system, MDKeeper can store and analyze patient data or transmit the data to a remote medical center for further analysis and care via its built-in Siemens wireless module. Fitbit accurately tracks burned calories, steps taken, distance traveled, and sleep quality. The Fitbit contains a 3D motion sensor by which it tracks the motion in three dimensions and converts this into useful information about user's daily activities. Bio-Harness BT couples the pioneering smart fabric sensor technology of garments with the power and ubiquity of Bluetooth. It enables the capture and transmission of comprehensive physiological data on the wearer and transmits the data via mobile and fixed data networks. The IntelesensVitalSens 100 (CE-approved) patch is based on the Bluetooth vital signs monitor. VitalSens provides wireless ECG, heart rate, skin surface temperature, and motion monitoring. It is designed to be rapidly integrated into medical devices or systems and can be configured for two different modes of data collection: periodic data or live data streaming via Bluetooth.

Continua-Compliant Devices The Continua Health Alliance (a nonprofit, open industry coalition of health care and technology companies) was formed to establish an ecosystem of interoperable personal health systems and medical devices that empower people to better manage their health and wellness. The goal was to ensure interoperability between components, systems, and subsystems incorporated within existing or emerging health systems. To achieve this aim, Continua established a technical working group (TWG) to select the standards and specifications and define design guidelines to further clarify interoperability specifications and standards. The Continua Health Alliance objectives include:

- Developing design guidelines that will enable vendors to build interoperable sensors, home networks, telehealth platforms, and health and wellness services
- Establishing a product certification program with a consumer-recognizable logo signifying the promise of interoperability across certified products
- Collaborating with government regulatory agencies to provide methods for safe and effective management of diverse vendor solutions
- Working with leaders in the health care industries to develop new ways to address the costs of providing personal telehealth systems

In order to build trust and ensure customer peace of mind, Continua Health Alliance has created a product certification program with a recognizable logo signifying interoperability with other certified products. The Continua Health Alliance's design guidelines contain references to the standards and specifications that Continua selected for ensuring interoperability of devices. It also contains additional design guidelines for interoperability that further clarify these standards and specifications by reducing options in the underlying standard or specification or by adding a feature missing in the underlying standard or specification. These guidelines were specifically written for device manufacturers that intend to go through the Continua certification process with their devices, companies that integrate Continua devices in systems and subsystems, and test labs that certify compliance to Continua specifications. Table 14.3 presents some Continua-certified medical devices available for physiological measurements.

TABLE 14.3 Examples of Available Continua Devices

Device Type	Manufacturer	Model	Transport
Blood pressure	A&D	UA-767PBT-C	Bluetooth
	Cypak	CPX186 Continua convertor cable	USB
	Omron	Home blood pressure	Bluetooth
Glucose meter	Roche	Accu-Check Smart Pix	USB
Pedometer	Omron	Pedometer with Bluetooth docking	Bluetooth
Pulse oximeter	Nonin	PalmSAT 2500	Bluetooth
	Nonin	Onyx II 9560 Fingertip	Bluetooth
Weighing scales	A&D	UC-321PBT-C	Bluetooth
	Omron	Weighing scales with body composition	Bluetooth

14.9 FUTURE RESEARCH DIRECTIONS AND CHALLENGES

Today's health information technologies (i.e., biomedicine, bioinformatics, electronic and mobile health) provide tools for health authorities, professionals, and patients that allow the realization and deployment of health information networks, telemedicine services, personal, wearable, and portable communication systems, and many other information and communication technology-based systems to assist in prevention, diagnosis, treatment, health monitoring, and lifestyle management of citizens. These systems, combined with organizational changes and the development of new skills for medical practice, promise to assist in achieving better and lower cost people-centered health delivery systems.

Nevertheless, a number of open issues need to be successfully addressed which today represent the focus of concerted international research efforts. Future developments in biomedical telemetry and telemedicine will focus on the provision of rapid response to critical medical care regardless of geographic barriers; flexible and swift access to expert opinion and advice at the point of care without delay; better management of medical resources; enabling fast communication links for high-rate data transmission; interactive medical consultation; increased empowerment and management of medical expertise, especially in rural and underserved areas; intelligent personal health monitoring systems generating early warnings for a variety of medical conditions; swift medical care in emergencies and management of medical data in catastrophes or natural disasters where conventional communication links may be disrupted; continuous health monitoring promoting healthy lifestyles and closed-loop reaction and monitoring; and synergy of information from individual sensors providing better insight into the physiological state and the type and level of activity. In subsequent sections, we specifically review current state-of-the-art and innovative research results in two challenging and innovative domains.

14.9.1 Biotelemetry Systems for High-Rate Biomedical Signals

Brain and cardiac biopotential signals in the form of EEG and ECG are two indicators appropriate for long-term monitoring, but despite the advancements of electronics and biomedical instrument, their use has been limited by the inconvenience and discomfort of conventional wet contact electrodes. Chi and Cauwenberghs (2010) presented a wireless EEG/ECG system using noncontact sensors expanding on previous presented work by building a sensor with much improved noise performance. The full design and schematics for this system are presented for practical body sensor networks. Gosselin et al (2006) present a wireless high-throughput data transmission system dedicated to interface implantable multichannel neural recording devices. Its main component is a low-end wireless IEEE 802.11g-compliant router, which makes it very cost effective and reliable. The most important feature of the presented design is its portability and wearability as a high-data-rate biomedical telemetric system. Chandler et al. (2009) performed an analysis of a wireless biomedical telemetry system. The results compare and cross reference key parameters of a biomedical telemetry system, such us the number of channel of useful transmitted information, power consumption, noise level, number of bits, and much more.

Technological challenges for the future must incorporate user acceptance, lightweight implementation, long battery life or batteryless sensors, biocompatibility, maintainability, usability and reliability, smart autonomous sensor design, reliable and unobstructed communications, new computational, networking environments, and protocols for biomedical telemetry medical networks. Additionally, if biomedical telemetry systems are to support or enable the future telemedicine services envisioned, a number of economical, business, regulatory, and societal issues must also be successfully addressed. These include appropriate regulatory approval from national and international agencies (FDA,[2] CEN,[3] CENELEC,[4] ETSI[5]), compliance with the European clinical trials directive (Directive 2001/20/EC of 4 April 2001),[6] and the Good Clinical Practice (GCP) Guidelines[7] if they are employed in the context of clinical trials and studies, standardization of protocols and interfaces that will significantly decrease cost, new business opportunities for inexpensive, small, and possibly disposable sensors, health care coverage and patient participation and reimbursement, liability issues, and lastly privacy and security of patient's records and transmissions.

14.9.2 EEG Portable Monitoring and Electrode Design

For biosignal recordings, electrodes are the initial elements which are used for converting biopotential signals due to biopotential sources into electrical signals, and thus it is clear that the EEG electrode technology is crucial for the acquisition of high-fidelity signals and the improvement of the system's measurement performance. Currently, several types of EEG electrodes are commercially available which are used for common neurophysiological applications (Sinclair et al., 2007). In order to reduce

[2] http://www.fda.gov/.
[3] https://www.cen.eu/cen/pages/default.aspx.
[4] http://www.cenelec.eu/.
[5] http://www.etsi.org/.
[6] http://eur-lex.europa.eu/LexUriServ/LexUriServ.do?uri=OJ:L:2001:121:0034:0044:en:PDF.
[7] http://ec.europa.eu/health/files/eudralex/vol-10/3cc1aen_en.pdf.

skin preparation time and increase measurement accuracy, multiarray thin-film electrodes are developed (Prohaska et al., 1986), while nitride-covered steel electrodes eliminate the need for EEG paste (Taheri et al., 1994). The active electrode, another type of electrode recently developed, promises noise reduction without the use of electrode gel but with much more skin preparation (Dunseath and Kelly, 1995). Apart from classical cup or disk-shaped electrodes and active electrodes, subdermal wire electrodes have been experimentally developed and have been found to be less susceptible to artefacts and more suitable for long-term monitoring, for example, in the ICU (Young et al., 2006). Another type of electrode has been also investigated in order to overcome the limitations of the conventional electrodes: the so-called dry electrode that is made of conductive rubber. With dry electrodes the preparation time of the patient prior to measurement is minimized, while no conductive gel or paste is required (Gargiulo et al., 2008). Ng et al. (2009) developed a dry EEG electrode with many microspikes of about 6 μm diameter and 110 μm length. Microspikes penetrate the stratum corneum of the epidermidis in a minimally invasive way and improve electrical conductivity at the skin–electrode interface; therefore they are a worthy candidate for the measurement of neurophysiological signals. The proper functioning of such electrodes for EEG has been proved, which suggests that the concept is valid (Ruffini et al., 2008). However, a final product that complies with the requirements of the clinical environment is required. Therefore, we aim to design a wearable long-term monitoring platform that will record the necessary biosignals for the task of seizure prediction (EEG/ECG) by introducing the microspike dry EEG electrode technology in the state-of-the-art clinical practice. Figure 14.4 shows initial microfabrication results for the proposed advanced microelectrodes implemented at the Foundation for Research and Technology—Hellas (FORTH). This approach will extend to the implementation of not only one single electrode but also the "whole picture" consisting of optimum configurations of such electrodes, for example, in the form of suitably designed arrays as well as the circuitry and the algorithms that are applicable to this specific approach, with a final product in mind capable of delivering

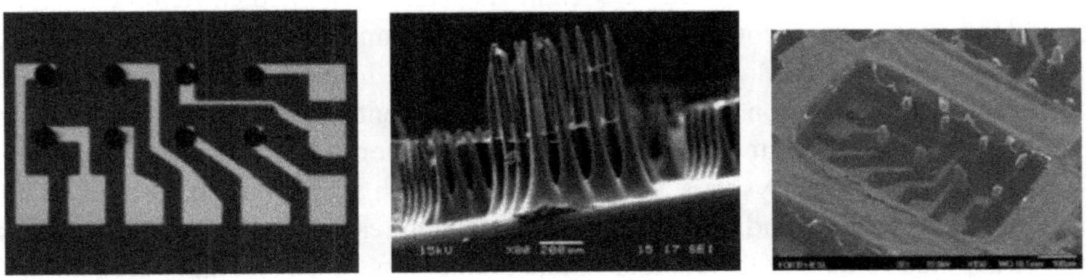

(*a*) (*b*) (*c*)

Figure 14.4 (*a*) Optical microscope image of out-of-plane, 100-μm-high polymer microspikes (unmetallized), fabricated onto planar addressing gold conductors. The diameter at the base is 30 μm. (*b*) Scanning electron micrograph (SEM) of 700-μm-high polymer microspikes. (*c*) SEM showing results of first attempt to selectively metallize microspikes, shown in (*a*), in order to electrically connect them to their respective planar conductor lines.

high throughput in given clinical scenarios. Furthermore, due to the nature of the proposed electrodes, some of which could be potentially functionalized as biosensors, we will investigate if other physiological parameters/readings, at the very immediate vicinity of the EEG recording sites, could be reliably monitored within a known time frame using the same underlying platform and an appropriate processing algorithm, thus yielding more relevant information as compared to conventional EEG. The utilization of these advanced electrodes will be integrated in the monitoring system, which will also include the appropriate devices to continuously record and amplify the acquired signals.

14.9.3 Bioinspired Approaches

Lastly a bioinspired approach for effective, smooth, and safe navigation in the human body and, in particular, in the gastrointestinal tract is an idea originating from the medical need to develop more powerful tools for microendoscopy, which is one of the most challenging frontiers of advanced medicine. Understanding motion and perception systems of lower animal forms can help to design and fabricate bioinspired robots able to navigate in tortuous, slippery, and difficult-to-access cavities of the human body. A preliminary study of a biomimetic adhesion system for human tissues, together with some technological implementations, is presented by Menciassi and Dario (2003). The authors present a bioinspired approach to effective, smooth, and safe navigation in the human body and, in particular, in the gastrointestinal tract.

14.10 CONCLUSION

An emerging approach to improve the wearability of continuous biosignal monitoring systems is the development of body-attached sensors with built-in wireless telemetry, thus freeing the user from having to carry personal data units for recording and retransmitting the medical data. Such biomedical telemetry systems involve a large number of wireless links coexisting in the same area, around a patient's body, sharing the same electromagnetic environment. Hence, biomedical telemetry must provide QoS guarantee, compensation of interference and surrounding thermal noise effects for all active medical acquisition devices and formed networks, ensuring a successful communication environment between patients and medical experts using intelligent health-related information management systems. Many tools and solutions today include products, systems, and services for health authorities and professionals as well as for patients and citizens. Literature examples include health information networks, electronic health records, telemedicine services, personal wearable and portable systems, health portals, and many other information and communication technology-based tools assisting prevention, diagnosis, treatment, health monitoring, and lifestyle management.

Recent advances in wireless sensor technology facilitate the development of smart ubiquitous health care systems, which combined with a life-long electronic health record can become the cornerstone for supporting continuity of care and cost-effective health care for every citizen (Darwish and Hassanienen, 2011). Despite the initial promising results, many obstacles remain to successfully deploy this technology in practical medical health care systems. Such issues include wearability, reliable and robust communication even in resource-constrained application environments, security, interoperability, self-configured medical devices, intelligent agent technology and ambient awareness, decision support systems, wireless resource and interference management, and more. The aim, for the future, is to create integrated systems to support long-term management of diseases based on biotelemetry intelligent wearable sensors and ubiquitous communication technologies (interoperable peer-to-peer communication platform, cloud-enabling middleware, model-driven application development, dynamic ontologies, etc.).

The new possibilities enabled by innovative developments in the domain of biomedical devices are provided at four levels:

(a) Microsensors offer the possibility of small size, as well as intelligent, active devices, working with low energy, wireless, and noninvasive or minimally invasive.
(b) Wrist devices are particularly user friendly and combine sensors, circuits, supply, display, and wireless transmission in a single box, making them very convenient for common physical activities.
(c) Health-related smart clothes make contact with 90% of the skin and offer many possibilities for the location of sensors. These sensors have to be thin, flexible, and compatible with textiles or be made using textile technologies, such as new fibers with specific (mechanical, electrical, and optical) properties.
(d) Health-oriented smart home platforms assist in improving the patient's living conditions and avoiding the cost of long hospitalization. Wearable devices and smart exosensors are very promising solutions providing a safer and much more comfortable environment for home health care, illness prevention, and citizen medicine (Axisa et al., 2005). The term *exosensor* is used to describe sensing equipment not directly placed on a patient's body but embedded within the surrounding ambient smart monitoring environment (home, room, place of work, vehicle, etc.) posing no physical constraints for the users (Axisa et al., 2005). Such sensing equipment is used for monitoring and measuring activity and behavioral patterns of people and are foundational elements of the smart home environments in the future.

In this chapter we discussed issues regarding computational medicine, biomedical telemetry, and telemedicine to describe the way smart networking environments can deliver seamless, personalized, and nonobtrusive health care services. We presented a number of key technologies on telecommunication and body area networks and technologies for device connectivity and raised issues on the use of the communication resources of miniaturized smart wireless medical devices and sensors concerning interference management. Lastly we presented state-of-the-art biomedical telemetry devices certified and available today to be used for advanced health care services.

REFERENCES

Akyildiz I. F, Su W, Sankarasubramaniam Y, Cayirci E. 2002. Wireless sensor networks: A survey. *Comput. Netw.*, 38(4), 15:393–422.

Axisa F, Schmitt PM, Gehin C, Delhomme G, McAdams E, Dittmar A. 2005. Flexible technologies and smart clothing for citizen medicine, home healthcare, and disease prevention. *IEEE Trans. Inf. Techn. Biomed.*, 9(3):325–336.

Bashshur RL. 2002. Telemedicine and health care. *Telemed. J. e-Health*, 8(1):5–12.

Breen G-M., Matusitz J. 2010. An evolutionary examination of telemedicine: A health and computer-mediated communication perspective, *Social Work Public Health*, 25(1):59–71.

Brunelli D, et al. 2006. Bio-feedback system for rehabilitation based on a wireless body area network. Proc. of the Fourth Annual IEEE International Conference on Pervasive Computing and Communications Workshops (PERCOMW06).

Cerny M, et al. (2008). Design and implementation of textile sensors for biotelemetry applications. 14th Nordic-Baltic Conference an Biomedical Engineering and Medical Physics, Riga, Latvia, vol. 20, pp. 194–197.

Chandler R. J, Gibson S, Karkare V, Farshchi S, Markovíc D, Judy JW. 2009. A system-level view of optimizing high-channel-count wireless biosignal telemetry. 31st Annual International Conference of the IEEE EMBS Minneapolis, MN, September.

Chi YM, Cauwenberghs G. 2010. Wireless non-contact EEG/ECG electrodes for body sensor networks. International Conference Body Sensor Networks, June 7–9.

Chiarugi F, Spanakis E, Lees PJ, Chronaki C, Tsiknakis M, Orphanoudakis SC. 2003. ECG in your hands: A multi-vendor ECG viewer for personal digital assistants. *Computers in Cardiology 2003*. Thessaloniki Chalkidiki, Greece.

Darwish A, Hassanien AE. 2011. Wearable and implantable wireless sensor network solutions for healthcare monitoring. *Sensors*, 11(6):5561–5595.

Dunseath WJR, Kelly EF. 1995. Multichannel PC-based data-acquisition system for high-resolution EEG. *IEEE Trans. Biomed. Eng.*, 42:1212–1217.

Eysenbach G. 2008. Medicine 2.0: Social networking, collaboration, participation, apomediation, and openness. *J. Med. Internet Res.*, 10(3):e22.

Fassbender H, et al. 2008. Fully implantable blood pressure sensor for hypertonic patients. IEEE SENSORS Conference.

Fernanda Cabrera-Umpíerrez M, et al. 2006. Communication platform for biosensor-based sleep management applications. Proc. of the 28th IEEE EMBS Annual International Conference, Aug.

Galarraga, M, Serrano L, Martinez I, de Toledo P, Reynolds M. 2007. Telemonitoring systems interoperability challenge: An updated review of the applicability of ISO/IEEE 11073 standards for interoperability in telemonitoring. 29th Annual International Conference of the IEEE EMBS, Lyon, France, August 23–26.

Gargiulo G et al. 2008. A mobile EEG system with dry electrodes. *IEEE Biomed. Circuits Syst. Conf.*, 273–276.

Gosselin B, Faniel L, Sawan M. 2006. A high data rate telemetry system for multi-channel biosignal recording. Biomedical Circuits and Systems Conference, IEEE BioCAS, Nov.

Hughes B, Joshi I, Wareham J. 2008. Health 2.0 and Medicine 2.0: Tensions and controversies in the field. *J. Med. Internet Res.*, 10(3):e23.

Jurik AD, Bolus JF, Weaver AC, Calhoun BH, Blalock TN. 2009. Mobile health monitoring through biotelemetry. International Conference on Body Area Networks, Los Angeles, April.

Khoor S, Nieberl K, Fugedi K, Kail E. 2001. Telemedicine ECG-telemetry with Bluetooth technology. *Comput. Cardiol.*, 28:585–588.

MacKay RS. 1993. *Biomedical Telemetry, Sensing and Transmitting Biological Information from Animals and Man*, 2nd ed. Piscataway, NJ: IEEE Press.

Matusitz J, Breen GM. 2007. Telemedicine: Its effects on health communication. *Health Commun.*, 21(1):73–83.

Menciassi A. Dario P. 2003. Bio-inspired solutions for locomotion in the gastrointestinal tract: Background and perspectives. *Philos Trans. R. Soc. Series A*, 361:2287–2298.

Meystre S. 2005. The current state of telemonitoring: A comment on the literature telemedicine and e-Health. *Telemed. J. e-Health*, 11(1):63–69.

Ng WC, et al. 2009. Micro-spike EEG electrode and the vacuum-casting technology for mass production. *J. Mater Process. Technol.*, 209:4434–4438.

Nguyen T-T, Cho M-Ch, Lee T-S. 2009. Automatic fall detection using wearable biomedical signal measurement terminal. 31st Annual International Conference of the IEEE EMBS Minneapolis, MN, September 2–6.

Panescu D. 2006. MEMS in medicine and biology. *IEEE Eng. Med. Biol. Mag.*, 25(5):19–28.

Phunchongharn P, Niyato D, Hossain E, Camorlinga S. 2010. An EMI-aware prioritized wireless access scheme for e-Health applications in hospital environments. *IEEE Trans. Inf. Techn. Biomed.*, 14(5):September.

Prohaska OJ, Olcaytug F, Pfundner P, Dragaun H. 1986. Thin-film multiple electrode probes: Possibilities and limitations. *IEEE Trans. Biomed. Eng.*, 33:223–229.

Ruffini G et al. 2008. First human trials of a dry electrophysiology sensor using a carbon nanotube array interface. *Sens. Actuators A*, 144:275–279.

Sakkalis V, Chiarugi F, Kostomanolakis S, Chronaki C, Tsiknakis M, Orphanoudakis SC. 2003. A gateway between the SCP-ECG and the DICOM 3.0 Supplement 30 Waveform Standards. *Proc. IEEE Comput. Cardiol.*, 30:25–28.

Schwiebert L, Gupta S, Auner P, Abrams G, Lezzi R, McAllister P. 2002. A biomedical smart sensor for visually impaired. *IEEE Sensors*, June.

Sinclair CM, Gaspair MC, Blum AS. 2007. Basic electronics in clinical neurophysiology. In Blum AS, Rutkove SB (eds). *The Physical Neurophysiology Primer*. 3–18. Humana Press.

Spanakis EG, Chiarugi F. 2011. Diabetes management: Devices, ICT technologies and future perspectives. 2nd International ICST Conference on Wireless Mobile Communication and Healthcare—MobiHealth 2011, Kos Island, Greece, October 5–7.

Spanakis E, Lelis P, Chiarugi F, Chronaki C. 2005. R&D challenges in developing an ambient intelligence eHealth platform, EMBEC 2005, Prague, Czech Republic, pp. 1727–1983, Nov.

Spanakis EG, Sakkalis V, Marias K, Traganitis A. 2014. Cross Layer Interference Management in Wireless Biomedical Networks. Entropy, 16(4), 2085–2104; doi:10.3390/e16042085

Spanakis E, Traganitis A, Ephremides A. 2009. Rate region and power considerations in a simple 2x2 interference channel. Information Theory Workshop on Networking and Information Theory (ITW 2009), Volos, Greece.

Taheri BA, Knight RT, Smith RL. 1994. A dry electrode for EEG recording. *Electroenceph. Clin. Neurophysiol.*, 90:376–383.

Timmons N, Scanlon W. 2004. Analysis of the performance of IEEE 802.15.4 for medical sensor body area networking. Proc. of the First Annual IEEE Communications Society Conference on Sensor and Ad Hoc Communications and Networks (SECON).

Traganitis A, Trypakis D, Spanakis E, Condos S, Stamkopoulos T, Tsiknakis M, Orphanoudakis SC. 2001. Home monitoring and personal health management services in a regional health telematics network. 23rd Annual International Conference of the IEEE Engineering in Medicine and Biology Society. Istanbul, Turkey, October.

Tsiknakis M, Spanakis E. 2010. Adoption of innovative eHealth services in prehospital emergency management: A case study. 10th IEEE International Conference on Information Technology and Applications in Biomedicine (ITAB 2010), Corfu, Greece.

Übeyli, ED. 2010. Telemedicine and biotelemetry for e-Health systems: Theory and applications. In Pease W, Cooper M, Gururajan R. (eds.) *Biomedical Knowledge Management: Infrastructures and Processes for E-Health Systems*, (pp. 1–17). Hershey, PA: Medical Information Science Reference.

Webster G, John. 2006. Encyclopedia of Medical Devices and Instrumentation, 6 Volume Set, 2nd Edition, April 2006, Wiley, ISBN: 978-0-471-26358-6

Young GB, Ives JR, Chapman MG, Mirsattari SM. 2006. A comparison of subdermal wire electrodes with collodion-applied disk electrodes in long-term EEG recordings in ICU. *Clin. Neurophysiol.*, 117:1376–1379.

15

SAFETY ISSUES IN BIOMEDICAL TELEMETRY

Konstantinos A. Psathas, Asimina Kiourti, and Konstantina S. Nikita

School of Electrical and Computer Engineering, National Technical University of Athens, Athens, Greece

15.1 INTRODUCTION

Achievements in wireless communications and microtechnologies combined with the growing need for advanced health care delivery systems have created a whole new field of interest in biomedicine. This includes the design of small biomedical devices with real-time biomedical telemetry functionalities for wearable, implantable, or ingestible applications. Wireless biomedical telemetry, performed by means of antennas, receives significant scientific interest. In the most complicated scenario, such devices not only communicate with a remote base station for telemetry purposes but also get powered wirelessly.

Even though biomedical telemetry is offering unique opportunities and enabling great advancements in prognosis, diagnosis, and treatment, there have been raised many questions regarding the safety of the aforementioned medical devices due to their proximity to the human body. Apart from all those risks and safety concerns that any medical device may have as a device that is responsible for the health of human beings, wireless medical devices have one further safety-related aspect that creates concerns to engineers and physicians. Since electromagnetic fields are used for communication with the exterior world, patients are exposed to electromagnetic energy.

Handbook of Biomedical Telemetry, First Edition. Edited by Konstantina S. Nikita.

In order for these devices to be marked as safe for use, special safety guidelines need to be followed.

This chapter analyzes all possible risks and safety concerns related to medical devices with biomedical telemetry functionalities. Apart from the usual operational and patient safety issues such as product hazard, device integrity and malfunction, side effects, human factors, and erroneous use, guidelines which need to be applied in order to ensure patient safety against electromagnetic field exposure are being extensively analyzed. Future research directions related to patient safety and the understanding of how electromagnetic fields interact with the human tissues are, finally, provided.

15.2 OPERATIONAL SAFETY

15.2.1 Electrical Hazards

Most of us have experienced some form of "electric shock," where electricity causes our body to experience pain or trauma. For a physiological effect to happen, the human body should, for some reason, become part of an electric circuit (Webster, 2009). The most basic and easy-to-understand effect of electricity on living tissues is when the current flows through the tissue, causing a dissipation of energy, usually in the form of heat. The effect is similar to that caused by an open flame to human tissues, except that the electricity also has the ability to further burn those tissues which are located well beneath the skin. Another effect of electricity on the human body, which can probably be considered as even more hazardous, is related to the electric stimulation of the excitable tissues, that is, nerves and muscles. Nerves all over the human body create a network of special cells called "nerve cells" or "neurons," which process and conduct those signals responsible for the regulation of several body functions (Brodal, 2010). Neurons communicate with each other by acting as "transducers" which create electrical signals (very small voltages and currents) in response to the input provided by certain chemical compounds, known as neurotransmitters. If electric current of sufficient magnitude is conducted to the living tissues, its effect will be to override the electric impulses generated by the neurons, overloading, in turn, the nervous system. Similar effects apply to muscle tissues, in which case current flow inside the human body may result in loss of full control of the muscular activity.

Figure 15.1 shows the magnitude of the current required to cause a number of physiological effects on the human body. A current with a frequency of 60 Hz is considered to be applied on a 70-kg human body for a time period ranging from 1 to 3 sec. The frequency of the current as well as the duration of the exposure and the weight of the individual are only indicated here as a reference. The threshold at which all the aforementioned effects happen is determined by the "let-go current," which is indicated in Figure 15.1 and defined as the maximum current at which a subject (e.g., a human) can withdraw voluntarily.

In the field of biomedical telemetry, use of all kinds of available equipment to-date in a way which deviates from the designer-set specifications may cause either

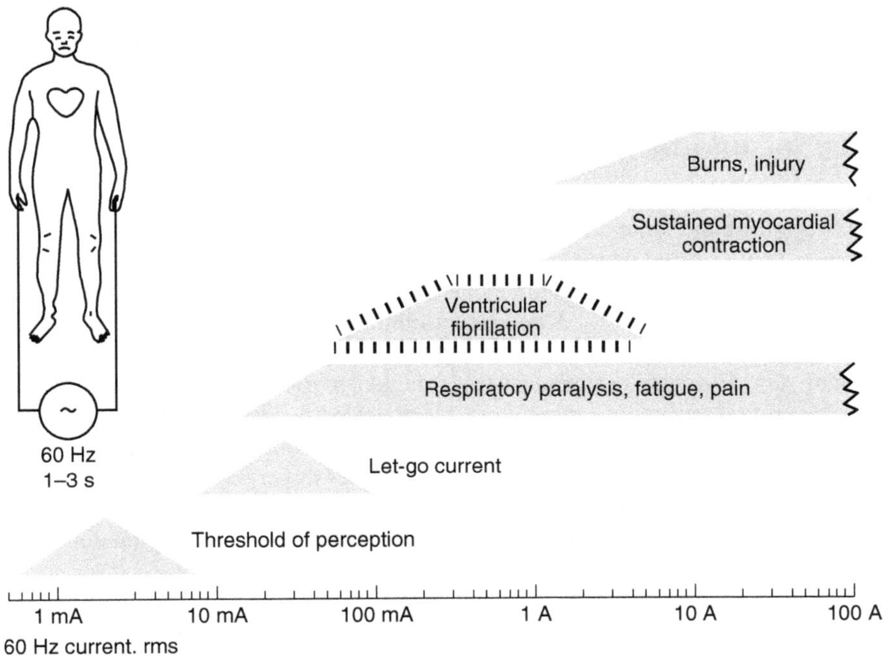

Figure 15.1 Physiological effects of electricity. Estimated thresholds are given for each effect. The current applied has a frequency of 60 Hz and is assumed to be applied on a 70-kg human for 1–3 sec via copper wires grasped by the hands (Webster, 2009).

macro- or microshocks. Macroshocks refer to those shocks which occur when the human body becomes a conductor of electric current without, however, the functions of the human's heart being affected. During a macroshock, current travels between two different areas of the skin. (Jacobson and Murray, 2007). For example, macroshock hazards may happen when on-body electrodes used to measure some physiological parameters induce a leakage current inside the human body as attributed to a malfunction or a failure (Anandanatarajan, 2011). On the other hand, microshocks are related to electric currents which are directly applied to the humans heart (Chan, 2008; Nokes et al., 1995) and could, for example, be a hazard in patients with intracardiac electrical conductors. Despite the fact that advanced medical devices are designed to prevent such phenomena by limiting any circuit currents to safe levels, microshocks continue to be a concern in cases where conductors are introduced into the human body and placed in close proximity to the heart.

To summarize, the physiological effects of electricity could be hazardous to patients' health. In the field of biomedical telemetry and in cases of biomedical devices which require electricity to operate, such hazards need to be addressed. Furthermore, due to the latest advances in biomedical telemetry, invasive devices that are either implanted or ingested inside the human body are used, making such hazards more apparent and crucial than ever before. Careful approaches on the design and completion of devices and systems used in biomedical telemetry are thus required in order to end up with medical equipment which minimizes the risks related to current induction inside human tissues.

15.2.2 Heat-Related Risks

Operation of medical devices placed on or inside the human body raises several questions regarding the amounts of heat produced. Causes of temperature rise can be associated with several operations of the medical device, such as the stimulation and sensing functions, the power requirements of the electronics, and the communication functionalities (Lazzi, 2005). A device with limited electronics and communication functionalities which operates on demand can rarely be a problem from a thermal point of view. However, recent advances in microelectronics and antenna design provide the ground toward designing more sophisticated devices that can stimulate the human body and its neural tissues, allow contiguous communication with the exterior world, and provide vital physiological information 24 hr per day (Akay, 2008; Beach et al., 2005; Karacolak et al., 2008). Under these circumstances, heat generated by the device could be significant and should be carefully taken into account during the design process.

Figure 15.2 summarizes the potential causes of temperature increase in the surrounding human tissues produced by the operation of an invasive biomedical device. As mentioned earlier, several parameters can be the cause of temperature increase. If, for example, the invasive device uses a transceiver for exchanging data with the exterior world and/or uses an advanced wireless power system, then electromagnetic

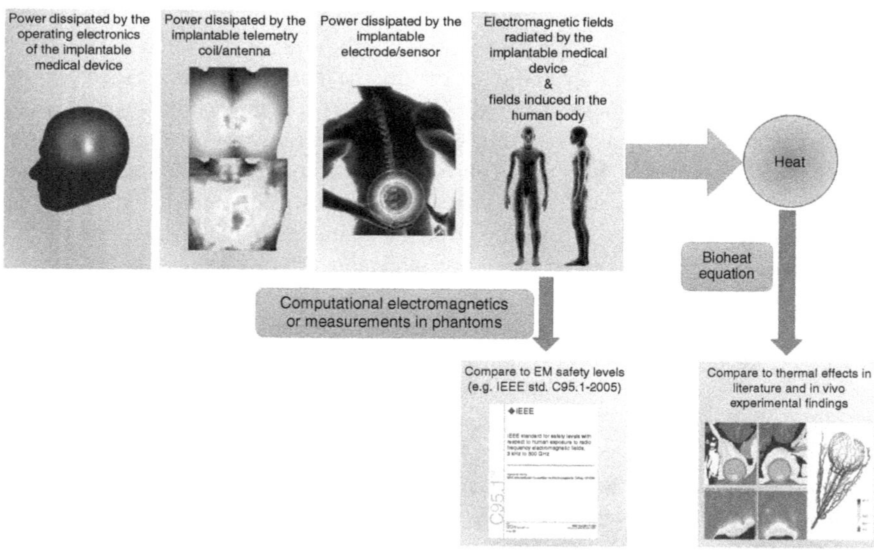

Figure 15.2 Schematic of potential causes of temperature increase in human body associated with operation of invasive devices. (*See insert for color representation of the figure.*)

power will be absorbed by the surrounding tissues and will in turn cause the temperature to increase. The aforementioned phenomenon can be quantified in terms of the specific absorption rate (SAR), which is expressed in watts per kilogram. SAR describes, in essence, the power induced per unit mass of tissue and is mathematically expressed as

$$\text{SAR}\ (x, y, z) = \frac{\sigma(x, y, z)E^2(x, y, z)}{2\rho(x, y, z)} \tag{15.1}$$

where ρ is the tissue density (in kg/m^3), σ is the conductivity (S/m), and E is the electric field amplitude (V/m) at the point of coordinates x, y, and z.

Apart from the temperature increase caused by the electromagnetic fields (EMFs) induced in the human body, temperature can also be raised due to the contiguous operation of the electronics of the invasive device. Contiguous use of batteries to power the device can further increase the temperature locally. To quantify the thermal impact of the operation of the device, the bioheat equation can be used and adjusted to include all the aforementioned effects (Lazzi, 2005):

$$C\rho\frac{\partial T}{dt} = \nabla \cdot (K\,\nabla\,\mathrm{T}) + A_0 - B_0(T - T_B) + \rho SAR + P^{\text{Density}}_{\text{Electronics}} \qquad (^{\text{W}}/_{\text{m}^3}) \tag{15.2}$$

where T is the temperature (°C), C is the specific heat [J/(kg °C)], ρ is the tissue density (kg/m^3), K is the thermal conductivity [J/(m s °C)], A_0 is the basic metabolic rate [J / (m^3 s)], B_0 is the blood perfusion coefficient [J / (m^3 s °C)], T_B is the temperature of blood (°C), and $P^{\text{Density}}_{\text{Electronics}}$ is the power dissipated by the implanted electronics (W/m^3). In the bioheat equation, the left side of the equation represents the temperature elevation. The right side consists of all possible reasons that can cause this temperature elevation, including thermal spatial diffusion, tissue metabolism, blood perfusion, and external heat sources.

15.2.3 Failure/Malfunction of Devices

Due to the fact that electronic equipment comes with a constant failure rate most of the time, there is a common belief that medical devices used in biomedical telemetry fail independent of their age (Dhillon, 2000; Fries, 2005). However, this assumption has been criticized as inaccurate, especially when it comes to invasive medical devices. This is attributed to the fact that invasive medical devices and their reliability and failure patterns could be seriously affected by external factors that are mostly random and cannot be easily predicted or categorized during the design and development stages (Taghipour et al., 2011). Therefore, in 1990, the U.S. Congress enacted the Safe Medical Devices Act (SMDA), which aims to increase the amount of information that the Food and Drug Administration (FDA) and the manufacturers receive about serious problems related to medical devices.

A failure or malfunction takes place when a medical device does not meet its performance specifications. The causes of failure can vary, including electrical, mechanical, thermal, software or materials problems (Fig. 15.3). Analyzing and understanding

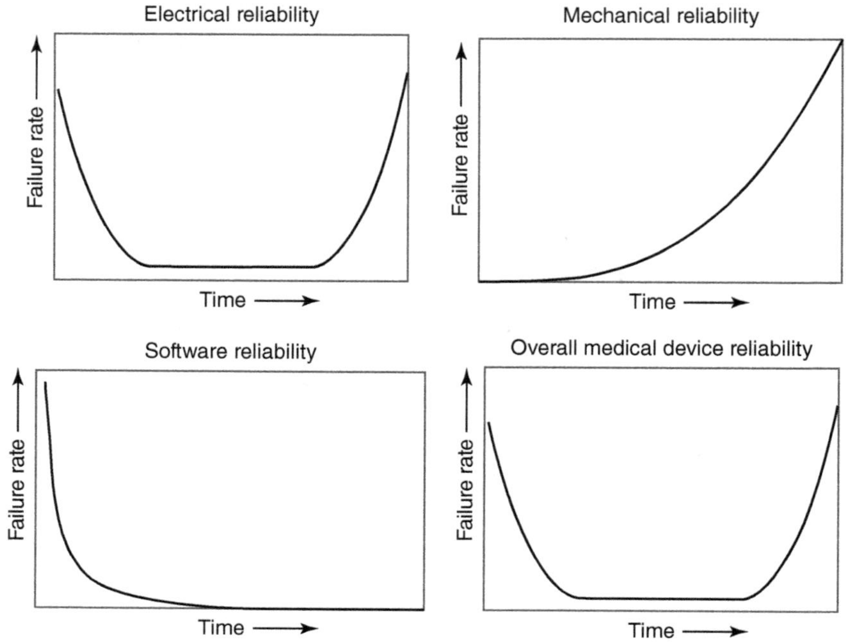

Figure 15.3 Reliability issues on medical devices, divided and analyzed by category.

the actual causes of malfunctions or failures of medical devices are considered crucial, because they can provide significant feedback to the designers, thus resulting in more fail-safe and advanced devices.

Due to their significant impact on human lives, medical devices used for biomedical telemetry must be evaluated in terms of reliability. Every aspect of their development—from early stages that cover design and prototyping through manufacture, distribution, and disposal—must adhere to quality standards that are well documented to specific functional and safety requirements set by the applications' standards. The standards should apply not only to manufacturers but also to all others involved in product development and distribution in order to ensure the quality of the devices and the indispensable safety and reliability. The process of identifying, analyzing, and eliminating or controlling the risks associated with medical devices is known as risk management and should address potential risks throughout the entire lifecycle of such products.

15.3 PRODUCT AND DEVICE HAZARDS

15.3.1 Adverse Tissue Reaction and Immune System Rejection Risks

Adverse reaction means a noxious and unintended response to any medical device that is implanted inside or ingested by the human body which in turn triggers the response of the human immune system. This type of reaction is well known and widely studied in the field of metal-on-metal hip implants. Such implants are used to relieve arthritis pain or fix severe physical joint damage as part of a hip fracture treatment. Similar effects can take place with any invasive medical device which gets implanted inside the human tissues or ingested by the human body for diagnosis or treatment purposes. Adverse reactions can be mild, such as a simple rash, or more severe and even life threatening. They can occur immediately or develop over time. During an adverse reaction, the human immune system "identifies" the invasive device as a threat to the organism. In such a case, the defensive mechanisms would subsequently turn against the implant and probably cause a rejection of the device. Adverse reaction could cause prolonged pain and discomfort to the patient. Furthermore, a possible rejection of the device will inevitably require the removal of the device by means of another, most likely painful, surgery procedure.

A solution to the possible rejection of a medical device by the human immune system lies in the use of biomaterials, that is, materials which have the ability to coexist with the tissues of the human body without causing any nonacceptable harm (Onuki et al., 2008). Biocompatibility is, thus, becoming a critical factor toward the development of medical devices for biomedical applications such as tissue engineering, invasive monitoring, drug delivery, and gene transfection. There are several ways in which materials and tissues can be brought together. The search for biomaterials which guarantee the optimum performance of such devices has been based upon the understanding of the interactions between these materials and human tissues, as studied in the literature during the last century (Thull, 2002). Nowadays, biocompatible materials or encapsulation layers are used to ensure the biocompatibility of invasive devices used in biomedical telemetry (Prodromakis et al., 2009; Silay et al., 2010; Soontornpipit et al., 2005).

15.3.2 Migration

Migration occurs when an implantable device tunnels through the flesh to a different part of the body than its original implanted location. Implantable device migration is a serious ongoing problem for animals with implants, where microchips are implanted in pets for identification purposes, but it could be generalized to any application of implantable devices. Despite the use of a special coating designed to anchor the implant to the original implanted location, migration still occurs in many cases. The migration issues get more imminent when it comes to nerve stimulators or other devices intended to interact with the muscles of the human body. In these cases, the position of the device is crucial to its functionality and a small migration could cause a failure of the system.

15.3.3 Security Risks

Apart from the well-known questions regarding computer security, concerns also extend to medical devices with biomedical telemetry functionalities. Since microprocessors are used as part of these devices to perform a number of tasks, questions related to security issues are coming up. Despite the recent advances in medical devices and technologies, security and privacy have not been in the top priorities during the design process. Security of medical devices should ensure reliable, secure data transmission and contiguous functionality, while, at the same time, preserving the patients' safety, confidentiality, and data integrity. For example, a recent research work has demonstrated that wireless connectivity can be exploited to gain access to confidential data transmitted by implantable medical devices (IMDs) or to send unauthorized commands to the IMDs—even commands that can cause the device to deliver an electric shock to the patient (Gollakota et al., 2011). As medical devices continue to evolve, balancing security and privacy with safety and efficacy will be one of the major challenges.

Adding a cryptography layer to existing medical devices in an attempt to solve the problem of security is difficult and ineffective (Gollakota et al., 2011). In terms of inalterability, many medical devices are already in use by humans serving several purposes. This issue becomes even more crucial in the case of invasive devices, where replacement with other devices which are enhanced with cryptographic abilities would require a surgery with all the risks, complications, and patient discomfort that this may cause. Overall, replacement of medical devices, except for the operational cost, also carries many risks that could directly affect the health of the patients. Another concern lies in the accessibility of the device. Adding a security cryptographic layer to a medical device could result in limited accessibility of the device in terms of communication. For example, if the device can be used only with proper credentials, then those credentials may not be available in specific scenarios, such as when the patient is unconscious or is examined by another doctor in another hospital. Last but not least, cryptography should enclose a software layer in the medical device that would manage the whole process. This software layer could be proved problematic due to possible software bugs. In this case, security of the device would be in question; in addition, the device could end up being inaccessible or nonfunctional.

Given the aforementioned difficulties, other ways should be found to ensure the safety, security, and privacy of biomedical telemetry devices. Gollakota et al. (2011) have proposed an interesting and promising advanced technique to surpass the problems mentioned earlier. In their study, the authors suggest the use of an external device, called a shield, which is interposed between an implantable device and potential counter parties—worn on the body near an implantable device. The shield acts as a gateway that relays messages between the IMD and authorized endpoints. Even though the method is promising, there is still ground to cover in the years to come in order to solve the security problems which come up with the growing needs of biomedical telemetry.

15.3.4 Development of Cancer

Eleven journal articles published between 1990 and 2006 have mentioned tissue reactions because of devices implanted into laboratory animals and dogs. A brief summary of the results is presented in Table 15.1. As perceived from the table, in some of these articles, animals developed malignant tumors around or adjacent to the

TABLE 15.1 In Vivo Animal Studies in Which Microchip-Induced Cancer Was Found

Authors	Species	Number of Animals	Length of Exposure	Developed Cancer
Johnson et al., 1996	Mice	2000	2 years	~1%
Tillmann et al., 1997	Mice	4279	Lifespan	0.8%
Palmer, et al., 1998	Mice	800	2 years	2.0%
Blanchard et al., 1999	Mice	177	6 months	10.2%
Elcock et al., 2001	Rats	1040	2 years	0.8%
Vascellari et al., 2004	Dog	N/A	18 months (at age 11)	1 dog
Vascellari et al., 2006	Dog	N/A	7 months (at age 9)	1 dog
Le Calvez et al., 2006	Mice	1260	2 years	4.1%

implanted devices. The percentage of the development was between 0.8 and 10.2% depending on the research. The devices used in at least 10 of those 11 studies were industry-standardized, passive implantable RF identification (RFID) transponders encapsulated in medical-grade glass and coated in an antimigration polymer sheath. Possible explanations behind the induced tumors have been proposed in these studies, which are summarized as follows:

- Foreign-body tumor genesis
- Postinjection sarcoma
- Possible genotoxic properties of the implant
- RF energy emissions

Even though malignant tumors were found to be developed in animals, the results cannot be generalized to humans, and scientists need to be really careful in criticizing the results. What happened to dogs and mice in response to an implantable device does not mean that it will be the same in the case of human beings. The human organism has a more advanced recovery and immune system and it can possibly assimilate the implant. It is important to highlight that, apart from the studies which are summarized in Table 15.1, there are a number of similar studies where no induced cancer effects were reported (Ball et al., 1991; Murasugi et al., 2003; Rao and Edmondson, 1990).

Furthermore, several medical devices, such as pacemakers, are already currently in use by humans and significantly enhance the quality of their lives without creating such phenomena. So, even though there is evidence about possible carcinogenetic reactions to animals, these results should be analyzed and criticized properly in order to get accurate results which could lead to a further improvement of the design of implantable medical devices.

15.3.5 Magnetic Resonance Imaging Incompatibility

Magnetic resonance imaging (MRI) is nowadays used to obtain highly detailed images of organs and tissues throughout the body without the need for x-rays or "ionizing" radiation. MRI uses rapidly changing magnetic fields to create images which indicate the existence of an injury, disease process, or another abnormal condition. The major advantage of MRI over the other traditional methods of diagnosis is that it can characterize and discriminate among tissues by making use of their physical and biochemical properties. Moreover, its ability to obtain images in multiple planes adds to its versatility and diagnostic utility. Among all other capabilities, MRI can be used to image the cardiac anatomy and function, study the rest and stress myocardial perfusion, perform contrast-enhanced angiograms, and assess and quantify regurgitant lesions and intracardiac shunts.

MRI is, thus, a great diagnostic tool in modern medicine and is used in several ways to obtain useful medical information from patients. Given the powerful magnetic fields which are required for the operation of MRI, great care should be given to ensure that objects such as ferromagnetic materials are not brought into the MRI system. If a patient with implantable or ingestible devices in his or her body needs, for some reason, to be examined in an MRI system, then problems multiply. Invasive biomedical telemetry devices use materials that can generally interact with the magnetic fields produced by the MRI systems. Magnetic fields may cause traumas to the patient by pulling on the devices or may induce electric currents to the implantable devices which could, in turn, cause electric shocks. Moreover, a metallic implant or other object may cause signal loss or distort the MRI images. In other words, there exists an incompatibility between MRI and invasive devices used in biomedical telemetry that forbids the use of MRI systems to patients with such devices in their body.

Several studies have tried to suggest solutions to the aforementioned problem. In 1999, Schueler et al. proposed a testing protocol for the evaluation of the compatibility of the magnetic fields produced by MRI systems with implantable devices. In 2007, Buchler et al. studied and proposed a numerical analysis methodology to evaluate the MRI safety of active medical implants. At the same time, industry is trying to provide implantable devices that use some kind of coating which can ensure their compatibility with MRI systems. Even though there is a significant ongoing work, there is still much ground to cover in order to deal with the incompatibilities between MRI systems and invasive devices used in biomedical telemetry.

15.4 PATIENT AND CLINICAL SAFETY

15.4.1 Patient Safety

"Freedom from accidental injury" is the definition of patient safety used by the Institute of Medicine in its landmark 1999 publication, "To Err is Human: Building a Safer Health System" (Kohn et al., 2000). The authors estimated that as many as 44,000–98,000 people die in U.S. hospitals each year as a result of lapses in patient safety. The study characterizes medical errors as the eighth leading cause of death, more frequent than motor vehicle accidents (43,458), breast cancer (42,458), and AIDS (16,516). Recent studies in an increasing number of countries of the European

Union show that health care errors occur in around 10% of the hospitalizations, while adverse events may take place in all settings where health care can be delivered. In the field of biomedical telemetry, where several sophisticated medical devices and equipment are used for a number of medical applications, from diagnosis to targeted treatment or drug delivery, patient safety could be severely threatened without proper consideration.

Broadly speaking, according to Health EU (2005), patient safety is defined as the freedom of a patient from unnecessary harm or potential harm associated with health care. Errors that may threaten patient safety can be summarized in the following two basic categories:

- Medical complexity
- System failures

As far as medical complexity is concerned, and especially in the field of biomedical telemetry, errors and problems may occur due to the advanced technologies used or the intensive care provided. Moreover, other complicated technologies, powerful drugs, or even prolonged hospital stays could further result in several issues associated with patient safety. Regarding system failures, several issues associated with patient safety could occur. Some of these problems have been discussed earlier as part of operational or devices hazards. Apart from these, poor communication throughout the health care system, devices and systems that are not properly interconnected, and infrastructure failure issues could also threaten patient safety.

Patient safety can only be considered in relative terms [World Health Organization] (WHO), 2003]. Similar to any electronic device, medical devices carry a certain degree of risk and could cause problems under specific circumstances. Apart from device failure, part of a device or a component can also fail in an unpredictable or random way, thus resulting in a possible threat to human health. The current approach to medical device safety is to estimate the potential of a device to become a hazard that could result in safety issues. This estimate is often referred to as risk assessment (Wilkins and Holley, 1998). Risk is a measure of the combination of

- the hazard,
- the likelihood of occurrence of the adverse event, and
- the severity or overall impact.

In general, risk assessment significantly depends on experience, evidence, computation, or even guesswork. In practice, risk assessment of medical devices is based on the experience of the health care professionals and the degree of device safety as provided by engineers.

One response to the concern for patient safety has been an effort to improve the systems which report medical errors by including those errors which could have caused injuries or death due to malfunctions or other reasons but, fortunately, did not (Hughes, 2008). The philosophy behind this idea is to take advantage of the nonhazardous events and failures and use this knowledge to find ways to prevent specific errors from recurring. This approach can be associated with safety design engineering which comes from the health care professionals under realistic conditions.

Another recent approach for securing patient safety has been the identification of existing knowledge and resources which come from nonrelated scientific fields, that is, lying outside the health care delivery system, and which may be applied to the problem under consideration. An example is the use of the extensive knowledge that industry has gained in the last 50 years regarding microelectronics devices in developing biomedical telemetry devices used for diagnosis and treatment. This approach can give a significant boost to patient safety.

World Health Organization The WHO was established on April 7, 1948, with headquarters in Geneva, Switzerland, and is a member of the United Nations Development Group. The constitution of WHO had been signed by all 61 countries of the United Nations (Burci and Vignes, 2004; Lee, 2008). The WHO's constitution states that its objective "is the attainment by all people of the highest possible level of health" (WHO, 2006). In October 2004, WHO launched a patient safety program in response to a World Health Assembly Resolution (2002) urging WHO and member states to pay the closest possible attention to the problem of patient safety (WHO, 2004). The mission of the WHO Patient Safety Program is to coordinate, facilitate, and accelerate patient safety improvements around the world by:

- Being a leader and advocating for change
- Generating and sharing knowledge and expertise
- Supporting member states in their implementation of patient safety actions

All these years, WHO, in participation with other organizations, has contributed to many projects related to public health and patient safety.

15.4.2 Clinical Safety

In general, pursuing safety is broader in scope than eliminating a risk. However, minimizing clinical risks that are closely related to applications of biomedical telemetry under specific circumstances and personnel will always be part of a clinical safety program. A rather practical and realistic definition of clinical risk is the avoidable increase in the likelihood of loss, damage, complaint, or harm to a patient, family member, staff member, or a member of the public during the treatment process in a clinical environment (e.g., a hospital). Clinical risks may result in harm, loss, or damage or they may lead to near-miss incidents that could have, but fortunately did not, threaten patient health or safety. Even though clinical risks may exist for some time without causing any harm or damage, that does not mean that the clinical risk does not exist or is insignificant. Clinical risk management tries to improve the quality and safety of health care by identifying the causes which may put patients at risk of harm and taking action to prevent or control those risks (Anderson, 2003). Clinical risk management is a continuous process as shown in Figure 15.4. The goals of risk management can be summarized as follows:

Figure 15.4 Clinical risk management process that ensures safety of medical devices in clinical environment.

- Identification of possible risks
- Prevention of harm, injury, and loss
- Control of systems and processes in a way to avoid possible risks

As shown in Figure 15.4, risk management does not stop after the identification, analysis, and control of possible risks. It is crucial to monitor the whole process for possible future risks or risks that could not be identified or solved in the first place.

In biomedical telemetry, where several medical devices are used for different reasons, for example, diagnosis, monitoring, and treatment, clinical risks are becoming more feasible. Implantable or ingestible devices which are placed inside the human body as well as simplified wearable devices can be used for monitoring and transmitting physiological parameters and vital data to other devices. Medical devices can interact, interfere, or malfunction, and the possibility of such effects further increases inside a clinical environment where several other electronic devices and equipment are being used at the same time. This means that clinical safety should be evaluated with a serious perspective while designing such systems in a way to prevent, minimize, or even eliminate clinical risks that can lead to threats for patient health and safety.

Some examples of harm related to clinical safety in biomedical telemetry are given below.

- The patient does not receive the right dose of a drug due to malfunction of implanted drug delivery system or due to problems of the telemetry link and/or remote control.
- Invasive diagnostic medical devices miss or cannot identify a possible threat to human health.
- Treatment devices or drug delivery systems deliver, by mistake, a medication that the patient does not need, thus resulting in adverse drug reaction.

- Operational device problems or malfunctions could result to incorrect or misleading diagnosis. An erroneous diagnosis could lead to mistakenly prescribing medication that either does not treat the disease or aggravates the disease symptoms.
- Interference issues could cause a medical device to operate in an improper way and cause several adverse reactions to patient health and safety.

15.4.3 Establishing Clinical Safety

Kohn et al. (2000) characterized medical errors as responsible for a higher number of annual deaths as compared to those caused by breast cancer, HIV, and road traffic accidents combined. Moreover, Nash (2012) shows that only a small fraction (less than 10%) of physicians are providing the best possible health care conditions for their patients in hospitals. Given the significance of clinical safety and its impact on patient health, as well as the growing need for more advanced medical devices that can be placed even inside the human body, ensuring clinical safety and avoiding clinical risks become more vital than ever before. In the same study, a robust, comprehensive, and integrated data infrastructure is proposed to continuously track the quality-of-care metrics and clinical safety issues in an attempt to minimize the clinical risks and the problems related to them (Nash, 2012). Web-based tools, rather than traditional methods, are suggested for real-time data collection, aggregation, statistical analysis, and report/graph generation, giving a significant boost to the effectiveness of current systems. Systems proposed try to deal with critical elements highlighted by the Institute of Medicine (IOM) (Kohn et al., 2000).

15.5 HUMAN FACTOR AND USE ISSUES

Human factor is a discipline that focuses on all those variables that affect the performance, reliability, and effectiveness of a medical device in terms of its interaction with the end user (Carayon, 2011). Errors in the use of medical devices are often attributed to the design of the user interface of the device or to the lack of knowledge regarding its proper use or some of its functionalities. Errors taking place while the medical device is operating could, in some cases, lead to injury or even death. Such errors may cause erroneous diagnosis, misleading monitoring, or ineffective patient treatment. Therefore, it becomes obvious that medical devices should be designed by taking into account the impact of the design in terms of safe use.

Use-related hazards are best identified and addressed by human factors engineering (HFE) (Lin et al., 1996). The ultimate goal of HFE is to minimize, ideally eliminate, any use-related hazards and assure that users are able to use the medical devices safely and effectively, as intended, throughout the product life cycle (FDA, 2000). Considering how and by whom a medical device will be used is highly significant in addressing use-related hazards. Substantial components of such considerations include:

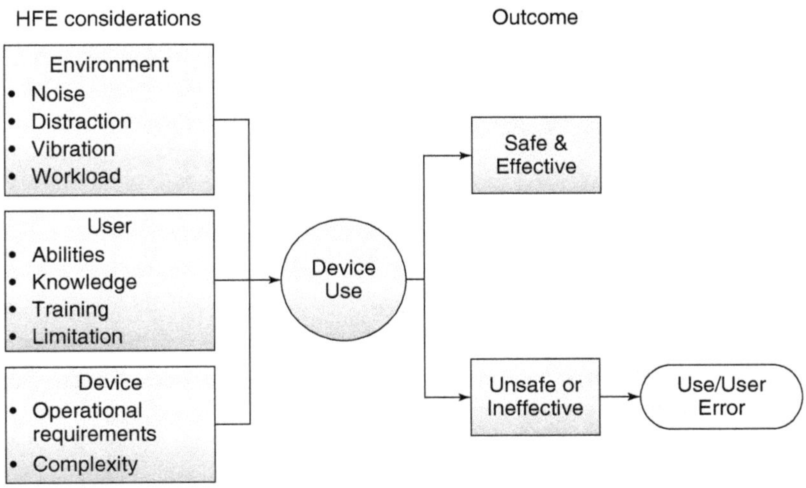

Figure 15.5 Interaction of HFE considerations could result in either safe and effective use or use error.

- User profile (e.g., patient, family member, qualified users)
- Typical and atypical uses of the device
- Device characteristics and particularities
- Environmental use characteristics
- Interaction between users, medical device, and use environment

In Figure 15.5, all the aforementioned components of HFE and the possible outcome of the interaction of the user with the medical device are being presented. The ineffective or unsafe use of the device could cause a use/user error, possibly affecting the patient's health or treatment process.

Most commonly, analysis and testing of prototype devices are carried out with actual users in an attempt to identify and investigate all possible usage scenarios that may lead to unanticipated hazards. This step is highly important because it is extremely difficult to identify and foresee all possible use-related problems in advance. After identifying any possible use-related hazards, hazards are mitigated or controlled by modifying the device–user interface or the abilities of potential users to use the specific device (e.g., training users, limiting use to qualified users).

15.5.1 Use-Related Hazards

Evidence from researchers suggests that the frequency and consequences of hazards resulting from medical device use might, by far, exceed those arising from device failures. A report by the IOM released in November 1999 estimated that as many as 98,000 people die every year from medical errors that occur in hospitals, a number

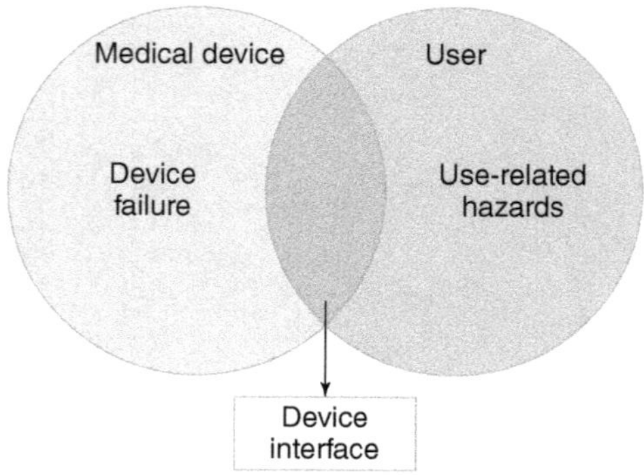

Figure 15.6 Medical device–interface–end user: How use-related hazards can affect device use and cause failure.

which is much higher than the number of people who die from vehicle accidents, breast cancer, or AIDS.

A medical device can be used safely and effectively if the device features, the operating environment, and the user capabilities are considered by the manufacturer as parameters during the design process. Operational hazards may cause a medical device to fail or malfunction and, thus, threaten the patient's health. Apart from operational safety or device failure, an erroneous user interaction with the device can also cause the device to fail. Use-related hazards and device failure hazards are connected as shown in Figure 15.6.

Use-related hazards can occur for one or more of the following reasons:

- Devices are used in ways that were not anticipated by the designer.
- Devices are used in ways that were anticipated but inadequately controlled for.
- Device use requires physical, perceptual, or cognitive abilities that exceed those of the user.
- Device use is inconsistent with the user's expectations or intuition about the operation of the device.
- The use environment affects device operation and this effect is not understood by the user (patient or qualified user).
- The user's physical, perceptual, or cognitive capacities are exceeded when using the medical device in a particular environment.

Use-related issues can become hazards or threats for human health under specific circumstances. For example, a user could calibrate a medical device intended for home use in a wrong way, use the device, and then act on the results that the device provides. In this case, inaccurate results are obtained because the device was not properly calibrated, and decisions involving a patient's health are made on these inaccurate results.

As home health care is becoming increasingly common due to the superior advantages (great improvement in quality of life, minimizing costs) and the unique challenges and opportunities it provides compared to traditional health care systems inside hospitals, use-related issues in home health care should be considered so that human health and safety are secured. Home health care has been enabled by the use of medical devices that can remotely connect to a hospital or a base station and transmit real-time physiological data. As a result, doctors are given the ability to diagnose or monitor diseases without the need of being physically close to the patient. In an environment where patients themselves rather than qualified users use the medical devices which are responsible for their disease monitoring or treatment, use-related hazards are becoming more apparent. Doctors should ensure that these devices are being used as intended by the patients or their family so that they can be sure about the validity of the results they get, remotely.

Due to the significance of home health care in modern medicine, in 2010 the FDA kicked off its Medical Device Home Use Initiative to assure the safety, quality, and usability of such medical devices (FDA, 2010). The Initiative's five actions can be summarized as follows:

- Developing formal guidance documents for home health care devices.
- Creating a labeling repository which provides online access for care recipients and caregivers on the proper use of these devices.
- Partnering with home health accrediting bodies to assess and provide training for home health care providers.
- Increasing postmarket surveillance of approved devices to understand their functionality and alert caregivers to specific risks identified by real-world usage
- Increasing public awareness on using the devices as directed

Apart from the aforementioned initiative, the FDA has also organized a Human Factors Program. In this program, the FDA works along with manufacturers to help ensure the application of human factors engineering to the design of new medical devices as well as to postmarket surveillance of currently marketed devices.

15.6 ELECTROMAGNETIC COMPATIBILITY AND INTERFERENCE ISSUES

The frequency bands which are of main interest to biomedical telemetry systems are the Medical Implant Communications Service (MICS) band and the industrial, scientific, and medical (ISM) bands. Each of these bands has advantages and disadvantages and is, thus, applicable to different medical scenarios.

The MICS band was adopted by the U.S. Federal Communications Commission (FCC) on October 10, 1999, and has been developed for use in medical implant communication systems [European Telecommunications Standards Institute (ETSI),

2002; FCC, 1999]. Example applications include heart pacemakers and defibrillators, while, in the future, wireless medical implant systems might include automated drug delivery devices and 24-hr real-time patient monitoring. The maximum available bandwidth of the MICS band has been set by the standard to 300 kHz. The MICS band uses the frequency range between 402 and 405 MHz and gives an implantable medical device the ability to be controlled and monitored from outside. Due to SAR limitations and interference issues, MICS band systems have an upper limit on the effective radiated power (ERP) that equals 25 μW. The MICS band enabled additional flexibility as compared to older inductive technologies by giving a range of a couple of meters in the communication link. However, due to the limited bandwidth available, MICS systems exhibit low bit rates as compared to systems using the higher frequency WiFi or Bluetooth technologies.

The ISM bands are portions of the radio spectrum which are internationally reserved for industrial, scientific, and medical purposes other than communications. In the early 1990s the FCC allowed the use of ISM bands for unlicensed communication equipment, which were defined in the following frequency ranges:

- 902–928 MHz
- 2400–2483.5 MHz
- 5725–5875 MHz

In the United States, use of these bands for communication purposes is regulated by part 15 of the FCC rules. The maximum output power fed into the transmitting antenna is defined as 30 dBm (1 W), while the maximum effective isotropic radiated power is set to 36 dBm (4 W). As compared to the MICS band, the ISM bands provide (a) easier antenna design due to the higher operation frequency and the expanded expertise obtained through the use of WiFi and Bluetooth systems in recent years, (b) better SAR distributions, and (c) wider bandwidth. However, signals at higher frequencies are attenuated considerably faster inside the human body, thus making the band difficult to adopt, especially for invasive (implantable and ingestible) devices.

In 1993, the International Electrotechnical Commission (IEC) published the first edition of the international electromagnetic compatibility standard for medical devices (60601-1-2) (IEC, 1993). Soon after, the European Committee for Standardization (CENELEC) adopted it as a European Norm (EN) standard with only minor changes. Electromagnetic compatibility (EMC) means that a device is compatible with its electromagnetic environment and does not emit EM energy of levels that could cause electromagnetic interference (EMI) to other devices in its close vicinity. EMC is a major concern in medical electronics and devices given the fact that functioning anomalies may be life threatening.

With all those medical devices used today, design approaches to face the EMC problems are straightforward and mainly include shielding and filtering techniques to protect the device. Pacemakers are typical examples of medical devices which are susceptible to EMC issues. However, pacemakers have greatly improved over the years with respect to compactness and immunity to RF interference so that current designs can foresee the electromagnetic environment where the device will be placed.

As a result, nowadays, pacemakers are very reliable to use and may only fail under extreme conditions.

EMI is defined as the disturbance that affects an electrical circuit as attributed to either electromagnetic induction or electromagnetic radiation emitted from an external source placed in the vicinity of the circuit. Problems caused by EMI can be minimized or eliminated by ensuring that all electronic equipment of the device is operated with a good electrical ground system. However, this can be hard to implement when it comes to invasive medical devices. Given the fact that the world's industrialized nations are becoming increasingly dependent upon the use of electrical and electronic equipment, EMI issues for medical devices are becoming more feasible than ever before.

EMI of devices can be classified under the two headings of "internal" and "external" EMI. Internal EMI is associated with the devices and components which are present within the device, such as resistors and transistors. External EMI is generated by sources outside the system and is normally the kind of EMI which causes most of the problems to medical devices. The effects of the EMI should be of great concern for designers/manufacturers of medical devices, qualified personnel, and end users of those devices.

For electromagnetic disturbances to cause interference, they must propagate in some way. The three main physical mechanisms of electromagnetic propagation are (Figure 15.7):

- Conduction Electromagnetic energy may be conducted in either the common mode or differential mode. This can be via power cables, earth conductors, signal cables, antenna feeders, or other low-impedance paths.
- Reactive Coupling Electromagnetic energy may also propagate by reactive coupling, either inductive or capacitive.
- Radiation For frequencies above approximately 30 MHz radiation tends to be the dominant propagation mechanism.

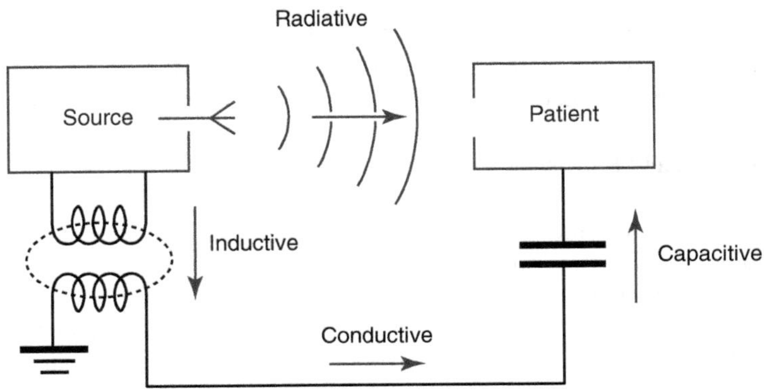

Figure 15.7 Schematic description of EMI coupling paths.

As in the case of EMC, EMI is a major concern for the design, fabrication, and testing of medical devices. Interference phenomena may cause operational problems, which can, in turn, result in patient life-threatening events.

It is important to highlight that interference issues are more intense in the ISM band, as more and more products use the 2.4-GHz portion of the radio spectrum. For example, interference issues are most commonly produced by WiFi networks or Bluetooth links which operate in the vicinity of the medical device under consideration. Due to its global availability, the ISM band constitutes a popular frequency band suitable to low-cost radios. New proposed solutions such as the Institute of Electrical and Electronics Engineers (IEEE) 802.15 technologies and Bluetooth as well as the IEEE 802.11 technologies and microwave ovens operate in the ISM band and are expected to cause interference issues to the medical devices since they operate in the same environment and frequency space. Designers should include all kind of interference right from the start and design the devices in a way to minimize or even eliminate the EMI effects.

At this point, it is worth noting that special mention is made on both IEEE and International Commission on Non-Ionizing Radiation Protection (ICNIRP) standards (see applicable guidelines for more info) concerning interference issues between mobile phones or any other electromagnetic device with pacemakers, a widely used medical device, which uses electrical impulses to regulate the beating of the heart. In the IEEE C95.1-2005 standard a suggestion of keeping a minimal separation distance between mobile phones and implantable cardiac devices of 15–22 cm is made, without though any specific restrictions or further recommendations (IEEE, 2005).

15.7 APPLICABLE GUIDELINES

The rapid development of technologies and systems that use RF EMFs to provide two-way radio communication between medical devices and exterior equipment poses several questions related to human exposure to EMFs.

The first standard of the American National Standards Institute (ANSI) that covered issues related to exposure to RF radiation was issued in 1966 and recommended that exposure should be limited to field levels no higher than $10\,mW/cm^2$ with a 0.1-hr (6-min) averaging time (ANSI, 1966). In 1982, IEEE developed the C95.1-1982 standard, which described the recommended limits for human exposure to RF fields (IEEE, 1982). Industry, engineers, and medical researchers contributed to the development of this standard as the first SAR-based human exposure standard in the world. It is noteworthy that the FCC wrote a set of regulations which asked RF services to comply with the limits proposed in C95.1-1982. Nine years later, in 1991, IEEE published a revised standard, C95.1-1991 (IEEE, 1991). The new standard decreased the maximum recommended RF exposure levels and extended the frequency range covered by the original standard. In 1996, upon the request of the U.S. Congress, a mandatory standard was adopted by the FCC for evaluating the radiation effects on human beings (FCC, 1996). Three years later, in 1999, the European Union Health Council, agreed on a similar recommendation for limiting exposure to electromagnetic fields. The recommendation adopted was based on exposure limits that had been earlier proposed by the ICNIRP. At the same time a new revision of IEEE C95.1-1999 was introduced. The latest revised standard available is the IEEE C95.1-2005 (IEEE, 2005). Today, any of the standards which are adopted worldwide are based on the C95.1 standard developed by IEEE and the recommendations of ICNIRP, and quantify RF exposure in terms of the SAR.

15.7.1 Development of IEEE C95.1-1991 Standard

The C95.1-1991 standard was developed by the IEEE Standards Coordinating Committee 28, Non-Ionizing Radiation Hazards (SCC-28). The C95.1-1991 standard is so important because it was the first two-tiered standard. It was based on SAR values as this approach had already been incorporated into the previous ANSI C95.1-1982 standard and it was developed after an extensive and serious evaluation of all available scientific literature. At that time, SCC-28 consisted of researchers (in a percentage of about 70%) as well as people from various organizations and the industry. While developing the standard, SCC-28 considered a large number of input sources and research studies and evaluated them against several scientific criteria. After excluding papers that did not include measured EMF levels, SCC-28 gathered around 120 papers for consideration. Following an extensive analysis of the topics covered and the conclusions derived in these papers, SCC-28 reached a consensus that a standard for exposure could be set.

The SAR value of 4 W/kg, which is temporally and spatially averaged over the whole body mass, was the final threshold that was adopted by the committee. Several adverse biological effects were demonstrated above this threshold, such as disruption of work schedules in trained animals. For example, a monkey trained to push a button six times in order to get a banana decided that it did not want a banana when exposed to a 4-W/kg field. As soon as the field was removed, the monkey soon decided that it was hungry after all. The committee deems these results to be thermal effects of the application of electromagnetic fields. Recognizing and understanding the scientific uncertainty and the biological variability in the human population, the committee applied a safety factor of 10, setting a SAR value of 0.4 W/kg for controlled/occupational exposure and an additional safety factor of 5 for the general population, resulting in a total margin of 50 and a SAR value of 0.08 W/kg. Since, for a given exposure, the SAR distribution inside the human body can vary from point to point, a partial-body limit was also introduced. The local SAR limit in the IEEE C95.1-1991 standard was set to 1.6 W/kg for the uncontrolled environment in the body or 8 W/kg for the controlled environment and is averaged in any 1 g of tissue (defined as a tissue volume in the shape of a cube) (IEEE, 1991). Clearly, the C95.1-1991 standard provides recommendations in an attempt to prevent adverse thermal effects that could result in limited functionality of the human body. It is worth noting, however, that some scientists consider this whole topic as unfounded because the assessment criteria used in deriving the reports of biological effects did not take the mechanisms of interaction into account.

TABLE 15.2 MPE Limits for Uncontrolled Environments (IEEE C95.1-1991)

Frequency (MHz)	Power Density (W/m^2)
0.1–1.34	1000
1.34–30	$1800/f^2$
30–400	2.0
400–2000	$f/200$
2000–100,000	10
100,000–300,000	Increases from 10 to 100

TABLE 15.3 MPE Limits for Controlled Environments (IEEE C95.1-1991)

Frequency (MHz)	Power Density (W/m^2)
0.1–1	9000
1–30	$9000/f^2$
30–300	10
300–3000	$f/30$
3000–300,000	100

As mentioned in the previous paragraph, the IEEE standard sets exposure limits for electric fields and magnetic fields that refer to the whole body and are averaged over time. Limits can be expressed in terms of the maximum permissible exposure (MPE). The IEEE's MPE limits for both general public and occupational environment are shown in Tables 15.2 and 15.3. Limits indicated are spatially averaged over the whole body and time averaged over 6 and 30 min for controlled and uncontrolled environments, respectively.

15.7.2 International Commission on Non-Ionizing Radiation Protection and Its Role

The ICNIRP is an international commission specializing in the protection from non-ionizing radiation and consists of an independent group of experts. The activities of the organization include:

- Responsibility to evaluate the current state of knowledge regarding the effects of nonionizing radiation (NIR) on humans
- Responsibility to provide science-based advice and recommendations related to human protection against the effects of nonionizing radiation

ICNIRP was founded in 1992 by the International Radiation Protection Association (IRPA), located in Geneva. It consists of a main commission with 14 members,

4 standing committees of 8 members, and a large group of consulting experts. The 4 standing committees are the following: (a) SCI on epidemiology, (b) SCII on biology and medicine, (c) SCIII on physics and engineering, and (d) SCIV on optical radiation. ICNIRP collaborates with a number of regional and international partners and organizations such as the WHO, the International Labor Organization (ILO), the European Council (EC), DG SANCO (Health), the European Society for Skin Cancer Illumination (CIE), and the International Commission for Occupational Hygiene (ICOH).

In order to fulfill its aim and provide information and advice on the potential health hazards due to EMF exposure, ICNIRP periodically publishes scientific reviews and reports and arranges relative meetings. The results combined with conclusions from studies carried out by other organizations form the ICNIRP guidelines. Examples are the guidelines for limiting exposure to EMF, lasers, and ultraviolet radiation.

Despite the wide use of the ICNIRP guidelines and guidance, the commission has already been criticized. The European Council has stated: "It is most curious, to say the least, that the applicable official threshold values for limiting the health impact of extremely low frequency electromagnetic fields and high frequency waves were drawn up and proposed to international political institutions (WHO, European Commission, governments) by the ICNIRP, an NGO [nongovernmental organization] whose origin and structure are none too clear and which is furthermore suspected of having rather close links with the industries whose expansion is shaped by recommendations for maximum threshold values for the different frequencies of electromagnetic fields" (Europe Council, 2011).

15.7.3 Issues on Developing Safety Standards

Developing a standard related to human health is of great importance due to the severity of the process itself and becomes even more crucial at a time when the actual interaction mechanisms between RF radiation and human tissues are still under consideration (Lin, 2000). There are two major issues related to the development of human exposure standards. The first issue is related to the procedure and process invoked to get the results and set the standards. The second issue has to do with the philosophy adopted in the process of developing the guideline or standard.

Typically, development of guidelines or standards should comprise (a) review of the available scientific literature, (b) accumulation of pertinent databases, and (c) identification of possible hazards. Following the considerations mentioned, the process continues to the drafting step, which often includes a review of the existing guidelines and standards. The process is completed when the exposure guideline have been finalized, published, and adopted by a recognized body. It is important to highlight that acceptance of the resulting standard is enhanced when the procedure invoked is as transparent as possible.

The identification of a hazard should be accompanied by a way or at least a philosophy of protection. This also applies to the possible hazards of EMF radiation on human health. The philosophical approaches to setting permissible exposure limits for RF radiation can be classified into the following categories (Lin, 2000):

- No demonstrable effect
- Observable effects with no known physiological consequence
- Minimal physiological consequence
- No adverse effect

It becomes obvious that the adoption of any no-effect philosophy would result in a non-wireless-technologies society, similar to the one of the nineteenth century. Given the aforementioned issues, it is apparent that setting, accepting the validity, and finally adopting a guideline or a standard could become really tricky.

15.7.4 Evolution and Comparison of Guidelines

Difficulties and different approaches in adopting the standards, combined with the publicly expressed lack of confidence pertaining to the maximum permissible MPE levels, have given rise to the development of two different sets of guidelines for limiting human exposure to RF radiation worldwide (Lin, 2003). IEEE and ICNIRP adopted different biological rationales to propose the RF exposure guidelines, resulting in two sets of distinct MPEs, which differ in terms of SAR levels and averaging tissue volume. Apart from the differentiation of standards across the two organizations, as already mentioned, there have been changes over time on the IEEE C95.1 standard to include available scientific data. Several scientific groups worldwide are trying to examine the existing guidelines or MPEs for human exposure in order to decide on the safety of the existing guidelines and come up with an explicit philosophy and procedure invoked in the process of setting the standards. The use of SAR was initiated as a recommendation of the National Council of Radiation Protection and Measurements (NCRP) and was adopted by the American National Standards Institute (ANSI) in 1982.

For example, the IEEE C95.1-1991 standard limited human exposure to RF radiation based on SAR dosimetric quantities and set the maximum SAR level to 1.6 W/kg averaged in 1 g of body tissue (IEEE, 1991). The guidelines proposed by ICNIRP restricted RF exposure by setting the maximum SAR level to 2 W/kg averaged in any 10 g of contiguous tissue (ICNIRP, 1998). For reference, both SAR levels mentioned earlier are referred to head tissues. Aside from the quantitative difference between the IEEE and ICNIRP standards (1.6 or 2 W/kg), there is also a difference in the definition of the tissue mass used to define the SAR standards (1 g for IEEE standard or 10 g for ICNIRP standard). This difference could have a great impact on the actual quantity of RF energy allowed to be deposited in human tissues. Due to the fact that the distribution of absorbed microwave energy varies from point to point inside the human body, a larger averaging mass (like the 10-g mass used in the ICNIRP standard) would tend to artificially smooth out the SAR distribution. Apart from the distribution smoothing of SAR, a larger averaging mass also tends to lower the value of the measured SAR by a factor of 2 or more (Lin, 2003). This means that the ICNIRP SAR limit of 2 W/kg averaged over 10 g of tissue could be equivalent to 4 W/kg or higher averaged over 1 g of tissue.

In many cases, such as in the example of the human eye mentioned in Lin (2003, 2006) or the muscle tissue, the 10-g tissue mass used to compute the average SAR distribution completely ignores the wide variation of SAR distribution and masks the intensity of the absorption of RF energy at the superficial layer of tissues. Another example of the distraction that a larger SAR averaging volume could cause is related to the anatomy of the human brain. Types and populations of cells and neurons inside the human brain are notably different even in 1 g of tissue. Absorption of RF radiation highly depends on the electric properties of the tissues involved, so that a larger averaging volume could mask the absorption of RF energy by neighbor cells or neurons inside the human brain.

As clearly seen by the examples mentioned above, SAR is a localized quantity and its value significantly depends on the averaging volume or mass. The larger the averaging volume or mass, the lower the resolution and sensitivity of SAR get, and the less useful the SAR becomes in terms of a metric which is used for quantifying the localized exposure to RF energy and the biological response to radiation.

As pointed out earlier, there are two different sets of standards worldwide for EM radiation safety. The first one has been developed, proposed, and published by IEEE and the second one by ICNIRP. Although in the past there were basic differences between the standards, most of them analyzed in the previous paragraphs, the latest revision of the IEEE standard (IEEE C.95.1-2005) is more similar than ever before with the ICNIRP standard, as indicated in Tables 15.4 and 15.5.

In Table 15.4 the whole-body average and localized SAR basic restrictions are presented according to the ICNIRP 1997 standard. There are two main parameters that should be well defined for these limits to get their actual meaning, the averaging time and the averaging tissue volume. All SAR values in Table 15.4 are to be averaged over any 6-min period. According to ICNIRP standards, the SAR averaging mass is defined as any 10 g of contiguous tissue without restrictions in terms of shape of the averaging mass.

In Table 15.5, the whole-body average and localized SAR basic restrictions are presented for the IEEE C95.1-2005 standard. Consistent with the approach used in the previous version of the IEEE C95.1 standard, a traditional safety factor of 10

TABLE 15.4 Basic Restrictions for Time-Varying Electric and Magnetic Fields for Frequencies from 100 kHz to 10 GHz According to the Latest ICNIRP Standards (1997)

Exposure Characteristics	Frequency Range	Whole-body Average SAR (W/kg)	Localized SAR (Head and Trunk) (W/kg)	Localized SAR (Limbs) (W/kg)
Occupational exposure	100 kHz–10 GHz	0.4	10	20
General public exposure	100 kHz–10 GHz	0.08	2	4

TABLE 15.5 Basic Restrictions for Time Varying Electric and Magnetic Fields for Frequencies from 100 kHz to 3 GHz According to Latest IEEE C95.1-2005 Standard

Exposure Characteristics	Frequency Range	Whole-body Average SAR (W/kg)	Localized SAR (Peak Spatial Average) (W/kg)	Localized SAR (Limbs) (W/kg)
Occupational exposure	100 kHz–10 GHz	0.4	10	20
General public exposure	100 kHz–10 GHz	0.08	2	4

has been applied to the established SAR threshold for such effects, yielding a SAR of 0.4 W/kg averaged over the whole body. As mentioned earlier, the averaging time and tissue volume should be defined. Contrary to the previous versions of IEEE C95.1, the averaging volume on the IEEE C95.1-2005 standard has been changed. The SAR is now averaged over any 10 g of tissue, defined as a tissue in the shape of a cube with volume of approximately 10 cm^3. In cases where the cubical volume rule cannot be satisfied (e.g., at the surface of the body), special rules apply for setting the SAR value in a given voxel. The averaging time for fields and power density are presented in Table 15.6. It is noteworthy that for the first time in its history the IEEE C95.1-2005 standard instituted exclusion for the pinnae or the external ears by relaxation of the above-mentioned basic SAR restriction from 2 to 4 W/kg. This relaxation by IEEE segregated the tissues of pinnae from all other of the human's head.

As clearly seen, the differences between the standards after the latest revision of IEEE C95.1 include the averaging tissue mass and the averaging time period that are being used for SAR calculation.

TABLE 15.6 Averaging Time of SAR Calculation for General Public in IEEE C95.1-2005 Standard

Frequency Range (MHz)	Averaging Time	
0.1–1.34	6	6
1.34–3	$f_M^2/0.3$	6
3–30	30	6
30–100	30	$0.0636 f_M^{1.337}$
100–400	30	30
400–2000	30	
2000–5000	30	
5000–30,000	$150/f_G$	
30,000–100,000	$25.24/f_G^{0.476}$	
100,000–300,000	$5048/[(9f_G-700)f_G^{0.476}]$	

Note: f_M is the frequency in MHz, f_G is the frequency in GHz.
Source: IEEE (2005).

TABLE 15.7 Ranges of Threshold Currents for General Public (ICNIRP, 1998)

	Threshold Current (mA)			
Effect	50/60 Hz	1 kHz	100 kHz	1 MHz
Touch perception	0.2–0.4	0.4–0.8	25–40	24–40
Pain on finger contact	0.9–1.8	1.6–3.3	33–55	28–50
Painful shock	8–16	12–24	112–224	
Severe shock	12–23	21–41	160–320	

Comparative evaluations of the IEEE C95.1-1999 and IEEE C95.1-2005 safety standards have been performed for implantable antennas (Kiourti and Nikita, 2012a, 2013). In these studies, numerical investigations were carried out for antennas implanted inside the skin tissue of the human head for applications such as intracranial pressure monitoring, stroke rehabilitation, brain wave sensing, and brain edema evolution monitoring. The IEEE C95.1-1999 standard was found to be much stricter, limiting the maximum allowable net input power to the antenna to more than six (Kiourti and Nikita, 2012a) and four (Kiourti and Nikita, 2013) times lower than that imposed by the IEEE C95.1-2005 standard. Furthermore, compliance with the recent IEEE C95.1-2005 standard was found to be almost insensitive to head properties (anatomical features and dielectric parameters), in contrast with the IEEE C95.1-1999 standard (Kiourti and Nikita, 2013).

Finally, it is worth noting that the new ICNIRP guidelines for limiting exposure to time-varying electric, magnetic, and electromagnetic fields have been developed to take into account safety issues regarding intrabody communications (currents propagating through the human body). Threshold currents that produce perception and pain or even shocks are presented in Table 15.7. It has been shown that threshold currents vary only a little over the frequency range of 100 kHz–1 MHz, while they are unlikely to vary significantly over the frequency range up to about 110 MHz.

15.8 OCCUPATIONAL SAFETY

Occupational safety for biomedical telemetry can be divided into two major categories: (a) safety that is related to patients who use wearable, implantable, or ingestible medical devices that could interact or interfere with other devices in the occupational environment of the patient and (b) safety that is related to the qualified users of the medical devices who are responsible for the patient monitoring and treatment. Either way, safety concerns arise for both patients who use a biomedical device to maintain their good health and qualified users who use medical equipment as part of their work.

As far as interaction or interference of the medical devices with their surrounding occupational environment is concerned, great progress has been made regarding the design of biomedical devices. Having in mind all those risks and possible interactions of the medical devices that have been analyzed and discussed earlier in this chapter, designers try to foresee and solve/eliminate any problems that could become to hazards or life-threatening events. Gustrau et al. (2002) presented a study about the possible interference of low-frequency electric and magnetic fields with implanted cardiac pacemakers.

Safety of workers with biomedical devices is currently based on general and conservative estimations. These estimations introduce great uncertainties and raise many questions about the safety of workers. On the other hand, workers and their employers demand a more realistic estimation of the risks and related hazards.

In order to improve the assessment of the occupational risk for workers with medical devices, the Federal Institute of Occupational Safety and Health in Germany has launched two projects. The first project investigates the voltage induced in the medical device due to the electric and magnetic fields emanating from electrical devices in the workplace, whereas the second project investigates the immunity of medical devices to low-frequency electromagnetic fields. Combining the results of both projects will enable a more detailed analysis on how medical devices interact with occupational equipment and electromagnetic fields and will provide valuable information toward reducing the risks and hazardous events for workers with medical devices.

As long as workers and qualified users of biomedical equipment are concerned, safety issues are strongly related to the extensive exposure to electromagnetic fields generated for telemetry communication. Despite the fact that such fields are not so intense due to safety issues discussed earlier, workers are exposed to electromagnetic fields for a much longer period of time as compared to patients, so that specific guidelines and safety standards should apply to them.

15.9 FUTURE RESEARCH DIRECTIONS

Safety of medical devices with telemetry functionalities is a field of growing scientific interest. The field has attracted significant research efforts, especially during the last 10 years as attributed to the technology and microelectronics advances which have enabled the use of devices with telemetry abilities to be implanted or ingested within the human body for diagnosis and therapy applications. Implanted medical devices have been tested and are being used as diagnostic tools for heart or neural diseases. Proposed implanted systems for contiguous monitoring of physiological parameters are waiting for testing in real conditions. Ingested devices have already been used to investigate diseases and disorders of the small intestine where traditional methods failed to deliver reliable results. All those innovative medical devices have the potential to greatly improve the patient's quality of life. Despite the remarkable progress in the last years, there is still ground for research and several engineering problems to be solved.

As far as operational issues, product and device hazards, patient and clinical safety, and human factors are concerned, advanced design processes, engineering, experience, and user training are expected to minimize or even eliminate the possible hazards. It is important to highlight, however, that, regardless of the recent technological advances, some of these risks cannot be eliminated completely because of their nature. Thanks to the experience gained and the statistical data collected by the current health care system and medical devices, safer devices are expected to be designed and be available in the market soon. In other words, engineers are expected to be able to design fail-safe devices based on the causes of failures or malfunction of the current devices. As a result, from an operational point of view, medical devices will be safer to use over time.

The real challenge, however, lies in the full understanding of the interaction of electromagnetic radiation with human tissues. Biological effects related to tissue exposure to repeated, prolonged, or lifelong RF energy emitted by low-power medical devices have only been very recently investigated. The existing scientific results are, therefore, equivocal and arguable in many respects, and scientists' understanding of biological effects of exposure to RF radiation is still evolving over time. As a result, as new research findings become available, it becomes necessary to periodically evaluate and possibly revise any criteria set forth for human safety due to exposure to RF radiation.

Last but not least, global harmonization of RF exposure standards would be a very desirable goal for consumers, manufacturers, and operators. At this time, two standards (IEEE C.95.1-2005 and ICNIRP) are available which are based on SAR limitations (ICNIRP, 1998; IEEE, 2005). However, the harmonization process should take place with respect to the overall improvement of the standards in terms of higher precision in the specification of the SAR, lower uncertainties in exposure assessment, more accurate biological results, and greater reliability in health status data. Advances and findings in bioelectromagnetics research will continue to facilitate this process. A global scientifically based and recognized exposure standard would help to address the existing widespread public concern related to the adequacy of the current guidelines.

15.10 CONCLUSION

In this chapter, a summary was provided regarding the safety issues encountered in biomedical telemetry. Safety issues related to device integrity, device use, clinical treatment, and RF exposure were presented. Compatibility and interference issues as well as applicable RF exposure guidelines were also discussed. Future research directions were finally proposed to improve safety in biomedical telemetry.

Regarding the device itself, operational and occupational safety, as well as user issues, human factors and device hazards were discussed. Due to the nature of biomedical devices used in biomedical telemetry, there exist several risks which could threaten the health of the patient during their operation. In addition, as for any electronic device, medical devices carry a risk of failure or malfunction. Moreover, due to human interaction with medical devices, user issues and additional hazards might appear. Finally, the clinical environment of the patient could result in a problematic operation of medical devices used for biomedical telemetry.

In addition to those common hazards, there are serious concerns related to the safety of patients exposed to repeated and prolonged exposure to RF energy. Different approaches are followed for the derivation and adoption of standards, and different sets of guidelines are developed for limiting human exposure to RF radiation, so that public concern about their adequacy is, in turn, spreading. Two standards are currently in use [IEEE C.95.1-2005 and the ICNIRP standard (1998)] which use the SAR distribution to quantify and limit the RF exposure to human tissues. Their major differences are related to the tissue mass used for averaging and the average time period used in the SAR measurement.

Finally, an attempt was made in this chapter to provide some general directions for future research. For example, hazards related to the design of medical devices and their interaction with the end user (patient or qualified user) could be minimized by using the data obtained by medical devices which are currently in use. Furthermore, advanced scientific findings about the biological effects of RF radiation are expected to come over time, which should be used to develop global standards, that is, standards that would be adopted by every organization worldwide.

REFERENCES

Akay M. 2008. Advances in neural implants and prostheses. 5th International Summer School and Symposium on Medical Devices and Biosensors.

American National Standards Institute (ANSI). 1966. C95.1-1966 standard.

Anandanatarajan R. 2011. *Biomedical Instrumentation and Measurements*. PHI Learning, New Delhi.

Anderson P. 2003. Improving patient safety: Insights from American, Australian and British healthcare. *Qual. Saf. Health Care*, 12(3): 235.

Ball DJ, Argentieri G, Krause R, Lipinski M, Robison RL, Stoll RE, Visscher GE. 1991. Evaluation of a microchip implant system used for animal identification in rats. *Lab. Anim. Sci.*, 41(2):185–186.

Beach RD, Conlan RW, Godwin MC, Moussy F. 2005. Towards a miniature implantable in vivo telemetry monitoring system dynamically configurable as a potentiostat or galvanostat for two- and three-electrode biosensors. *IEEE Trans. Instrum Meas.*, 54(1):61–72.

Blanchard KT, Barthel C, French JE, Holden HE, Moretz R, Pack FD, Stoll RE. 1999. Transponder-induced sarcoma in the heterozygous p53+/– mouse. *Toxicol. Pathol.*, 27(5):519–527.

Brodal P. 2010. *The Central Nervous System*. Oxford University Press, New York.

Buchler P, Simon A, Burger J, Ginggen A, Crivelli R, Tardy Y, Olsen S. 2007. Safety of active implantable devices during MRI examinations: A finite element analysis of an implantable pump. *IEEE Trans. Biomed. Eng.*, 54(4):726–733.

Burci GL, Vignes CH. 2004. *World Health Organization*. Kluwer Law International, The Hague, The Netherlands.

Carayon P. 2011. *Handbook of Human Factors and Ergonomics in Health Care and Patient Safety*. Taylor & Francis, Boca Raton, Florida.

Chan AYK. 2008. *Biomedical Device Technology: Principles and Design*. Charles C. Thomas, Illinois.

Dhillon BS. 2000. *Medical Device Reliability and Associated Areas*. Taylor & Francis, Boca Raton, Florida.

Elcock LE, Stuart BP, Wahle BS, Hoss HE, Crabb K, Millard DM, Lake SG. 2001. Tumors in long-term rat studies associated with microchip animal identification devices. *Exp. Toxicol. Pathol.*, 52(6):483–491.

European Telecommunications Standards Institute (ETSI). 2002. Electromagnetic compatibility and radio spectrum matters (ERM); Radio equipment in the frequency range 402 MHz to 405 MHz for ultra low power active medical implants and accessories; Part 1: Technical characteristics, including electromagnetic compatibility requirements, and test methods: European Standard (Telecommunications series).

Europe Council. 2011. The potential dangers of electromagnetic fields and their effect on the environment. Available: http://assembly.coe.int/ASP/Doc/XrefViewPDF.asp?FileID=13137&Language=EN.

Federal Communications Commission (FCC). 1996. Guidelines for Evaluating the Environmental Effects of Radionfrequency Radiation.

Federal Communications Commission (FCC). 1996. 96–326. Washington, DC.

Federal Communications Commission (FCC). 1999. Rules and Regulations. MICS Band Plan.

Food and Drug Administration. (FDA) 2000. Medical Device Use—Safety: Incorporating Human Factors Engineering into Risk Management. Guidance for Industry and FDA Premarket and Design Control Reviewers.

Food and Drug Administration. (FDA) 2010. Medical Device Home Use Initiative. Available: http://www.fda.gov/downloads/MedicalDevices/ProductsandMedicalProcedures/Home HealthandConsumer/HomeUseDevices/UCM209056.pdf.

Fries RC. 2005. *Reliable Design of Medical Devices*, 2nd ed. Taylor & Francis, Boca Raton, Florida.

Gollakota S, Hassanieh H, Ransford B, Katabi D, Fu K. 2011. They can hear your heartbeats: Non-invasive security for implantable medical devices. *SIGCOMM Comput. Commun. Rev.*, 41(4):2–13.

Gustrau F, Bahr A, Goltz S, Eggert S. 2002. Active medical implants and occupational safety—measurement and numerical calculation of interference voltage. *Biomed Tech*, 2:656–659.

Health EU. 2005. Patient Safety. Available: http://ec.europa.eu/health-eu/care_for_me/patient_safety/index_en.htm.

Hughes R, Research USAfH, & Quality. 2008. *Patient Safety and Quality: An Evidence-Based Handbook for Nurses*. Washington, DC: Agency for Healthcare Research and Quality, U.S. Department of Health and Human Services.

Institute of Electrical and Electronics Engineers (IEEE). 1982. IEEE Standard for Safety Levels with Respect to Human Exposure to Radio Frequency Electromagnetic Fields, 3 kHz to 300 GHz.

Institute of Electrical and Electronics Engineers (IEEE). 1991. IEEE Standard for Safety Levels with Respect to Human Exposure to Radio Frequency Electromagnetic Fields, 3 kHz to 300 GHz.

Institute of Electrical and Electronics Engineers (IEEE). 2005. Safety levels with respect to human exposure to radio frequency electromagnetic fields, 3 kHZ to 300 GHz.

International Commission on Non-Ionizing Radiation Protection (ICNIRP). 1998. Guidelines for Limiting Exposure to Time-Varying Electric, Magnetic, and Electromagnetic Fields (up to 300 GHz).

International Electrotechnical Commision (IEC). 1993. Medical equipment|medical electrical equipment—Part 1: General requirements for basic safety and essential performance.

Jacobson B, Murray A. 2007. *Medical Devices: Use and Safety*. Churchill Livingstone, Edinburgh.

Johnson K. 1996. Foreign-body tumorigenesis: Sarcomas induced in mice by subcutaneously Implanted transponders. *Toxicol. Pathol.*, 33:619.

Karacolak T, Hood AZ, Topsakal E. 2008. Design of a dual-band implantable antenna and development of skin mimicking gels for continuous glucose monitoring. *IEEE Trans. Microw. Theory Techn.*, 56(4):1001–1008.

Kiourti A, Nikita KS. 2012a. Miniature scalp-implantable antennas for telemetry in the MICS and ISM bands: Design, safety considerations and link budget analysis. *IEEE Trans. Antennas Propag.*, 60(6):3568–3575.

Kiourti A, Nikita KS. 2013. Numerical assessment of the performance of a scalp-implantable antenna: Effects of head anatomy and dielectric parameters. *Wiley Bioelectromagnetics*, 34(3):167–179.

Kohn LT, Corrigan JM, Donaldson MS, America CQHC, Medicine I. 2000. *To Err Is Human: Building a Safer Health System*. Washington, DC: National Academies Press.

Lazzi G. 2005. Thermal effects of bioimplants. *IEEE Eng.Med. Biol. Mag.*, 24(5):75–81.

Le Calvez S, Perron-Lepage MF, Burnett R. 2006. Subcutaneous microchip-associated tumours in B6C3F1 mice: A retrospective study to attempt to determine their histogenesis. *Exp. Toxicol. Pathol.*, 57(4):255–265.

Lee K. 2008. World Health Organization. *Taylor & Francis*, New York.

Lin J, Chen J, Bai J. 1996. Using human factors engineering as the basis for developing medical human-computer systems. Presented at the IEEE International Conference on Systems, Man, and Cybernetics.

Lin JC. 2000. Perspectives on guidelines for human exposure to radio-frequency radiation. *IEEE Trans. Antennas Propag. Mag.*, 42(1).

Lin JC. 2003. Biological bases of current guidelines for human exposure to radio-frequency radiation. *IEEE Trans. Antennas Propag. Mag.*, 45(3).

Lin JC. 2006. A new IEEE standard for safety levels with respect to human exposure to radio-frequency radiation. *IEEE Trans.* Antennas and Propag. Mag., 48(1).

Murasugi E, Koie H, Okano M, Watanabe T, Asano R. 2003. Histological reactions to microchip implants in dogs. *Vet. Rec.*, 153(11):328–330.

Nash DB. 2012. Population health and health reform—Inseparable concepts. *Prescript. Excell. Health Care*, 16.

Nokes L, Jennings D, Flint T, Turton B. 1995. *Introduction to Medical Electronics Applications*. New York: Elsevier Science.

Onuki Y, Bhardwaj U, Papadimitrakopoulos F, Burgess DJ. 2008. A review of the biocompatibility of implantable devices: Current challenges to overcome foreign body response. *J. Diabetes Sci. Technol.*, 2(6):1003–1015.

Palmer JN, Palazzolo M, Ryan T. 1998. Fibrosarcomas associated with passive integrated transponder implants. *Toxicol. Pathol.*, 26: 170.

Prodromakis T, Michelakis K, Zoumpoulidis T, Dekker R, Toumazou C. 2009. Biocompatible encapsulation of CMOS based chemical sensors. Presented at the IEEE Sensors.

Rao GN, Edmondson J. 1990. Tissue reaction to an implantable identification device in mice. *Toxicol Pathol*, 18(3):412–416.

Schueler BA, Parrish TB, Lin JC, Hammer BE, Pangrle BJ, Ritenour ER, Truwit CL. 1999. MRI compatibility and visibility assessment of implantable medical devices. *J. Magn. Reson. Imaging*, 9(4):596–603.

Silay KM, Dehollain C, Declercq M. 2010. Inductive power link for a wireless cortical implant with biocompatible packaging. Presented at the IEEE Sensors.

Soontornpipit P, Furse CM, You Chung C. 2005. Miniaturized biocompatible microstrip antenna using genetic algorithm. *IEEE Trans. Antennas Propag.*, 53(6):1939–1945.

Taghipour S, Banjevic D, Jardine AKS. 2011. Reliability analysis of maintenance data for complex medical devices. *Qual. Reliability Eng. Int.*, 27(1):71–84.

Thull R. 2002. Physicochemical principles of tissue material interactions. *Biomol. Eng.*, 19(2–6):43–50.

Tillmann T, Kamino K, Dasenbrock C, Ernst H, Kohler M, Morawietz G, Mohr U. 1997. Subcutaneous soft tissue tumours at the site of implanted microchips in mice. *Exp. Toxicol. Pathol.*, 49(3–4):197–200.

Vascellari M, Melchiotti E, Mutinelli F. 2006. Fibrosarcoma with typical features of postinjection sarcoma at site of microchip implant in a dog: Histologic and immunohistochemical study [Case Reports]. *Vet. Pathol.*, 43(4):545–548.

Vascellari M, Mutinelli F, Cossettini R, Altinier E. 2004. Liposarcoma at the site of an implanted microchip in a dog [Case Reports]. *Vet. J.*, 168(2):188–190.

Webster JG. 2009. *Medical Instrumentation Application and Design*. Hoboken, NJ: Wiley.

Wilkins RD, Holley LK. 1998. Risk management in medical equipment management. Presented at the Engineering in Medicine and Biology Society, Proceedings of the 20th Annual International Conference of the IEEE.

World Health Organization (WHO). 2003. Medical device regulations: Global overview and guiding principles. Available: http://www.who.int/medical_devices/publications/en/MD_Regulations.pdf.

World Health Organization (WHO). 2004. Patient safety. Available: http://www.who.int/patientsafety/about/en/index.html.

World Health Organization (WHO). 2006. Constitution of the world health organization. Available: http://www.who.int/governance/eb/who_constitution_en.pdf.

PART III

EXAMPLE APPLICATIONS OF BIOMEDICAL TELEMETRY

16

CLINICAL APPLICATIONS OF BODY SENSOR NETWORKS

RICHARD M. KWASNICKI AND GUANG-ZHONG YANG

Imperial College London, United Kingdom

16.1 INTRODUCTION

Using the body as the medium, a source of inspiration and energy, body sensor networks (BSN) may facilitate the provision of high-quality, personalized healthcare through continuous monitoring of human physiological parameters in natural, unrestricting environments (Yang, 2006). A BSN is not a specific system but is in fact an architecture in which one or more sensors connected to the body, either wearable or implantable, may seamlessly communicate with each other to provide coordinated monitoring and intervention. The simplest embodiment of each sensor consists of a microprocessing unit, battery pack, and wireless transmitter. Future development is focused on the use of bioinspired design to avoid this stereotypical architecture and to rely on the body as the transmission medium and power source instead (through power harvesting) as well as to explore cellular level signaling and smart particles (a wireless network of minisensors) combined with advanced sensing and imaging techniques. The current generation of BSN components may use the same or different input measurements, including temperature, motion, and surface conductance. Effective "on-node" processing and storage are key to the wider deployment of the technology, although it is common to use a local processing unit (LPU) for data aggregation and then transmit to the central database for potential intervention. At this

Handbook of Biomedical Telemetry, First Edition. Edited by Konstantina S. Nikita.

point, data may undergo further processing and fusion before being wirelessly transmitted to a central server.

Advances in technology and high-quality research programs continue to reshape the standards of healthcare. Meanwhile, patient expectations have increased, and the general public expects to receive the latest treatment options available. With a growing population and rising healthcare demands, it is vital that health technologies and healthcare reform are developed with sustainability in mind. Pervasive sensing with BSN offers high-quality care that is potentially cost and resource effective. Part of the aim is to safely relocate appropriate healthcare services to the community setting, allowing patients to take a more active role in their own management. Additionally, continuous patient monitoring strategies are likely to improve conventional clinic-based recordings of patient physiology, where time and resources are often wasted trying to capture transient pathological episodes.

Platforms for BSN include wearable devices (such as a ringlike finger-worn sensor for measuring heart rate and blood pressure), textile-incorporated sensors [such as intelligent vests with integrated electrocardiography (ECG) electrodes], and implantable or ingestible devices (including subcutaneous glucose sensors or engineered pills that transmit core body temperature). Within the BSN architecture, these modalities may be combined to produce a hybrid platform capable of multimodal human physiology monitoring. For example, data from an ingested pill for core temperature monitoring may be combined with that of a wearable motion sensor to produce an early-warning system for heat exhaustion in athletic performance. Similarly, BSN may be combined with ambient sensors from wireless sensor networks (WSN) to facilitate context-aware sensing, such as relating body temperature to that of the local environment.

Communication plays a vital role in allowing the evolution of a BSN to that of a functional biotelemetric platform. Data transmission must be reliable and secure and must have low energy cost to preserve battery life. On-node processing may reduce the demand for data transmission. The destination of the data depends on the function of the platform. A BSN designed to monitor vital signs of a patient associated with heart disease would be suited to real-time communication with the local emergency department. Patients at high risk of fatal cardiac events, such as a myocardial infarction (MI) ("heart attack"), are often treatable as long as the interval between the emergency call and the intervention is brief (often described as the "golden hour"). A coronary occluding thrombus (clot) can be dissolved rapidly with medication, but the longer the heart muscle is starved of oxygen and nutrients, the more widespread and irreversible the damage becomes. Early detection of event-preceding vital signs may facilitate a more timely intervention with improved patient outcomes.

Alternatively, data collected from a BSN assessing postoperative gait in knee arthroplasty patients may be more suitably sent to a central server for kinematic modeling. In this case it is less likely that there will be a need for emergency intervention, yet supervised rehabilitation, including goal setting and strengthening regimes, is essential for optimal recovery. A personalized follow-up strategy is best achieved with supporting functional data gained from regular, objective mobility assessments.

Another important usage of BSN is that of a "closed-loop" system such as the Artificial Pancreas. Patients with diabetes lack the important homeostatic mechanism required to control their blood sugar effectively. Normophysiologically, blood sugar levels are detected by the pancreas, which initiates the corresponding response of secreting insulin to reduce sugar levels or glucagon to increase them. In diabetes, the failing pancreas is replaced by a BSN which first allows glucose levels in the interstitial fluid (between cells) to be monitored by a subcutaneous sensor and then uses those levels to titrate the dose of exogenous insulin or glucagon infused into the patient. Frequent, accurate adjustments of blood sugar not only reduces user demand (e.g., the need for self-testing and injecting) but also may improve outcomes as a result of finer regulation of the blood sugar profile. A self-sustainable platform such as the artificial pancreas is the desirable outcome for many biotelemetry systems.

In this chapter, we will first identify healthcare challenges. We will then show a variety of case reports, in a range of healthcare settings, where BSN are used in order to meet these challenges. Further, we will consider the bigger picture with regard to BSN in healthcare, including the practical challenges of implementation, data interpretation, and the necessary steps needed to translate research devices into everyday healthcare.

16.2 HEALTHCARE PARADIGM SHIFT FOR PERVASIVE SENSING

Advances in medical care and human hygiene are causing continuous evolution of the population and a corresponding change in healthcare demands. The population is not only growing but aging as well (Schofield and Heath, 1999). In addition to the rise in age-related diseases, including cancer, heart failure, arthritis, and dementia, we are also experiencing an obesity epidemic.

Consequently, 29% of the U.K. population are living with a long-term medical condition, of which type 2 diabetes mellitus (DM) forms a significant proportion. Today, 80% of the healthcare budget is spent on chronic disease (Saxby, 2007).

In the U.K., for example, when the National Health Service (NHS) was created in 1946, it was designed to manage acute illness. In the early twentieth century, infectious diseases were more prevalent, and conditions often associated with long-term health support such as cancer and severe trauma had vastly higher mortality rates. Patients were mostly managed through short-term concentrated inpatient care where they would either be cured and discharged or pass away in hospital. Modern hospitals now have relatively small departments dedicated to managing this type of patient and instead have facilities for a wide variety of specialist services, including health screening, elective surgery, psychiatric care, interventional radiology, and hemofiltration for those waiting for renal transplants. Ironically, it is increasingly unusual for one to see a sick (critically unwell) patient in hospital due to improved access to care and early detection of disease and advances in monitoring and supporting patient physiology.

The quality of patient care is improving in line with research advances in pharmacology, technology, and surgery. Structured research programs allow valuable data collection and organization, which initiates and supports the progression of many specialties toward evidence-based patient management. When deciding how to manage a patient, the most effective and reliable plans will often be those based on high-quality patient information. Information can be collected in a variety of forms, ranging from simple tasks such as taking a patient's history and clinical examination to more advanced methods including three-dimensional imaging or obtaining a tissue biopsy for histological assessment. Often, intermediate assessments are performed in the clinic setting, such as blood pressure readings or capturing an electrocardiogram to map the electrical activity of the heart. These data, give the doctor a snapshot impression of the patient's physiological state. However, human physiology is dynamic, and pathology has a tendency to be sporadic and as such can be missed. This particular situation can arise on many levels within the healthcare setting, with important recognized sequelae. The awareness of this low diagnostic sensitivity phenomenon is such that when a snapshot test result returns negative, the healthcare professional may interpret this as (1). true negative, there is no pathology, and the patient is healthy and can go home or (2). false negative, we cannot confirm pathology, but neither can we allow the patient to go home because we are not sure. Situation 1 may result in a missed diagnosis, which would compromise patient safety, whereas situation 2 is likely to cause a number of inappropriate hospital admissions and unnecessary further investigations. This is one situation where a continuous, and potentially remote, data collection method might provide a solution.

There are many resource-demanding scenarios associated with the management of chronic disease and of the elderly population. In each scenario, whether in community care, diagnostics, rehabilitation, or any other, two key components are, first, the efficient collection of valuable patient information and, second, the accurate processing of the data in order to determine further management strategies. To reiterate the scope of these long-term healthcare demands, three typical case examples are the typical follow-up and rehabilitation of a patient with arthritis who has undergone a hip replacement, monitoring and advising an overweight child who is at risk of developing DM, and titrating (optimizing) the dose of medication given to a patient with Parkinson's disease to manage their symptoms. A number of these scenarios may be more effectively managed through the use of BSN, either in the clinical setting or at the patient's home.

16.3 USAGE SCENARIOS

This section provides a nonexhaustive review of BSN applications to meet the current healthcare challenges and is organized according to the clinical setting (Table 16.1).

TABLE 16.1 Clinical Applications of BSN

Community	Diagnostics
Promoting healthy lifestyle choices and improving chronic disease management • Remote patient surveillance for cost-effective, appropriate medical social support, including detection of acute deterioration: Bespoke BSN with wireless communication capabilities facilitate a medically supervised home care package • Raising awareness and modifying lifestyle behavior through user-friendly interventions at primary-care level: Real-time lifestyle assessment and feedback through activity pattern detection and motivational reaction	Replacing snapshot clinical data with continuous pervasive monitoring to reflect true physiology in unrestricted, natural environments • Encouraging ambulatory patient assessment in order to optimize diagnostic sensitivity in more realistic environments: Streamlined cardiac arrhythmia detection through autonomic data organization and user-controlled electronic symptom tagging • Investigating dynamic patient symptomatology in response to fatigue, natural diurnal variation, and interventions to refine diagnostic accuracy and titrate pharmacological therapy (a) Supplementing subjective, qualitative data with objective measures such as motion/location estimation alongside patient activity diaries (b) Remote ad hoc consultations using a Web interface with integrated home-worn sensor inputs
Next: Safely shifting the responsibility and setting of appropriate healthcare scenarios to the patient in their home	Next: Reducing resource demand through autonomic sensing protocols and closed-loop systems combining monitoring with therapeutics
Perioperative	**Extreme Environments**
High-quality patient data, collected remotely and stored electronically, which can be continued following discharge in a home environment • Utilizing wireless BSN to collect detailed bioinformatics, including during the ward-to-home transition period, with secure integration with electronic medical records: Combined sensor inputs allow inference of several vital signs, with mobile phone or ward-based data relay stations to allow raw/processed signals to be sent to a central server	Pervasive monitoring of vital signs in hazardous, unpredictable environments • Remote surveillance of vital signs in combat/hazardous environments: Simulated environments and events such as high humidity, extreme temperatures, and hemorrhage allow training of BSN systems to recognize and report notable incidents • Smart sensor integration with personal apparel (such as smart textiles), including rescue medication delivery apparatus: Pervasive biochemical sample interrogation, skin surface

(*continued*)

TABLE 16.1 (*Continued*)

Perioperative	Extreme Environments
• Pervasive motion sensors provide objective kinematic data, allowing pervasive monitoring of functional rehabilitation: Postoperative analysis of patient biomechanics with BSN allows early detection of surgical complications characterized by movement abnormalities	conductance, and salivary hormone profiling; minimally invasive tissue and blood evaluation with microneedles for oxygen, CO_2, lactate, electrolytes
Next: Pervasive sensing to indicate need for surgery, prediction of outcome, and personalization of postoperative follow-up	Next: Combining implantable and wearable sensors for continuous performance monitoring, with retrospective data analysis for training, evaluation, and psychophysical event preparation.

16.3.1 In the Community

One day in a standard hospital bed, without an operation or specialist tests, can cost up to £400. Although cost alone provides a strong motive to reduce the length of patient stay in hospital, there are added risks, including hospital-acquired infections (HAIs) and the potential consequences of immobility such as muscle wasting and blood clotting conditions.

This is particularly important when considering the elderly population, who are more at risk of HAI and losing their independence following prolonged bed rest. Monitoring the elderly and other high-risk population groups such as the immune deficient and the disabled, especially during winter, may allow healthcare professionals to be alerted should their attention be required (the elderly often fail to seek medical attention) but also reduce unnecessary hospital appointments through remote consultations.

BSN in the community should improve overall healthcare standards as well as change the location of treatment. For example, the treatment of patients with DM requiring exogenous insulin has made considerable advances through BSN technology. DM is a chronic metabolic disorder in which the patient either fails to produce enough insulin or becomes resistant to its effects. This results in a dangerous inability to control blood glucose autonomously, and as such the patient must manage the balance actively with multiple daily finger prick glucose tests and appropriate insulin injections. Not only does this method of management reduce quality of life, in the majority of cases the blood glucose control is vastly inferior to that of healthy subjects. The consequence of poor glycemic control is an increased risk of micro- and macrovascular disease, including ischemic heart disease, stroke, renal failure, blindness, and ischemic limbs often requiring amputation. A miniature, continuous subcutaneous insulin infusion (CSII) device, known as the patch pump, is revolutionizing the management of diabetes. Through simple algorithms using minimal information entered by the patient, the device calculates the necessary insulin dose, including compensatory increases during mealtime, and provides a natural delivery throughout the day through an indwelling catheter (which only needs replacing twice a week). When combined with a continuous glucose monitoring device, even less manual data input is needed. In addition, some devices are able to deliver both insulin and glucagon, allowing control of both high and low blood sugar levels. CSII, also known as the Artificial Pancreas, claims the responsibility for glycemic control and also has been shown to significantly improve the glycosylated hemoglobin levels (HbA1c) which cause a marked reduction in the risk of diabetic complications and therefore increase cost-effectiveness (Pickup, 2012).

Central to the natural history of many diseases are often lifestyle choices, including diet, exercise, alcohol use, and smoking. Desiderius Erasmus was the first to say that "prevention is better than a cure", and today we actively target those who are making unhealthy lifestyle choices with media campaigns and new technology. One cost-effective, scalable approach to promoting healthy living is through utilization of smart phones, which already contain useful inbuilt hardware and communication capabilities. The key to tackling obesity is finding an effective method of rebalancing the food intake to energy output ratio. As such, multiple smart phone applications are combining food intake diaries with activity monitors in order to make the user aware of why they might be gaining unwanted weight (Denning et al., 2009). The creative approach of the UbiFit wellness system (Klasnja et al., 2009) projects activity levels onto the condition of a virtual garden, providing the user with regular, graphical feedback. Other products such as Nike+ and Map My Run allow the user to share workout data with friends on social networking sites.

Regarding the efficacy of these system, no randomized controlled trials (RCTs) were found (May, 2012). However, some low-level evidence shows that during a case–control study the intervention arm, using a mobile phone activity and diet application, incurred significant weight loss whereas the control arm did not (Lee et al., 2010). The next step in this research is to implement a robust smart phone application in a large-scale RCT to assess the value of this technology in weight management. A simple tool with the ability to influence behavioral economics, particularly considering cues such as incentives, norming, and ego (Dolan et al., 2010), may play an important role in dealing with the obesity epidemic. The phone application ActiveMiles, under development at Imperial College London, is one example of such technology (Figure 16.1). Paying particular attention to the beneficial exploitation of human psychology, ActiveMiles allows the primary user to compare their activity to others within a chosen cohort (e.g., Classmates, weight management group) as well as directly competing against a chosen friend or rival—the so-called nemesis. The potential for competition, extending the access of personal activity levels and raising awareness regarding the variation in performance of contemporaries, is likely to act as an activity promoting behavioral intervention.

16.3.2 Diagnostics

When evidence-based medical (EBM) practice is used wisely, without discounting the unique features of the case at hand, it has the potential to improve patient

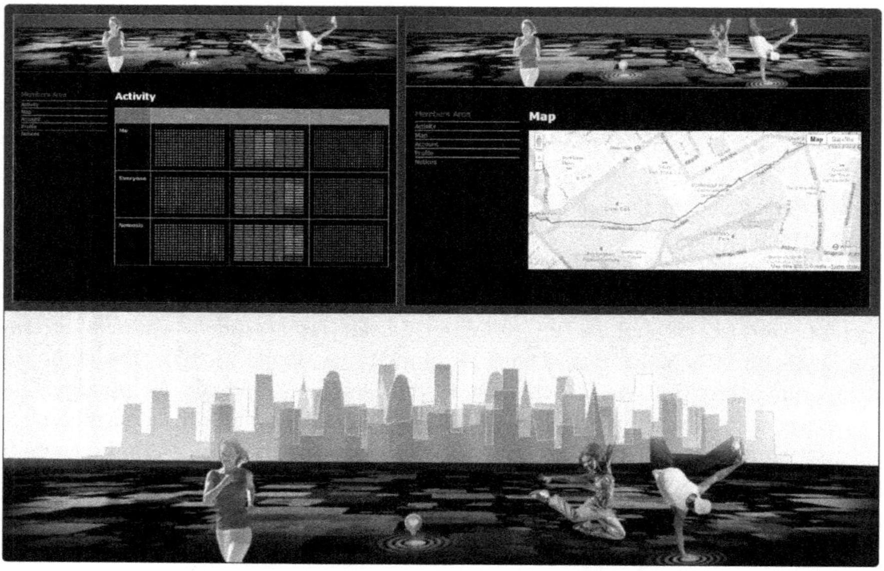

Figure 16.1 Example screenshots of ActiveMiles (Imperial College London, U.K.) showing activity levels, including that of their associated group and chosen nemesis (left) as well as geographical activity review (right).

outcomes by utilizing the results of one or more research studies. Diagnostics, and EBM in general, requires high-quality patient information, that is, reliable, reproducible, accurate, representative, specific, and contextualized. BSN begin to play a more central role in improving the quality of patient information through continuous, pervasive monitoring. There are several examples of this in everyday healthcare.

Considering the management of cardiac arrhythmias (CAs), often events are sporadic and as such are often difficult to capture in a snapshot diagnostic investigation. Misdiagnosis of CA may result in life-threatening consequences such as stroke and cardiac arrest. Although there are a variation of CAs, each with different etiology and symptomatology, current medical practice often involves clinical examination and electrocardiography (ECG). ECG provides a short cardiac trace that allows the rhythm to be analyzed for diagnostic purposes. This is often inadequate for diagnosis, and as such several technological devices and techniques have been built and implemented (Kamel and Smith, 2011).

In a scenario where the initial ECG fails to show pathology, patients are either admitted to a ward for continuous inpatient cardiac assessment over 24–48 hr or alternatively sent home for 24-hr ambulatory cardiac monitoring using a Holter monitor. A Holter monitor records electrical signals from the heart via a set of wired electrodes attached to the chest. The electrodes are connected to a base station unit worn on the patient's belt or hung around the neck. The device improves the management process through the facilitation of continuous monitoring in the patient's natural environment. However, the device is quite bulky and therefore likely to affect the patient's behavior. Additionally, the entire sequence of raw data collected by the device (limited by memory and battery constraints) must be reviewed by a physician at the end of the collection period, which is time consuming and does not allow warning of the healthcare team at the time of a cardiac event.

A further improvement to the management of CAs is the external loop recorder (ELR), which avoids the memory and battery life limitations of the Holter monitor by data collection being user initiated upon experiencing symptoms (Wallmann et al., 2007). ELRs also offer software that automatically detects arrhythmias such as atrial fibrillation (AF). ELR over four to seven days detects arrhythmias in 6–8% of patients that had a negative 24-hr Holter test, hence showing greater diagnostic sensitivity (Barthelemy et al., 2003). Further, in patients who are at high risk of unsafe CAs, an implantable loop recorder (ILR) may be used. This subcutaneous device also features software capable of detecting arrhythmias, which conveniently prompts data saving immediately before, during, and for a short duration after the event. This significantly reduces the amount of data passed on to the physician for review, and energy consumption, allowing the implant to function for up to two years postimplantation (Ziegler et al., 2010).

Mobile Cardiac Outpatient Telemetry (MCOT) combines the Holter monitor and ILR techniques to produce a wearable device capable of continuous data collection, but with daily data transmission to a central library for processing and review. The data are electronically tagged by the patient at the onset of symptoms, complemented by feature detection software to maximize the cost effectiveness of the service and the diagnostic sensitivity (Tayal et al., 2008). A future development of MCOT to close the feedback loop, as with the Artificial Pancreas, may be to link therapeutic action in response to data collected such as rescue medication delivery or cardiac defibrillation in appropriate circumstances (Figure 16.2).

Many clinical conditions are dynamic in the sense that the associated symptom severity and character can vary depending on factors such as a medication duration of action, systemic illness, and environmental changes. Also, many conditions deteriorate over time, requiring reassessment and occasionally rediagnosis. Parkinson's disease (PD), for example, is a neurological disorder characterized by tremor, bradykinesia (slow movements), rigidity, and impairment of postural balance (Chen et al., 2011). Current diagnostic strategies mainly involve clinical examination and exclusion of other differential diagnoses that may be seen on 3D imaging, neurological tests, and blood tests. BSN in the form of motion sensing has been trialed in characterizing the movement patterns of patients with PD for potential use in diagnostics but also to assist in titration of medication regimes and monitoring disease progression.

In a simple but effective study, Cavanaugh et al. (2012) collected accelerometry data from a small cohort of PD patients for one week at baseline and at one year follow-up in order to explore long-term changes in daily ambulatory activity and how these measurements compare to conventional clinical measurements of gait and disease severity. Clinical measurements included questionnaires regarding limitations of activities of daily living and physician-led movement analysis during a variety of short activities. Although the tests provide a score on the Unified Parkinson's Disease Rating Scale (UPDRS), they are limited, as they remain subjective, snapshot

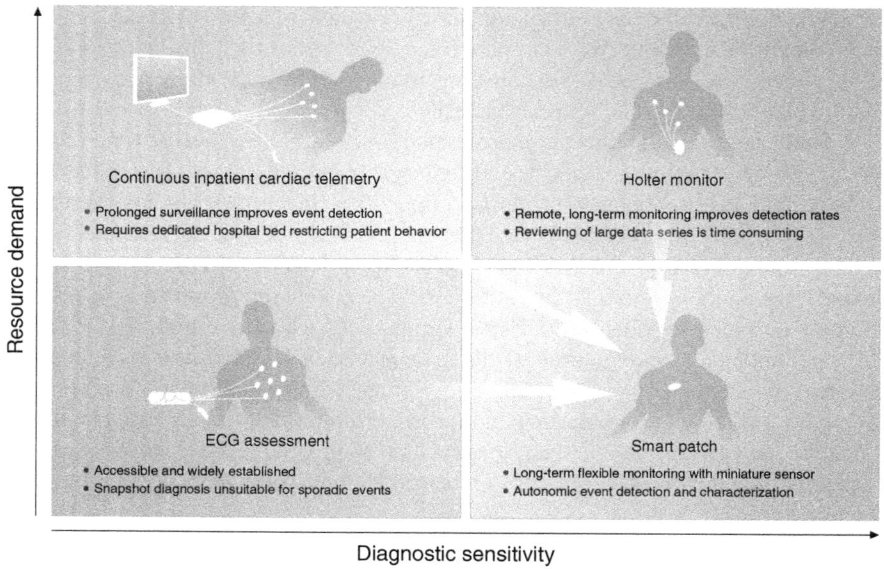

Figure 16.2 Advances in cardiac monitoring show a trend of increased resource demand alongside with improvements in sensitivity. Developing a highly sensitive diagnostic tool with low resource demand is one target for BSN.

assessments. Accelerometer data collected over days or weeks can be analyzed using a variety of methods, such as machine learning techniques for activity classification and pattern recognition, specific detailed analysis of signal patterns for biomechanical profiling including gait parametrics, activity indexing whereby the magnitude of the motion data is taken and subdivided into variable activity intensities, and many more. In this study, accelerometer data showed objective reductions in the amount and intensity of daily ambulatory activity but not in its frequency or duration.

Moving from observational studies to clinical translation in the form of telemetry platforms in diagnosis and disease monitoring, Chen et al. (2011) developed MercuryLive, a system that connects wearable sensors to a laptop-based data collection engine and a Web-based graphical user interface with video conferencing capabilities. MercuryLive is programmed to conduct automatic data analyses, which could provide the user (and their care team) with estimates of clinical scores reflecting disease severity at appropriate intervals, avoiding a formal visit to the hospital. Based on preliminary test results, the authors speculate that frequent longitudinal observations will allow clinicians to adjust patients' medication regimen more effectively.

16.3.3 Perioperative

The scope for perioperative BSN applications is vast, as by definition this time period involves the time before, during, and after an operation. Every operation has a risk of complications which may arise from the anaesthetic, from the surgery itself, or just as a result of being in a hospital. As many operations are elective, as opposed to emergencies, it is essential that the risk of perioperative morbidity and mortality is minimized. Although this section will focus on BSN for inpatient care and postoperative recovery, uses of biotelemetry in the preoperative setting must not be overlooked. Preoperative continuous pervasive monitoring may be useful in developing an evidence-based approach to determining the need for and the timing of a particular operation. Similarly, assessment of preoperative fitness allows prediction of postoperative morbidity (Struthers et al., 2008), which may allow services to plan resource allocation and personalize patient follow-up strategies.

Patients in hospital are monitored according to their individual needs. Monitoring is conducted through many strategies, including continuous data collection from worn devices (e.g., pulse oximetry), observations from nurses and other healthcare staff (e.g., respiratory rate and temperature), and through specific investigations such as blood tests, urinalysis, and imaging. Results are most often inserted into the patient's paper-based notes, though some will be available on a computer database, depending on the hospital. Each patient in the intensive care unit has their own nurse and hourly observations are recorded, whereas a patient in a standard hospital ward most likely will have observations recorded every 4 hr. Patient monitoring is resource demanding, subject to human error and noncontinuous.

The LOBIN platform (locating and biomonitoring by means of wireless networks in hospitals) combines e-textiles and wearable technologies to provide remote location data and healthcare monitoring (Lopez et al., 2010). This advanced IT platform is an excellent example of a BSN, capable of monitoring physiological parameters such as ECG, heart rate, angle of inclination, activity index, and body temperature as well as allowing location tracking of patients within the hospital. Additionally, software designed to identify abnormal ECG traces can trigger an alarm. This wealth of data is collected using a smart e-textile shirt and is transmitted through a low-cost wireless sensor network infrastructure based at the hospital. This boasts the advantage of automatic electronic documentation of patient information, which could then be transferred to relevant displays and settings such as a tablet PC on a ward round or to an on-call surgeon's home PC when considering an "out-of-hours" intervention.

The system hardware consists of two subsystems, the physiological sensors and a wearable data acquisition device (WDAD). The physiological parameters are calculated using data from three sensor inputs: e-textile electrodes, triaxial accelerometer, and thermometer. The WDAD contains a data acquisition and processing board (DAPB) and a wireless transmission board (WTB). Preliminary tests to assess the feasibility of using the device in a clinical setting were conducted, allowing validation of the sensor measurements and providing feedback from the patients and healthcare staff. Appropriate post-feedback modifications of the software and smart shirt dimensions will soon result in larger scale application of this technology.

Following an operation, or time in hospital, it is important to undergo a period of functional rehabilitation. A focused effort serves to ensure that the patient may resume standard activities of daily living and, accordingly, regain optimum quality of life. In many cases, patients are discharged from hospital with a set of rehabilitation exercises to perform at home, for example, following a knee replacement. A structured and timely recovery to a high functional level is likely to reduce the amount of time taken off work and prevent future complications and healthcare resource requirements. Accurate, reliable home-based assessment may facilitate improved recovery and allow those patients who are struggling to be identified for additional support.

The Gait Evaluation Differential Entropy Method (GEDEM) has been used to analyze gait data obtained from a triaxial accelerometer in patients following anterior cruciate ligament (knee) reconstruction (Tsivgoulis et al., 2011). Trunk gait accelerometry is able to show minor changes in normal walking caused by pathological biomechanics and learn injury behavior. In this study, patients were fitted with an accelerometer and asked to walk along a straight 40-m walkway. GEDEM analysis revealed highly significant reductions in mean spectral differential entropy postoperatively in the mediolateral and anteroposterior axes of the signal, implying decreased gait variability after surgical repair. At six to eight months after surgery, there was still a significant difference in mediolateral spectral differential entropy compared to the control group, suggesting that some rotational instability remained.

Objective gait analysis is a luxury that is often unavailable as part of standard surgical follow-up, mostly due to centers lacking expensive gait laboratories and trained technicians. Current clinical evaluation of knee stability is limited to questionnaires and simple examination techniques, which often do not correlate well with functional outcome. Use of BSN in an efficient, low-cost gait analysis setting would facilitate high-value data collection which could be used to improve clinical practice through personalization of postoperative care and meaningful evaluation of surgical techniques.

16.3.4 Extreme Environments

When a healthy body is put under unusual stress, there may be obvious or not so obvious effects. Human physiology tends to adapt to meet external challenges, but often due to the extreme environment or circumstances we are not aware of exactly how the body is changing. Paradoxically, it is a time where human monitoring is vital yet very rarely carried out. Capturing and investigating said physiological coping mechanisms may give more insight into how the body is responding, what the threshold of human adaptation is, and at what stage the extreme conditions are causing irreversible damage and there becomes a necessity for medical support. Furthermore, answering these questions is undoubtedly important to permit optimal design specifications for specialized BSN in a variety of settings.

Extreme environments that the human body may encounter include elite sport, expeditions, and military pursuits. In all of these settings, peak physical and mental condition must be maintained throughout to support top-level performance. A subject's control of body temperature, water and electrolyte balance, and energy levels and avoidance of unnecessary external insults will determine their relative success. It is not unusual for dehydration of up to 8% body mass to occur during physical activity (Sharp, 2006) yet just a 5% loss is enough to cause a significant increase in heart rate, core temperature, and perceived exertion ratings, with concomitant decrease in blood volume, stroke volume, cardiac output, and skin blood flow (Coyle, 1998). Impaired cardiovascular and thermoregulatory function is likely to have a negative effect on physical and mental performance. However, verifying the above-mentioned parameters outside of a controlled environment is technically challenging.

Climbing Mount Everest is universally considered extreme, and it is one scenario in which a telemetric BSN platform was applied to monitor the health status of three mountaineers. A combination of wearable and ingested sensors were used to provide heart rate, three-lead ECG, accelerometry, global positioning system (GPS), surface body temperature, and core body temperature. Time and geostamped data collected by the user's local receiver were aggregated every minute and transmitted to a laptop computer at Everest Base Camp (EBC). Data from EBC was sent via satellite (Imarsat) using transmission control protocol/Internet protocol to Yale University School of Medicine, United States, rerouted from Malaysia. Vital signs and global positioning were monitored for 24 hr. Two of the three sensors functioned 100% of the time, with the other functioned at around 75%. Although the sensing only covered 24 hr, it is commendable that even through periods of intense and arduous activity at 17,800 ft, none of the subjects were ever without a full set of data for more than 25 min.

Another extreme setting in which improved physiological monitoring would be of great benefit is in field-based military personnel. First, the physical strain of noncontact operations in many battle climates (e.g., Middle East) is of a similar magnitude to that of many elite sporting activities. Second, to add the risk of injuries related to a war zone, one might argue that the soldiers should be monitored as if they were about to undergo an invasive procedure. As pervasive vital signs monitoring has already been discussed in this chapter, we have chosen an example of how manipulation of bioinformatics can be used to predict the extent of hemorrhage, which is the leading cause of death on the battlefield. The need for an accurate predictor of hemorrhage comes from the phenomenon observed whereby during the early compensatory phases of haemorrhage no clinical alterations occur in mental status, pulse character, blood pressure, oxygen saturation, respiratory rate, or blood pH. This results in patients who appear stable to suddenly and irreversibly "crash" (decompensate) due to massive blood loss.

In order to investigate the effects of mass hemorrhage in human subjects, a safe and realistic model was designed (Convertino et al., 2001). This involves applying negative pressure to the lower body, which simulates controlled hypovolemic hypotension by the redistribution of blood away from the head and heart. Lower body negative pressure (LBNP) was shown to accurately mimic hemodynamic, autonomic, respiratory, and metabolic compensatory responses when compared to the early stages of hemorrhage in anesthetized animal models and computational models (Summers et al., 2009). Using this model, feature extraction and machine learning methods were applied to recognize and predict hypovolemic cardiovascular collapse from complex, multi-input data sequences. Continuous data were collected from a pulse oximeter, transthoracic impedance plethysmography, a near-infrared tissue perfusion monitor, a transcranial Doppler flow probe, and a finger-worn infrared blood pressure monitor.

This provided respiratory rate, heart rate, arterial blood pressure, stroke volume, and cardiac output. In 28 subjects, LBNP was gradually increased to simulate a gradual reduction in central blood volume. The algorithm was 96.5% accurate in predicting the level of LBNP. When the data set was expanded using computer-generated model subjects, the refined algorithm was capable of tracking central blood volume reduction in real time at 95% accuracy. If military personnel were able to wear similar sensors, this method may be used to immediately, remotely, and continuously assess the physiological status of the subject, enabling efficient resource allocation and critical casualty prioritization in multitrauma scenarios.

16.4 OPPORTUNITIES AND FUTURE CHALLENGES

In this emerging field, there are many opportunities for sensor development and application, all with their own specific challenges and questions to explore. Although the relationship between technical and clinical development is entwined, the matters discussed, in this chapter come from a practical, clinical perspective. Readers from a variety of backgrounds should, at a minimum, have an awareness of the holistic approach to BSN and biotelemetry, as we all share the common goal of improving user outcome. In healthcare, this may be extended to improving patient outcome, increasing quality of life, and improving the socioeconomic efficacy of healthcare provision.

16.4.1 User Preferences

Biotelemetry projects, as with all product developments, are often approached from two different angles: top down, in which a clinical problem is identified and a system is developed to solve the problem, or bottom up, when an existing technology is applied to various clinical problems in an attempt to solve them. Of course, a mixed approach may be employed. It should be noted, as stated at the beginning of the chapter, that BSN are not specific systems for solving clinical problems (or otherwise); rather an the architectural backbone that may be manipulated and applied to various scenarios. Whichever approach is taken, recording and considering user preferences are fundamental in ensuring successful uptake of the technology, optimal data collection, efficient and meaningful data processing, and consequently improvements in patient outcomes and cost benefit. In a bottom-up scenario, if the sensing apparatus cannot be modified, the study or usage protocol and the manner in which it is introduced to the user should be appropriately designed.

The user is not only the subject wearing the device but also the healthcare professional prescribing it as well as any supporting staff who may be involved in data analysis and further management. The user preferences of patients and their doctors do not always match. This was highlighted by a recent systematic review by, Bergmann and McGregor (2011), who reported that patients preferred small and compact sensors that are easy to operate, do not affect normal daily activities, and reduce the required travel to clinics and hospitals, whereas clinicians want sensors that have a real-time function, are capable of adapting dynamically to different situations, and are focused in assisting patient training, education, and self-management. Additionally, it should be noted that in most cultures patients have a great deal of respect for their doctor, leading to two important points: first, if the doctors trust the device, it is likely that they will be able to convince the patient to wear it, and, second, patients do not want to replace their doctor with a sensor and so it may be prudent to describe biotelemetry as a communication portal from the patient to the doctor. Finally, if the sensor is well received by all users, it is likely to have a positive effect on patient outcome, thus improving commercial viability.

"Disappearance" is a concept of increasing popularity in the BSN community, referring to the level of pervasiveness of a sensing system. Miniaturization is not the only method of reducing the awareness of sensor usage; furthermore, size restrictions often impact negatively on the functionality of the device. People wear a variety of garments and accessories for fashion and functional purposes. It is not farfetched to foresee one or more sensors disappearing within the daily wearable apparel. In fact, this is already occurring. For example, many sports watches function as heart rate monitors, and training shoes may be fitted with off-the-shelf GPS devices and accelerometers for exercise monitoring. There must be a balance however between disappearance and functionality. Most users consider the burden–benefit ratio, whether or not this is verbalized. For instance, should a patient suffer from sporadic cardiac arrhythmia, they are likely to prefer a slightly larger, more apparent device that was highly effective at alerting themselves and appropriate healthcare professionals when a dangerous rhythm is detected, rather than a less effective miniaturized sensor disguised as a bracelet. Conversely, if a device monitors food intake or everyday health parameters, the health benefits are perceptually smaller and this must be offset with a more discrete sensor design. Possibly the increasing feasibility of implantable or ingestible sensors will indisputably swing the balance of modern BSN implementation in a favorable direction (Figure 16.3).

Although few studies have assessed the ability for BSN to improve patient outcome, the theory is closely linked to behavioral economics. In other words, how can BSN change the way a subject behaves or feels in order to improve their outcome. Considering the concept of competition, for example if a group of school children were taught the benefits of exercise and were then fitted with activity monitors, it is expected that the children would exercise more in an attempt to score the highest and "win" the competition. Or similarly with motivation, if someone is given a set of exercises to perform every day as part of a rehabilitation program, they are more likely to complete this if they have the extrinsic motivation of being supervised by their physiotherapist or doctor or even the immediate feedback from a motion sensor. Therefore the sensor's ability to direct behavior must be evaluated.

16.4.2 Clinical Translation

Implementation of healthcare technologies is largely dependent on cost. More specifically, a cost–utility analysis must be performed in order to provide a monetary value on the impact of the technology or intervention on a patient. This is often calculated using quality-adjusted life years (QALYs), which are decided by multiplying the resultant quality-of-life change by the number of years where that quality of life will

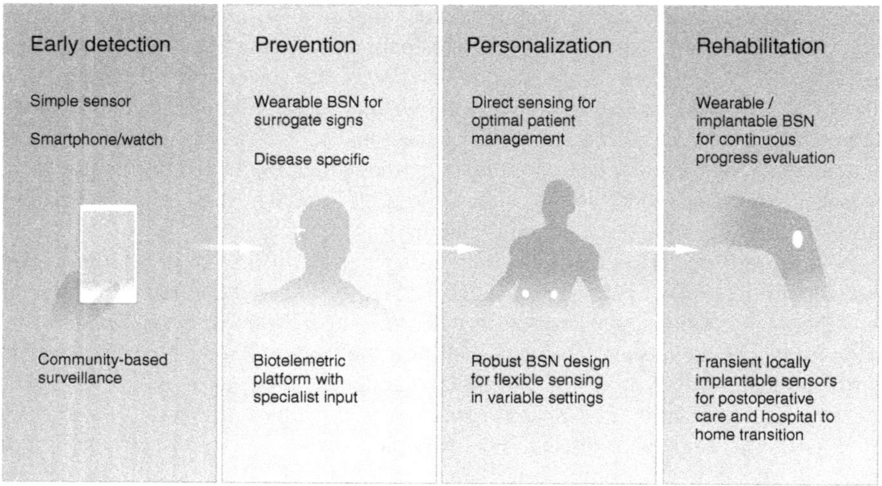

Figure 16.3 Examples of BSN in healthcare pathway. In all user scenarios, the BSN must have a positive benefit–demand ratio. A healthy user requires appropriately pervasive monitoring, preferably through a device already carried. Implementation of a BSN with higher user demand is more acceptable in high-risk subjects as their potential benefit is likely to be greater. (*See insert for color representation of the figure.*)

be experienced. It has been suggested that in the UK if the technology does not gain 1 QALY for less than around £30,000, it is unlikely that the National Health Service (NHS) will implement it. For example, if the artificial pancreas (continuous subcutaneous insulin infusion pump) improved quality of life by 20% (0.2) and increased life expectancy by five years ($0.2 \times 5.0 = 1.0$ QALY), it would be implemented should it cost £30,000 or less in total. However, should the implementation of a device provide an essential step in refining another aspect of healthcare or reduce the number of clinic appointments in a patient group, it may also be adopted.

To maximize the chances of meeting QALY targets, one must consider the scalability and versatility of the device. Many mobile phone healthcare applications are used which are highly scalable and low cost due to downloadable software and versatile as different software can be downloaded for different applications. Moreover, this software can gain data from any combination of inbuilt, or additional, phone sensors. BSN nodes may be considered as more specialized devices, but a simple user interface may allow convenient, wireless programming to facilitate a variety of functions. For example, a patient may wear a small number of BSN nodes during the day for general activity and vital signs monitoring, but during their rehabilitation exercises reprogram the nodes to start high-frequency data collection for more detailed biomechnical analysis.

16.4.3 Practical Considerations

There are still many unknowns in biotelemetry. Important questions exist including who will benefit the most from these technologies, when they need to be implemented, what needs to be measured, how much information is needed, and what will be the optimum balance between sensor and clinician input. Answering these questions may be specific to each study, or technology, or clinical problem, but as more high-quality trials are carried out, we expect a pattern to emerge which is likely to provide some direction to the development of the field in general. Regarding sensor inputs, several groups are investigating the importance of sensor placement to provide useful user metrics as well as exploring the possibilities of reducing the sensor number and using mathematical models to reconstruct missing data (Lai et al., 2011, Atallah et al., 2010, Zhang et al., 2013) (Table 16.2).

On a similar line of questioning, what happens to the data when it is collected is up for debate. On the postoperative ward, for example, if a patient is wearing a BSN allowing continuous collection of vital signs, how often should the data be sent to the patient files? Current patient management is based on four hourly observations, so what benefit would result from reducing that time interval to once hourly or every minute? Potentially, vast amounts of physiological data postoperatively could be analyzed to give an early predictor of recovery or complications, but it is also possible that a problem could arise where there is too much data, causing handling difficulties with little or unknown benefit. Continuing with the hospital theme, the data have many potential recipients. First, it is important for an appropriate amount of data to be stored in the patient's medical records. Next, it might be useful for the medical team, including the nurses, to have a continuous flow of vital signs data from each patient to monitor their condition. This could be accessed on a tablet computer on the ward round, perhaps. Also, the status of the patient might be of interest to the family, in a more general sense, perhaps with a graphical representation or description of their physical condition.

Collecting and transmitting large volumes of data also creates a security challenge. On the matter of patient confidentiality, the U.K. Data Protection Act 1998 describes clearly how data must be handled in order to preserve one's privacy and human rights. Part of this legislation includes that data should be used for the specific purposes for which it was collected. As such, detailed data from sensing devices such as global positioning, motion, and vital signs must be treated sensitively. In addition, patients must consent to wearing these devices, and they should be informed of the intended use of the data collected. On-node data processing to derive physiological measures before transmitting might make it more acceptable, ethically and legally as there are clear outcomes with predetermined uses.

Healthcare technology innovation and adoption can be unpredictable. A careful balance must be found between providing supplementary patient information to healthcare staff and overloading them (Wilson, 2006). As mentioned previously, devices such as the Holter monitor are able to provide clinicians with continuous, ambulatory data but analyzing the data is time consuming. Implementing autonomic data processing techniques is an essential step toward producing a BSN that is feasible for use outside the research environment. As a matter of safety, this is best limited to highlighting areas of data needing expert review. Alternatively, early pattern recognition suggesting an immediate health risk may prompt further

TABLE 16.2 Technical and Clinical Considerations for BSN in Healthcare

	Technical Considerations
Nodes	Limited number[a] and size of nodes, miniaturized or hidden on user (e.g., bracelet); biocompatibility (and biodisposability in implantable sensors) essential to reduce patient adverse reactions
Software	Flexible, customizable, and lightweight software infrastructure required to cater for heterogeneous sensors, diverse data formats, application-specific metadata requirements and varied end-user visualization platforms; options for on-node data processing, designed to increase battery life without compromising data output quality; remote reprogramming
Fault tolerance	System should be robust against local failure, including hardware failure and software crashes; data normalization; audit trails to ensure efficient system recovery
Dynamics	Changing environment requires need for context awareness, possibly integrating ambient wireless sensor networks; motion artifact creates a challenge and should be actively addressed
Power	Miniaturized sensor creates battery limitations, energy scavenging potential from body heat and motion; reduced accessibility to implantable sensors
Data handling	Options for on-node data storage, otherwise low-power communications needed; data transfer between multiple nodes must be optimized to reduce packet loss and sensor drift; real-time data interrogation; flexible connectivity allowing LPU changeover in different locations
Incremental processing	Process and update statistics from sensor data online to increase abstraction and compression, enhancing scalability of the system
Security	High-level security required to maintain data integrity and authentication; ensuring patient information privacy by giving users the ability to control the information collected from them
Medical device	Tested to ensure safety for medical application; consider interaction with other medical equipment such as bandages, splints, catheters, blood pressure cuffs, and wireless interference of life support machines and other electronics
Electronic medical records (EMRs)	Consider integration of BSN data with national EMR system, facilitating data access to appropriate medical personnel through conventional means

(*continued*)

TABLE 16.2 (*Continued*)

	Clinical Considerations
Problem	Clinical problem must be identified and defined; theorize how BSN can explain or solve the problem
Population	Device choice based on user characteristics; e.g., elderly person may be unable to change small battery; patient location considered (hospital, care home, etc.); decide best method for delivering feedback, receiving used sensors; appropriately matched healthy comparison group may be required
Device	Sensor input, data sampling rate and battery life must satisfy trial requirements; if the sensor data are going to be analyzed alongside other clinical data, time-stamp calibration may be required
Study design	Observational studies may provide quantitative insight into a clinical problem, but to show value of BSN in improving patient outcome or the cost effectiveness of a service, a randomized controlled trial will often be required
Activity protocol	Is there an activity protocol for the patient to follow or is natural behavior being recorded? Does the data require labeling? Manual labeling requires easily recognizable signal events.
Time scale	What duration of data collection will provide an accurate representation of patient physiology? For activities of daily living, two weekdays and one weekend day is a sufficient duration. Does the sensor operate effectively over the chosen time schedule without replacement? Is longitudinal data collection needed?
Ethics	When working with healthy or sick subjects, performing simple or high-risk activities usually requires ethical approval from local and national committees; begin this process well in advance of study
Data analysis	Data often must undergo multiple layers of processing before formal analysis; what outcome measures are required and can the analysis be streamlined; expert assistance may be needed
Data output	Output must be suited to the user in terms of the detail and extent of information shown, the format (e.g., scatter plot, table, bar chart), and the available further options such as seeking help, contacting a patient, or conducting further data analysis
User interface and preferences	Bespoke packages for specific applications, a suitable user interface and phone/online technical support improve acceptability of BSN to patients and clinicians

[a]For external body sensors only; for implantable sensors the size of the sensor network may become extremely large.

investigations such as a phone call to the patient and possibly asking the patient to complete a short test such as home blood pressure monitoring or pulse oximetry.

16.4.4 Personalization

Although EBM is largely practiced worldwide, it is of utmost importance to treat every case on its own individual merits. An algorithm for treating a particular medical condition, however complex, must be used only as an adjunct to clinical judgment and also the patient's own wishes. Utilizing the qualities of a BSN in healthcare can allow more personalized care provision. The more data captured from a patient, from a variety of sensor inputs, the more the clinician knows about that patient. This includes their physiological character, their response to pathology, and their tolerance or reaction to pharmacological or physical interventions. Continuous, pervasive data collection throughout diagnosis and treatment provides a more reliable indication of health status, which in some circumstances can allow closing the loop through personalized medication dose titration, feedback including encouragement and motivation, and prompting patients to seek further medical care when necessary, rather than at arbitrarily timed follow-up appointments.

Clinical application of BSN should follow the forementioned practical concepts using the technological advances detailed in previous chapters. Sensor inputs must be chosen carefully, considering the role of the BSN. Gaining data from the environment to provide context awareness may be necessary to allow meaningful analysis of human data. Quality of service with regard to sensor communications should be a high priority. The complexity of the human body in its natural environment creates communication challenges, which often lead to unwanted data packet loss and reduced data quality. Sensor embodiment design, including material choices, must also be taken into account to optimize communications. In addition, a flexible radio protocol that allows secure communication between multiple devices such as mobile phones, other sensors, and hospital computer systems should be pursued. The sensor battery life should be maximized through intelligent conservation methods, that is, when data need not be collected or sent (or when data sampling rate can be reduced), the device enters a low-power mode. A BSN must be designed with usability in mind, including the needs of both the patient and the clinician. Sensors where possible should be dynamic and scalable. The ability to reprogram a sensor node for a different function rather than need to buy or wear another will improve usability and cost-effectiveness. Studies must be completed to prove the cost-effectiveness of the BSN, taking into account the limitations of an already stretched healthcare system.

16.4.5 Future

Although the potential for BSN to positively influence healthcare provision has been shown, the frequent reliance on surrogate signs rather than direct sensing creates an aspect of uncertainty. For example, it has been suggested that monitoring patient activity levels postoperatively may allow early detection of complications such as

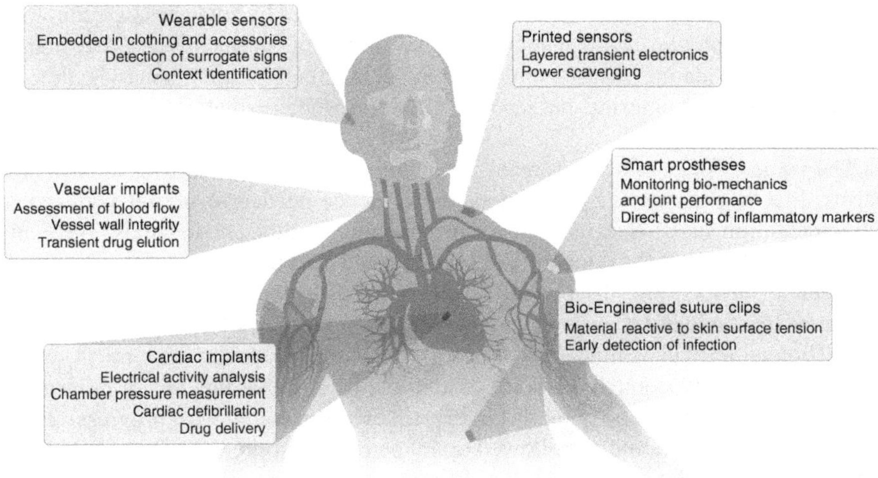

Figure 16.4 Future BSN apparatus: smart sensor placement and integration with surgical intervention to facilitate reliable, unrestricted, personalized health surveillance and dynamic interventions.

infection or deep-vein thrombosis (DVT) (Atallah et al., 2009). This theory relies on the correlation between reduced activity, or a change in gait, with the presence of aforementioned complications. A more robust approach would be to implant a device during the operation to directly assess the surgical site postoperatively for features such as inflammatory markers.

Although several healthcare settings are candidates for implantable sensors, postoperative application is a prime target as the skin has already been breached and a potential implant site accessed. Minimally invasive surgical techniques are reducing the amount of iatrogenic trauma sustained in the operating theatre, resulting in a drastic increase in rate of convalescence. Although patients are often discharged much earlier, it must be recognized that postoperative care should be augmented, not withdrawn. Smaller incisions reduce the need for inpatient care with frequent dressing changes and physiological support, but a similar internal operation is likely to be conducted and complications may still occur. If the patients recover at home, it is crucial that complications are detected promptly to alert the healthcare team and arrange readmission. Clinician-directed technical developments in BSN are essential to allow the realization of effective, transient biochemical sensing (Figure 16.4).

16.5 CONCLUSION

Health technologies such as BSN have the potential to improve patient outcomes, the efficiency of the service, and our understanding of medical conditions. In addition, they may have a positive effect on society as a whole by promoting healthy living as a disease prevention strategy, including psychosocial support.

There are clear changes in healthcare demands initiated by advances in technology and an aging population with an increasing prevalence of chronic disease. Healthcare services worldwide are unsustainable, and as such, a focused approach to healthcare reform through restructuring and resource-effective implementation of technology is required.

The value of BSN in healthcare has been demonstrated in numerous clinical scenarios. It is now a matter of providing the evidence needed to permit widespread implementation of BSN as an answer to changing healthcare demands. BSN and biotelemetry facilitate the relocation of several health services to the community environment. In this setting, the patient is able to take responsibility for their disease management, knowing that medical assistance is readily available should the need arise. Biotelemetry in healthcare is also bridging the gap between the clinical setting and the home. Postoperative patients should feel safe in that they are appropriately monitored upon prompt hospital discharge. Likewise, patients with progressive disorders, such as PD, should be followed up by periodic symptom assessment in order to maintain a most favorable quality of life.

Through appropriate resource allocation, personalization of care, and advanced bioinformatics for health research, BSN are well poised to meet current and future healthcare challenges.

ACKNOWLEDGMENT

Many thanks to Dr. Robert Merrifield for the design and production of the figures.

REFERENCES

Atallah L, Aziz, O, Lo B, Yang G-Z. 2009. Detecting walking gait impairment with an ear-worn sensor. Body Sensor Networks Conference Proceedings, pp. 177–182.

Atallah L, Lo B, King R, Yang GZ. 2010. Sensor placement for activity detection using wearable accelerometers. Proc of IEEE International Conference on Body Sensor Networks.

Barthelemy JC, Feasson-Gerard S, Garnier P, Gaspoz JM, DA Costa A, Michel D, Roche F. 2003. Automatic cardiac event recorders reveal paroxysmal atrial fibrillation after unexplained strokes or transient ischemic attacks. *Ann. Noninvasive. Electrocardiol.*, 8:194–199.

Bergmann JH, McGregor AH. 2011. Body-worn sensor design: What do patients and clinicians want? *Ann. Biomed. Eng.*, 39:2299–2312.

Cavanaugh JT, Ellis TD, Earhart GM, Ford MP, Foreman KB, Dibble LE. 2012. Capturing ambulatory activity decline in Parkinson's disease. *J. Neurol. Phys. Ther.*, 36:51–57.

Chen BR, Patel S, Buckley T, Rednic R, McClure DJ, Shih L, Tarsy D, Welsh M, Bonato P. 2011. A web-based system for home monitoring of patients with Parkinson's disease using wearable sensors. *IEEE Trans. Biomed. Eng.*, 58:831–836.

Convertino VA. 2001. Lower body negative pressure as a tool for research in aerospace physiology and military medicine. *J. Gravit. Physiol.*, 8:1–14.

Convertino VA, Moulton SL, Grudic GZ, Rickards CA, Hinojosa-Laborde C, Gerhardt RT, Blackbourne LH, Ryan KL. 2011. Use of advanced machine-learning techniques for non-invasive monitoring of hemorrhage. *J. Trauma*, 71:S25–32.

Coyle EF. 1998. Cardiovascular drift during prolonged exercise and the effects of dehydration. *Int. J. Sports Med.*, 19 (Suppl 2): S121–124.

Denning T, Andrew A, Chaudhri R, Hartung C, Lester J, Borriello G, Duncan G. 2009. BALANCE: Towards a usable pervasive wellness application with accurate activity inference. Proc IEEE Workshop Mob. Comput. Syst. Appl., 2009, p. 5.

Dolan P, Hallsworth M, Halpern, K D, Vlaev I. 2010. MINDSPACE—Influencing behaviour through public policy. Available: http://www.instituteforgovernment.org.uk.

Kamel H, Smith WS. 2011. Detection of atrial fibrillation and secondary stroke prevention using telemetry and ambulatory cardiac monitoring. *Curr. Atheroscler. Rep.*, 13:338–343.

Klasnja P, Consolvo S, McDonald DW, Landay JA, Pratt W. 2009. Using mobile & personal sensing technologies to support health behavior change in everyday life: lessons learned. *AMIA Annu. Symp. Proc.*, 2009, pp. 338–342.

Lai Y-X, SHU L, Vasilakos A, Rodrigues JJPC, LAI C-F, Huang, Y-M. 2011. 3D Adaptive reconstruction of human motion from multi-sensors. Proc. of IEEE 3rd International Workshop on Wireless Sensor, Actuator and Robot Networks, pp. 531–536.

Lee W, Chae YM, Kim S, Ho SH, Choi I. 2010. Evaluation of a mobile phone-based diet game for weight control. *J. Telemed. Telecare*, 16:270–275.

Lopez G, Custodio V, Moreno JI. 2010. LOBIN: E-textile and wireless-sensor-network-based platform for healthcare monitoring in future hospital environments. *IEEE Trans. Inf. Technol. Biomed.*, 14:1446–1458.

Pickup JC. 2012. Management of diabetes mellitus: Is the pump mightier than the pen? *Nat. Rev. Endocrinol*, 100:1628–36.

Saxby R. 2007. How silicon will transform healthcare. 3rd IET Int. Conf. on Medical Electrical Devices & Technology, London, 2–3 Oct. 2007.

Schofield I, Heath H. 1999. *Healthy ageing: Nursing older people*. Mosby, UK.

Sharp RL. 2006. Role of sodium in fluid homeostasis with exercise. *J. Am. Coll. Nutr.*, 25:231S–239S.

Struthers R, Erasmus P, Holmes K, Warman P, Collingwood A, Sneyd JR. 2008. Assessing fitness for surgery: A comparison of questionnaire, incremental shuttle walk, and cardiopulmonary exercise testing in general surgical patients. *Br. J. Anaesth.*, 101:774–780.

Summers RL, Ward KR, Witten T, Convertino VA, Ryan KL, Coleman TG, Hester RL. 2009. Validation of a computational platform for the analysis of the physiologic mechanisms of a human experimental model of hemorrhage. *Resuscitation*, 80:1405–1410.

Tayal AH, Tian M, Kelly KM, Jones SC, Wright DG, Singh D, Jarouse J, Brillman J, Murali S, Gupta R. 2008. Atrial fibrillation detected by mobile cardiac outpatient telemetry in cryptogenic TIA or stroke. *Neurology*, 71:1696–1701.

Tsivgoulis SD, Tzagarakis GN, Papagelopoulos PJ, Koulalis D, Sakellariou VI, Kampanis NA, Chlouverakis GI, Alpantaki KI, Nikolaou PK, Katonis PG. 2011. Pre-operative versus post-operative gait variability in patients with acute anterior cruciate ligament deficiency. *J. Int. Med. Res.*, 39:580–593.

Wallmann D, Tuller D, Wustmann K, Meier P, Isenegger J, Arnold M, Mattle HP, Delacretaz E. 2007. Frequent atrial premature beats predict paroxysmal atrial fibrillation in stroke patients: an opportunity for a new diagnostic strategy. *Stroke*, 38:2292–2294.

Wilson CB. 2006. Adoption of new surgical technology. *BMJ*, 332:112–114.

Yang, G-Z. 2006. *Body Sensor Networks*. New York: Springer.

Zhang Z, Wong C, Yang GZ. 2013. Forearm functional movements recognition using spare channel surface electromyography. Proc. of IEEE International Conference on Body Sensor Networks, pp. 1–6.

Ziegler PD, Glotzer TV, Daoud EG, Wyse DG, Singer DE, Ezekowitz MD, Koehler JL, Hilker CE. 2010. Incidence of newly detected atrial arrhythmias via implantable devices in patients with a history of thromboembolic events. *Stroke*, 41:256–260.

17

WEARABLE HEALTH CARE SYSTEM PARADIGM

YANG HAO AND ROBERT FOSTER
Queen Mary University of London, London, United Kingdom

17.1 INTRODUCTION

Future health care is expected to manage a large increase in chronic diseases with limited resources, caused in part by population growth and population aging. Globally, there has been an increase in the size of the elderly population since the middle of the twentieth century (Fass, 2007); in addition, the incidence of chronic diseases, including cardiac disease and diabetes, has increased during the same period, caused by changes in lifestyle (such as the increase in obesity levels), particularly in the West (Saxby, 2007). This is placing greater pressure on the health care resources of countries around the world (Feied et al., 2006) and is expected only to get worse, as the demographic trends are likely to continue for the foreseeable future.

However, there has also been a tremendous growth in information and communication technology, with growth in wireless technologies particularly notable over the past decades or so. These changes have enabled new services, as well as providing new ways of delivering services. Mobile health care technology promises to enhance the quality of life for all people, both healthy and ill, while simultaneously reducing costs in health care provision (Department of Health, 2007).

Wearable physiological measurement systems have many potential applications, in areas such as medicine, sports, and security. Health applications, whether civilian or military, are arguably the strongest drivers for the development of these devices.

Handbook of Biomedical Telemetry, First Edition. Edited by Konstantina S. Nikita.

Section 17.2 provides some background for the rest of the chapter, focusing on the motivation for wearable wireless technology in health care and the challenges that emerge; much of this has already been covered in more depth in previous chapters, but it is included here for completeness. Section 17.3 discusses the methods available for wearable wireless devices, with an emphasis on available wireless communication protocols. Some example applications are discussed in Section 17.4, with such things as the implications on node locations, radio channel performance and antennas considered. Finally, a summary of the chapter appears in Section 17.5. Some of this work first appeared in Hao and Foster (2008); this is a revised and updated version.

17.2 WIRELESS WEARABLE TECHNOLOGY IN HEALTH CARE

A WPMS can be defined as a system using a wireless medium to carry (potentially real-time) physiological measurement data from wearable (or implantable) medical sensors to a central processing unit. The primary purpose of the WPMS is to improve the quality and efficiency of health care (GE Healthcare, 2007). Wireless physiological measurements hold a number of advantages over wired alternatives, including (Townsend et al., 2005):

- Ease of use
- Reduced risk of infection
- Reduced risk of failure
- Reduced user discomfort
- Enhanced mobility
- Lower cost of care delivery

Wireless physiological measurements have not only found applications in health care, but have also been applied in the areas of the military, security, sports, and fitness monitoring.

A number of functions fulfilled by WPMSs can be identified for the improvement of health care delivery services. The first function of the WPMS is to alert its wearer of the approach or development of a potential medical emergency, so that precautionary action can be taken. Moreover, the use of wireless physiological measurement systems outside the hospital could also help reduce overall health care costs, especially among patients with chronic diseases capable of receiving care at home. A good example of this is to alert a patient with chronic cardiac diseases when an increase in blood pressure or abnormal electrocardiography (ECG) measurement occurs, potentially allowing him to seek assistance before the onset of a cardiac event.

Although multiparameter patient monitoring systems are available today, they operate in a cost-inefficient way, as a high-cost monitoring unit is assigned to each patient. Moreover, established patient monitoring solutions typically involve a web of cables connecting the on-body sensors to nearby monitoring equipment. This system can restrict the patient's mobility and comfort and can dangerously result in unattached sensors when the tangle of cables causes the patient, either intentionally or accidentally, to become disconnected (GE Healthcare, 2007). Much work has been invested in the development of wireless versions of patient monitoring systems, but relatively few have reached commercial availability, most of which may still be bulky and not well integrated into the patients' daily lives.

The second function of a wireless physiological measurement system is to alert the medical emergency system if vital signs drop below a certain threshold. In this scenario, the exact location of the patient needs to be transmitted, along with any useful medical information that could assist the emergency team. For example, the U.K.'s Office for National Statistics (ONS) reported that (Hao and Foster, 2008) "heart disease (including heart attacks) was the leading cause of death for both sexes in England and Wales in 2005, accounting for one in five male deaths and around one in six female." In the United States, "each year, about 1.1 million Americans suffer a heart attack. About 460,000 of those heart attacks are fatal. About half of those deaths occur within 1 hour of the start of symptoms and before the person reaches the hospital" (Hao and Foster, 2008). It seems reasonable, therefore, to assume that the use of wireless physiological measurement systems could help save lives, in that they could detect and warn for early symptoms of impending cardiac (or other) problems, enabling the patient to receive potentially life-saving treatment sooner.

The third function of a wireless physiological measurement system is to measure a real-time biosignal for local processing. A good example is an automatic internal cardiac defibrillator (AICD, also known as an implantable cardioverter defibrillator, or ICD), which aims to restore the regular heart rhythm by delivering an electrical shock if abnormal behavior is detected, potentially averting sudden cardiac death. Another example is an implantable drug delivery system: this delivers medication more efficiently for such applications as chemotherapy, pain management, diabetic insulin delivery, and AIDS therapy, by locally processing wireless physiological measurements (Jones et al., 2006). The wireless physiological measurement system can also provide real-time biosignal information for postprocessing.

In the military, a WPMS implementation can facilitate remote noninvasive monitoring of vital signs of soldiers during training exercises and combat. For example, it can be used to remotely determine a casualty's condition by medics in combat, without exposing first-responders to increased risks; or to quickly identify the severity of injuries and continuously track the injured condition until the casualty arrives safely at a medical care facility. In safety and emergency scenarios, wireless physiological measurement information can also be useful for firefighters, hazardous material workers, mountain climbers, or emergency first-responders operating in harsh and hazardous environments. It can keep track of an injured person's vital signs, allowing rapid distribution of the information to medical providers and assisting emergency responders in making critical, and often life-saving, decisions in order to expedite rescue operations (Mendelson et al., 2006). Wireless physiological measurement systems can be classified into critical and non-critical applications, as shown in Table 17.1.

It has been recognized that "the health care industry has reached the beginning of a wave of breakthroughs in providing care and rehabilitation that will use radio communication technologies in a variety of ways" (Hao and Foster, 2008).

TABLE 17.1 Applications of Wireless Physiological Measurement Systems and Their Classification

Critical Monitoring	Noncritical Monitoring
Monitoring chronically ill patients with heart disease, diabetes, epileptic (those that need alarms)	Monitoring of physical conditions and efficiency of sport athlete during exercises
Monitoring at home and nursing home for the elderly and patients with dementia	Control and feedback during athlete training
Monitoring vital signs of soldiers in battle	Crime investigation with wireless lie detectors
Vehicles such as the ambulance when transporting patients	Monitoring the consciousness of drivers, pilots, and operators of heavy machinery
Medical research team can carry out unobtrusive patient study and clinical field trials over an extensive period	In the hospital to reduce discomfort and restriction of wires
Remote telemedicine	Monitoring employees to identify those that are engaging in unlawful activities

With recent advancements in wireless technology, it is possible to use wireless technology in the delivery of health care services as never before. One of the most promising concepts on the horizon is that of the wireless body sensor network (WBSN), a sub-type of the wireless body area network (WBAN). WBSNs typically consist of multiple sensor nodes worn on the body, each capable of sampling, processing, and communicating wirelessly one or more physiological measurements or environmental parameters (Otto et al., 2006). These physiological parameters are typically blood glucose and oxygen levels, pulse rate, blood pressure, circadian rhythm, and wake–sleep patterns through actigraphy. Antenna components embedded in the sensor nodes make it possible for the data generated to be transmitted wirelessly to a body-worn or closely located hub device, eliminating the need for cables. The hub device, in turn, receives the data generated from the various sensor nodes on the body and may process the data locally, or transmit it wirelessly via an appropriate radio link for centralized processing, display, and storage, or both (GE Healthcare, 2007).

Another benefit of WBSNs is the greater spectrum efficiency that can be achieved. Traditional telemetry systems generally create a separate radio frequency (RF) link to the remote monitoring station for each patient sensor. WBSNs utilize network bandwidth more efficiently, by creating a very short-range point-to-point wireless network, thereby enabling greater spectrum reuse. A common approach in the past was to use a mesh topology for WSBNs, primarily due to the common use of the IEEE 802.15.4 standard (see Section 17.3). This means that there are typically many links between nodes and each node chooses the best path through the network to the destination. Data transfer thus occurs in multiple "hops" between nodes.

Mesh networks are generally self-forming: each node builds up a routing table of the network. This may, however, be calculated by a network controller node in some cases. One advantage of self-forming is that limited configuration is required, a particular need in environmental sensing over relatively wide geographic areas (e.g., in military applications) or if the nodes are mobile, making the network topology dynamic. Mesh networks are also relatively fault tolerant, as the availability of multiple paths enables the network to handle individual node failures gracefully. However, it has been recognised that, in general, WBANs will have a fixed architecture and, as a result, they may efficiently use a star or tree topology. Hence, the IEEE 802.15.6 standard for WBANs does not include mesh capability; it does, however, include provision for a two-hop star topology in most of its operating bands.

Wireless sensor nodes, also known as motes, are tiny microcomputers capable of doing small amounts of processing, sensing and using wireless media to communicate with other wireless sensor nodes in the sensor network. The motes combine processing, sensing, and wireless networking into a single tiny package. The main components of a wireless sensor node are a radio transceiver with antenna for wireless communication, a microprocessor, memory, analog or biochemical sensors (e.g., ECG, temperature, glucose), and a battery for electrical power. These motes typically use an operating system with a compact footprint, in terms of memory requirements and system overhead; a commonly used example in the literature is the open-source TinyOS software, coupled with the nesC programming language. Each wireless sensor node is strategically placed on the human body as a tiny patch or implant, or hidden in users' cloths, allowing ubiquitous physiological measurements in the natural environment over an extended period of time.

To permit prolonged ubiquitous monitoring and seamless integration into a WBSN, wireless sensor nodes have to meet requirements, such as being low profile, lightweight, and low cost and having low power consumption and a high degree of integration and packaging with the sensor. Consideration should be given to wireless powering methods, such as inductive, capacitive, ultrasonic, and light. Wireless sensors of the future may need energy harvesting methods like vibration (such as piezoelectric) or temperature gradient (such as thermopile), or use an alternative power supply from body fluid, such as glucose (Lekkala, 2006).

17.3 METHODS AND DESIGN APPROACH FOR WIRELESS WEARABLE SYSTEMS

17.3.1 Design Goal and Considerations

A number of parameters need to be considered when designing a miniature wireless sensor device. The application sets constraints and requirements for the device. The volume, shape and weight are important factors, particularly for wearable sensors. For a given application, aimed at wireless physiological measurements, operational lifetime, duty cycle, accuracy, and stability of sensor output need to be considered. The most challenging scenario is 24-hr, round-the-clock monitoring, where even the incorporation of sensors into clothing may fall short of the goal, making implanted or skin-mounted devices necessary. Implantable devices have a level of invasiveness exceeding current mainstream acceptance. However, the necessity of skin-worn devices or skin-contacting devices is unavoidable in some cases, such as galvanic skin response. Devices embedded into clothing are certainly champions of unobtrusiveness, as long as their weight, shape, and volume stay within limits (Ouwerkerk et al., 2007). To meet the design requirements, an investigation into various applicable wireless technologies must be performed.

17.3.2 Wireless Technologies Available for Wearable Systems

Wireless communication within a WBSN may be based on infrared, light, microwave radio, and even near-field coupling through skin conductivity. A number of protocols exist that have been used in networks of these types. A protocol, for the purposes of this discussion, is a set of rules that define how devices exchange information. It is usually broken into seven layers, to simplify discussions (Leon-Garcia and Widjaja, 2004). Microwave radio communication, especially using ZigBee wireless technology, is a popular approach. WBSNs may communicate externally with other networks (which may themselves be WBSNs) using one of the many available wireless technologies. For short-, medium-, and long-range communications, Bluetooth, ultrawide-band, wireless local area network (LAN) (WiFi), WiMAX, Global Systems for Mobile Communications (GSM), General Packet Radio Service (GPRS), Universal Mobile Telecommunications System (UMTS), and satellite communication are available, allowing a wide coverage area and offering the possibility of ubiquitous worldwide wireless mobility (GE Healthcare, 2007). Some of these technologies will be discussed in the following sections, in the context of their applicability to WBSNs.

Bluetooth and Bluetooth Low Energy One of the most recent versions of Bluetooth wireless technology is Bluetooth low energy (BLE), also called Bluetooth version 4.0 BLE and now marketed as Bluetooth Smart. It introduces a new communication method, which is optimized for small packet sizes, integrating low-energy and low-latency features to the Bluetooth core specification. It targets short-range communications for devices that do not need streaming data or high data throughput; thus, battery-operated sensor nodes could last for months or even years (Bluetooth Special Interest Group, 2010).

Although the BLE specification uses a frequency hopping transceiver, as in the core Bluetooth specification, in order to combat interference and fading [i.e., many frequency-hopping spread-spectrum (FHSS) carriers are available, albeit fewer than in the standard modes], the lack of direct-sequence spread spectrum (DSSS) on the modulation scheme results in a radio lacking in robustness and vulnerable to jamming and interference; thus, mesh network topologies are unavailable. Table 17.2 highlights the main differences of Bluetooth wireless communication core specification and BLE.

The new technological features of BLE (e.g., for infrequent sending of small data packets to a mobile phone and/or to a web service) are creating new opportunities for developers, manufacturers, and applications, bringing to life entirely new markets for devices that are low-cost and operate with low-power wireless connectivity.

TABLE 17.2 Comparison of Bluetooth (Core Specification) and BLE

Technical Specification	Core Bluetooth Technology	BLE Technology
Distance/range	100 m	50 m
Over the air data rate	1–3 Mbps	1 Mbps
Application throughput	0.7–2.1 Mbps	0.26 Mbps
Active slaves	7	Not defined; implementation dependent
Security	56/128-bit and application layer user defined	128-bit AES with counter mode CBC-MAC and application layer user defined
Robustness	Adaptive fast frequency hopping, FEC, fast ACK	Adaptive frequency hopping, 24-bit CRC, 32-bit message integrity check
Latency	Typically 100 msec	6 ms
Total time to send data	100 msec	6 ms, <3 ms
Voice capable	Yes	No
Network topology	Piconet, Scatternet	Star-bus
Power consumption	1 as the reference	0.01–0.5 (depending on use case)
Peak current consumption	<30 mA	<20 mA (max. 15 mA to run on coin cell battery)
Service discovery	Yes	Yes

Note: AES = advanced encryption standard, CBC-MAC = cipher block chaining media access layer, FEC = forward error correction, CRC = cyclic redundancy check.

IEEE 802.15.4, ZigBee and Related Standards IEEE 802.15.4 is a popular low-power, low-data-rate standard that defines the physical (PHY) and media access (MAC) layers of the communications protocol. The physical layer operates in a number of unlicensed RF bands, the main ones being at 2.4 GHz (global), 915 MHz (America), and 868 MHz (Europe). The IEEE 802.15.4 physical layer uses spreading techniques to minimize data loss due to noise and interference.

The original version of IEEE 802.15.4 (IEEE 802.15 Task Group 4, 2006) supported two PHY layer modulation options. The 868/915 MHz PHY, known as the low band, used binary phase shift keying (BPSK) modulation, whereas the 2.4 GHz PHY, called the high band, used offset quadrature phase shift keying (OQPSK) (IEEE 802.15 Task, Group 4 2006). The data rate was 250 kbps at 2.4 GHz (16 channels), 40 kbps at 915 MHz (10 channels), and 20 kbps at 868 MHz (1 channel). There have been a number of PHY layer amendments and additions, and as a result there are now seven different PHY layers allowed, with new modulation schemes and operating bands defined. Table 17.3 summarizes some of these PHY layers (IEEE 802.15 Task Group 4, 2011); UWB PHY layers are not considered due to their complexity (e.g., support of multiple data rates). Other PHY layers are being considered; the most significant for health care applications is one that uses the 2.36–2.4-GHz band for medical devices (USA only), which was approved in February 2013.

TABLE 17.3 Frequency Bands and Data Rates

PHY (MHz)	Channel Parameters			Spreading Parameters		Data Parameters		
	Frequency Band (MHz)	Number of Channels	Channel Bandwidth (MHz)	Chip Rate (kchips/sec)	Modulation	Bit Rate (kbps)	Symbol Rate (ksymbols/sec)	Symbols
780	779–787	7	2	1000	OQPSK	250	62.5	16-ary orthogonal
780	779–787	7	2	1000	MPSK	250	62.5	16-ary orthogonal
868/915	868–868.6	1	0.6	300	BPSK	20	20	Binary
	902–928	10	2	600	BPSK	40	40	Binary
868/915 (optional)	868–868.6	1	0.6	400	ASK	250	12.5	20-bit PSSS
	902–928	10	2	1600	ASK	250	50	5-bit PSSS
868/915 (optional)	868–868.6	1	0.6	400	OQPSK	100	25	16-ary orthogonal
	902–928	10	2	1000	OQPSK	250	62.5	16-ary orthogonal
950	950–956	11	0.4	N/A	GFSK	100	100	Binary
950	950–956	11	0.2/0.6	300	BPSK	20	20	Binary
2450 DSSS	2400–2483.5	16	5	2000	OQPSK	250	62.5	16-ary orthogonal
2450 CSS (optional)	2400–2483.5	14	5		DQPSK	250	166.667	64-ary bi-orthogonal
2450 CSS (optional)	2400–2483.5	14	5		DQPSK	1000	166.667	8-ary bi-orthogonal

Source: IEEE 802.15 Task Group 4, 2011.

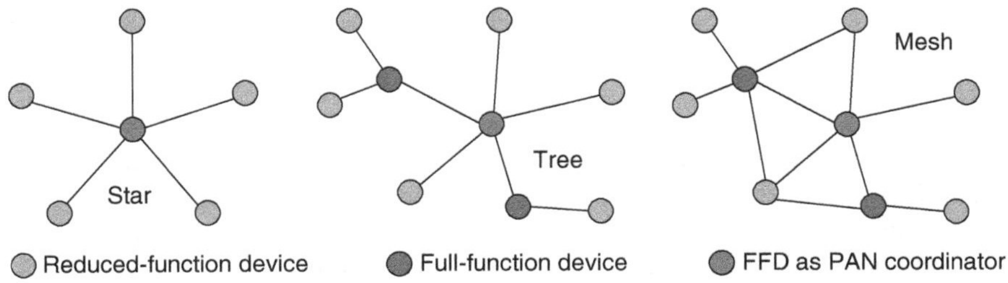

Figure 17.1 Network topologies supported by IEEE 802.15.4 2011 (IEEE 802.15 Task Group 4, 2011).

The MAC layer controls access to the radio channels using the CSMA-CA (carrier sense multiple-access collision avoidance) mechanism, together with an optional time slot structure and security functionality. The MAC layer supports several network architectures (Fig. 17.1), including the star, cluster-tree, and mesh [called peer-to-peer in the 2011 edition of the IEEE 802.15.4 standard (IEEE 802.15 Task Group 4, 2011)] topologies, allowing protocols built on IEEE 802.15.4 to create a scalable, reliable, and self-healing network. These protocols define the functionality of the higher layers of the protocol stack and include ZigBee (ZigBee Alliance), JenNET (NXP Semiconductor), MiWi (Microchip Technology, Inc.), WirelessHART (Hart Communication Foundation), and 6LoWPAN [Internet Engineering Task Force (IETF)], among others.

There are two categories of IEEE 802.15.4 devices (IEEE 802.15 Task Group 4, 2011):

- IEEE 802.15.4 Full-Function Device (FFD). This is a smart node that can operate as a personal area network (PAN) coordinator (i.e., controls the formation of a network) or coordinator (i.e., includes relay or router functionality). This type of node links nodes and, potentially, groups together PANs. It can communicate with both types of IEEE 802.15.4 device and provides multihopping for messages.
- IEEE 802.15.4 Reduced-Function Device (RFD). This node can only communicate with one full-function device at a time and never with other reduced-function devices. It carries limited functionality to control cost and complexity. Typical applications might be light switches or passive infrared sensors, where only small amounts of data are transmitted.

Somewhat confusingly, ZigBee defines three types of device: "co-ordinator," "router," and "device" [ZigBee Alliance (Ergen, 2004)]. The co-ordinator is the equivalent of the PAN co-ordinator in the IEEE standard, while the router is the co-ordinator of the IEEE standard; both are full-function devices. The device is always a reduced function device.

IEEE 802.15.4 provides authentication, encryption, and integrity services for wireless systems through the security layer, with 128-bit AES encryption and authentication available. The transmission range is from 10 to 75 m, depending on the power output and environmental characteristics. IEEE 802.15.4 devices are expected to have a battery life ranging from months to years.

IEEE 82.15.6 Perhaps the most significant development for wearable WBANs for health care so far has been the publication of the IEEE 802.15.6 standard, in February 2012 (IEEE 802.15 Task Group 6, 2012). This standard has been developed to "provide an international standard for a short range ... , low power and highly reliable wireless communication for use in close proximity to, or inside, a human body" (Astrin, 2011). The new standard has a single MAC layer but supports multiple physical layers, grouped into three classes (IEEE 802.15 Task Group 6, 2012; Astrin, 2011):

- Narrow band (NB)
- Ultrawide band (UWB)
- Human body communication (HBC)

Table 17.4 shows some key features of the NB physical layers (IEEE 802.15 Task Group 6, 2012; Astrin, 2011).

The NB layers are designed primarily for medical applications (Astrin, 2011); in fact, the 2.36–2.4-GHz band is a medical BAN (MBAN, or MedBAN) band for the United States. This band is used by aeronautical mobile telemetry and radio astronomy users; thus, MBAN use would be on a secondary-user basis. Its key benefit is that the band is electromagnetically quiet, with relatively few users, compared to the adjacent spectrally congested 2.45-GHz ISM (Industrial, Scientific and Medical) band, implying that high reliability would be easier to maintain. It would also benefit from the commercial operations in the 2.45-GHz band, potentially allowing reductions in cost. Each frequency band supports multiple data rates, allowing a trade-off between data rate and operating range. The IEEE 802.15.6-NB system has even lower power consumption than BLE, permitting longer battery lifetimes to be achieved (Astrin, 2011). The main trade-off is the reduced operating range for 802.15.6.

TABLE 17.4 Key Features of IEEE 802.15.6 Narrow-Band Physical Layers

Frequency Band (MHz)	Number of Channels	Channel Bandwidth	Supported Data Rates (kbps)	Operation
402–405	10	300 kHz	75.9/151.8/303.6/455.4	Implantable
420–450	12	320 kHz	75.9/151.8/187.5	Wearable
863–870	14	400 kHz		Wearable
902–928	60	400 kHz	101.2/202.4/404.8/607.1	Wearable
950–958	16	400 kHz		Wearable
2360–2400	39	1 MHz	121.4/242.9/485.7/971.4	Wearable
2400–2483.5	79	1 MHz		Wearable

Source: IEEE 802.15 Task Group 6, 2012; Astrin 2011.
Note: The 2.36-2.4 GHz band is a medical BAN (MBAN) band for the USA.

The UWB layers are designed for high-bandwidth applications, both medical and nonmedical. UWB systems combine potentially high data rates with low power consumption, albeit over relatively short ranges. However, range is not an issue in WBANs. As the signal levels are low, interference with other medical devices and equipment can also be kept to low levels. Impulse radio (IR-UWB) and wide-band frequency modulation (FM-UWB) layers are available, with the FM-UWB elements being optional and the IR-UWB mandatory (IEEE 802.15 Task Group 6, 2012; Astrin, 2011). There are two operational modes: the high quality-of-service mode is used for high-priority medical applications, while the default mode is used in nonmedical applications and lower priority medical scenarios (Astrin, 2011). Table 17.5 gives the channel allocation.

The FM-UWB layer supports a raw data rate of 250 kbps, using a form of Gaussian frequency shift keying combined with wide-band FM (IEEE 802.15 Task Group 6, 2012; Astrin, 2011); use of FEC to improve reliability reduces this to 202.5 kbps (IEEE 802.15 Task Group 6, 2012; Astrin, 2011). The IR-UWB layer supports raw data rates from 0.487 to 15.6 Mbps. Three modulations are supported by the IR-UWB layer, with different data rates available for each:

- On–off signaling (OOK)
- Differential binary-phase shift keying (DBPSK)
- Differential quadrature-phase shift keying (DQPSK)

The use of FEC reduces the raw data rate differently for each modulation.

The HBC physical layer operates at 21 MHz and supports the following data rates: 164 kbps, 328 kbps, 656 kbps, and 1.3125 Mbps. It is designed for the exchange of data through body channels; in other words, it requires the person to touch the

TABLE 17.5 Channel Allocation for Proposed IEEE 802.15.6 UWB Physical Layers

Band Group	Channel Number	Center Frequency (MHz)	Bandwidth (MHz)	Channel Attribute
Low band	1	3494.4	499.2	Optional
	2	3993.6	499.2	Mandatory
	3	4492.8	499.2	Optional
High band	4	6489.6	499.2	Optional
	5	6988.8	499.2	Optional
	6	7488	499.2	Optional
	7	7987.2	499.2	Mandatory
	8	8486.4	499.2	Optional
	9	8985.6	499.2	Optional
	10	9484.8	499.2	Optional
	11	9984	499.2	Optional

Source: IEEE 802.15 Task Group 6, 2012; Astrin 2011.

device or other person, in order to establish a communications link (Astrin, 2011). Its principle of operation requires an electrode in contact with the body, which "is used for transmitting or receiving an electrical signal through the body to a device (e.g., smart-phone)" (Astrin, 2011). Application examples include e-payment mechanisms and the exchange of electronic business cards, among others.

The next decade will, therefore, benefit from a new standard for interoperability of wearable, and implantable, devices and systems, helping the vision of wearable sensor networks for health, and other, applications to become a reality.

17.4 EXAMPLE WIRELESS BODY AREA NETWORK (WBAN) APPLICATIONS IN HEALTH CARE

This section discusses some examples of WBAN applications in health care. For each application, some background is first provided; then, the application is discussed with regard to implications for wireless propagation channels, antennas, and other elements of the communications system.

17.4.1 Wearable Artificial Pancreas

An artificial pancreas refers to technology that mimics the role of the pancreas in controlling blood sugar levels. This concept dates back to the late 1950s (Nishida et al., 2009) in implantable and fixed installation forms. Wearable artificial pancreases utilizing wireless communication links date back at least as far as 1984, when a patent (Shichiri et al., 1984) was granted for a "portable artificial pancreas comprising a blood sugar detecting unit including electrode means to be inserted into ... the living body, an injection unit including a container for a blood sugar control agent and a feed pump ... and a ... microcomputer for calculating the blood sugar value from an output ... from the detecting unit and the dose of the control agent ... and controlling the pump [accordingly]. ... The output signal of the detecting unit is transmitted to the control unit by radio." This early description echoes the functionality of advanced wearable continuous glucose monitors (CGMs) and closed-loop systems including an insulin (control agent) pump. Indeed, a closer reading of the patent reveals that even the use of a second pump and counteragent (either glucagon or glucose) was considered.

Modern systems, such as Medtronic's Guardian Real-Time (no insulin pump) and MiniMed Paradigm Real-Time Revel (with insulin pump) systems (Medtronic, Inc.), obviously benefit from advances in technology, such as miniaturization of the electronics involved. CGMs are typically two-node networks, with the biosensor in one unit on one part of the torso and the user interface (UI) unit—and insulin pump, where present—on another part of the torso, although a three-node network could be utilized with separate UI and pump units.

Despite the early origins of the artificial pancreas concept, there are still a number of research challenges to overcome, including such areas as the control algorithms for safe blood sugar level management, accurate and long-life biosensors, and wearable systems that are safe, comfortable, and secure. An alternative to biosensors attempts to detect changes in blood glucose level by monitoring the changes in the effective permittivity of the body tissue, through the use of an RF or microwave resonator and appropriate signal processing. Examples in the literature are Topsakal et al. (2008), Freer and Venkataraman (2010), Hancock and Chaudhry (2007), Jean et al. (2008), and Yilmaz and Hao (2011). There remains much to be done to refine this technique, however.

From a radio propagation perspective, a two-node network requires a single channel along the torso; a three-node network may include a second channel along the torso or, possibly, a torso-to-wrist channel. It is understood that the vast majority of WBANs, including those considered for IEEE 802.15.6, are in star network topologies; the advantages of multiple-hop network topologies are rarely worth the extra complexities involved, given the relatively small distances between WBAN nodes.

Torso–torso channels are relatively static channels during many human activities, with a relatively small variance around the mean. To put this into context, a motionless subject may produce variations about the mean of ± 3 dB; this scenario can approximate various real-life scenarios, such as working at an office desk or sitting at home watching television. On the other hand, a jogging person may produce variations of the order of ± 15 dB (Munoz et al., 2012). Small variations are predominantly related to breathing and heart activity, and larger variations are related to movement and changes in the surrounding environment (Munoz et al., 2012). The variations can also be assessed according to their time period: fast variations are attributed to the multipath effect, where a signal travels by multiple paths and arrives with different magnitudes and phases, causing interference. Slow (long-term) variations are related to changes in the path length (or absorption), caused by the relative motion of the transmitter and receiver or the presence of an obstruction.

Torso–wrist channels are more dynamic, due to the relative motion of the wrist with respect to the torso. This can introduce additional losses in the propagation path. For example, an arm swinging during walking will periodically move behind the torso, increasing the loss through the shadowing effect at that point.

As the distances between nodes are relatively small, the use of a higher frequency operation is less of an issue, as absorption by the body is less. An ideal scenario might involve subsecond resolution in the control of blood glucose levels, but the data rate requirements would still be fairly small between sensor unit and pump unit. Even including multiple sensor units, in order to improve the accuracy of the glucose reading, would not imply large increases in data rate requirements.

The antennas used would ideally launch waves along the body surface, and not away from the body (for efficiency and security reasons) or into the body (efficiency and health/safety reasons). Antennas would obviously need to be compact and conformal. Chip antennas are one solution but may not be optimized for the on-body environment. Hence, if not protected by a ground plane from the effects of the body, nonoptimal performance may occur. The use of a microstrip patch antenna in its fundamental mode would not be appropriate, due to the fact that the radiation is radiated away from (normal to) the body; however, a higher order mode (e.g., TM_{21})

can give radiation that is tangential to the body. Besides that, standard square or rectangular patches are relatively large. More compact antennas, such as planar inverted-F antennas (PIFAs), can be used, but often sacrifice performance for compact size.

The addition of an off-body link for telemedical applications adds complications for the network controller node (usually the one with the UI). The data rate requirements may not change significantly, but the frequency of operation could potentially change and the operating mode of the antenna is normal to the body, implying either the use of two separate antennas or the use of a dual-mode, and possibly dual-frequency, antenna.

17.4.2 Functional Electrical Stimulation

Functional electrical stimulation (FES) refers to the technique of applying controlled pulses of current to muscles, in order to stimulate contraction and achieve movement. In this example, the focus is on the use of FES to assist walking; as in other areas, there has been a shift toward wireless technology in these systems, as in Lawrence and Schmidt (1997), Toussaint et al. (2010), and Mecheraoui et al. (2012). This technique is used to treat a number of conditions, including (Toussaint et al. 2010):

- Spinal cord injuries
- Stroke
- Multiple sclerosis
- Cerebral palsy
- Parkinson's disease

Wireless FES devices can be used in clinical settings, for therapy and research, or in nonclinical settings, where they "have the potential to become practical assistive devices that will provide a level of home and workplace mobility beyond that provided by the wheelchair or braces" (Lawrence and Schmidt, 1997). A historical review of various experimental and commercial FES systems can be found in Thrasher and Popovic (2008).

The basic FES system considered uses a sensor in one (or both) shoes to detect the step events and, depending on the sensor, provide the data required for gait analysis. The electrodes are attached to the thigh muscle and are either controlled via a waist-worn unit (implying the use of relatively long wires) or with a local controller attached to the leg and a waist-worn device acting as the overall controller and UI unit. The use of wires is inhibitory to most patients, so the second version, with three nodes, is preferred. When a step is detected (e.g., by a foot switch detecting the impact of a foot on the ground), the electrodes are triggered to stimulate muscle contraction, resulting in the lifting of the foot for the next step.

Usage of this system implies that data rate requirements would be very small, being essentially event triggering. However, this presupposes the system operates in "stand-alone" mode: If used in a clinical setting, and particularly when a user is first becoming accustomed to the system, it is possible that additional "telemetry" data is required off-body for analysis and fine-tuning of the system, increasing the data rate requirements to some extent, both on-body and from the gateway controller node (most likely the waist-worn unit) off-body.

The node positions are significantly different to those in the first example (Section 17.4.1). Here, there is only one waist node, which is relatively stationary; that is, it does not contribute significantly to the dynamic variation in the channel, compared to the contribution from the other two nodes. The other two nodes are both on the leg, and hence experience more motion than that on the waist, but there are still significant differences between them. In particular, the node at the foot can experience more relative motion than the node on the thigh, as well as experience a different electromagnetic environment, due to its proximity to the ground. Hence, significant fading can occur on the foot–waist channel.

Most examples in the literature have the antenna attached to the outside of the shoe; however, real-world versions must integrate the antenna (and all electronics and power source) into the shoe, for aesthetic reasons. This again implies that the foot–waist channel will experience a reduction in performance, due to the added absorption of the RF signal through the foot and due to the additional constraints on antenna design, which will likely be manifested in reduced efficiency and gain.

17.4.3 Multiparameter Monitoring

Ideally, monitoring systems for mobile health should be unobtrusive, incorporate multiple sensors, give real-time feedback, and provide wireless communication with online data evaluation. Much research has been conducted in this area, with some successful commercial real-time systems produced, particularly devices for performance monitoring, such as the Nike iPod kit (Apple, Inc.) and "adidas miCoach" (Adidas). However, these devices are limited to measuring one or two parameters and are not suitable for monitoring chronic patients. Combining sensors for vital signal monitoring has been studied by many research groups in academia, as well as in industry, often integrated into textiles. A common application are the emergency first responders, as well as the military, but there are potential applications in chronic disease management, as well as general "wellness" monitoring.

A common set of physiological signals for multiparameter monitoring might include some or all of (Yilmaz et al., 2010):

- Blood glucose level
- Blood pressure
- Pulse rate
- ECG patterns
- Electroencephalography (EEG) patterns
- Electromyography (EMG) patterns
- Respiration rate

- Respiration effectiveness (e.g., blood oxygen saturation)
- Temperature
- Skin pH level (sweat monitoring)

These use a range of sensing techniques and, in many cases, require the sensors to be placed on different parts of the body. It is also worth noting that multiparameter systems can make use of the context of the user in interpreting the results; context can be obtained by additional sensors, particularly accelerometers for motion and posture detection, but also other devices close to the user, including a mobile phone and devices in the environment.

The sensors may also have different data rate requirements. In particular, ECG (and EEG/EMG) waveforms could require much higher rates than the other physiological signals. There are four potential solutions to this. First, the ECG device could be the network controller, user interface, and gateway off the body, meaning on-body transmissions are avoided. However, this may either involve long electrode leads or positioning the UI in a nonoptimal location.

Second, the ECG node can perform local processing and feature extraction (e.g., the QRS width, RR distance, and QT interval parameters) to reduce the on-body data rate requirements. This adds computational and memory requirements to the device specification, which may adversely impact the cost, device battery lifetime, or device volume (shape, weight, etc.).

Third, the ECG node could implement a PHY layer from IEEE 802.15.6 with the required data rate, while other nodes use a PHY with a data rate appropriate to their needs. This means that the network controller/UI unit is potentially more complex in that it may need to support multiple PHY layers, but it reduces the complexity of the other nodes in the network. Although the PHY layer may be different, it has the advantage that the MAC layers are identical; hence, interoperability is not compromised.

Finally, all nodes could use the same PHY, with the data rate chosen to meet the requirements for the ECG; this makes the hardware and software requirements for the radio simpler, but may imply inefficiency, particularly for those nodes needing a smaller data rate. Ultimately, the specific application must inform the implementation and the choice between the above options, which will be affected by cost, performance, comfort, and even aesthetic considerations. This choice also affects other elements of the design, such as the type(s) of antenna used.

Most of the devices are expected to be located on the torso of the body, but some, such as pulse rate sensors, EMG sensors, and posture/activity sensors (e.g., accelerometers), could be located on the limbs and even the head (e.g., the ear). Hence, there will be significant variation in the on-body channels, with differing levels of fading between any two links (a link is a pair of nodes). There is also some evidence in the literature that the statistics describing the radio propagation channel will differ between links, and even that a given link is nonreciprocal (Heaney et al., 2010).

Analyses of on-body radio channel characteristics for the extraction of kinesthetic and physiological signals have already been conducted [e.g., Munoz et al. (2012) and

Guraliuc et al. (2010, 2011)]. Although more work is required, it is possible to use these techniques in multiparameter systems, either to augment or complement existing sensors, or even replace them. This latter option is attractive, if it minimizes the number (hence, cost) of nodes, but its validity as a solution depends on the performance such techniques provide, as well as the significance of the data, particularly safety-critical data.

17.5 CONCLUSION

This chapter has looked at wearable wireless health care devices, with the objective of enhancing understanding of some of the design choices and trade-offs. It began by presenting some background information on the use of wireless technology and sensors to develop a wireless physiological measurement system, before discussing the relevant wireless communications protocols and standards, including the IEEE 802.15.6 medical BAN standard.

The chapter concluded with examples of wearable wireless physiological measurement systems (WPMSs) in specific application areas. For each application, some background was first provided; the application was then discussed with regard to implications for wireless propagation channels, antennas, and other elements of the communications system.

REFERENCES

Adidas. Undated. Product webpage for adidas miCoach system. Available: http://www.adidas.com/uk/micoach/; last accessed September 2012.

Apple, Inc. Undated. Product webpage for Nike + iPod system. Available: http://www.apple.com/ipod/nike/; last accessed September 2012.

Astrin A. 2011. IEEE-802.15.6-tutorial. Available: https://mentor.ieee.org/802.15/documents?is_group=0006; last accessed December 12, 2011.

Bluetooth Special Interest Group (SIG). 2010. Bluetooth Specification Version 4.0 Vol 0. Available: https://www.bluetooth.org/Technical/Specifications/adopted.htm; last accessed September 2012.

Department of Health. 2007. What is physiological measurement? Available: http://www.dh.gov.uk/en/Publicationsandstatistics/Publications/PublicationsPolicyAndGuidance/DH_074448; last accessed September 2012.

Ergen SC. 2004. ZigBee/IEEE 802.15.4 Summary. Technical Report, Advanced Technology Lab of National Semiconductor. Available: http://pages.cs.wisc.edu/~suman/courses/838/papers/zigbee.pdf; last accessed September 2012.

Fass L. 2007. Patient centric healthcare. The 3rd IET International Conference on Medical Electrical Devices & Technology, 2–3 October 2007, London.

Feied C, Jordan N, Kanhouwa M, Kavanagh J. 2006. The New World of Healthcare Work: A Microsoft White Paper. UK Focus International Lecture, The Royal Academy of Engineering, 7 November 2006.

Freer B, Venkataraman J. 2010. Feasibility study for non-invasive blood glucose monitoring. In Proceedings of IEEE Antennas and Propagation Society International Symposium (APSURSI), pp. 1–4.

GE Healthcare. 2007. Investigation of the spectrum requirements for advanced medical technologies. Private communication.

Guraliuc AR, Barsocchi P, Potorti F, Nepa P. 2011. Limb movements classification using wearable wireless transceivers. *IEEE Trans. Inform. Tech. Biomed.* 15(3):474–480.

Guraliuc AR, Serra AA, Nepa P, Manara G, Potorti F. 2010. Detection and classification of human arm movements for physical rehabilitation. In IEEE Antennas and Propagation Society International Symposium (APSURSI), pp. 1–4.

Hancock C, Chaudhry S. 2007. A non-invasive monitoring system. In Proceedings of European Microwave Conference, pp. 313–316.

Hao Y, Foster R. 2008. Wireless body sensor networks for health-monitoring applications. *Physiol. Meas.* 29(11):R27–56.

Hart Communication Foundation. Undated. WirelessHART protocol website. Available: http://www.hartcomm.org; last accessed September 2012.

Heaney SF, Scanlon WG, Garcia-Palacios E, Cotton SL. 2010. Fading characterization for Context Aware Body Area Networks (CABAN) in interactive smart environments. In Proceedings of Loughborough Antennas and Propagation Conference, pp. 501–504.

IEEE 802.15 Task Group 4. 2006. IEEE 802.15.4 Standard. Available: http://standards.ieee.org/getieee802/download/802.15.4-2006.pdf; last accessed September 2012.

IEEE 802.15 Task Group 4. 2011. IEEE 802.15.4 Standard. Available: http://standards.ieee.org/getieee802/download/802.15.4-2011.pdf; last accessed September 2012.

IEEE 802.15 Task Group 6. 2012. IEEE Standard for local and metropolitan area networks Part 15.6: Wireless Body Area Networks. Available: http://dx.doi.org/10.1109/IEEESTD.2012.6161600; Last accessed September 2012.

Internet Engineering Task Force (IETF). Undated. 6LoWPAN, an IETF Working Group and protocol. Available: http://datatracker.ietf.org/wg/6lowpan/charter/; last accessed September 2012. See also the Wikepedia page at http://en.wikipedia.org/wiki/6LoWPAN; last accessed September 2012.

Jean BR, Green EC, McClung MJ. 2008. A microwave frequency sensor for non-invasive blood-glucose measurement. In Proceedings of IEEE Sensors Applications Symposium, pp. 4–7.

Jones VM, Shashar N, Ben Shaphrut O, Lavigne K, Rienks R, Bults RGA, Konstantas D, Vierhout P, Peuscher J, van Halteren AT, Herzog R, Widya IA. 2006. Remote monitoring for healthcare and for safety in extreme environments. In: *M-Health: Emerging Mobile Health Systems*. Berlin: Springer, pp. 561–574.

Lawrence TL, Schmidt RN. 1997. Wireless in-shoe force system. In Proceedings—19th International Conference IEEE/EMBS, pp. 2238–2241.

Lekkala J. 2006. Wireless biomedical sensing challenges and opportunities. Biomedical Sensors—Foresight Workshop IV, June 7, 2006, Tampere University of Technology (TUT).

Leon-Garcia A, Widjaja I. 2004. *Communication Networks: Fundamental Concepts and Key Architectures*. New York: McGraw-Hill.

Mecheraoui C, Cobb J, Swain I. 2012. Evaluation of a wireless in-shoe sensor based on ZigBee used for drop foot stimulation. In Proceedings of 2012 IEEE Radio and Wireless Symposium, pp. 423–426.

Medtronic, Inc. Undated. Diabetes-related services webpage. Available: http://www.medtronicdiabetes.com/AboutDiabetes/ManagingDiabetes; last accessed September 2012.

Mendelson Y, Duckworth RJ, Comtois G. 2006. A wearable reflectance pulse oximeter for remote physiological monitoring. In Proceedings of the 28th IEEE EMBS Annual International Conference, Vol. 1. Institute of Electrical and Electronics Engineers, Piscataway, NJ, pp. 912–915.

Microchip Technology, Inc. Undated. MiWi Homepage. Available: http://www.microchip.com/stellent/idcplg?IdcService=SS_GET_PAGE&nodeId=2664&param=en520414; last accessed September 2012. See also http://www.microchip.com/stellent/idcplg?IdcService=SS_GET_PAGE&nodeId=2664&param=en535761; last accessed September 2012.

Munoz MO, Foster R, Hao Y. 2012. Physiological features from an on-body radio propagation channel. In Proceedings of Wearable and Implantable Body Sensor Networks, pp. 52–57.

Nishida K, Shimoda S, Ichinose K, Araki E, Shichiri M. 2009. What is artificial endocrine pancreas? Mechanism and history. *World J. Gastroenterol.*, 15(33):4105–4110.

NXP Semiconductors. Undated. JenNET. Available: http://www.jennic.com/products/protocol_stacks/jennet; last accessed September 2012.

Otto C, Milenkovic A, Sanders C, Jovanov E. 2006. System architecture of a wireless body area sensor network for ubiquitous health monitoring. *J. Mobile Multimedia* 1:307–326.

Ouwerkerk M, Pasveer F, Langereis G. 2007. Unobtrusive sensing of psychophysiological parameters: Some examples of non-invasive sensing technologies. Private communications.

Saxby R. 2007. How silicon will transform healthcare. The 3rd IET International Conference on Medical Electrical Devices & Technology, October 2–3, 2007, London.

Shichiri M, Hattori K, Nishiyama T, Singuh Y, Takagi K, Higuchi M, Okamoto M, Okudaira T. 1984. Portable artificial pancreas. European Patent, no. EP0098592. Available: http://www.freepatentsonline.com/EP0098592A2.html, last accessed September 2012.

Thrasher TA, Popovic MR. 2008. Functional electrical stimulation of walking: Function, exercise and rehabilitation. *Annales de réadaptation et de médecine physique: revue scientifique de la Société française de rééducation fonctionnelle de réadaptation et de médecine physique*, 51(6):452–460.

Topsakal E, Moreland EC, Karacolak T, Acar M. 2008. The impact of glucose concentration in blood plasma on relative dielectric constant and conductivity. In Proceedings of National Science Meeting URSI.

Toussaint M, Andreuy D, Fraisse P, Guiraudy D. 2010. Wireless distributed architecture for therapeutic functional electrical stimulation: A technology to design network-based muscle control. In Proceedings of Annual International Conference of the IEEE Engineering in Medicine and Biology Society, pp. 6218–6221.

Townsend KA, Haslett, JW, Tsang TKK, El-Gamal MN, Iniewski, K. 2005. Recent advances and future trends in low power wireless systems for medical applications. In Proceedings of the 9th International Database Engineering & Application Symposium (IDEAS'05). Institute of Electrical and Electronics Engineers Computer Society, Piscataway, NJ.

Yilmaz T, Foster R, Hao Y. 2010. Detecting Vital Signs with Wearable Wireless Sensors. *Sensors*, 10(12):10837–10862.

Yilmaz T, Hao H. 2011. Sensing of dielectric property alterations in biological tissues at microwave frequencies. In Proceedings of Loughborough Antennas and Propagation Conference, pp. 1–4.

ZigBee Alliance. Undated. ZigBee web site. Available: http://www.zigbee.org/en/index.asp; last accessed September 2012.

18

EPIDERMAL SENSOR PARADIGM: INNER LAYER TISSUE MONITORING

DIMITRIS PSYCHOUDAKIS, CHI-CHIH CHEN, GIL-YOUNG LEE, AND JOHN L. VOLAKIS

ElectroScience Laboratory, Ohio State University, Columbus, Ohio

18.1 INTRODUCTION

Continuous real-time monitoring of the dielectric properties of embedded tissues and organs is critical for early detection of common medical conditions such as pulmonary edema. The next generation of tumor or organ failure detection equipment to be used on the human body will identify the mass based on its dielectric characteristics without the need for surgery. Unique for each material, the permittivity serves as an identifier in detecting changes in internal organs such as the lung. Therefore, it is important to understand the body's electromagnetic interaction properties when designing these type of implanted wireless sensors or other body-worn antennas. As these properties also depend on signal frequency and tissue types, it becomes necessary to consider different human models and properties for each frequency band and application.

In this regard, an epidermal sensor capable of in vivo monitoring of the inner layer permittivity is analyzed. To this extent, the relevant properties of the human body from the very high frequency (VHF) to the X band are studied and dominant propagation modes around the body are discussed. Based on these data, approximate human torso models are built for each propagation mode using multiple concentric cylinders emulating the skin, muscle, organ, and the like. After this study, a simpler two-layer model of the human body is proposed for efficient calculations without loss of accuracy and used in the analysis of the epidermal sensor.

Handbook of Biomedical Telemetry, First Edition. Edited by Konstantina S. Nikita.

18.2 REVIEW OF ELECTROMAGNETIC PROPERTIES OF HUMAN BODY

The radio frequency (RF) properties of human body tissues affect propagation, reflection, attenuation, and other behaviors of electromagnetic fields around the body. These properties depend strongly on the types of tissue and frequency. Temperature and blood or fluid perfusions also affect these properties, but these secondary effects are normally not considered. In this chapter, focus will be given to the relative permittivity (ϵ_r) and conductivity (σ) of the body tissues. Further, because the body is nonmagnetic, the relative magnetic permeability (μ_r) is assumed to be unity ($\mu_r = 1$), except for special applications such as magnetic resonance imaging (MRI).

18.2.1 Numerical Expression of Dielectric Properties for Human Tissues

The dielectric properties of body tissues have been studied by several authors (Cook, 1951; Duck, 1990; Dumey et al., 1986; England, 1950; Foster and schwan, 1989; Lee, 2012; Yamamoto and Yamamoto, 1976). The most comprehensive and referenced study is that in the report by Gabriel (1996). Gabriel measured over 25 tissue types in the frequency range from 1 MHz to 20 GHz using an open-ended coaxial probe technique (Gabriel et al., 1994). Most of these measurements were carried out with excised animal tissue from freshly killed sheep at 20 and 37°C. In addition, accessible parts of the human body such as palm, sole, and forearm skin were measured in vivo. To describe the frequency dependence of these tissues' dielectric properties, a model based on the 4-Cole–Cole expression is used (Gabriel, 1996) with the relative dielectric constant expressed as

$$\epsilon_r(\omega) = \epsilon_\infty + \sum_{m=1}^{4} \left[\frac{\Delta\epsilon_m}{1 + \left(j\omega\tau_m\right)^{(1-\alpha_m)}} \right] + \frac{\sigma_i}{j\omega\epsilon_0} \tag{18.1}$$

where ϵ_∞ is the material's relative permittivity at terahertz frequencies; ϵ_0 is the free-space permittivity; σ_i is the ionic conductivity; and ϵ_m, T_m, and α_m are material parameters for each dispersion region. Since the dielectric constant is generally complex, it consists of real and imaginary parts:

$$\epsilon_r(\omega) = \epsilon_r'(\omega) - j\epsilon_r''(\omega) \tag{18.2}$$

with the tissue conductivity and skin depth given by

$$\sigma(\omega) = \epsilon_r''(\omega) \cdot \epsilon_0 \cdot \omega \tag{18.3}$$

and

$$\delta(\omega) = \frac{1}{\omega} \left\{ \frac{\mu' \epsilon_0 \epsilon_r'(\omega)}{2} \left[\left(1 + \left(\frac{\sigma(\omega)}{\omega\epsilon_0\epsilon_r'(\omega)} \right)^2 \right)^{1/2} - 1 \right] \right\}^{-1/2} \tag{18.4}$$

respectively. As usual, δ is defined as the depth at which the radiation intensity becomes $1/e$ of the original value it has at the surface.

The required parameters to determine the dielectric properties of the selected human tissues are tabulated in Table 18.1. Using these, along with equations (18.1)–(18.4), the important dielectric properties of the human tissues of interest can be found.

18.2.2 Human Tissue Dielectric Properties

Figure 18.1 plots the relative permittivity, conductivity, and penetration depth of selected tissues of the human torso from 10 MHz to 20 GHz. Also, Figure 18.2 shows the same properties of relatively large organs inside the human torso (heart, liver, and lungs). From these, we observe several important tissue properties:

1. Human body tissue can be classified into two types: the first has low water content (i.e., fat, bone, etc.), low dielectric constant, and low loss. The second has high water content (i.e., skin, muscle, heart, etc.), high dielectric constant, and high loss (see Figure 18.1).
2. Almost all internal organ tissues with high water content have similar electromagnetic (EM) properties. We also notice that the inflated lung has substantially lower dielectric constant and conductivity. In fact, the inflated lung has values similar to those of the low-water-content body parts. The differences in dielectric constants can, of course, be used to identify whether the lung is inflated or deflated and how much liquid it contains (see Figure 18.2). Specifically, at 10 MHz the $100 < \epsilon_r' < 200$ depending on the lung being inflated or deflated.
3. The conductivities of all tissues increase with frequency. In other words, the human body becomes lossy and any incident wave rapidly attenuates at higher frequencies (see Figures 18.1 and 18.2). From Figure 18.2, the maximum penetration for any body part is 3–4 cm at about 10 MHz.

The above observations lead to some insights for body-worn antenna applications. Some of these are:

1. The human torso can be modeled using multilayers (skin, fat, muscle, organs, and bone) in the same order as they appear in the human body. However, since many organs have EM properties similar to muscle, they can be grouped, with the exception of inflated lungs.
2. Variations in the lung's electrical properties can be exploited to sense the human's respiratory rate and vital signs.
3. In-body communication (between on-body antenna and implanted sensors) should be carried out at low frequencies due to high losses experienced by the wave once it penetrates the body. Thus, in-body communications should be performed at 13.56 MHz via inductive coupling and at 402–405 MHz, both allocated for medical implanted communication systems.

TABLE 18.1 Parameters Required for Finding Human Tissue Dielectric Constant and Conductivity

	ϵ_∞	σ_i	$\Delta\epsilon_1$	τ_1 (psec)	α_1	$\Delta\epsilon_2$	τ_2 (nsec)	α_2	$\Delta\epsilon_3$	τ_3 (μsec)	α_3	$\Delta\epsilon_4$	τ_4 (msec)	α_4
Bone — cancellous	2.5	0.07	18	13.263	0.22	300	79.577	0.25	2.0e+4	159.155	0.2	2.0e+7	15.915	0
Bone — cortical	2.5	0.02	10	13.263	0.2	180	79.577	0.2	5.0e+3	159.155	0.2	1.0e+5	15.915	0
Fat	2.5	0.01	3	7.958	0.2	15	15.915	0.1	3.3e+4	159.155	0.05	1.0e+7	7.958	0.01
Heart	4	0.05	50	7.958	0.1	1200	159.155	0.05	4.5e+5	72.343	0.22	2.5e+7	4.547	0
Liver	4	0.02	39	8.842	0.1	6000	530.516	0.2	5.0e+4	22.736	0.2	3.0e+7	15.915	0.05
Lung — deflated	4	0.2	45	7.958	0.1	1000	159.155	0.1	5.0e+5	159.155	0.2	1.0e+7	15.915	0
Lung — inflated	2.5	0.03	18	7.958	0.1	500	63.662	0.1	2.5e+5	159.155	0.2	4.0e+7	7.958	0
Muscle	4	0.2	50	7.234	0.1	7000	353.678	0.1	1.2e+6	318.31	0.1	2.5e+7	2.274	0
Skin — dry	4	0	32	7.234	0	1100	32.481	0.2	0	159.155	0.2	0	15.915	0.2
Skin — wet	4	0	39	7.958	0.1	280	79.577	0	3.0e+4	1.592	0.16	3.0e+4	1.592	0.2
									$e+4 = \times 10^4$			$e+4 = \times 10^4$		
									$e+3 = \times 10^3$			$e+3 = \times 10^3$		
									$e+5 = \times 10^5$			$e+5 = \times 10^5$		
									$e+6 = \times 10^6$			$e+6 = \times 10^6$		
									$e+7 = \times 10^7$			$e+7 = \times 10^7$		

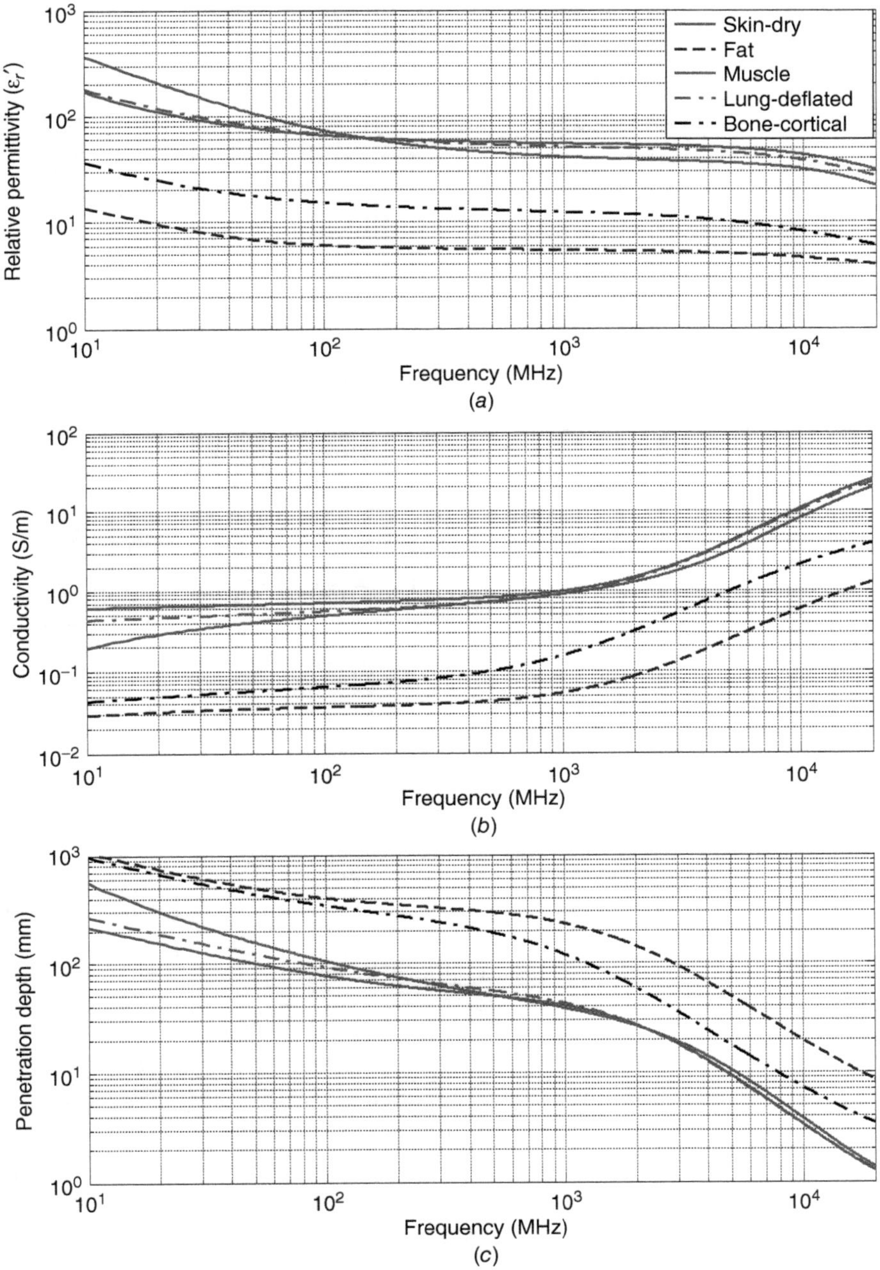

Figure 18.1 Electromagnetic properties of various tissues: skin, fat, muscle, and lung (deflated): (*a*) relative permittivity (real value), (*b*) conductivity, and (*c*) penetration depth in millimeters. (*See insert for color representation of the figure.*)

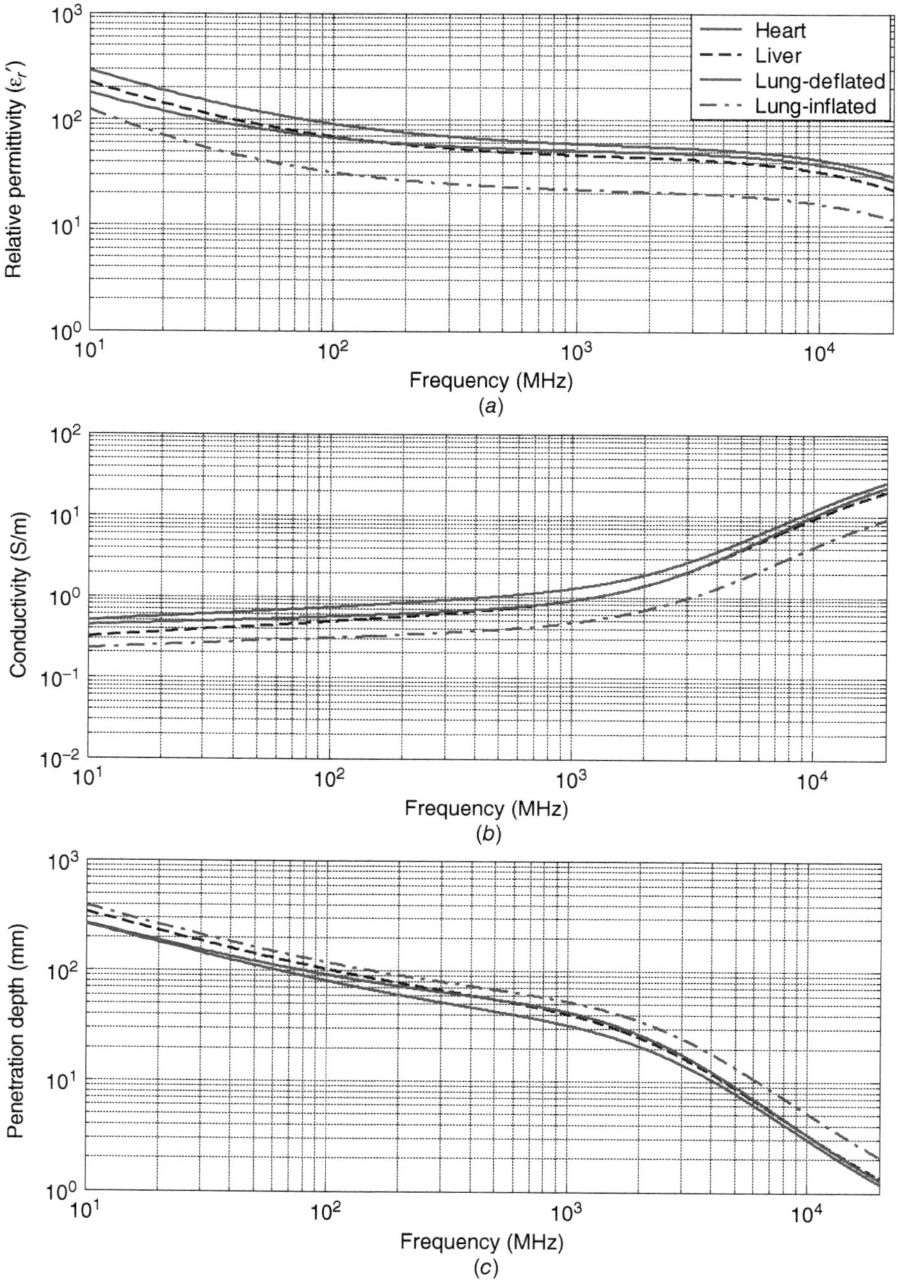

Figure 18.2 Electromagnetic properties of various internal organs: heart, liver, lung (deflated), and lung (inflated): (*a*) relative permittivity (real value); (*b*) conductivity, and (c) penetration depth in millimeters. (*See insert for color representation of the figure.*)

The EM properties of the skin, muscle, and fat at different frequencies are also tabulated in Table 18.2. These values are based on equations (18.1)–(18.4) and can also be extracted from Figures 18.1 and 18.2.

18.3 PROPAGATION MODES FOR BODY-CENTRIC WIRELESS COMMUNICATIONS

Body-centric wireless communications can be divided into three types based on the interaction between the body and the EM field. In each case, different propagation modes are dominant. The types of body-centric communications are:

1. *Off-Body Communication* Human-to-human or human-to-base-station communication. In this case, waves propagating away from the body are the dominant mode of communication.
2. *On-Body Communication* Multiple antennas placed on the same body surface communicate with each other using mostly creeping waves that propagate along the body surface.
3. *In-Body Communication* Implanted sensors interact with antennas placed on the body's surface. In this case, losses within the tissues dominate.

A simple approach to analyze propagation within the human tissue is to consider an infinitely planar layered model (emulating the human torso geometry). Figure 18.3 shows such a layered geometry with separate EM sources/excitations relating to each propagation mode. For off-body communications, a plane wave (far-field) interaction is assumed as depicted in Figure 18.3*a*. Alternatively, for on-body and in-body communication, vertical and horizontal dipole excitations are considered, respectively, as shown in Figures 18.3*b–c*. The thickness of each tissue layer is $t_1 = 2.4$ mm, $t_2 = 13.5$ mm, and $t_3 = 168$ mm (Gordon et al., 1976), implying a total thickness of 200 mm (see Figure 18.3). The EM properties of these tissues are specified in Table 18.2.

When considering a plane wave incidence, of most interest is the reflected wave off the body's surface. Using the model in Figure 18.3 and FEKO, the plane wave excitation $E = \hat{x}E_0 e^{+jk_0 z}$ (with $E_0 = 1$ V/m) was considered to calculate the electric field within the human body. Figure 18.4 plots these fields. Also, Table 18.3 gives the voltage standing-wave ratio (VSWR) and reflection coefficient, $|\Gamma|$:

$$|\Gamma| = \frac{\text{VSWR} - 1}{\text{VSWR} + 1} \tag{18.5}$$

The transmitted E field within the human tissue is also given. Note that most of the incoming waves are reflected at the boundary. Actually, a very small percentage of the wave penetrates the body. In other words, the interference caused by the reflected and incident fields near the body surface reduces field penetration, disallowing any communication with the interior layers. Specifically, at 30 MHz only 10% of the

TABLE 18.2 Electromagnetic Properties of Skin (Dry), Muscle, and Fat at Selected Frequencies (30 MHz, 433 MHz, 915 MHz, 2.45 GHz, and 12 GHz)

Tissues	EM Properties	$f = 30$ MHz	$f = 433$ MHz	$f = 915$ MHz	$f = 2.45$ GHz	$f = 12$ GHz
Skin—dry	Relative permittivity (ϵ_r')	152.9	46.08	41.33	38.01	29.33
	Conductivity (σ) (S/m)	0.3415	0.7017	0.8715	1.464	10.33
	Penetration depth (δ) (mm)	222	54	40	23	3
Fat	Relative permittivity (ϵ_r')	8.11	5.57	5.46	5.28	4.46
	Conductivity (σ) (S/m)	0.033	0.042	0.051	0.105	0.728
	Penetration depth (δ) (mm)	614	304	242	117	16
Muscle	Relative permittivity (ϵ_r')	91.81	56.87	55.0	52.73	40.1
	Conductivity (σ) (S/m)	0.658	0.805	0.948	1.739	13.540
	Penetration depth (δ) (mm)	127	52	42	22	3

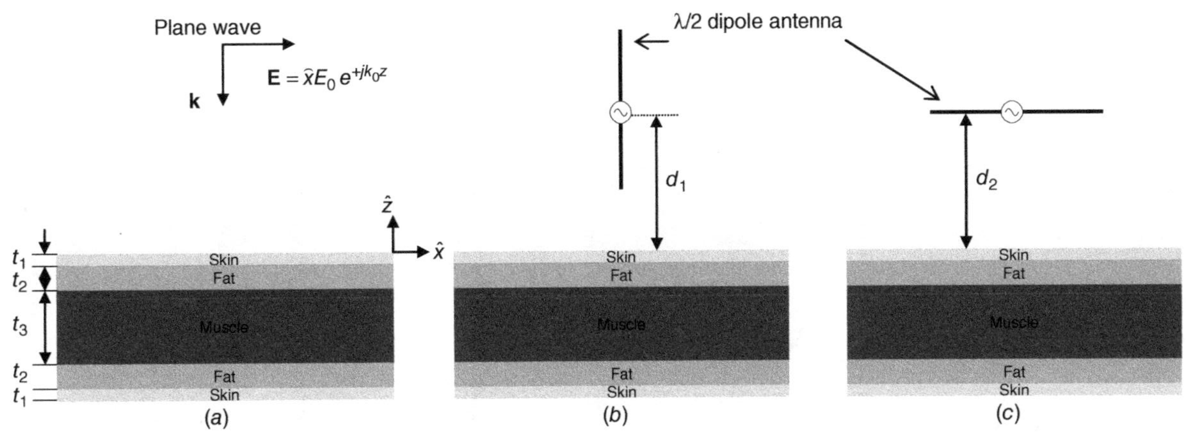

Figure 18.3 Infinite human tissue layers (skin, fat, muscle, fat, skin) and radiating sources near the body surface: (*a*) plane wave incidence, (*b*) vertical dipole antenna illumination, and (*c*) horizontal dipole antenna excitation.

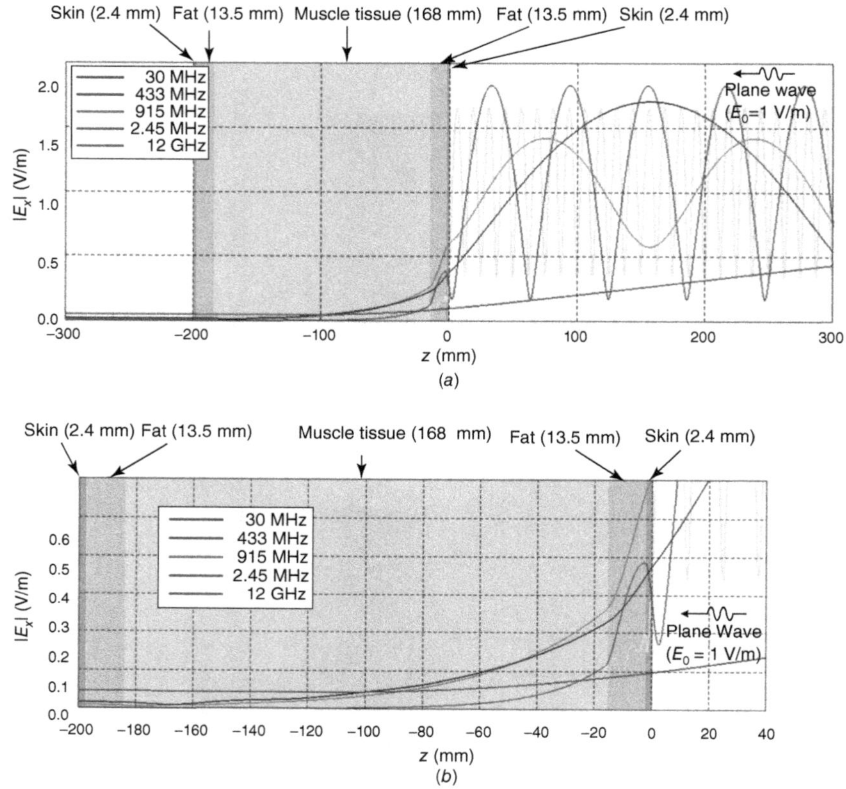

Figure 18.4 Electric field magnitude due to plane wave incidence on the layered tissue model depicted in Figure 18.3 (rotated clockwise by 90°). The electric field, generated using FEKO is sampled along the direction of propagation, normal to the tissue surface: (*a*) holistic view and (*b*) zoomed-in view focusing on human tissue layers. (*See insert for color representation of the figure.*)

TABLE 18.3 Calculated VSWR Reflection Coefficient and Transmitted E Field at Each Frequency

	f = 30 MHz	f = 433 MHz	f = 915 MHz	f = 2.45 GHz	f = 12 GHz
SWR	29.29	5.84	2.45	10.89	4.94
$\lvert\Gamma\rvert$	0.934	0.707	0.421	0.832	0.663
$\lvert E_t\rvert$ (V/m)	0.0458	0.016	0.001	0.00005	0

incoming field penetrates into the body. Moreover, the 30-MHz wave has the least decay, as expected.

The above analysis suggests some design tips for in-body communication and off-body communication antennas. Specifically, for in-body communication the antenna should be mounted close to the body and if possible directly on the body (without any gaps) to reduce reflections at the boundary due to the high dielectric contrast between air and skin. Alternatively, for off-body communication any gap between antenna and body should be carefully chosen to avoid the severe gain drop caused by reflection cancellation. This will be shown later.

Dipole radiation next to the layered body model gives a better insight into propagation modes for off-body and on-body communication modes. Figure 18.5 shows the behavior of the fields generated by a resonant dipole near the planar human body interface. Clearly, the decision to use a vertical versus a horizontal dipole depends on the communication scenario. In the case of the vertical dipole, the fields along the interface remain at normal power levels, whereas the fields from the horizontal dipole vanish in presence of the body surface. Therefore, for on-body communication or body area networks (BAN), it is more desirable to use vertical dipole or monopole type of radiators. On the other hand, for off-body communication it would be more prudent to use broadside radiators, like the horizontal dipole or the microstrip patch antenna. Other considerations such as operating frequency and bandwidth play a major role in the selection of the antenna for such applications. Patch antennas, due to their ground plane requirement are ideal for off-body applications since they are also conformal. However, they have an inherently low bandwidth performance, making them less desirable for applications with high bandwidth requirements. Although monopole configurations may seem inappropriate for body-worn applications due to their profusion, they could be attractive for high-frequency applications where the wavelength is small and body surface profusion is minimal.

18.3.1 Space Wave Analysis for Off-Body Communication

Space waves emitted by body-worn antennas radiating away from the body consist of a direct radiating field from the antenna and a reflected wave from the body surface. In this case, the gap between the antenna and human body (reflecting surface) plays a critical role. To investigate the influence of the gap, a horizontal dipole antenna is considered as the excitation for off-body communication as shown in Figure 18.3 (*c*). The dipole is placed at a distance, d_2, from the body surface. Figure 18.6 shows the far-field magnitude pattern at 433 MHz for various antenna distances from the body surface. The pattern in absence of the body is also given for comparison.

From Figure 18.6, the following can be concluded:

1. The radiation patterns are similar for distances up to $d_2 = 0.2\lambda$. When the distance is greater than 0.3λ, the main beam broadens or splits in the horizontal direction. This is due to cancellations from the dipole's image at $z = -d_2$.
2. The strongest radiated field is observed at $d_2 = 0.2\lambda$ along the main beam direction. This field is even stronger than that of an antenna in free space, implying constructive interference from the image dipole. This is consistent with Figure 18.4*a* in which the first maximum of the total field occurs at approximately 150 mm ($\approx 0.22\lambda$ at 433 MHz).
3. The radiated field is weaker when the antenna is close to the body. This is expected due to antenna shorting by the body's conductive surface.

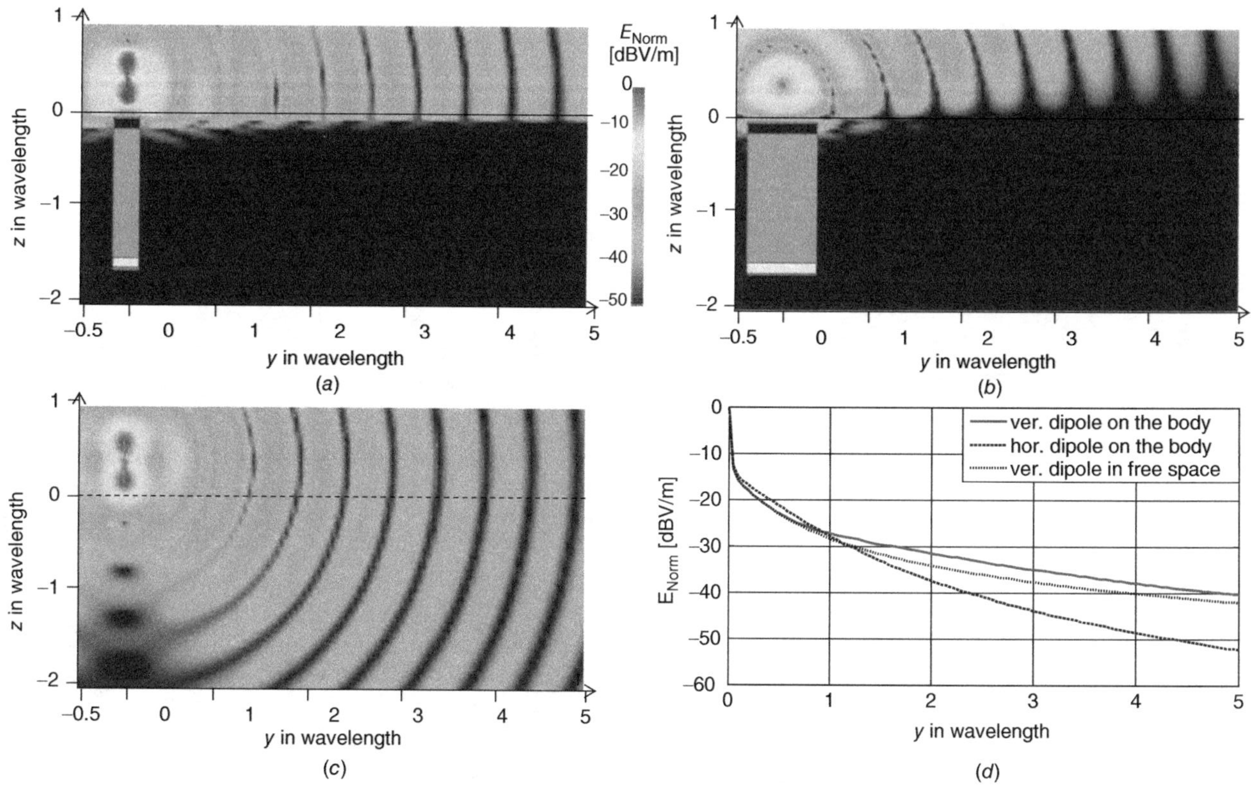

Figure 18.5 Dipole radiation near the infinite layered human tissue model as depicted in Figure 18.3: (*a*) vertical resonant dipole fields, (*b*) horizontal resonant dipole fields, (*c*) vertical dipole fields in free space, and (*d*) normalized field magnitude along a line ($z = 0.4\lambda$) parallel to the body surface. (*See insert for color representation of the figure.*)

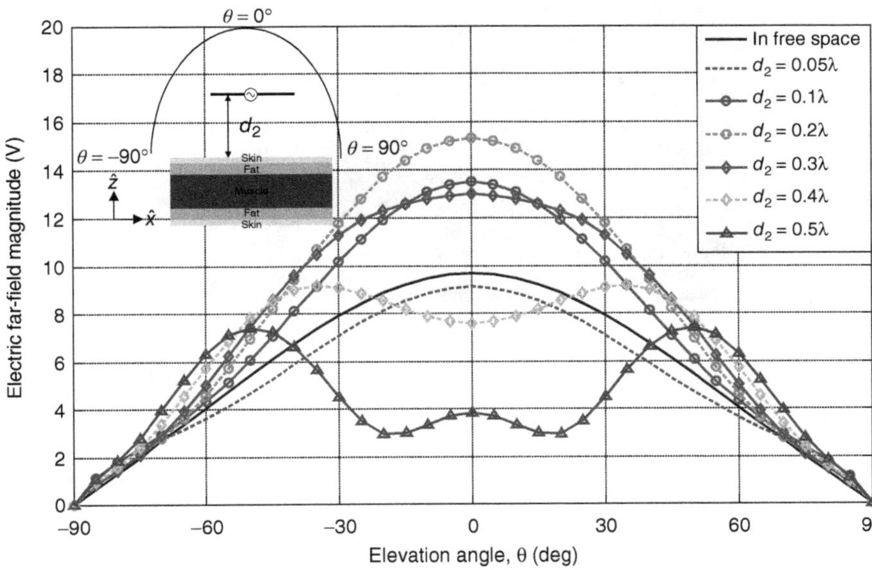

Figure 18.6 Elevation angle far-field pattern for different distances between antenna and human body. (*See insert for color representation of the figure.*)

From the above, one can conclude that the antenna should be placed at a certain electrical distance off the body, depending on the reflection coefficient of the human body. Since the human body can be treated as a metallic surface (large reflection with negative polarity), a distance of $\lambda/4$ is usually selected. However, this is as relevant for wide-band antennas as for low-frequency antennas. For the latter, the physical distance corresponding to $\lambda/4$ is too large for practical realization. An approach is to reduce the reflection from the body surface using wide-band electromagnetic band gap (EBG) surfaces (Zhu and Langley, 2009) or magnetic material coatings (Bell et al., 2007; Erkmen et al., 2008, 2009; Tzanidis et al., 2010).

18.4 HUMAN TORSO MODEL FOR BODY-CENTRIC WIRELESS COMMUNICATION

For accurate RF analysis, an anatomically realistic human model composed of a set of voxels is desired. A human voxel model for the entire body is available from Dimbylow (1996). This model was based on MRIs of an adult male and was segmented into 37 different tissues. Higher resolution human models have also been developed by other researchers (Mason et al., 2000; Nagaoka et al., 2004; Xu et al., 2000).

An important consideration is that computational analyses for the human body are time consuming and can even be impractical. In this light, several approximate human torso models are proposed herewith for in-body, on-body, and off-body communication. Specifically, a concentric elliptical cylinder composed of multiple tissue layers

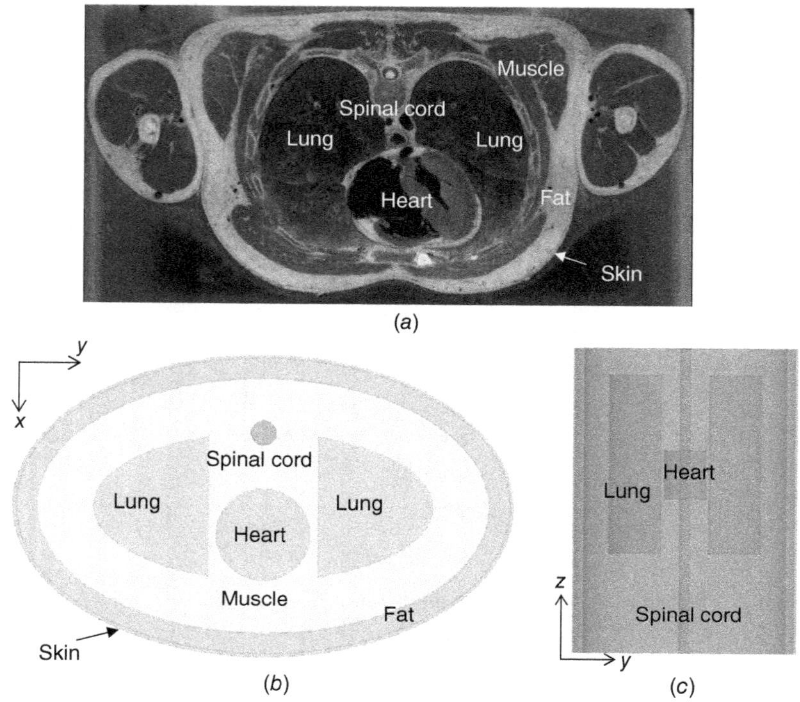

Figure 18.7 Reference human torso model: (*a*) thorax cut image from visible human project, (*b*) top view, and (*c*) front view of torso model.

is proposed, and the simplest model giving reasonably small error in EM analysis is selected.

18.4.1 Human Torso Model for In-Body Communication

To establish a human torso model, communication between on-body antenna and a sensor implanted on the heart is considered (see Figure 18.7). The calculation is performed at 433 MHz because the antenna size is realizable and the penetration depth at this frequency is also suitable for implanted sensor applications.

A reference human model was first constructed based on the thoracic cut image from the visible human project (Ackerman, 1998), as shown in Figure 18.7. This reference model consists of skin, fat, muscle, lung, heart, and spinal cord and is depicted in Figure 18.7*b*. The outer diameter of this model is 20 cm along the short axis and 32 cm along the long axis. To evaluate RF performance, the penetrated field was calculated inside the human model with the plane wave incident from the +*x* direction (from the top as depicted in Figure 18.7). The calculated *E* fields from the reference model are given in Figure 18.8. An approximate four-layer model comprised of skin, fat, muscle, and lung is also included. Surprisingly, the simplified four-layer model is in good agreement with the more exact one. Only a small difference is observed

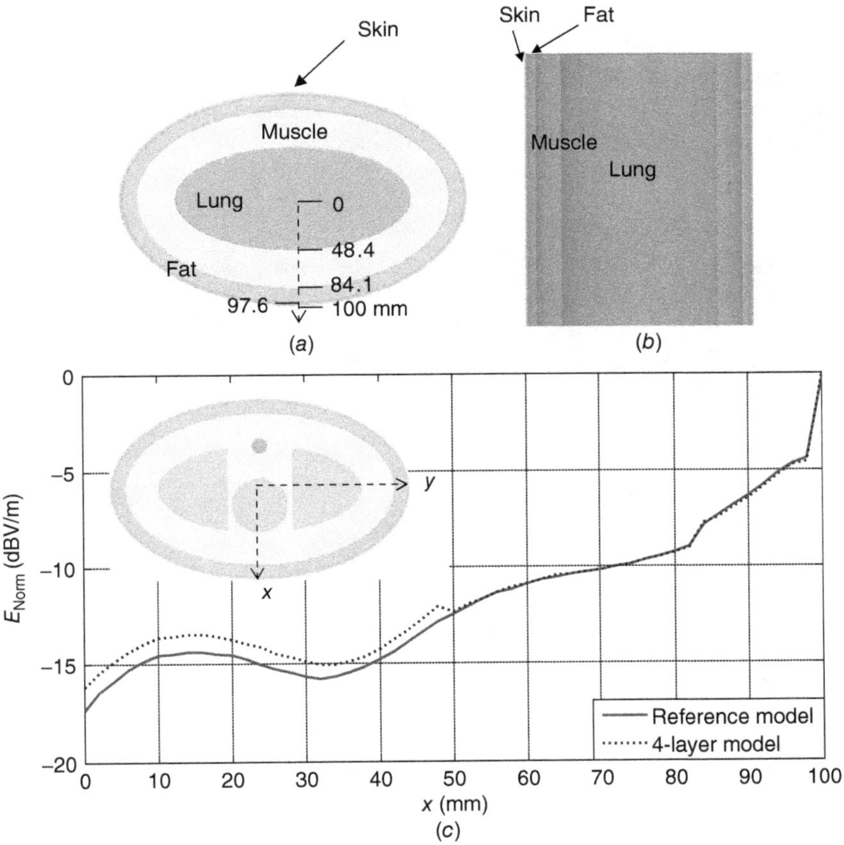

Figure 18.8 Comparison of the propagated fields within a human torso using a detailed reference human model and a more approximate four-layer model. (*a*) Top view and (*b*) front view of the concentric four-layer model approximating propagation into the human torso. (*c*) Comparison of the normalized *E*-field magnitude inside the human body using the detailed and four-layer human models.

between the models. This difference is caused by RF property variations between the heart and lung (deflated). Nonetheless, the field curves show that the four-layer model (skin–fat–muscle–lung) is sufficient for in-body propagation analysis.

18.4.2 Human Torso Model for On-Body Communication

The main propagation mode for on-body communication (Alomainy and Hao, 2006; Alomainy et al., 2005a, 2005b; 2006; Hall and Hao, 2006) is the creeping waves. These are not significantly affected by the core medium (body organs in this case). This is because the creeping wave cannot penetrate into the core medium at higher frequency as shown in Figure 18.5*a*. To demonstrate this, three human torso models (homogeneous and three- and four-layer concentric cylinders) were modeled and the radiation/transmission between two antennas close to the body (0.25λ away from the surface) was computed. The homogeneous model is a phantom whose relative permittivity and conductivity are equivalent to two/three times that of muscle. The four-layer model is the same as that shown in Figure.18.8, and the three-layer model is similar to the four-layer model except that the lung is replaced with muscle. Figure 18.9*d* shows

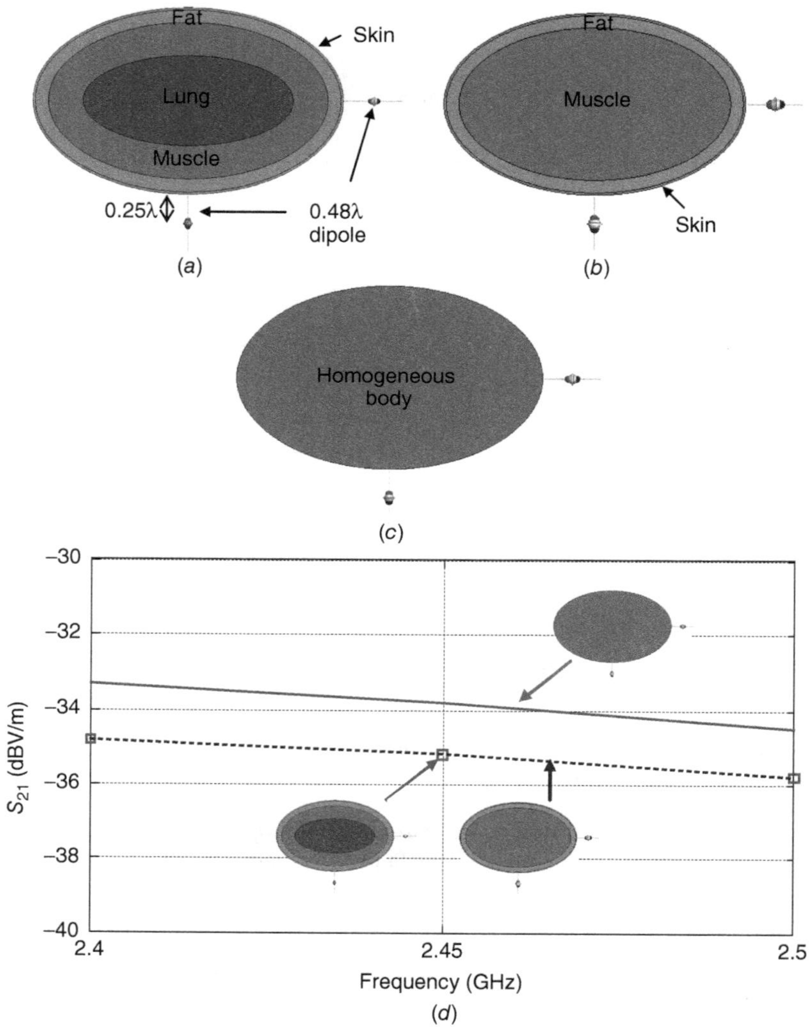

Figure 18.9 Reception (S_{21}) between two horizontal dipoles placed 0.2λ off the body's surface: (*a*) top or cross-sectional view of the four-layer model, (*b*) top view of the three-layer model, (*c*) top view of the homogeneous model with dielectric constant two or three-that of muscle, and (*d*) transmission coefficients (2.4–2.5 GHz) between the shown antenna pair for each model.

the transmission coefficients from 2.4 to 2.5 GHz between two antennas. As depicted, the four- and three-layer models are in excellent agreement, and the homogeneous model has about 2 dB higher values but follows the same behavior. We remark that the fat layer is critical in our models. Actually, RF power does not propagate beyond the fat layer. Therefore, it is necessary to include skin, fat, and muscle layers in an approximate human model for on-body communication. However, the organ tissues are not as important, especially at higher frequencies. Given its simplicity, we can conclude that the model is quite accurate for on-body communication.

18.4.3 Human Torso Model for Off-Body Communication

Space waves are the main propagation mode for off-body communication. As already noted, the reflected field is critical when considering off-body communication. To evaluate the human model for off-body communication, radiation patterns using the homogeneous and three-layer models at 433 MHz were calculated and compared as shown in Figure 18.10. Both models use the same material layers as those in Figure 18.9 *b*, *c*. Also, in this case, the $\lambda/2$ dipole antennas are located in front of

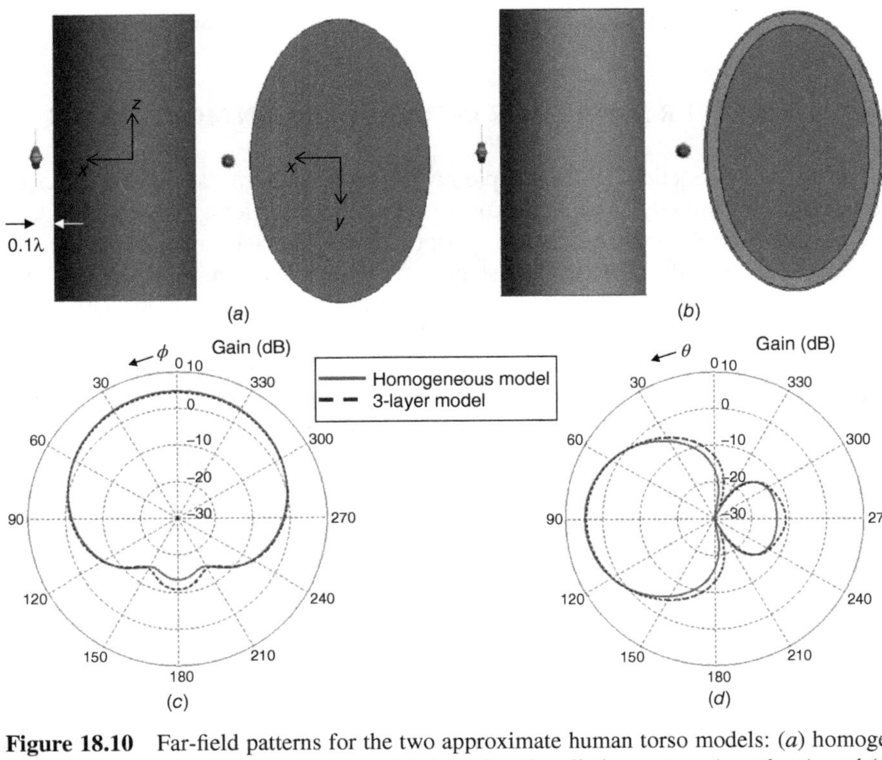

Figure 18.10 Far-field patterns for the two approximate human torso models: (*a*) homogeneous human model, (*b*) three-layer model, (*c*) azimuth radiation pattern (*x*-*y* plane), and (*d*) elevation radiation pattern (*x*-*z* plane).

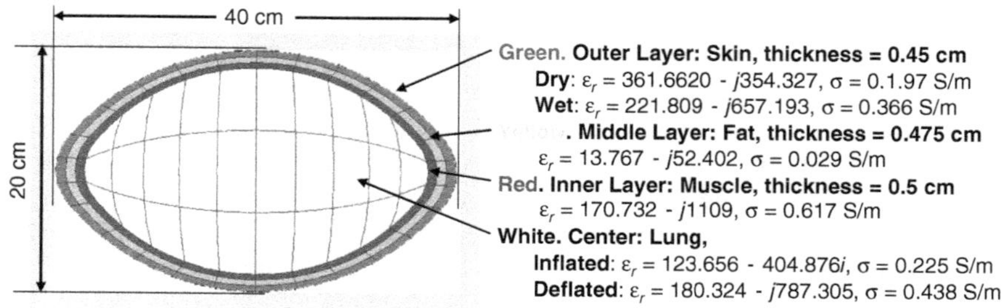

Figure 18.11 Depiction of four-layer torso model and corresponding tissue values for investigating field penetration at 10 MHz. Dielectric constant and conductivity values are from Salman et al. (2012). (*See insert for color representation of the figure.*)

the human model at a distance of 0.1λ and parallel to the body surface. The resulting azimuth (horizontal to the body) and elevation patterns are shown in Figure 18.10*c*, *d*. Overall, they are in good agreement except for back radiation and a small difference in gain (less than 0.5 dB) in the main beam direction. This is likely due to a difference in the RF property. In conclusion, the homogeneous model with proper EM properties is sufficient to represent the human body model for off-body communication.

18.5 TWO-LAYER MODEL FOR INTERNAL ORGAN MONITORING

As mentioned in Section 18.4.1, simple multilayer models can accurately describe propagation for in-body communication with implanted devices. These models can also be used for developing biomedical sensors capable of determining the inner layer properties, such as the dielectric constant and conductivity. Typical dielectric constant values at 10 MHz associated with tissues prevalent in the torso are outlined in Figure 18.11. This four-layer model assumes three external "shell" layers consisting of skin, fat, and muscle tissue and can be used to develop an epidermal RF sensor array. However, the main objective of this array is to continuously monitor long-term changes occurring within *internal* organs (liver, lungs, pancreas, spleen, etc.). Therefore, the accurate description of the shell layers is superfluous in our analysis. The mere existence of a shell with a range of thicknesses and effective dielectric values suffices to develop such a sensor. Hence, a unique feature of the sensor is required to remove influence from the skin and fat layers. We will return to discussing a simplified two-layer model after discussing the operation of the epidermal RF sensor.

18.6 EPIDERMAL RF SENSOR FOR INNER LAYER TISSUE MONITORING

To facilitate our discussion a depiction of the sensor is given in Figure 18.12. It consists of a plurality of conductive strips or electrodes (Salman et al., Submitted). Each

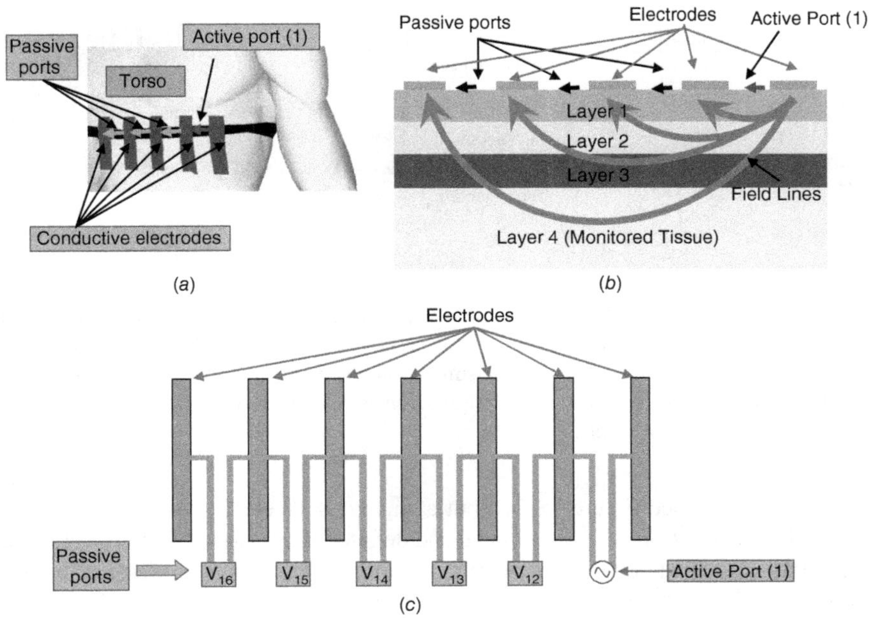

Figure 18.12 Epidermal RF sensor for deep-tissue water content monitoring in lungs (Salman et al., 2012): (*a*) sensor mounted on the body, (*b*) Side view of the sensor mounted on the four-layer torso model, and (*c*) schematic of the sensor's layout.

strip or electrode is 10 cm long (and less than 1 cm wide) and placed parallel to the adjacent ones as depicted in Figure 18.12*c*. These strips are connected via a horizontal conducting line with port insertions to measure the voltage between strips, referred to as V_{1i}. As shown in Figure 18.12*c*, only the rightmost probe is excited, giving rise to the voltage V_{1i} at the *i*th port. Alternatively, if we choose to excite, say, the third port, the voltage at the *i*th port will be V_{3i}. The key operational principle of the sensor is depicted in Figure 18.12*b*. Specifically, the fields propagating from port 1 to port 2 or 3 stay mostly within the skin and fat tissue regions. For the subsequent, further out ports, the muscle layer becomes part of the propagation region. Eventually, the distant electrodes receive mostly the fields propagating through the lungs. As shown in Figure 18.11, there is significant dielectric constant difference between the inflated and deflated lungs ($\varepsilon_r = 123-180$) for reliable detection of any changes in dielectric constant and in breathing rates.

The dielectric constant change manifests itself as a variation in the voltages V_{1i} or transmission coefficients S_{1i} at the electrode ports. It is graphically depicted in Figure 18.13*a*. It is seen that the far away port voltages show significant variation in their values as the dielectric constant of the mass changes. In fact, as shown in Figure 18.13*b*, certain sensor designs can lead to more than 20 dB in probe voltage changes between the extremes of an inflated and deflated lung. Verification measurements confirmed the sensor design as well. The measurement setup used

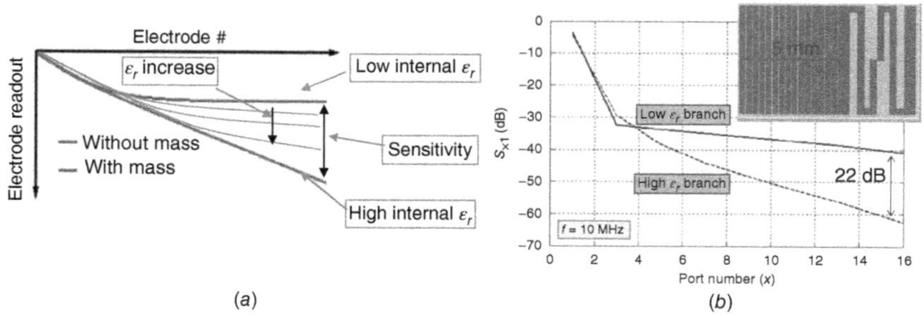

Figure 18.13 (*a*) Graphical representation of the voltage readout changes due to changes in the dielectric of the innermost layer in the four-layer model. The outer layers are assumed to be unchanged. (*b*) Lung sensor sensitivity simulation results when the lung (mass) is inflated or deflated (no mass) using the sensor design given in Salman et al., (2012). (*See insert for color representation of the figure.*)

only 9 electrodes (see Figure 18.14), but as many as 17 electrodes were needed to achieve 22 dB variation between inflated and deflated lungs, shown in Figure 18.13*b*. Indeed the agreement between measurements and calculations is impressive (bottom of Figure 18.14), verifying the functionality of this advanced sensor for deep-tissue dielectric constant extraction and even imaging.

18.7 EXTRACTION OF DIELECTRIC CONSTANT

Having presented the sensor setup, mode of operation, and geometry, we return to our description of the human torso model to demonstrate how the extraction of the dielectric values is achieved from the electrode readouts. The simplification in Figure 18.8 is quite useful for RF calculations and remains more accurate as the frequency decreases. However, the overall goal is to determine the lung's dielectric constant in the presence of variation in the outer three layers (skin, fat, and muscle). Nevertheless, as we alluded previously in Section 18.5, the actual composition of the three external layers is of little significance since the interest lies on the inner tissue. Therefore, an even more simplified two-layer model can be considered as shown in Figure 18.15. In this model and without loss of accurate description of the physics, the three outer shell layers are grouped into an effective layer that can now be described by two parameters: its thickness and its dielectric constant (as opposed to six parameters needed for the four-layer model). By narrowing down the degrees of freedom needed to describe the shell, it can readily be used to introduce perturbations to the model for calibrating the sensor, as discussed below.

Dielectric constant extraction begins by expressing the dielectric constant as a weighted sum of the voltages, V_{1i}, measured from the multiprobe epidermal textile sensor (Zhang et al., 2012) (see Figure 18.12, 18.13, and 18.14) at the ports $i = 1,2,3, \ldots, N$, where N is the number of ports. A possible form of this expression is

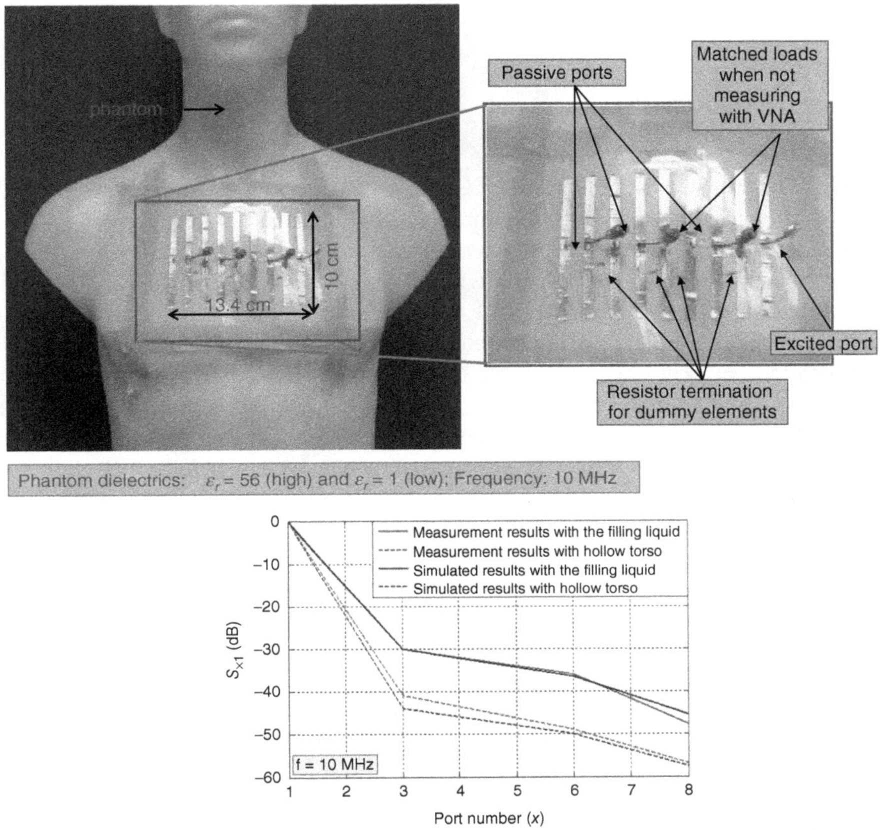

Figure 18.14 (*Top*) The experimental setup on a plastic mannequin phantom using nine electrodes and four ports to demonstrate the validity of the simulated results. (*Bottom*) Graphic representation of electrode outputs (S_{x1}) showing agreement between simulations and experiment. (*See insert for color representation of the figure.*)

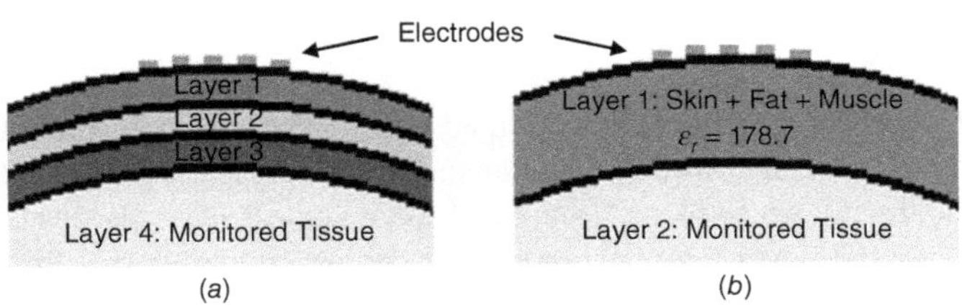

Figure 18.15 Differences between (*a*) four-layer torso model and (*b*) two-layer model used in demonstrating the epidermal RF dielectric sensor.

(Salman et al., submitted)

$$\epsilon_r = \sum_{i=1}^{N} w_i V_{1i} \tag{18.6}$$

where i is the port number and w_i are the weights to be calculated for each port. Normally, and for constant body outer layers, the weights, w_i, would be determined by carrying out a set of reference/calibration measurements (at least as many as the weights) using various inner layer materials with different $\epsilon_{r,j}$ values ($j = 1,2, \ldots , M$, where M is the number of different reference ϵ_r values). However, this is hardly the case for the subject application. Instead, the weights are determined via a least-square method after generating a system of reference equations for a variety of outer layer combinations of skin, fat, and muscle (various thicknesses and dielectric constant variations), in addition to the reference ϵ_r cases. For the kth combination of these layers, the voltages $V_{1i,j,k}$ are computed and the kth equation is generated to form part of the linear system of equations for the weights. After including all the variation contributions into equation (18.6), the system of calibration equations can be given by

$$\epsilon_{r,j} = \sum_{i=1}^{N} w_i V_{1i,j,k} \quad j = 1, 2, \ \ldots , \ M \quad k = 1, 2, 3, \ \ldots , \ M_t \tag{18.7}$$

where M_t is the number of variations of outer layer combinations of skin, fat, and muscle, giving a total of $M \times M_t$ calibration equations (note that M_t equations correspond to a single $\epsilon_{r,j}$ value).

From a generic point of view, the above process does nothing more but create an average of the outer three layers as depicted in Figure 18.11. As such, the extracted permittivity is insensitive to outer layer variations as much as possible. The latter is accomplished by increasing the number of electrodes/number of ports and the number of calibration measurements.

18.8 CONCLUSION

This chapter analyzed the paradigm of an advanced epidermal sensor, which can be used for in vivo determination of the human tissue dielectric properties, and thus monitoring of the inner layer tissues. A review of tissue dielectric properties was initially performed, and the operation of the proposed epidermal sensor was further carried out. The sensor consists of several probes and, equivalently, ports, which are used for measuring the scattering parameters when a single port is excited. Postprocessing of the measured scattering parameters helps extract the dielectric constant at a specific depth inside the tissues. Therefore, the proposed sensor can be used for continuous monitoring of organs placed deep inside the human body, for applications such as drug delivery and medical diagnosis (e.g., lung fluid monitoring).

REFERENCES

Ackerman MJ. 1998. The visible human project. *Proc. IEEE*, 86:504–511.

Alomainy A, Hao Y. 2006. Radio channel models for UWB body-centric networks with compact planar antenna. IEEE Int. Symposium on Antennas and Propagation Society; pp. 2173–2176.

Alomainy A, Hao Y, Hu X, Parini C, Hall PS. 2006. UWB on-body radio propagation and system modelling for wireless body-centric networks. *IEE Proc.- Commun.*, 153:107–114.

Alomainy A, Hao Y, Parini C, Hall PS. 2005a. Comparison between two different antennas for UWB on-body propagation measurements. *IEEE Antennas Wireless Propag. Lett.*, 4:31–34.

Alomainy A, Hao Y, Parini C, Hall PS. 2005b. On-body propagation channel char-acterisation for UWB wireless body-centric networks. IEEE Int. Symposium on Antennas and Propagation Society; pp. 694–697.

Bell JM, Iskander MF, Lee JJ. 2007. Ultrawideband hybrid EBG/ferrite ground plane for low-profile array antennas. *IEEE Trans. Antennas Propag.*, 55:4–12.

Cook HF. 1951. The dielectric behaviour of some types of human tissues at microwave frequencies. *Br. J. Appl. Phys.* 2:295–300.

Dimbylow P. 1996. The development of realistic voxel phantoms for electromagnetic field dosimetry. Proc. Int. Workshop on Voxel Phantom Development; pp. 1–7.

Duck F. 1990. *Physical properties of tissue: A comprehensive reference book*. San Diego, CA: Academic Press.

Durney CH, Massoudi H, Iskander MF. 1986. Radiofrequency Radiation Dosimetry Handbook, 4th ed. DTIC Document, Tech. Rep.

England TS. 1950. Dielectric properties of the human body for wave-lengths in the 1–10 cm. range. *Nature*, 166:480–481.

Erkmen F, Chen CC, Volakis JL. 2008. UWB magneto-dielectric ground plane for low-profile antenna applications. *IEEE Trans. Antennas Propag.*, 50:211–216.

Erkmen F, Chen CC, Volakis JL. 2009. Impedance matched ferrite layers as ground plane treatments to improve antenna wide-band performance. *IEEE Trans. Antennas Propag.*, 57:263–266.

Foster KR, Schwan HP. 1989. Dielectric properties of tissues and biological materials: A critical review. *Crit. Rev. Biomed. Eng.*, 17:25–104.

Gabriel C. 1996. Compilation of the dielectric properties of body tissues at RF and microwave frequencies. Brooks Air Force Technical Report, AL/OE-TR-1996-0037.

Gabriel C, Chan TYA, Grant EH. 1994. Admittance models for open ended coaxial probes and their place in dielectric spectroscopy. *Phys Med. Biol.*, 39:2183–2200.

Gordon RG, Roemer RB, Horvath SM. 1976. A mathematical model of the human temperature regulatory system—transient cold exposure response. *IEEE Trans. Bio. Med. Eng.*, 23:434–444.

Hall PS, Hao Y. 2006. Antennas and propagation for body centric communications. European Conference on Antennas and Propagation, pp. 1–7.

Lee GY. 2012. Conformal body-worn smart antenna system for wideband UHF operation. Ph.D. Dissertation. The Ohio State University.

Mason PA, Hurt WD, Walters TJ, D'Andrea JA, Gajsek P, Ryan KL, Nelson DA, Smith KI, Ziriax JM. 2000. Effects of frequency, permittivity, and voxel size on predicted specific absorption rate values in biological tissue during electromagnetic-field exposure. *IEEE Trans. Microw. Theory. Tech.*, 48:2050–2058.

Nagaoka T, Watanabe S, Sakurai K, Kunieda E, Watanabe S, Taki M, Yamanaka Y. 2004. Development of realistic high-resolution whole-body voxel models of Japanese adult males and females of average height and weight, and application of models to radio-frequency electromagnetic-field dosimetry. *Phys. Med. Biol.* 49:1–15.

Salman S, Psychoudakis D, Volakis JL. 2012. Determining the relative permittivity of deep embedded biological tissues. *IEEE Antennas Wireless Propag.* Let., 11:1694–1697.

Tzanidis I, Chen CC, Volakis JL. 2010. Low profile spiral on a thin ferrite ground plane for 220–500 MHz operation. *IEEE Trans. Antennas Propag.*, 58:3715–3720.

Xu X, Chao T, Bozkurt A. 2000. VIP-man: An image-based whole-body adult male model constructed from color photographs of the visible human project for multiparticle monte carlo calculations. *Health Phys.* 78:476–486.

Yamamoto T, Yamamoto Y. 1976. Electrical properties of the epidermal stratum corneum. *Med. Biol. Eng.*, 14:151–158.11

Zhang L, Wang J, Volakis JL. 2012. Textile antennas and sensors for body-worn applications. *IEEE Antennas Wireless Propag. Lett.*, 11:1690–1693.

Zhu S, Langley R. 2009. Dual-band wearable textile antenna on an EBG substrate. *IEEE Trans. Antennas Propag.*, 57:926–935.

19

IMPLANTABLE HEALTH CARE SYSTEM PARADIGM

MASAHARU TAKAHASHI AND KOICHI ITO
Chiba University, Chiba, Japan

19.1 INTRODUCTION

Small implantable devices have been investigated with great interest for wireless medical applications because they are promising different clinical usages in order to promote patient's independence (Hall and Hao, 2006; Steinhaus et al., 1994). These implantable devices are embedded into the human or pet body. The vital information can be transmitted from implantable devices to external equipment using a wireless communication link. It is expected that in this way reducing the pressure on overburdened health care systems and replacing the cable connectivity implanted into the human body for biological telemetry monitoring will be achieved. Nowadays, the in-body technology is an important matter for the implanted cardiac pacemaker and the animal chip for tracking and tagging. Moreover, it is desired to be applied in health care to transfer information on temperature, blood pressure, cardiac beat, and so on (Malasri and Wang, 2009; Sani et al., 2010). However, the human body is not an attractive environment for wireless signals because of the mutual influence between the human body and the implanted device's antennas, namely, implanted antennas. According to studies of the prior arts, antenna gain and efficiency will decrease due to the high dielectric loss caused by the human body. Moreover, as the implanted device should be embedded into the human body, the implanted antenna must have small or thin size. Therefore, the research on antennas for implantable devices is very important. In addition, when designing the implanted antenna, it is necessary to

Handbook of Biomedical Telemetry, First Edition. Edited by Konstantina S. Nikita.

consider various constraints because of being buried into the body, including power consumption, size, frequency, and conformance to the biological.

An example of an implantable device is Verichip, which was used in the United States (Foster et al., 2012). Verichip was the Food and Drug Administration (FDA)–approved human-implantable radio frequency identification (RFID) microchip. Verichip was typically implanted between the shoulder and elbow area of an arm by use of syringe. The device was used, for example, for age verification and significant institutional admission. The operating frequency of Verichip is 134 kHz. Therefore, when you read the information of the microchip, you need to close the dedicated reader because communication distance is very short, about a few centimeters. If implantable devices are applied to medical situations as described above, the communication distance of Verichip is too short. To solve this problem, it is necessary to increase the operation frequency (Chen et al., 2009; Merli et al., 2008; Scanlon et al., 2000; See et al., 2010). Until now, microstrip antennas or PIFAs were proposed for implantable devices which were covered with a dielectric material in the frequency band at 402–405 MHz (CEPT, 1997; Gosalia et al., 2005; Kim and Rahmat-Samii, 2004; Soontornpipit et al., 2004). In these studies, the characteristics of the antennas, such as input impedance, radiation pattern, and specific absorption rate (SAR), around the antennas were presented. However, these antennas are not small enough to be embedded into the human body. Therefore, one of the ultrahigh-frequency (UHF) bands is suggested to be used for in-body wireless communication. A higher frequency band will cause higher path loss, but it has the advantage of more compact antenna size and the UHF band can also be applied for RFID application. In order to reduce the size of the antenna, some implanted antennas were investigated for 2.45-GHz applications (Collin et al., 2008; Huang et al., 2011; Karacolak et al., 2008; Merli et al., 2011; Mizuno et al., 2011; Usui et al., 2006; Xia et al., 2009). These antennas are small enough to be embedded into the human body.

In this chapter, an implanted antenna that can be placed into a small implantable device as a capsule is examined. We used the finite-difference time-domain (FDTD) method for simulation and some experimental approaches to investigate the antenna performances, including reflection coefficients and radiation patterns. In addition, we investigated the difference of several numerical human models (a cylindrical, rectangular, and layered phantom) according to the strength of the electric field and radiation performance.

19.2 MULTILAYERED MODEL SIMULATING HUMAN BODY

Since implantable devices are embedded into the human body, which consists of lossy dielectric materials, including skin, fat, and muscle, the performances of implantable devices are supposed to be affected by them. Therefore, how the human body is modeled becomes very important. We should consider a nonuniform model but a multilayered one. Since the antenna is embedded in the layer of fat in general, the

TABLE 19.1 Electrical Properties of Tissues

	400 MHz		950 MHz		2.45 GHz	
	ϵ_r	σ (S/m)	ϵ_r	σ (S/m)	ϵ_r	σ (S/m)
Skin	46.8	0.69	41.1	0.88	38.0	1.5
Fat	5.6	0.04	5.5	0.05	5.3	0.1
Muscle	57.1	0.80	54.9	0.96	52.7	1.7

Source: Gabriel et al. (1996).

effects of bone can be ignored because of a high dielectric constant in the muscle between them. The electrical properties are defined as the value of each tissue in Table 19.1.

Depending on the part of the body, the model can be simulated as a multilayered structure like that shown in Figure 19.1. Limbs are replaced by a cylinder and a rectangular structure and the torso can be replaced with the layered structure.

In the measurement of the implanted antenna, electrical and thermal characteristics of human phantoms depend on phantom materials. It is therefore necessary to know the characteristics of materials. In addition, various shaped phantoms are used in the measurement (e.g., head, abdomen, torso, whole body). The shape of phantoms is very important because the radiation characteristics of implanted devices depend on the shape of phantoms (Ogawa et al., 2001). In the experimental investigations, different tissue-equivalent phantoms are used for their purposes. Okano et al. (2000) proposed the high-water-content tissue-equivalent semihard phantom. It can simulate the dielectric properties of high-water-content tissues over frequency ranges of 100 MHz–10 GHz (Ishido et al., 2004) The other features of the phantom are summarized as follows:

- Easy adjustment of dielectric properties
- Easy to make arbitrary shaps and multilayers (Okano et al., 2000)

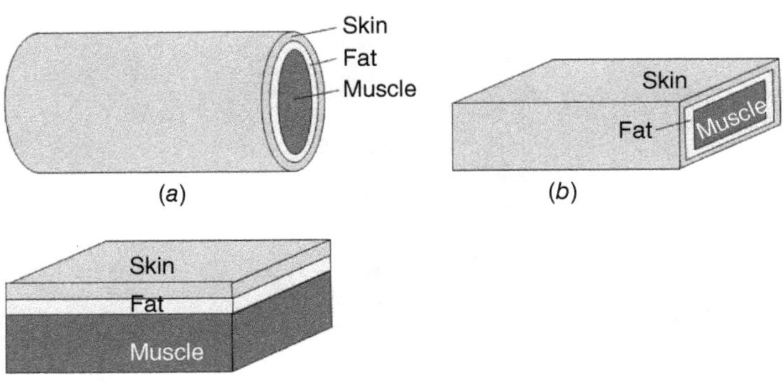

Figure 19.1 Calculation models of human body:(*a*) cylindrical phantom; (*b*) rectangular phantom; (*c*) layered phantom.

TABLE 19.2 Example Compositions of Semihard Phantoms at 2.45 GHz

	Ingredient	Amount (g)
Muscle	Deionized water	3375
	Polyethylene powder	340
	Agar	104.6
	Thickner:TX-151	82.9
	Sodium chloride	10
	Sodium dehydroacetate	2.0
Skin	Deionized water	55
	Silicon emulsion	215
	Gelatin	35
Fat	Silicon	200
	Glycerin	70
	Aluminum powder	40

- Not necessary to use a container (or shell)
- Inexpensive and popular ingredients

Table 19.2 shows an example of the composition of a semihard phantom at 2.45 GHz. Here, the dielectric properties are adjusted by the amount of polyethylene powder (relative permittivity) and that of sodium chloride (conductivity), respectively. Sodium dehydroacetate is a preservative and agar is a coagulant. As a result, the dielectric properties of the phantom agree with those of target values within 5%. Figure 19.2 show photographs of the fabricated phantom.

Realistic high-resolution whole-body models of the Japanese male and female are used for numerical calculation. Nagaoka et al., (2004) have developed a realistic

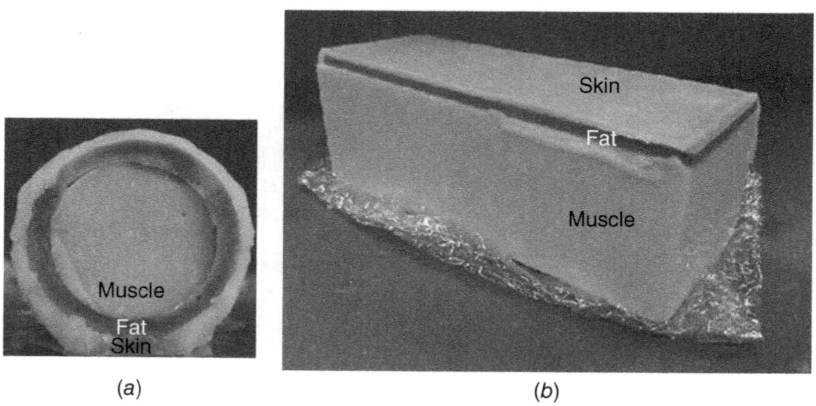

Figure 19.2 Photographs of fabricated phantoms: (*a*) cylinder phantom (Uno et al., 2010); (*b*) layered phantom (Mizuno et al., 2011).

high-resolution whole-body voxel model by collecting magnetic resonance imaging (MRI) data from volunteer Japanese adult males and females with average body size. The average height and weight of Japanese adults between the ages of 19 and 30 years are 171.4 cm and 63.3 kg for males and 159.1 cm and 52.6 kg for females, respectively. On the basis of these data, a male and a female subject whose height and weight were close to the Japanese average values were chosen for the model. It is impossible to automatically segment the voxels with sufficient accuracy using available imageprocessing technologies. All tissue and organidentification processing was therefore performed manually by medical personnel. Researchers checked the three-dimensional images and the three orthogonal planes and smoothed the boundaries of the following tissues and organs: axial, sagittal, and coronal. The final positioning and shaping of the tissues and organs were performed under the supervision of medical doctors using three-dimensional visualization software. Figure 19.3 shows the developed male and female models that are assumed to stand in the upright position with their hands at both sides of their bodies. These models are composed of voxels of $2 \times 2 \times 2\,\text{mm}^3$ and are divided into 51 different tissues and organs.

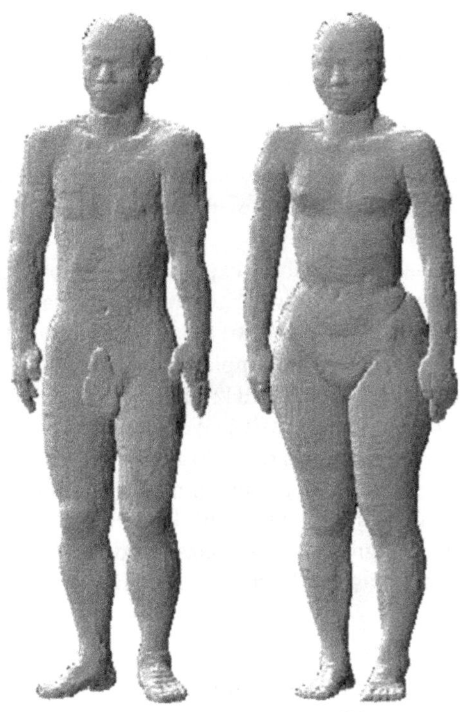

Figure 19.3 Realistic high-resolution whole-body models of Japanese (*a*) male and (*b*) female.

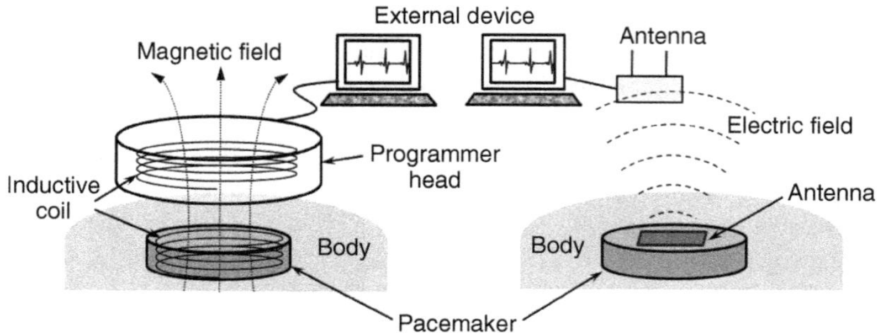

Figure 19.4 Images of wireless communication systems for implantable cardiac pacemaker.

19.3 CARDIAC PACEMAKER EMBEDDED IN MULTILAYERED MODELS

An implantable cardiac pacemaker, which forms pulses and senses a heart, is one medical implanted device (Mond, 2001). Implantable cardiac pacemakers have a need for communication, both to rewrite new programs and to transmit vital information. Today, communication between the implanted cardiac pacemaker and external devices is performed by inductive coupling. Therefore, data transmission speed is not enough for normal communication and the communication distance is only about a few centimeters. These factors cause the following two problems: long binding hour of a medical doctor and a patient and a risk for the patient to acquire an infection caused by the embedment of a cardiac pacemaker. In order to solve these problems, a wireless communication system with a small antenna has been investigated (Figure 19.4).

In this section, a cardiac pacemaker is expected to have an implantable antenna as a new tool for the communication between a cardiac pacemaker and external devices instead of a coil which is used for inductive link. A PIFA loaded onto a cardiac pacemaker was proposed (Figure 19.5). An implantable antenna for a cardiac pacemaker is designed to operate in the 402–405 MHz Medical Implant Communications Service (MICS) band (Takahashi et al., 2010). We analyzed the antenna characteristics embedded in a multilayer medium model. In addition, we implanted a pacemaker loading-type antenna in the numerical human model and investigated the effectiveness of this antenna and the validity of the multilayered structure model by numerical analysis. As a result, the design of the pacemaker loadings-type antenna that used the multilayer medium model was valid, and the effectiveness of the designed antenna was confirmed.

19.3.1 Modeling and Analytical Method

Figure 19.5 shows the model of the antenna loaded onto a cardiac pacemaker, including its housing, pacing lead, and electrode. In order to load the antenna onto a pacemaker, it is necessary to design an antenna adequately small and lowprofile. As an

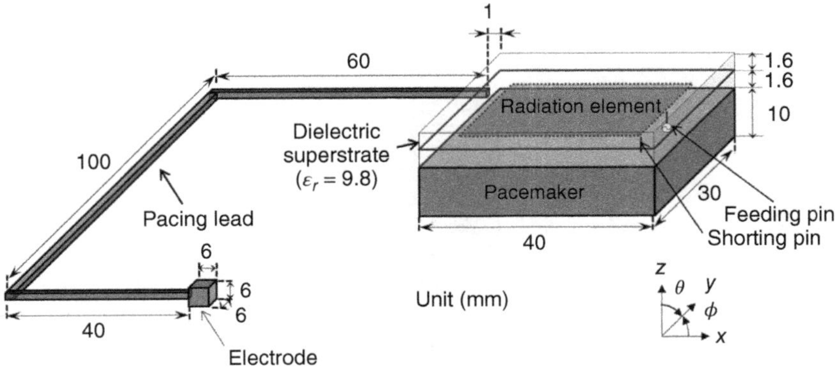

Figure 19.5 Structure of proposed antenna and cardiac pacemaker model. Copyright © 2010 IEICE.

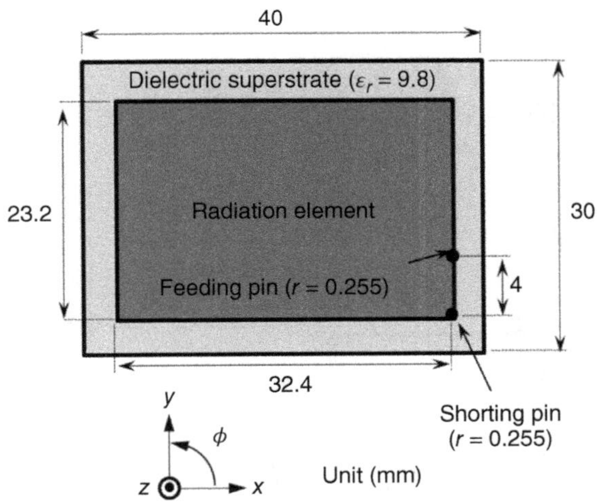

Figure 19.6 Structural details of antenna. Copyright © 2010 IEICE.

antenna which fulfills the demands, PIFA is proposed and designed. The antenna is designed to operate in the 402–405 MHz MICS band. The pacemaker is 40 mm × 30 mm × 7.5 mm, the pacing lead has a diameter of 2 mm and a length of 200 mm, and the electrode is a cube 6 mm on a side (Wang et al., 2000). Figure 19.6 shows a detail configuration of the PIFA. The radiation element has a dimension of 30 mm × 21 mm and is covered with dielectric substrate ($\varepsilon_r = 9.8$) so as to shorten wavelength and miniaturize the antenna. In order to match the antenna to 50 Ω, the distance between the feeding and shorting pins is adjusted.

Figure 19.7 shows the size of tissue-equivalent phantom and the position of the implanted antenna. In this section, two types of torso phantoms with dimensions 290 mm × 450 mm × 190 mm are used. One is a multilayered phantom which simulates the skin, fat, and muscle. The electric constants are shown in Table 19.3. An

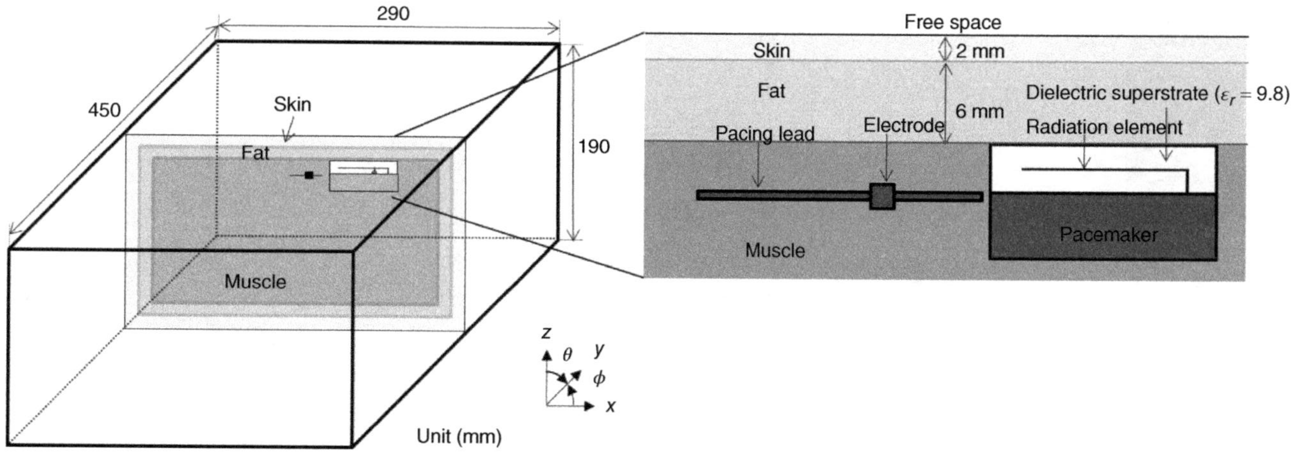

Figure 19.7 Position of implanted antenna in tissue-equivalent phantom. Copyright © 2010 IEICE.

TABLE 19.3 Electric Constants of Tissues

	Relative Permittivity ε_r	Conductivity σ(S/m)
Skin	46.8	0.69
Fat	5.6	0.04
Muscle	57.1	0.80
Two-thirds muscle	38.1	0.53

input characteristic of the implantable antenna is a significant impact on the thickness of the layers of tissue. Therefore, we consider that the average thickness of the skin of the human body is 2 mm and the skin layer is to a depth of 2 mm from the surface of all six sides of the box. The thickness of the fat layer is 6 mm, and inside it is full of muscle. The other type of phantom has a uniform structure. Electric constants of the uniform phantom mimics two-thirds human muscle, which is commonly used to represent average body properties (Gabriel et al., 1996). The pacemaker model is implanted 6 mm deep into the torso surface.

The FDTD method is used for the simulation of the proposed PIFA, and analytical conditions are as follows. The cells are separated nonuniformly: minimum cell size is $\Delta x = \Delta y = 0.5$ mm, $\Delta z = 0.4$ mm and maximum cell size is $\Delta x = \Delta y = \Delta z = 5$ mm. Twelve perfectly matched layers (PMLs) are employed as the absorbing boundary conditions and the time step is 0.785 psecs. The feeding pin and the shorting pin are modeled by subcell wire approximation: The feeding pin has a diameter of 0.51 mm and the shorting pin has a diameter of 0.286 mm. All of the shielded housing, pacing lead, and electrode are modeled with a perfect electric conductor (PEC).

19.3.2 Link Budget

Table 19.4 shows a link budget, as an example, between an implanted antenna and a receiving antenna (Usui et al., 2006). Since the purpose of the communication system is transmitting information between an implanted device and an external programmer which is kept in a hospital or clinic, it is assumed that the antenna is used in a square sickroom with a side about 6 m. Considering the space of the room, the distance between the transmitter and the receiver is 8 m. When the bit rate is 800 kbps for transmitting vital information and new programs of a pacemaker and the input power is 25 μW, the implanted antenna gain should be kept more than −33.0 dBi. Based on the link budget, the availability of the proposed antenna is judged by its gain: whether it exceeds −33.0 dBi or not.

19.3.3 Antenna Characteristics

Figure 19.8 shows the reflection coefficient of the implanted antenna. As shown in the figure, the range in which the reflection coefficient is less than −10 dB is 398–407 MHz and is also sufficient to meet the 402–405 MHz desired frequency band. Moreover, the input characteristics when this antenna is set up in the uniform

TABLE 19.4 Link Budget

Frequency	403.5 MHz
Transmission power	−46.0 dBW
Tx antenna gain	−33.0 dBi
Effective Isotropic Radiated Power (EIRP)	−80.0 dBW
Distance	8.0 m
Path loss	42.6 dB
Rx antenna gain	0.0 dBi
Bit rate	800 kbps
Link C/N_0	75.5 dB/Hz
Required C/N_0	71.1 dB/Hz
Margin	4.4 dB

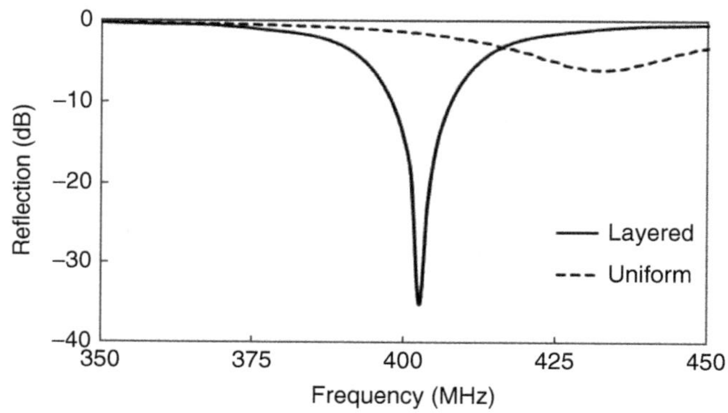

Figure 19.8 Reflection coefficient of multilayered model. Copyright © 2010 IEICE.

phantom are shown. In a uniform phantom, the resonant frequency of this antenna will be as high as 431 MHz, so it is not possible to match the desired frequency. As a result, the human body is treated not as a uniform medium but as multilayered media in the design of implantable devices.

The far-field radiation patterns are illustrated in Figure 19.9. It secures more than the gain −33 dBi within the range of ±120° in the x–z plane. Thus, it is possible to communicate in the front direction of the human body.

The pacemaker model for the experiment is shown in Figure 19.10. This model is embedded in a uniform phantom for experiment, and the radiation characteristics were measured. Figure 19.11 shows the measurement results with the calculated result. Both results are in good agreement. The antenna gain of the front direction is −30.7 dBi by the uniform model.

19.3.4 Verification by Human Body Model

In order to verify the validity of the design of the antenna loading a cardiac pacemaker in the model rectangular multilayer medium presented in Section 19.3.3, in

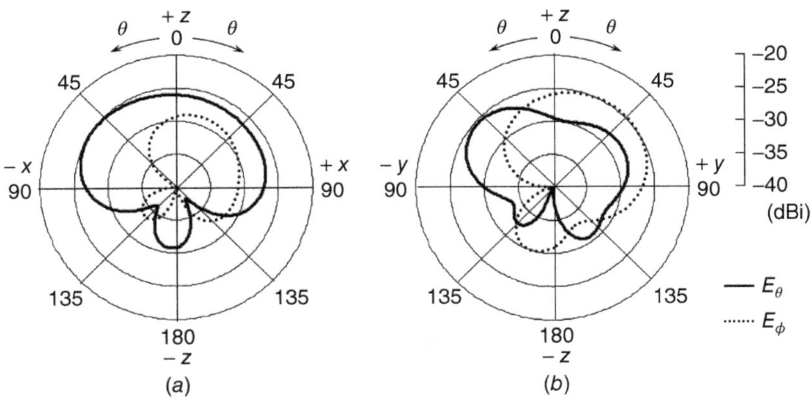

Figure 19.9 Radiation patterns of multilayered model: (*a*) *x*–*z* plane; (*b*) *y*–*z* plane. Copyright © 2010 IEICE.

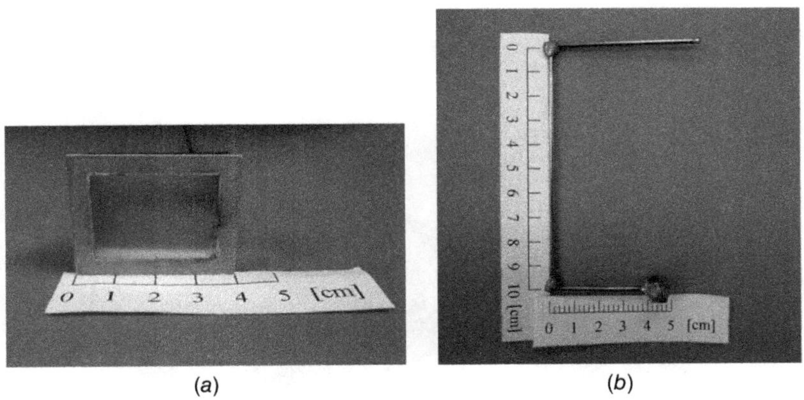

Figure 19.10 Fabricated models: (*a*) antenna; (*b*) lead and electrode.

this section, analysis was performed using the high-resolution human body model (Nagaoka et al., 2004). Figure 19.12 shows the state of the male model in whom a pacemaker was implanted. In general, the pacemaker is implanted between the fat layer and muscle layer of the human breast. For this reason, at the patient's chest where the pacemaker was implanted a bulge occurs due to the pacemaker. Assuming the implantation of a pacemaker in the chest of the human body model which was modified to swell, Figure 19.13 shows the cross-sectional view of the *x*–*z* plane that passes through the center of the antenna when the pacemaker was installed in the male model. The pacemaker was placed in a way that guarantees a distance of up to 6 mm between the dielectric surface of the antenna and the inner surface of the skin layer. In general, pacemaker leads, were attached via the subclavian vein, the superior vena cava from the pacemaker, to reach the inner wall of the right ventricle. Therefore, the

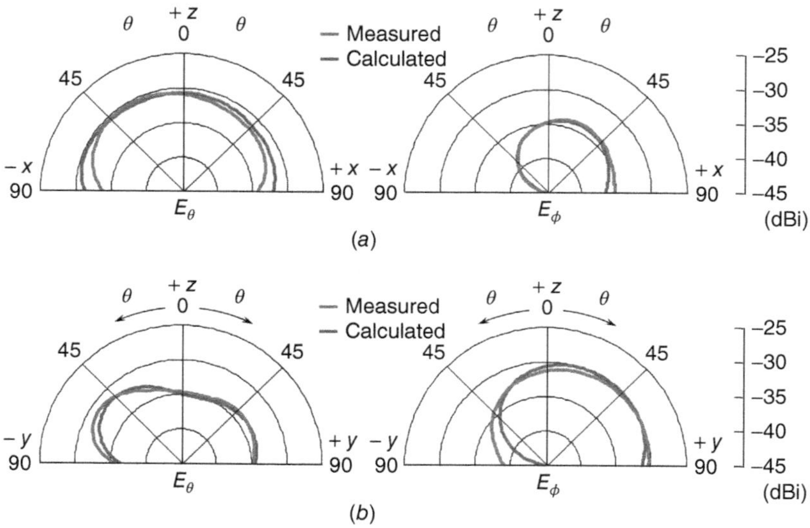

Figure 19.11 Measured radiation pattern: (*a*) x–z plane; (*b*) y–z plane.

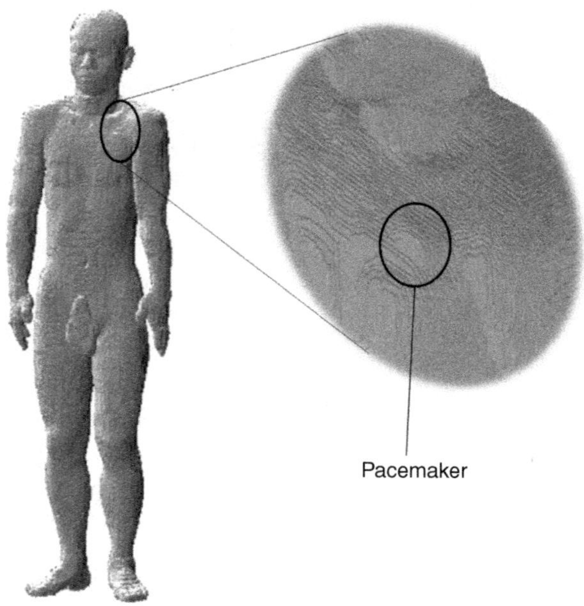

Figure 19.12 Implanted pacemaker in a high-resolution human model. Copyright © 2010 IEICE.

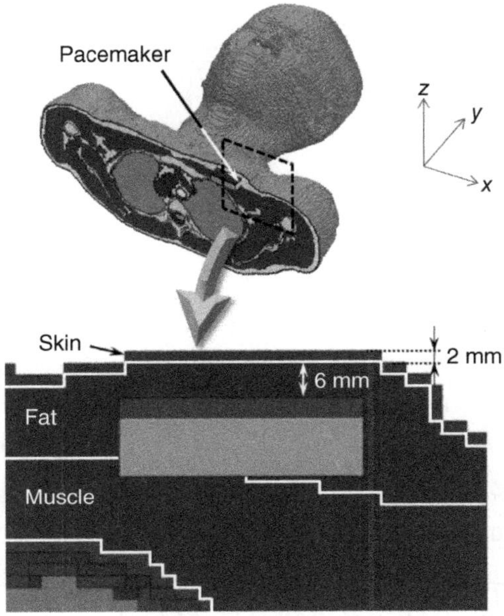

Figure 19.13 Position of implanted pacemaker. Copyright © 2010 IEICE.

lead wire simulated by a perfect conductor was set up in this way. The same was done for female models.

When the antenna loading pacemaker was planted in the human body model, the distance between the feeding pin and the short pin shown in Figure.19.6 was changed from 4 to 4.6 mm to improve the matching. Figure 19.14 shows the reflection characteristics of the antenna which was installed in each model for the male and female. In addition, the results show that the antenna implanted in a male model was designed using the multilayer model. The frequency ranges for less than −10 dB reflection on all models are almost identical. From these results, the reflection in the male Uersus the female model does not differ. In addition, the effectiveness of the design by a multilayer model has been shown.

Figure 19.15 shows radiation patterns of the male and female models and multilayer model. The radiation characteristics of these are almost the same and the fact that they radiate strongly in the direction of the front of the body can be confirmed. Antenna gain in this direction ($\theta = 0°$) is −25.1 dBi for both human models. When $\theta \leq 127°$ in the x–z plane, the antenna gain is greater than −33 dBi. In the y–z plane, it became $\theta \leq 93°$ for the male model and $\theta \leq 106°$ for the female model. Moreover, the gain of the antenna is −30.7 dBi for the uniform model. By an antenna implanted in the human body model, the antenna gain is increased, and the range of the communication angle is also broaded. From this result, the characteristics of the antennas designed using the multilayer model and the high-resolution human model are almost the same, and the effectiveness of the design that used the multilayer model

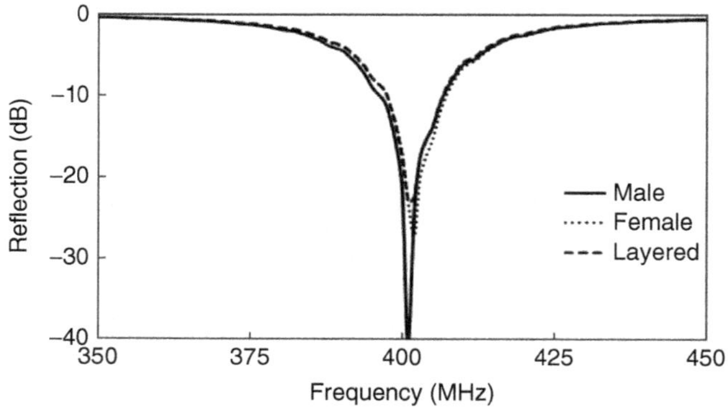

Figure 19.14 Reflection in human models. Copyright © 2010 IEICE.

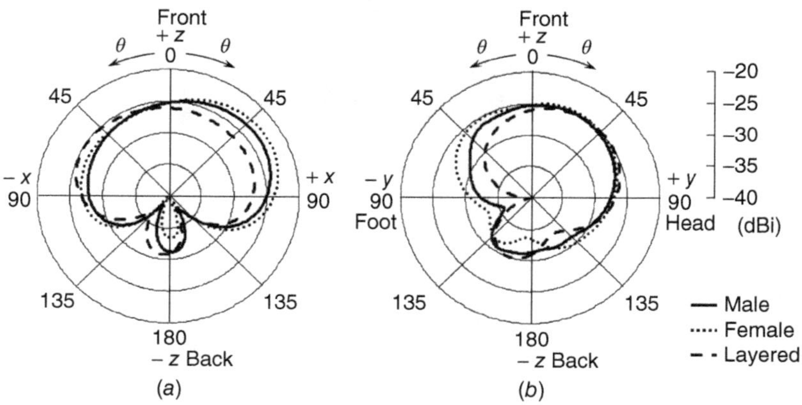

Figure 19.15 Radiation pattern of human models: (*a*) *x*–*z* plane (E_θ); (*b*) *y*–*z* plane (E_ϕ). Copyright © 2010 IEICE.

was shown. In addition, it was confirmed that it is able to secure a gain that is enough for the communication as MICS.

19.4 IMPLANTABLE HEALTH CARE SYSTEM PARADIGM

In this section, an implantable antenna with miniaturized structure was proposed which is designed to operate at the industrial scientific medical (ISM) bands (2.45 GHz) (Collin et al., 2008; Karacolak et al., 2008; Mizuno et al., 2011; Usui et al., 2006). This antenna has a size that is ideal for it to be inserted with the injection syringe. An implanted antenna must match with an integrated circuit (IC) chip whose input impedance is not 50 Ω. In this section, the helical dipole antenna which can match with an IC chip was proposed (Mizuno et al., 2011). It is not suitable to embed metallic materials directly into the human body. Therefore, the implantable device is covered with fluororesin, which has a cylindrical shape of 1.2 mm diameter and 12.0 mm length and the antenna element is buried into this fluororesin in order to be able to be inserted into the human body using a syringe. This material is resistant to heat. In addition, to confirm the validity of the analyzed results of the input impedance and radiation patterns, the proposed antenna is fabricated and measured using a biological tissue-equivalent phantom.

19.4.1 Link Budget for Wireless Communication

The proposed antenna is assumed to be used in a room for medical data transmission. Therefore, it is necessary to know some key parameters related to the link budget for communication between the implanted antenna and the receiver. The operating frequency is set to 2.45 GHz and the input power to the implanted antenna is 25 μW [from the European Research Council (ERC) limitation (CEPT, 1997)]. The receiving antenna is assumed to be a monopole antenna. The required Rx power (RP) is calculated as

$$\mathrm{RP} = \frac{E_b}{N_0} + kT_0 + B_r \qquad (\mathrm{dBW}) \tag{19.1}$$

where E_b/N_0 is the ideal phase shift keying (PSK) (dB), k is the Boltzmann constant (J/K), T_0 is the ambient temperature (K), and B_r is the required bit rate for transmitting vital information (kbps). The available Rx power (AP) is calculated as

$$\mathrm{AP} = P_t + G_t + G_r - L_f \qquad (\mathrm{dBW}) \tag{19.2}$$

where P_t is the required power of the implanted antenna (dBW), G_t is the gain of the implanted antenna (dBi), G_r is the gain of the receiving antenna (dBi), and L_f is free-space path loss [dB]. Here, it is supposed to be $B_r = 10.0$ kbps, $L_f = 52.26$ dB. If the AP is larger than the RP by 20 dB, the wireless communication between the implanted antenna and external equipment will be possible. Table 19.5 shows the parameters used to calculate the RP and the AP. The calculated results are shown in Table 19.6. If the gain of the proposed antenna is more than −18.0 dBi, the margin exceeds 40 dB. This result is enough to guarantee reliable communication between the implanted antenna and external equipment. Communication will be made possible if the gain of the proposed antenna is −18.0 dBi.

19.4.2 Calculation of Helical Dipole Antenna

The implantable device is small enough to be buried into the human body. Therefore, it is necessary that the implanted antenna be very small too. Figure 19.16 shows the designed antenna structure. The antenna element is in the form of a 1.0 mm diameter helix to extend the distance of the element. In addition, the antenna provides a shorting pin in order to match with an IC chip. The number of turns of the helix is 6 and the conducting strip is 0.4 mm. The thickness of the conducting strip is 0.04 mm.

TABLE 19.5 Parameters of Link Budget

Required Rx power	
Noise power density N_0 (dB/Hz)	−199.95
E_b/N_0 (ideal PSK) (dB)	9.6
Boltzmann constant k	1.38×10^{-23}
Ambient temperature T_0 (K)	293
Bit rate B_r (kbps)	10
Available Rx power	
Rx power P_t (dBW)	−46.02
Rx antenna gain G_t (dBi)	−18.0
Tx antenna gain G_r (dBi)	2.15
Free-space loss	
Frequency (GHz)	2.45
Free-space loss L_f (dB)	52.26

TABLE 19.6 Calculated Results of Link Budget

Required Rx power (dBW)	−154.3
Available Rx power (dBW)	−114.1
Margin (dB)	40.2

Moreover, the implantable device is covered with fluororesin ($\varepsilon_r = 2.1$) which has a cylindrical shape of 1.2 mm diameter and 12.0 mm length and the antenna element is buried into this fluororesin in order to be able to be inserted into the human body using a syringe. This size is almost equivalent to Verichip. The operating frequency is 2.45 GHz. The antenna is assumed to be a match with a μ-Chip (Usami and Ohki, 2003) that has conjugate impedance 9.3−j55.2 Ω (Yamada et al., 2007).

For the analysis the layered phantom shown in Figure. 19.1 was used. The dimensions of the model are 60 mm × 60 mm × 60 mm. In this research, the human-body-equivalent phantom is simulated as a human arm, which consists of three tissues, dry skin, fat, and muscle. The thickness of each is 2 mm, 4 mm, and 54 mm. The electrical properties are defined as the value of each tissue at 2.45 GHz (skin: $\varepsilon_r = 38.0$, $\sigma = 1.5$ S/m; fat: $\varepsilon_r = 5.3$, $\sigma = 0.1$ S/m; muscle: $\varepsilon_r = 52.7$, $\sigma = 1.7$ S/m)[1]. The designed antenna is buried into the fat layer and positioned just beneath the skin layer. In this condition, the antenna characteristics for each of them were analyzed by the FDTD method.

19.4.3 Experiment of Helical Dipole Antenna

Figure. 19.17 is a photograph of the fabricated antenna. The antenna is fed by a coaxial cable, and the $\lambda/4$ balun, which is embedded in order to change the unbalanced

[1]FCC website: http://www.fcc.gov/fcc-bin/dielec.sh.

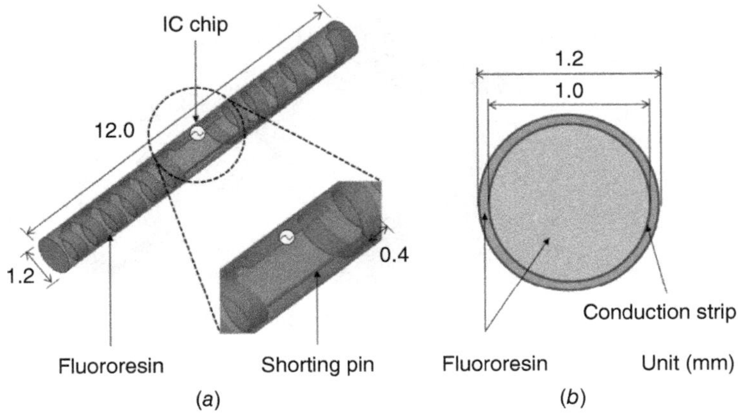

Figure 19.16 Configuration of the helical dipole antenna: (*a*) overview; (*b*) cross-sectional view.

Figure 19.17 Fabricated antenna.

line to a balanced line, is also used to improve the matching. The input impedance and radiation pattern of the antenna with the phantom are measured. Figure. 19.18 is a photograph of the phantom. The phantom consists of three tissues which are the same as the calculation model.

Figure. 19.19 shows calculated and measured input impedances of the proposed antenna. The calculated input impedance and reflection coefficient of each antenna at 2.45 GHz are 11.9+j53.3 Ω and −16.6 dB. The reflection coefficient is normalized by the impedance of the μ-Chip. The measured input resistance is 12.3 Ω, and the input reactance is 52.7 Ω. The measured reflection coefficient at 2.45 GHz is −15.2 dB. As a result, the fabricated antenna can match with a μ-Chip and the antenna operates well at 2.45 GHz.

Figure. 19.20 indicates the radiation pattern in the x–z and y–z planes when the antenna is embedded into the human-body-equivalent phantom. It is important that the radio wave passes through the skin because there is an external equipment in that direction. Therefore, the gain of the +z direction is important. It is found that the

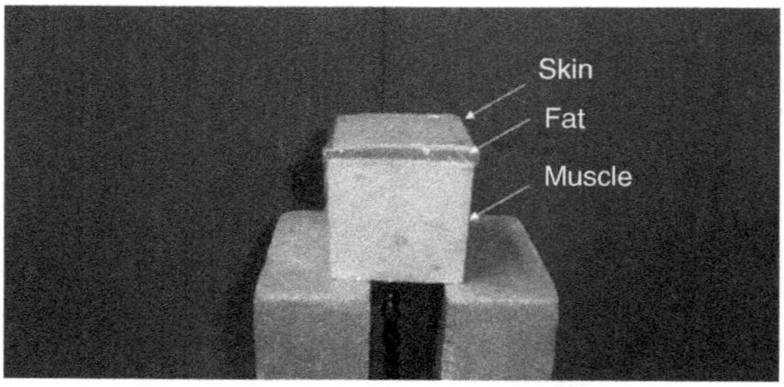

Figure 19.18 Measurement model.

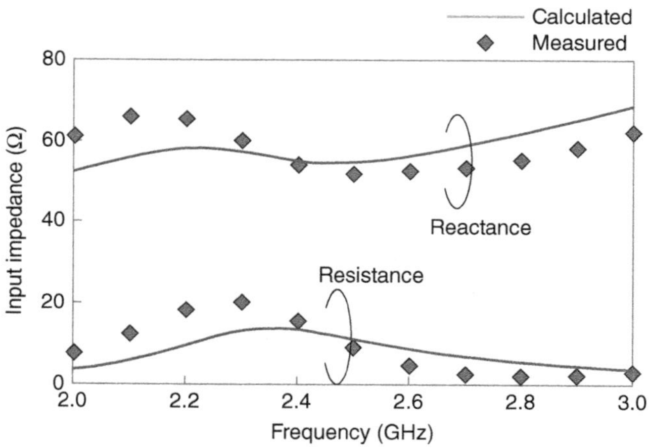

Figure 19.19 Input impedance of fabricated antenna.

direction of maximum radiation of the calculated results in both planes is in the $+z$ direction. The maximum calculated and measured gains are -14.5 dBi and -15.3 dBi, respectively. According to the link budget for wireless communications if the gain of the antenna is more than -18.0 dBi, the antenna can communicate with external equipment. From the above, the calculated and measured results of the antenna are higher than this target centered at $+z$ in both planes, and it is found that the antenna is suitable for implantable devices.

19.4.4 Analysis Using High-Resolution Model

The characteristics of the proposed antenna were investigated with the three-layer rectangular phantom. However, the structure of the real human body is complicated. In order to confirm the validity of the phantom, the proposed antenna was buried into the high-resolution whole-body model (Nagaoka et al., 2004) and analyzed by the

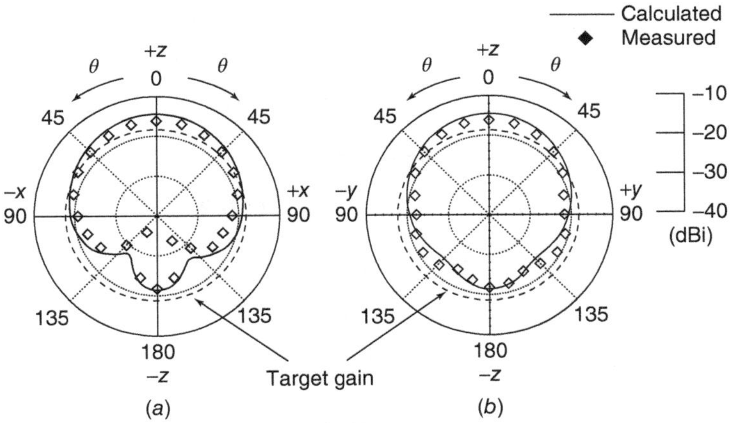

Figure 19.20 Radiation pattern: (*a*) *x*–*z* plane; (*b*) *y*–*z* plane.

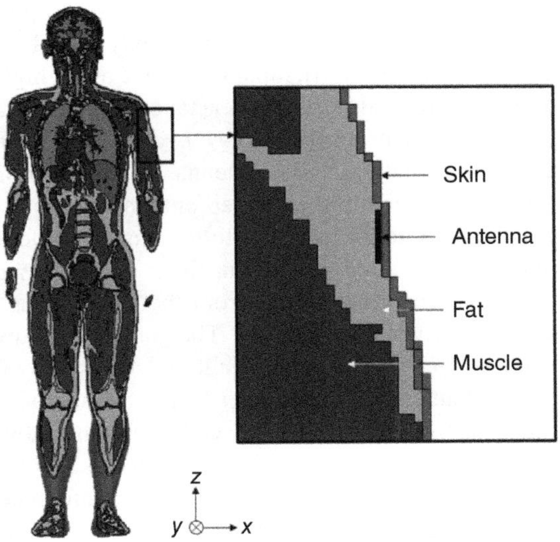

Figure 19.21 Position of embedded antenna.

FDTD method. Figure. 19.21 shows the position where the antenna was embedded. The proposed antenna is implanted into the upper arm of the high-resolution model. The antenna is embedded into the fat layer and positioned just beneath the skin layer, which is the same as the calculation using the rectangular phantom. The relative permittivity and conductivity of each tissue are equal to the values at 2.45 GHz (see FCC website).

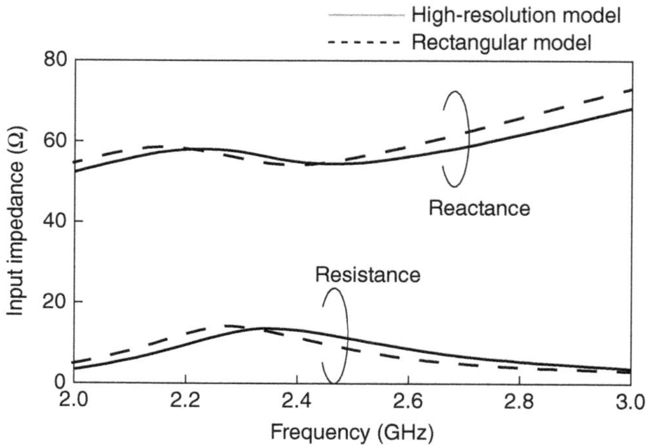

Figure 19.22 Input impedance of arm-buried antenna.

Figure. 19.22 shows the calculated input impedance of the proposed antenna when the antenna is buried into the high-resolution whole body model and the three-layer rectangular phantom for comparison. An input resistance of the high-resolution body model at 2.45 GHz is 12.2 Ω, and the input reactance is 54.7 Ω. According to the results, the reflection coefficient normalized by μ-Chip at 2.45 GHz is −17.4 dB. This means that the tissues that touch the antenna are closely related to the input impedance. This result shows that the proposed antenna can match with a μ-Chip even if the antenna is buried into the high-resolution human model.

Figure. 19.23 indicates radiation patterns in the x–z and x–y planes. The gain of the $+x$ direction is important and differs from the earlier ones because the distance to the outside of the body is the nearest. The gain of the $+x$ direction in both the x–y and x–z planes is −14.7 dBi. This result is decreased by 0.2 dB compared with the rectangle-buried antenna. On the other hand, the gain of the -x direction is decreased significantly compared with the above. This is caused by the human body existing in the -x direction and the fact that the radio wave is absorbed by the human body. However, considering the link budget for wireless communications, the gain is enough to communicate with external equipment. Moreover, the result indicates that the communication range of the antenna that is buried into high-resolution model is wider than that of the rectangle-buried one. It shows that the antenna can be used for implantable devices even if the device is buried into human body.

19.5 CONCLUSION AND FUTURE RESEARCH DIRECTIONS

In this chapter, implanted antennas to be applied to medical situations were described. One is the PIFA loaded onto a cardiac pacemaker for use in the 400 MHz MICS band; another one is the helical dipole antenna for the ISM band. Implanted antenna models

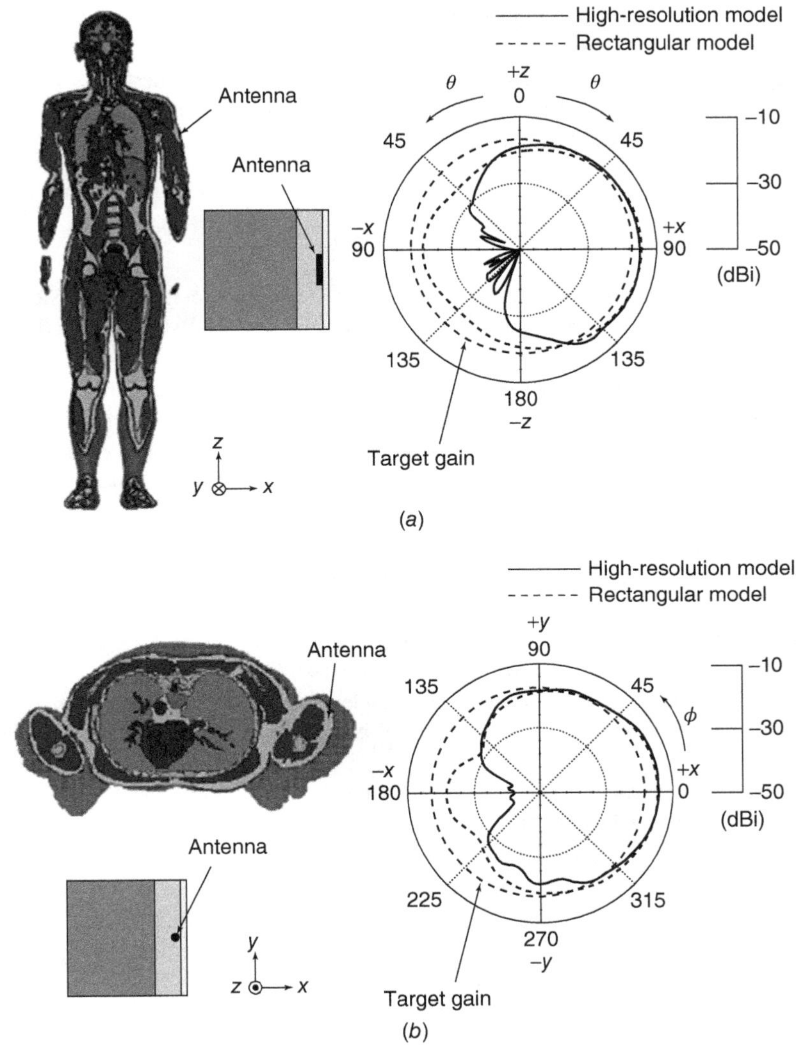

Figure 19.23 Radiation pattern: (*a*) *x*–*z* plane; (*b*) *x*–*y* plane.

were numerically analyzed by the FDTD method and the antenna characteristics were evaluated. We analyzed the antenna characteristics of an antenna implanted in a multilayer medium model. Moreover, implanted antennas were buried into high-resolution numerical human models and the effectiveness of these antennas and the validity of the multilayered structure model were investigated by numerical analysis.

In addition, the antenna was fabricated and the input impedance and radiation patterns were measured with phantoms in order to confirm the validity of the calculations. As a result, it is possible to use these implantable antennas for health care wireless communication systems.

As future research, it is necessary to measure antennas embedded into animal and human bodies. Moreover, it is important to develop a device with a sensor that measures not only ID information but also biological data.

REFERENCES

CEPT. 1997. ERC recommendation 70–3 relating to the use of short range devices (SRD). Eur. Postal Telecommunications Administration Conf., Tromsø, Norway, CEPT/ERC70-03, Annex 12.

Chen ZN, et al. (2009). Transmission of RF signals between MICS loop antennas in free space and implanted in the human head. *IEEE Trans. Antennas Propag.*, 57(6):1850–1853.

Collin G, et al. 2008. Human implanted spiral antenna for a 2.45 GHz wireless temperature and pressure SAW sensor system. IEEE Int. Symp., Antennas and Propagation, July.

Federal communications commission (FCC).

Foster K.R, Jaeger J. 2012. RFID Inside. *IEEE Spectrum.* 44(3): 24–29.

Medical Implant Communication Service (MICS) Rules and Regulations, *Fed. Reg.*, 64(240):69926–69934.

Gabriel S, et al. 1996. The dielectric properties of biological tissues: II. Measurements in the frequency range 10 Hz to 20 GHz. *Phys. Med. Biol.*, 41:2251–2269.

Gosalia K, et al. 2005. Impedance matching and implementation of planar space-filling dipoles as intraocular implanted antennas in a retinal prosthesis. *IEEE Trans. Antenna Propag.*, 53(8):2365–2373.

Hall PS, Hao Y. 2006. Antennas and propagation for body centric communications. Proc. European Conference on Antenna and Propagation (EuCAP).

Huang FJ, et al. 2011. Rectenna application of miniaturized implantable antenna design for triple-band biotelemetry communication. *IEEE Antenna Propag.* 59: 2646–2653.

Ishido R, et al. 2004 A study on the solid phantoms for 3–6 GHz and evaluation of SAR distributions based on the thermographic method. Proc. 2004 International Symposium on Electromagnetic Compatibility, EMC'04, Sendai, Japan, Vol. 3, B3-2, pp. 577–580, June.

Karacolak T, et al. 2008. Design of a dual-band implantable antenna and development of skin mimicking gels for continuous glucose monitoring. *IEEE Trans. Microw. Theory Tech.*, 56.(4):1001–1008, April.

Kim J. Rahmat-Samii Y. 2004. Implanted antennas inside a human body: Simulations, designs, and characterizations. *IEEE Trans. Microw. Theory Tech.*, 52(8):1934–1943, Aug.

Malasri K, Wang L. 2009. Securing wireless implantable devices for healthcare: Ideas and challenges. *IEEE, Commun. Mag.*, vol.47, pp. 74–80, July.

Merli F, et al. 2008. Implanted antenna for biomedical applications. IEEE Antennas Propag. Soc. International Sym., pp. 1–4, July.

Merli F, et al. 2011. Versatility and tunability of an implantable antenna for telemedicine. Proc. European Conference on Antennas and Propagation (EuCAP), pp. 2487–2491, April.

Mizuno H, et al. 2011. Design of a helical folded dipole antenna for biomedical implants, European Conference on Antenna and Propagation (EuCAP), pp. 3640–3643, April.

Mond HG. 2001. The World Survey of Cardiac Pacing and Cardioverter Defibrillators: Calender Year 1997—Asian Pacific, Middle East, South America, and Canada. *J. Pacing Clin.* Electrophysiol. 24(5):856–862, May.

Nagaoka T, et al. 2004. Development of realistic high-resolution whole-body voxel models of Japanese adult males and females of average height and weight, and application of models to radio-frequency electromagnetic-field dosimetry. *Phys. Med. Biol.* 49: 1–15, Jan.

Ogawa K, et al. 2001. A high-precision real human phantom for EM evaluation of handheld terminals in a talk situation. *Proc. Int. IEEE Antennas Propag. Symp.*, 2: 68–71.

Okano Y, et al. 2000. The SAR evaluation method by a combination of thermographic experiments and biological tissue-equivalent phantoms. *IEEE Trans. Microw. Theory Tech.*, 48(11):2094–2103.

Sani A, et al. 2010. Antennas and propagation of implanted RFIDs for pervasive healthcare applications. *Proc. IEEE*, 98(9):1648–1655.

Scanlon WG, et al. 2000. Radiowave propagation from a tissue-implanted source at 418 MHz and 916.5 MHz. *IEEE Trans. Bio. Eng.*, 47(4):527–534.

See TSP, et al. 2010. RF transmission in/through the human body at 915 MHz. Antennas and Propagation Society International Symposium (SPSURSI), pp. 1–4, April.

Soontornpipit P, et al. 2004. Design of implanted microstrip antenna for communication with medical implants. *IEEE Trans. Microwa. Theory Tech.*, 52(8):1944–1951.

Steinhaus BM, et al. 1994. The role of telecommunications in future implantable device systems. *Proc. IEEE Conf. Med. Biol*, 2: 1013–1014.

Takahashi M, et al. 2010. Characteristics of antenna on cardiac pacemaker embedded in multilayered models. *Trans. IEICE*, J93-B (12):1636–1643 (in Japanese).

Uno Y, et al. 2010. Structure of cylindrical tissue-equivalent phantom for medical applications. International Conference on Electromagnetics in Advanced Applications 2010, pp. 406–409, Sept.

Usami M, Ohki M. 2003. The μ-chip: An ultra-small 2.45 GHz RFID chip applications. *IEICE Trans.Electron.*, E86-C (4):521–528.

Usui H, et al. 2006 Radiation characteristics of an implanted cavity slot antenna into the human body. Proc. IEEE Antennas and Propagation Soc. Int. Symp., pp. 1095–1098, July.

Wang J, et al. 2000. A model for predicting electromagnetic interference of implanted cardiac pacemakers by mobile telephones. *IEEE Trans. Microw. Theory Tech.*, 48(11):2121–2125.

Xia W, et al. 2009. Performances of an implanted cavity slot antenna embedded in the human arm, *IEEE Trans. Antennas Propag.*, 57(4):894–899.

Yamada S, et al. 2007. Study on a mutual coupling and maximum read range in RFID. Proceedings of the 2007 IEICE Society Conference, p. S-6, Sept. (in Japanese).

20

INGESTIBLE HEALTH CARE SYSTEM PARADIGM FOR WIRELESS CAPSULE ENDOSCOPY

NIKOLAOS BOURBAKIS AND ALEXANDROS KARARGYRIS
ATR Center, Wright State University, Dayton, Ohio

20.1 INTRODUCTION

It is estimated that 20 million people in the United States suffer from diseases of the small intestine, such as obscure bleeding, Crohn's disease, chronic diarrhea, or cancer. Wireless capsule endoscopy is a technology that offers medical doctors (gastroenterologists) the ability to view the interior of the small intestine with a noninvasive procedure [International Conference on Consumer Electronics (ICCE), 2005; Bourbakis 2005; Chan Y, et al. 2005; Fireman and Kopelman 2007; Miftahof RN. 2005; Valdastri P, et al. 2008; Karargyris and Bourbakis, 2010; Food and Drug Administration (FDA), 2008]. WCE was invented by a group of researchers in Baltimore in 1989 and later introduced by Given Imaging Incorporated as a commercial tool. Given Imaging has developed a swallowable wireless capsule called PillCam™ SB2 that has been the only available product on the market since 2002 (Figure 20.1). A competitor, Olympus Corporation, has been selling the EndoCapsule™ in Europe since 2005 and received FDA approval in September 2007. The differences between the two products are minor from an operational point of view.

At one end of the PillCam capsule is a miniaturized color camera and an optical dome with four white-light light-emitting diodes (LEDs). The camera can operate up to 8 hr and capture more than 50,000 images which are transmitted wirelessly to a storage device worn by the patient. After the 8-hr procedure, the video feedback is

Handbook of Biomedical Telemetry, First Edition. Edited by Konstantina S. Nikita.

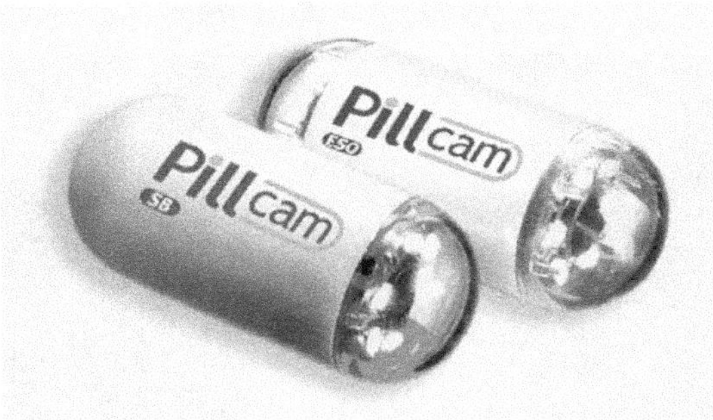

Figure 20.1 Common capsule endoscopy products on the market. Left: PillCam distributed by Given Imaging. Right: EndoCapsule distributed by Olympus.

uploaded to a workstation where the medical personnel can review and examine the endoscopic images.

The capsule endoscopy market is rapidly emerging and in the United States it was above $370 million in 2011.

20.1.1 Wireless Capsule Endoscopy and Other Technologies

Besides WCE, a small number of other endoscopic techniques are used for identifying small-bowel diseases.

Push endoscopy is a procedure that allows investigation and treatment of diseases in the upper small intestine. Push endoscopy reaches further into the small intestine than the standard esophagogastroduodenoscopy (EGD).

Sonde enteroscopy involves the use of a long, flexible, fiber-optic instrument propelled through the small bowel by peristalsis; this procedure may allow for viewing the remainder of the small bowel. Sonde instruments rely on a balloon placed at the instrument's tip. Peristalsis then advances the long flexible endoscope to the distal small bowel, and the endoscopic examination is performed during withdrawal.

Intraoperative endoscopy is performed in an operating room with the patient under general anesthesia. An endoscope is inserted into the bowel by the surgeon and passes through the small bowel manually while the gastroenterologist views the image. Intraoperative endoscopy is the last resort procedure to be considered in situations of severe recurrent bleeding.

20.1.2 Need for Computer-Aided Diagnostic System

The patient swallows a capsule and as it traverses through the digestive tract a computer-aided system captures approximately three frames per second. When the procedure finishes, over 50,000 and sometimes 80,000 screenshots are wirelessly transmitted to a receiver that is worn by the patient. The video is uploaded to a workstation so that physicians may examine it for any intestinal diseases.

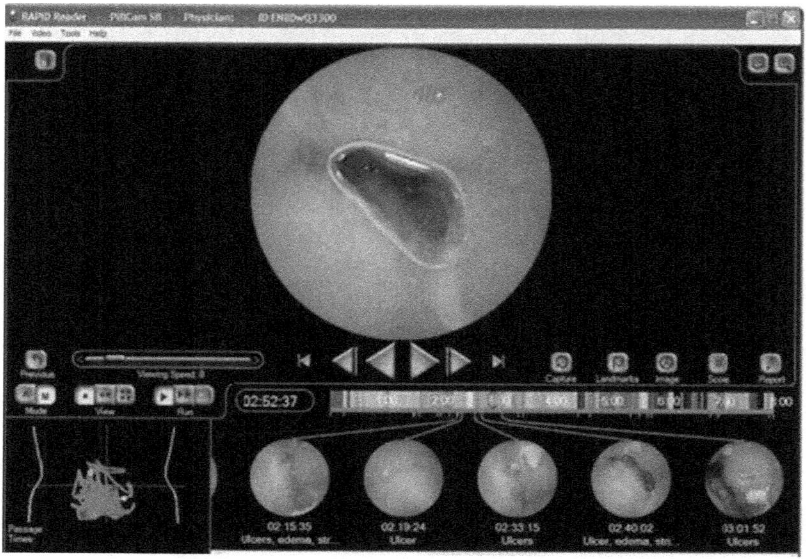

Figure 20.2 RAPID Reader software. (Picture Courtesy of Given Imaging Inc.). (*See insert for color representation of the figure.*)

In the case of the PillCam capsule, Given Imaging provides gastroenterologists with RAPID Reader software to examine the uploaded video. Although the video is usually 8 hr long, an experienced gastroenterologist can view the video in about 2 hr depending on the case. RAPID Reader offers several features that help the medical personnel with the video examination (Figure 20.2).

Although all the above features are very useful, they are mainly tools for video navigation. Additionally, RAPID Reader offers a detection tool—a blood detector. The blood detector tool is an artificial intelligence feature that goes through the video frames searching for red color. Thus, in case a frame exceeds a certain value of the red component, it assumes there is bleeding and saves the frame for future reference. Although the success rate of the bleeding detector is conflicting and gastroenterologists are not convinced about the robustness of the software tool, credit should be given since it is a computer-aided diagnostic tool.

Therefore, it from the above the need for the development of computer-aided diagnostic methodologies in the form of synergies to solve these open research challenges is summarized as follows: (a) educe viewing time and (b) detect intestinal abnormalities.

In particular, for the first challenge the synergistic methodology has to search the whole video and, based on various criteria, will remove redundant frames, thus decreasing the number of frames to be viewed by the gastroenterologist. For the second challenge the synergistic methodology has to search all the video frames for known diseases and save only suspicious frames for further evaluation. Therefore, for this challenge near automatic detection of abnormalities can be achieved.

Before moving on to the various research methods proposed and our synergistic methodologies on capsule endoscopy, we have to address the weak points of the current WCE that make the development of diagnostic computer-based methods quite difficult and challenging:

1. Video resolution is low—256 × 256 pixels per frame.
2. Capsule may be bouncing and flipping inside the digestive tract.
3. Frame rate is low.

In particular, the issue makes image interpretation more difficult, since a lot of visual information (details) is lost and creates the problem of a wrong location sequence because, in the case of bouncing or flipping, the next frame is taken not necessarily at a forward position but rather at an undetermined place. Finally, the issue may leave out important pieces of visual information which can help with the smart interpretation of the video.

20.1.3 Results from Recent WCE Methods

The successful operation of the WCE device depends on several software methods (modules) that detect abnormal patterns (bleedings, polyps, ulcers). Figure 20.3 presents a better WCE module and its capabilities to detect bleeding, polyps, and ulcers (Karargyris and Bourbakis, 2008, 2011).

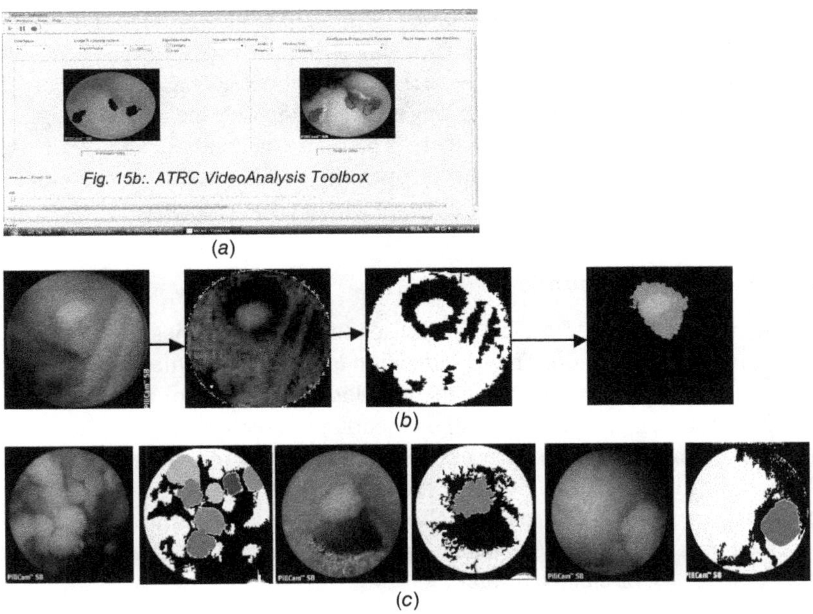

Figure 20.3 Modules for detecting (*a*) bleeding, (*b*) polyps, and (*c*) ulcers in WCE frames. (*See insert for color representation of the figure.*)

20.2 WCE AND ENDOSCOPIC IMAGING

In this section we offer a quick survey of WCE methodologies from Karargyris (2010). Computer-assisted technologies, especially image analysis and computer vision methods, can be used to automatically examine WCE videos and extract important features (here pathological) useful for an interpretation of the context of the video frames and significantly reduce the viewing time spent by physicians to examine these videos.

So far, research studies have been done on two key problems of endoscopic image analysis: *malicious tissue detection* and *tissue discrimination.*

The first problem is about the detection of malignant intestinal features like polyps, bleeding, and abnormal regions (tumors). The latter problem is about finding the boundaries of the digestive organs. This enables medical doctors to find areas of interest easily.

Although some significant research efforts have been proposed for the above problems, endoscopic analysis is still an important field with many open problems, such as recovery of missing visual information, accurate detection of pathological features, and removal of shadows and bubbles from the intestinal surface.

It is important to mention here that this chapter offers a survey of WCE and endoscopic imaging by evaluating some representative methodologies as well as that WCE technology is also used in colonoscopy (Given Imaging PillCam colon capsule). Thus, researchers have to deal with problems that are common to both the small intestine and the colon. That is why both endoscopy methodologies and colonoscopy methodologies are described in this survey. This chapter also provides an evaluation of each methodology individually for achieving its highest maturity. It does not offer a comparison among methodologies because each methodology has its own approaches and techniques to solve various endoscopy problems and very few methodologies deal with a specific problem. In addition, there is no common database with WCE videos for a fair comparison. Thus, this survey attempts to offer credit to methodologies based on that maturity and on those utilized techniques/approaches.

20.2.1 Methods Classification

In this section we provide a classification of the methodologies under evaluation before their full presentation. The classification is based on the theory or experimental techniques used to compose these methodologies for accomplishing their results. Thus, after a thorough evaluation of all methodologies, three major classification categories were created: (i) signal processing methods, (ii) color and image processing methods, and (iii) artificial intelligence methods.

In the signal processing category fall all methodologies that involve processing information in the frequency domain. These methods can be Fourier transforms, wavelet transforms, or bandbass filters.

The color and image processing category includes color processing methods like segmentation and color transformations, although these techniques can be viewed as computer vision as well.

The artificial intelligence category includes methods based on machine learning capabilities, such as clustering and artificial neural networks.

Automatic Detection of Intestinal Juices in Wireless Capsule Video Endoscopy This methodology deals with the good visibility of the intestinal lumen and wall, which is usually hindered by the presence of intestinal juices mixed in with the remains of food. These intestinal secretions are visualized as a semiopaque turbid liquid accompanied by the presence of bubbles and other artifacts related to the flux of the different fluids into the gut. As a result, this turbid liquid is interposed between the camera and the gut, obstructing visualization.

Thus, this methodology as it relates to the problem above can be simplified as follows: First, Gabor filters are applied for the characterization of the bubblelike shape of intestinal juices in fasting patients. The use of a bank of Gabor filters has been proven to be a very accurate characterization of turbid liquid. Also, a Gabor function can be viewed as a Gaussian function shifted in frequency to the position (u_0, v_0) and with orientation $\tan^{-1}\theta = u_0/v_0$. Thus, the Gabor filter has standard deviation of the Gaussian envelope σ_x and σ_y, as also the u_0 and v_0 parameters. The methodology uses a filter bank of 16 units with orientations 0°, 45°, 90°, 135° and standard deviation $\sigma = 1, 2, 4, 8$. The choice of these values was made through an empirical search. If the size of the filter bank increases, then the detection of false-positive areas (intestinal juices) increases respectively. This is undesirable because these false alarms accumulate as noise in the resulting image. Thus, at the start, all 16 Gabor filters are applied on the video frame and the resulting filtered images are added up according to the function $I_n = \sum_{i=1}^{4}\sum_{j=1}^{4} \text{abs}\{I_n^{i,j}\}$.

The medical specialists selected a set of 100 video frames in which the area containing the bubbles was delimited by hand, and a random set of 300 frames in which no bubbles existed was also selected. In the next step, an exhaustive search was applied over a threshold for I_n (the response images, after the Gabor implementation) to obtain a binary image by defining two exclusively complementary areas in the frame: (1) the region with intestinal juices and (2) the region with no evidence of intestinal juices. The final threshold was searched in an iteratively convergent algorithm which compared the areas obtained by the given threshold value with those described by the specialists. Finally, frames with a detected region of bubbles greater than 50% of the useful visualization area were excluded. The methodology was tested with 10 endoscopic videos. From those videos, 200 frames were chosen and characterized as valid, and 200 frames were characterized as invalid. The method has reached a very high percentage of invalid frames detection—98 and 95% of valid frame detection. Thus, by using this method, the examination time was reduced by 23% because the medical personnel could skip video frames with no information due to the existence of intestinal juices.

Neural-Network-Based Approach for Classification of WCE Images The purpose of this methodology is to implement a fusion-based methodology to make use of the advantages of color and texture features at the same time. As a first step, texture information is extracted from endoscopic images using nine statistical measures (standard deviation, variance, skew, kurtosis, entropy, energy, inverse difference moment, contrast, and covariance) for each vector of color space RGB and HSV. Thus, the result is 54 features. In this case, the histogram of the original image carries no information regarding the relative position of the pixels in the texture. This can fail to distinguish between textures with similar distributions of grey levels. Therefore, there are methods that recognize characteristic relative positions of pixels of given intensity levels. Thus, an additional scheme, NTU transformation (texture unit number transformation), is proposed to extract new textures from the texture spectra in the chromatic and achromatic domains for a selected region of interest from each color component histogram of the endoscopic images. The texture spectra of various image components {I(intensity), R(Red), G(Green), B(Blue), H(Hue), S(Saturation)} are obtained from their texture unit numbers.

In the second step, the statistical features (from step 1) are estimated on the histograms of the NTU transformations of the chromatic and achromatic planes of the image (R, G, B, H, S, V). The six subsystems (R, G, B, H, S, and V) that have been developed (steps 1 2) go through a fusion scheme in which the extended normalized radial basis function network (ENRBF) is used. Although a usual scheme chooses one best subsystem from the set of candidate subsystems based on a winner-takes-all strategy, the current proposed (ENRBF) approach runs all multiple subsystems with an appropriate collective decision strategy. The ENRBF network is incorporated into a multiple classifier scheme, where the structure of each individual (for R, G, B, H, S, V planes) classifier is composed of nine input nodes (nine statistical features) and two output nodes. In the final stage, the nine statistical measures for each individual image component–color space are calculated through the related texture spectra after applying the NTU transformation. The results were very good, giving a final 95.71% accuracy.

Deformable Rings Model for Aiding WCE Video Interpretation This methodology deals with the extraction of an estimate of the gastrointestinal tract as the capsule traverses it, so that a map can be created. This map helps physicians to easily go through the video feedback and find areas of interest. The result is a great reduction of the viewing time. In addition, the physician has an estimate of the whereabouts of the capsule. This methodology creates a 2D map of the internal surface of the digestive system and computes a rough estimate of capsule velocity. The velocity of the capsule provides sufficient data for localization of video sequences that show narrow sections of the gastrointestinal (GI) tract, where the capsule stopped or considerably slowed down. The method here uses a model of deformable rings (MDR). The MDR consists of nodes connected to form a mesh. The mesh forms concentric rings surrounding the center of the frame. MDR nodes store information on local image properties, such as RGB color components. The main steps of the method are:

Step 1 The nodes search within their vicinity of the current frame for locations having similar properties using Euclidean distance d. Each node is pushed to such a location.

Step 2 However, the arrangement of the nodes within the MDR mesh has to be retained. This is possible through the elastic matching method that gives a balance between these two effects (steps 2 and 1).

Step 3 The elastic matching, which is essentially the combination of the first two steps, is repeated until a state of equilibrium (defined by a threshold) is reached.

The process of elastic matching is repeated for every frame. It must be noted that as the capsule moves forward, the MDR expands while matching consecutive video frames. If the capsule moves backward, the model shrinks.

Automatic Classification of Digestive Organs in WCE Videos As mentioned in the presentation of the previous methodology, one of the main problems in wireless capsule imaging is the creation of a map of the digestive tract. A similar method is the discrimination of the digestive organs such as the esophagus, stomach, duodenum, small intestine, and colon. In this case, the physician also has a rough estimate of the digestive tract of the patient. When contractions occur periodically in the digestive tube, they distort the color values. When a capsule enters the next digestive organ, the corresponding color signal has a short-term change, which is the suddenness of the signal change and the increase in energy. Thus, this methodology uses energy functions in the frequency domain to characterize contractions extracted from the WCE video. In particular, a video frame is converted from the RGB to the HSI color space. The reason for this choice is because HSI uses components (hue, saturation, and intensity) that are close to the human eye interpretation.

Blood Detection in WCE Using Expectation–Maximization Clustering The Given Imaging M2A newest software version has provided a feature of detecting the luminal blood called suspected blood indicator (SBI), but its accuracy is not high enough to replace human examination. The reported sensitivity and the specificity of SBI were 72 and 85%, respectively. In response to this problem, a technique of detecting the bleeding regions automatically utilizing the expectation–maximization (EM) clustering algorithm and the Bayesian information criterion (BIC) has been proposed.

For this method the key point is to know that there is a certain color range distinguishing blood pixels from nonblood pixels. However, most dark pixels in blood overlap with dark pixels in nonblood because dark pixels have no chrominance. Some bleeding types are red, whereas others are blue. Consequently, bleeding pixels with similar characteristics need to be grouped together to extract the unique visual characteristics. The EM clustering algorithm is used to train probability models for the blood pixels and nonblood pixels.

Bleeding pixels and nonbleeding pixels were selected to train the system to recognize bleeding. The system is EM clustering that categorizes bleeding pixels and nonbleeding pixels into different subcategories because, as mentioned before, the intensity of blood varies (red–dark, red–blue). Each subcategory is modeled by a set of parameters of Gaussian mixture density. Thus, the blood detection is summarized as follows: (i) remove dark pixels, (ii) conditional probability restriction for selecting the blood candidate pixels, and (iii) region size filtering to remove small areas of wrongly identified blood.

Using Color Distributions to Discriminate Tissues in WCE Images As mentioned in the introduction of this chapter, the detection of various organs inside the digestive tract is an important tool for the physician. Of the alternatives that manufacturers have provided, besides SBI, another software feature is the localization function, which shows a rough estimate of the route and the relative position of the capsule. In order for the software to work properly, the physician must detect and manually mark two locations, the pylorus and the IV. Thus, a methodology for an automatic discrimination of tissues in the GI tract based on WCE images was presented. The methodology can be used for the mouth/esophagus and stomach. Note that the same team previously published two other methodologies for discriminating stomach/intestine and stomach/intestine/colon.

For this method, images are initially converted into the HSI color space. Then, the intensity I component is removed to achieve intensity invariance and data size reduction. A color distribution histogram is used here because it is the feature that physicians need for their investigation. The key point is that the HS histogram of stomach images is shifted toward red compared to the intestine histogram. However, the colon tissue color distribution is very similar to that of the small intestine. Next, data are compressed by applying discrete cosine transform (DCT) followed by principal-component analysis (PCA) for better results. Finally, the system is trained to detect the mouth/esophagus and stomach, stomach/intestine, and intestine/colon using the k nearest-neighbor (kNN) classifier and support vector classifier (SVC). The two classifiers are compared. The results were very good for the mouth/esophagus and stomach but not good enough for the intestine/colon. The kNN classifier and the SVC gave very similar results. The SVC managed to provide a stomach/intestine median frame error around 50–70 frames.

Topographic Segmentation and Transit Time Estimation for WCE Exams Here, a different approach to digestive organ detection is proposed. This methodology detects the boundaries of the digestive organs, entrance–stomach (esogastric junction), stomach–small intestine (pylorus), and small intestine–large intestine (ileocecal valve).

The main tools used here are SVM classifiers and MPEG-7 descriptors. Support vector machines (SVMs) have an error surface that is free from local minima and have only a global optimum. The features of the images that seed the classifiers are extracted from the MPEG-7 descriptors, which are of scalable color and homogenous texture. The advantage of these two descriptors is that they are widely tested and offer color and a texture representation of the image. The main steps of the methodology can be synopsized as follows:

Step 1 Use the MPEG-7 descriptors scalable color and homogenous texture to extract image features.

Step 2 Seed the above features as inputs to SVMs for training. The SVM uses four different types of kernel functions: (a) linear, (b) polynomial (c) radial-base, and (d) sigmoid.

Step 3 Assign each frame, in the zone it belongs to, by selecting the SVM classifier with the largest positive distance to its corresponding hyperplane. For this method, the training set was 30,000 frames and the testing set was 60,000 images. Eight-six percent accuracy was reached using a Gaussian kernel as an SVM classifier. Scalable color outperforms homogenous texture. This is quite interesting, because scalable color relies on color information. The accuracy percentage cannot be overcome because global model fitting (SVM method) has an "inbuilt" low-pass filter that reduces accuracy. Another problem is that the automatic system, given the fact it is based on color and texture features, annotates as a valve the moment the capsule crosses an area with completely different characteristics, but when the capsule bounces, then it annotates these bounces as multiple valves.

Neuronal Networks and Textural Descriptors for Automated Tissue Classification Detecting malignant regions in endoscopy video sequences encounters several problems due to the time-varying nature of the process and variable perceptual conditions (shading, shadows, lighting conditions, reflections, etc.). This methodology deals with issues by extracting texture features from video frames using co-occurrence matrix 2D discrete wavelet transform (DWT). From the co-occurrence matrices were extracted 4 texture features: energy-angular second moment, correlation, inverse difference moment, and entropy at four different angles: 0°, 45°, 90°, 135°. Thus, this resulted in 16 features. Applying DWT to each frame provide four wavelet channels: one approximate and three detailed. The final features are $3 \times 16 = 48$ and consist of the input of the neural network. The method here uses various kinds of neural networks for training, testing, and comparison: (1) offline adaptive back-propagation (BP) algorithm with 100 multilayer perceptrons (MLPs), (2) offline adaptive BP with 1000 MLPs, (3) offline Levenberg–Marquardt (LM) with 1000 MLPs, (4) offline Rprop algorithm, scaled conjugate gradient (SCG) algorithm and LM algorithm with 1000 MLPs and 11 hidden neurons, (5) online Rprop algorithm, (6) online BP, (7) online Alternative Benefit Plan (ABP), (8) online evolution, and (9) ensemble based. The last algorithm—ensemble based—is the main proposed training method. It is a combination of several individual networks combining a variant of the stochastic gradient descent with a differential evolution strategy to readapt the neural network to modified environmental conditions.

Neural-Network-Based Colonoscopic Diagnosis Using Online Learning and Differential Evolution This methodology is also based on various methods of offline and online training neural networks. It gives more emphasis to online training with online BP learning and online differential evolution (DE) strategy. The authors claim that online training is more efficient because it helps the process to escape from local minima and provides an approach to learning in nonstationary environments. This positive view of online training can be considered as a disadvantage, because here an ideal trade-off must be found between the newly acquired knowledge and its memory. Thus, adaptation must be as smooth as possible. In this case, texture features are extracted from colonoscopy images using co-occurrence matrices (energy-angular second moment, correlation, inverse difference moment and entropy), and online BP learning is used to initialize the population of the DE, as BP provides an initial approximation of the solution.

The major steps of the methodology here are as follows:

1. An n-dimensional (n-D) weight vector is used as initial population offered by BP.
2. The weight vectors are generated by the combination of vectors randomly chosen from the current population (mutation).
3. The resulting vectors are mixed with a predetermined target vector producing the trial vector.
4. This trial vector is accepted for the next generation if and only if it reduces the value of the objective function.

The last operator is referred to as a selection.

Thus, the DE presentation offline training was presented and all the results from the various methods were compared. The best results came from the Rprop offline training algorithm, because offline training uses the true gradient of the error function as it exploits information from all the training patterns. The online BP with DE provided the second best results, because it works with an instantaneous approximation of the true gradient and uses information only from one pattern per iteration. On the other hand, the online method has the advantage of readapting to modified conditions, while Rprop requires all information a priori.

As was mentioned in the introduction, the purpose of this chapter is to offer a survey of endoscopy methodologies besides WCE methodologies because there exist similarities in the problems and objectives in both areas. Thus, the following research methods are used for identifying colorectal polyps in colonoscopy video frame sequences: Polyps are characterized from their color, appearance of their mucosal surface, presence of ulcers, bleeding tendency and above all presence of pedunculus (pedunculated or nonpedunculated).

Computer-Aided Tumor Detection in Endoscopic Video Using Color Wavelet Features The main concept of the methodology discussed here can be briefly described as follows: A novel feature extraction technique was used based on a discrete wavelet decomposition that was applied on different color spaces. Then statistical analysis of the wavelet coefficients associated with the color bands was used. The wavelet features, named color wavelet covariance (CWC), were based on second-order textural information estimated on the domain of the discrete wavelet decomposition of each color band of a video frame.

The methodology is as follows:

Step 1 RGB video frames are converted into a Karhunen–Loeve color space resulting in three color channels. This diagonalizes the covariance (or optionally the correlation) matrix of the colors. It performs very well for color texture characterization.

Step 2 Since the size of the lesions varies, a multiresolution approach plays an important role. Due to the fact that texture is essentially a multiscale phenomenon, the DWT works well here because it offers spatial/frequency representation. As is known, when the decomposition level decreases in the spatial domain, it increases in the frequency domain, giving zooming capabilities. So the DWT transformation consists of (1) convolution of the image with a low-pass filter H, (2) convolution of the image with a bandpass filter G, and (3) repetition of the previous convolutions to decompose the image at several scales. The result is a set of low-resolution image Ln along with detailed images $D_1, \ldots, D_{3n}$, where n is the scale.

Step 3 Co-occurrence matrices were used on images from the previous step to extract second-order statistical measures, such as (1) second moment, (2) correlation, (3) inverse difference moment, and (4) entropy. Two more features were used, namely contrast and dissimilarity, but their results were not satisfactory, so they were removed from the process.

Step 4 The covariance among these different statistical values was applied between the various color channels of the examined region in order to rank changes. In other words, using the covariance between two features, a measure of their "tendency" to vary together was given.

Step 5 Finally, linear discrimination analysis (LDA) was used for classification and evaluation of the proposed features.

In this methodology, very good sensitivity ($93.6 \pm 0.8\%$) and specificity ($99.3 \pm 0.3\%$) were achieved. It needs to be mentioned here that a 128×128-pixel window was used, because the smaller the window size set, the bigger the mean classification error (MCE) and the uncertainty. This means that a large population of pixels is required in order to characterize tumor regions using second-order statistical features in the wavelet domain.

CoLD: A Versatile Detection System for Colorectal Lesions in Endoscopy Video Frames This work is an incremental improvement of a methodology previously discussed using a graphical software implementation. This software can be used and trained by medical personnel to detect precancerous polyps in colonoscopy images. The new part here is only the classification scheme, an artificial neural network (ANN) that can handle nonlinear multidimensional input features that have the advantage of generalization. The neural network used in the software system consisted of an online BP with 48-X-2 MLPs, where X is the number of hidden neurons. The relationship between X and time was linear.

Thus, the modified methodology is presented below:

Step 1 Take a region of the image (square window).

Step 2 Apply DWT to this region resulting into three bands (bands 2, 3, and 4).

Step 3 Calculate the co-occurrence matrix for each band resulting in 16 statistical measures for each band ($3 \times 16 = 48$ textural value vector).

This procedure is applied for either training or testing. In the training case, the training set is calculated on a limited set of regions that are randomly selected from the image. In the testing case, overlapping windows are used. Any examined area is assigned to a class corresponding to the normal or abnormal case, and this information is stored as the 49th component of each 48th vector. Also isolated misclassified vectors or groups of vectors may appear as spots or small formations within the output processed image. A postprocessing stage is applied to cut out these "errors." Medical personnel train the neural network by tuning the MLPs. The software is user friendly.

Classification of Endoscopic Images Based on Texture and Neural Network This methodology proposes different ways of detecting malignant regions in endoscopy images, such as ulcer or coli. A local binary pattern (LBP) is used for texture feature extraction and then a log-likelihood ratio, called a G statistic, is used to evaluate LBP distributions. The basic idea of LBP is a binary code that describes the local texture pattern. This code is built by thresholding a neighborhood by the value of its center. Through its recent extensions, the LBP operator has been trasformed into a very powerful measure of image texture, showing excellent results in terms of accuracy and computational complexity in many empirical studies. This work uses, however, a modified LBP method combined with the intensity I of the image to generate more stable results for endoscopic images. Note that the original LBP method does not use the contrast and intensity of the texture.

The G statistic, which is a modification of Kullback's criterion, shows the probability that two sample regions are from the same population. The higher the G statistic, the less the probability they have. Also, a self-organizing map (SOM) is used for the classification. The SOM is an unsupervised process and this is a great advantage given the complexity of endoscopic images.

Methodology for Detecting Blood-Based Abnormalities in WCE Videos This methodology deals with bleeding detection in WCE videos in a more sophisticated way. Past research has focused only on identifying bleeding. The present methodology introduces an advanced categorization consisting of three cases of blood-based abnormalities: (a) bleeding, (b) angioectasia, and (c) erythema. It utilizes a mathematical transformation as a color space transformation, known as Karhunen–Loeve to make R, G, B values as uncorrelated as possible. More specifically, let S be the region to be segmented and Σ be the covariance matrix of the distributions of R, G, and B in S. Let λ_1, λ_2, and λ_3 be the eigenvalues of Σ and $\lambda_1 \geq \lambda_2 \geq \lambda_3$. Let $W_i = (W_{Ri} W_{Gi} W_{Bi})^t$ for i=1,2,3 be the eigenvectors of Σ corresponding to λ_i, respectively. The color features X_1, X_2, X_3 are defined as $X_i = W_{Ri} \cdot R + W_{Gi} \cdot G + W_{Bi} \cdot B$;

($\|W_i\| = 1$, i=1, 2, 3). After applying this transform onto WCE frames, blood-based cases appear more distinct from the rest of the image. A complex segmentation process based on fuzziness extracts the regions of a WCE frame. After this step, a merging procedure is carried out based on a local–global (LG) graph scheme to combine similar regions into bigger ones. In such a way, small blood-based segments are not discarded. This scheme uses color information, geometry information, and segment interrelation of each region to achieve the merging process. Finally, an experimental color threshold is applied to keep only blood-based regions.

Neural-Network-Based Detection of Bleeding in Sequences of WCE Images This methodology deals with the problem of bleeding detection in WCE images. As mentioned previously, Given Imaging has provided its software package with a bleeding detection feature. However, its detection rates vary from 21 to 41%. The present methodology detects bleeding successfully at 80% with potential for high rate of success. The main concept used in this methodology is color features. It applies a color adaptation approach by using a reference color value of the image. Color adaptation is carried out by the cone response transform and the reference color is detected using a neural network. Color histograms are produced first for the whole image and second for the designated blood area. Both histograms are input and output in the training stage of the neural network. Image color is detected in the testing phase after color adaptation and estimation of deviation from the gray value axis in the RGB cube. Color adaptation can be considered as a dynamic mechanism of the human visual system to optimize the visual response to a particular viewing condition. Chromatic adaptation is the ability of the human visual system to discount the color of the illumination and to approximately preserve the appearance of an object. In the bleeding detection problem, this helps visualize the bleeding areas better. However, the reference color used by the color technique needs to be extracted by the image. This is done by a neural network through the histograms of the whole image and the bleeding areas. The selection of bleeding areas is done manually by the physician so that the neural network can be trained.

20.3 DIAGNOSTIC METHODS AND CHALLENGES

In this section we present challenges related to the WCE approaches. In diagnostic endoscopy, based on a distributed perception of local changes, the medical expert interprets the physical surface properties of the tissue, such as the roughness or the smoothness, the regularity, and the shape, to detect abnormalities. Adjacent surfaces showing different surface properties are distinguished on the basis of the texture differences.

It is well known that the performance of human perception on this complex task is outstanding when there is no heavy workload to be completed under time constraints. The physician has the ability to view and associate patterns in an image and to interpret its context. The physician can also adapt to changes so that even he can recognize a distorted image can be recognized. There is no machine available today that can achieve the equivalent perceptive power of a human brain. The idea of using machine learning methodologies to help with the interpretation of endoscopic images is meritorious. This is due to the difficulty of a human to continuously observe the same scene for long periods and to accurately evaluate it. In addition, it is important for one to take into consideration the vast difficulties in physical attributes of the intestine: *no intestines are identical to each other. Also, within the same intestine, one section may have very similar characteristics with another.*

This fact introduces limitations in the use of traditional computer-assisted endoscopy for interpreting endoscopic images. Given an endoscopic image, the "true" features that are associated with it are not exactly known to us. Usually, one or more feature extraction models are used to provide values for each feature's parameters. One effort in this direction—to partially automate the observation of the WCE videos—was undertaken by Given Imaging by incorporating into its video viewer the capability to simply detect "red spots" (Red Blood Identification System, RBIS). Unfortunately the RBIS has been proven to have very low sensitivity (20.5%) and specificity (41.6%) [ICC-2005 (Karargyris, 2010)] and it does not shorten the review time of the recordings.

It is also important to mention here advances in WCE that are related to the design of capsules with additional capabilities (Gono, 2008, Fireman and Kopelman, 2007) and new semi-invasive systems with controllable capsules for the entire digestive system (Menciasi et al., 2007). These efforts are directly related to our research proposed here, and our work could be useful to those efforts as a noninvasive approach that offers flexibility to patients in their daily activities. Our research effort is based on a synergy of techniques for robust and reliable automation of the WCE video observation.

20.4 FUTURE DIRECTIONS: DESIGN NEW GENERATION OF WCE

According to Fireman and Kopelman (2010): "Researchers have proposed the idea that endoscopists will be able to 'control and steer' the Capsule Endoscopy (CE), as they are able to do in standard endoscopy. This means being able to maintain the capsule steady in a selected area and hold the view in order to have more time to examine the opposite wall of the bowel."

Based on these thoughts, several endoscopic capsule designs have been proposed and some are shown in Figure 20.4.

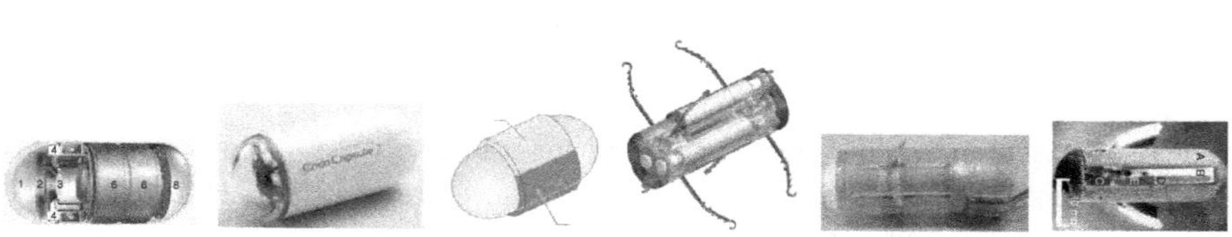

(a) (b) (c) (d) (e) (f)

Figure 20.4 Various endoscopy capsule designs.

These designs are all very promising with different potentials. Some, like the one from Given Imaging, look simple but have limited capabilities. However, the capsule by Given Imaging offers better safety than the other designs in the case of blockages in the intestine. The one from Olympus has more capabilities due to its magnetic fields for possible movement forward and backward guided by the gastroenterologist. The last two present capabilities of movements forward and backward due to their robotic legs (Gono, 2008, Glass et al., 2008). These two capsule designs present, however, low safety features due to high risk of a possible grabbing problem inside the intestinal wall and at the same time they may damage the wall, causing bleeding. The last capsule (Figure 20.4*f*) (Glass et al., 2008), which is passive, offers better grasping and smooth legs, minimizing bleeding accidents.

20.4.1 Design of New Robotic WCE

There are research efforts to address the detection of abnormalities in WCE videos as well as the reduction of examination time (Bourbakis Giakos Karargyris 2010: Karargyris, 2010). Because of the diversity of the abnormalities, the design of the system's architecture was carefully planned to have the ability to be extended to address different cases.

Using the knowledge we have extracted throughout our research, an ideal capsule endoscopy for diagnostic purposes can be constructed as follows:

Depth cameras (*time-of-flight cameras*) have reached a mature level of resolution and detail. The problem is the size of these cameras, which has been reduced to 1.5 inches. Given the current trend, the miniaturization of this camera will be achieved.

Accelerometers and *magnetometers* offer direction and acceleration information. Fortunately these sensors are so small that their size is less than half of a fingertip.

Narrow-band light filters can be utilized in existing cameras to reveal additional information on the surface of the digestive tract. Such sensors are used in other endoscopy devices.

Existing *miniature cameras* offer a viewing field of more than 100°. The use of such cameras could increase the existing viewing field. Using two cameras placed on opposite sides of the capsule could tremendously increase the viewing field. Ideally, if cameras achieved a viewing field of nearly 180°, then using two of those cameras would reach a viewing field of 360°.

Power consumption is probably the most important issue in capsule endoscopy. Some researchers have proposed use of an external magnetic field to power the capsule. However, studies have to be carried out to examine health risks. We believe that redesigning the way current technology transmits information to the patient's receiver is a potential solution: Since the capsule constantly transmits video frames to the receiver, a lot of energy is wasted. A different approach would be making the capsule store all the information on a memory bank, and at the end of the procedure the video could be uploaded to a workstation. In this way even battery technologies could be used to increase the resolution of the video frame and more importantly increase the frame rate, which is really low.

Thus, it is a challenge for researchers to propose a new design with as many features as discussed above. Having thoroughly examined these WCE problems the design has to achieve two goals: function as both a diagnostic capsule and a robotic therapeutic capsule. In this way, all state-of-the-art features in each capsule to serve their purpose with maximum efficiency must be efficiently combined. Additionally, features that are redundant on either capsule were excluded to reduce the cost. Finally, both proposed designs are as safe as the Given Imaging and Olympus capsules.

In summary, after discussion with gastroenterologists, the diagnostic capsule should be able to:

(a) Capture video at much larger video frame
(b) Have the potential of grabbing 3D information of the gastric structures
(c) Provide position information while traversing the intestine

Each position sensor is attached to a soft spiral spring that has the ability to extend as much as possible so that it can reach the intestinal wall (this depends on how the intestine is deformed) and can let the precision sensor (wheel) rotate due to frictional forces. More specifically, the two position sensors basically follow the same principle as odometers. They are composed of frictional rubber materials so that when they are in contact with the soft tissue of the intestinal wall they are able to rotate. This rotation gives information about how far the capsule has moved inside the intestine by emitting electromagnetic peaks that are received by precision sensors on the control board of the capsule.

The two position sensors not only offer accurate and robust information (in comparison with a single-position sensor) about the position of the capsule but also work as guidance for the capsule to keep as steady as possible on its traversing direction. Finally, due to its extension capabilities the capsule can successfully record video from the large bowel whereas with current capsule technology its small size does not allow for a robust, smooth video of the large bowel.

Researchers are conducting experiments on the type of springs and the material of the rubber wheels to be used for the aforementioned purposes. Such materials exist and serve these goals. By knowing the position of the captured frame inside the intestine, gastroenterologists will be able to know where a potential abnormal case is inside the gastric tract. This information is extremely important for future reference or for a possible surgery.

It must be mentioned that the proposed capsule has been carefully planned to be the same size as the current commercial capsules (27 mm × 13 mm). Additionally, the capsule does not require a wearable receiver since it does not transmit the video wirelessly but rather saves the video to an onboard memory storage unit. When the process is over, the front part of the capsule, which contains the memory storage, can be easily removed and connected to a memory reader to upload the video to a workstation.

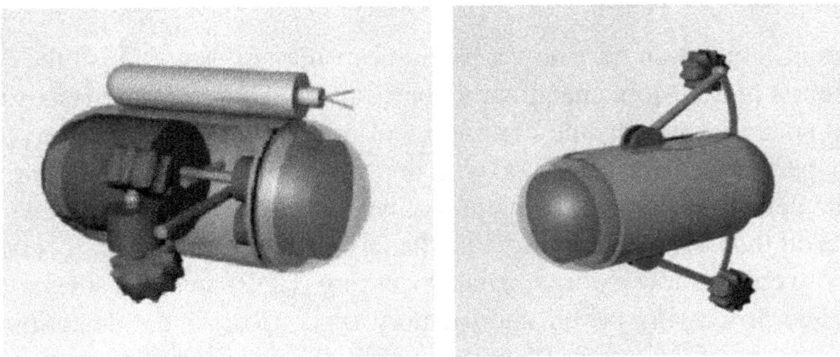

Figure 20.5 Views of robotic WRC capsule; (*a*) capsule opening of extended wheels; (*b*) capsule internal view and injection/biopsy selection arm coming out from capsule interior.

Whereas the diagnostic capsule is a passive device, the therapeutic capsule is an active and fully controllable robotic device (see Figure 20.5). Additionally, the capsule barely exceeds the size of current commercial capsules.

If the capsule has to stop at a certain position, the wings need to extend to neutralize the peristalsis force and the gravity force applied to the capsule. The extended wings will touch and deform the intestinal wall through their attached rubber wheels. Consequently, this will generate reaction forces from the intestine to the capsule and these reaction forces will cause friction forces on the wheels. Finally, the motor will have to provide sufficient torque to overcome the torque of the friction forces. There are miniature motors that have this capability. A quick research in the miniature motor industry (i.e., Namiki Co., New Scale Technologies) showed that there are multiple micromotor models that offer sufficient power, torque, and size capabilities. Careful planning must be taken to find materials that provide the right friction coefficient, which can halt the capsule and allow the motor's torque to rotate them. After examining the literature (Glass et al., 2008) we are confident that these materials exist. Additionally, the wheels can be designed to have soft rubber cut teeth to enable the better rotating momentum.

The robotic WCE includes an advance complex mechanical mechanism that:

1. Expands the wings that carry four wheels to hold the capsule at the current position in the intestine.
2. Offers the ability to expand the size of the capsule's width three times. This enables the capsule to fully function in the large bowel too.
3. Offers the ability to rotate the wheels in both directions and therefore move the capsule forward/backward.
4. Offers the ability to carry a Micro-Gripper in order to take biopsy of the intestine. The gripper can be replaced with a needle or a cauterizing device.
5. Offers the ability to rotate the Micro-Gripper to some degree of angle in order to have better access.

20.4.2 Alternative Design

It is important to mention here how a robotic capsule will accomplish the diagnostic and therapeutic tasks. More specifically, there are several important issues regarding the limited space in the capsule compartment and how we are going to efficiently utilize it. In particular, in order to deal with the storage issue (storing into the capsule's memory the detection synergistic methods), we will use an external receiving device more powerful than the existing one (like the Given Imaging external receiver). More specifically, we use a new external data processing (EDP) device as the external unit with sufficient processing power and memory space to store the detection methods and the sequences of the image frames transmitted by the capsule and execute these methods. Thus, the EDP device will receive image frames, process them using the detection methods, detect and recognize the existence of tissue abnormalities, and determine the action that has to be performed by the robotic capsule. In addition, EDP will transmit to robotic capsule the necessary tissue information and the commands of actions. Thus, this collaboration between EDP and WRC will eliminate the total video storage issue and the process issue as well, since the EDP device will perform these tasks in real time.

The WRC robotic capsule will perform the following main subtasks in collaboration with the EDP unit:

1. *Capsule* *Capturing image frames* (on the average three images per second) from the interior wall of the small intestine and transmitting them to the external receiving EDP device. The EDP device will process these frames for detecting possible abnormalities (polyps, bleeding, ulcers), decide what actions should be taken by the capsule, and send an information package (commands and tissue information) to the capsule.
2. *Capsule* ***Receiving commands*** (actions to be performed) ***and tissue information*** (location of the abnormality and its main characteristics) from the EDP device. The robotic capsule will use this information package to perform either (i) a *diagnostic task*—capturing or ***recapturing image frames from a certain region*** (by going backward or forward to reach the location of the region) ***and sending*** new frames to EDP)—or (ii) *a therapeutic task*—after locating the candidate region, by using the information package, the capsule ***opens its compartment, extends its wheels*** to control its movements ***(stops, moves forward, moves backward), extends the injection arm*** to release medicine in a certain region (if needed), or ***cuts a piece of tissue*** and saves it in its compartment for a later biopsy). After completion of any of these actions the capsule closes its compartment, captures new frames from that region, sends them to EDP, and moves on.
3. *Capsule* ***Closing the compartment*** and following the movement due to peristalsis or leaving it open and moving backward or forward according to the EDP commands. Figure 20.6 graphically shows (a) the EDP device tied to the human body; (b) the general view of the stomach, small intestine, and colon; and (c) the capsule inside the small intestine.

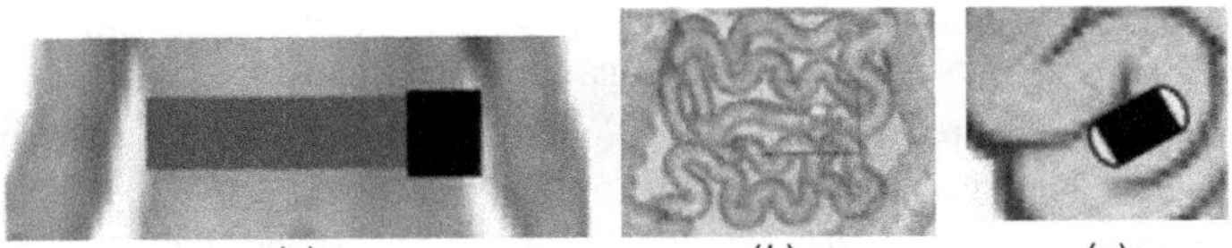

Figure 20.6 (*a*) Location of EDP device (black box) attached to human body; (*b*) human stomach–intestine–colon; (*c*) capsule inside small intestine.

20.5 CONCLUSION AND WCE GLOBAL HEALTH CARE

It is known that surgical robotic has become a routine-based procedure for surgeons in many medical centers (Rosen et al., 2011). In addition, the technological progress in microrobotics (Bellouard, 2010) offers a new dimension to robotic surgery. In this chapter the design of a WCE robotic capsule for diagnostic/therapeutic endoscopy was proposed. The main idea behind this research effort is to investigate opportunities that the microrobotics technology may offer to the medical field. The WCE design offers to the capsule capabilities of backward and forward movements either under gastroenterologist control or under autonomous control by the capsule under the assumption that the capsule has a continuous access to the real-time processing of the WCE video, which will take place in the EDP device. The design of the WCE robotic capsule and its EDP was carefully planned to have the ability to deal with different cases of abnormalities.

An important feature of the WCE design can be related to the global health care system. The WCE will have the capability to detect in real time pathological abnormalities and contact the gastroenterologist for online evaluation of the patient's condition and for the further actions to be taken. This type of communication between patient (via capsule) and physician implicates all the security and privacy issues addressed in Chapter 14. It becomes part of a global health care system, in which health data will be transferred from one place to many and additional health records can be generated or transferred for better evaluation of patients' conditions and more reliable decisions to be made. Thus, all generations of iPhones, iPads, and others will become part of the technology to assist physicians in offering better and faster health care—of course with a "cost."

REFERENCES

Bellouard Y. 2010. *Micro-robotics: Methods and Applications*, Boca Raton, FL: CRC Press.

Bourbakis N. 2005. Detecting abnormal patterns in WCE images. 5th Int. IEEE Bioinformatics and Bioengineering, BIBE-05, Oct. 20–21, pp. 232–238.

Bourbakis N, Giakos G, Karargyris A. 2010. Design a new generation of robotic capsule in wireless endoscopy. Proc. IEEE Conf. on Imaging Techniques and Systems, ITS-10, Thessaloniki, Greece, pp. 1–6.

Chan Y, et al. 2005. A prototype design of wireless capsule endoscope. *IEEE Conf. Mechatronics & Automation*, Niagara Falls, Canada, July.

Food and Drug Administration (FDA). 2008. Available: http://www.fda.gov/fdac/features/2005/205_pillcam.html. January.

Fireman Z, Kopelman Y. 2007. New frontiers in capsule Endoscopy. *J. Gastroenterol.* Hepatolo., 22: 1174–1177.

Glass P, et al. 2008. A legged anchoring mechanism for capsule endoscopes using micropatterned adhesives. *IEEE Trans., Biomed. Eng.*, 55 (12): 2759–2767.

Gono K. 2008. Multifunctional endoscopic imaging system for support of early cancer diagnosis. *IEEE J. Sel. Topics Quantum Electron.*, 14(1): 62–69.

International Conference on Consumer Electronics (ICCE). 2005. Proceedings of the 4th Int. Conference on Capsule Endoscopy, March 7–8, Miami, FL.

Karargyris A. 2010. A synergistic methodology for detecting abnormal patterns in WCE videos. Ph.D. Dissertation, ATRC, Wright State University.

Karargyris A, Bourbakis N. 2008. Detecting blood based abnormalities in WCE video-images. Proc. IEEE Int. Conf. on BIBE08, Athens, Greece, Oct., pp. BE5.4.4/1-5.

Karargyris A, Bourbakis N. 2010. A survey on WCE and endoscopic imaging. *IEEE Eng. Med. Biol.*, 29(1).

Karargyris A, Bourbakis N. 2011. Detecting polyps and ulcers on WCE images. *IEEE Trans. Biomed. Eng.*, 58(10): 2777–2786.

Menciassi A, Quirini M, Dario P. 2007. Microrobotics for future gastrointestinal endoscopy. *Minim. Invasive Ther.*, 16(2): 91–100.

Miftahof RN. 2005. The wave phenomena in smooth muscle syncytia. *In Silico Biol.*, 5(5–6): 479–498.

Rosen J, Hannaford B, Satava R. (eds). 2011. *Surgical Robotics: Systems Applications and Visions*, 1 vol. XXII, Springer.

Valdastri P, et al. 2008. Wireless therapeutic endoscopic capsule: In vivo experiment. *Endoscopy*, 40: 979–982.

21

STIMULATOR PARADIGM: ARTIFICIAL RETINA

CARLOS J. CELA AND KEYOOR C. GOSALIA
North Carolina State University, Raleigh, North Carolina
ANIL KUMAR RAMRAKHYANI AND GIANLUCA LAZZI
University of Utah, Salt Lake City, Utah
SHRUTHI SOORA, GERARD J. HAYES, MICHAEL D. DICKEY
North Carolina State University, Raleigh, North Carolina

21.1 INTRODUCTION

Degenerative diseases of the eye, such as retinitis pigmentosa and age-related macular degeneration, progressively disable photoreceptors while leaving the rest of the optical path relatively intact. The AR partially replaces the optical portion of the vision path, including the nonfunctional photoreceptors, with a combination of external and implanted electronic devices.

The AR device is comprised of one extraocular unit containing a data processor, a video camera, the transmitter coil, the external antenna and a battery, and one intraocular unit, containing the receiver coil, implanted antenna, electronic circuits for power, data recovery, and retinal stimulation, and thin microelectrode array.

Both power and data transfer between the external and implanted units are through wireless links. Power transmission is done using a set of coils coupled by an inductive link operating at low-megahertz frequencies. Data transmission for low electrode

Handbook of Biomedical Telemetry, First Edition. Edited by Konstantina S. Nikita.

count and low frame rate implants can be achieved with the inductive link; however, for larger bandwidth configurations, miniature implantable antennas operating at microwave frequencies may be necessary. The electrode array electrically stimulates ganglion and bipolar cells in the retina, bypassing the nonfunctioning photoreceptors. Data communication on the implant is ideally bidirectional, with an upstream link providing image data to the neurostimulator and a downstream link providing diagnostics and configuration information to the external processor.

21.2 TELEMETRY FOR ARTIFICIAL RETINA

Implanted telemetry devices used for medical implants are affected by their surrounding environment since tissues are frequency-dependent lossy dielectric materials. The extent of this influence depends on the location and depth of the antenna within the body. As the implanted patient's body absorbs the radiated power, the relatively high conductivity of the body tissue results in high transmission losses, and the high dielectric permeability narrows the impedance bandwidth, detuning the antenna with the consequent change in its operating frequency. In a retinal implant, the telemetry circuitry will reside within the eyeball, and the dielectric properties of the vitreous humor will have a significant impact on the antenna's performance. For vitreous humor, the dielectric constant reduces and the conductivity of the tissue increases as the operating frequency becomes larger (Gabriel et al., 1996).

For low-bandwidth configurations of the AR, the power and data transmission can be accomplished inductively between a pair of external and implanted coils at low frequency (Buckett and Peckham, 1988). In this configuration, the primary coil is located externally to the body, and the secondary coil is implanted in the body tissue. In order to minimize absorption loss in the body tissue, low frequencies are utilized for the power transfer link. The design of biotelemetry coils depends on multiple parameters, including the implant size, the maximum power requirement, the bandwidth of the data to be transmitted to the implant, and safety considerations such as maximum allowed power absorbed by the body and temperature increase in the tissues due to Joule heating in the implanted device. Wireless telemetry for medical implants can also be implemented using radio frequency or microwave frequency coupling mechanisms between the external unit and implanted unit. Such coupling is generally accomplished by a pair of antennas, one implanted and one external.

A hybrid configuration can be used in which small antennas are used for high-speed data transfer in conjunction with paired coils for power transfer to the implanted device. In such configurations the inductive coil coupling operates at low frequencies (1–50 MHz), while the antenna link can be used at microwave frequencies (1–3 GHz). Recently, several implantable antennas for similar applications have been reported (Merli et al., 2011; Izdebski et al., 2009; Karacolak et al., 2008). A conceptual sketch showing the relative positions of the extraocular and the implanted antenna is shown in Figure 21.1.

Since the tendency is for implanted devices to require increasingly larger volumes of telemetry data with a corresponding increase in required bandwidth, in the

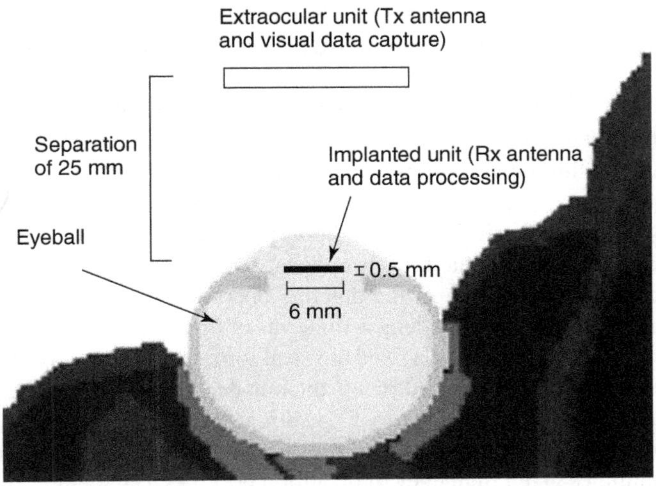

Figure 21.1 Numerical model of transverse section of head showing inductive telemetry link. The position of the internal and external coils is marked in the plot.

remainder of this chapter we will focus on the considerations and methodologies for the design of high-performance and small-size antennas appropriate for intraocular implants.

21.3 INTRAOCULAR TELEMETRY ANTENNAS

Antenna design for the implanted ocular unit is a challenging task due to several reasons:

- The space available is limited, so size/form factor is constrained.
- The antenna has to be an efficient passive radiator of energy.
- The antenna has to be able to efficiently transmit energy through the lossy environment.

The resonant frequency (and, in most cases, the operating frequency) of an antenna is inversely proportional to the electrical length of the radiating element. Because of this, the physical size of the antenna is generally decided based on the frequency of operation. The physical length of a simple antenna such as a half-wave dipole is $\lambda/2$ (where λ is the wavelength) at its frequency of operation. Such a dipole is also $\lambda/2$ long in terms of its electrical length.

Highly efficient passive antennas have to satisfy the following constraints:

- Exhibit high radiation efficiency (based on the electrical loss in the structure)

- Have optimal impedance matching to the source [low voltage standing-wave ratio (VSWR), maximal power transfer between the source and antenna]

At a physical length of $\lambda/2$, the simple dipole satisfies all of these important characteristics for acceptable antenna performance: It is an efficient low-loss radiator, as it transmits a significant portion of the energy it receives from the source and exhibits a reasonable impedance bandwidth (low VSWR over a specific frequency range). However, the physical size of a dipole at the considered frequencies makes this configuration inappropriate for this application.

To reduce the physical size of the antenna operating at a given frequency, it is possible to increase the electrical length (in terms of the operating wavelength) of the radiating element while retaining a fixed physical form factor. However, the efficiency and impedance bandwidth attained by an antenna depend on the volume it occupies expressed in terms of the wavelength of operation. Because of these considerations, any efficient antenna design for a retinal prosthesis needing physical miniaturization while maintaining a given frequency of operation becomes a challenging proposition.

In general, due to the limited form factor necessary for a retinal implant, most implanted antenna designs will be classified as an electrically small antenna. The conventional definition classifies any antenna which fits within a radiansphere as a small antenna. A radiansphere is a spherical volume having a radius of $\lambda/2\pi$. The electrical size or volume of the antenna can be expressed as the value of ka where k is the wavenumber $2\pi/\lambda$ and a is the radius of an imaginary sphere which includes the maximum dimension of the antenna structure, as shown in Figure 21.2. An antenna with $ka < 0.5$ is classified as an electrically small antenna.

Electrically small antennas designed to operate in the lossy environment of the vitreous humor of the eyeball present multiple design challenges. Because of the size reduction, any reduction in total realized efficiency adds to the already significant amount of energy loss that takes place in the tissues. So it is critical that implanted antennas are impedance matched to the system impedance—this is a challenging aspect for heavily loaded electrically small antennas.

For a retinal prosthesis which is designed to be implanted in the area of the lens or the central portion of the eyeball, the available volume for the implanted circuitry is

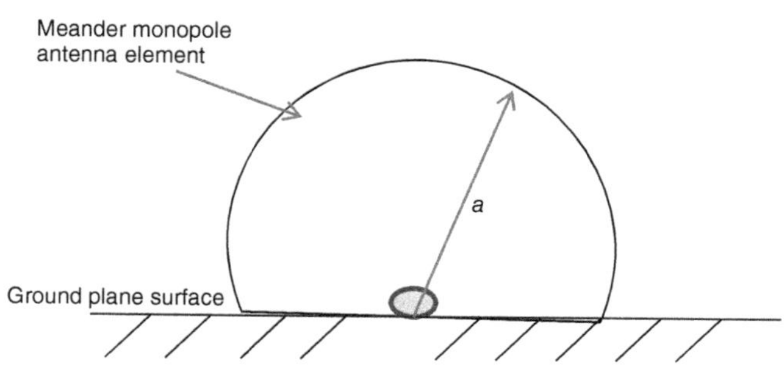

Figure 21.2 Electrically small antenna.

estimated to be within $6 \times 6 \times 6\,mm^3$ (Gosalia et al., 2004a). This includes the receiver architecture as well as any power transfer mechanism (coil/antenna), which leaves approximately $6 \times 6 \times 2$ mm for the antenna structure itself. To put the size constraints in perspective, at the industrial, scientific, and medical (ISM) band of 2.45 GHz, the wavelength is 122 mm and the radiansphere has a radius of approximately 20 mm. An implanted antenna designed to operate within the available volume ($6 \times 6 \times 2$ mm) will exhibit $ka \sim 0.2$ (based on the radius $a \sim 4$ mm). Thus, the implanted antenna will be classified as an electrically small antenna. Such a radiating structure will need significant size reduction to operate at 2.45 GHz within a form factor of $6 \times 6 \times 2$ mm.

Several different types of antenna structures can be utilized for such an implant. Microstrip patch antennas and wire antennas have been demonstrated before as suitable options for a retinal prosthesis (Gosalia et al., 2005). Microstrip patch antennas can be miniaturized and offer a lightweight, reliable, and conformal radiating structure which can be easily integrated in monolithic microwave integrated circuits (MMICs) and printed board designs. Several techniques can be employed to operate them as electrically small antennas where their overall size may be such that $ka < 0.5$. Increasing the electrical length of the radiating element is the most frequently adopted technique for size reduction (Wong, 2002). Another method to achieve this is by the reactive loading of a patch antenna element which involves presenting inductive and capacitive loads at specific locations along the radiating structure. The strategically located loading mechanism can take the form of slots cut out on the patch surface or it can involve introducing electrical shorting posts between the patch and ground plane.

Inductive reactive loading can be implemented by etching out slots or creating notches or insets at the edges of the patch antenna. These features can increase the surface current path (and thereby the electrical length of the radiating element) for a given physical size of the antenna. Several instances of reactive loading have been employed for compactness and are summarized in Wong (2002). Shorting pins can be used between the patch and the ground plane as in Waterhouse (2003) to achieve size reduction for a given frequency. These concepts have been utilized to design electrically small antennas for a retinal implant at 1.45 and 2.45 GHz, which can be accommodated within the required 6 x 6 x 2-mm form factor. Figure 21.13 shows the picture of the antennas for both the external and implanted units from Gosalia et al. (2004a).

Another category of antennas that can be efficient compact radiators is comprised of different types of wire antenna configurations. Several techniques to reduce the size of wire antennas (and make them electrically small) have been reported with the most popular being fractal antennas and meander line dipoles. Figure 21.14 shows a miniature implantable meander antenna prototype (Soora, 2005). Both these types of wire antennas achieve size reduction by increasing the electrical length of the radiating element within a fixed antenna volume. For wire antennas, this is accomplished by bending the wire in either two or three dimensions to accommodate a long section within a given volume. The amount of size reduction possible and the electromagnetic behavior of the wire antenna will be a function of several parameters: total length of wire, volume of antenna, wire diameter, and details of the geometry of the antenna.

As noted earlier, it is important to be able to impedance match any compact wire antenna to the system impedance.

In general, electrically small fractal or meander antennas are inherently capacitive at the operating frequency. The antennas are made resonant by the self-inductance developed at the feed point due to the closely coupled sections of wire within the antenna. Effective impedance-matching techniques for such wire antennas are discussed elsewhere (Gosalia et al., 2005; Zhu et al., 2003; Best and Morrow, 2002). For wire antennas at a given operating frequency, a high degree of current vector alignment (current vectors in closely coupled parallel sections of wire antennas in phase) can lead to more size reduction (Best and Morrow, 2003). These principles were utilized by Gosalia et al. (2005) to design ultracompact ($ka \sim 0.2$) wire antennas for a retina. These antennas also utilize a novel way of impedance, matching the feed to the system impedance, which leads to a higher directivity broadside to the structure than can be achieved with fractal antennas (Gosalia et al., 2005).

21.3.1 Fractal Antennas

Fractal geometries are self-replicating space-filling curves which are traditionally used for representing complex objects found in nature such as clouds and coastlines. These iterative fractal curves can be generated mathematically and the contours can add significant amount of wire length in a smaller volume. Such efficient packing abilities of fractals can be utilized for antenna miniaturization. Fractal curves have been extensively researched along with their application in antenna design (Gianvittorio and Rahmat-Samii, 2002; Werner and Mittra, 2000; Puente et al., 1998). The first four iterations of both the Hilbert curve are shown in Figure 21.3 and the first four iterations of the Sierpienski curve are depicted in Figure 21.4.

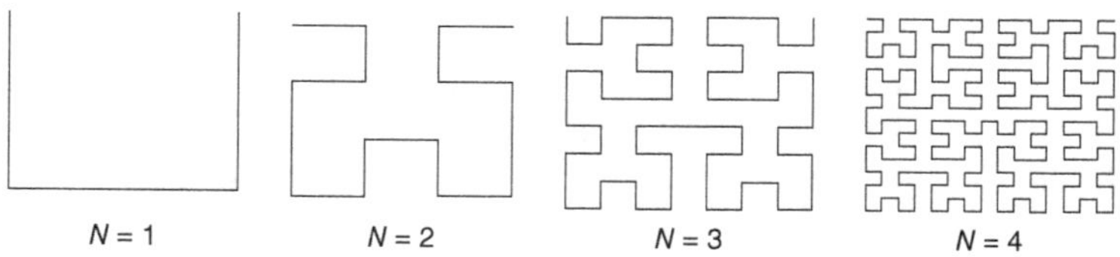

Figure 21.3 Hilbert curve, N=1 to N=4 (Soora, 2005).

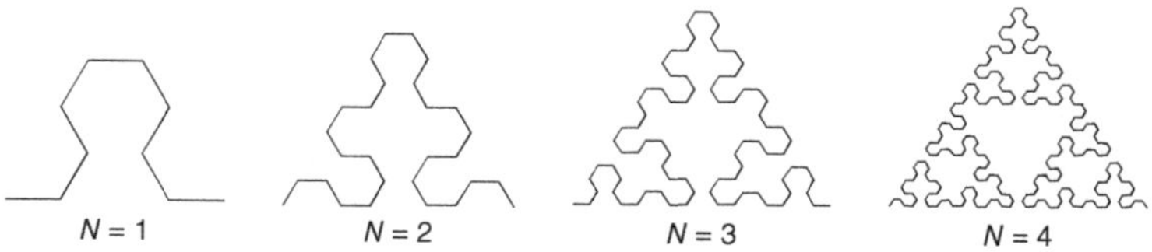

Figure 21.4 Sierpienski curve, N=1 to N=4.

21.3.2 Meander Antennas

Another manner in which wire antennas can be utilized for a compact application is by meandering the wire length for greater electrical length within a fixed volume. The meander antennas shown in Figures 21.5–21.9 were simulated using the NEC software (Voors, 2005) in free space with a footprint size constrained to 5.25 × 5.25 mm and a wire radius of 10 μm. Two- and three-dimensional designs were explored to find the smallest footprint size while achieving a strong directivity in a single direction. The 3D antennas were simulated by splitting the 2D antenna at the feedpoint and offsetting the two layers by 4 mm on the z axis.

The 2D and 3D antennas implemented the matching technique presented in Gosalia et al., (2005) where the source point remains constant and the match is achieved by creating an offset dipole. The antenna has very narrow bandwidth and high directivity due to the strong current vector alignment at the central portion of the antenna, as shown in Figure 21.5. In an effort to increase the bandwidth and compress the footprint size to fit inside the ocular globe, the structure was folded to create the 3D structure shown in Figure 21.6. With the 3D folded version, there is a higher proportion of current vector alignment throughout the structure, which results in a more omnidirectional radiation pattern and a wider bandwidth. By rotating one of the arms in the 3D structure as shown in Figure 21.7, the bandwidth and radiation pattern change due to the different current vector alignment between the two dipole arms. The radiation pattern becomes even more directive with the arm rotated. These simulations demonstrate the feasibility of attaining a range of variation for bandwidth and directivity while honoring the form factor restrictions discussed. The numerical results for these simulations are presented in Table 21.1.

Other radiator geometries are possible using meander lines. As an example, 2D and 3D triangular shaped spiral dipoles were also investigated. Triangular shapes have the ability to increase bandwidth in structures such as bowtie antennas (Balanis, 1996). In Figure 21.8, we have a 2D triangular spiral structure. This geometry has a wide bandwidth due to the higher proportion of aligned current vectors across the antenna. When the 2D triangular structure is folded across the center to create a 3D version, the current vectors align vertically as shown in Figure 21.9. Due to the angular orientation of the different wire segments, it is difficult to completely align the planar x–y currents resulting in a somewhat less directive antenna. The 3D orientation is not as beneficial as the 2D version in this specific case. Simulation results for triangular spiral meander antennas are presented in Table 21.2.

During this analysis, when simulating different cases for meander antennas, it was found that in wire antennas, when currents on one section of the antenna are strongly aligned in one direction compared to the rest of the antenna, the antenna has a higher directivity but smaller bandwidth with respect to 50 Ω. When the current vectors on the antenna are more aligned with each other (where the currents are more or less symmetric across the antenna structure), the antenna has a wider bandwidth and a more omnidirectional pattern.

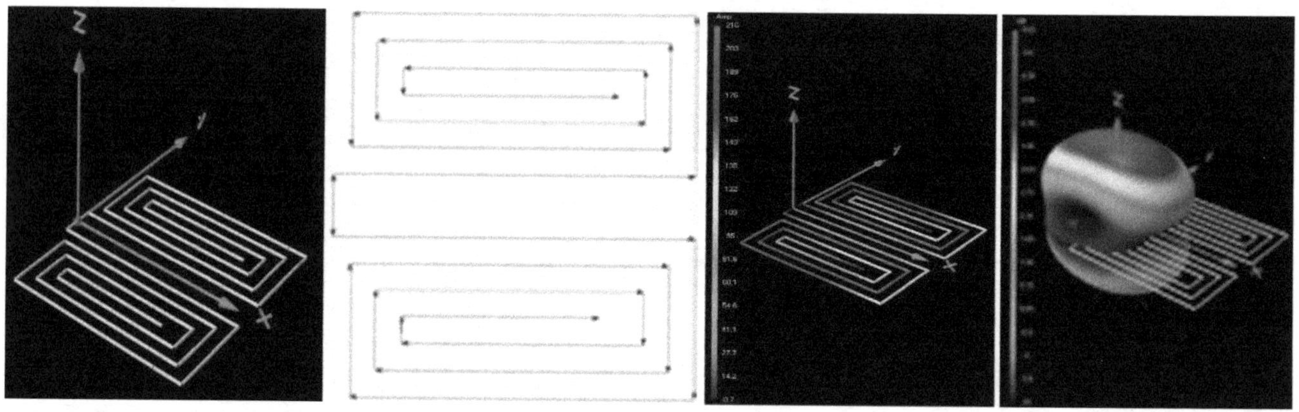

Figure 21.5 Meander antenna structure (2D spiral), current vector alignment, and radiation pattern (Soora, 2005; Soora et al., 2008). (*See insert for color representation of the figure.*)

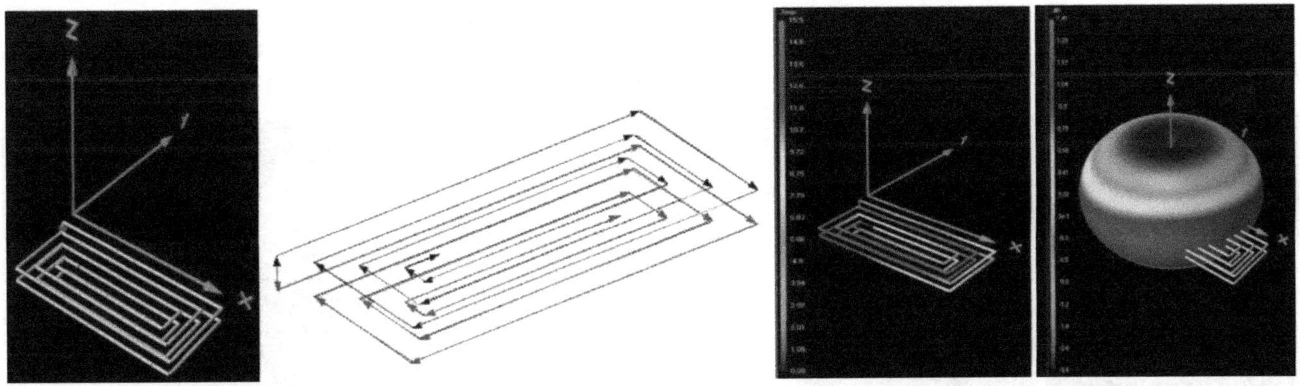

Figure 21.6 Meander antenna structure (3D spiral, folded), current vector alignment, and radiation pattern (Soora, 2005; Soora et al., 2008). (*See insert for color representation of the figure.*)

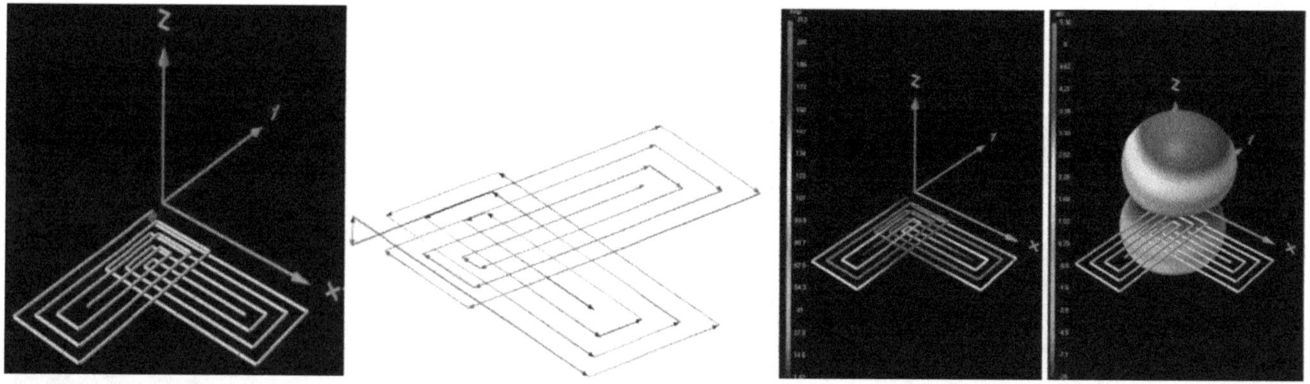

Figure 21.7 Meander antenna structure (3D rotated), current vector alignment, and radiation pattern (Soora, 2005; Soora et al., 2008). (*See insert for color representation of the figure.*)

TABLE 21.1 Meander Antenna Performance Simulation Results

Antenna Type	Resonant Frequency	Bandwidth (VSWR = 3)[a]	Maximum Gain (dBi)
Spiral 2D	3.6 GHz	8.8 kHz	4.9
Spiral 3D	3.5 GHz	1.3 MHz	1.4
Spiral 3D rotate	3.9 GHz	9.7 kHz	5.4

Source: Soora et al. (2008).

21.3.3 Prototypes and Experimental Results

A number of electrically small high-performance antennas have been designed and fabricated on copper-clad boards using photolithography and copper etching. These antennas, and in particular the wire antennas, required very fine pitch copper traces. While several fabrication techniques were investigated, the photolithography process provided the most accurate and consistent results. The prototype antennas were built and measured in our lab using two different experimental apparatus, one considering free space, shown in Figure 21.10, and the other simulating the dielectric properties of the eye, shown in Figures 21.11 and 21.12.

To fully test the intraocular antennas in an accurate environment using the apparatus described in Figure 21.11, the eye phantom was filled with a solution that mimicked the electrical properties of the vitreous humor of the eye. The actual dielectric properties of the vitreous humor of the eye were taken from Gabriel et al., (1996). Considering frequencies between 2 and 2.6 GHz, the relative permittivity for vitreous humor ranges between 68.1 and 68.5, and the conductivity varies from 2.15 to 2.6 S/m (Gabriel et al., 1996).

Creating a chemical solution to mimic these values is challenging. At frequencies above 100 MHz, the electrical properties of biological tissues are usually lower than water. Therefore, materials such as sucrose, ethanol, and glycerol have been used to lower the dielectric constant when added to a water base solution. Salt is used to increase the conductivity parameter. Various solutions of salt, water, and sugar were prepared to mimic the properties of the vitreous humor at 2.0–2.6 GHz. The solutions were measured using an open-ended coaxial probe and a network analyzer based on the technique presented in Wei and Sridhar (1989). This method uses data from experiments using open and shorted configurations in both known liquids (distilled water or acetone) and the unknown liquid to calculate the dielectric constant of the unknown liquid

Gosalia et al., (2004a, 2005), fabricated and measured experimental prototypes for both implanted and external antennas operating at 1.45 and 2.45 GHz (Figure 21.13). The performance of these designs was characterized in terms of the S parameter S_{21} between the external and implanted antennas and is reported in Table 21.3.

Measurements using the experimental eye phantom apparatus showed an improvement in the coupling performance between the external and implanted antennas as compared to coupling at the same distance in free space. This enhancement was of the

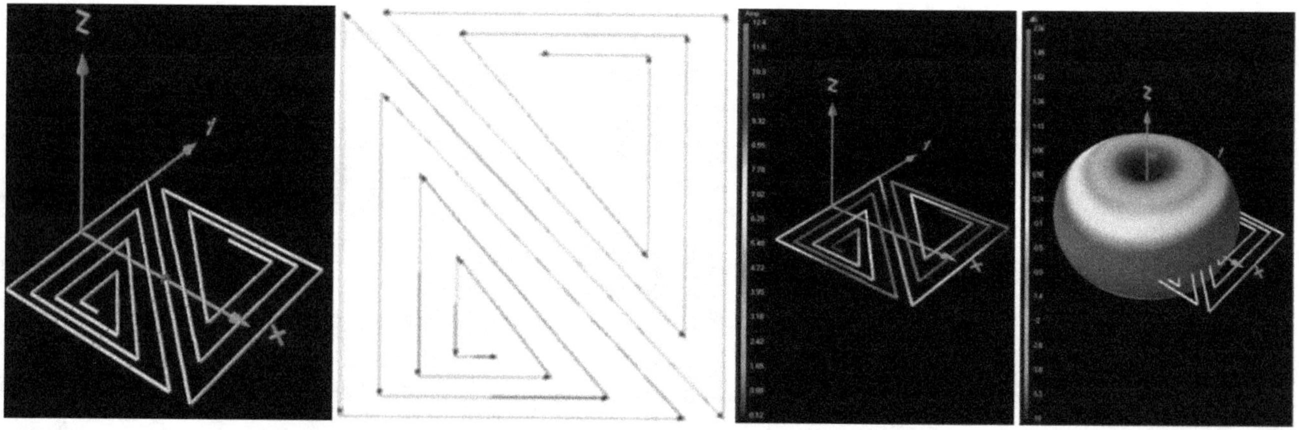

Figure 21.8 Meander antenna structure (2D triangular spiral), current vector alignment, and radiation pattern (Soora, 2005; Soora, et al., 2008). (*See insert for color representation of the figure.*)

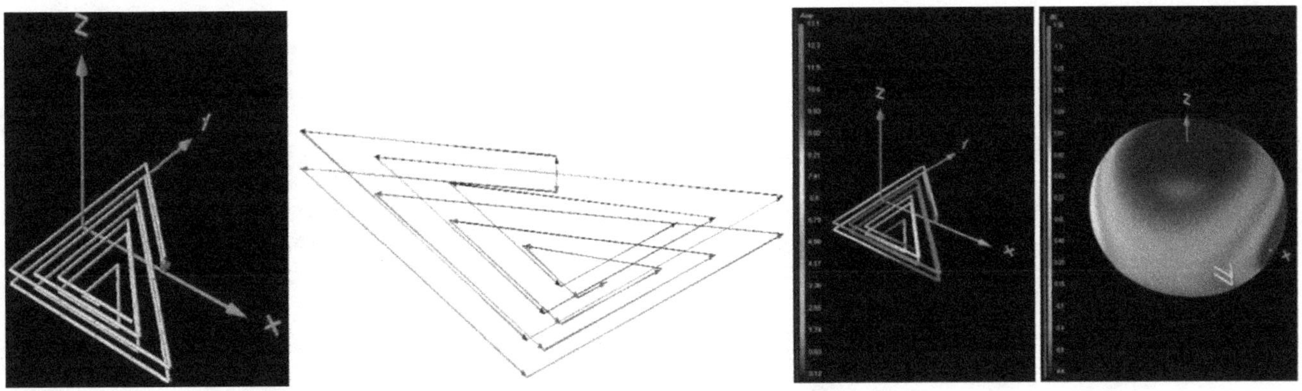

Figure 21.9 Meander antenna structure (3D triangular spiral, folded), current vector alignment, and radiation pattern (Soora, 2005; Soora et al., 2008). (*See insert for color representation of the figure.*)

TABLE 21.2 Triangular Meander Antenna Performance Simulation Results

Antenna Type	Resonant Frequency	Bandwidth (VSWR = 3)	Maximum Gain (dBi)
Tri-Spiral 2D	3.8 GHz	3.2 MHz	2.1
Tri-Spiral 2D	3.5 GHz	2.2 MHz	1.4

Source: Soora et al., (2008).

order of 4–6 dB at 1.45 GHz and 1–2 dB at 2.45 GHz for the intraocular patch antennas; for the intraocular wire dipoles, the coupling enhancement was in the 1–2-dB range.

For measuring using the eye phantom experimental apparatus, the implanted antennas were immersed to a depth that will locate their position at the region of the lens in the eyeball. A distance of 25 mm was maintained between the external and the implanted antenna. It was observed that the eye phantom acted as a radome or a curved dielectric lens for microwave energy received by the implanted antenna.

During the experimental measurements, it was noticed that when the encapsulated intraocular antennas (both microstrip patch and the meander line dipole) were immersed in the eye phantom filled with vitreous humor simulant fluid, the antennas exhibited severe detuning due to the dielectric loading effect. The resonant frequency decreases, and the return loss degrades due to an inductive effect in its impedance characteristics. The dielectric loading effect is more pronounced at the higher frequency band of 2.45 GHz.

The 2D and 3D wire meander antennas had a planar footprint size of 5.25 × 5.25 mm on a substrate with a relative dielectric constant of 4.8. An extra copper pad was included to solder the antenna element to the SMA connector. In addition, corresponding external (extraocular) antennas were designed and built.

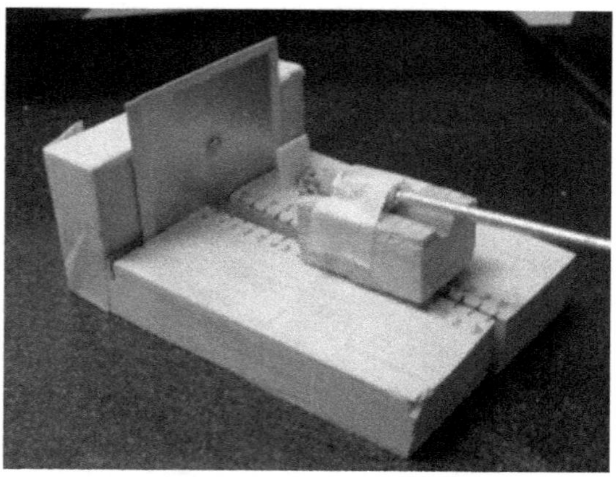

Figure 21.10 Free-space measurement apparatus (Soora, 2005).

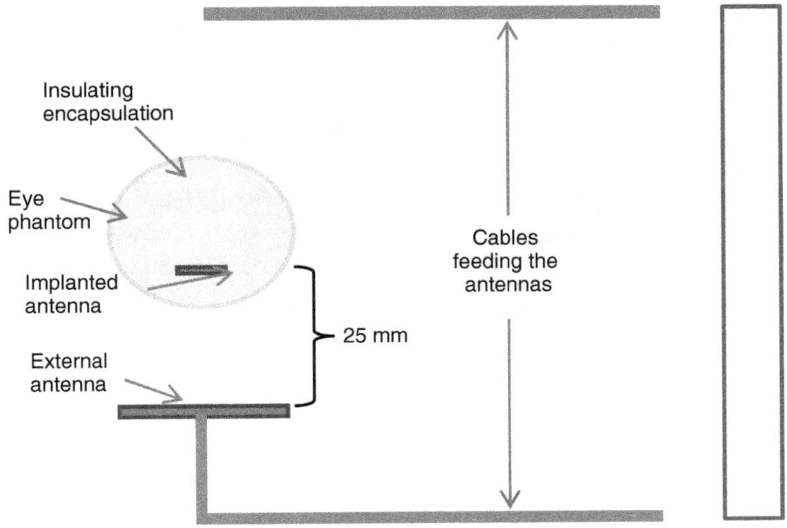

Figure 21.11 Eye phantom measurement apparatus diagram.

The external antennas were all patch antennas because of the directive nature and ease of tuning of that configuration.

Some of the 2D antennas were measured in free space to determine the performance at various points in space. The location of the test points relative to the

Figure 21.12 Eye phantom measurement experimental apparatus (Soora, 2005).

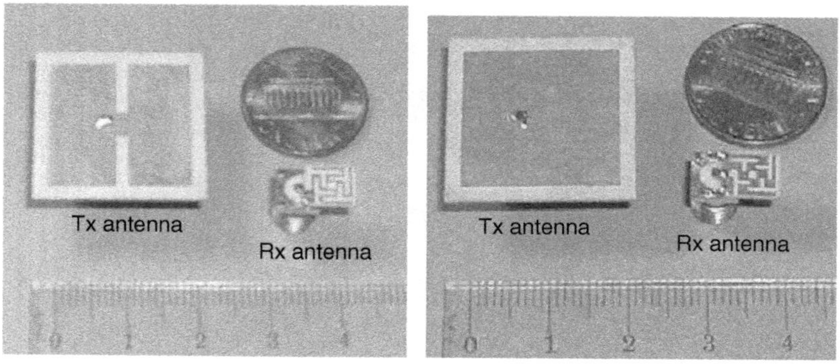

Figure 21.13 Implantable antenna prototypes for 1.45 and 2.45 GHz (Gosalia et al., 2004a, 2005).

TABLE 21.3 Measured Antenna Coupling for Radiators Shown in Figure 21.13

	Implanted Microstrip Antenna		Implanted Wire Antenna	
Coupling (S_{21}) at Separation of 25 mm	Free Space	Eye Phantom	Free Space	Eye Phantom
1.45 GHz	−42.5 dB	−37.8 dB	−36.8 dB	−35.8 dB
2.45 GHz	−32.0 dB	−31.1 dB		

implanted antenna are shown in Figure 21.15. The antenna was placed parallel to the xy plane. Points 1 and 3 are located at a 25-mm distance on either side of the x axis. Points 2 and 4 are 25 mm on either side of the y axis and point 5 is 25 mm above the antenna on the z axis. In addition, the uniformity of the coupling between the two antennas was measured at different rotation angles to mimic the rotation of the eyeball movement.

Figure 21.16 shows the measured coupling (S_{21}) between the 2D spiral intraocular antennas and the extraocular antennas at the selected test points. These results are measured at 25 mm, which is within the near field of the antenna. The far-field radiation pattern and the near-field directivity may not be the same. The results show that the measured coupling at point 5 for the 2D spiral is higher compared to the other test points. This means that the near-field pattern is more directive in the broadside direction, making it a good structure for implantable devices. The change in S_{21} at 25 mm at different rotation angles is minimal (Figure 21.17), making this radiator robust to eye movement and a good candidate for an intraocular implantable device.

21.3.4 Biocompatibility and Safety Considerations

Chronically implanted antennas need to be encapsulated in a biocompatible material. The biocompatible insulation helps prevent infections and maintains the isolation

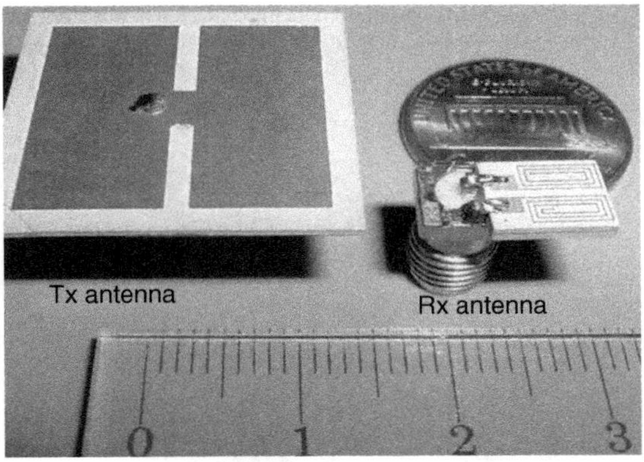

Figure 21.14 Implantable meander wire antenna. (Gosalia et al., 2005)

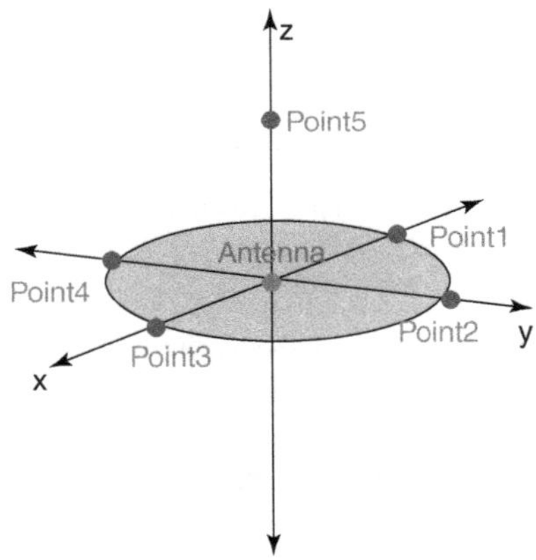

Figure 21.15 Meander antenna test points (Soora, 2005, Soora et al., 2008).

of the antenna from the surrounding conductive medium. It has been reported (Bahl et al., 1982) that a properly designed insulating biocompatible encapsulation can help in lowering the amount of loading on the embedded antennas and also provide a smoother transition to the electromagnetic energy from the antenna surface in to the dissipative medium. Suitable insulators can improve the impedance match, and previously reported implanted antennas have used superstrates or substrates with high dielectric constant in the range of 5–10 to achieve acceptable impedance matching. Silicone superstrates have been used for biocompatibility (Warty et al., 2008), and it has been shown that the thickness and dielectric properties of superstrates affect the miniaturization of the antenna (Djaiz et al., 2008).

Biomedical devices have to adhere to regulations concerning the use of specific frequencies and safety due to operation of the devices either externally, periodically, or as implants. The ISM band (433, 915, and 2450 MHz) is stipulated to be used for

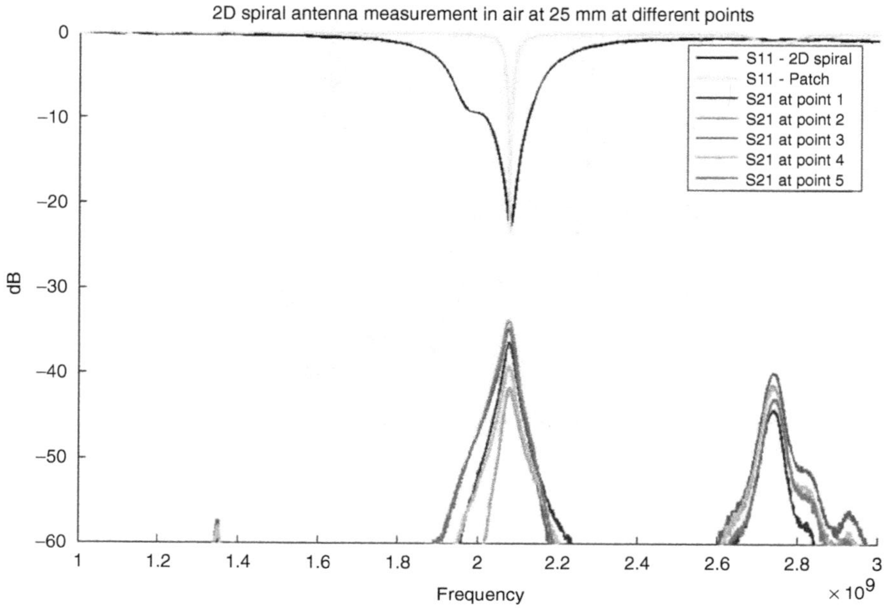

Figure 21.16 2D spiral meander antenna performance (coupling) at test points (Soora, 2005). (*See insert for color representation of the figure.*)

devices placed external to the body in the United States and Europe. Implantable medical devices that chronically stay within the body are allocated the Medical Implant Communication Services (MICS) band, ranging from 402 to 405 MHz. The safety regulation that medical device antennas have to comply with is the maximum absorbed power by the body. The limits on localized power absorption are critical and are specified in terms of the specific absorption rate (SAR), which is calculated as

$$\text{SAR (W/kg)} = \frac{\sigma |\mathbf{E}_{\text{peak}}|^2}{2\delta}$$

where σ is the conductivity of the tissue (in S/m), $\mathbf{E}_{\text{peak}}$ is the peak value of the electric field in the specific tissue at that location, and δ is the mass density of the tissue (kg/m^3).

As depicted in Figure 21.1, due to the proximity of the transmitting antenna to the head of the implanted patient, electromagnetic energy is deposited in the eye and surrounding head tissues. Computations to quantify the power deposited in terms of SAR were performed at both the frequency bands. Currently available sensitive receiver architectures will enable the Tx antenna to limit radiated power to 50 mW. Thus, the SAR values were normalized to a maximum transmitted power of 50 mW. Peak 1-g SAR values were calculated at both frequency bands to determine compliance with the IEEE/ANSI stipulated guidelines. As reported in Gosalia et al. (2004a), using the configurations discussed, the maximum power deposition takes place in the forehead region for both the frequency bands due to the presence of the external transmitting

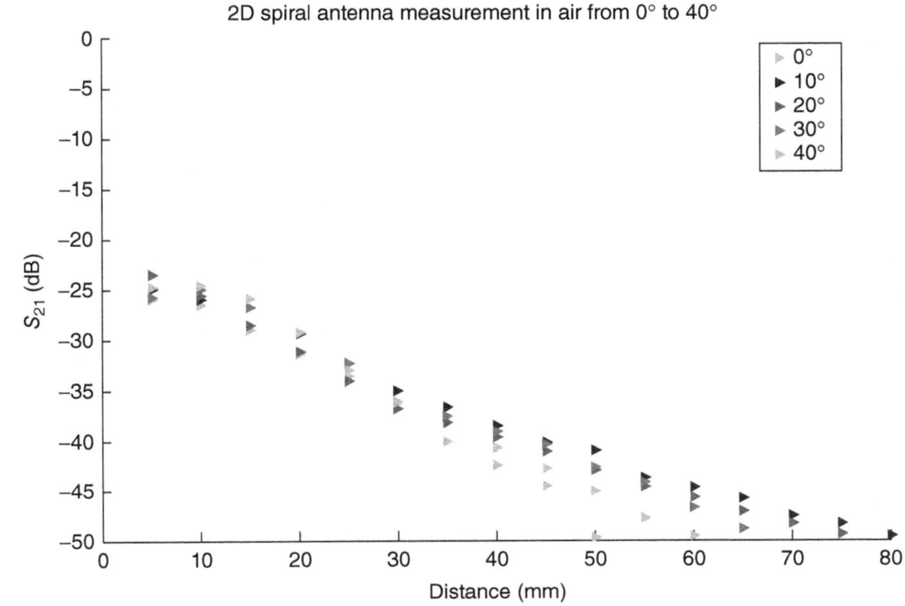

Figure 21.17 2D spiral meander antenna coupling as function of eye rotation (Soora, 2005). *(See insert for color representation of the figure.)*

antenna's proximity. At 1.45 GHz, a computed peak 1-g SAR was 0.985 W/kg, and at 2.45 GHz, it was computed at 1.158 W/kg. Both these values are within the maximum allowed under IEEE/ANSI guidelines [IEEE Standard for Safety Levels with Respect to Human Exposure to Radiofrequency Electromagnetic Fields (IEEE, 1999)].

Physiological hazards due to electromagnetic power deposition in tissues are known to be largely thermal in nature (Adair, 2002). In addition to the deposited energy due to the wireless telemetry link (quantified in terms of SAR above), the operation of the implanted unit itself will also cause a localized thermal increase due to power dissipation in the receiver electronics. This will impact the natural basal steady-state temperature of the eye and surrounding tissues of the head. The thermal increase effect due to power dissipation of the implanted electronics was investigated in detail in DeMarco et al., (2003).

Gosalia et al., (2004b) report a detailed investigation on the thermal effects of a retinal implant due to power dissipation in the implanted electronics. This computational study presents a comprehensive account of temperature elevations in different tissues surrounding the eye and head under different operational conditions.

21.4 MULTICOIL TELEMETRY

Inductive coupling can be used for wireless power transfer (WPT) and wireless data transfer when operating distances are small, on the order of 1–2 cm; this is the case with many biomedical implants (Chestek et al., 2009). In these designs, a power amplifier is used to drive an external coil (driver), causing a time-varying magnetic field in close proximity of an implanted coil (load).

For wireless data transfer, data are modulated using a number of modulation schemes; these include frequency shift keying (FSK), on–off keying (OOK), and amplitude shift keying (ASK) (Sawan et al., 2005). Power transfer efficiency (PTE) and frequency bandwidth are performance parameters to be considered for efficient biomedical implant telemetry links. Optimum PTE is obtained when both the external coil and the implanted coil are resonating at the same operating frequency. In that situation, the PTE is limited by the resistive losses and the magnetic coupling between the external and implanted coils. Magnetic coupling k is determined by the distance between the coils and the dimensions of each coil (Smith et al., 2007).

A two-coil inductive link as depicted in Figure 21.18*a* offers moderate data rates (10 kbps–1 Mbps) depending on the available frequency bandwidth and the modulation scheme of choice. Due to finite source and load resistances, most two-coil systems have relatively low PTE, up to a maximum of 40%, with a frequency bandwidth of about 10% of the carrier frequency (Catrysse and Hermans, 2005).

Multicoil-based power and data transfer links are relatively new configurations that use three (Figure 21.18*b*), four (Figure 21.18c), or more coils, resulting in higher

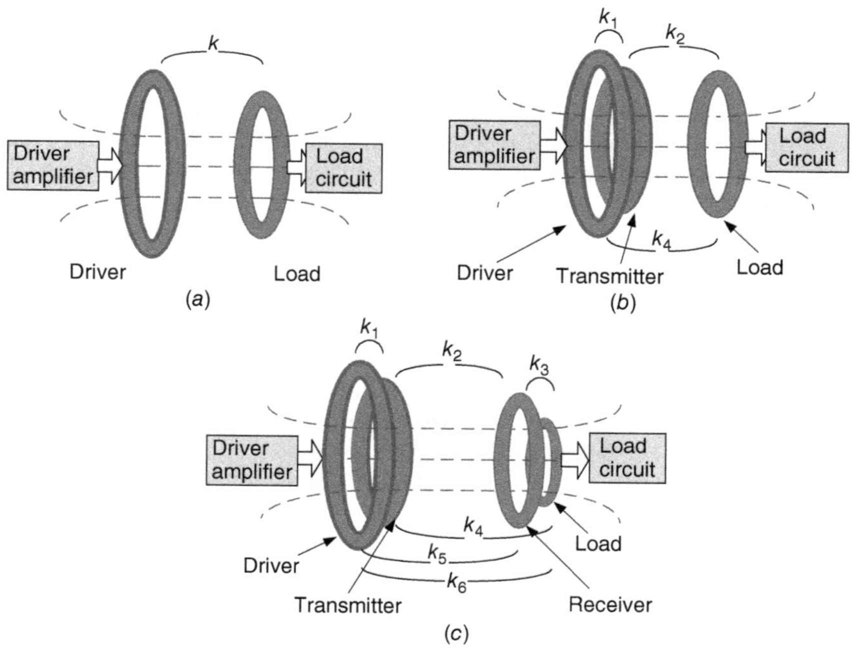

Figure 21.18 Block diagram for (*a*) two-coil, (*b*) three-coil, and (*c*) four-coil telemetry systems. Load and receiver coils (if present) are implanted. Driver and transmitter coils (if present) are part of the external device [from RamRakhyani and Lazzi (2012)].

overall PTE and frequency bandwidth when compared with similarly sized two-coil links (RamRakhyani and Lazzi, 2012; RamRakhyani and Mirabbasi, 2011). Using three or more for the wireless link allows the driver and load resistances to be decoupled from the Q factor of the coils, making it possible to use high-Q-factor transmitter and receiver coils, which leads to improved PTE. Moreover, because of the strong coupling between the driver and the transmitter coils and between the receiver and the load coil (in a four-coil configuration), it is possible to achieve higher tunable frequency bandwidth when compared to similarly sized two-coil links.

21.4.1 Power Transfer Efficiency

Power transfer efficiency is one of the key performance parameters for a telemetry link. Table 21.4 provides an analytical formulation to calculate the PTE at the resonance (operating) frequency for telemetry systems using two, three, and four coils (RamRakhyani and Lazzi, 2012).

For two-coil telemetry systems, the PTE is a strong function of the magnetic coupling k and the Q factors of the driver coil, Q_d and the load coil, Q_l. In general, the load resistance value is in the 100–100-Ω range, and the source resistance value is between 3 and 6 Ω. Within these ranges, the loaded Q factors of the driver coil and the load coil are on the order of ~30 and ~10, respectively.

Using the multicoil approach, the Q factor of the transmitter coil is independent of the driver resistance. Moreover, in the four-coil case, the Q factor of the receiver coil is independent from the load resistance. This allows to minimize the effect of low coupling between external and implanted coils, making it possible to achieve a high PTE. The PTE of the three-coil system can be modeled considering a two-step power transfer from the driver electronics to the load circuit. Due to high coupling k_1 between the driver and the transmitter coil ($k_1 \sim 0.3$–0.5) and the high Q factor of the

TABLE 21.4 Power Transfer Efficiency for Two-, Three-, and Four-Coil Telemetry Systems

	Efficiency (η)
Two coils	$\eta = \dfrac{k^2 Q_d Q_l}{1 + k^2 Q_d Q_l}$
Three coils	$\eta \approx \dfrac{k_1^2 Q_d Q_t}{1 + k_1^2 Q_d Q_t} \dfrac{k_2^2 Q_t Q_l}{1 + k_2^2 Q_t Q_l} = \eta_{dt}\eta_{tl}$
Four coils	$\eta \approx \dfrac{k_1^2 Q_d Q_t}{1 + k_1^2 Q_d Q_t} \dfrac{k_2^2 Q_t Q_r}{1 + k_2^2 Q_t Q_r} \dfrac{k_3^2 Q_r Q_l}{1 + k_3^2 Q_r Q_l} = \eta_{dt}\eta_{tr}\eta_{rl}$

Source: RamRakhyani and Lazzi (2012).

transmitter coil ($Q_t \sim 100-150$), three-coil systems can achieve more than twice the efficiency of two-coil telemetry systems using coils of the same size (RamRakhyani and Lazzi, 2012). An additional advantage of three-coil systems is that they can be used to upgrade implanted two-coil telemetry systems by modifying the external unit without having to resort to surgery. Similarly, for the same overall dimensions of external and implanted coils, four-coil systems can achieve high-Q-factor transmitter and receiver coils to achieve efficient wireless power transfer link.

21.4.2 Voltage Gain

The output voltage to operate the implanted electronics can be provided relatively independently of the voltage of the power supply by controlling voltage gain. The voltage gain is a strong function of the coupling between the individual coils, the Q factor of each coil, and the source and load resistances.

In two-coil telemetry systems, given a fixed value of k, source resistance R_d, and load resistance R_L, the voltage gain is constant.

Three- and four-coil systems can control the voltage gain by modifying the coupling k_1 between the driver and transmitter coil and the coupling k_3 between the receiver and load coil. Voltage gains for systems operating at the resonance frequency and having two, three, and four coils can be calculated using the expressions in Table 21.5.

Figure 21.19*a* shows that voltage gain of a three-coil system can be controlled by varying coupling k_1 based on design example 2 in Table 21.6. For all value of coupling $k_1 > 0.15$, the power transfer efficiency of the three-coil system is higher than the same-size two-coil telemetry system and voltage gain can be controlled linearly with the variation of k_1.

TABLE 21.5 Voltage Gain for Telemetry Systems having Two, Three, and Four Coils

	Voltage Gain ($G = V_{out}/V_{in}$)
Two coils	$G_2 = -j\sqrt{\frac{R_L}{R_d}}\frac{k\sqrt{Q_dQ_l}}{1+k^2Q_dQ_l}$
Three coils	$G_3 = -\sqrt{\frac{R_L}{R_d}}\frac{(k_1\sqrt{Q_dQ_t})(k_2\sqrt{Q_tQ_l})}{1+k_1^2Q_dQ_t+k_2^2Q_tQ_l}$
Four coils	$G_4 = j\sqrt{\frac{R_L}{R_d}}\frac{(k_1\sqrt{Q_dQ_t})(k_2\sqrt{Q_tQ_r})(k_3\sqrt{Q_rQ_l})}{(1+k_1^2Q_dQ_t)(1+k_3^2Q_rQ_l)k_2^2Q_tQ_r}$

Source: RamRakhyani and Lazzi (2012).

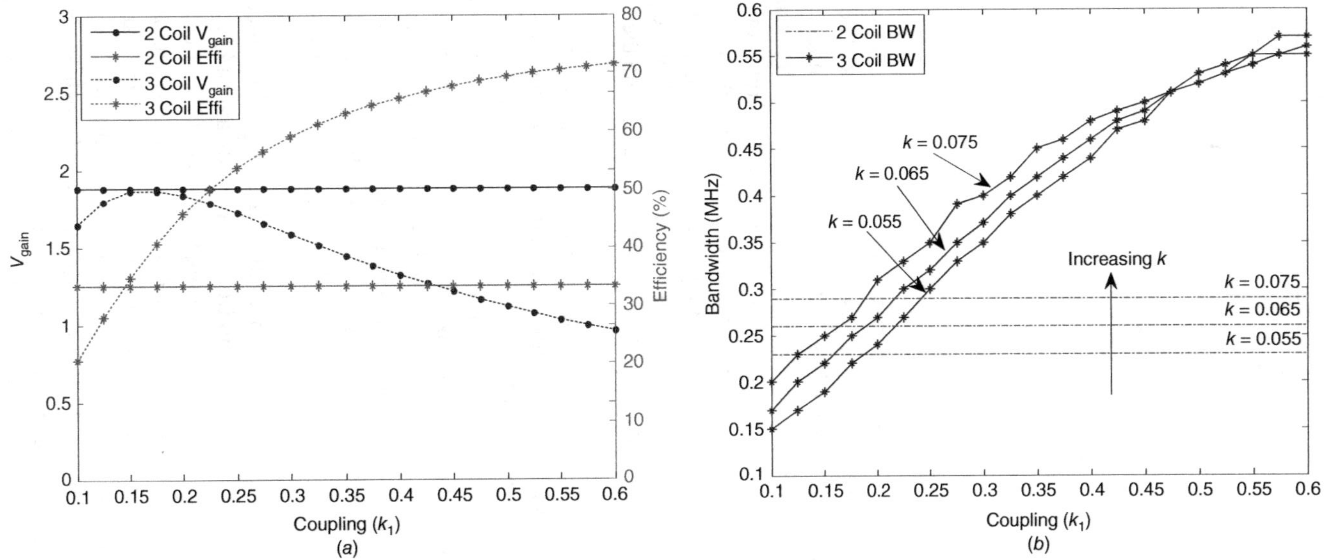

Figure 21.19 (*a*) Voltage gain and PTE of systems with two and three coils (same size) with change in coupling k_1. Both systems use the same operating distance between the external and the implanted coils. (*b*) Channel bandwidth variation of two-coil system and its equivalent three-coil system with change in coupling k_1, and $k_2 = k$ (RamRakhyani and Lazzi, 2012).

TABLE 21.6 Experimental Results Comparing Two- and Three-Coil Telemetry System Performance

	Design 1		Design 2	
	Two Coils	Three Coils	Two Coils	Three Coils
Efficiency, η (%)	3.12	6.3	35	65
BW (kHz)	60	332	240	580
Gain	0.42	0.125	1.9	1.07
FOM	−34.28	−28.8	−18.94	−14.91

Note: The size of the coils was the same for the compared cases.
FOM: Figure of merit.
Source: RamRakhyani and Lazzi (2012).

21.4.3 Frequency Bandwidth

The frequency bandwidth (BW) of the inductive link is an important performance parameter as it affects the available data rate over the inductive link. The inductive link behaves as a bandpass filter around its operating frequency, limiting bandwidth. The BW of the inductive link is determined by computing the 3-dB frequency points on the voltage gain versus frequency curve of the system. In a two-coil inductive link, the BW is a function of the Q factor of the coils and their mutual coupling k. The inductive link uses two or more resonating coils for power and data transfer. Since the frequency bandwidth is directly proportional to the coupling between resonating oscillators (Haus, 1991), the system bandwidth increases with the increase in coupling.

For a given set of system constraints, the BW is usually limited to 10% of the carrier frequency. Multicoil WPT systems utilize the high coupling between the driver and transmitter coil k_1 (and between the receiver and load coil k_3) to improve the frequency bandwidth (RamRakhyani and Lazzi, 2012). As shown in Figure 21.19*b*, the frequency bandwidth of the inductive link can be improved by increasing the coupling k_1. For a three-coil system, k_1 can be controlled independently of the distance between the transmitter and the implant coil, just by properly designing the driver and transmitter coils. As expected from coupled mode theory, the frequency bandwidth on the inductive link is linearly proportional to the coupling k_1 (RamRakhyani and Lazzi, 2012), which can be seen in Figure 21.19*b*.

Figure 21.20*a* shows the simulated and measured efficiency of a two-coil and its three-coil (same-size coils) telemetry system from (RamRakhyani and Lazzi, 2012). Simulation was done with and without considering the effect of secondary coupling k_4. It can be observed that coupling between driver and load coil has negligible effect on the system performance and that efficiency of the three-coil system is at least double that of the two-coil equivalent. Figure 21.20*b* shows that, due to strong coupling between driver and transmitter coil, the frequency bandwidth of the three−coil system is considerably higher than the traditional two-coil system.

The multicoil approach for inductive coupled telemetry offers high power transfer efficiency for biomedical telemetry. Having multiple control parameters to tune

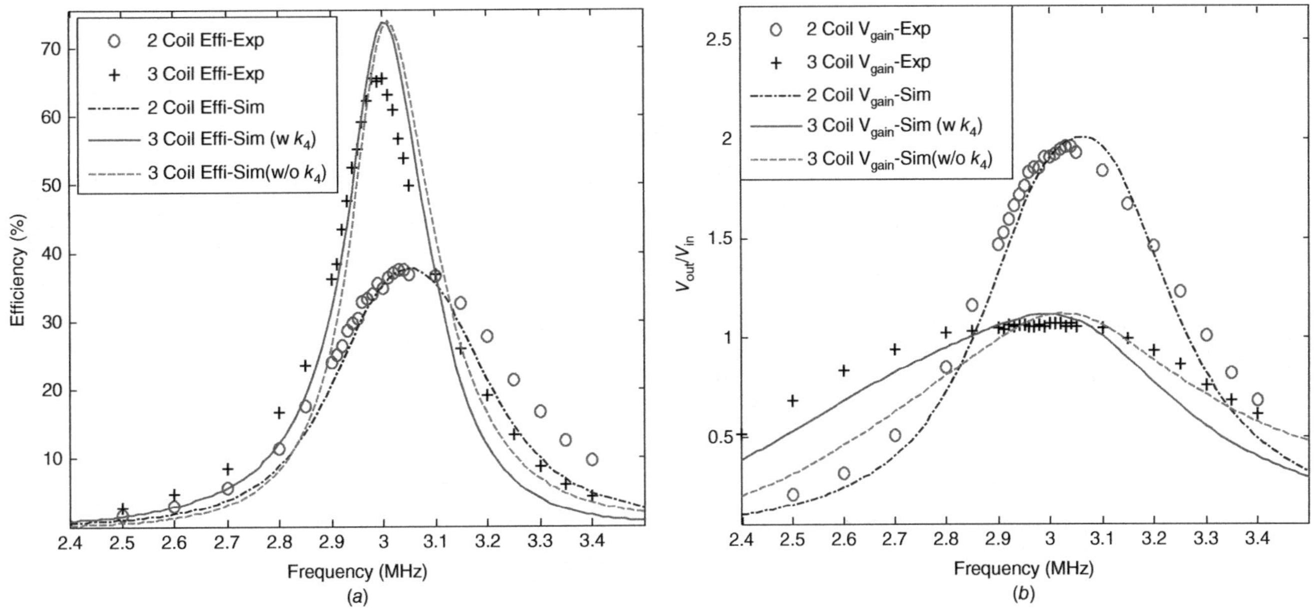

Figure 21.20 Simulated (Sim) and experimental (Exp) (*a*) power transfer efficiency and (*b*) voltage gain of two-coil system and its three-coil equivalent system with respect to frequency for design 2 in Table 21.6. Simulation is done with (w k_4) and without (w/o k_4) the consideration of coupling between driver and load coils k_4. [(From RamRakhyani and Lazzi (2012).] (*See insert for color representation of the figure.*)

efficiency, voltage gain, and bandwidth, multicoil telemetry systems are in general more flexible than dual-coil systems.

21.5 FUTURE RESEARCH DIRECTIONS: FLEXIBLE AND LIQUID ANTENNAS

For implanted-device telemetry applications, the mechanical, biological, and environmental parameters, including size, weight, biocompatibility, and reliability of the system components, constrain the material selection and fabrication process. Often, as understanding of the application increases and associated technology (i.e., microelectronics, electrode arrays) advances, novel combinations of antenna configurations and materials are required.

Many of the early conformal small-antenna systems consisted of etched copper on flexible substrate materials, such as polyimide and thin printed circuit board materials, including 0.008-in. G10 or FR-4. These materials are flexible due to their thin geometry. Electrically, these antennas performed well. The conductor and substrate losses were minimized through the use of copper and low-loss dielectric substrates. Mechanically, these antennas could be curved in one dimension or folded in multiple dimensions, provided that there is a rigid support structure available to affix them. That is usually not the case for intraocular systems, where minimizing the device volume and the size of the incision needed to perform the implant surgery is critical.

The use of rigid metals (such as etched copper or stamped stainless steel) limits the dynamic flexing capabilities of the antennas. Furthermore, when metal antennas are subjected to repetitive flexure, they eventually fatigue and/or crack the conductive elements with catastrophic impact to their performance. This is a cause of concern in implanted devices, while the structure cannot be easily accessed.

Flexible conductive elements can be incorporated into on-body applications to overcome mechanical limitations. Possible technologies include conductive threads and conductive fabrics properly encapsulated in biocompatible materials. Conductive thread applications typically consist of woven dielectric fibers (including yarn and glass fibers) with small interwoven metallic wires. Through a sufficiently dense weave pattern, the wires can be arranged to form patch elements, microstrip lines, and ground planes as needed. Similarly, conductive fabrics (copper- and nickel-coated nylon or polyester fabrics) can be stitched or glued to a dielectric fabric substrate to form patch antennas and planar inverted-F antenna (PIFA) systems. Electrically, these antennas perform well compared to conventional microstrip elements. The substrate losses are minimized through the use of low-loss dielectric substrates. Mechanically, these antenna systems can conform to complex, irregular surfaces (such as clothing). Although these antennas are more flexible than a rigid conductor construction, the metallic components (thin wires or coated plastics) are still vulnerable to fatigue and wear over time. Furthermore, these constructions may not allow the antenna system to stretch beyond the inherent strain permitted by the weave pattern.

Antenna systems incorporating the thin-film metallization of the antenna conductor on elastomeric substrates and the formation of liquid metal radiating elements

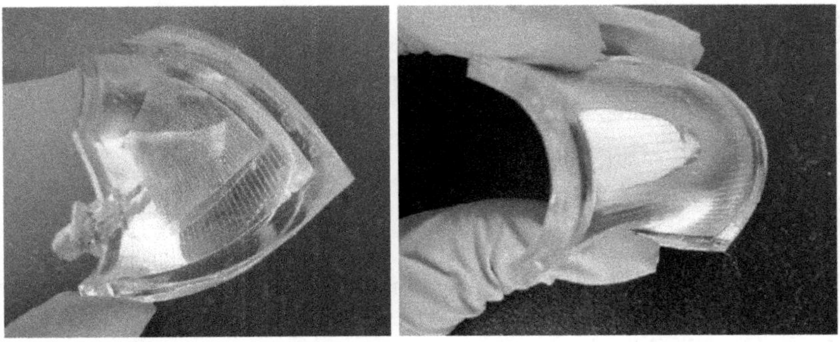

Figure 21.21 Photographs of multilayer, flexible microstrip patch antenna composed of liquid metal encased in elastomer. (*See insert for color representation of the figure.*)

in microfluidic channels within an elastomeric substrate can be designed to overcome the limitations with respect to flexibility and stretchability. Thin metal films (such as <100-nm-thick gold films) deposited on elastomeric substrates [such as polydimethylsiloxane (PDMS)] can stretch reversibly without losing electrical conduction. The presence of dense arrays of microcracks in the substrate allows for the thin film to twist and flex when strained up to 100% strain without losing conductivity. Up to 20% elongation has been reported with acceptable electrical performance after hundreds of thousands of cycles. However, when subjected to excessive stress, microcracks can be formed within the gold film, which may result in higher resistance and lower antenna efficiency. In the absence of microcracks, these antennas perform well electrically. The conductor and substrate losses are minimized through the use of gold and low-loss dielectric substrates.

Fluidic antennas represent the latest combination of antenna systems that are ideally suited for in-body applications. An example of antennas using fluidic technology is shown in Figure 21.21. These antennas are extremely flexible and highly stretchable and maintain acceptable electrical performance (Hayes et al., 2012). These antennas fully adopt the mechanical properties of the encasing substrate material, such as PDMS or selectively hydrogenated poly(styrene-block-butadiene-block-styrene) (SEBS), and are therefore flexible and mechanically durable. Unlike solid metal structures or metallized surfaces, liquid metal conductors do not crack or fatigue over time.

Mechanically, the typical embedded system is extremely small. As a result, the antenna design has to function in close proximity to the electronics and its associated ground plane. These designs can be highly flexible to accommodate a movable prosthetic implant.

Intrabody systems must also be biocompatible to minimize any potential hazards to the user. With respect to antenna systems, the mechanical and biocompatibility requirements significantly limit the selection of applicable materials. Commonly used biocompatible substrates include PDMS and SEBS. Gold is the most common biocompatible conductor. As previously mentioned, gold wire embedded in PDMS can be very flexible and desirable for many applications. Furthermore, the metal is typically composed of gallium or gallium alloys. Since gallium has low toxicity and negligible vapor pressure, gallium alloys encapsulated in PDMS or SEBS demonstrate potential for intrabody applications.

21.6 CONCLUSION

The artificial retina is undergoing clinical medical trials in several countries, and it can be available to patients worldwide. The device is evolving from a concept to a sophisticated medical implant which performs exceptionally well. Each progressive generation of the implant refines on the previous iteration and increases the comfort for the patient as well as the resolution of the electrode array.

As the density of the implanted electrode array increases and the device capabilities grow in complexity, higher data transfer rates and higher power transfer efficiency will be needed between the external electronics and the implanted stimulator. This implies a larger bandwidth requirement for data transfer. There are promising technologies available to fulfill these needs: Electrically small antennas, including the approaches presented in this chapter, can deliver the data bandwidth needed for the foreseeable future. Multicoil inductive link also provides an alternate approach for high-data-rate communication and efficient wireless power transfer. The implementation of flexible small antennas using liquid metals holds the promise of better fitting implantable conformal devices, which is important in the proximity of delicate biological structures such as the eye's retina and smaller incisions for implant surgery.

REFERENCES

Adair ER, Petersen RC. 2002. Biological effects of radiofrequency microwave radiation. *IEEE Trans. Microw. Theory Techn.* 50(3):953–962.

Bahl IJ, Bhartia P, Stuchly SS. 1982. Design of microstrip antennas covered with a dielectric layer. *IEEE Trans. Antennas Propag.*, 30(2):214–318.

Balanis CA. 1996. *Antenna Theory: Analysis and Design*. New York: Wiley.

Best SR, Morrow JD. 2002. The effectiveness of space filling fractal geometries in lowering resonant frequency. *IEEE Wireless Antennas Propag. Lett.*, (1):112–115.

Best SR, Morrow JD. 2003. On the significance of current vector alignment in establishing the resonant frequency of small space filling wire antennas. *IEEE Antennas Wireless Propag. Lett.* 2:201–204.

Buckett R, Peckham PH. 1988. An externally powered, multichannel, implantable stimulator-telemeter for control of paralyzed muscle. *IEEE Trans. Biomed. Eng.*, 45(4):463–475.

Catrysse M, Hermans B. 2005. An inductive power system with integrated bidirectional data transmission. Proc. 17th Eurosensors Conf.

Chestek C, Gilja V, Nuyujukian P, Kier R, Solzbacher F, Ryu S, Harrison R, Shenoy K. 2009. Hermesc: Low-power wireless neural recording system for freely moving primates. *IEEE Trans. Neural Syst. Rehabil. Eng.*, 17(4):330–338.

DeMarco SC, Lazzi G, Liu W, Weiland JD, Humayun MS. 2003. Computed SAR and thermal elevation in a 0.25 mm 2D model of the human eye and head in response to an implanted retinal stimulator: Parts I and II. *IEEE Trans. Antennas Propag.*, 51(9): 2274–2295.

Djaiz A, Nedil M, Habib MA, Talbi L, Denidni TA. 2008. A new CPW-feed miniaturized dual-band monopole antenna using spiral matching resonators. IEEE Int. Antennas Propag. Symp., San Diego, CA.

Gabriel S, Lau RW, Gabriel C. 1996. The dielectric properties of biological tissues: III. Parametric models for the dielectric spectrum of tissues. *Phys. Med. Biol.*, 41(11).

Gianvittorio JP, Rahmat-Samii Y. 2002. Fractal antennas: A novel antenna miniaturization technique and applications. *IEEE Antennas Propag.*, 40(1):20–36.

Gosalia K, Humayun M, Lazzi G. 2005. Impedance matching and implementation of planar space-filling dipoles as intraocular implanted antennas in a retinal prosthesis. *IEEE Trans. Antennas Propag.*, 53(8):2365–2373.

Gosalia K, Lazzi G, Humayun M. 2004a. Investigation of a microwave data telemetry link for a retinal prosthesis. *IEEE Trans. Microw. Theory Techn.*, 52(8):1925–1933.

Gosalia K, Weiland J, Humayun M, Lazzi G. 2004b Thermal elevation in the human eye and head due to the operation of a retinal prosthesis. *IEEE Trans. Biomed. Eng.*, 51(8):1469–1477.

Haus H, Huang W. 1991. Coupled-mode theory. *Proc. IEEE*, 79:1505–1518.

Hayes GJ, Qusba A, Dickey MD, Lazzi G. 2012. Flexible liquid metal alloy (EGaIn) microstrip patch. *IEEE Trans. Antennas Propag.*, 60(5):2012.

Institute of Electrical and Electronics Engineers (IEEE). 1999. Standard for Safety Levels with Respect to Human Exposure to Radiofrequency Electromagnetic Fields, 3 kHz to 300 GHz. IEEE Standard C95.1.

Izdebski PM, Rajagopalan H, Rahmat-Samii Y. 2009. Conformal ingestible capsule antenna: A novel chandelier meandered design. *IEEE Trans. Antennas Propag.*, 57(4):900–909.

Karacolak AZ, Hood AZ, Topsakal E. 2008. Design of a dual-band implantable antenna and development of skin mimicking gels for continuous glucose monitoring. *IEEE Trans. Microw. Theory Techn.*, 56(4):1001–1008.

Merli F, Bolomey L, Zurcher J, Corradini G, Meurville E, Skriverkik A. 2011. Design, realization and measurements of a miniature antenna for implantable wireless communication systems, *IEEE Trans. Antennas Propag.*, 59(10):3544–3555.

Puente C, Romeu J, Ramis J, Hijazo A. 1998 Small but long Koch fractal monopole. *IEEE Electron. Lett.*, 34(9).

RamRakhyani AK, Lazzi G. 2012 On the design of efficient multicoil telemetry system for biomedical implants, *IEEE Trans*. Biomed. Circuits Syst., 5(6).

RamRakhyani AK, Mirabbasi S, Chiao M. 2011. Design and optimization of resonance-based efficient wireless power delivery systems for biomedical implants. *IEEE Trans. Biomed. Circuits Syst.*, 5(1):48–63.

Sawan M, Hu Y, Coulombe J. 2005. Wireless smart implants dedicated to multichannel monitoring and microstimulation. *IEEE Circuits Syst. Mag.*, 5(1):21–39.

Smith S, Tang T, Terry J. 2007. Development of a miniaturised drug delivery system with wireless power transfer and communication, *IET Nanobiotech.*, 1:80–86.

Soora S. (2005). A Comparison of two and three dimensional wire antennas for biomedical applications, Thesis, NC State University.

Soora S, Gosalia K, Humayun MS, Lazzi G. 2008. A comparison of two and three dimensional dipole antennas for an implantable retinal prosthesis. *IEEE Tans. Antennas Propag.*, 56(3):622–629.

Voors A. 2005. Numerical Electromagnetic Code (NEC). Available: www.home.ict.nl/~arivoors.

Warty R, Tofighi M, Kawoos RU, Rosen A. 2008. Characterization of implantable antennas for intracranial pressure monitoring: Reflection by and transmission through a scalp phantom. *IEEE Trans. Microw. Theory Tech.*

Waterhouse R. 2003. *Microstrip Patch Antennas: A Designer's Guide*. Springer, 56(10): 2366–2376.

Wei YZ, Sridhar S. 1989. Technique for measuring the frequency dependent complex dielectric constants of liquids up to 20 GHz. *Rev. Scient. Instrum.*, 60(9):3041–3046.

Werner DH, Mittra R. 2000. *Frontiers in Electromagnetics*. New York: IEEE Press.

Wong KL. 2002. *Compact and Broadband Microstrip Antennas*. Hoboken, NJ: John Wiley.

Zhu J, Hoorfar A, Engheta N. 2003. Bandwidth, cross-polarization and feed point characteristics of matched Hilbert antennas. *IEEE Antennas Wireless Propag. Lett.*, 2:2–5.

22

mHEALTH-INTEGRATED SYSTEM PARADIGM: DIABETES MANAGEMENT

ALESSIO FIORAVANTI, GIUSEPPE FICO, AND ALEJANDRO GONZÁLEZ PATÓN

Life Supporting Technologies, Universidad Politécnica de Madrid, Madrid, Spain

JAN-PAUL LEUTERITZ

Institute for Human Factors and Technology, Management at the University of Stuttgart, Stuttgart, Germany

ALEJANDRA GUILLÉN ARREDONDO

Medtronic Ibérica, Madrid, Spain

MARÍA TERESA ARREDONDO WALDMEYER

Life Supporting Technologies, Universidad Politécnica de Madrid, Madrid, Spain

22.1 CLINICAL TREATMENT

Diabetes mellitus is a commonl disease, affecting millions of people in Europe and throughout the world. According to the International Diabetes Federation (IDF) in fact, currently 35 million people in Europe have diabetes and 42 million have prediabetes [European Diabetes Leadership Forum (EDLF), 2012]. Any age-related changes affect the clinical presentation of diabetes. It is said that at least half of the diabetic elderly population do not even know they have the disease (Chau et al., 2001). Current treatment of diabetes includes lifestyle modification, oral hypoglycemic agents to stimulate insulin secretion and perhaps improve insulin

Handbook of Biomedical Telemetry, First Edition. Edited by Konstantina S. Nikita.

resistance, and insulin replacement therapy, if necessary [American Diabetes Association (ADA) 2010].

22.1.1 Blood Glycemic Variability

Glucose levels are generally stable in healthy people; however, they are quite variable in patients with diabetes. This glycemic variability in patients with diabetes involves interprandial hypoglycemia and periods of postprandial and acute hyperglycemia. Figures 22.1 and 22.2 show graphs with a comparison between diabetes patients, T1DM and T2DM.

Factors influencing glycemic variability include deficiency of endogenous insulin and amylin secretion, lack of appropriate suppression of glucagon upon eating, poor compliance with diet and exercise therapy, and inappropriate use of exogenous insulin and other hypoglycemic agents.

22.2 DIABETES TREATMENT THROUGH TELEMETRY

People suffering from chronic diseases like diabetes are ideal candidates for telemonitoring. In the last decade, technological progress in telemedicine has had a positive impact on health care management (Norris et al., 2002). It is now possible to provide telemetric services to a large number of patients using relatively simple

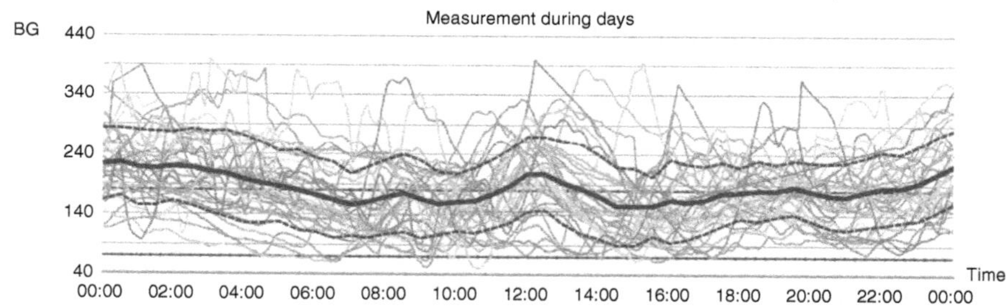

Figure 22.1 Glycemic variation during day in type 1 patient (*Source*: METABO).

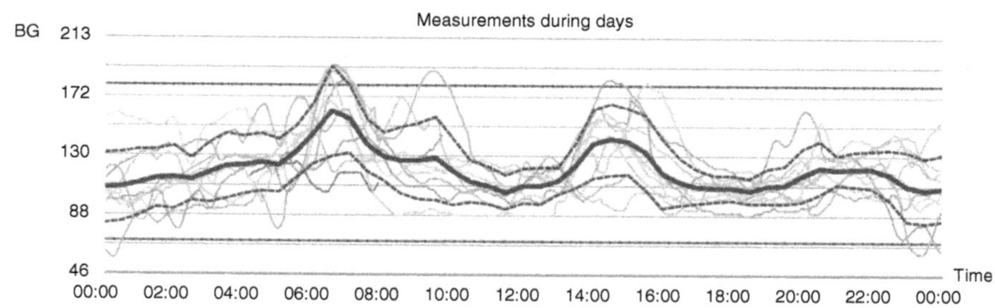

Figure 22.2 Glycemic variation during day in type 2 patient (*Source*: METABO).

telemetric applications (Mezger, 2012). Nowadays there are many products available that already allow telemetric monitoring of chronic diseases like diabetes. A clear example is evident in patients with heart failure, where weight gain caused by increased water retention indicates deterioration in the patient's condition. Thanks to the use of a telemetric scale for daily measuring the weight, it is possible to keep this condition under control and avoid expensive hospital stays.

Actually, the current telemedicine systems are based on the transmission and storage of biomedical data to an electronic medical record (EMR). Then, an application allows the health care providers to communicate with patients for checking their status and providing any modification of therapy (Harvey et al., 2011). Thanks to engineering advances in telecommunications, software, and human factors, the core technologies for the practice of telemedicine for diabetes and other diseases are rapidly advancing. Increasingly robust biomedical sensors for collecting several types of physiologic data are continually developed. Implanted medical devices are common among the therapies of diabetes through telemetric systems. These devices (e.g., infusion pump) are used to communicate with an external device via telemetric messages during listening periods, typically kept small to minimize power consumption. The external device can be then connected with a patient monitoring device (PMD) such as a Smartphone and manage a variety of data such as the patient's history and other biomedical signals.

Recently, in order to increase the control on telemetry, decision support systems (DSSs) have been incorporated into the telemedicine platforms to help clinicians on providing recommendations for acute care in a remote way avoiding a face-to-face visit.

22.3 PROBLEMS RELATED TO CURRENT TREATMENTS

There is a lack of accuracy of the data that patients collect in their diaries concerning habits. Patients are usually asked to record information of their food intake and glycemic levels: five times a day for Type 1 and insulin-dependent patients and once or twice a week for Type 2 non-insulin-dependent patients. (ADA, 1994). These diaries often lack information on accurate timing of the measurements and food intake, the correct contents of each meal, some (supposed) hypoglycemic events that were not measured, or the insulin units correspondent to each insulin shot and related to the time when it was given (not to mention the intake of other type of drug, which is missing in most of the cases) (Fontaine et al., 2003).

22.4 ASSESSMENT: STATE OF THE ART

The ongoing improvement in electronic technology, computing, and communications during the last century has led to an ideal scenario for the development of information and communication technology (ICT) applied to medicine as meant telemedicine. Recently, industrialized and developing countries have shown a growing interest in this area and have started the reform and modernization of their health sector aiming at greater equity in service delivery, effectiveness, and usefulness. The contribution of ICTs has been essential, allowing a great simplification of the devices used, lower costs, and a breakthrough in performance and capabilities. Diagnostic methods have been computerized, and the effectiveness of the equipment is much higher when they are combined together. One example is telemedicine, which allows remote diagnosis, monitoring, therapeutic and surgical activity, and consultation with outside groups (Helmrich et al., 1991). This technology makes possible the solution of extreme situations, such as treatment of people who are isolated, at sea or in remote locations. It is also useful in most common situations, such as monitoring of vital parameters of a diabetic and providing a controlled delivery of insulin. The use of bionic implants such as pacemakers, cochlear implants, eye implants, smart orthopedic prostheses to replace parts or to correct functions of the human body organs is advancing steadily day by day thanks to the convergence between ICT and medicine. In a recent study conducted the United States the researchers have successfully tested a fully implantable glucose monitoring device in pigs for nearly two years (Valdastri et al., 2008). If these tests lead to positive results, these devices will be used in order to show trends in blood sugar levels, helping patients to better tailor their next dose of insulin (Gough et al., 2010).

22.5 TECHNOLOGICAL SOLUTION

Telemetric care is based on providing the patients with devices that also transmit the collected data to their doctor or medical service center. Yet telemetric monitoring can save money in the health care sector by having beneficial effects on the cost–efficiency of the health care system. A telemetry glucose monitoring system consists of a transmitter connected to a needle-type glucose sensor and also a receiver. Among the available systems in the market/research, there are several products produced by certain research groups or commercially available for the assessment of telemetry glucose monitoring and diabetes management. Some examples follow.

22.5.1 Sensors for Medicine and Science

Sensors for Medicine and Science Company (S4MS) is developing transformative glucose monitoring products that are intended to enable people with diabetes to confidently live their lives with ease. Utilizing breakthrough fluorescence sensing technology, the S4MS system is designed to be the first fully implantable continuous glucose monitoring system that is highly accurate and stable throughout its long sensor life.

22.5.2 Philips IntelliVue MX40 Patient Monitoring

The MX40 patient monitoring system gives doctors, nurses, and other clinicians the ability to view patients' vital information right at their bedsides rather than from a central location on a hospital floor. This highly innovative telemetry technology system can be used on any type of patient but is primarily used on cardiac patients.

The mobile monitors, which are about the size of a deck of cards, help clinicians spend more time with patients because the monitor's display allows them to quickly check electrocardiography (ECG) rhythms, heart rate, blood oxygen levels, trends, and monitoring status.

22.5.3 GlucoBand

The GlucoBand is a compact electronic scanning device that uses a resonance phenomenon of bioelectromagnetic noninvasive way to measure blood glucose levels continuously.

The GlucoBand is based on a biotechnology called electromagnetic resonance (BEMR) that detects a change in electrical impedance in the human body caused by an electromagnetic wave of glucose.

Each concentration of glucose solution has its unique electromagnetic molecular self-oscillation wave called glucose signature.

As shown in the previous paragraph, several applications exist for monitoring diabetic patient parameters. However, they usually lack some solid scientific background, and they are sold as standalone products and being oriented more to the wellness market instead of the medical one: The platform presented in this chapter has the ambition to be certified as a software medical device, being part of a holistic and comprehensive monitoring platform that guarantees the closing of the loop between the caregiver and patient (long loop) and between the patient and the monitoring device (short loop).

22.6 METABO SYSTEM

METABO is a research and development (R&D) project cofunded by the European Commission (EU, 2010). The METABO project consists of 21 partners from 9 EU member states. The project started on January 2007 and ended on the end of 2012. The aim of the METABO project is the improvement of the clinical management of diabetes by using the latest available technologies to define new patient follow-up pathways inside the health care systems and to help the diabetic patients improve their own disease management. Real-time glucose-level data, combined with automated system suggestions are used by patients to adjust their lifestyle and therapy. The system offers existing and validated services for the assessment of diabetes, such as a clinical decision support system (CDSS) based on clinical rules and guidelines. The CDSS consists of an inference engine for data analysis and the refined modeling procedures. The final goal is to achieve a closed loop among data acquisition, data interpretation, and clinical decision to enhance knowledge of the disease and, therefore, its treatment.

22.6.1 METABO Challenges

The METABO system serves the overall purpose of supporting the diabetes mellitus (DM) patient—no matter whether he or she is type 1 or type 2—in following the prescriptions given by the physician or health care specialist. On the basis of this purpose, user requirements can be collected and challenges defined. METABO addresses two different main categories of actors: the DM patient and the patient's health care specialist. In this project, user requirements were collected by reviewing the literature on the management of DM and by acquiring information from patients and health care professionals, using questionnaires and focus groups. The challenges arising from the needs of the METABO users can be summarized as follows:

1. Different Treatment Standard: METABO is meant to cater to DM patients in different European countries. However, there is no established standard of treatment of this condition.
2. Different Types of DM Type 1 and type 2 users are characterized by differences in their lifestyles, their aim in using METABO, and the way they prefer to use it.
3. Inter- and intraindividual differences Apart from differences between users in one type, such as those related to age, comorbidities, and the like, there are also relevant mood changes within the users to be respected.
4. Diabetes mellitus is a very complex disease, the management of which requires the control of a lot of different variables, which in turn requires extensive data input by the users.

22.6.2 METABO Medical and Technological Vision

The METABO platform has been developed to enable the collection of physiological parameters from patients with, or at risk of, diabetes and other associated metabolic disorders, like obesity, retinopathy, heart diseases, and kidney diseases. The platform intends to provide full connectivity in order to be integrated in existing information systems, for example, in hospitals. The interoperability with other information systems, such as hospital information system (HIS) and already developed data transmission system, is guaranteed through the HL7 protocol messages and the Institute of Electrical and Electronics Engineers (IEEE) 11073 storage data. Contacts with Continua Health Alliance (http://www.continuaalliance.org/) have been also maintained to keep up with standardization and interoperability initiatives.

At the same time METABO wants to improve communication between patient and caregivers, making it customizable to the patient's personal preferences, lifestyle, and knowledge deficits.

22.6.3 System Overview

The METABO platform deals with a series of elements that relate to each other through a flow of data that is interpreted and displayed in different ways depending on the final user of the system. The technological approach followed for the development of the platform offers a scalable solution capable to accommodate more sensors and devices and thus fulfill the requirements of a future scenario.

The elements consist of a set of mobile and compact wearable sensors, a table-top blood pressure monitor, a weight scale, a PDA (personal digital assistant), and a UMPC (ultramobile PC).

The METABO application hides the heterogeneity of the various sensors and provides a uniform way for the user to interact with the sensors and update the platform with newly acquired measurements. The different clinical situations of the diabetic patients require different monitoring scenarios that are defined by the clinician for each patient separately. Personalized medical observation of some patients by their clinicians may also require specific adjustments on the use of the METABO monitoring platform. To put it differently, the clinician needs to be able to decide which sensors and devices will be used and assign them to each patient accordingly. The platform contains six different sensors, such as glucose sensors (Medtronic CGMS and Bayer glucometer), physical activity sensors (Sensewear armband and Omron pedometer), a weight scale, and a sphygmomanometer. In order to facilitate the efficient exchange of vital signs and medical device data in the health care environments, it has been decided to comply with the IEEE 11073 standard (ISO/IEEE, 2004). The challenge is to set up communication requirements automatically and receive the measurements without the interaction of the user (see Figure 22.3). The standard covers the different aspects of communicating the semantics of medical data from the different devices. In this way the information coming from sensors can be mapped and stored to a proper database based on the specific standard nomenclature. This solution allows an efficient data exchange with the medical devices used as well as with other external health care information systems.

22.7 EVALUATION METHODOLOGY: DATA COLLECTION AND SYSTEM TESTING

Two main pilot tests and two additional empirical studies have been conducted from 2009 to 2012 in order to collect data for the improvement of the METABO system and to test whether it meets its quality criteria. The aim of these trials is to provide an exploratory analysis of the usability of the METABO system, its acceptability by the patients and the impact of clinical relevant parameters in comparison with the standard care. Fifty-four patients have been enrolled for the study and allocated 1:1 to the two treatment groups for 6 weeks. The first pilot test served to collect data from diabetes patients, using a continuous blood glucose measuring device. Additionally, behavior data was collected by reviewing the metabolic diaries of these participants. This data was taken as a primary input for the development of the metabolic models. More precisely, the additional data collected referred to food intake, physical activity, motivation, comorbidities, and details on the personal health care. The second and final pilot test was meant to test the fully integrated METABO system under realistic working conditions. For the telemetric monitoring diabetic's pilot tests, a set of tools has been adopted, comprising a variety of telemedical devices, diabetes training materials, digital medical records, access to a medical service center, and a network of doctors involved in patient monitoring. With all these in place, a 6-week training

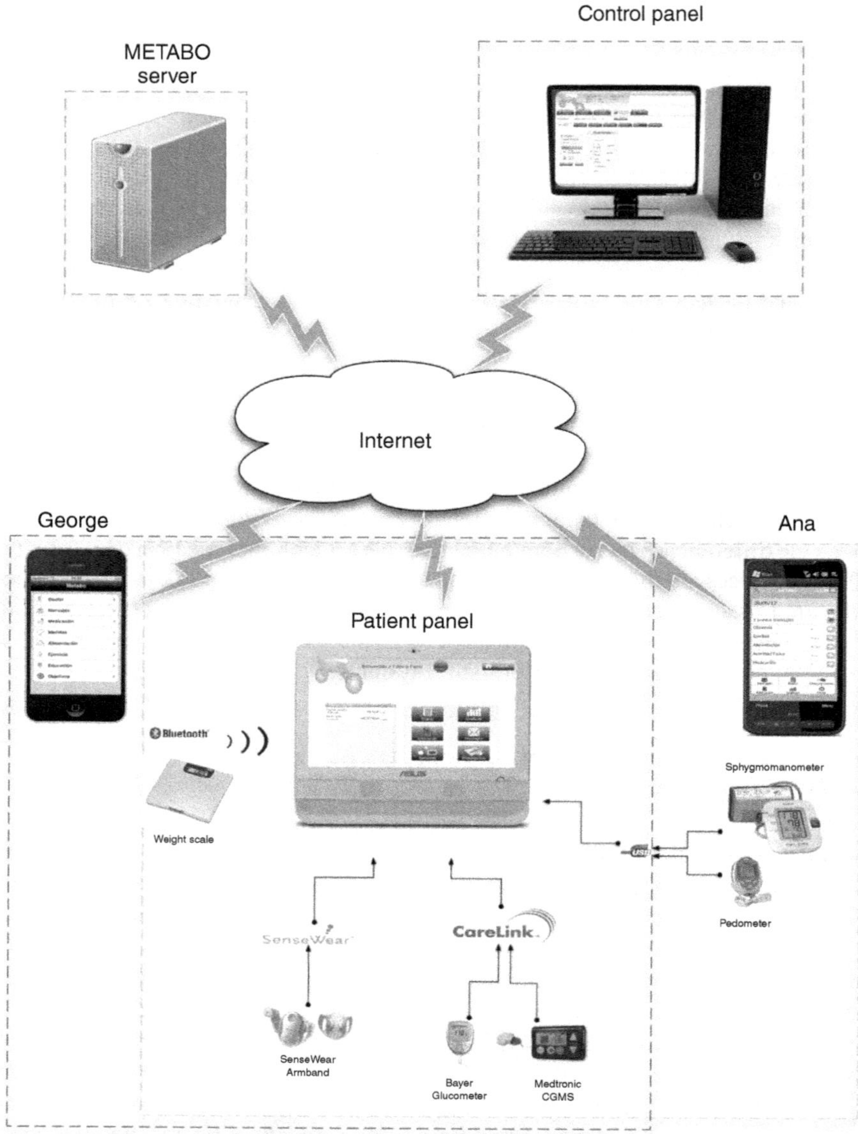

Figure 22.3 METABO platform elements architecture.

program has been set up for patients with type 1 and type 2 diabetes. In order to demonstrate the effectiveness of the system, patients have been divided in two groups, one that use the METABO system and one that is treated without the system. During the test, METABO group patients learn about diabetes and a complete set of medical telemetry system is set up record blood sugar values and other metabolic parameters. The objective is to evaluate if patients using the METABO system show a higher adherence to their prescriptions, an improved glycemic condition, or even an increased quality of life. Compared to patients following the traditional treatment, this approach aims to prevent the patient's health condition from progressing to a worse stage, by improving the patient's ability to manage his or her condition.

22.8 RESULTS

By this writing, the final pilot tests of the METABO system have been conducted. The results have shown the successfully developed applications and the METABO platform, which have been used to collect data during five weeks in four european clinical centers. Additionally, results from the above reported usability pilots show that the modules are rated positively by the test users, which indicates that user requirements, both for patients (lifestyle and medical information, treatment, real-time alerts, personalized recommendations through the DDS, for hypo/hyperglycemic events), and for physicians (sets information and recommendations of patients' treatment, check patient profile and treatment compliance, use simulation tool) have been taken into account satisfactory.

22.9 CONCLUSION

In this chapter, METABO system has been presented, focusing particularly on the context of its implementation, the user requirements, and problems faced and solved during the development process. The METABO system has been contrasted to the state of the art in telemedical diabetes management in order to underline its innovativeness. The main innovation is the ability to analyze and interpret the context-sensitive fusion of information from both body and environmental monitoring noninvasive devices and depict a complete framework of patient's status, enabling behavior prediction and consequent feedback generation.

Despite the extensive empirical work completed, further clinical studies would be necessary in order to validate the system as an effective health management tool and treatment aid. If these issues are addressed in time, we can expect telemedical diabetes treatment to be adopted on a large scale. This is desirable, as telemedicine promises to become one of the main twenty-first-century tools for diabetes health, increasing the well-being of millions of people and decreasing national health costs a substantial extent.

ACKNOWLEDGMENTS

The authors thank METABO project Consortium for their valuable contribution to this work. METABO project (Contract Nr. E080945117) is partially funded by the European Commission under the 7th Framework Programme.

REFERENCES

American Diabetes Association (ADA). 1994. Nutrition recommendations and principles for people with diabetes mellitus. *Diabetes Care*, 17:519– 522.

American Diabetes Association (ADA). 2010. Standards of Medical Care in Diabetes.

Chau D, et al. 2001. Clinical management of diabetes in the elderly. *Clin. Diabetes*, 19(4): 172–175.

European Commission (EU). Information Society Technologies Program. 2010. METABO project. Chronic diseases related to metabolic disorders. ICT-26270.

European Diabetes Leadership Forum (EDLF), Copenhagen. 2012.

Fontaine KR., et al. 2003. Years of life lost due to obesity. *JAMA*, 289(2):187–193.

Gough D, et al. 2010. Function of an implanted tissue glucose sensor for more than 1 year in animals. *Sci. Transl. Med.* 2:42–53.

Harvey P, et al. 2011. Data acquisition in a wireless diabetic and cardiac monitoring. In Proceedings of the Engineering in Medicine and Biology Society, pp. 3154–3157.

Helmrich SP, et al. 1991. Physical activity and reduced occurrence of non-insulin-dependent diabetes mellitus, New England Journal of Medicine. 325:147–152.

ISO/IEEE 11073. 2004. Point of Care Medical Device Communication Standards.

Mezger P. 2012. A helping hand from telemetry. Available: http://www.steinbeis.de/en/publications/transfer-magazine/transfer-032010/wie-telemetrie-helfen-kann.html (Accessed: 24th November 2011).

Norris S. et al. 2002. Increasing diabetes self-management education in community settings. *American J. of Preventive Med.* 22(4), pp. 39–66.

Valdastri P, et al. 2008. An implantable ZigBee ready telemetric platform for in vivo monitoring of physiological parameters. *Sensors Actuators. A*, 142: 369–378.

23

ADVANCED MATERIAL-BASED SENSING STRUCTURES

Manos M. Tentzeris, Sangkil Kim, Vasileios Lakafosis, Hoseon Lee, Taoran Le, Rushi Vyas, Sagar Mukala, and Anya Traille

Georgia Institute of Technology, Atlanta, Georgia

23.1 INTRODUCTION

Currently, there are many problems in the medical industry such as increased patient acuity, increased device/technology complexity, increasing volume of information, and increasing workloads on clinical staff. This increase in information volume and workload on the clinical staff has led to medical errors such as:

7000–98,000 deaths annually

54–61% errors in administration

Device operation errors

15–20% identification device failures

As we move on to the ubiquitous cognition era, the capability of automatically and reliably extracting, gathering, storing, and processing large amounts of human vital signal data collected from spatially distributed sensors becomes more and more critical.

There are many types of wireless body area networks for biomedical applications. Intrabody communication networks use the human body as a communication channel, while on-body communication networks operate in the air–body interface. Both

Handbook of Biomedical Telemetry, First Edition. Edited by Konstantina S. Nikita.

on-body and implanted sensors acquire biomedical information and transmit data through the human body to a central monitoring unit such as a network controller. For convenience, many systems include a wireless network system on the body communicating with a network controller and the sensor modules while making use of the Institute of Electrical and Electronics Engineers (IEEE) 802.15.4 (wireless personal area network, WPAN) and IEEE 802.11 (WiFi) standards typically operating in the 2.4–2.5-GHz frequency band. Especially for the machine-to-machine (MCM) communication depicted in Figure 23.1, the efficiency of wireless module on the body is important.

In this chapter, we are presenting research work the goal of which is to tackle all the subtle but essential problems that hinder the performance or even the realization of the abovementioned vision of automatic and remote human vital signal extraction. In particular, the following are presented: (i) different designs of electromagnetic band gap structures for human-body-wearable antennas that help eliminate the effects of the lossy body, which are appropriate for any typical antenna, (ii) an advanced liquid ionic antenna that can be mounted on the human wrist based on the earlier finding that dielectric filled tubes become resonant at specific frequencies, (iii) inkjet-printed substrate-integrated waveguide prototypes, (iv) carbon-nanotube-based and graphene-based passive and very low-profile sensing structures for ammonia and other hazardous gas detection for medical diagnosis, and (v) wireless integrated modules including two ZigBee-based low-power, low-cost wireless EKG prototypes that can be deployed for continuous monitoring of the three-lead EKG data of a patient at 2.4 GHz for a range of 150 m in open space and 50 m in a closed area, the "smart wireless integrated module," which is totally hosted by a regular photo paper substrate, and an EPC-Gen2 RFID gas sensor prototype platform.

23.2 HUMAN-BODY-WEARABLE ANTENNAS

23.2.1 Challenges of Wearable Wireless Device

There are lots of things that should be taken into account to design wearable wireless devices. First of all, the elimination or minimization of the interaction between the human body and the wireless device is one of the biggest challenges in the design of wearable wireless devices such as an antenna, because it is inevitable for a wireless device to be exposed to the human body effect when the device is mounted on the body. The human body consists of a large portion of organic matter and some inorganic matter. In the high-frequency case, the organic matter is very lossy material with high dielectric constant (ϵ_r), and this results in it having a dielectric constant of 52.5 and conductivity of 1.78 S/m on average (Sangkil et al., 2012). Most of the radiated waves from wireless devices propagate through the human body and are dissipated as heat due to high loss by the human body. This happens because the electromagnetic wave tends to propagate through the material with the higher dielectric constant. This issue results in a decrease of the wireless device's efficiency since the radiation efficiency is drastically decreased due to the absorption by the human body. For example, let's assume that a simple dipole antenna is mounted on the human body.

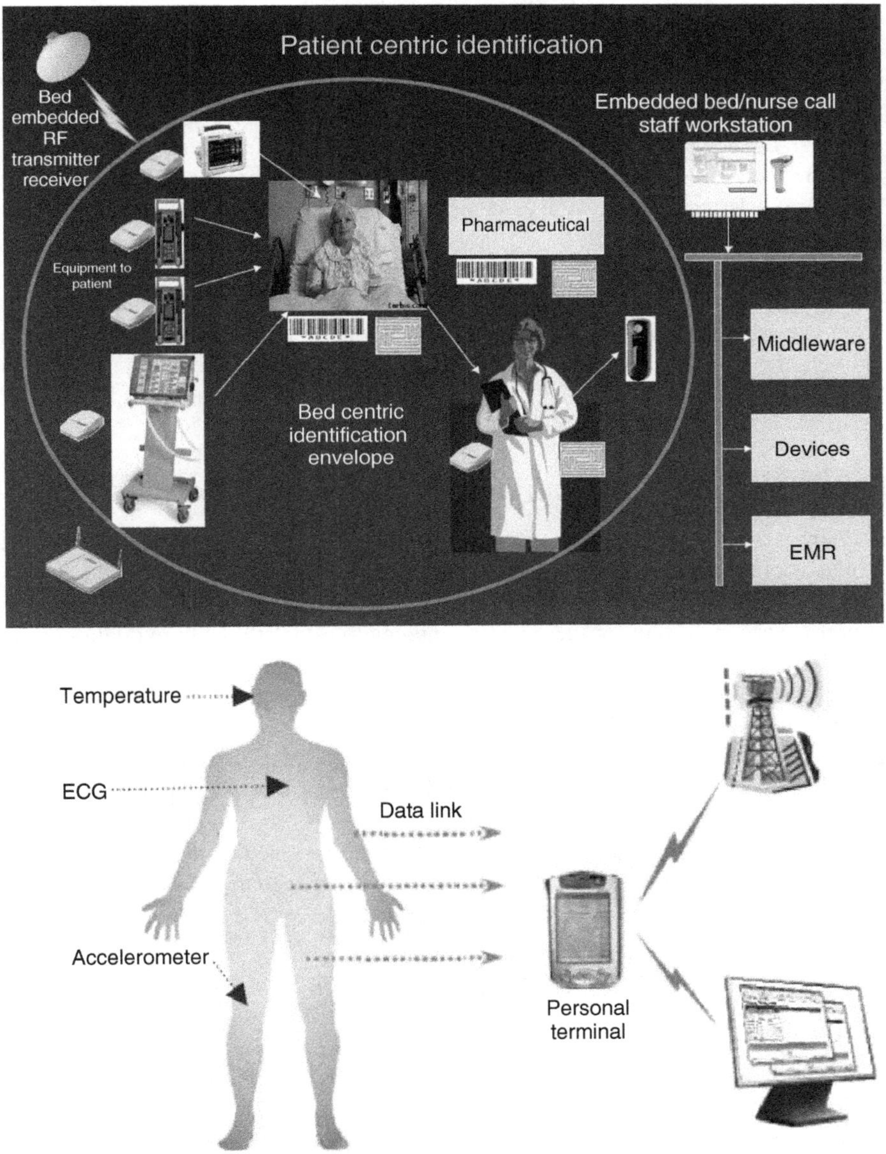

Figure 23.1 Machine-to-machine (MCM) communication for health monitoring applications.

At first glance, its frequency response seems to be improved since it has the wider -10 dB bandwidth and the lower magnitude of return loss (S_{11}). However, actual antenna parameters are worse than those of the antenna in the free space when we examine them more thoroughly. The wider and deeper return loss plot is caused by loss by the human body because the human body absorbs most of power throughout the frequency range. In addition to this, the resonance frequency of the antenna shifts to a lower frequency because of the high dielectric constant of the human body. Therefore, it is undesirable to expose wireless devices to the human body without shielding or isolating methods.

Besides that, biocompatible and flexibility should be taken into account to accommodate for the curvature of the human body and for typical body motions. It is difficult to recommend a wearable wireless device if it is brittle or causes health problems. In addition, the cost issues in terms of the fabrication process or power efficiency should be as low as possible to allow for the easy fabrication of a large number of radiating elements and efficient power use by the devices.

23.2.2 Role of Antenna in Wireless Body Area Networks (WBANs)

A wireless on-body mobile sensor system, as in Figure 23.2, acquires and modulates the data before sending them to the network controller, such as a PC. The modulated data then is radiated and collected/processed by the network controller. In this case, it is very important for the on-body antenna to establish the high quality of the wireless link.

This is because the received power at the network controller is directly proportional to the gain of the transmitting antenna according to the Friis equation. The system's required link can be estimated through the Friis equation. In the Friis equation listed below, P_r is the delivered power to the receiving antenna, G_t and G_r are the gain of the transmitting and receiving antenna, respectively, λ is the wavelength at the frequency of interest in the free space, R is the distance between the transmitting and receiving antenna, and P_t is the delivered power to the transmit antenna. If the transmitting antenna radiates all the delivered power, P_t can be considered the radiated power:

$$P_r = G_t G_r \left(\frac{\lambda}{4\pi R}\right)^2 P_t \tag{23.1}$$

In this kind of on-body wireless link, P_r and G_r are fixed because of the receiver's sensitivity, while the communication distance, R, can also be fixed or constrained due to system specifications. Therefore, the transmitting antenna's gain (G_t) and the delivered power to the transmitting antenna (P_t) are critical to set up a stable wireless communication link with a small amount of power. However, in normal antenna designs without the consideration of human body effect, the human body absorbs most of radiated power since the waves tend to propagate through the high dielectric material, namely the human body ($\epsilon_r \approx 52.5$) as we have discussed above, thus leading to the deterioration of the antenna characteristics, such as gain and return loss (S_{11}), which is going to be discussed in the following.

23.2.3 Inkjet Printing on Paper Substrate

As it is the most common method of printing in commercially sold printers, inkjet printing technology is very familiar to most people. The commonality of inkjet printing comes from its very fast and relatively cheap printing technique. This printing method has been widely used because it can be applied to many different types of substrates beyond just paper, and the ink can be replaced with conductive inks that form conductive traces when cured.

Inkjet printing has a lot of advantages over current methods of electronic device fabrication such as etching since it has smaller environmental impact. A circuit fabricated by inkjet printing only uses as much ink as needed since the method places droplets of ink where desired. Thus, this method does not produce any by-products as a result of etching away the unwanted metal and does not use hazardous chemicals that produce extra waste not found in inkjet printing (Lakafosis et al., 2010).

In microwave applications, the use of paper as a substrate also has many advantages. It allows for cheaper production cost of electronics because paper is one of the cheapest materials, and a large-scale production system can be easily established when it is integrated with inkjet printing technology. For wearable electronics, paper can become hydrophobic when it is coated by a water-resistant chemical. The properties of a paper substrate were studied and characterized using the microstrip ring resonator method (Li et al., 2007). The dielectric constant of a 0.23-mm-thick paper substrate is 3.2 at 2.4–2.5 GHz, and the loss tangent (tan δ) is 0.07 throughout the

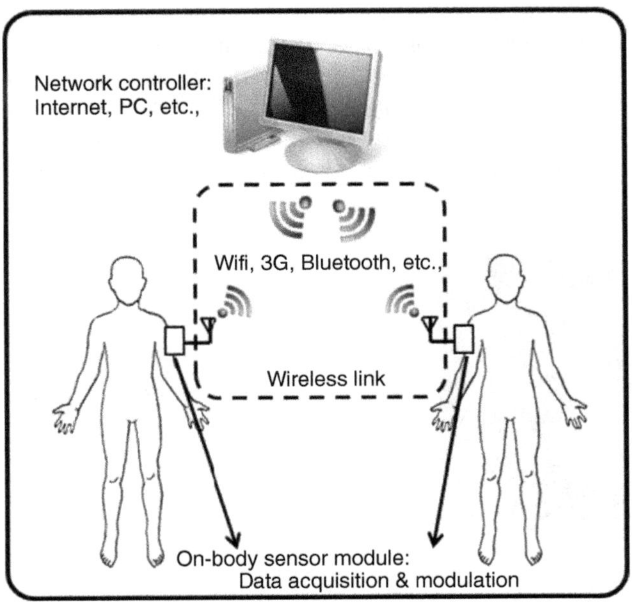

Figure 23.2 Wireless on-body mobile system.

frequency range. The resistivity of a printed silver pattern on a paper substrate after sintering at 130°C for 4 hrs is 9.09×10^{-8} Ω/m to 1.11×10^{-7} Ω/m with a printing resolution of 1270 dpi (Sangkil et al., 2012). The sintering process is used to increase conductivity by removing the nonconductive solvent of silver ink. Figure 23.3a shows the surface of an inkjet-printed pattern on paper substrate.

23.2.4 Antenna on Electromagnetic Band Gap Structure for Wearable Applications

One of the biggest challenges for RFIDs real-world applications is to overcome the effect of metal and lossy substrates onto which a RFID tag needs to be commonly placed. The degradation of gain due to the antenna proximity to such metal and lossy materials seriously limits the effective read range. To resolve this issue, there have been numerous previous efforts utilizing electromagnetic band gap (EBG) or high-impedance surfaces. EBG designs are considered by Kong et al. (2000) and Collins et al. (2000). However, they require the use of vias, which limit the flexibility and ruggedness of the derived wearable RFID tag. Another example of EBG-based design focuses on the improvement of return loss rather than gain and is based on a patch antenna on top of an EBG surface (Fujiwara et al., 2001). Since the patch antenna already has a ground plane, this EBG surface does not provide improvement in gain. Moreover, no measurements to validate improvement in range are provided. Marliere et al. (2000), Chopra et al. (2002), and Varghese et al. (2001) proposed different designs to mitigate the effect of metal, but they lack the flexibility needed for a truly wearable RFID tag.

Our work considers an ultrathin completely via-less, split-ring resonator-based EBG design for both on-metal and on-body mounted rugged applications featuring low cost, low fabrication complexity, and flexibility characteristics. In detail, our proposed "rugged" RFID configuration consists of an inkjet-printed dipole antenna laid on top of an inkjet-printed EBG surface, matched to commercially available RFID chips. Its significantly improved performance was verified by measurements

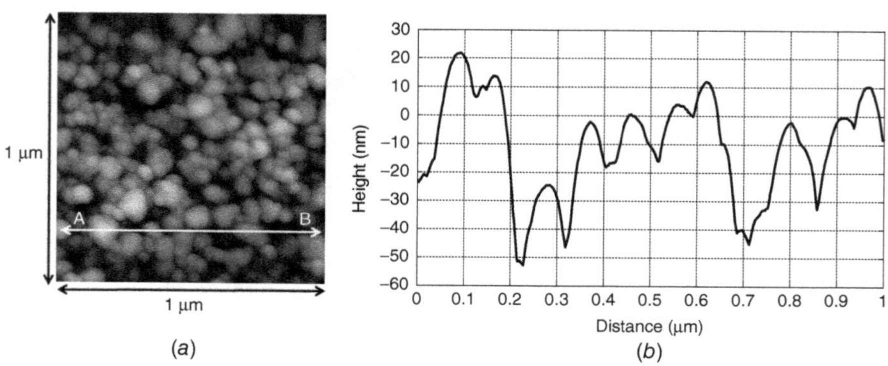

Figure 23.3 (*a*) Surface of inkjet-printed nanosilver ink; (*b*) cross section of line. AB [8].

performed both inside a anechoic chamber and in real indoor environments by using a commercial tool for tag performance characterization and a standard RFID reader. Both on-metal and on-body measurements were performed.

It has to be noticed that due to the design of the EBG surface with a ground plane on the bottom, this prototype can theoretically be "universally" mounted on any (e.g., metal, water, body, etc.) surface. Applications range from wearable antennas for factory workers in industry and patients in the hospital to plastic water bottles and metal containers, to name a few. In the following sections, the EBG design and fabrication is introduced followed by measurement results and a conclusion.

EBG Structure As discussed above, it is important to avoid effects by the human body on wireless devices for on-body or biomedical applications due to the detrimental effects it causes. Therefore, it is necessary to develop effective isolating structures. As an isolator, utilizing an EBG structure is one of the most popular and effective ways to avoid detrimental effects of the human body on wearable applications (Rais et al., 2009; Shaozhen and Langley, 2009). The artificial magnetic conductor (AMC) is the most popular EBG structure, and it is a sort of metamaterial based on a frequency-selective surface (FSS) and is sometimes called high-impedance surface (HIS) (Salonen and Rahmat-Samii, 2006). There are a lot of structure types for FSS design. Like the AMC or HIS, FSS can also be considered an EBG structure due to its band gap property of frequency response. The FSS is acting like a band stop or band pass filter for plane waves and plays an important role in the isolation of the wireless device, namely for the antenna from the effects of the human body. EBG structures such as FSS or AMC consist of periodic metallic structures that stop or pass the plane waves at desired frequency bands simlar to the microwave filters.

There are many methods to design AMC, HIS, and FSS, but the reflection phase method is one of the most popular design method because of its ease (Sievenpiper et al., 1999). This method identifies the band gap of the EBG structure based on the reflection phase diagram, assumes the illumination of a plane wave onto the infinite FSS backed by a ground plane, and monitors the phase of reflection coefficient (S_{11}) on the FSS structure. The reflected waves from the FSS structure will be constructively added to the radiated waves from the antenna when the reflection phase from the surface is in the range of $-90° \sim +90°$. In this way, the on-body wireless devices can be effectively isolated from the effect of the lossy human body. For example, Figure 23.4 depicts design geometry of AMC ground plane for wearable antenna for biomonitoring applications. The AMC ground plane has a band gap property at 2.4–2.5 GHz because reflection phase of this frequency range is in the range of $-90°$ to $+90°$.

In addition, there are more advantages to utilizing EBG structures. Any type of antenna can be mounted on the EBG structures if operation frequency is matching. For instance, meandered monopole, bow-tie, or dipole antenna can be mounted on the same EBG structure without any performance degradation such as return loss (S_{11}). In addition, EBG structure is scalable to many different frequency bands. If an EBG structure is designed for a 2.4-GHz band application, it can easily scaled down to a 5.8-GHz one because the operation of the EBG structure is determined by the relative

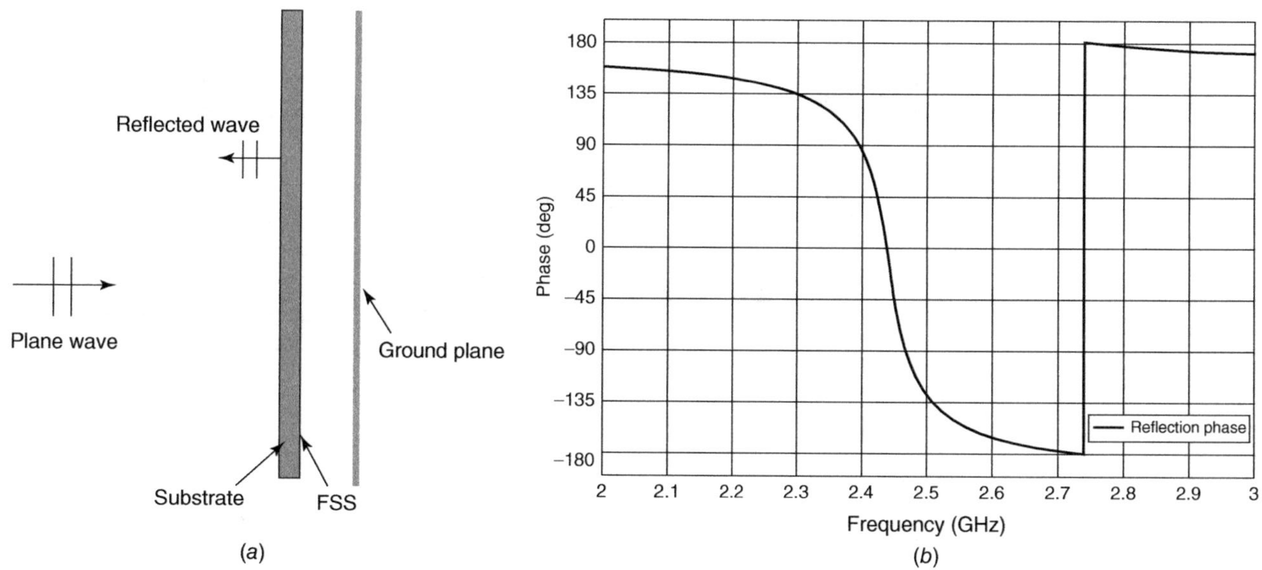

Figure 23.4 Reflection phase method: (*a*) AMC and plane wave geometry and (*b*) phase of reflected wave.

size of the EBG structure to the wavelength of the operation frequency. Therefore, it has a broad range of applications due to this property.

Antenna on AMC Ground Plane As proof of the concept, a simple antenna is placed on the AMC ground plane and its antenna defining factors are investigated, as shown in Figure 23.5 (Li et al., 2007). A microstrip-fed monopole antenna is placed on the AMC ground plane, which operates at 2.4–2.5 GHz band. The split ring resonators are inkjet printed on a paper substrate in order to reduce production cost. The single split ring resonator is chosen for an EBG structure since it has the smaller printing area compared to a conventional double split-ring resonator. Thus it uses less ink, and it is less affected by errors in printing such as splatter or inherited positioning error. The size of ring resonator is decided based on the wavelength of the operation frequency. One wavelength (λ_0) at the central frequency of 2.45 GHz, in free space, corresponds to the circumference of the ring resonator. For ease of design, one-tenth of the edge length of the resonator is chosen as the thickness of the resonator, and the thickness of each side is adjusted to optimize the resonance frequency at 2.45 GHz. The EBG characteristic is identified by the reflection phase method described in Section 23.4, and the array size of 4×3 is chosen in order to maintain its band gap property after placing the microstrip-fed monopole antenna because it is impossible to realize the infinite resonator array. A small size is generally preferable, but the AMC ground plane loses its band gap property when the EBG

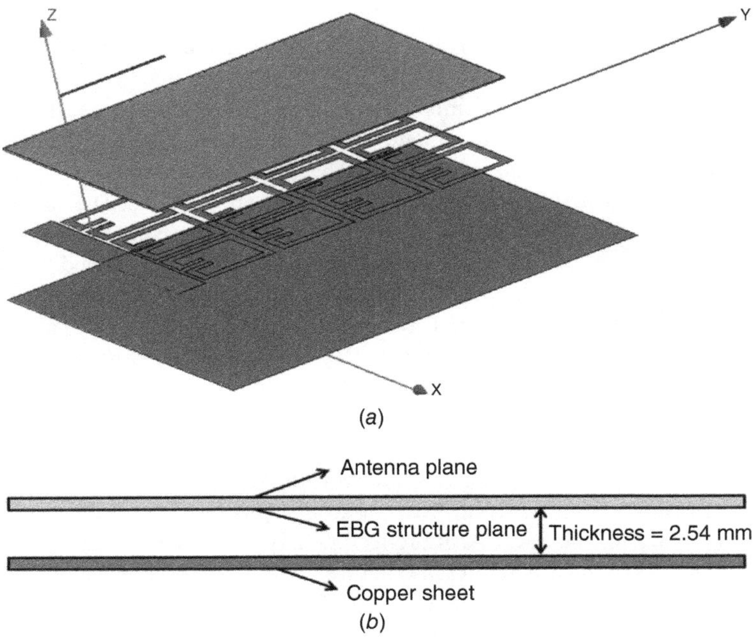

Figure 23.5 (*a*) Antenna plane and (*b*) copper sheet.

array is too small because of the parasitic reactance between the antenna and AMC ground.

Figure 23.6 shows measured return losses (S_{11}) of the antenna with AMC ground plane. The return loss is measured inside an anechoic chamber, and the measurements are in very good agreement with the simulated results. The EBG structure effectively isolates the human body's detrimental effects by comparing the return loss of the antenna with/without an EBG structure. In this measurement, only the ring resonator array is removed in order to verify the effectiveness of EBG structures, namely the return loss is measured on the ground plane with/without ring resonator array (EBG structure). The antenna without EBG structure does not properly resonate at the desired frequency while -10 dB bandwidth of the antenna with EBG structure covers the frequency band of interest (2.36–2.61 GHz). Figure 23.6*b* clearly shows the usefulness of the AMC ground plane for wearable application because the frequency response of the antenna doesn't change even if the antenna is placed on a human body phantom, which is shaped like the human torso. Figure 23.7 shows the radiation pattern of the antenna, which is consistent in the whole frequency band from 2.4 to 2.5 GHz. The gain of antennas with or without an EBG structure in different environments is measured to verify the performance of the antenna with the EBG structure in Figure 23.7. The phantom deteriorates the gain of the antenna when the AMC ground plane is not implemented. The gain difference of the antenna with the EBG structure and without the EBG structure on the phantom is about 8 dBi. On top of that, the AMC ground plane causes the gain of the antenna on the phantom to increase more than 0 dBi. Therefore, the AMC ground plane effectively isolates the antenna from the surrounding environment because the gain remains the same in both the presence and the absence of the phantom. Thus, the detrimental effect of lossy materials like the human body on the antenna gain is effectively suppressed due to the AMC ground plane, which allows using the maximum communication range. These results support the statement that the AMC ground plane isolates the antenna from the human body.

The flexibility of the antenna is also studied because conformity with the curvature of the human body is an important factor for wearable applications. The antenna with the AMC ground plane was bent along a cylindrical surface, as shown in Figure 23.8. Figure 23.8*a,b* illustrate radii of the cylindrical surface of 84, 125, and 249 mm to make bending angles 5°, 10°, and 15°. There are almost no changes in the antenna's return loss (S_{11}), even if the antenna is bent along the cylindrical surface as shown in Figure 23.8*c*. These results support the fact that the proposed antenna doesn't experience performance degradation when it is attached to body parts such as chest, upper arm, and thigh.

EBG Design In this section, the well-known split-ring resonator (SRR) is adopted as a unit cell of the EBG structure. Next, a ground plane is placed below the EBG structure in order to obtain an AMC reflector. It is well known that an RFID tag antenna achieves the maximum increase in directivity when it is placed at a distance of about 80 mm ($\lambda/4$ in free space at 915 MHz) from a conductive ground plane. Such a distance can be significantly reduced if an AMC reflector is considered as the ground plane for the RFID tag.

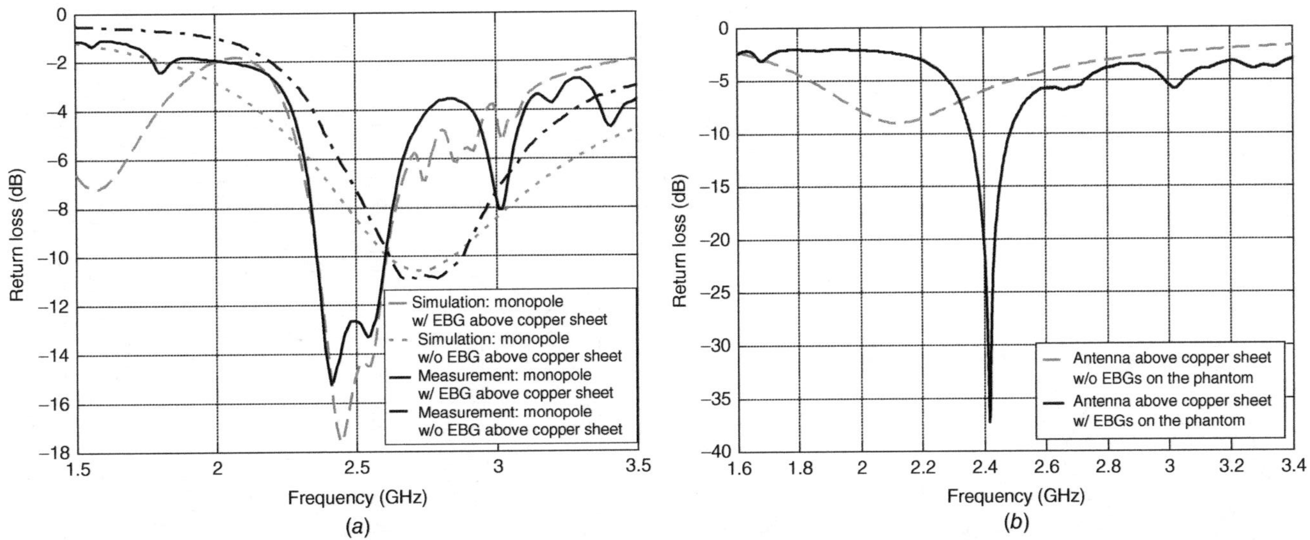

Figure 23.6 Measured return loss (S_{11}): (*a*) in free space and (*b*) on the phantom.

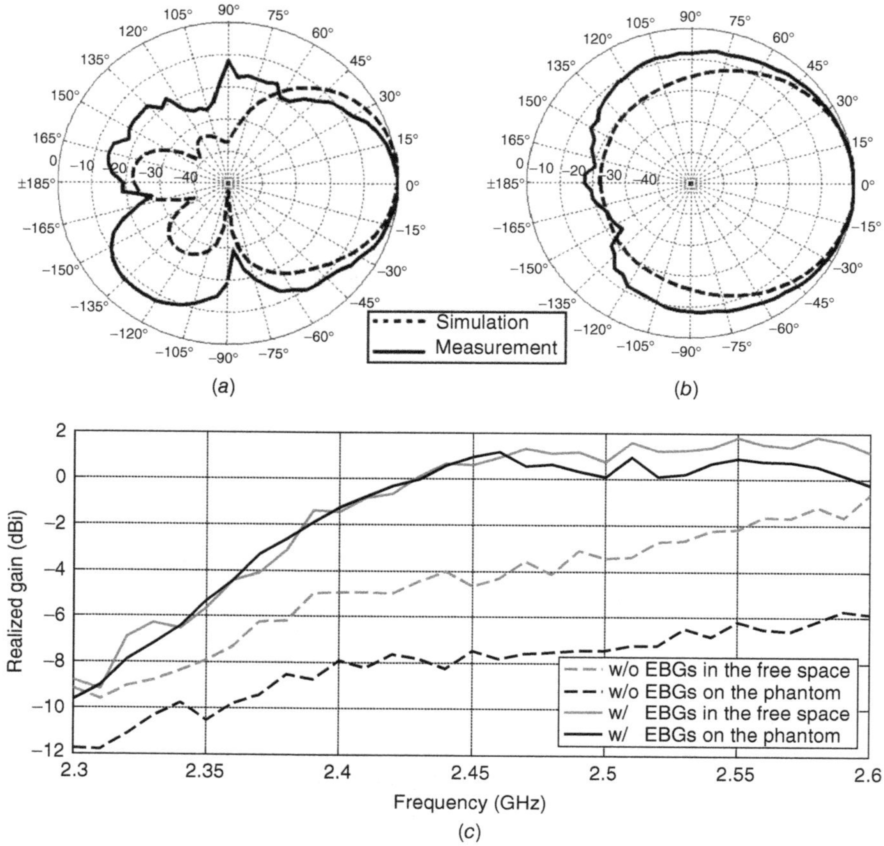

Figure 23.7 Copolarized (E_φ) radiation pattern: (*a*) *YZ* plane (*E* plane), (*b*) *XZ* plane (*H* plane), and (*c*) measured antenna gain.

Based on this, we have designed an AMC with a 3-mm gap between the EBG and the ground plane. The distance between the RFID tag and the EBG is always kept to 3 mm because it is an optimum value for flexibility and gain improvement. The AMC reflector is designed by means of reflection phase characterization (Fan and Rahmat-Samii, 2003). A plane wave is illuminated onto the infinite EBG surface, and the phase of reflection coefficient on the surface of the EBG is observed.

The EBG surface can be considered as an AMC reflector if the reflection phase is within range of −90° to +90° since the reflected waves will be constructively added to the radiated waves. The top of Figure 23.9*a* shows the unit cell of the EBG structure. The gap between EBG and ground plane is kept by inserting microwave foam, whose dielectric constant is close to that of the air. This provides more flexibility, low losses, and reduced difference between paper substrate and air dielectric constants. The reflection coefficient at the paper–foam interface is 0.5, a small value because the dielectric constants of paper and foam are 3 and 1, respectively. In other words, 75%

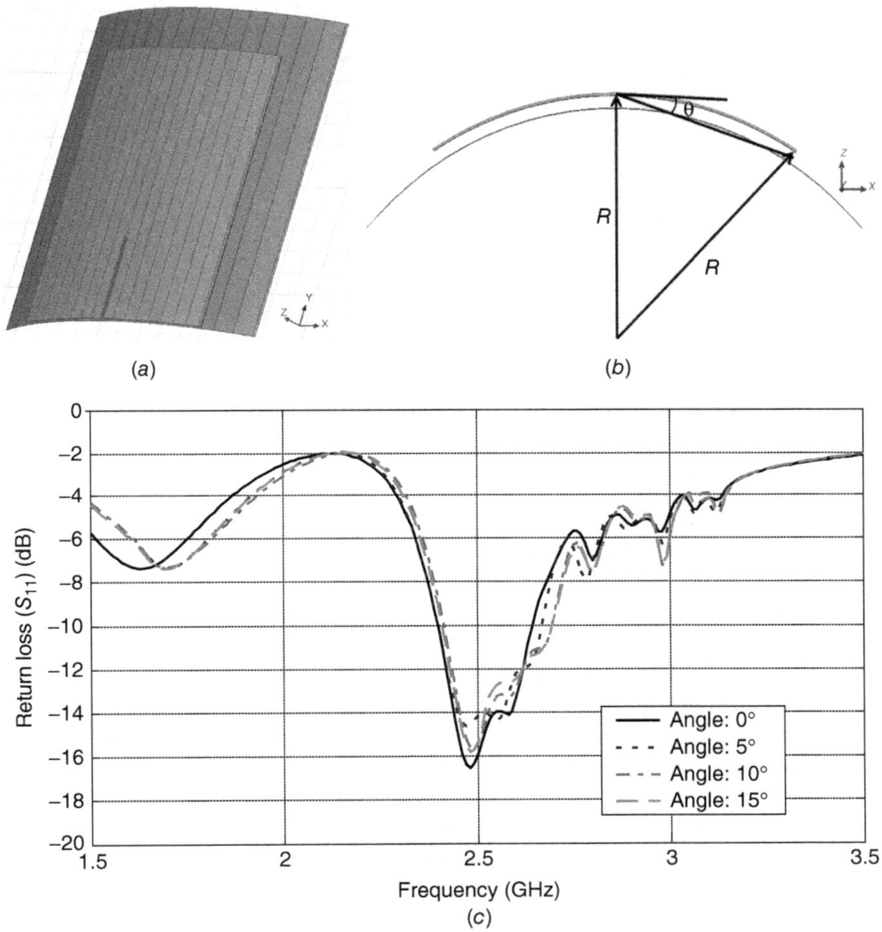

Figure 23.8 (*a*) Bent antenna geometry, (*b*) cross section of bent antenna (*R*: radius of the cylindrical surface for antenna bending, θ, bending degree), and (*c*) return loss of the bent antenna.

of power can be transmitted to EBGs to make in-phase reflection. Different materials can be considered to fill the EBG–ground plane gap. For example, with high dielectric materials like ceramics, the tag loses its benefits in terms of flexibility and weight. With paper substrates, maximum power transfer is achieved, but the EBG loses its property due to high loss tangent, and the overall flexibility is degraded by the structure thickness. The bottom part of Figure 23.9*b* shows the phase of reflection coefficient on the EBG structure. The reflection phase has a value between −90° and +90° in the RFID frequency band.

Based on the single-cell EBG design, we have built a two-dimensional (2D) array made up of 3 × 2 EBG unit cells. This is the minimum dimension required to cover the

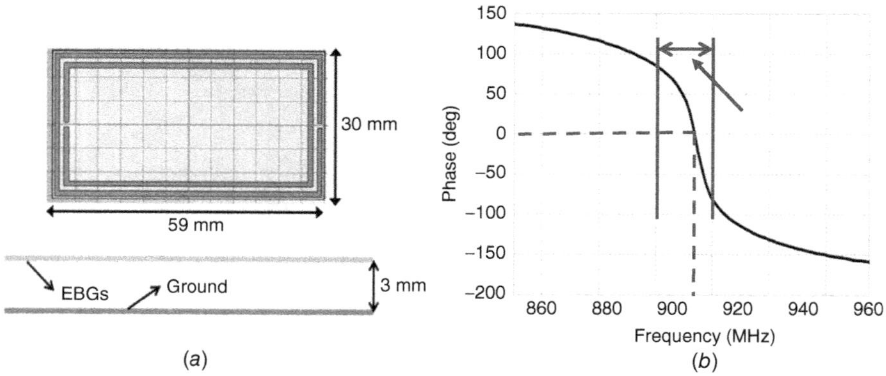

Figure 23.9 (*a*) EBG unit cell tuned to 915 MHz simulated on paper substrate and (*b*) phase reflection graph.

area of the RFID tag. Figure 23.10 shows the fabricated EBG structure by means of inkjet printing on commercially available photo paper, while the designed EBG-based RFID tag made up of the three-layer structure (ground plane + EBG + dipole) is shown in Figure 23.11.

The EBG-based RFID tag was extensively tested in order to assess its capability to work regardless of the surface where it is mounted. In the first experiment, we performed a sensitivity analysis in an anechoic chamber varying the interrogation frequency. The EBG-based tag was mounted on a polystyrene sheet and placed 1 m away from a circularly polarized antenna (8 dBi gain) connected to the Voyantic Tagformance Reader. The minimum power required to communicate with the Voyantic tag was recorded at different frequencies in the 840–1000-MHz band with 1-MHz steps. Figure 23.12 shows the comparison between the EBG-based tag and an inkjet-printed dipole tag. Both tags exhibit a broad-band behavior with the best performance achieved in the U.S. ultra high-frequency (UHF) RFID band (902–928 MHz). Here, the EBG-based tag outperforms significantly the dipole, which requires about

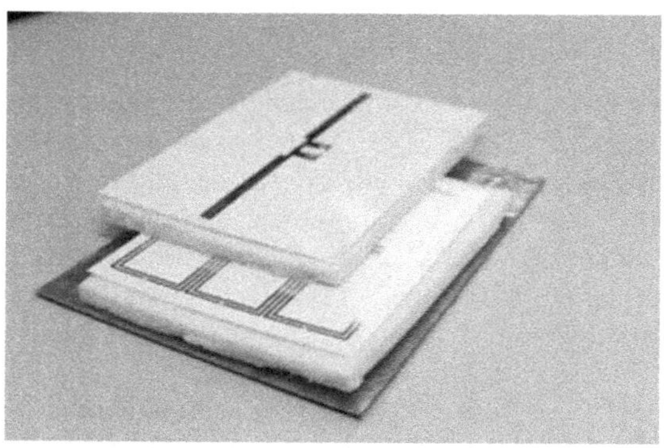

Figure 23.10 Fabricated EBG structure by means of inkjet printing.

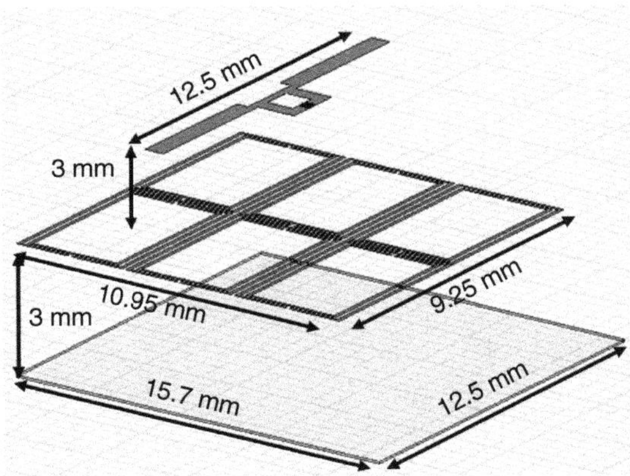

Figure 23.11 Dipole anten na matched to the RFID chip on top of the designed SRR-based EBG surface.

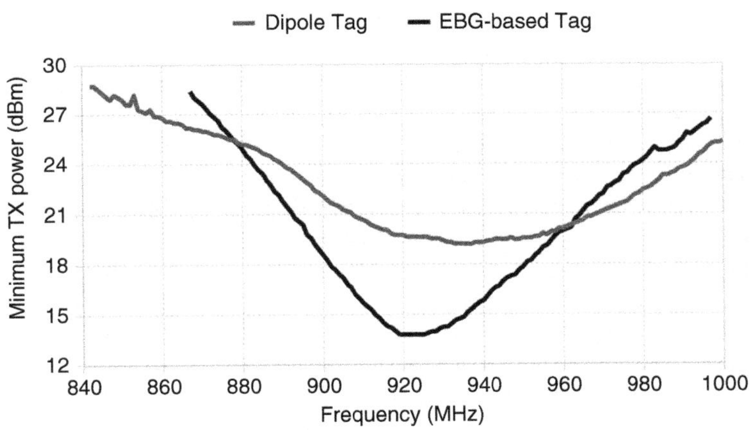

Figure 23.12 Sensitivity analysis in anechoic chamber: comparison between EBG-based tag and dipole tag.

6 dB more power to communicate with the reader. Since current RFID systems are forward-link limited ("Super-sensitive and small: New MIT detector uses nanotubes to sense deadly gases," 2008) and therefore the maximum read range exclusively depends on the reader Effective Isotropic Radiated Power (EIRP), a sensitivity gain of 6 dB in free space allows to double the read distance of the EBG-based tag over the dipole.

In the second experiment, we evaluated the reliability of the tag–reader communication in real application scenarios. The EBG-based tag was attached both to a metal surface and to a human body for future applications like container tagging and biomonitoring. The Impinj Speedway Reader (operating frequencies 902–928 MHz) was used to interrogate the tag in a standard office room. Figure 23.13 shows the

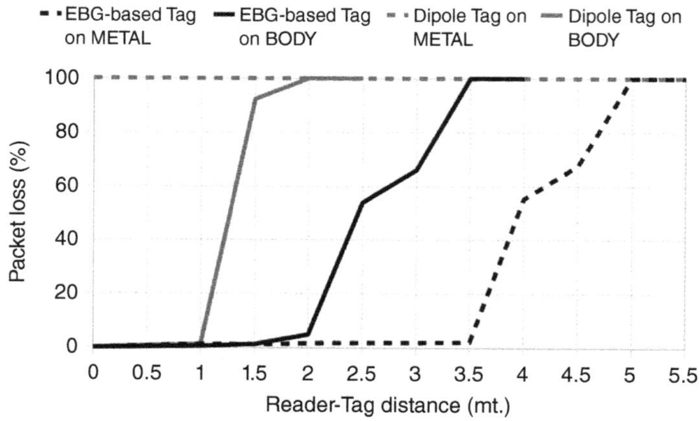

Figure 23.13 Indoor read-range analysis: comparison between EBG-based tag and dipole tag attached to different surfaces.

measured packet loss, that is, the ratio between the number of lost tag replies to the total number of performed interrogations, at the reader. The EBG-based tag can be reliably read up to a distance of 4.5 and 2.5 m when attached to metal and human body, respectively. The difference in range between metal and body due to the lossy dielectric properties of the human body is close to 0.3 for the loss tangent (Someya et al., 2003). There is less constructive reflection off the body, which contributes less to the gain compared to metal. The dipole tag instead exhibits extremely poor performance on both surfaces: It is never read on metal and can communicate with the reader up to 1 m away when attached to the human body. The measurement results are in perfect agreement with simulations performed by HFSS: The maximum gain of a dipole tag is extremely poor both on metal and on body, while the EBG-based tag exhibits excellent performance on both surfaces (see Figure 23.7). This clearly demonstrates the inappropriateness of commonly adopted RFID tags in harsh application environments. In such situations, ad hoc tags like the EBG-based one presented here are required in order to ensure reliable communication with the reader.

Communication Range on Human Body The antenna with an AMC ground plane was integrated with a commercial wireless temperature sensor module in order to demonstrate its performance on the human body. The on-body communication is setup as shown in Figure 23.14 to measure communication distance of the wearable wireless sensor modules. The circuit's original chip antenna of the module is replaced by a female SMA connector, and the proposed antenna is connected. For the comparison, a conventional chip antenna and inkjet-printed meandered antenna were also mounted on the sensor module and their communication range was measured. Figure 23.15 shows the designed wearable temperature sensors. The measurement took place on the top of a large parking deck in order to minimize multipath effects. For the communication range measurement, the sensor (transmitter) was mounted on the chest of a human. The transmitter was set at close distance to the receiver to ensure

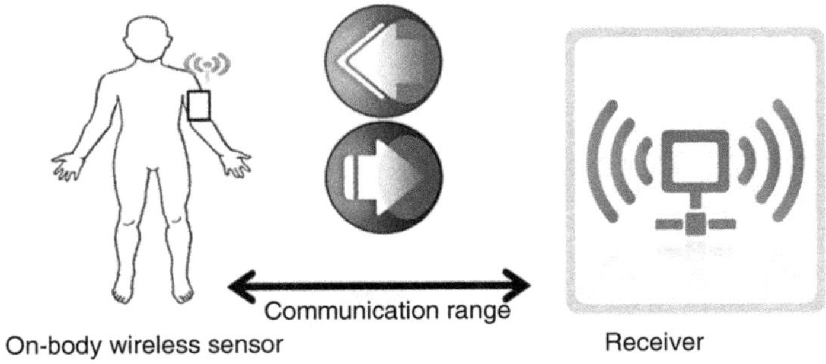

Figure 23.14 Communication range measurement concept.

Figure 23.15 Communication range measurement setup. (*See insert for color representation of the figure.*)

their communication while mounted on the human body. Afterward the distance from the receiver was increased until the receiver lost signal from the transmitter. In this way, the distance from the receiver to the transmitter was measured when the receiver failed to get a signal from the transmitter, and the antennas' alignment was kept in their general maximum direction. The measured communication range of the on-body sensor and the conventional chip antenna was 18.3 m (Fig. 23.16), and that of the sensor and the inkjet-printed meandered monopole antenna was 29.5 m. However, the sensor with a simple monopole antenna on the AMC ground plane has a communication range of 82.8 m, which is almost triple that of the sensor and the conventional chip antenna on the human body. This suggests that the proposed antenna can be easily integrated with previously proposed biomonitoring systems to significantly improve the performance of body area networks (BANs) or mobile health monitoring systems.

The increased communication range on the body means that the same amount of power level at the receiver can be achieved using a small amount of power at the transmitter while keeping the same communication range. In other words, it is possible to save power for transmitting and increase battery lifetime because we can use the power efficiently without dissipation of the radiated power in the human body. For example, the sensor module with the proposed antenna with AMC ground plane has about four times longer communication range compared to the conventional chip

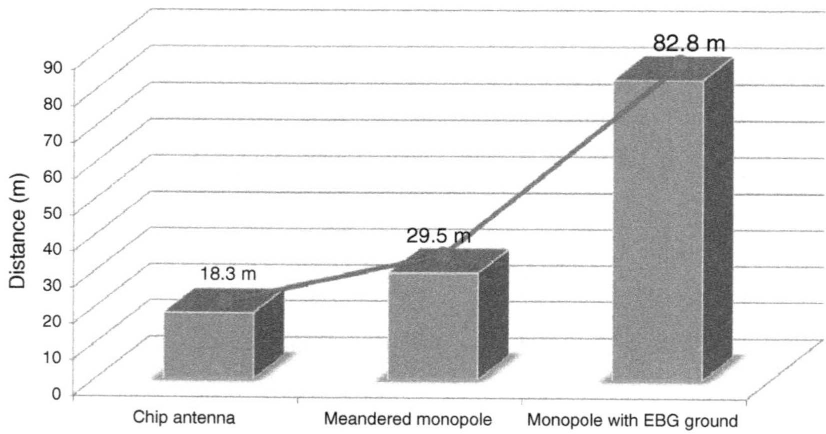

Figure 23.16 Communication range measurement.

antenna. This implies that only $\frac{1}{16}$ of power is required for the sensor with the proposed antenna to maintain the same communication range with the sensor with the chip antenna. This is because the radiated power from the chip antenna is absorbed by the human body, which restricts communication range.

23.2.5 Liquid Ionic Antenna for Biosignal Monitoring Applications

Introduction The explosive growth of the biosensors and health-related wearable monitoring devices has accentuated the need for miniaturized, high-efficiency conformal materials that can operate over a wide range of frequencies, while they can be integrated in wearable and lightweight configurations. One of the major issues for the implementation of wireless body area networks (WBANs) of these devices is the very limited range of commonly used metal antennas. Due to the high dielectric constant (Table 23.1) between the metal antenna material (as well as the metal-based circuitry) and the human body tissues consisting mostly of "ionized-water," the near field is significantly disturbed, while local reflections due to the dielectric mismatch cause unwanted distortion in the reading image, which may require extensive reconstruction

TABLE 23.1 Dielectric Properties at 1.5–2.0 GHz

Measurement Results Conc. Ammonia (ppm)	Exposure Flow Rate (L/min)	Reversion Flow Rate (L/min)	Δf_{res} of S_{11}
50	0.46	0.5	43.125
75	0.31	5.0	46.875
100	0.23	0.5	48.750

Source: Andreuccetti et al. (2002).

to produce a sufficient quality for accurate analysis. Current forms of biomonitoring are invasive because they involve the application of stimulation currents by attaching electrodes directly to the body or skin, posing the risk of harming the immediate and surrounding tissue due to electrolysis at the electrode–tissue interface.

The goal of the liquid antenna is the same. Since the electrical properties of the body vary with tissue type, temperature, water content, organ movement, and blood flow, it is possible that the resulting variation of a liquid antenna's radiation pattern with inhomogeneous dielectric properties can be used to communicate information about a person' health condition such as body temperature, tissue damage, tumors, inflammation, and breathing. Like other biosensing devices, to detect internal bioactivities based on the dielectric and conductive properties of tissue, the microwave signal must be able to penetrate into the tissue posing the need to utilize liquids with a high dielectric constant in order to better focus and direct the electromagnetic energy.

The idea of a liquid antenna was explored as early as 1970 when Ting and King (1970) proved theoretically that dielectric-filled tubes become resonant at specific frequencies. Ida et al. (2002) demonstrated the dependence of the efficiency bandwidth on the permittivity of the dielectric surrounding a metal conductor. This dielectric material slows the velocity of the electromagnetic wave and leads to physically smaller antenna size than the purely metallic thin-wire antennas at the same frequency. In this work, we propose for the first time the development of liquid antennas, based on aqueous salt solutions, operating in microwave frequencies (around 1.7 GHz) with significantly improved range and efficiency characteristics, as well as a much simpler and flexible fabrication mechanism. Encapsulating the proposed liquid antennas in flexible plastic containers makes them quite easily wearable or implantable in the proximity of human organs or biomonitoring devices (e.g., pacemakers) in order to enable short-range wireless communication/control. In addition to this, corrosion resistance is another advantage of the ionic-liquid, glass/plastic-enclosed antennas, while the easy elimination of air gaps permits shape manipulation and an improved electromagnetic coupling between the probe and the probed dielectric. Since the dielectric property and the conductivity of these solutions are a function of the salinity level as well as of the type of dissolved salt, these antennas can be easily reconfigured for different areas of mounting/observation.

Electrical Properties of Water Solutions Liquid water is a dielectric and a conductor that self-ionizes ($H_2O \leftrightarrow H^+ + OH^-$). As a dielectric, the finite charge displacement (rotation) of the polar molecules occurring under an electric field is referred to as Maxwell displacement current. The electrical conduction (collective translational charge movement) occurring in water is not very well understood due to the very complex structural properties of water, having to do with the transfer of hydrated water ions (hydronium H_3O, and hydroxide OH^-) between various water clusters or chains held together by hydrogen bonds (Sadkowski, 2000).

Adding salt to water is believed to break the structure of water clusters, freeing molecules, which in tum have a lower relaxation time, thus causing an overall depolarizing effect in the bulk solution. As the salt concentration increases, the salt ions reduce the amount of free molecules, eventually causing the relaxation time to decrease again. The overall effect is that adding salt increases the loss due to the drift of ions (conductive loss) and decreases the dielectric constant of water.

When the water (or electrolyte) is biased with electrodes of opposite polarity, the negative electrode attracts the hydrogen atoms while the positive electrode attracts the oxygen atoms. This accelerates the ion transport mechanisms and results in the electrolysis process at the solid–solution interface where O_2 is produced at the anode and H_2 is produced at the cathode. These processes occur at a potential limit that depends on the electrode materials (Abdur Rahman et al., 2007; Bera et al., 2004; Brüesch and Christen, 2004; Gabriel et al., 1998; Somaraju and Trumpf, 2006; Whalen III et al., 2006; Zhou et al., 2005).

At low frequencies, within a particular potential range, the method of charge injection can be a nonfaradic process that involves an increase or decrease in water ion concentration at the liquid–electrode (adsorption and desorption) and charge polarization (electrical double-layer capacitance) at the liquid–electrode interface but no charge transfer between the liquid and electrode. This is undesirable since the electrical double layer blocks the electric field from the bulk solution and prevents the electric field from coupling to the bulk liquid, which is to serve as the conducting antenna element. At higher frequencies, a reversible faradic process occurs via charge transfer at the interface (oxide formation and reduction) in addition to the adsorption and desorption of water ions. Since the polarity alteration of the electrodes must cycle rapidly at some resonant frequency, a reversible electrode process is necessary, whereas, in the case of an irreversible process, the alternating potential would result in the accumulation of solids or gasses onto the electrode surface overtime (Abdur Rahman et al., 2007; Bera et al., 2004; Brüesch and Christen, 2004; Gabriel et al., 1998; Somaraju and Trumpf, 2006; Whalen III et al., 2006; Zhou et al., 2005).

It is desirable to use electrode materials, which are noncorrosive, chemically stable, biocompatible, and nontoxic so that they will not induce an immune response on the skin or tissue. Additional material-specific characteristics relate to surface morphology and microstructure, extrinsic exchange density, and adsorption behavior (which can be controlled using coatings). Examples of such materials are platinum, gold, and titanium nitride.

The liquid must be encapsulated in a flexible, noncorrosive, biocompatible, nontoxic container. Kynar polyvinylidene fluoride (PVDF) is a tough, flexible, heat- and chemical-resistant plastic with ϵ_r of around 8.4. It is available in two grades, the homopolymer Kynar, which is the semirigid grade, and the copolymer Kynar, which is the flexible grade.

Benchmarking Structure Simulations To characterize the performance of a liquid antenna, we performed extensive simulations for the structure shown in Figure 23.17. The bracelet-type liquid antenna consists of a 2.8*M* aqueous salt solution ($\epsilon_r = 38$) inside a very thin plastic tubing container having a tube radius of 2.5 mm. The bracelet was excited using two electrodes on both sides of the slot and mounted on various positions of an inhomogeneous multilayer cylinder emulating the human body. This cylinder with a length of 250 mm (similar to the lower arm) was composed by two

Figure 23.17 Wearable liquid antenna bracelet around the lower-arm phantom.

materials: for the inner cylinder with radius $r = 24$ mm, the material was blood with dielectric constant 58 and conductivity 1.2 S/m, while for the outer cylindrical ring the material was skin with dielectric constant 37.5 and conductivity 0.6 S/m. The simulation was performed at the central frequency of 1.7 GHz.

The simulated radiation patterns of the antenna made of a metal (copper) and the liquid were shown in Figure 23.18. The designed antennas had the same dimensions and mounted at the same position while utilizing the transmission line matrix (TLM) method. It can be clearly seen that the liquid antenna has a maximum to the direction transverse to the arm with 59% efficiency and a directivity of 5.64 dBi. On the other hand, the metal antenna has the maximum of its radiation pattern along the direction of the arm, thus minimizing its communication range away from the body. The simulated directivity was 3.36 dBi and the efficiency was 26%. Comparison of the range to the direction that is transverse to the body, which is usually the most critical dimension for the communication of bracelet-mounted wearable biosensors, demonstrated that the liquid antenna led to an improvement by a factor of 12.

Changing the position of the bracelet on the lower-arm phantom can easily tilt the radiation pattern, further improving the directivity and the range of reach. As the bracelet is close to the edge of the arm, the radiation pattern slightly tilts vertically but

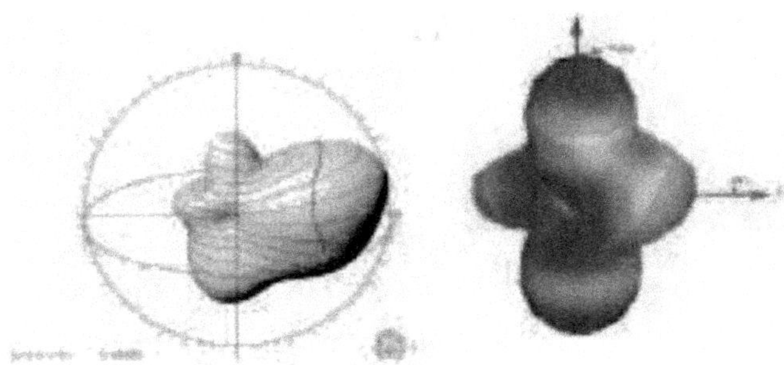

(*a*) (*b*)

Figure 23.18 Radiation patterns for wearable bracelet: (*a*) liquid antenna and (*b*) copper antenna arm phantom.

still maintains the main pattern characteristics, having the maximum in the direction of the slot as shown in Figures 23.19*a*,*b*. The directivity is 5.86 and 6.44 dBi and the efficiency 61 and 63%, respectively, for the position of the liquid antenna at 60 mm from the bottom edge and at 125 mm from the bottom edge (middle of the arm).

One of the most important features of the proposed advanced antenna is its very easy reconfigurability by simply changing the salinity levels (or modifying the diluted salt). Time-domain simulations for two different values of salinity (2.853 mol/liter and 4.643 mol/liter), shown in Figures 23.20*a*,*b* demonstrate that as the salinity increases the pattern gets more isotropic but maintains the radiation characteristics with a peak in the direction that is transverse to the arm. For Figure 23.20*a* the dielectric constant was 38 with a conductivity value of 12.4 S/m leading to a directivity of 5.64 dBi, while for Figure 23.20*b* the dielectric constant was 23 with a conductivity of 21 S/m. It is clear that the dielectric constant and the conductivity values of the saltwater can be very easily modified by changing the salinity. This allows the easy reconfigurability of this type of structure in order for them to readjust for different types of tissues, such as internal organs, skin, and blood. In addition to

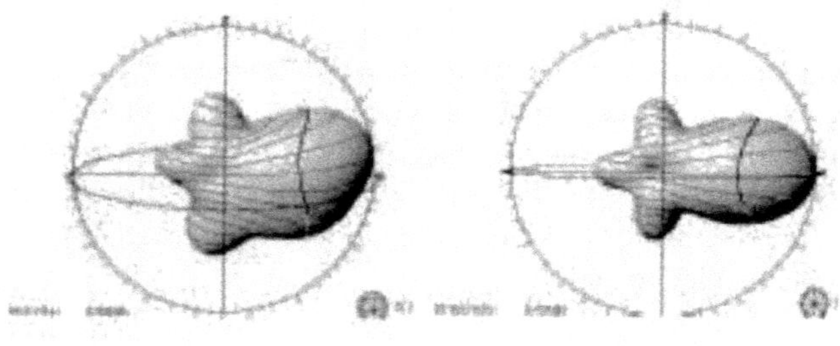

(*a*) (*b*)

Figure 23.19 Radiation patterns for wearable bracelet liquid antenna: (*a*) 60 mm from bottom and (*b*) middle (125 mm from bottom).

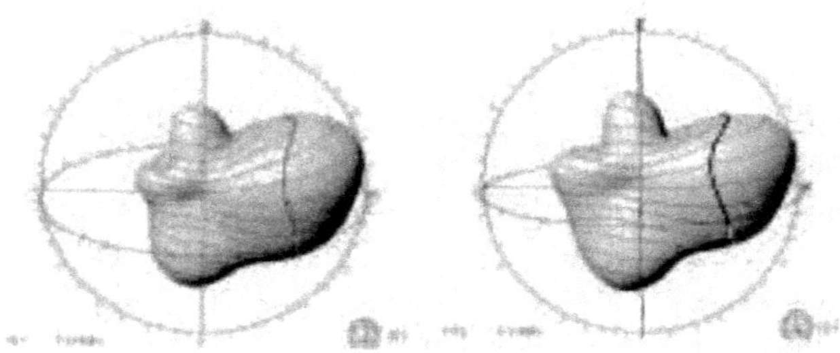

(*a*) (*b*)

Figure 23.20 Radiation patterns for liquid antenna salinities: (*a*) 2.853 mol/Liter, (*b*) 4.634 mol/Liter this, that feature would enable the easy readjustment and operability of this device in different ambient conditions (desert, polar, and space environments) when utilized in wearable applications.

23.2.6 Inkjet-Printed Substrate-Integrated Waveguide

Inkjet-printed substrate-integrated waveguide (SIW) on paper substrate is going to be discussed in this section. The SIWs are comparable to traditional dielectric-filled rectangular waveguides utilizing two metalized rows of vias (Bozzi et al., 2011). Microstrip line structures are not effective structures for high-frequency applications since the short wavelength of the high frequency has narrow fabrication tolerance. Fabrication issues such as fabrication difficulty and cost should be taken into account to use waveguide structures for high-frequency applications. However, SIW structures have advantages so that they can be considered as a transition between microstrip and dielectric-filled waveguide. SIW technology is a promising approach for the microwave and millimeter wave components in planar form, which enables to integrate all components on the same substrate including passive and active components (Bozzi et al., 2011). Implementation of SIW structures on paper using inkjet printing technology has many advantages over traditional SIW technology. The total cost can be greatly reduced by introducing inkjet-printed SIW on paper, and it allows for arbitrary geometry. In addition, paper substrate enhances the advantages of SIW components like flexibility and environmentally friendly properties. Moreover, the possibility of multilayered structures allows miniaturization of the size of the components.

A simple SIW on paper has been implemented in order to investigate its feasibility in Figure 23.21. It has been designed for operation frequency of 5 GHz, and its cut-off frequency is 3.65 GHz. The width (w) of the SIW is 23.2 mm, the diameter of vias is 0.8 mm, and their pitch spacing is 1.0 mm. The via holes have been drilled using

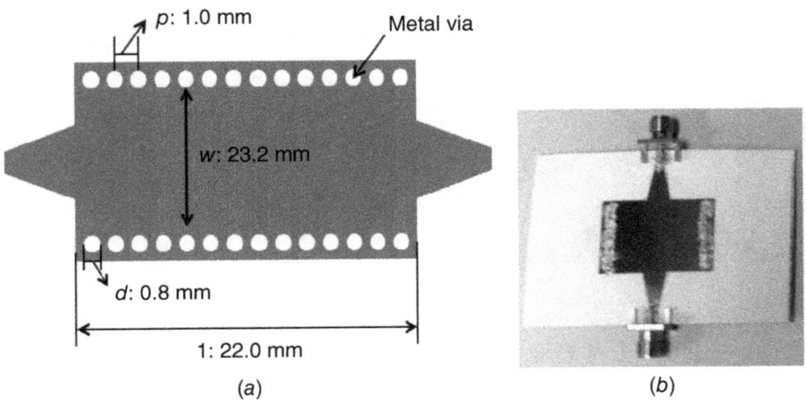

Figure 23.21 (a) Geometry and (b) fabricated inkjet-printed SIW interconnect on paper.

a mechanical drill, and a cylinder-type copper rivet whose diameter is corresponding to that of via hole is inserted in order to metalize the via. For the substrate, three layers of paper were stacked, which resulted in a substrate thickness of 0.69 mm. This thickness has been selected in order to minimize conductor losses by increasing the substrate thickness since the inkjet-printed silver nanoparticle ink has relatively low conductivity (Bozzi et al., 2011). The scattering parameters such as S_{11} and S_{21} are simulated and measured in Figure 23.22. The high insertion loss above 7 GHz is due to relatively large via diameter, spacing, and high loss of the substrate at high frequency. The wider bandwidth can be obtained if the diameter of via holes and the distance between the vias were smaller than those of the proposed design. This result suggests the possibility of many other inkjet-printed SIW applications on paper, for example, cavity filter, ridge substrate-integrated slab waveguide (RSISW), radar, and antenna.

23.3 CARBON-NANOTUBE-BASED AMMONIA DETECTION FOR MEDICAL DIAGNOSIS

23.3.1 Introduction

The accurate and real-time detection of chemical and biological agents is advantageous for biomedical applications and is of benefit to both the public and private sectors. The technology can be used to monitor patients and to predict and avoid dangerous occurrences. Current analysis solutions, such as gas and liquid chromatography and multiple biological methods, exist for the detection of various biological agents (Joshi et al., 2005). These techniques are very sensitive and reliable and can detect and decipher a range of known substances, but they are unsuitable for the rugged field applications and frequently suffer from long detection time limitations, which preclude real-time detection. Portable sensors have been developed using thin-film sensing devices optimized to detect a particular

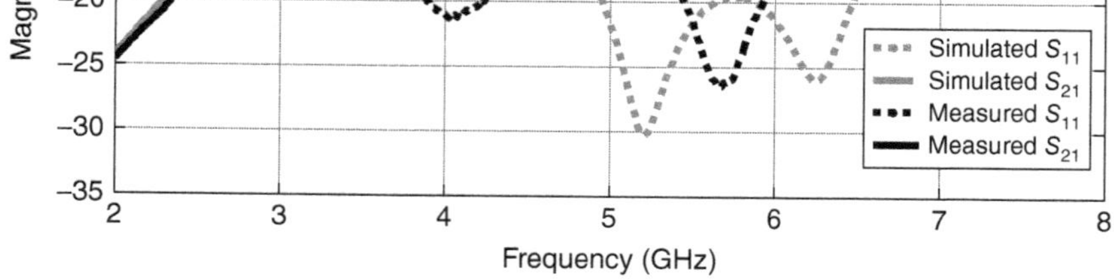

Figure 23.22 Scattering parameters of the SIW on paper.

substance and utilizing wireless transmission of the sensed data. These solutions offer real-time monitoring in a low-cost, power-efficient packaging. One such technology, which is heavily explored at the research level for these purposes, is the use of nano-carbon-based thin films made from carbon nanotubes and graphene. Excellent electrical conductivity and the ability to be functionalized for a wide range of chemicals make these advanced materials ideal candidates for the development of a broad spectrum of portable and wearable sensors (Balandin et al., 2008; Geim and Novoselov, 2007; Li and Kaner, 2008). By direct write methods, they can be easily integrated with existing wireless/wired technologies to create mass-producible, low-cost, high-performance wireless/wired sensor devices that enable immediate detection of various substances with a high degree of accuracy and, subsequently, transmit the information wirelessly to a nearby monitoring facility. The films can be incorporated into rugged wireless gas sensors with low form factor packaging, which can be easily printed/attached to uniforms and worn on the body.

Carbon nanotubes (CNTs) are long, thin cylinders of carbon (typical diameter ranging from 1 to 100 nm) that are unique for their size, shape, and physical properties. Due to their large tubular surface area comprising a hollow inner core and thin outside cylindrical walls, CNTs exhibit moderately high surface adsorption and produce appreciable changes in electrical conductivity at room temperature in the presence of various gases (Marliere et al., 2000; Chopra et al., 2002; Varghese et al., 2001; Zhang et al., 2007). The research reported in this section pertains to the development of a low-cost CNT-based sensor for ammonia detection. Ammonia is found in household cleaning agents, traces of explosives, fertilizers, and as an environmental contaminant in water runoff from streams and estuaries. For the sensor element, we employ a thin film comprising commercially available single-wall nanotubes (SWNTs) coated with a conductive polymer (at the molecular level) to enhance its chemical reaction with ammonia. In comparison with multiwalled CNTs (MWNTs), SWNTs offer higher sensitivity and produce larger variation in electrical conductivity. For example, by measuring the resonance frequency shift upon exposure to ammonia of a microstrip resonator coated separately with SWNTs and MWNTs, the authors (Pham, 2003; Huang et al., 2008) demonstrated a larger shift for the former and argued that SWNTs provide better surface adsorption of ammonia. SWNTs also depict smaller DC resistances (of the order of 20–200 Ω) than MWNTs, characterized by high resistances, on the order of kilomhs (Varghese et al., 2001) for the same thicknesses. Therefore, in RF applications, where we would like to match the sensor impedance to a transmission line or an antenna with typical impedance values below 100 Ω, it is preferable to use SWNT-based sensors. We have also observed that MWNTs require significantly thicker films than SWNTs of comparable resistance, thereby adding to the sensor cost and complexity. Also, SWNTs can be coated with a polymer that can enhance the adsorption of ammonia molecules onto the SWNT (Zhang et al., 2007). For these reasons, we have chosen functionalized SWNTs as the sensing medium, applied as a film of a few micron thickness between electrodes fabricated on paper substrate.

Chemical gas sensors use the property of binding the gas to some form of electrodes and detecting the change in both the DC and RF impedance of the electrodes.

For example, using SWNTs, a study ("Super-sensitive and small: New MIT detector uses nanotubes to sense deadly gases," 2008) introduced a sensitive electronic detector for sensing deadly gases, such as the nerve agent sarin and mustard gas, utilizing an array of carbon nanotubes epitaxially grown over a substrate across interdigitated gold electrodes. Each tube consists of a single-layer lattice of carbon atoms, rolled into a long cylinder with a diameter about 1/50,000 of the width of a human hair, which acts as a molecular wire. When a particular gas molecule binds to this SWNT, the tube's electrical conductivity changes. Each gas affects conductivity differently, so gases can be identified by measuring the conductivity change after binding. The authors in Someya et al. (2003) grow CNT film on a silicon substrate between the drain and source electrodes of an Field-effect transistor (FET) and observe the change in drain current when the sensor is exposed to alcohol vapor. In the work reported in Zhou et al. (2010), MWNTs are vertically grown out of a polymer surface at a very high density (about 3×10^9 nanofibers per cm^2) to form a CNT sheet with high flexural strength. Semiconductor growth requires expensive fabrication processes and precludes the development of printable sensors in the form of thin films for low-cost applications.

All these sensors based on the growth of CNTs ("Super-sensitive and small: New MIT detector uses nanotubes to sense deadly gases," 2008; Someya et al., 2003; Zhou et al., 2010) are designed for stand-alone operation and need to be integrated with a device that communicates the detection information remotely upon interrogation. Thus, although the gas detector itself is very small and quite sensitive, integration of antenna and sensor electronics with the detector poses serious challenges: (a) impedance mismatch between the antenna and the electrodes, (b) difficulty in interfacing the microelectrodes inside the integrated circuit (IC) with the antenna terminals, and (c) limiting the power consumption with active devices present. For rapid deployment in buildings, highways, bridges, and other infrastructure, it is desirable to have low-cost sensors that can be powered from thin-film batteries or by power scavenging, and *incorporate integration of the sensor and the RF communication device on the same substrate*.

There has been significant work done in wireless CNT-based gas sensors that rely on detecting either change in the resonant frequency or change in the amplitude upon exposure to the analyte of interest. As an example in the first class of sensors, Chopra et al. (2002) discussed the design of a patch antenna, designed at 3.9 GHz, fully coated with a mixture of SWNT (in powder form) and a conductive epoxy, for ammonia detection. When the CNT coating is exposed to ammonia, it changes the effective permittivity of the antenna and shifts its resonant frequency. Only a 5-MHz shift has been detected even with a high ammonia concentration (1000 ppm). A similar sensor tracking the change in the resonant frequency of a CNT-coated patch resonator is discussed by Zhou et al. (2010), reporting a shift of 7 MHz when the resonator is immersed in methanol. In Pham (2003) and Zhou et al. (2010) the sensor is used in a laboratory setting to describe phenomenology, and communication of the detected information is not addressed. In a remote-sensing mode, however, such a small shift can lead to false alarms (in fact, small detuning due to antenna manufacturing tolerances may cause such small changes). Ong and Grimes (2002) placed a sensor element comprising of a composite of MWNTs and SiO_2 (which simply acts to bind the MWNTs) over a planar *LC* resonator, fabricated by photolithography on a printed circuit board substrate. This sensor detects the change in effective dielectric constant caused by surface interaction with the gas. Due to very small variation observed in the dielectric constant upon exposure to gas, this sensor suffers from low sensitivity too, thus being limited to its practical utility.

Another class of CNT-based gas sensors relies on the change in amplitude of the return loss or transmitted signal for detection. Dragoman et al. (2006) reported experimental results to demonstrate change in the modulus and phase of transmitted signal when a micromachined coplanar waveguide (CPW), filled with a mixture of SWNT, is exposed to nitrogen gas. The attenuation constant of the CPW line changes by only a few decibels per centimeter—too small to be of any utility as a wireless sensor node due to potential multipath interference. Yang et al. (2009) demonstrated the sensing capability of a printed SWCNT film by printing it at the feeding-gap of a bow-tie antenna. The amplitude of the return loss/back-scattered power varies when the SWCNT loaded bow-tie antenna is exposed to a gas, such as ammonia. This design requires a very high concentration of gas to get a large variation in the loading resistance, and thus suffers from limited sensitivity and slow detection response due to sampling over a wide bandwidth.

Among these different detection mechanisms, the method relying on the shift of the resonant frequency is the most effective for remote sensing because the amplitude is susceptible to interference and noise, leading to false positives (ammonia detection), while the frequency shift is relatively insensitive to these detrimental factors.

This section discusses an ammonia gas sensor integrated with a microstrip patch antenna on a thin paper substrate and utilizes the shift in the resonant frequency as a means of gas detection. The salient advanced features of the proposed wireless gas sensor design include (a) incorporation of functionalized SWNT to improve sensitivity, (b) an advanced characterization methodology, (c) integration of sensor and communication device (antenna) on the same platform to facilitate wireless detection, and (d) inkjet printing fabrication on paper substrate for low-cost, light-weight wireless sensor nodes.

23.3.2 Functionalized CNTs

In this work, a poly(*m*-aminobenzene sulfonic acid) (PABS) SWNT is utilized for the gas detection surface. PABS SWNT is a water-soluble nanotube–polymer compound, formed by covalently bonding the polymer (PABS) to SWNTs via amide functionalization (Zhang et al., 2007). The utilized functionalized PABS SWNTs are 1.1 nm in diameter and 0.5–1 μm in length.

Chemical functionalization is a method to enhance both processability and sensing performance of SWNTs. First, it allows the unique properties of SWNTs to be coupled to other materials, such as conducting polymers, metals, and metal oxides, to create hybrid sensing materials with enhanced sensitivity, selectivity, and faster response time. Second, it can improve dissolution and dispersion of SWNTs in various solvents, including water, which enables cost-effective methods to fabricate sensors by simple dispensing or printing techniques. Chemical surface modification of nanotubes by attaching functional groups has been utilized as an important step for preparing suspensions and stable solutions (Gracia-Espino et al., 2010). However, the original electrical properties of the nanotubes are not left intact, which should be taken into account when designing a new device (Geng et al., 2007). It is recognized that covalent sidewall functionalization is one of the promising methods of modifying electric properties of SWNT for many applications (Ciraci et al., 2004). Sidewall chemical functionalization disrupts the π*-bonding* system and breaks the translational symmetry of SWNTs by introducing saturated sp^3 carbon atoms. As a result, electronic and transport properties of SWNTs are significantly altered (Hyoungki and Lu, 2006). By introducing different chemical groups to the nanotube wall, it is possible to change the interaction between nanotubes and the surrounding environment. This reactivity could enhance the sensitivity and selectivity of CNT-based gas sensor devices (Bekyarova et al., 2005; Kim et al., 2007).

The preparation of well-defined materials through covalent bond formation constitutes the only controlled route for the introduction of chemical receptors into SWNTs that allows for the isolation and characterization of specific compounds of known electrical conductivity (Bekyarova et al., 2007). The well-defined composition and properties of covalently functionalized SWNTs provide an opportunity to dynamically characterize the nanotube electronic structure and electrical conductivity in the presence of analyte molecules (Shaker et al., 2010).

For this study, the PABS SWNT in powder form was purchased from Carbon Solutions Inc. (www.carbonsolution.com) to prepare our ammonia (NH_3) gas sensing film. Aqueous dispersions of the PABS SWNT with a concentration of ~5 mg/mL in water were prepared by ultra-sonication for 2 hr, followed by deposition onto a paper substrate using a paint brush technique. About 20 layers of PABS SWNT were applied for each fabrication to assure the reproducibility of the gas sensor.

The molecular structure and the interaction mechanism of PABS SWNT with NH_3, proposed by Haddon et al. (Shaker et al., 2010), is shown in Figure 23.23. The PABS functional group that is attached to the SWNT is chemically sensitive to NH_3. NH_3 sensing occurs by removing a proton (H^+) from the side-chain oligomers (the PABS functional group) as shown in the grey box in Figure 23.23, which in turn induces electron transfer between the side chain and the SWNTs. In effect, conductivity of the PABS SWNT molecule is decreased because addition of electrons to the semiconducting SWNTs refills the valence band. Our results have shown that the PABS SWNT sensing film offers great promise for NH_3 sensing.

23.3.3 Material Properties and Characterization

Printing Setup Constructing low-cost environmentally friendly sensors using PABS SWNTs, we propose the utilization of inkjet printing technologies of conductive inks on paper sheets. Paper possesses a number of intriguing attributes that makes it amenable for low-cost "green" electronics. It is cellulose in nature, thus considered as a renewable resource. Additionally, it can be easily processed in a reel-to-reel fashion enabling low-cost manufacturing solutions. Driven by the fact that it is process and environmentally challenging and cost inefficient to apply photolithography techniques to paper, inkjet printing of conductive particles provides a promising

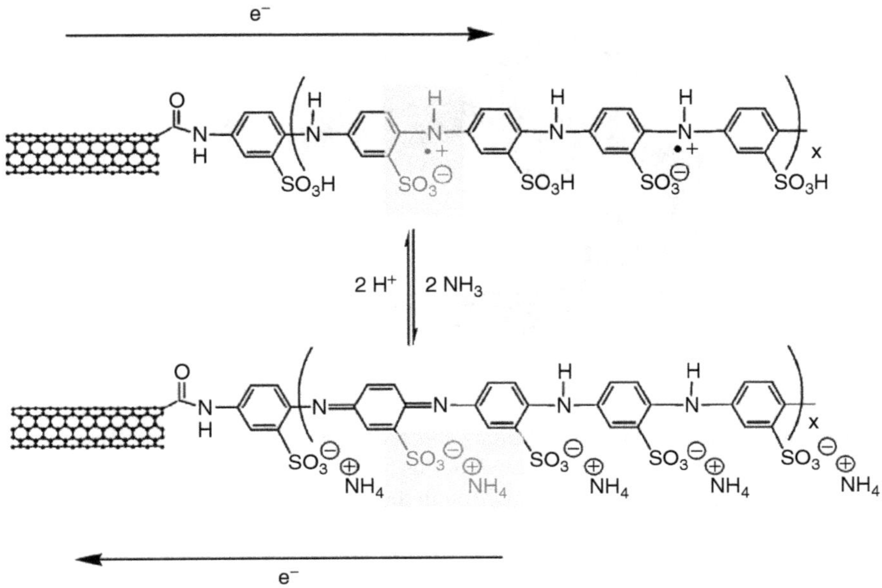

Figure 23.23 Mechanism of interaction of PABS SWNT with NH_3. Arrows indicate charge transfer between SWNT and PABS (Shaker et al., 2010).

solution (Jun et al., 2007; Subramanian et al., 2005). It is widely known that one of the most common methods of digital printing is through inkjet printers, almost ubiquitous in home or office applications. Notably, inkjet printing has been recently enhanced with the capability of printing new functional materials, such as conductive and CNT-based inks, leading to an increased deployment in printed electronics, such as flexible displays, RFIDs, sensors, solar panels, fuel cells, batteries, and most recently in antennas (Kupka et al., 2000;). Inkjet printing for RF applications is a challenging endeavour, where precise control of the achieved conductivity and surface roughness is required. In general, there are two critical factors that affect the print quality: the ink properties and the settings of the printing system itself. The most notable ink properties to observe are viscosity, surface tension, and dispersion stability (printing with an ink of high viscosity and high surface tension at a high contact angle produces smaller sized dot patterns). The most crucial settings of the printing system include the volume of the jetted ink, the traveling velocity of the ejected droplet, the gap distance between each droplet, the printing frequency, the temperature of the jetted ink, the temperature of the substrate, and the sintering/curing mechanism performed on the printed structures. In this work, we utilized a DMP-2800 ink-jet printer shown in Figure 23.24, a table-top printer available from Dimatix Inc. (www.dimatix.com). To ensure good RF properties of the printed device, an in-house recipe has been developed considering all of the aforementioned parameters. For all devices, the utilized Dimatix 1-pL silver nanoink cartridges (DMC-11601) are kept at a distance of 0.5 mm from the surface of the paper. The printer head is first

Figure 23.24 Table-top ink-jet printer used in this work.

adjusted to achieve a print resolution of 2540 dpi, which ensures good RF conductivity of above 9×10^6 S/m, up to several gigahertz. Cabot conductive ink CCI-300 (www.cabot-corp.com) is then jetted through the cartridges at a temperature of 40°C, with the paper substrate maintained at 60°C. Each printed device is then cured in a thermal oven for 2 hr at 120°C.

CNT Characterization The RF characterization of the deposited CNT layers was performed using a two-tier calibration process: First, a short-open-load-through (SOLT) calibration process was applied to the coaxial feed of the SMA connector. Next, a custom-designed multiline through-reflect-line (TRL) calibration set using silver coplanar strip lines was deployed. Figure 23.25 depicts a schematic of such a custom set, where W1 denotes the strip width, L1 denotes the length up to the calibration reference plane, LL denotes the physical lengths of the short line, and of the load termination. L2 and L3 are different line lengths used in the calibration procedure. TRL is utilized to extract the bulk RF behavior of the CNT thin film. The mathematical procedure behind TRL calibration is described in Ukkonen (2004). The silver printed lines form a set of coupled lines, with the CNT layer deposited as a thin film at the TRL reference plane. Using the TRL calibration set, it is possible to embed the silver lines up to the reference plane and extract the input admittance at that plane.

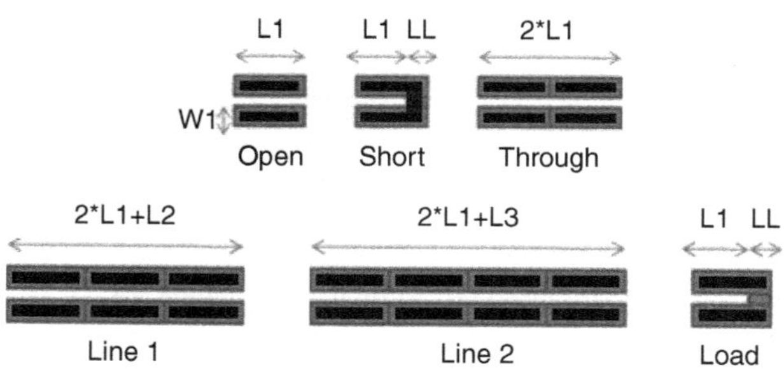

Figure 23.25 Schematic for the TRL calibration set used in this work.

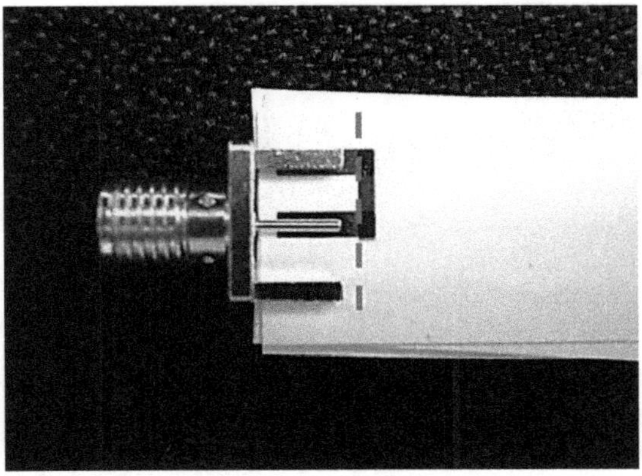

Figure 23.26 CNT film placed at the edge of printed silver lines.

A sample of the CNT film deposited between two electrodes on the paper substrate is shown in Figure 23.26. The fabrication process consists of printing silver nanoink to form a transmission line coupon of coplanar strip lines on photo paper, followed by the application of SWNT at its edges, as outlined in Figure 23.25. Multiple layers of paper were stacked to ensure good contact of the transmission line to the connector center pin as well as ground. Note that the dashed red line in Figure 23.26 defines the reference plane used in the TRL extraction algorithm.

The extraction for various samples with different thicknesses of the SWNT film revealed that the RF behavior of CNT layers can be adequately modeled through a parallel *RC* tank. Such *R* and *C* values are functions of the frequency, the geometrical configuration, as well as the concentration and material properties of the deposited CNT film. Figures 23.27 and 23.28 demonstrate sample results for the extracted resistance and capacitance of a CNT film of dimensions 1 mm × 4 mm deposited at the gap shown in Figure 23.26. The total thickness of the CNT film (10 layers) is ~11 μm. At higher frequencies, the losses in the SWNT increase as shown in Figure 23.27. Also, the capacitance decreases as frequency increases. It is important to note that the resulting equivalent resistance and capacitance can be designed to a desired value by altering the dimensions of CNT thin film (width and length) and its concentration (or thickness).

The test fixture shown in Figure 23.26 was subjected to ammonia of 50-ppm concentration in a closed-cell system. Figure 23.29 shows the reflection coefficient at the coaxial feed of the SMA connector for three distinct cases: without CNT (open-ended transmission line), with CNT before gas exposure, and with CNT after the exposure. It is clear that the RF properties of the CNT thin film are changed upon exposure, resulting in noticeable change in the input reflection coefficient. It should be noted that Figure 23.29 is indicative of the overall behavior of the printed silver ink and the CNT thin film on the paper substrate.

The equivalent admittance of the exposed CNT thin film was extracted using the outlined TRL scheme. The extracted resistance and capacitance values are compared

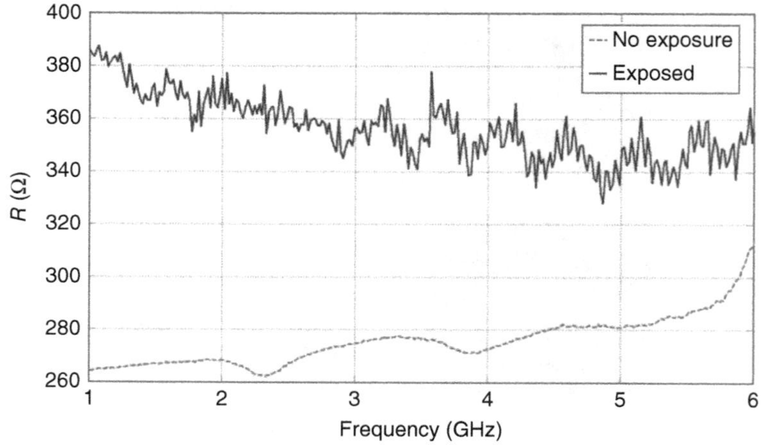

Figure 23.27 Equivalent resistance of CNT thin film (before and after gas exposure).

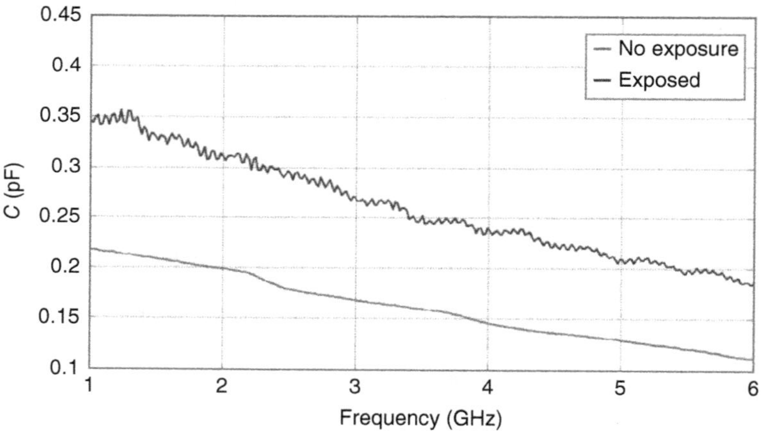

Figure 23.28 Equivalent capacitance of CNT thin film (before and after gas exposure).

and overlaid with and without ammonia exposure in Figures 23.27 and 23.28, respectively. Both the resistance and the capacitance of the exposed film increase due to the adsorption of ammonia molecules on the PABS SWNT thin film. The losses in the CNT film due to ammonia exposure are higher but not significantly dependent on frequency. The trend in capacitance variation with frequency is similar with and without ammonia exposure.

23.3.4 Sensor Design

Given the impedance model of the SWNT film, the sensor was designed with two main objectives: to obtain high RF sensitivity through a high *Q*/narrow bandwidth

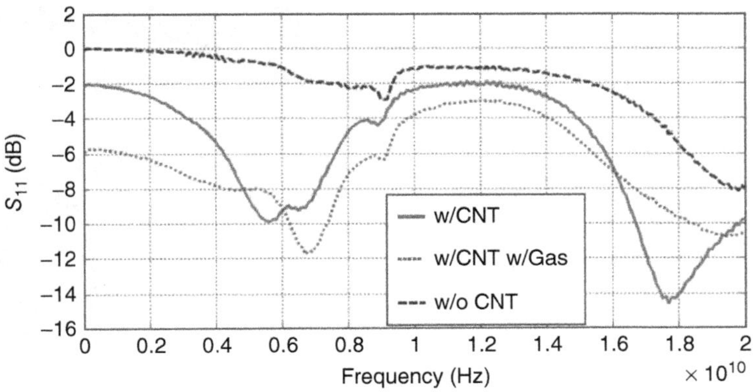

Figure 23.29 Input reflection coefficient at connecter coaxial feed in different scenarios (50-ppm ammonia concentration).

design and moderate gain using a patch antenna. This is achieved by separating the sensing and radiation functions as will be described shortly. Sensitivity from the RF design context means narrow-band sampling of the detection phenomena and does not refer to the chemical sensitivity of the CNT film to ammonia exposure. In contrast to previous efforts [cf. Chopra et al. (2002)], where resonators were utilized without incorporating remote sensing, our design focuses on integrating both high sensitivity and remote-sensing capabilities within the same structure.

Previously, patch resonators have been designed to sense ammonia by covering the entire patch surface with CNT (Chopra et al., 2002). However, SWNT retains high conductivity only in single tubular form, and the conductivity decreases significantly when mass deposited in a random configuration. Therefore, the key to attain high RF detection sensitivity and simultaneously maintain moderate antenna gain, which is required for remote communication, is to minimize the losses in the antenna structure.

Here, we propose using silver ink for the patch antenna realization, with only a small area needed for the CNT film at the tip of the stub (see Figure 23.30). Thus, silver covers a large part of the antenna surface (namely, the patch), ensuring moderate efficiency and good radiation pattern, while the high Q of the sensor element results from minimum losses occurring due to a small CNT film. The idea is that when the CNT impedance changes due to chemical reaction with ammonia, the matching conditions using the loading stub will change, thereby altering the input reflection coefficient at the feed.

To elaborate on the principle of operation in a simplified manner, let us consider the sensor design case where the equivalent capacitance of the CNT thin film before exposure is 1 pF and resistance R is fixed at 1 Ω. Using this lumped-element combination to load a parallel stub as shown in Figure 23.30, a patch antenna was designed [using SPEAG's SEMCAD and Ansoft's high-frequency structural simulator (HFSS) software] to resonate around 4.5 GHz. Based on Figures 23.27 and 23.7, we know that the equivalent circuit of the thin film will change upon exposure to ammonia gas. So, we numerically investigate the cases where the equivalent resistance is fixed at 1 Ω, and

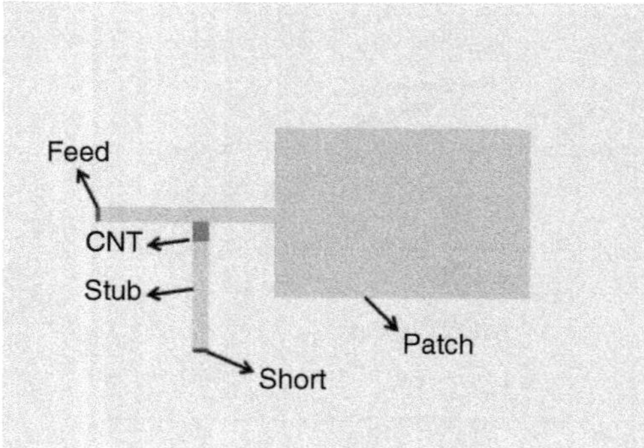

Figure 23.30 Schematic for proposed sensor design (*left*) and fabricated sensor (*right*)

the capacitance changes from 1 pF to 1.1, 1.5, and 2 pF. The resulting reflection coefficient at the antenna feed terminal is shown in Figure 23.31. It is clear that a significant frequency shift is observed, even for a 10% change in the nominal capacitance.

Now, let us consider the case where the nominal (before exposure) equivalent circuit model is a parallel combination of $R = 10\ \Omega$ and $C = 1$ pF. Assuming an exposure level that changes R to 20 Ω and C to 2 pF, we get the resonance shift shown in Figure 23.32. It should be noted that the antenna efficiency in the nominal case will be degraded due to the CNT losses represented by the equivalent resistance. A theoretical calculation shows that the typical antenna efficiency of such design is around 30%.

Based on the outlined principle of operation, a loaded patch antenna was designed and printed on 10-mil-thick paper as shown in Figure 23.30. The ground plane is made of a thin copper sheet. Conductive epoxy is used to solder the connectors onto

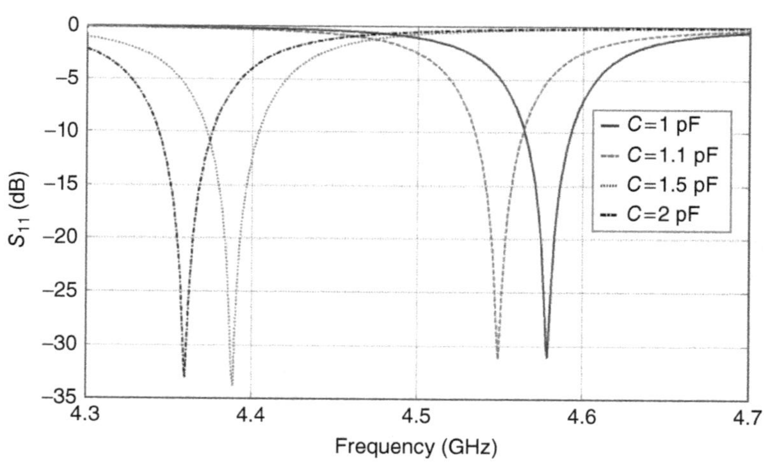

Figure 23.31 Numerical predictions of shift in response to changes in equivalent capacitance of CNT layers.

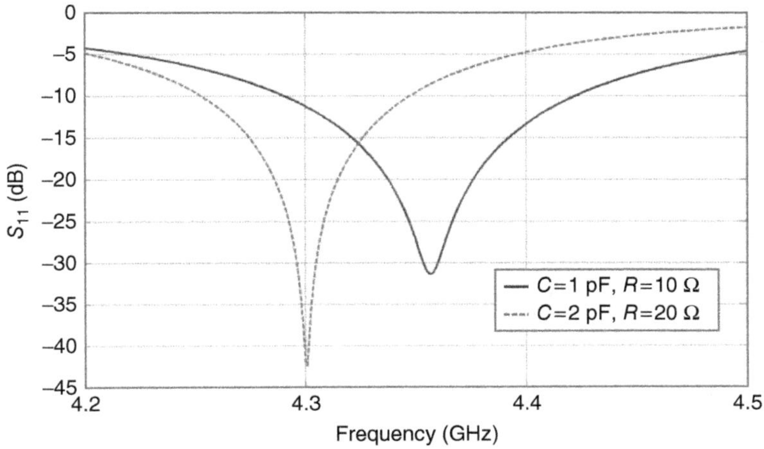

Figure 23.32 Numerical predictions of shift in response to changes in equivalent admittance of CNT layers.

silver traces on the paper substrate. The ground pins of an edge-mount connector are connected with the stub end to create the RF short circuit.

To evaluate the sensor performance, we used household ammonia (10% ammonium hydroxide) in a small plastic container, with the antenna placed nearby (Figure 23.33). The input reflection coefficient was recorded before and after filling the container with ammonia, and the resulting response for these two cases is shown in Figure 23.34. Resonance frequency shift of about 120 MHz was observed. Notably, the response time was quick (less than 30 sec).

23.3.5 Controlled Sensor Measurement

The measurements agree with the circuit model presented in Section 23.3.4 and the setup for the measurement is shown in Figure 23.33. The resonant frequency of the

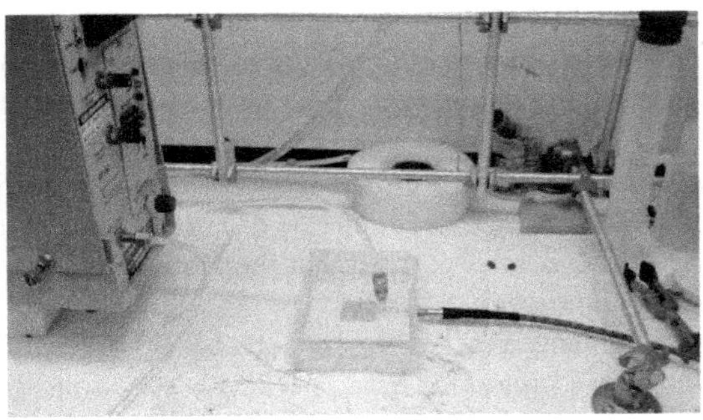

Figure 23.33 Closed test cell connected to gas generator and PNA.

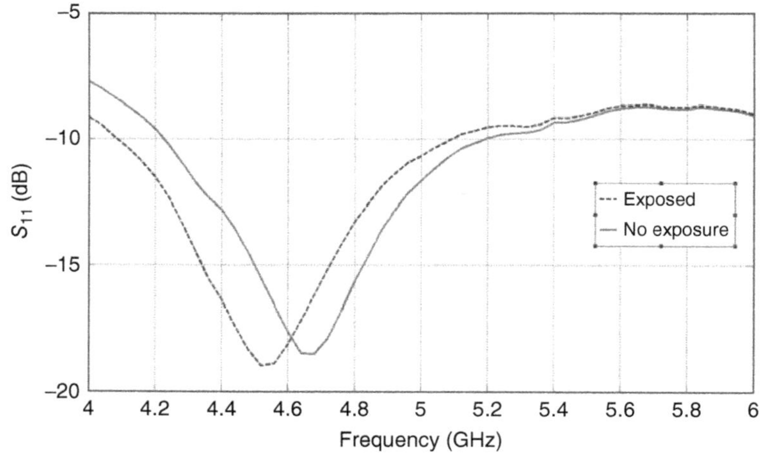

Figure 23.34 Frequency shift in response after exposure to household ammonia.

sensor has shifted to lower frequencies due to the increased resistance and capacitance due to interaction between NH3 and CNT. Such measurements verify the principle of operation of the sensor. However, to evaluate the sensitivity, it is desired to determine the gas concentration needed to trigger a noticeable shift in the resonant frequency. To this end, a controlled measurement system is deployed.

Our controlled measurement system consists of five major components: gas generator, NH_3 permeation tube, closed-system sensor cell, Agilent PNA-L N5230A programmable network analyzer (PNA), and data acquisition system. We used the 491M gas generator and the NH_3 permeation tube, from Kin-Tek Company (La Marque, TX). At all times, this system was used under a chemical fume hood.

The NH_3 permeation tube has a calibrated emission rate under controlled temperature. Dry nitrogen was delivered from an ultrahigh-purity (grade 5) compressed nitrogen gas cylinder from Airgas. By controlling the ammonia emission rate and the nitrogen flow rate, desired concentrations (in parts-per-million range) of ammonia were generated from the gas generator. Before each ammonia exposure test, the NH_3 permeation tube was installed in the gas generator and heated to a predetermined temperature for around 4 hr in order to obtain a stabilized ammonia concentration. The accuracy of gas concentration output is $\pm 4\%$ according to the calibration results from Kin-Tek.

The sensor was installed inside a closed test cell, depicted in Figure 23.33. The concentrated NH_3 stream was delivered into the test cell via a Teflon tube. The CNT film was placed near the gas inlet port, so that the ammonia stream was allowed to interact with the sensor surface first.

For the RF measurement, expendable connectors and a precision-phase steady cable assembly was used to interface between the PNA and the antenna under test. An SOL one-port calibration was conducted with coaxial standards.

Measurements were made at three different concentrations: 50, 75, and 100 ppm. For each measurement, the data was saved automatically every 10 sec. Also, each measurement consisted of several minutes of baseline recording (exposure to pure nitrogen), followed by exposure to ammonia, after which the system was purged with nitrogen gas to bring the sensor back to the baseline.

Based on the measured data for the three concentrations, we conclude that the resonance frequency shift increases with ammonia concentration until the sensor eventually saturates. The resonant frequency shift is summarized in Table 23.2 along with the nitrogen flow rate required to maintain appropriate ammonia concentration from the 491M gas generator during the exposure and reversion (back to ambient) phases. Figure 23.35 plots the frequency shift as a function of time during these two phases.

From Figure 23.35, it can be observed that the gas concentration affects two things: the time it takes to reach saturation and the saturation level itself (or the maximum frequency shift). The 100-ppm concentration showed the steepest slope to saturation as well as highest shift at 48.75 MHz, while the 50 ppm showed the slowest response as well as the lowest shift at 43.125 MHz.

After exposing the sensor to ammonia for approximately 18 min, the closed system was purged of ammonia using nitrogen gas. In the third column of Table 23.2,

TABLE 23.2 Measurement Results

Conc. Ammonia (ppm)	Exposure Flow Rate (L/min)	Reversion Flow Rate (L/min)	Δf_{res} of S_{11} (MHz)
50	0.46	0.5	43.125
75	0.31	6.0	46.875
100	0.23	0.5	48.750

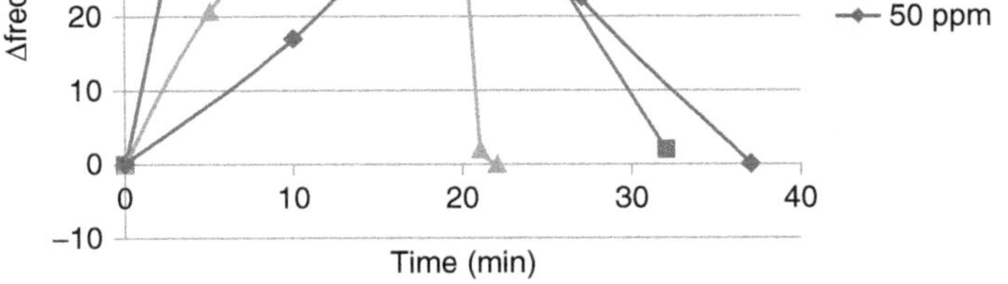

Figure 23.35 Measurement of frequency shift versus time for three different concentrations.

reversion flow rate indicates the flow rate of nitrogen gas to flush out the closed cell system so that the input reflection coefficient of the antenna returns to its baseline. Two different flow rates, 0.5 and 5 L/min, were used to observe the difference in reversion. From this test (see Figure 23.35), it is clear that faster purging rate of nitrogen gas flow (5 L/min for 75 ppm) allows the sensor to return to baseline faster.

In Figure 23.36, the maximum shift in frequency is plotted against ammonia concentration, showing that higher concentration yields a larger resonant frequency shift. Also, the plot shows a decreasing amount of shift for incrementally higher concentrations past the 80-ppm level, indicating a trend toward saturation of the sensor with ammonia.

23.4 GRAPHENE-BASED AMMONIA DETECTION FOR MEDICAL DIAGNOSIS

23.4.1 Introduction

In this section, we demonstrate the use of graphene as the basis for the design and development of low-cost, self-powered, batteryless, wireless sensor solutions utilizing thin films produced from environmentally friendly, water-based, inkjet-printed graphene oxide (GO) ink. The advanced sensor material demonstrates good response to ammonia gas (NH_3), yielding a 6% normalized resistance change within 15 min of exposure to a concentration of 500 ppm. In addition to this, excellent recovery time is achieved using the reduced GO (RGO) thin films, with over 30% of material recovery observed within 5 min without exposure to high temperature or any ultraviolet (UV) treatments. Finally, we present important distinctive characteristics of the behavior of the RGO sensor when exposed to different types of gases, including the hard-to-detect CO gas, which can be exploited in order to further enhance the applicability of the material. The introduction of mass-producible, stable, environmentally friendly,

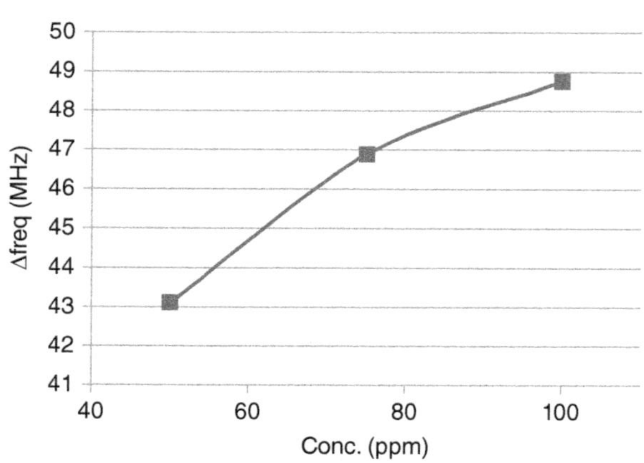

Figure 23.36 Measurement plot of resonant frequency shift versus concentration of NH_3 gas.

inkjet-printable GO on organic paper/Kapton substrates lays the foundation for the development of a wide range of new low-cost, high-performance graphene-based devices, such as inkjet-printed diodes, capacitors, and transistors.

23.4.2 Principle of Operation

Carbon-based biosensors have a promising future as molecular detectors due to their unique electronic and structural characteristics. These materials interact with a given analyte (chemical agent) by surface absorption (binding), which induces charge transfer between carbon atoms and the analyte molecules. Depending on the type of substance which comes into contact with the sensing thin film, either electron (n-type) or hole (or p-type) doping of the semiconducting thin films occurs. The resulting increase or decrease in charge carrier density (electrons or holes) creates a corresponding change in conductance of the material. The change in electronic structure of the thin films can yield alterations in properties such as real and imaginary impedance, DC resistance, and effective dielectric constant (Lakafosis et al., 2011). These changes in material properties in the presence of the analyte can be converted into measureable changes in electrical properties such as voltage, current, circuit resonant frequency, and backscatter power amplitude in appropriately designed electrical systems. *This comprises the fundamental principle of gas detection using nano-carbon thin films.*

23.4.3 Design Example

To illustrate the concepts of gas detection using nano-carbon thin films, we will present here a state of the art design using graphene as the sensing material for wireless detection of carbon- and nitrogen-based compounds, two universal constituents of biological systems. The text will take the reader through all of the stages of prototype development, from creation of the thin films through validation testing, to final wireless system build.

Previous studies have demonstrated excellent gas detection for nitrogen-based gases using a multiwalled CNT (MWCNT) (Lakafosis et al., 2011). At 10 ppm NO_2 we achieved 21% sensitivity, and at 4 ppm NH_3 9% sensitivity was obtained at 864 MHz using a functionalized MWCNT dissolved in water and inkjet printed on paper substrate.

Graphene is a single planar layer of carbon atoms with exotic physical properties including high mobility of charge carriers (200,000 cm^2 V^{-1} s^{-1}), and high thermal conductivity (~5000 Wm^{-1} K^{-1}). Most importantly, as a zero band gap semiconductor, graphene's high metallic conductivity and low charge carrier density enable extreme sensitivity because small variations in charge carrier density yield notable changes in conductivity. Additionally, due to the fact that graphene is an extremely low-noise material electronically, graphene offers clear advantages in terms of minimum detection levels to other solid-state gas sensors, such as those created from CNT. Single-molecule detection has been achieved in the lab (Schedin et al., 2007).

The designs discussed below illustrate the principles of wireless gas sensor design utilizing graphene thin films and combining both analog (passive) and digital (batteryless semiactive) wireless remote transmission principles with graphene-based enhanced selectivity, featuring advanced inkjet printing of aqueous graphene solution on environmentally friendly paper and Kapton substrates. Regarding the digital transmission scenario, in particular, the coupling of gas sensing with power-efficient wireless data transmission is achieved by deploying Intel's wireless integrated sensing platform (WISP). The platform adds microcontroller functionality to traditional RFID tags, which operate as the sensor's integration platform. The WISP can be programmed to sample a variety of different responses and produce gas detection results, which take advantage of additional processing, such as time-averaged stabilized or normalized difference gas-induced material changes.

23.4.4 Inkjet Printing of Graphene Sensor Prototype

The core of the wireless gas sensor is a prototype board made from graphene deposited onto a flexible substrate, in this case paper and Kapton. Cabot conductive nanoparticle silver ink was used to create the traces connecting the graphene pad to the external circuitry. We used Cabot's CCI-300 ink here, which contains surface-modified ultrafine silver nanoparticles, providing reliable inkjet printing of high-resolution and low-resistivity conductive features on a variety of substrates ("CCI-300 Data Sheet"). Microstrip topology was used since the silver microstrip trace can be easily integrated with a large range of inkjet-printed antennas as noted in Li et al. (2009).

Creation of Graphene Oxide The first step in the sensor development process was the creation of stable, long-life, inkjet-printable graphene-based inks. This was accomplished by first converting the graphene into graphene oxide powder. Unlike pristine graphene, which has very poor dispersion in common solvents, GO exhibits excellent solubility in water due to the existence of hydrophilic functional groups on the surface (Lin et al., 2010), rendering it an excellent candidate for development of environmental friendly water-based inks. After deposition, pure graphene was obtained by the reduction of GO, which reverts the conjugated basal plane and restores the electrical properties of the material. The reduction of GO is considered as one of most promising methods for low-cost, high-yield, and scalable preparation of graphene materials (Li et al., 2010).

The GO was produced by chemical oxidation of graphite, which introduces oxygen-containing functional groups to exfoliate pristine graphite into individual GO sheets. GO was prepared with Hummers' method (Hummers and Offeman, 1958). Graphite flake was placed into a $NaNO_3$ /concentrated H_2SO_4 solution in an ice bath. Subsequently, $KMnO_4$ was slowly added to the solution while maintaining the temperature at $< 20°C$. The mixture was stirred in the ice bath for 2 hr and for an additional 30 min in a 35°C water bath. After that, 70 mL of 70°C water was added slowly into the flask. The heat generated via exothermic reaction raised the solution temperature up to 98°C. Then an additional 140 mL of 70°C water was added, followed by hydrogen peroxide solution to terminate the reaction. The mixture was filtrated and washed with water to remove excess acid and inorganic salts. The resulting GO was dried overnight at 55°C to produce the GO powder. To prepare the GO ink, dry GO powder was dispersed in a water/glycerol solution and sonicated to form a homogenous dispersion.

Fabrication via Inkjet Printing A Dimatix Materials Printer (DMP-2800) series material deposition system was used to print both the silver and GO inks. Before printing, the Kapton substrate was treated by UV ozone to improve its hydrophilicity. The treated substrate demonstrated improved performance during printing compared to the native Kapton substrate, enabling higher resolution prints and more layers to be printed during a given cycle.

The fabrication process involved deposition of 10 layers of conductive silver onto the Kapton substrate, followed by 12–42 layers of graphene oxide. First, the 10 layers of silver ink were deposited and cured at 120°C for 8 hr. Afterward, 2 layers of GO were deposited and cured at 80°C for 4 hr to serve as a foundation for further prints. These two foundational layers overlapped the silver pads by 0.5 mm to ensure optimum connectivity. Next, the GO was deposited in 5-layer increments with curing in between to ensure the highest consistency between samples. Figure 23.37 illustrates the printing of GO ink. Figure 23.38 shows the completed sensor elements prior to reduction of the GO thin films.

Reduction of Graphene Oxide After printing and curing, each sample was reduced to obtain the desired graphene from the GO. Due to the reduction process of graphene oxide, we fabricated the samples on a Kapton substrate. After thermal reduction,

Figure 23.37 Graphene drop ejected from inkjet printer onto Kapton substrate.

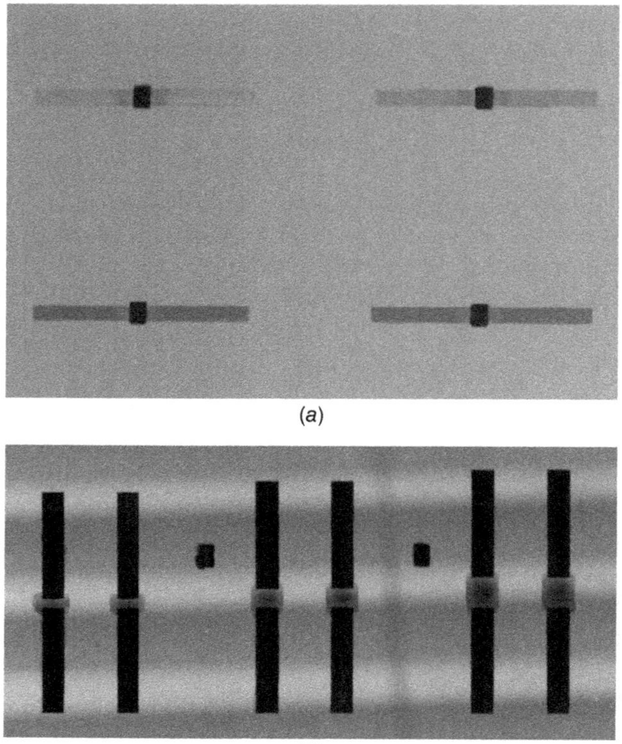

(a)

(b)

Figure 23.38 (*a*) Graphene-based thin films inkjet printed in between and overlapping silver conductive inkjet-printed trace on paper substrate. (*b*) Graphene oxide thin films inkjet-printed in between and overlapping silver conductive inkjet-printed trace on Kapton substrate.

the resistance became as low as 150 Ω, achieving approximately 1000 times better performance than that mentioned in the current literature (Le et al., 2011). Reduction was achieved by placing the printed GO thin films in elevated temperature in a hydrogen and argon atmosphere. The samples were reduced at 200°C for 30 min and 300°C for another 30 min. The heating rate was 5°C/min. After reduction, the samples were left to cool down to room temperature naturally. Scanning electron microscopy (SEM) was used to determine the overall quality of the graphene thin films (i.e., presence of defects in the print, etc.) and to make adjustments in processing to further prints. Figure 23.39 provides SEM images of both the initial GO and (reduced) RGO thin films. The wrinkles on the surface of the samples in Figure 23.39 are typical characteristics of GO and RGO thin films, which are likely indicative of minor defects in the structure.

The examination on the whole patterned area revealed a uniform RGO layer with nearly 100% coverage, indicating the high quality of the GO layer obtained from the

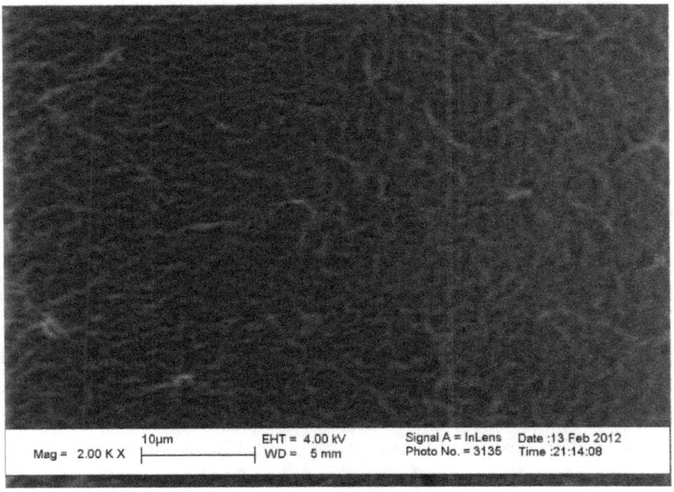

(a)

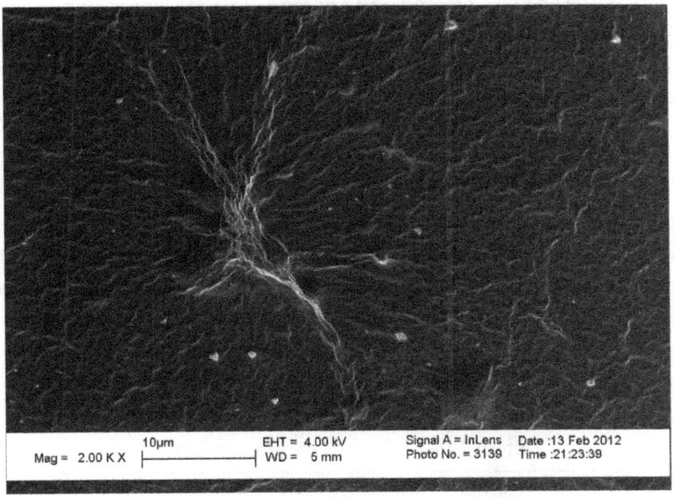

(b)

Figure 23.39 (*a*) SEM image of inkjet-printed GO thin film and (*b*) SEM image of inkjet-printed RGO thin film.

inkjet printing process and the fact that the reduction process doesn't bring significant change in the morphology of printed graphene.

Final Packaging Steps One important fabrication factor, which was explored in the framework of this work, was the impact of the contact between silver and graphene on the system performance. Based on experimentation, we determined that applyning

10 additional layers of silver on top of the graphene guarantees better connectivity between the graphene and silver traces and helps to de-embed any additional impedance that would offset the measured change of impedance unevenly among different samples of the similar dimensions. We perform this important step as a means to reduce the variability between samples.

23.4.5 Optimization of RGO Thin Film

The RGO thin films were optimized to obtain the proper sensor area to ensure the maximum sensitivity while maintaining the requisite intrinsic impedance for the wireless sensor platforms presented in Section 23.

To determine the optimum sensor area, three patterns of differing lengths of RGO material were produced, as shown in Figure 23.38*b*. It was expected that the larger surface area of the last pattern would yield the highest sensitivity. Based on previous efforts on thin-film design, it was estimated that between 15 and 25 layers of the material would be necessary to achieve the desired resistance for the given dimensions of the RGO pads. To validate this assertion, samples of both 15- and 25-layer versions of the patterns were printed and reduced. The measured results can be found in Table 23.3. These values represent the average resistance values for a given combination of RGO pad dimension and thickness (number of layers) and the reported works are summarized in Figure 23.4.

23.4.6 Gas Sensor Experimentation

Experiment Setup The gas test setup is shown in Figure 23.40. The sensors were tested using an Environics S4000 gas mixing system. The setup was capable of producing reliable mixtures up to 500 ppm of ammonia gas diluted in air and delivered at

TABLE 23.3 Average Intrinsic Resistance Values of RGO Thin Films

Number of Layers	Pattern 1 (Ω)	Pattern 2 (Ω)	Pattern 3 (Ω)
15	501.9	1123.8	2014.8
25	248.2	488.0	881.8

TABLE 23.4 Comparison of Our Designs with Previous Work

Design	*Power (mW)*	*Transmit Range (m)*
Fulford-Jones et al., 2004	60	20–30 m
Ekström, 2006	170	20
Design 1	774	150
Design 2	87	432

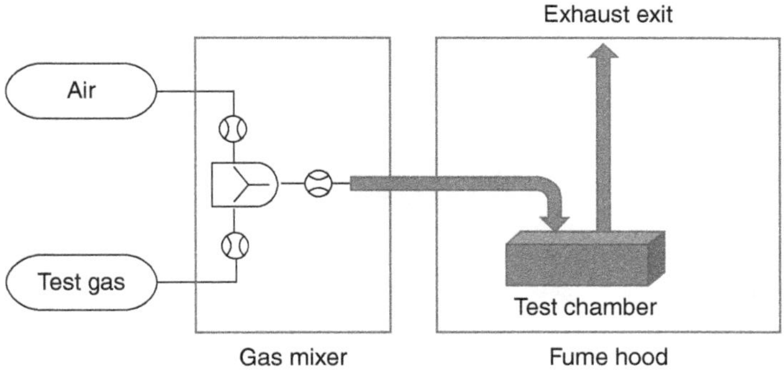

Figure 23.40 Measurement setup to test RGO thin films.

a rate of 50 mL. To calibrate the sensor, the graphene samples were placed into a semi-closed glass chamber to direct the flow of gas across the sensor. The custom-designed gas flowing chamber, shown in Figure 23.41, has internal dimensions of 4 cm × 1.5 cm × 1.5 cm.

Air flowed through the system for 15 min to establish a system baseline. Next, the gas mixture (500 ppm of ammonia/air or 0.1% CO/air) was introduced into the system, and measurements of the resistance were taken at 1-min internals for 15 min until near-steady-state condition in the material was achieved. Finally, the gas source was removed from the device while measurements continued to be taken for another 15-min interval in order to measure the recovery time.

Experimental Results The results of 500 ppm NH_3 gas testing are shown in Figure 23.42. The inkjet-printed RGO thin films of different dimensions show similar responses to 500 ppm NH_3. The electrical resistance rapidly increases in the first few minutes after the introduction of NH_3, indicating a fast detection rate. With the continued supply of NH_3, the resistance change begins to diminish after 10 min, which occurs as the material enters the saturation region.

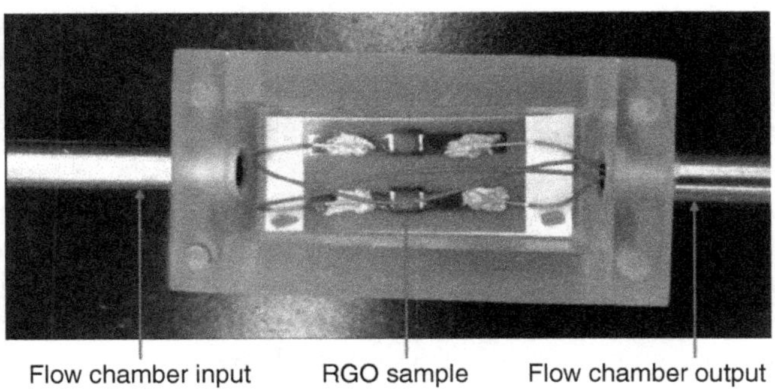

Figure 23.41 Test chamber with two samples inside.

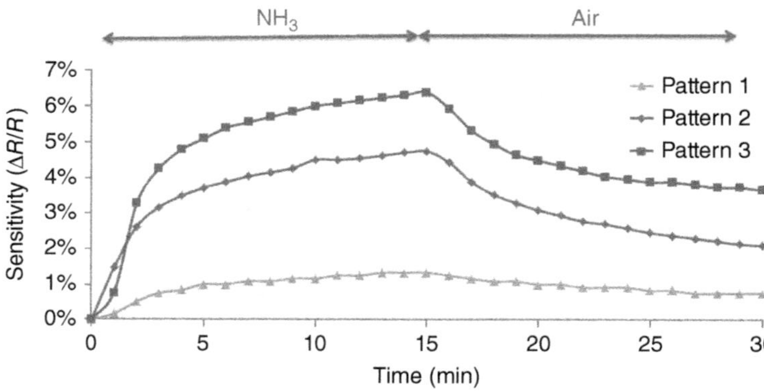

Figure 23.42 Measured response of RGO thin films in presence of 500 NH_3.

The maximum sensitivity is observed in the case of pattern 3, which shows greater than 6% increase in normalized resistance. The effect of the RGO pattern dimension on the sensor performance is still under investigation. Here, it was observed that the pattern with the largest area provided the best result. After the introduction of NH_3, pure air was introduced into the system, and it was found that the resistance quickly recovered a large portion of its original value (~30%) within 5 min.

The increase in resistance upon exposure to NH_3 is due to the electron donation of NH_3 to the p-type RGO film, and is consistent with the literature. Compared to other works on graphene-based gas sensors, our inkjet-printed RGO demonstrates superior performance, including high sensitivity, fast response, and quick recovery. In particular, the short recovery time in natural environmental conditions is extremely advantageous over other reports that used UV and heat treatment to assist the recovery and is also of great importance for practical applications (Schedin et al., 2007). The sensor exhibits an observable (~10 Ω) change in resistance within 1 min after introducing NH_3, a measure that is certainly in the detectable range of the backend circuitry of the WISP platform.

The measured response of the material to CO gas is shown in Figure 23.43. Here, the RGO thin film was exposed to 0.1% CO. This concentration was necessary to invoke response of the material. For the comparison between CO and NH_3 testing results, we plot the results of the pattern 3 samples together. The sensitivity for CO gas is substantially lower than that of NH_3, which is similar to previous reports (Schedin et al., 2007). However, the important difference, which may potentially be used to differentiate between the two gases, is the unique response given by the CO gas. Note that the exposure to CO gas does not provide the same monotonic response as for NH_3. Although the samples had some variance as noted above, each of the eight samples tested under CO gas conditions contained the same rapid rise and fall while being under the influence of the gas. As we have found only a few works in the literature concerning testing of inkjet-printed RGO thin films with CO gas, we have little basis for comparison. However, the relevant sample size obtained during the experiments demonstrates that the effect is repeatable enough to be considered

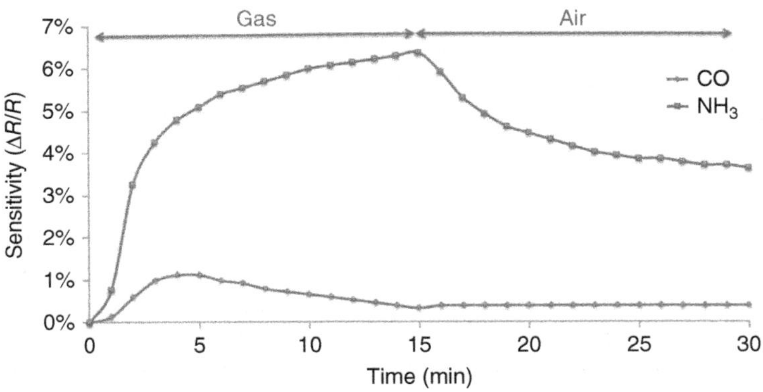

Figure 23.43 Measured response of RGO thin films to NH_3 and CO.

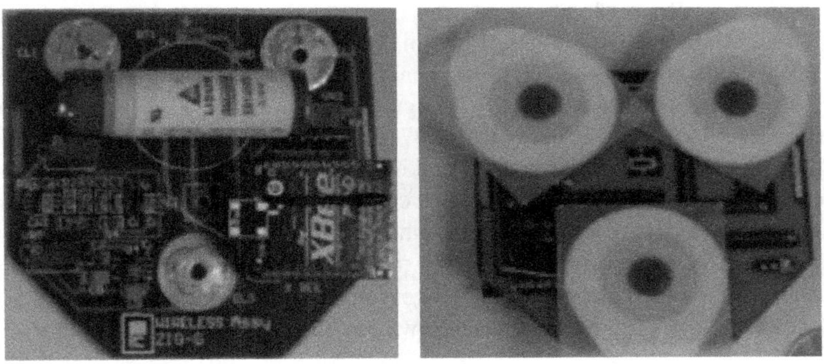

Figure 23.44 Wireless EKG transmitter. (*See insert for color representation of the figure.*)

useful for the potential selectivity between different gases. Further experiments are ongoing to determine the significance of the observed phenomenon.

23.5 INTEGRATED WIRELESS MODULES

23.5.1 Wireless EKG System Utilizing Low-Power ZigBee Standard

Introduction Electrocardiography has been in clinical use for the diagnosis and monitoring of heart abnormalities for more than a century. Moreover, EKG remains the best and least invasive method for the intended task. Eliminating the use of wires in medical operating theaters, as well as for elderly people that need to wear monitoring devices for a continuous tracking of their heart condition is, undoubtedly, a highly desirable feature and a big step forward for the technological advancement of EKG. A low form factor is very important as it can further increase the use of such a device in pediatrics.

The electrical impulses within the heart act as a source of voltage, which generates a current flow in the torso and corresponding potentials on the skin. The frontal projection of the cardiac vector can be captured by three standard limb leads/electrodes, which are bipolar leads that measure the potential difference between the right arm, left arm, and left leg. Here we focus on the standard limb leads and not the precordial lead configuration.

Wireless Module Prototypes The two wireless modules proposed (Mukala et al., 2010) support multihopping over the popular ZigBee protocol, and their size is 7 cm × 7 cm × 1 cm. The first design is powered by an Atmega32 microcontroller that is interfaced with the Xbee Pro ZigBee module for the wireless transmission (Fig. 23.44). The relatively high current demand of this Xbee module has necessitated the proposal of a new design that includes a single system on the chip that houses both the microcontroller and the RF transmitter. The second design is based on the TI CC2530. Both designs operate at a bit rate of 9600 bits/sec for transmission of a single lead II signal. The resolution of the analog-to-digital converter (ADC) was 8 bits in the first design and 9 bits in the second one while the sampling rate was 500 samples/sec. Both designs are analyzed with electrodes placed close to the heart, which provide a better quality of signal, and the EKG data obtained from the tested circuit is shown later in Figure 23.47.

Power consumption is calculated by taking an average of the worst- and best-case power consumption. As the best case we consider placing the source very close to the receiver, whereas the worst case is when the source is at the maximum possible distance from the receiver. The higher power consumption in our designs can be attributed to the higher transmission range. Power consumption of design 2 is low because of the less transmitter current drawn.

Testing and Validation The received signal is validated using a logged EKG signal with Biopack (www.biopac.com) as reference. Frequency analysis is done, and the received signal frequency plot is compared to the reference signal. Figure 23.45 shows the logged EKG data for 3 min. Frequency analysis is shown in Figure 23.46. The frequency of each peak is about 0.8 Hz.

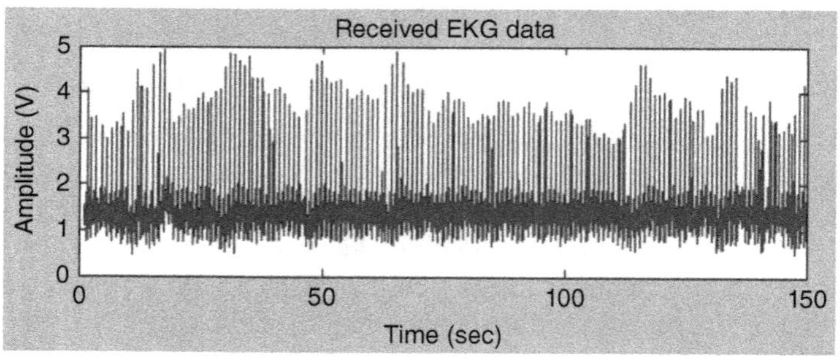

Figure 23.45 Logged EKG data.

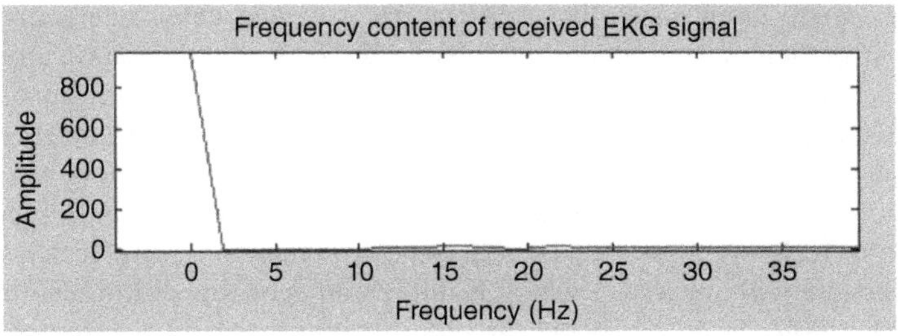

Figure 23.46 Frequency analysis of received EKG signal.

Fast Fourier transform (FFT) is performed to determine the signal-to-noise ratio (SNR). The SNR is calculated by finding 20 log [(amplitude of 0.8 Hz component)/(amplitude of other frequency components)], that is found to be equal to 15.2 dB.

Analysis showed that closer placement of the electrodes to the heart results in a better quality of signal obtained. Initial placement of electrodes was 30 cm apart; however, a better signal was observed when the electrodes were placed 7 cm apart. Both design 1 and design 2 are analyzed with electrodes placed close to the heart. Figure 23.47 shows EKG data obtained from the tested circuit on the oscilloscope. The experimental setup is conducted by having a fixed receiver connected to the computer and moving source, which bears the EKG device and moves away from the receiver. It is observed that a good signal (SNR > 10 dB) is obtained while the source is within a certain distance from the transmitter beyond which the signal is distorted. We fix this point as the range of transmission. Power consumption is calculated by taking an average of the worst- and best-case power consumption.

23.5.2 Smart Wireless Integrated Module

Efficient communication protocol, low power, and low cost are important design factors for realizing the wireless sensor networks (WSNs) (Akyildiz and Vuran, 2010; Mahalik, 2007). Furthermore, sensor nodes should be small enough to deploy in large

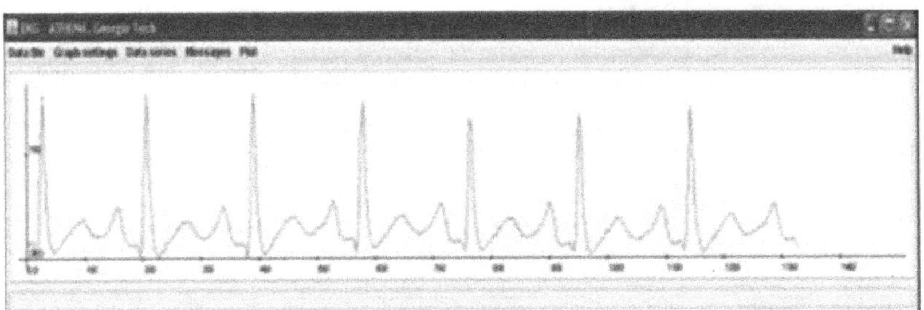

Figure 23.47 Plot of obtained EKG signal.

quantities, which emphasizes the need for low-cost and environmentally friendly technology for their mass production (Mahalik, 2007). As discussed previously, inkjet printing technology on organic paper substrate is one of the most promising candidate for this kind of RF device (Yang et al., 2009). Paper is one of the cheapest and most widely used organic material in the world. In addition, it is very environmentally friendly because it is recyclable (Bozzi et al., 2011). Moreover, inkjet printing is a growing technology of interest that deposit conductive nanoparticles directly on the substrate without waste, which results in an economical fabrication process (Yang et al., 2009). Lots of papers have reported inkjet-printed electronics on paper such as sensor-enabled RFID tags (Chung et al., 2009). Based on these results, a smart wireless integrated module (SWIM) is designed on paper using inkjet printing technology for IEEE 802.14.4 and ZigBee-compliant wireless sensor node.

A system-on-chip (SoC) solution was chosen to implement the sensor module as can be seen in Figure 23.48. The sensor circuit was powered by 3.3 V, and the SoC voltage regulator is decoupled with external circuit components to constant 1.8 V to the RF transceiver. The output from the analog temperature sensor is connected to a microcontroller unit (MCU) ADC multiplexer in order to read and broadcast the information. Finally, a matching network that matches differential feeding of the RF transceiver to a 50-Ω antenna is designed and connected to the antenna. First, the sensor node was implemented on a conventional FR4 board using swivel antenna, and its feasibility is tested as shown in Figure 23.49*a*. The sensor on FR4 successfully communicates with the receiver so that the designed circuit footprint has been modified to inkjet print on paper substrate as shown in Figure 23.49*b*. In this phase, a planar inverted-F antenna (PIFA) is designed in order to make the sensor planar structure. Inkjet-printed circuit on paper is shown in Figure 23.49c, and the other circuit components such as resistors, capacitors, and inductors are mounted on the inkjet-printed circuit board using conductive epoxy.

In summary, inkjet printing technology has been utilized to materialize wireless sensor module and IEEE 801.14.4 and ZigBee-compliant sensor module on paper. The sensor module is designed and tested on FR4 and paper, and data packets are successfully transmitted. This technology is promising in the future applications due to its ease of mass production, low cost, and environmentally friendly properties.

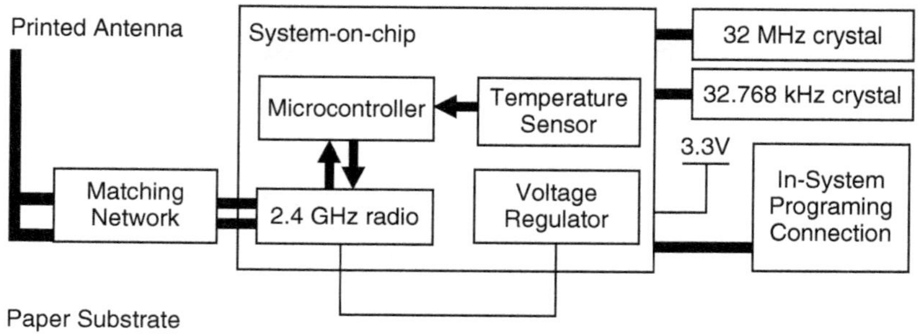

Figure 23.48 Block diagram of wireless sensor.

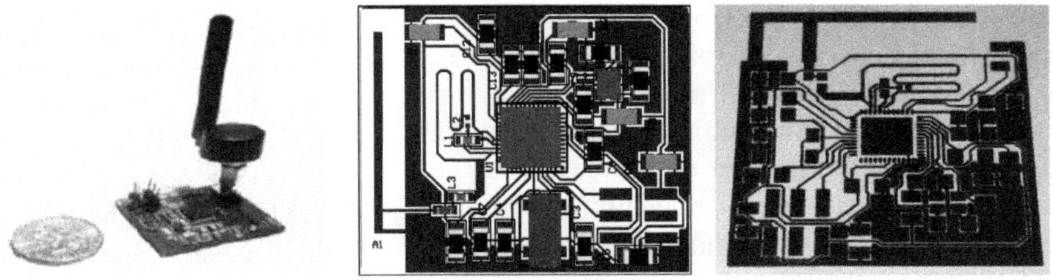

(*a*) (*b*) (*c*)

Figure 23.49 (*a*) Implemented sensor node on FR4, (*b*) schematic of wireless sensor node on paper, and (*c*) inkjet-printed footprint of wireless sensor on paper.

23.5.3 Wireless Transmission of CNT-based Sensed Information

Analog Transmission The incorporation of our graphene-based sensor thin films with microstrip lines at this early development stage only serves as a proof of concept. Of course, these transmission lines can be thought of as part of a wide range of antenna and microwave structures. For instance, the graphene microstrip lines can be part of the meander line of a bowtie antenna or even part of short lines of microstrip patch antennas (Lakafosis et al., 2011), such as that shown in Figure 23.50. Under this approach of analog transmission, the wireless remote-sensing mechanism is actually the detection of the shift of the resonant frequency and/or the magnitude of the backscattered power as a direct result of the change of the impedance of the graphene stripes and, thus, of the whole structure as well. Calibration may be needed in order to

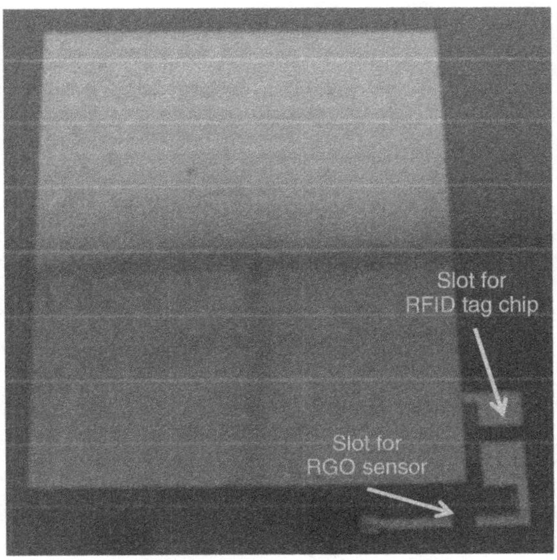

Figure 23.50 Inkjet-printed antenna for development of analog gas sensor.

map particular amounts of shift in the resonant frequency or change in backscattered power to the concentration of toxic gases under test.

Regarding the range of frequencies that can be supported through these types of antenna structures, this ultimately depends on both the manufacturing accuracy of these transmission lines and, on the other hand, on how thin the graphene lines can be fabricated while still maintaining high levels of sensitivity. Preliminary results toward that direction demonstrate that it is the surface area of the inkjet-printed graphene and not the number of its inkjet-printed layers that determines the sensitivity to the presence of small concentrations of poisonous gases. In other words, as long as the surface area of the RGO exposed to the air remains constant, regardless of the width of the lines that the RGO is part of, the gas sensitivity remains the same.

Digital Transmission As opposed to the aforementioned analog-based transmission of the sensed information that might be subject to distortion due to the dynamic nature of the air medium, the WISP-GS (gas sensor) prototype that we have developed, shown in Figure 23.51, offers the reliability of wireless digital transmission.

The WISP is a fully passive, battery-free, and programmable RFID tag (Sample et al., 2008), which can be powered and read by off-the-shelf EPC Gen2 UHF RFID readers and has an on-board microcontroller for sensing and computing functions. In this work, we modify (hardware- and software-wise) the WISP to create a sensing and communication platform, which enables for the first time the introduction of gas sensing capabilities to RFID tags. The WISP-GS is solely powered by the RF energy illuminated by any commercial RFID reader. This 915-MHz RF energy is rectified with a charge pump topology of diodes and capacitors to charge the on-board capacitor. Whenever located within the interrogation zone of an RFID reader, the WISP-GS is automatically detected and begins transmission of the sensed information within the EPC message of the RFID communication.

The external WISP-GS board consists, of course, of the graphene-based inkjet-printed RGO sensor, presented in the previous section, as well as of the analog interface of this sensor. It is due to this analog interface, consisting essentially of a

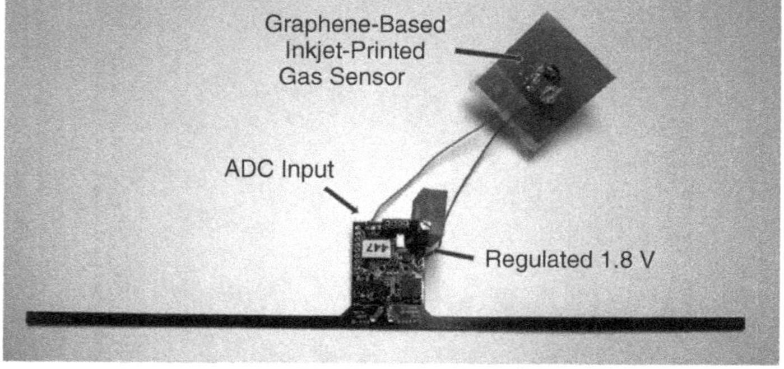

Figure 23.51 WISP-GS (gas sensor) prototype platform.

voltage divider, that the gas detection takes place locally at the end device level in terms of a simple impedance measurement and can be relayed digitally.

23.6 CONCLUSION

The human body can be considered a very lossy and high dielectric material, which causes detrimental effects on on-body wireless devices because radiated waves tend to propagate through high dielectric material. Most of the power radiated from the device is absorbed by the human body being dissipated as heat energy due to loss of human body, which may cause biomedical and power managing issues. It is critical to design a noble wearable electronic system to overcome the issues.

In this chapter, vialess, flexible, EBG-based antennas and a complete RFID tag have been introduced. Both in chamber and real environment measurements demonstrate the increase in gain and range due to the EBG structure. One of its main applications is a wearable RFID tag. However, this can be extended to any surface because any effect of the surface is isolated from the tag due to the ground plane.

We have also demonstrated the advanced concept of liquid antennas for wearable or implantable biomonitoring applications. Although conventional metal antennas suffer from low-range and oxidation problems when in proximity to the human body, the proposed topology features an improved range by up to an order of magnitude. In addition to this, it is very conformal allowing for an easy wearable WBAN implementation and can also be very easily reconfigured by adjusting the salinity of the salt solutions in order for it to operate in different ambient conditions.

Different designs of highly sensitive antenna-based CNT and graphene sensors, which respond to interaction with ammonia by a shift in resonant frequency, and their characterization have been reported.

Measurements have shown frequency shifts in the range of a few megahertz shift for ammonia concentrations ranging from 4 up to 100 ppm. The change in the resonant frequency with ammonia exposure qualitatively validated the characterization model, and the sensor demonstrated both high sensitivity at low concentrations and fast return to baseline. The proposed design can be used for remote sensing and can be integrated with RFID or WISP tags for low-cost, wireless gas sensing applications thanks to the ease of integration with existing wireless packaging technologies such as inkjet printing. These results can be further improved by optimization of the deposition and curing techniques, and with enhancements to the output circuitry of the final sensor design.

Among the three different types of integrated wireless medical modules, an advanced ZigBee-based low-power, low-cost wireless EKG design has been proposed. The resulting device can be deployed for continuous monitoring of the EKG signal of a patient. The transmission of three-lead EKG data was tested in a wireless system around 2.4 GHz for a range of 150 m in open space and 50 m in a closed area.

In conclusion, the presented work could set the foundation for the implementation of truly wireless biosensor networks with small-size, lightweight conformal nodes for wireless health monitoring in pharmaceuticals, hospital, ambulance, and home-based patient care, as well as in space/underwater missions.

REFERENCES

Abdur Rahman AR, Price DT, Bhansali S. 2007. Effect of electrode geometry on the impedance evaluation of tissue and cell culture. *Sensors Actuators B: Chem.*, 127(1):89–96.

Akyildiz I, Vuran MC. 2010. Wireless sensor networks. Vol. 4. John Wiley & Sons.

Andreuccetti D, Fossi R, Petrucci C. 2002. An Internet resource for the calculation of the dielectrtic properties of body tissues in the frequency range 10 Hz–100 GHz. http://niremf.ifac.cnr.it/tissprop/.

Balandin AA, Ghosh S, Bao W, Calizo I, Teweldebrhan D, Miao F, Lau CN. 2008. Superior thermal conductivity of single-layer graphene. *Nano Lett.*, 8(3):902–907.

Bekyarova E, Cabrera N, Zhao B, Yu A, Gao J, Haddon R.C. 2005. Electronic properties of single-walled carbon nanotube networks. *J. Am. Chem. Soc.*, 127:6.

Bekyarova E, Itkis ME, Beer L, Carbrera N, Haddon RC. 2007. Mechanism of ammonia detection by chemically functionalized single-walled carbon nanotubes: In situ electrical and optical study of gas analyte detection. *J. Am. Chem. Soc.*, 129:7.

Bera S, Chattopadhyay S, Chakraborty B. 2004. An experimental analysis of the non-linear behaviour of a bio-electrode polarisation impedance with excitation frequency. *Measurement*, 35(4):363–370.

Bozzi M, Georgiadis A, Wu K. 2011. Review of substrate-integrated waveguide circuits and antennas. *Microwaves, Antennas Propag., IET*, 5(8):909–920.

Brüesch P, Christen T. 2004. The electric double layer at a metal electrode in pure water. *J. Appl. Phys.*, 95:2846.

CCI-300 Data Sheet. Available: http://www.cabot-corp.com/.

Chopra S, Pham A, Gaillard J, Parker A, Rao A. 2002. Carbon-nanotube-based resonant-circuit sensor for ammonia. *Appl. Phys. Lett.*, 80:4632.

Chung DJ, Bhattacharya SK, Papapolymerou J. 2009. Low loss multilayer transitions using via technology on LCP from DC to 40 GHz. In Electronic Components and Technology Conference, ECTC 2009. 59th., pp. 2025–2029. IEEE.

Ciraci S, Yildirim T, Gülseren O, Senger RT. 2004. Functionalized carbon nanotubes and device applications. *J. Phys. Condensed Matt.*, 129:2.

Collins PG, Ishigami M, Zettl A. 2000. Extreme oxygen sensitivity of electronic properties of carbon nanotubes. 287(5459):1801–1804.

Dragoman M, Dubuc D, Bary L, Fourn E, Plana R, Flahaut E. 2006. The experimental determination of microwave attenuation and electrical permittivity of double walled carbon nanotubes. *Appl. Phys. Lett.*, 88.

Ekström M. 2006. Small wireless ECG with Bluetooth™ communication to a PDA. M.A. Thesis, *Mälardalen University*.

Fan Y, Rahmat-Samii Y. 2003. Reflection phase characterizations of the EBG ground plane for low profile wire antenna applications. *IEEE Trans. Antennas Propag.*, 51(10):2691–2703.

Fujiwara A, Suematsu H, Kataura K, Maniwa Y, Suzuki S, Achiba Y. 2001. Gas adsorption in the inside and outside of single-walled carbon nanotubes. *Chemical Physics Letters*, 336(3):205–211.

Fulford-Jones TR, Wei GY, Welsh M. 2004. A portable, low-power, wireless two-lead EKG system. In Engineering in Medicine and Biology Society, IEMBS'04. 26th Annual International Conference of the IEEE. 1:2141–2144. IEEE.

Gabriel C, Gabriel S, Grant EH, Halstead BSJ, Mingos DMP. 1998. Dielectric parameters relevant to microwave dielectric heating. *Chem. Soc. Rev.*, 27(3):213–224.

Geim AK, Novoselov KS. 2007. The rise of graphene. *Nat Mater*, 6(3):183–191.

Geng HZ, So KP, Lee YS, Chang Y, Lee YH. 2007. Effect of acid treatment on carbon nanotube-based flexible transparency conducting films. *J. Am. Chem. Soc.*, 129.

Gracia-Espino E, Pino GS, Halonen N, Luomahaara J, Maklin J, Toth G, Kordas K, Jantunen H, Terrones M, Helisto P, Seppa H, Ajayan PM, Vajtai R. 2010. Electrical transport and field-effect transistors using inkjet-printed SWCNT films having different functional side groups. *J. Am. Chem. Soc.*, 4(6).

Huang Y, Sarkar TK, Carlo J. 2008. Design of a low profile end-fire antenna using split-ring resonators. Presented at the IEEE Antennas and Propagation Society International Symposium, San Diego, CA.

Hummers WS, Offeman RE. 1958. Preparation of graphitic oxide. *J. Am. Chem. Soc.*, 80(6), 1339–1339.

Hyoungki P, Lu JP. 2006. Effects of sidewall functionalization on conducting properties of single wall carbon nanotubes. *Nano. Lett.*, 6(5):4.

Ida I, Sato J, Sekizawa T, Yoshimura H, Ito K. 2002. Dependence of the efficiency-bandwidth product on electrical volume of small dielectric loaded antennas. *IEEE Trans. Antennas Propag.*, 50(6):821–826.

Joshi KA, Prouza M, Kum M, Wang J, Tang J, Haddon R, Mulchandani A. 2005. V-Type nerve agent detection using a carbon nanotube-based amperometric enzyme electrode. *Anal. Chem.*, 78(1):331–336.

Jun SC, Cha SN, Baik CW, Lee SH, Jin Kim H, Hone J, Kim JM. 2007. Radio-frequency transmission characteristics of a multi-walled carbon nanotube. *Nanotechnology*, 18(25).

Kim S, Yun YJ, Ji S, Yoo K, Yun WS, Koo JY, Ha DH. 2007. Effect of polymer coating on the adsorption of gas molecules on carbon nanotube networks. *Appl. Phys. Lett.*, 91.

Kong J, Zhou C, Chapline MG, Peng S, Cho K, Dai H. 2000. Nanotube molecular wires as chemical sensors. *Science*, 287.

Krupka J, Clarke RN, Rochard OC, Gregory AP. 2000. Split post dielectric resonator technique for precise measurements of laminar dielectric specimens-measurement uncertainties. In Microwaves, Radar and Wireless Communications. MIKON–2000. 13th International Conference on Vol. 1, pp. 305–308. IEEE.

Kupka J, Rochard OC, Gregory AP. 2000. Split post dielectric resonator technique for precise measurements of laminar dielectric specimens-measurement uncertainties. *Microw. Radar Wireless Commun.*, 1:4.

Lakafosis V, Rida A, Vyas R, Li Y, Nikolaou S, Tentzeris MM. 2010. Progress towards the first wireless sensor networks consisting of inkjet-printed, paper-based RFID-enabled sensor tags. *Proc. IEEE*, 98(9):1601–1609.

Lakafosis V, Xiaohua Y, Taoran L, Gebara E, Yang W, Tentzeris MM. 2011. Wireless sensing with smart skins. Presented at the Sensors, October 28–31, 2011 IEEE.

Le LT, Ervin MH, Qiu H, Fuchs BE, Lee WY. 2011. Graphene supercapacitor electrodes fabricated by inkjet printing and thermal reduction of graphene oxide. *Electrochem. Commun.*, 13(4):355–358.

Li D, Kaner RB. 2008. Materials science. Graphene-based materials. *Science (New York)*, 320(5880):1170–1171.

Li Y, Rida A, Vyas R, Tentzeris MM. 2007. RFID tag and RF structures on a paper substrate using inkjet-printing technology. *IEEE Trans. Microw. Theory Techn.*, 55(12):2894–2901.

Li Y, Rongwei Z, Staiculescu D, Wong CP, Tentzeris MM. 2009. A novel conformal RFID-enabled module utilizing inkjet-printed antennas and carbon nanotubes for gas-detection applications. *Antennas Wireless Propag. Lett. IEEE*, 8, 653–656.

Li Z, Yao Y, Lin Z, Moon K-S, Lin W, Wong C. 2010. Ultrafast, dry microwave synthesis of graphene sheets. *J. Mat. Chem.*, 20(23):4781–4783.

Lin Z, Yao Y, Li Z, Liu Y, Li Z, Wong C-P. 2010. Solvent-assisted thermal reduction of graphite oxide. *J. Phys. Chem. C*, 114(35):14819–14825.

Mahalik NP. 2007. Sensor networks and configuration: Fundamentals, standards, platforms, and applications. *Recherche*, 67:2.

Marliere C, Poncharal P, Vaccarini L, Zahab A. 2000. Effect of gas adsorption on the electrical properties of single-walled carbon nanotube mats. In MRS Proceedings, 593(1). Cambridge University Press.

Mukala V, Lakafosis V, Traille A, Tentzeris MM. 2010. A novel Zigbee-based low-cost, low-power wireless EKG system. In Microwave Symposium Digest (MTT), 2010 IEEE MTT-S International, pp. 624–627. IEEE.

Ong KG, Grimes CA. 2002. A wireless passive carbon nanotube-based gas sensor. *IEEE Sensors*, 2(2):7.

Pham A. 2003. Carbon nanotube resonator sensors for remote sensing systems. *Proc. IEEE Topical Conf. Wireless Comm. Techn.*, 4.

Rais NHM, Soh PJ, Malek F, Ahmad S, Hashim NBM, Hall PS. 2009. A review of wearable antenna. Presented at the Antennas & Propagation Conference, Nov. 16–17 2009. LAPC 2009. Loughborough.

Sadkowski A. 2000. On the ideal polarisability of electrodes displaying cpe-type capacitance dispersion. *J. Electroanal. Chem.*, 481(2):222–226.

Salonen P, Rahmat-Samii Y. 2006. Textile antennas: Effects of antenna bending on input matching and impedance bandwidth. Presented at the First European Conference on Antennas and Propagation, Nov. 6–10, 2006. EuCAP 2006.

Sample AP, Yeager DJ, Powledge PS, Mamishev AV, Smith JR. 2008. Design of an RFID-based battery-free programmable sensing platform. *IEEE Trans. Instrum. Meas.*, 57(11):2608–2615.

Sangkil K, Tentzeris MM, Nikolaou S. 2012. Wearable biomonitoring monopole antennas using inkjet printed electromagnetic band gap structures. Presented at the March 26–30 2012 6th European Conference on Antennas and Propagation (EUCAP).

Schedin F, Geim AK, Morozov SV, Hill EW, Blake P, Katsnelson MI, Novoselov KS. 2007. Detection of individual gas molecules adsorbed on graphene. *Nat Mater*, 6(9):652–655. Available: http://www.nature.com/nmat/journal/v6/n9/suppinfo/nmat1967_S1.html.

Shaker G, Duncan K, Tentzeris M. 2010. Integrated antenna with inkjet-printed compact artificial magnetic surface for UHF applications. Presented at the IEEE International Conf. on Wireless Information Tech.

Shaozhen Z, Langley R. 2009. Dual-band wearable textile antenna on an EBG substrate. *IEEE Trans. Antennas Propag.*, 57(4):926–935.

Sievenpiper D, Lijun Z, Broas RFJ, Alexopolous NG, Yablonovitch E. 1999. High-impedance electromagnetic surfaces with a forbidden frequency band. *IEEE Trans. Microw. Theory Techn.*, 47(11):2059–2074.

Somaraju R, Trumpf J. 2006. Frequency, temperature and salinity variation of the permittivity of seawater. *IEEE Trans. Antennas Propag.*, 54(11):3441–3448.

Someya T, Kim P, Nuckolls C, Yardley JT. 2003. Alcohol vapor sensors based on single-walled carbon nanotube field effect transistors. *Nano Lett.*, 3(7):4.

Subramanian V, Lee JB, Molesa SE, Volkman SK. 2005. Printed organic transistors for ultra-low-cost RFID applications. *IEEE Trans. Compon. Packag. Technol.*, 28(4):6.

Super-sensitive and small: New MIT detector uses nanotubes to sense deadly gases, MIT, Cambridge, MA, MIT Press Release, Jun. 2008. Available at: http://web.mit.edu/newsoffice/2008/nanotube-0605.html.

Ting CY, King R. 1970. The dielectric-filled tubular monopole. *IEEE Trans. Antennas Propag.*, 18(5):604–610.

Ukkonen LEA. 2004. Patch antenna with EBG ground plane and two-layer substrate for passive RFID of metallic objects. Presented at the IEEE Antennas and Propagation Society International Symposium.

Varghese OK, Gong D, Ong KG, Dickey EC, Grimes CA. 2001. Gas sensing characteristics of multi-walled carbon nanotubes. *J. Sensors Actuators B Chem.*, 81.

Whalen III JJ, Young J, Weiland JD, Searson PC. 2006. Electrochemical characterization of charge injection at electrodeposited platinum electrodes in phosphate buffered saline. *J. Electrochem. Soc.*, 153:C834.

Yang L, Rida A, Tentzeris MM. 2009. Design and development of radio frequency identification (RFID) and RFID-enabled sensors on flexible low cost substrates. *Synthesis Lect. RF/Microw.*, 1(1):1–89.

Zhang T, Bekyarova E, Yoo BY, Haddon RC, Myung NV, Deshusses MA. 2007. Poly(m-aminobenzene sulfonic acid) functionalized single-walled carbon nanotube based gas sensor. *Nanotechnology*, 18:6.

Zhou H, Preston MA, Tilton RD, White LR. 2005. Calculation of the electric polarizability of a charged spherical dielectric particle by the theory of colloidal electrokinetics. *J. Colloid Interf. Sci.*, 285(2):845–856.

Zhou Y, Du F, Dai L, Volakis JL. 2010. Polymer-carbon nanotube sheets for conformal load bearing antennas. *IEEE Trans. Antennas Propag.*, 58(7):6.

INDEX

Handbook of Biomedical Telemetry, First Edition. Edited by Konstantina S. Nikita.

IEEE Press Series in Biomedical Engineering

The focus of our series is to introduce current and emerging technologies to biomedical and electrical engineering practitioners, researchers, and students. This series seeks to foster interdisciplinary biomedical engineering education to satisfy the needs of the industrial and academic areas. This requires an innovative approach that overcomes the difficulties associated with the traditional textbooks and edited collections.

Series Editor: Metin Akay, University of Houston, Houston, Texas

1. *Time Frequency and Wavelets in Biomedical Signal Processing*
Metin Akay

2. *Neural Networks and Artificial Intelligence for Biomedical Engineering*
Donna L. Hudson, Maurice E. Cohen

3. *Physiological Control Systems: Analysis, Simulation, and Estimation*
Michael C. K. Khoo

4. *Principles of Magnetic Resonance Imaging: A Signal Processing Perspective*
Zhi-Pei Liang, Paul C. Lauterbur

5. *Nonlinear Biomedical Signal Processing, Volume 1, Fuzzy Logic, Neural Networks, and New Algorithms*
Metin Akay

6. *Fuzzy Control and Modeling: Analytical Foundations and Applications*
Hao Ying

7. *Nonlinear Biomedical Signal Processing, Volume 2, Dynamic Analysis and Modeling*
Metin Akay

8. *Biomedical Signal Analysis: A Case-Study Approach*
Rangaraj M. Rangayyan

9. *System Theory and Practical Applications of Biomedical Signals*
Gail D. Baura

10. *Introduction to Biomedical Imaging*
Andrew G. Webb

11. *Medical Image Analysis*
Atam P. Dhawan

12. *Identification of Nonlinear Physiological Systems*
David T. Westwick, Robert E. Kearney

13. *Electromyography: Physiology, Engineering, and Non-Invasive Applications*
Roberto Merletti, Philip Parker

14. *Nonlinear Dynamic Modeling of Physiological Systems*
Vasilis Z. Marmarelis